Estimation and Control of Systems

Estimation and Control of Systems

Theodore F. Elbert

The University of West Florida
Pensacola, Florida

VAN NOSTRAND REINHOLD COMPANY
NEW YORK CINCINNATI TORONTO LONDON MELBOURNE

Library of Congress Catalog Card Number: 83-6685
ISBN: 0-442-22285-8

Manufactured in the United States of America

Published by Van Nostrand Reinhold Company Inc.
135 West 50th Street
New York, New York 10020

Van Nostrand Reinhold Company Limited
Molly Millars Lane
Wokingham, Berkshire RG11 2PY, England

Van Nostrand Reinhold
480 Latrobe Street
Melbourne, Victoria 3000, Australia

Macmillan of Canada
Division of Gage Publishing Limited
164 Commander Boulevard
Agincourt, Ontario M1S 3C7, Canada

15 14 13 12 11 10 9 8 7 6 5 4 3 2 1

Library of Congress Cataloging in Publication Data

Elbert, Theodore F.
Estimation and control of systems.

Includes bibliographical references and index.
1. Control theory. 2. Estimation theory. I. Title.
QA402.3.E38 1984 629.8′312 83-6685
ISBN 0-442-22285-8

To the memory of my brother Donnie

"When I see the world around me
It marvels my very soul
To see what God has created
To see how little we know.

The vastness of the universe
The intricacy of the mind
The new things, new ways, new knowledge
That we never cease to find."

Don Elbert
1932-1980

Preface

The material in this book was developed in support of a program oriented toward the use of computers in system control. The two main topics are the control of systems by state variable feedback, and estimation theory as it applies to system state variables. The former topic is considered in Chapters 4, 5 and 6, with supporting material in Appendix D. The latter is covered in Chapters 2, 3 and 7, with a brief introduction to system concepts in Chapter 4. The developments in Chapter 8, which relate to dynamic programming, expand on both topics. There is sufficient material for a comprehensive one-semester course in either topic.

While students of these subjects usually have backgrounds in science or engineering, experience has shown that many suffer from inadequate recall in the critical areas of random variable theory and fundamental optimization theory. This problem is particularly noted in students returning to formal education after a period of industrial experience. The format of this text is influenced heavily by this observation. The basic intent is to provide enough background material in these critical areas, using the same nomenclature and terminology as the major portions of the text, to allow the reader to reaquaint himself with the subject, at the same time introducing new concepts which blend in with the main topics of the text. For example, the discussion of optimization techniques in Appendix B includes an introduction to the calculus of variations as a natural extension of the mathematical techniques of optimization, even though this topic serves as the basis for the development of optimal control theory in Chapter 6. By use of the supporting materials in these appendices, the reader may enter the development at any point at which he feels comfortable.

While the discussion of basic probability theory presented in Chapter 2 could also have been placed in an appendix, it is so closely aligned with the discussion of estimation theory, which eventually leads to the Kalman estimation technique, that it was included in the main flow of the text. This feature makes the transition from random variable theory to a discussion of random processes, and eventually to the theory of Markov random processes, relatively easy to accomplish. Such an arrangement also sets the standard for notation early in the text and, since a meaningful notational standard is critical to a lucid presentation of multivariable concepts, this feature is considered an important part of the overall presentation.

In keeping within the constraint of reasonable length, the presentation of a large number of solved example exercises has been sacrificed in order to include the basic background material contained in Chapter 2 and the appendices. While some may argue that an improper choice was made here, it is felt that any shortcoming is offset by the inclusion of a large number of problems, which permits an instructor to develop as many examples as he deems necessary. The inclusion of basic review materials, using the notational standards of the main flow of the text, should serve to lessen the burden on the instructor for review of those topics prerequisite to the course. There are, in general, two kinds of exercises at the end of chapters and appendices. The Developmental Exercises are designed to take the reader further into specific areas than does the textual material. Some of these exercises are simple and straightforward, but others may require reference to other publications and considerable developmental effort. By judicious use of these exercises, the instructor can augment the textual material to any degree he wishes. On the other hand, the Problems at the end of each chapter or appendix can be worked using the techniques developed in the text. Most of the problems result in fairly straightforward solution; some result in complex mathematics, even though the basic concept involved is obtained easily from the material in the text. A few of the problems are nebulously stated design problems which could be used for an end-of-term project of some kind. Whatever approach is used, the problem set should be considered an integral part of the text.

It is assumed that the reader has a background in classical control theory, and preferably has had some exposure to statistical analysis. The former may not be necessary to an understanding of most of the material in the text, but it is certainly mandatory if a complete grasp of the system control field is desired. Appendix D, which relates modern control concepts to those of the classical theory, cannot be understood without such a background. There is sufficient material in Chapter 2 to provide the necessary foundation in statistical concepts, but it is of necessity rather condensed and oriented specifically toward the developments of Chapters 3 and 7.

Several people were instrumental in the development of this manuscript, and to them I owe a debt of gratitude. To Ruby Sistrunk and Curt Jones, who helped proof the manuscript and offered helpful suggestions; to Kathy Landua, who not only offered technical suggestions but also proofed the complete text for proper use of the English language; to my son Mark for the countless hours spent seated at the word-processing computer; and to my wife Barbara for her assistance in arranging the entire manuscript, I am forever grateful. Professor A. K. Mahalanabis provided a most critical review of the original manuscript and offered many helpful suggestions. His ideas bear heavily on the organization and content.

Theodore F. Elbert

Pensacola, Florida

Contents

Estimation and Control of Systems

Chapter 1
Introduction

1.1. AUTOMATIC CONTROL

Automatic control is a 20th Century development, finding its origins in the engineering applications of the early 1900s, maturing rapidly through the development pressures of a world at war, and finally, beginning in the late 1950s, transitioning from the art of designing a satisfactory system to the modern science that it is today. The increased complexity of modern systems, coupled with the advent and rapid development of the modern digital computer, provided the impetus and the means for generalizing control theory to the broader concepts of the theory of organized systems. Frequency domain techniques such as the Nyquist, Bode, and root locus methods, ingeniously developed to provide ways around the computational complexities of stability analysis, were complemented by computer programs which made some of the finesse of the original frequency domain approach unnecessary. Then the advantages of digital hardware began to influence system design, first through the use of sampled-data systems, and then through developments leading to the all-digital systems which are becoming commonplace today.

This change of emphasis from analog to digital systems was accompanied by a requirement to express system dynamics by methods more amenable to yielding solutions of a digital nature. At the same time, rapid development of the time domain approach to system analysis, previously prohibited because of the vast amount of computation required, was made possible by the expanding digital computer technology. With the exploitation of the time domain approach came such innovations as the state space concepts, the Kalman filter, and control concepts based upon the work of Lyapunov and Pontryagin. It is a fortunate circumstance that the discrete nature of practical solutions to the time domain problem produce the very digitally oriented results needed for efficient use of the emerging digital hardware. The result is a natural marriage of analysis techniques and modern hardware technology, in that the numerical solutions so often required in analysis produce discrete time results amenable to implementation by modern digital hardware.

1.1.1. Classical Control Theory

Early workers in the emerging field of automatic control concentrated on the primary problem of the time–the single input-single output system. They developed ingenious concepts leading to techniques which were appropriate for the computational tools of the day. Heavy emphasis was placed on graphical construction and adequate but less than perfect description of feedback system characteristics. It was recognized that the differential equations which described the dynamics of feedback systems were often intractable, but that transformation to the frequency domain through application of the Laplace transform produced algebraic results from which system characteristics could be inferred. Design techniques based upon the frequency domain representation were developed, and these techniques made possible the design of linear feedback control systems that satisfied performance requirements. The Bode and Nyquist methods were the basis of these techniques. Later development of the root locus method provided a great step forward in frequency domain analysis, and, together with the Bode and Nyquist techniques, provided for design of systems which were stable and exhibited satisfactory performance in terms of such parameters as rise time, settling time, overshoot, steady-state error coefficients, and others. Today, classical control theory remains the foundation of feedback control. Most first courses in control theory still rely on classical methods because they are much more closely aligned with the students' perspective of differential equation solution by operational techniques.

1.1.2. Modern Control Theory

The complexity of the feedback control problem grew with developing technology. Systems with multiple inputs and multiple outputs began to tax the capabilities of available control system design techniques. System modeling in terms of more than one dependent variable became necessary, and the concept of a system state described by a set of state variables, the conglomerate of which was termed the state vector, came into being. The idea of generalized coordinates, used by theoretical physicists for over a hundred years, was applied to the feedback control problem and termed the modern approach. In essence, the control engineering community had come to realize that there are not that many basic mathematical problems arising in the physical world, but rather that one finds the same problem packaged in many different ways. At the same time, the time domain representation of a system by means of the state transition matrix, as opposed to the frequency domain representation by means of the system transfer function, became the more advantageous approach to the representation of system dynamics. This result was attributable to two effects: the first related to the natural biases of the state space representation of the system, and the second to the fact that it provides a concise, systematic approach to the digital computer modeling of systems. Now the richness of matrix theory can be brought to bear on the system analysis problem, and one speaks in terms of dominant eigenvalues and definiteness of a matrix when addressing system performance characteristics.

One additional advantage has arisen from the modern control concept. Just as the control problem was recognized to be a new form of an older theoretical physics problem, so also are classical problems in nonengineering fields such as biology, economics, medicine, and others. This recognition brings together all these problems under the general category of systems theory, and makes possible useful interchange of ideas between investigators and practitioners in these various fields.

1.2. OPTIMIZATION OF SYSTEMS

While classical control theory provided for analysis and design of systems which satisfied a more or less arbitrary set of performance requirements, the representation of the system in state space form made possible the optimization of system performance measures by the classical techniques of the calculus of variations. The methods of Lagrange and Hamilton apply almost directly to the feedback control problem, and the result is an optimization based upon some specified performance index. While the general result of the application of the calculus of variations to the optimal control problem is a set of two-point boundary value, nonlinear differential equations—solvable only by iterative numerical techniques—the linear problem results in a generally solvable set of equations. Solutions to the minimum time, minimum energy, and minimum control effort problems can be obtained from this general approach. The solution to the optimal linear regulator problem can also be obtained in a rather straightforward manner by taking the optimal control problem to the steady-state limit.

The use of optimal control concepts in system design has two primary advantages. First, it provides a systematic approach to the problem and specifically defines the performance measure to be optimized. Second, the solution of the optimal problem sets a standard by which any other design can be gauged. The latter feature is very useful in obtaining a measure of the degradation produced by use of suboptimal control schemes.

1.2.1. The Role of Estimation in System Control

In most practical instances, system control involves the concept of feedback, whereby some function produced by the system is measured or observed and used to determine the control function. Classical control theory is based upon the feedback of a single output function. Modern control theory, on the other hand, depends upon the ability to feed back all the state variables of a system.

In principle, this concept eliminates the need for dynamic compensation in the controller itself, a topic which constitutes a major portion of the classical theory. In most practical systems, however, the state variables themselves are not available for observation, but instead, some function of the state variables must be used for system control. What is required in such a situation is some way of estimating the system state variables, using the available output as the measured or observed function. A device which accomplishes this is called an *observer*, and the design of observers is an integral part of modern control theory. As might be imagined, there is a direct connection between dynamic compensation in the classical theory and the design of observers in the modern theory. This relationship is discussed in detail in Appendix D.

All systems are stochastic in nature to some extent. Input functions are never completely noise free, and perfect measurement of output functions is not possible in the real world. The whole question here is one of degree. In those systems in which input and measurement noise levels are negligible, both the classical and modern theory apply directly. If these noise levels are appreciable and must be considered, then estimation from a statistical viewpoint becomes a requirement of the controller. In the classical theory, the work of Wiener provided the design criteria for single input-single output filters with optimal statistical characteristics. Generalization of the Wiener filter to the multiple input-multiple output system was developed by Kalman in 1960. This general result is known as the *Kalman filter*, and serves as a basis for most of the theoretical and practical developments in system state estimation today. As might be expected, there is a close relationship between optimal control and optimal estimation by way of the Kalman filter. In fact, the problems are duals of one another, as initially pointed out by Kalman.

1.2.2. Random Process and Statistical Inference

When random effects are present to an appreciable degree, the logical approach is to consider the system state and output function to be random processes. The filtering process then becomes one of statistical inference, similar in every respect to estimation as it is applied in other fields. In fact, the derivation of the Kalman filter equations developed in Chapter 7 follows directly from basic principles of statistical inference. For these reasons, a clear understanding of random processes and statistical inference is necessary for full understanding of the Kalman filter and the implications of its use. Also, because multivariable systems are the norm, multivariate forms of probability distributions are an integral part of the theoretical development. This makes nomenclature critical, since a clear, concise, and consistent terminology is essential to a lucid presentation. In Chapter 2, statistical inference is developed using terminology which sets the standard for the remainder of the text. This is further extended in Chapter 3 to permit the systematic characterization of random processes.

A particular kind of random process, the *Markov process*, plays an important role in the stochastic control problem. The reason for this lies in the fact that linear stochastic systems tend to produce output functions which are Markov random processes. Markov processes are also used to model the input and measurement noise processes in certain Kalman filter applications. Chapter 3 provides an introduction to the Markov random process and serves to illustrate the correlation between Markov processes and linear dynamic systems.

1.2.3. Dynamic Programming

An important development of the 1950s, made practically possible by the computational power of the large-scale digital computer, was the introduction by Bellman of the technique of dynamic programming. Whereas the classical optimization techniques provided by the calculus of variations produce a generally intractable set of equations which must be solved by iterative numerical techniques, dynamic programming approaches the optimization problem from a more or less microscopic point of view by considering the basic model to be a multi-stage decision process. The overall optimization is achieved by dealing with the individual stage optimizations. Dynamic programming often provides a practical alternative to the calculus of variations approach to system optimization. It also can be used to develop alternative forms of the necessary conditions for system optimization.

Because dynamic programming is extremely versatile in the modeling of physical reality, it can be used in a variety of situations in which the classical calculus of variations approach fails. This feature has led to use of the technique in many areas of general optimization. These concepts are considered in Chapter 8.

1.3. CONTINUOUS AND DISCRETE SYSTEMS

Some of the theoretical developments in estimation and control are more easily explained when dealing with continuous systems. Others, particularly those relating to complex probability distributions, are more clearly expressible using discrete systems as the basic model. In most cases, conversion of continuous system results to corresponding discrete system results, and the converse, are fairly straightforward. Throughout this book, the easiest approach is taken. That is, if the continuous system version is more easily understood, the development is carried out in terms of the continuous system and later generalized to the discrete system version. The converse is true for those developments more appropriately carried out for discrete systems. In all cases, the discrete version should be considered the most practical, simply because modern systems tend to be discrete in nature.

1.4. NOTATIONAL CONVENTION

In the study of random processes, notation plays an important role in the development of a lucid presentation. Commonly encountered concepts, such as probability density functions, cumulative distribution functions, mean value functions, and covariance kernels, should be represented by symbols common to all developments. Random variables should be easily distinguishable from nonrandom quantities by the notation used, rather than by the specific content of the expression containing the variable.

In the development of multivariate concepts, good notation is even more critical. Differentiation among scalar, vector, and matrix quantities is fundamental to a clear presentation of the concepts involved. In treating vector-matrix operations, certain formalisms have been developed which are quite efficient and are compatible with the univariate specialization of the concepts involved. These formalisms are summarized in Appendix A; they depend heavily upon standardized notation in the representation of scalar, vector, and matrix quantities.

When the representation of random vector quantities is considered, the problems are compounded. Care must be taken to ensure that notational distinction between random and nonrandom quantities does not become confused with that distinguishing scalar, vector, and matrix quantities. In this book, most notations follow a specific set of standards, outlined below.

- Nonrandom quantities, whether constants, time functions, or vectors, are represented by lower case letters. Examples are:

  c—a scalar constant
  $x(t)$—a scalar time function
  $f_{\mathrm{X}}(x)$—a function of x.

- Scalar random quantities are represented by roman capital letters. Examples are:

  C—a scalar random variable
  $\mathrm{X}(t)$—a scalar random process.

- Vector quantities are represented by boldface letters—lower case italics for nonrandom vectors, and upright capitals for random vector quantities. Examples are:

  $\boldsymbol{c}$—a nonrandom vector
  $\boldsymbol{x}(t)$—a nonrandom time function
  $\mathbf{X}(t)$—a vector random process.

- Matrix quantities are represented by italic boldface capital letters. Examples are:

$\boldsymbol{F}$–a matrix constant
$\boldsymbol{F}(t)$–a matrix time function.

There are some departures from these standards, notably in the use of Greek letters, but for the most part the reader can feel comfortable using the notational policy outlined here. For example, the state equation

$$\dot{\boldsymbol{x}}(t) = \boldsymbol{F}(t)\boldsymbol{x}(t) + \boldsymbol{G}(t)\boldsymbol{u}(t)$$

is readily recognized as a vector-matrix differential equation for the deterministic vector time function $\boldsymbol{x}(t)$, characterized by time-varying matrices $\boldsymbol{F}(t)$ and $\boldsymbol{G}(t)$, and driven by the vector time function $\boldsymbol{u}(t)$. On the other hand, the equation

$$\dot{\mathbf{X}}(t) = \boldsymbol{F}(t)\mathbf{X}(t) + \boldsymbol{D}(t)\mathbf{W}(t)$$

describes the generation of the vector random process $\mathbf{X}(t)$ by a linear system driven by the vector random process $\mathbf{W}(t)$. Similarly, the function

$$f_{\mathbf{X}}(\boldsymbol{x})$$

is the scalar probability density function describing the distribution of the random vector $\mathbf{X}$. The argument is the deterministic vector $\boldsymbol{x}$. On the other hand, the function

$$g(\boldsymbol{x})$$

is a vector function of the deterministic vector $\boldsymbol{x}$, while

$$g(x)$$

is a vector function of the deterministic scalar x. Finally,

$$g(\mathbf{X})$$

is a scalar function of the random vector $\mathbf{X}$, and

$$g(\mathbf{X})$$

is a vector function of the random vector $\mathbf{X}$. Other combinations are possible; they are readily identifiable using the notational convention outlined here.

Chapter 2
Review of Probability and Statistical Inference

2.1. INTRODUCTION

While the reader is assumed to have knowledge of basic probability and statistical inference, this chapter is devoted to a review of some important concepts and the establishment of a notational convention to be used throughout the remainder of the text. The use of a *consistent* notation, even though it may not be the simplest, is important to a clear understanding of the principles to be developed here.

The *probability* of an event is defined as the limiting value to which the relative frequency of an event converges (if indeed it does converge) as the number of trials gets very large. Thus, if N is the number of trials, the expression

$$P[E] = \lim_{N\to\infty}\left[\frac{\text{number of occurrences of } E}{\text{number of trials}}\right] \tag{2.1}$$

serves as the definition of the probability of event E. If two events E_1 and E_2 are related in such a manner that knowledge of the occurrence of one event alters one's assessment of the probability of occurrence of the other event, then the two events are said to be *statistically dependent*. Conversely, if knowledge of the occurrence of one event does not change the assessment of the probability of the other event, then the events are said to be statistically independent. The probability of one event E_1, given knowledge of the occurrence of the second event E_2, is denoted as

$$P[E_1|E_2] = \text{probability of event } E_1\text{, given that } E_2 \text{ has occurred.} \tag{2.2}$$

Then, for two independent events E_1 and E_2,

$$P[E_1|E_2] = P[E_1]$$

$$P[E_2|E_1] = P[E_2]. \tag{2.3}$$

These equations represent both necessary and sufficient conditions. From the basic tenets of relative frequency, it can easily be shown that

$$P[E_1|E_2] = \frac{P[E_1 \cap E_2]}{P[E_2]} \tag{2.4}$$

where the event $E_1 \cap E_2$ is the intersection of events E_1 and E_2—that is, the simultaneous occurrence of the two events E_1 and E_2. There follows from Equations (2.3) and (2.4) the third necessary and sufficient condition for two events E_1 and E_2 to be statistically independent:

$$P[E_1 \cap E_2] = P[E_1]P[E_2]. \tag{2.5}$$

It can also be shown that if the first expression of Equation (2.3) holds, then the second expression also holds, and vice versa.

Using the fact that Equation (2.4) is symmetric with respect to events E_1 and E_2, one obtains the very useful relationship known as *Bayes' rule*:

$$P[E_1|E_2] = \frac{P[E_2|E_1]P[E_1]}{P[E_2]}. \tag{2.6}$$

Bayes' rule, of course, is also symmetric with respect to the events E_1 and E_2.

It is important to note that, for two events E_1 and E_2 to be statistically dependent, there need not be any readily identifiable physical relationship between the two events. What statistical dependency implies is that knowledge of the occurrence of one event alters the assessment of the probability of occurrence of the other, where the meaning of the term *probability* must be taken in the sense of the limiting value of relative frequency.

2.2. RANDOM VARIABLES AND THEIR PROPERTIES

A random variable can best be described as a functional relationship which assigns some real number to each possible random event that can be produced by an experiment. Thus, the outcome of the experiment determines the value of the random variable. Since random variables are associated with random events, and since random events can be described in terms of the probabilities of their occurrence, it follows that random variables can be described by the probabilities that they take on certain values.

2.2.1. The Cumulative Distribution Function

In general, random variables can be divided into two different types–discretely distributed and continuously distributed random variables. The discretely distributed random variable can assume only discrete values, although the number of discrete values may be infinite. On the other hand, the continuously distributed random variable may assume any of a continuum of values, although the range of values may be finite. In order to describe the manner in which the possible values of a random variable are distributed, a functional notation is defined:

$$F_X(x) = P[X \leqslant x]. \tag{2.7}$$

The function $F_X(x)$ is called the *cumulative distribution function* (CDF) of the random variable X. The variable x is an argument which spans the possible values of the random variable X. Use of the subscript notation may at first seem cumbersome, but the systematic use of F to imply a CDF and subscripting to identify the particular random variable whose distribution is being described is a very useful concept in what follows. In the connotation of the previous section, $F_X(x)$ is the probability of the event $X \leqslant x$.

For two numbers a and b, where $a \leqslant b$, the probability

$$P[a < X \leqslant b] = P[(X > a) \cap (X \leqslant b)] \tag{2.8}$$

can be expressed in terms of the CDF as

$$P[a < X \leqslant b] = F_X(b) - F_X(a). \tag{2.9}$$

This expression is valid for both discretely and continuously distributed random variables. From the above considerations, it can be seen that a CDF has certain characteristic properties:

$$\text{I. } F_X(x) \text{ is a nondecreasing function of } x.$$

$$\text{II. } \lim_{x \to \infty} F_X(x) = 1.$$

$$\text{III. } \lim_{x \to -\infty} F_X(x) = 0. \tag{2.10}$$

A *continuously distributed random variable* will be defined as one whose CDF is a continuous function of its argument. That is, it has a finite first derivative everywhere, so that there are no jumps in the CDF, $F_X(x)$. An example of the CDF of a continuously distributed random variable is shown in Figure 2-1(a). A *discretely distributed random variable* will be defined as one whose CDF con-

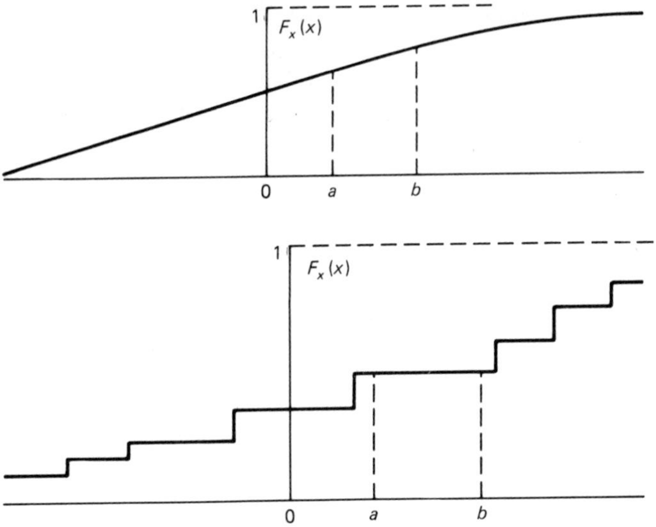

Figure 2-1. (a) CDF of a continuously distributed random variable. (b) CDF of a discretely distributed random variable.

sists only of a finite (or at most a countably infinite) number of jumps, as depicted in Figure 2-1(b). From Equation (2.9), the probability that a random variable whose CDF is depicted in Figure 2-1 lies between the values a and b is given by the difference in $F_X(x)$ evaluated at these two points a and b. In the case of the continuously distributed random variable, this difference is generally a nonzero number. In the case of the discretely distributed random variable, $F_X(a)$ and $F_X(b)$ may have the same value, in which event the probability that X lies in the range $(a < X \leqslant b)$ is zero. Such will always be the case for discretely distributed random variables unless the interval (a, b) includes one or more of the jumps in the function $F_X(x)$. This conclusion implies that X can take on only those values of x for which jumps occur, and, furthermore, that the probability of the random variable assuming this value is proportional to the size of the jump.

It is also possible to encounter random variables with distributions which have both discrete and continuous properties, in that one or more jumps may be superimposed on a generally continuous distribution. This representation is permissible so long as the CDF has the characteristics of Equation (2.10).

2.2.2. The Probability Density Function

The cumulative distribution function of a continuously distributed random variable, as exemplified by Figure 2-1(a), can be differentiated with respect to the variable x. In most cases, such differentiation results in a well defined function which is called the *probability density function* (PDF), and is denoted by the symbol $f_X(x)$. The probability density function is identified by the lower case letter f; the subscript indicates that it is the PDF of the random variable X; the argument is x. The integral relation between the PDF and CDF of the random variable x is

$$F_X(x) = \int_{-\infty}^{x} f_X(\sigma)\, d\sigma \tag{2.11}$$

where σ is a variable of integration introduced to avoid confusion with the upper limit of the integral. From the preceding discussion, the following properties of the PDF may be inferred:

$$\text{I. } f_X(x) \geqslant 0$$

$$\text{II. } \int_{-\infty}^{\infty} f_X(x)\, dx = 1$$

$$\text{III. } P[a < X \leqslant b] = \int_a^b f_X(x)\, dx \tag{2.12}$$

Note that the third property relates the probability that the random variable X will assume values between a and b to the area under the PDF $f_X(x)$ between the points $x = a$ and $x = b$. The relationship between $F_X(x)$ and $f_X(x)$ is depicted in Figure 2-2.

When the random variable under consideration is discretely distributed—implying a CDF of the form depicted in Figure 2-1(b)—then the concept of a probability density function as used above does not apply, since the CDF is not differentiable. However, the concept of a probability density function can be generalized by application of the mathematical formalism of the Dirac delta function. A delta function of strength C, designated by

$$C\delta(x - x_1) \tag{2.13}$$

is defined as a function of x which is zero everywhere except at the point $x = x_1$, where it assumes an infinite value in such a manner that the area under the function is C, the strength of the function. Such a function can easily be visualized as the limiting result of a rectangular pulse of height h and width C/h, as h gets very large. One of the useful properties of the Dirac delta function is the sam-

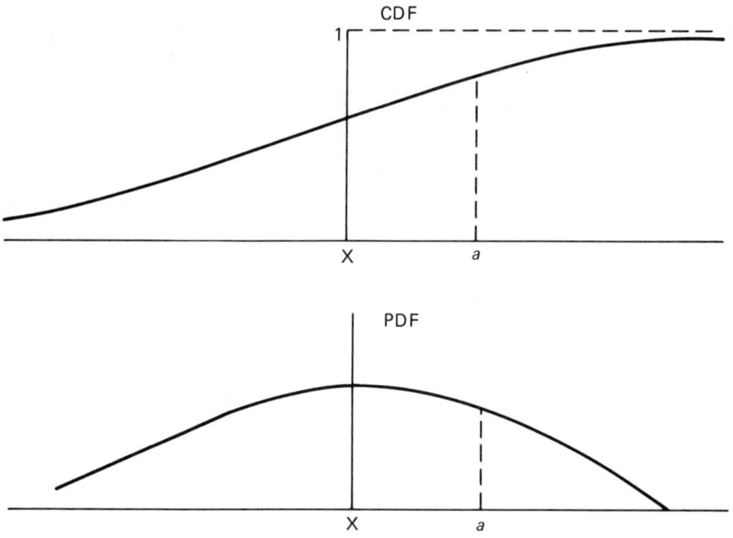

Figure 2-2. CDF and PDF of a continuously distributed random variable.

pling property, which states that, for $a \leqslant x_1 \leqslant b$,

$$\int_a^b g(x)\delta(x - x_1)\,dx = g(x_1) \tag{2.14}$$

where $g(x)$ is any function continuous at the point $x = x_1$. By use of the Dirac delta function, a function analogous to the probability density function—defined for a continuously distributed random variable—can be considered. The problem previously encountered was that the derivative of the CDF could not be defined by use of ordinary mathematical functions. However, the consideration of the Dirac delta function as an acceptable mathematical entity permits the derivative of such functions to be defined as a series of delta functions located at those points at which the CDF exhibits a discontinuity, or jump; each delta function has a strength $P_X(x_i)$ equal to the magnitude of the jump at its associated point x_i. Such consideration leads to the relationship shown in Figure 2-3—which is analogous to Figure 2-2—in which the height of the delta function has been used to indicate its strength. Some reflection on the properties of the Dirac delta function will verify that the PDF is actually the formal derivative of the CDF. Analytically, the PDF of a discretely distributed random variable X can be expressed as the sum of the individual delta functions of strength $P_X(x_i)$:

$$f_X(x) = \sum_i P_X(x_i)\delta(x - x_i) \tag{2.15}$$

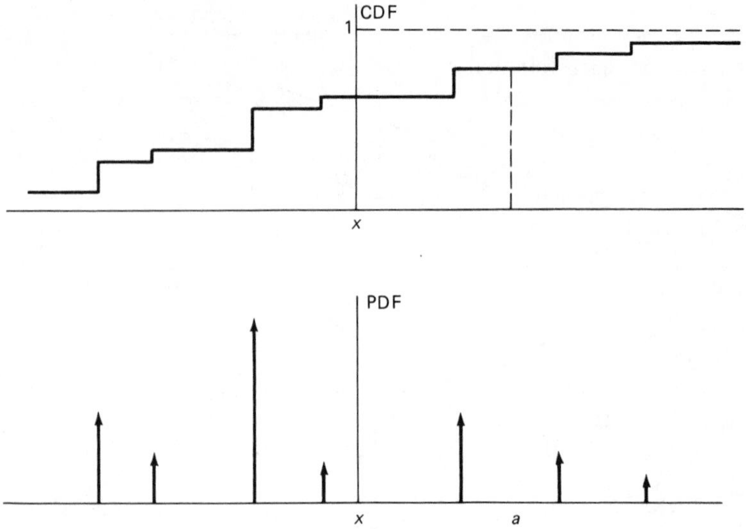

Figure 2-3. CDF and PDF of a discretely distributed random variable.

where

$$P_X(x_i) = P[X = x_i] \tag{2.16}$$

is the strength of the delta function at x_i and is called the *probability function* of the discretely distributed random variable X. Basic considerations will show that the PDF defined in this manner will satisfy the basic integral relationship of Equation (2.11). That is,

$$F_X(x) = \int_{-\infty}^{x} f_X(\sigma)\, d\sigma = \int_{-\infty}^{x} \sum_i P_X(x_i)\delta(\sigma - x_i)\, d\sigma \tag{2.17}$$

which, by virtue of the sampling property of the delta function, is

$$F_X(x) = \sum_i P_X(x_i) \tag{2.18}$$

where the sum is over all values of $x_i \leqslant x$. A simple extension of these principles permits consideration of CDF's with a combination of continuous and discrete characteristics.

The distribution of a discretely distributed random variable can be adequately described in terms of the probability function itself; this is the method used in most elementary texts. Use of the probability function has the disadvantage that it does not produce the same integral relationship with the CDF which results from the delta function approach, and which is completely analogous to the relationship obtained for continuously distributed random variables. Consequently, the use of the probability function does not permit the uniform definition of important operators, such as the expectation operator, while the delta function approach has been developed specifically for this purpose.

In general, the distribution of a random variable can be described by either the PDF or CDF. Normally, the PDF is the preferred method, because most operations of practical importance are defined in terms of the PDF. Also, it happens that some distributions are defined in terms of a PDF which can be integrated only by numerical techniques. That is, an analytical form for the CDF cannot be obtained. The most important continuous distribution—the normal or Gaussian distribution—is of this type. The physical concepts leading to the definition of the normal distribution produce a description in terms of the PDF. The CDF can only be obtained by use of numerical techniques or from tables resulting from numerical computation. This limitation results from the fact that the PDF given by

$$f_X(x) = \frac{1}{\sqrt{2\pi}\,\sigma} e^{-1/2(x-\mu/\sigma)^2} \tag{2.19}$$

is not an integrable function. That is, there is no analytic function of x which, when differentiated, yields the PDF of Equation (2.19). This is the reason why computations based upon the normal probability distribution make use of tables.

2.2.3. The Expectation Operator

An important operator in the mathematics of probability is the expectation operator. This operator is defined in terms of the probability density function and is given by the following expression:

$$E[g(X)] = \int_{-\infty}^{\infty} g(x) f_X(x)\, dx. \tag{2.20}$$

Here, the symbol $E[\cdot]$ is the expectation operator, $g(X)$ is some function of the random variable X, and $f_X(x)$ is the PDF of the random variable X. Since the expectation operator is defined with respect to the distribution of X, it will sometimes be denoted as

$$E_X[g(X)]. \tag{2.21}$$

Such notation will be used when there may be some confusion as to the exact implication of the symbol. When no doubt exists, the subscript will be omitted.

The expression of Equation (2.20) serves to define the expectation for both continuously distributed and discretely distributed random variables, if the discrete distributions are defined in terms of Dirac delta functions. Thus, if X is discretely distributed, then the PDF $f_X(x)$ consists of a sum of delta functions, each with its characteristic strength $P_X(x_i)$. Use of the sampling property of Equation (2.14) then yields

$$
\begin{aligned}
E[g(X)] &= \int_{-\infty}^{\infty} g(x) \sum_i P_X(x_i)\delta(x - x_i)\,dx \\
&= \sum_i \int_{-\infty}^{\infty} g(x)P_X(x_i)\delta(x - x_i)\,dx \\
&= \sum_i g(x_i)P_X(x_i)
\end{aligned}
\tag{2.22}
$$

where the x_i are those values of x at which the discrete jumps in the CDF occur, or, equivalently, the x_i are those values of x at which $f_X(x)$ exhibits a delta function.

The expectation operator is linear, so that

$$E[g_1(X) + g_2(X)] = E[g_1(X)] + E[g_2(X)] \tag{2.23}$$

and

$$E[cg(X)] = cE[g(X)] \tag{2.24}$$

where c is a nonrandom constant. Also note that, if c is a nonrandom constant, then

$$E[c] = c \tag{2.25}$$

where the expectation operator is taken with respect to any random variable X.

When the function $g(X)$ is the random variable X itself,

$$g(X) = X \tag{2.26}$$

so the expectation of X is

$$E[\mathrm{X}] = \int_{-\infty}^{\infty} x f_{\mathrm{X}}(x)\, dx. \tag{2.27}$$

This quantity is of fundamental importance in the description of the distribution of a random variable and is known as the *mean* of the distribution of X. It is usually represented by the Greek letter μ or, as required, by μ_{X}. The precise significance of the mean of a distribution will have to be postponed until discussion of the Law of Large Numbers, but it should be noted that the mean is a mathematical property of the PDF $f_{\mathrm{X}}(x)$.

The mean of the distribution is one of a class of descriptive parameters called the *moments* of the distribution. The nth moment of the distribution of the random variable X is defined as

$$M_n = E[\mathrm{X}^n]. \tag{2.28}$$

Thus, the mean is the first moment. When second or higher moments are considered, it is useful to define moments about the mean, or *central moments*. The nth central moment is defined as

$$\bar{M}_n = E[(\mathrm{X} - E[\mathrm{X}])^n]. \tag{2.29}$$

It can be shown that the moments of a distribution serve to define the distribution itself.

Many important distributions are completely characterized by the first one or two moments. The second central moment is also of considerable interest, so much so that it defines an operator called the *variance* of the distribution:

$$\mathrm{VAR}[\mathrm{X}] = E[(\mathrm{X} - E[\mathrm{X}])^2]. \tag{2.30}$$

An alternative and sometimes more useful form of Equation (2.30) is

$$\mathrm{VAR}[\mathrm{X}] = E[\mathrm{X}^2] - (E[\mathrm{X}])^2 = E[\mathrm{X}^2] - \mu^2 \tag{2.31}$$

The normal distribution described by Equation (2.19) has mean value and variance given by

$$\begin{aligned} E[\mathrm{X}] &= \mu \\ \mathrm{VAR}[\mathrm{X}] &= \sigma^2. \end{aligned} \tag{2.32}$$

2.3. DISTRIBUTION OF MORE THAN ONE RANDOM VARIABLE

As in most areas of mathematics, theories and procedures developed for dealing with single variables must be generalized to include the multivariate case. In the mathematics of random variables, this generalization constitutes expansion of the theory to consider the case in which more than one event can result from a particular experiment, and therefore more than one random variable can be associated with various events. This consideration necessitates some way of describing the distribution of related random variables. In order to develop the required techniques, it is necessary to return to consideration of the fundamental descriptor of the distribution of random variables, the cumulative distribution function.

2.3.1. Bivariate Distributions

For simplicity, consider the case of two random variables, X_1 and X_2, which then have a *joint cumulative distribution function*, defined as follows:

$$\begin{aligned}\text{joint CDF of } X_1 \text{ and } X_2 &= F_{X_1X_2}(x_1, x_2)\\ &= P[X_1 \leqslant x_1 \cap X_2 \leqslant x_2]. \qquad (2.33)\end{aligned}$$

Note the agreement here with previously established convention. The CDF is denoted by upper case letter F, subscripted by the upper case letters denoting the random variables whose distribution it describes. In this case, there are two arguments denoted by the lower case letters x_1 and x_2, which are the possible values of X_1 and X_2, respectively. Where previously the CDF was a function of one variable and could therefore be represented by a two-dimensional curve, now it is a function of the two arguments x_1 and x_2 and must be considered as defining a surface in three dimensions. The basic definition gives the CDF as the probability of the simultaneous occurrence of two events, $X_1 \leqslant x_1$ and $X_2 \leqslant x_2$.

There are certain properties of the bivariate distribution of X_1 and X_2 which can be readily deduced. These are as follows:

$$F_{X_1X_2}(x_1, \infty) = P[X_1 \leqslant x_1 \cap X_2 < \infty] = P[X_1 \leqslant x_1] = F_{X_1}(x_1). \qquad (2.34)$$

The cumulative distribution function of X_1 determined in this manner is called a *marginal CDF*. It is determined as the probability that $X_1 \leqslant x_1$, without regard to the value which X_2 assumes. Such a marginal function reflects all that is known concerning the distribution of X_1, considered without regard to X_2. Similarly, the marginal CDF of X_2 is

$$F_{X_2}(x_2) = P[X_1 < \infty \cap X_2 \leqslant x_2]. \qquad (2.35)$$

Further results may be obtained from basic principles. For example, the probability that a point (X_1, X_2) lies within the rectangular region in the (X_1, X_2) plane bounded by the values

$$a_1 < X_1 \leqslant b_1$$

$$a_2 < X_2 \leqslant b_2 \tag{2.36}$$

is given by the expression

$$\begin{aligned} P[a_1 < X_1 \leqslant b_1 \cap a_2 < X_2 \leqslant b_2] \\ &= P[X_1 \leqslant b_1 \cap X_2 \leqslant b_2] - P[X_1 \leqslant b_1 \cap X_2 \leqslant a_2] \\ &\quad - P[X_1 \leqslant a_1 \cap X_2 \leqslant b_2] + P[X_1 \leqslant a_1 \cap X_2 \leqslant a_2] \\ &= F_{X_1X_2}(b_1, b_2) - F_{X_1X_2}(b_1, a_2) - F_{X_1X_2}(a_1, b_2) + F_{X_1X_2}(a_1, a_2). \end{aligned} \tag{2.37}$$

In addition, the bivariate CDF also exhibits properties analogous to those of the univariate CDF's:

I. $F_{X_1X_2}(x_1, x_2)$ is a nondecreasing function of x_1 and x_2.

II. $\lim_{x_1, x_2 \to \infty} F_{X_1X_2}(x_1, x_2) = 1.$

III. $\lim_{x_1, x_2 \to -\infty} F_{X_1X_2}(x_1, x_2) = 0.$ (2.38)

Generalization of these properties to distributions of more than two variables is straightforward. Also, the concept of both continuously distributed and discretely distributed random variables applies directly in the multivariate case.

Analogous to the development of the univariate distribution–if the CDF is differentiable with respect to both x_1 and x_2–is the definition of a function $f_{X_1X_2}(x_1, x_2)$ such that

$$F_{X_1X_2}(x_1, x_2) = \int_{-\infty}^{x_1} \int_{-\infty}^{x_2} f_{X_1X_2}(\sigma_1, \sigma_2)\, d\sigma_2\, d\sigma_1 \tag{2.39}$$

where σ_1 and σ_2 are variables of integration used to avoid confusion with the limits of the integral. This expression serves as the definition of the bivariate probability density function $f_{X_1X_2}(x_1, x_2)$. In the case of discretely distributed random variables X_1 and X_2, this function takes the form of Dirac delta functions located at specified positions in the (x_1, x_2) plane; while for continuously distributed random variables, it is merely a function of x_1 and x_2. Note that the delta function approach to the discrete distribution permits a unified description of mixed distributions. For example, if X_1 has a continuous distribution and X_2 a discrete distribution, the PDF $f_{X_1X_2}(x_1, x_2)$ can still be defined in terms of delta functions with respect to x_2, whose strengths depend upon the continuous variable x_1. With the bivariate PDF defined as in Equation (2.39), it follows from Equation (2.34) that

$$F_{X_1}(x_1) = F_{X_1X_2}(x_1, \infty) = \int_{-\infty}^{x_1} \int_{-\infty}^{\infty} f_{X_1X_2}(\sigma_1, \sigma_2)\, d\sigma_2\, d\sigma_1. \tag{2.40}$$

If the inner integral—with respect to σ_2—is called the *marginal probability density function of* X_1 and denoted by

$$f_{X_1}(x_1) = \int_{-\infty}^{\infty} f_{X_1X_2}(x_1, x_2)\, dx_2 \tag{2.41}$$

then the expression of Equation (2.40) is

$$F_{X_1}(x_1) = \int_{-\infty}^{x_1} f_{X_1}(\sigma_1)\, d\sigma_1 \tag{2.42}$$

which is identical to Equation (2.11) describing the relationship between the univariate PDF and CDF. Thus, the marginal distribution of X_1, as given by either the marginal PDF or marginal CDF, merely describes the distribution of X_1 without regard to the fact that it has a joint distribution with X_2. There is no practical difference between the marginal distribution of X_1 as derived from the multivariate distribution of X_1 and X_2, and the univariate distribution obtained by neglect or ignorance of the joint distribution. These considerations are, of course, symmetric with respect to X_1 and X_2, and they are easily generalized to the multivariate case involving three or more variables.

If the random variables are discretely distributed, then the joint PDF is described by a two-dimensional generalization of the Dirac delta function

$$\delta(x_1 - x_{1i}, x_2 - x_{2j})$$

which has a zero value everywhere except at the point (x_{1i}, x_{2j}), where the value is infinite. Now, it is the volume under the delta function which is constrained to be one. Then, the joint PDF is given by

$$f_{X_1X_2}(x_1, x_2) = \sum_i \sum_j P_{X_1X_2}(x_{1i}, x_{2j})\delta(x_1 - x_{1i}, x_2 - x_{2j}). \tag{2.43}$$

From the sampling property of the delta function, it is seen that the marginal PDF of X_1 is

$$f_{X_1}(x_1) = \int_{-\infty}^{\infty} \sum_i \sum_j P_{X_1X_2}(x_{1i}, x_{2j})\delta(x_1 - x_{1i}, x_2 - x_{2j})\, dx_2$$

$$= \sum_i \sum_j P_{X_1X_2}(x_{1i}, x_{2j})\delta(x_1 - x_{1i}) \tag{2.44}$$

from which the marginal probability function is determined as

$$P_{X_1}(x_{1i}) = \sum_j P_{X_1X_2}(x_{1i}, x_{2j}) \tag{2.45}$$

where the j sum is over all possible values of x_2. Again, generalization to three or more dimensions is straightforward.

2.3.2. Conditional Distributions

Since the random variables X_1 and X_2 are defined by their association with certain random events, the observation of X_2 implies that the outcome of its associated random event is known. From the discussion of random events previously considered, it is known that observation of one event can affect the knowledge which the observer has of the probabilities associated with a related event. Hence, knowledge of X_2 may affect the distribution of X_1, and the cumulative distribution function of X_1 must be modified in order to reflect this effect. This modification defines the *conditional* CDF

$$F_{X_1|X_2=x_2}(x_1) = P[X_1 \leqslant x_1 | X_2 = x_2]. \tag{2.46}$$

Note that the upper case F implies a CDF, and the subscripts indicate that it is a CDF associated with X_1, with X_2 being fixed at the value x_2. The functional notation indicates that it is a function of the variable x_1. There are, of course, a family of functions depending upon x_2 as a parameter. The conditional CDF describes the distribution of the random variable X_1, just as does the marginal CDF $F_{X_1}(x_1)$, and as such it has a corresponding conditional probability density function defined by

$$F_{X_1|X_2=x_2}(x_1) = \int_{-\infty}^{x_1} f_{X_1|X_2=x_2}(\sigma)\, d\sigma \tag{2.47}$$

where again σ is merely a variable of integration.

The discrete versions of these relationships follow directly from previous discussions of the delta function interpretation of the discrete PDF's, and the conditional probability functions are thus defined. Specifically,

$$F_{X_1|X_2=x_2}(x_1) = \int_{-\infty}^{\infty} \sum_i P_{X_1|X_2=x_2}(x_{1i})\,\delta(x_1 - x_{1i})\,dx_1$$

$$= \sum_i P_{X_1|X_2=x_2}(x_{1i}) \tag{2.48}$$

where the sum is over values of $x_{1i} \leqslant x_1$, and the conditional probability function is

$$P_{X_1|X_2=x_2}(x_{1i}) = P[X_1 = x_{1i}|X_2 = x_2].$$

Again, there is a family of functions $P_{X_1|X_2=x_2}(x_{1i})$ with x_2 as a parameter.

The joint distribution of two random variables, one of which is continuously distributed and the other discretely distributed, can also be treated in this general manner. The joint PDF is described by

$$f_{X_1X_2}(x_1, x_2) = \sum_i P_{X_2}(x_{2i}) f_{X_1|X_2=x_2}(x_1)\,\delta(x_2 - x_{2i}) \tag{2.49}$$

indicating that there is a family of PDF's $f_{X_1|X_2=x_2}(x_1)$ with x_2 as a parameter. The delta function $\delta(x_2 - x_{2i})$ is a function having infinite values at prescribed values of the discrete variable x_2. The strength of the delta function is the product $P_{X_2}(x_{2i}) f_{X_1|X_2=x_2}(x_1)$. The joint CDF is, from the sampling property of the delta function,

$$F_{X_1X_2}(x_1, x_2) = \int_{-\infty}^{x_2} \int_{-\infty}^{x_1} \sum_i P_{X_2}(x_{2i}) f_{X_1|X_2=x_2}(\sigma_1)\,\delta(\sigma_2 - x_{2i})\,d\sigma_1\,d\sigma_2$$

$$= \sum_i P_X(x_{2i}) \int_{-\infty}^{x_1} f_{X_1|X_2=x_2}(\sigma_1)\,d\sigma_1 \tag{2.50}$$

where the summation is over $x_{2i} \leqslant x_2$.

From previous work in the conditional probability of events it is known that conditional, marginal, and joint probabilities are related by the expression

$$P[E_1|E_2] = \frac{P[E_1 \cap E_2]}{P[E_2]}. \tag{2.51}$$

This relationship can be used to show that the conditional, marginal, and joint probability density functions and probability functions are related by

$$f_{X_1|X_2=x_2}(x_1) = \frac{f_{X_1X_2}(x_1, x_2)}{f_{X_2}(x_2)}; \qquad P_{X_1|X_2=x_2}(x_1) = \frac{P_{X_1X_2}(x_1, x_2)}{P_{X_2}(x_2)} \tag{2.52}$$

These expressions are sometimes used as the definition of the conditional PDF, and follow directly from Equation (2.51) in the case of discretely distributed random variables, and through a limiting process for continuously distributed random variables. Just as in the previous section, a Bayes' rule for PDF's can be developed from interchanging x_1 and x_2 in Equation (2.52), and equating the expressions for the joint PDF $f_{X_1X_2}(x_1, x_2)$. This process yields

$$f_{X_1|X_2=x_2}(x_1) = \frac{f_{X_2|X_1=x_1}(x_2) f_{X_1}(x_1)}{f_{X_2}(x_2)} \tag{2.53}$$

for those values of x_2 for which $f_{X_2}(x_2)$ is not zero. The cumbersome notation associated with conditional distributions can be alleviated somewhat by using the symbol $f_{X_1|X_2}(x_1)$ to indicate the PDF of X_1 conditioned on X_2, in lieu of the left hand side of Equation (2.53). No generality is lost by this terminology, and the more precise form will be used in what follows only when it is deemed necessary for a clear understanding of the situation.

2.3.3. Independent Random Variables

The notion of statistically independent random events implies the concept of independent random variables. The usual definition states that the two random variables X_1 and X_2 are statistically independent if, and only if, the two events

$$[X_1 \leqslant x_1] \quad \text{and} \quad [X_2 \leqslant x_2]$$

are statistically independent events. If these events are independent, then their joint probability is

$$P[X_1 \leqslant x_1 \cap X_2 \leqslant x_2] = P[X_1 \leqslant x_1] P[X_2 \leqslant x_2] \tag{2.54}$$

which, when written in terms of the CDF's, is

$$F_{X_1X_2}(x_1, x_2) = F_{X_1}(x_1) F_{X_2}(x_2). \tag{2.55}$$

This expression is sometimes used as a definition for independence, but the definition based upon random events is more fundamental. Equation (2.55) also leads to the result

$$
\begin{aligned}
F_{X_1X_2}(x_1, x_2) &= \int_{-\infty}^{x_1} \int_{-\infty}^{x_2} f_{X_1X_2}(\sigma_1, \sigma_2)\, d\sigma_2\, d\sigma_1 \\
&= F_{X_1}(x_1) F_{X_2}(x_2) \\
&= \int_{-\infty}^{x_1} f_{X_1}(\sigma_1)\, d\sigma_1 \int_{-\infty}^{x_2} f_{X_2}(\sigma_2)\, d\sigma_2 \\
&= \int_{-\infty}^{x_1} \int_{-\infty}^{x_2} f_{X_1}(\sigma_1) f_{X_2}(\sigma_2)\, d\sigma_2\, d\sigma_1
\end{aligned} \tag{2.56}
$$

from which, for independent events,

$$f_{X_1X_2}(x_1, x_2) = f_{X_1}(x_1) f_{X_2}(x_2) \tag{2.57}$$

which is analogous to the expression of Equation (2.55). From Equation (2.52), if X_1 and X_2 are independent random variables,

$$f_{X_1|X_2}(x_1) = \frac{f_{X_1}(x_1) f_{X_2}(x_2)}{f_{X_2}(x_2)} = f_{X_1}(x_1) \tag{2.58}$$

and in a similar manner

$$f_{X_2|X_1}(x_2) = f_{X_2}(x_2). \tag{2.59}$$

These equations, which state that the conditional and marginal PDF's are identical for independent random variables, is also sometimes used as a definition for independent random variables. Note that equation (2.58) implies Equation (2.59)–and vice versa–as a result of the relationship of Equation (2.53).

2.3.4. The Multivariate Expectation Operator

The expectation operator, defined previously with respect to a univariate distribution described by the PDF $f_X(x)$, can now be generalized to apply in the multivariate case. Specifically, for a bivariate distribution of two random variables X_1 and X_2, the expectation operator is defined as

$$E_{X_1X_2}[g(X_1, X_2)] = \int_{-\infty}^{\infty} \int_{-\infty}^{\infty} f_{X_1X_2}(x_1, x_2) g(x_1, x_2)\, dx_1\, dx_2 \tag{2.60}$$

where the subscript notation has been used to indicate that the operation is performed with respect to the joint distribution of X_1 and X_2. In a similar manner, the expectation with respect to a conditional distribution is defined, and is termed the *conditional expectation*. For example

$$E_{X_1|X_2}[g(X_1, X_2)] = \int_{-\infty}^{\infty} f_{X_1|X_2}(x_1) g(x_1, x_2)\, dx_1 \tag{2.61}$$

is the conditional expectation of $g(X_1, X_2)$ given X_2. Equation (2.61) defines a family of expectations depending, of course, upon the specific value x_2 assumed by X_2. Together with those discussed previously, these concepts lead to some interesting relationships among conditional, marginal, and bivariate expectations. For example, if the expectation with respect to the distribution of the random variable X_2 of Equation (2.61) is taken, the result is

$$E_{X_2}[E_{X_1|X_2}[g(X_1, X_2)]] = \int_{-\infty}^{\infty} f_{X_2}(x_2) \int_{-\infty}^{\infty} f_{X_1|X_2}(x_1) g(x_1, x_2)\, dx_1\, dx_2. \tag{2.62}$$

By moving the $f_{X_2}(x_2)$ term inside the inner integral and then applying Equation (2.52), the following result is obtained:

$$\begin{aligned} E_{X_2}[E_{X_1|X_2}[g(X_1, X_2)]] &= \int_{-\infty}^{\infty} \int_{-\infty}^{\infty} f_{X_1X_2}(x_1, x_2) g(x_1, x_2)\, dx_1\, dx_2 \\ &= E_{X_1X_2}[g(X_1, X_2)]. \end{aligned} \tag{2.63}$$

This equation states that the expectation with respect to the bivariate distribution of X_1 and X_2 is the expectation, with respect to the marginal distribution of X_2, of the conditional expectation of X_1 given X_2.

Just as the expectation operator in the univariate case was used to define the useful properties of mean and variance of a distribution, so also can the expectation operator as defined for the bivariate distribution. The expectation of X_1, taken with respect to the bivariate distribution of X_1 and X_2 is

$$E_{X_1X_2}[X_1] = \int_{-\infty}^{\infty} \int_{-\infty}^{\infty} x_1 f_{X_1X_2}(x_1, x_2)\, dx_1\, dx_2. \tag{2.64}$$

Note, however, that this expression can be written as

$$E[X_1] = \int_{-\infty}^{\infty} x_1 \int_{-\infty}^{\infty} f_{X_1X_2}(x_1, x_2)\, dx_1\, dx_2 \tag{2.65}$$

where the inner integral is, from Equation (2.41), the marginal PDF $f_{X_1}(x_1)$. Then, the expected value of X_1 can be computed either from Equation (2.64) or from Equation (2.65), which is rewritten as

$$E_{X_1}[X_1] = \int_{-\infty}^{\infty} x_1 f_{X_1}(x_1)\, dx_1. \tag{2.66}$$

A similar relationship holds for the variance of X_1. Specifically,

$$\text{VAR}[X_1] = E_{X_1X_2}[(X_1 - E[X_1])^2] = E_{X_1}[(X_1 - E[X_1])^2]. \tag{2.67}$$

Note that a conditional variance can also be defined with respect to a conditional PDF as

$$\text{VAR}_{X_1|X_2}[X_1] = E_{X_1|X_2}[X_1^2] - (E_{X_1|X_2}[X_1])^2. \tag{2.68}$$

Use of this concept and the principles just developed leads to the following expression for the variance of X_1:

$$\text{VAR}[X_1] = E_{X_2}[\text{VAR}_{X_1|X_2}[X_1]] + \text{VAR}_{X_2}[E_{X_1|X_2}[X_1]]. \tag{2.69}$$

The quantity given by Equation (2.69) is that defined by Equation (2.67). It should be evident at this point that the mean and variance of a random variable are parameters associated with the distribution of that random variable, regardless of the observed values of any related random variables. When a related random variable is observed, the state of knowledge concerning the probabilities associated with the random variable of interest may be altered, and the mean and variances then change appropriately. The relationship between these conditional parameters and those relating to the basic distribution are given by expressions such as Equation (2.69) and the following expression for the mean, derived from Equation (2.63):

$$E[X_1] = E_{X_2}[E_{X_1|X_2}[X_1]]. \tag{2.70}$$

Similar expressions can be obtained for the mean and variance of X_2.

2.3.5. The Covariance of Jointly Distributed Random Variables

When a bivariate distribution is considered, a new property of the distribution is defined which yields some indication of the dependence of one variable upon the other. This property is called the *covariance* of X_1 and X_2, and is basically a cross-moment between the two:

$$\text{COV}[X_1, X_2] = E_{X_1X_2}[(X_1 - E[X_1])(X_2 - E[X_2])]. \tag{2.71}$$

The analogy to the variance operator is seen from examination of Equation (2.30), since

$$\text{COV}[X_1, X_1] = \text{VAR}[X_1]. \tag{2.72}$$

It can easily be shown that an alternative expression to Equation (2.71) is

$$\text{COV}[X_1, X_2] = E[X_1X_2] - E[X_1]E[X_2]. \tag{2.73}$$

Note that the subscripts on the expectation symbols have been dropped since the expectation of the product X_1X_2 is defined only in association with the joint distribution, while the expectations of X_1 and X_2 can be defined with respect to either the joint or marginal distributions. A commonly used quantity is the *correlation coefficient* of X_1 and X_2, represented by the symbol $\rho_{X_1X_2}$, and defined as follows:

$$\rho_{X_1X_2} = \frac{\text{COV}[X_1, X_2]}{\sqrt{\text{VAR}[X_1]\,\text{VAR}[X_2]}}. \tag{2.74}$$

This normalized representation of the covariance of X_1 and X_2 gives some indication of their dependency, without regard to the actual numbers describing the individual variances.

If the covariance of two independent random variables X_1 and X_2 is examined,

$$\begin{aligned}\text{COV}[X_1, X_2] &= E[X_1, X_2] - E[X_1]E[X_2]\\ &= \int_{-\infty}^{\infty}\int_{-\infty}^{\infty} x_1x_2 f_{X_1X_2}(x_1, x_2)\,dx_1\,dx_2\\ &\quad - \int_{-\infty}^{\infty} x_1 f_{X_1}(x_1)\,dx_1 \int_{-\infty}^{\infty} x_2 f_{X_2}(x_2)\,dx_2\end{aligned} \tag{2.75}$$

it can be seen that application of Equation (2.57) to the first integral reduces it to the same expression as the second integral, with the result,

$$\mathrm{COV}[\mathrm{X}_1, \mathrm{X}_2] = 0 \tag{2.76}$$

if X_1 and X_2 are independent random variables. (It should be noted that this is a necessary but not sufficient condition for independence). Two random variables satisfying Equation (2.76) are said to be *uncorrelated.*

A further useful result is obtained by examining the variance of the sum of two random variables X_1 and X_2,

$$\mathrm{VAR}[\mathrm{X}_1 + \mathrm{X}_2] = E[(\mathrm{X}_1 + \mathrm{X}_2)^2] - (E[\mathrm{X}_1 + \mathrm{X}_2])^2. \tag{2.77}$$

The second term on the right-hand side is equivalent to

$$(E[\mathrm{X}_1] + E[\mathrm{X}_2])^2$$

by virtue of the linearity properties of the expectation operator. By expanding the right-hand side of Equation (2.77) and collecting terms, one obtains

$$\mathrm{VAR}[\mathrm{X}_1 + \mathrm{X}_2] = \mathrm{VAR}[\mathrm{X}_1] + \mathrm{VAR}[\mathrm{X}_2] + 2\,\mathrm{COV}[\mathrm{X}_1\mathrm{X}_2]. \tag{2.78}$$

Since the covariance term may be positive or negative, the variance of the sum may be more or less than the sum of the variances. If X_1 and X_2 are uncorrelated random variables, then the covariance is necessarily zero, and Equation (2.78) reduces to

$$\mathrm{VAR}[\mathrm{X}_1 + \mathrm{X}_2] = \mathrm{VAR}[\mathrm{X}_1] + \mathrm{VAR}[\mathrm{X}_2] \tag{2.79}$$

for X_1 and X_2 independent or uncorrelated.

Finally, all of the above concepts can be generalized to the case of more than two variables. In general, the cumulative distribution function of a joint distribution of N random variables $X_1, X_2, \ldots, X_N$, is denoted by

$$F_{\mathrm{X}_1\mathrm{X}_2\ldots\mathrm{X}_N}(x_1, x_2, \ldots, x_N) = P[\mathrm{X}_1 \leqslant x_1 \cap \mathrm{X}_2 \leqslant x_2 \cap \ldots \cap \mathrm{X}_N \leqslant x_N]. \tag{2.80}$$

Notationally, considerable simplification is obtained by representing the multivariate CDF by

$$F_{\mathbf{X}}(\boldsymbol{x}) = F_{\mathrm{X}_1\mathrm{X}_2\ldots\mathrm{X}_N}(x_1, x_2, \ldots, x_N)$$

where vectors $\mathbf{X}$ and $\boldsymbol{x}$ are simply

$$\mathbf{X} = \begin{bmatrix} X_1 \\ X_2 \\ \vdots \\ X_N \end{bmatrix}, \quad \boldsymbol{x} = \begin{bmatrix} x_1 \\ x_2 \\ \vdots \\ x_N \end{bmatrix}.$$

This notation will be used exclusively for the general multivariate distribution. Similarly, the multivariate PDF is denoted by

$$f_{\mathbf{X}}(\boldsymbol{x}) = f_{X_1X_2\ldots X_N}(x_1, x_2, \ldots, x_N).$$

Marginal and conditional distributions, and the means, variances, and general expectation operations with respect to these distributions, are directly generalized to the multivariate case of more than two variables. Care must be taken, however, to recognize the existence of marginal distributions such as

$$f_{X_1X_2}(x_1, x_2) = \int_{-\infty}^{\infty} \int_{-\infty}^{\infty} f_{X_1X_2X_3X_4}(x_1, x_2, x_3, x_4)\, dx_3\, dx_4 \qquad (2.81)$$

and conditional distributions such as

$$f_{X_1X_2|X_3X_4}(x_1, x_2) \qquad (2.82)$$

and the expectation operations related to these distributions.

2.4. DISTRIBUTIONS OF FUNCTIONS OF RANDOM VARIABLES

A function of a random variable is also a random variable—for obvious reasons—and as such has its own distribution described by the appropriate CDF or PDF. A problem which arises quite often in the analysis of probabilistic systems concerns the determination of the distribution of some function of a random variable of known distribution. That is, if X is a random variable of known distribution described by

$$F_X(x) \quad \text{or} \quad f_X(x)$$

and if

$$Z = g(X) \qquad (2.83)$$

is some function of X, how can the distribution of Z, described by

$$F_Z(z) \quad \text{or} \quad f_Z(z),$$

be determined? The most basic approach to the problem is by way of the CDF, which is the fundamental descriptor of a distribution. From the basic definition of the CDF and from Equation (2.83), it can be seen that

$$F_Z(z) = P[Z \leqslant z] = P[g(X) \leqslant z]. \tag{2.84}$$

From this expression, the range of values of X for which $Z \leqslant z$ can be determined. The probability that X assumes a value within this range can then be obtained from the known distribution of X, and Equation (2.84) can be evaluated. If this determination is made for every value of Z, then the function $F_Z(z)$, and thus the distribution of Z, is defined.

2.4.1. One-to-One Transformations

When the transformation of Equation (2.83) is one-to-one and differentiable, then the inverse transformation can be defined, at least symbolically, as

$$X = X(Z) = g^{-1}(Z) \tag{2.85}$$

where $g^{-1}(Z)$ is the inverse transformation of Equation (2.83) expressing X as a function of Z. Let this equation also imply the relationship between the arguments x and z:

$$x = x(z) = g^{-1}(z).$$

From Equation (2.84), the probability that the random variable Z lies within a small interval $(z, z + \Delta z)$ is given by

$$\begin{aligned} P[z \leqslant Z \leqslant z + \Delta z] &= P[z \leqslant g(X) \leqslant z + \Delta z] \\ &= P[g^{-1}(z) \leqslant g^{-1}(Z) \leqslant g^{-1}(z + \Delta z)] \end{aligned} \tag{2.86}$$

where the last expression results from recognizing the corresponding values of X and Z for which the probability statement holds, and from using Equations (2.84) and (2.85) to describe these values. For small Δz, this expression can be written as

$$f_Z(z)\,\Delta z = f_X(g^{-1}(z))\,\Delta g^{-1}(z)$$
$$= f_X(x(z))\left|\frac{dg^{-1}}{dz}\right|\Delta z \tag{2.87}$$

where a first-order Taylor series expansion has been used to evaluate $\Delta g^{-1}(z)$. The absolute magnitude terms are required because the PDF's must be nonnegative. The Δz terms cancel, leaving the desired result.

$$f_Z(z) = f_X(x(z))\left|\frac{dg^{-1}}{dz}\right|. \tag{2.88}$$

This expression indicates that the PDF of Z is obtained by using $x(z)$ as the argument in the PDF of X, and multiplying the PDF by the term

$$\left|\frac{dg^{-1}}{dz}\right| = \left|\frac{dx}{dz}\right|.$$

In the more general case in which the transformation is not one-to-one—that is, more than one value of X can produce the same value of Z—a simple generalization of this development yields the result

$$f_Z(z) = \sum_i f_X(x(z)) \tag{2.89}$$

where the sum is over the multiple solutions of the inverse expression given by Equation (2.85).

2.4.2. Generalization to Multivariate Distributions

The generalization of the above procedure to the multivariate case involves a vector transformation

$$\mathbf{Z} = \boldsymbol{g}(\mathbf{X}) \tag{2.90}$$

where now $\mathbf{Z}$, $\mathbf{X}$, and $\boldsymbol{g}$ are all vectors. For the transformation to be one-to-one, $\mathbf{Z}$ and $\mathbf{X}$ must be of the same dimension. Then, if the inverse transformation exists,

$$\mathbf{X} = \mathbf{X}(\mathbf{Z}) = \boldsymbol{g}^{-1}(\mathbf{Z}). \tag{2.91}$$

The multivariate PDF of **Z** is given by

$$f_{\mathbf{Z}}(\mathbf{z}) = f_{\mathbf{X}}(\boldsymbol{x}(\mathbf{z})) \left| \det \frac{\partial \boldsymbol{g}^{-1}}{\partial \boldsymbol{z}} \right| \tag{2.92}$$

where the derivative $\partial \boldsymbol{g}^{-1}/\partial \boldsymbol{z}$ is a symbolic representation of the matrix

$$\frac{\partial \boldsymbol{g}^{-1}}{\partial \boldsymbol{z}} = \begin{bmatrix} \dfrac{\partial g_1^{-1}}{\partial z_1} & \dfrac{\partial g_1^{-1}}{\partial z_2} & \cdots & \dfrac{\partial g_1^{-1}}{\partial z_N} \\ \dfrac{\partial g_2^{-1}}{\partial z_1} & \dfrac{\partial g_2^{-1}}{\partial z_2} & \cdots & \dfrac{\partial g_2^{-1}}{\partial z_N} \\ \cdots & \cdots & \cdots & \cdots \\ \dfrac{\partial g_N^{-1}}{\partial z_1} & \dfrac{\partial g_N^{-1}}{\partial z_2} & \cdots & \dfrac{\partial g_N^{-1}}{\partial z_N} \end{bmatrix} \tag{2.93}$$

and g_i^{-1} is the ith element of $\boldsymbol{g}^{-1}(z)$. In the case of a linear transformation,

$$\mathbf{Z} = \boldsymbol{A}\mathbf{X} = \boldsymbol{g}(\mathbf{X}) \tag{2.94}$$

where $\boldsymbol{A}$ is an $N \times N$ nonsingular matrix, the inverse transformation is

$$\mathbf{X} = \boldsymbol{A}^{-1}\mathbf{Z} = \boldsymbol{g}^{-1}(\mathbf{Z}) \tag{2.95}$$

and, as indicated by the matrix formalism developed in Appendix A,

$$\frac{\partial \boldsymbol{g}^{-1}}{\partial \boldsymbol{z}} = \boldsymbol{A}^{-1}. \tag{2.96}$$

Thus, in this case,

$$f_{\mathbf{Z}}(\mathbf{z}) = f_{\mathbf{X}}(\boldsymbol{A}^{-1}\mathbf{z}) |\det \boldsymbol{A}^{-1}|. \tag{2.97}$$

This expression can be quite useful, since linear transformations occur often in analysis, either as simplifications of more complex relationships or as adequate representations of reality.

If Equation (2.90) does not represent a one-to-one relationship, then the multivariate generalization of the direct approach must be used, in which the range of values of transformed variables is determined from the nature of the transformation and the appropriate probabilities assigned.

2.5. THE NORMAL DISTRIBUTION

One of the most important distributions is the normal or Gaussian distribution in both its univariate and multivariate forms. The implications of the Central Limit Theorem make the normal distribution important from a physical viewpoint, and its transformation properties often offer analytical advantage. The form of the multivariate normal distribution is

$$f_{\mathbf{X}}(\boldsymbol{x}) = \frac{1}{(2\pi)^{N/2}\sqrt{\det \boldsymbol{K}}} \exp -\left\{\frac{1}{2}(\boldsymbol{x}-\boldsymbol{\mu})^T K^{-1}(\boldsymbol{x}-\boldsymbol{\mu})\right\} \tag{2.98}$$

where $\boldsymbol{K}$ is the covariance matrix and $\boldsymbol{\mu}$ is the vector of means. There are several necessary and sufficient conditions associated with this distribution, some of which are used as definitions of the distribution. In the interest of brevity, these conditions will be listed without proof.

2.5.1. Marginal Distributions

If the random variables $X_1, X_2, \ldots, X_N$ have a multivariate normal distribution, then any subset also has a multivariate normal distribution. Furthermore, the mean vector and covariance matrix of the distribution of the subset is obtained by taking the appropriate mean values and elements of the covariance matrix of the original distribution. Specifically, if the distribution of $X_1, X_2, \ldots, X_k$, where $k < N$, is desired, the mean vector and covariance matrix are obtained as $\boldsymbol{\mu}_1$ and $\boldsymbol{K}_{11}$ from the partitioned forms

$$\begin{bmatrix} \boldsymbol{\mu}_1 \\ \hline \boldsymbol{\mu}_2 \end{bmatrix}, \quad \left[\begin{array}{c|c} \boldsymbol{K}_{11} & \boldsymbol{K}_{12} \\ \hline \boldsymbol{K}_{21} & \boldsymbol{K}_{22} \end{array}\right] \tag{2.99}$$

where $\boldsymbol{\mu}_1$ is a $k \times 1$ vector representing the means of $X_1, X_2, \ldots, X_k$, and $\boldsymbol{K}_{11}$ is the $k \times k$ submatrix containing covariances of $X_1, X_2, \ldots, X_k$. Since the order of the variables is not important, Equation (2.99) is a general result.

2.5.2. Linear Operations

If the random variables $X_1, X_2, \ldots, X_N$ have a multivariate normal distribution, and if the new vector $\mathbf{Z}$ is formed by a linear operation on the vector $\mathbf{X}$, then $\mathbf{Z}$ also has a multivariate normal distribution. That is, if

$$\mathbf{Z} = \boldsymbol{A}\mathbf{X} + \boldsymbol{b} \tag{2.100}$$

then $\mathbf{Z}$ also has a multivariate normal distribution. Furthermore, the mean value and covariance matrix of **Z** are

$$\boldsymbol{\mu}_Z = A\boldsymbol{\mu}_X + \boldsymbol{b}$$

and

$$\boldsymbol{K}_Z = A\boldsymbol{K}_X A^T \tag{2.101}$$

where $\boldsymbol{\mu}_X$ and $\boldsymbol{K}_X$ are the mean value vector and covariance matrix of **X**, respectively. This result is quite useful, since it precludes the necessity of using the transformation procedures of the previous section. If the mean vector and covariance matrix of a multivariate normal distribution are known, then the PDF is specified.

2.5.3. Conditional Distributions

If the random variables $X_1, X_2, \ldots, X_k, X_{k+1}, \ldots, X_N$ have a multivariate normal distribution, with mean value vector

$$\boldsymbol{\mu} = \left[\begin{array}{c} \boldsymbol{\mu}_1 \\ \hline \boldsymbol{\mu}_2 \end{array}\right]$$

where $\boldsymbol{\mu}_1$ is the $k \times 1$ mean vector of $X_1, X_2, \ldots, X_k$ and $\boldsymbol{\mu}_2$ is the $(N - k) \times 1$ mean vector of $X_{k+1}, \ldots, X_N$, and covariance matrix

$$\left[\begin{array}{c|c} \boldsymbol{K}_{11} & \boldsymbol{K}_{12} \\ \hline \boldsymbol{K}_{21} & \boldsymbol{K}_{22} \end{array}\right]$$

where $\boldsymbol{K}_{11}$ is the $k \times k$ submatrix representing the covariances of $X_1, X_2, \ldots, X_k$, then the conditional distribution of $X_1, X_2, \ldots, X_k$ given $X_{k+1}, \ldots, X_N$ is also multivariate normal with mean vector

$$\boldsymbol{\mu}_1 + \boldsymbol{K}_{12}\boldsymbol{K}_{22}^{-1}(X_2 - \boldsymbol{\mu}_2) \tag{2.102}$$

and covariance matrix

$$\boldsymbol{K}_{11} - \boldsymbol{K}_{12}\boldsymbol{K}_{22}^{-1}\boldsymbol{K}_{12}^T \tag{2.103}$$

where the symmetry condition $\boldsymbol{K}_{21} = \boldsymbol{K}_{12}^T$ has been used. Again, since the original order of the terms is not important, this result is general.

Thus, linear operations on normally distributed variables produce other normally distributed variables, and both marginal and conditional distributions derived from multivariate normal distributions are also normal. These are extremely important properties of the normal distribution, which are quite useful in analysis.

Since each individual element of a random vector with a multivariate normal distribution has its own univariate normal distribution, its marginal probability density function is simply

$$f_X(x) = \frac{1}{\sqrt{2\pi}\,\sigma} e^{-(x-\mu)^2/2\sigma^2} \tag{2.104}$$

where σ^2 is the appropriate diagonal element of $\boldsymbol{K}$.

2.6. LIMIT THEOREMS

When initially introduced in Section 2.1, probability was defined as the number to which the relative frequency of an event tends as the number of trials becomes very large. The result of such a definition is to restrict interpretation of any of the subsequent results to those involved with limiting behavior. In this section, several theorems relating to limiting behavior are stated without proof; these theorems should be used as the basis for interpretation of all subsequent results involving probabilistic concepts.

2.6.1. The Law of Large Numbers

Previously, the mean value of a random variable was defined in terms of the expected value operator; specifically,

$$\mu = E[X] = \int_{-\infty}^{\infty} x f_X(x)\, dx. \tag{2.105}$$

It cannot be said that the mean is the value that X is most likely to assume, because it may be a physical impossibility for X to take on the specific value. For exmaple, if X is the variable obtained from the flip of a coin by assigning the number 1 to a head and the number 0 to a tail, it can be easily determined that

$$E[X] = \tfrac{1}{2}$$

which is not a possible value of X. The explanation, of course, is that the mean is a statistical parameter and has meaning only in terms of the basic premise of probability–relative frequency. The precise interpretation of the mean is given by the *Law of Large Numbers*, which is stated as follows: If $X_1, X_2, \ldots, X_N$ are independent, identically distributed random variables with common mean μ, and, if the sample mean is defined as

$$\bar{X} = \frac{1}{N}(X_1 + X_2 + \cdots + X_N) \tag{2.106}$$

then for any positive number ϵ,

$$\lim_{N \to \infty} P[|\bar{X} - \mu| \geqslant \epsilon] = 0. \tag{2.107}$$

The condition of Equation (2.107) is called *convergence in probability*, and from a practical standpoint indicates that the mean of the distribution, μ, and the sample mean, $\bar{X}$, are very close in value as N gets large.

2.6.2. The Central Limit Theorem

The *Central Limit Theorem* concerns the distribution of the sum of large numbers of independent random variables, regardless of their individual distributions. It is stated as follows: Let $X_1, X_2, \ldots, X_N$ be a sequence of independent random variables with means $\mu_1, \mu_2, \ldots, \mu_N$, and variances $\sigma_1^2, \sigma_2^2, \ldots, \sigma_N^2$. Then the random variable Z_N defined as

$$Z_N = \frac{\sum_{i=1}^{N} X_i - \sum_{i=1}^{N} \mu_i}{\sqrt{\sum_{i=1}^{N} \sigma_i^2}} \tag{2.108}$$

has a distribution which approaches a normal distribution with zero mean and unit variance as N gets large.

This amazing characteristic of the sum of random variables, which is independent of their individual distributions, is very important in justifying the use of the normal distribution as appropriate for many physical processes, since these processes are often the conglomerate effect of a large number of individual processes acting independently. This result is also quite fortunate, because the mathematical form of the normal PDF makes it amenable to analytical treatment in many cases of interest to the analyst.

2.7. STATISTICAL INFERENCE: ESTIMATION

The basic tenets of probability theory have now been reviewed, and the next logical step is to examine the way in which these principles are applied under the generic description of statistical inference. By virtue of the very nature of the broad category of problems falling within the realm of systems analysis, probabilistic concepts come into play generally in three areas: estimation, decision-making, and optimization related to random quantities. Primary interest here is in estimation, and it is important that an understanding of the basic principles involved be developed in relation to the theory of probability and random variables.

2.7.1. Point Estimation of Nonrandom Parameters

The problem of estimation arises when observations of some random quantity are obtained, from which an estimate of some other quantity is to be inferred. For the estimation process to make any sense, the quantity to be estimated must be related in some way to the random variable constituting the observation, and this can occur only in one of two ways. If the quantity to be estimated is itself random, then it must exhibit some dependency, in the statistical sense, on the observed variable. This dependency implies the existence of a joint probability distribution, which is the vehicle for the estimation process. On the other hand, if the quantity to be estimated is not random, then it must constitute a parameter of the probability distribution of the observed variable. If this were not the case, there would be no relationship between the values of the observed variable and the quantity to be estimated. Each of these two cases leads to particular estimation techniques. In this section, several techniques for estimating the parameters associated with the probability distribution of the observed variables will be introduced. First, however, certain desirable characteristics of estimates should be considered:

I. An estimate $\hat{\theta}$ of the parameter θ is said to be *unbiased* if $E[\hat{\theta}] = \theta$.

II. A sequence of estimates $\hat{\theta}_n$ is said to be *consistent* if $\hat{\theta}_n$ converges in probability to θ. That is,

$$\lim_{n\to\infty} P[|\hat{\theta}_n - \theta| \geqslant \epsilon] = 0.$$

III. Of all estimates within a given class (e.g., unbiased estimates), that possessing the smallest variance is termed the *minimum variance estimator* of the class.

These properties are often used in assessing the usefulness of a given estimation procedure.

2.7.1.1. The Method of Moments. Previous considerations have indicated that the distribution of a random variable is completely specified by the various moments. This circumstance implies that the parameters of the distribution can be

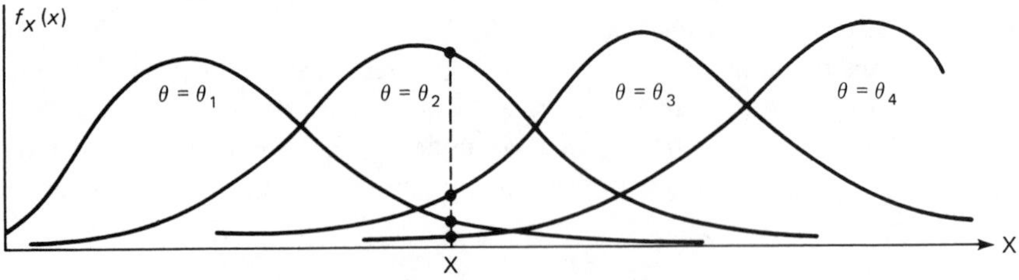

Figure 2-4. Family of probability density functions.

obtained from knowledge of the moments. In turn, this concept leads to an estimation process in which a number of sample moments are determined equal to the number of unknown parameters in the PDF of the observed variable. Sample moments are equated to the moments of the distribution, and the resulting equations solved for the desired parameters. Under certain usually met requirements, estimates derived by this method are consistent.

2.7.1.2. The Method of Maximum Likelihood. Since the probability distribution of the observed variable X is dependent upon the parameter or parameters to be estimated, there is defined a family of PDF's, with each set of possible parameters determining a member of the family. Since the concepts involved are most easily understood when there is but a single parameter, this case will be examined. The ensuing family of PDF's is

$$f_X(x, \theta), \qquad \theta = \theta_1, \theta_2, \ldots \tag{2.109}$$

where θ_i are the possible values of the parameter θ. This family of curves might be sketched as in Figure 2-4, where representative curves for θ_1, θ_2, θ_3, and θ_4 are shown. Now suppose that the random variable X has been observed, and from this observation an estimate of θ is to be formed. One possible approach is to select the value of θ which makes the observed value X most probable. That is, if the observed value of X is shown in Figure 2-4, then the distribution produced when $\theta = \theta_2$ is the distribution which yields the maximum probability that the indicated observed value would occur. This conclusion follows from the fact that the probability of a random variable lying within a region around some particular point is proportional to the height of the PDF at that point. The PDF produced when $\theta = \theta_2$ has a height at the point $x = \mathrm{X}$ which is larger than that produced by any other value of θ. Since this procedure consists of determining the value of the parameter θ which makes the actual observed value of X also the most probable value, it is called the method of maximum likelihood.

In the above development, only discrete values of θ were considered. In the general case θ may take on continuous values, and in this event the likelihood function is defined as

$$\ell(X, \theta) = f_X(X, \theta) \tag{2.110}$$

where, it should be noted, the arguments are θ and the actual observed value of X. Then, in direct analogy to the development above, the maximum likelihood estimate is obtained from the equation for maximizing $\ell(X, \theta)$,

$$\frac{\partial \ell}{\partial \theta} = 0. \tag{2.111}$$

Often the measurement consists of a sequence of observations, $\mathbf{X} = (X_1, X_2, \ldots, X_N)$ in which case the PDF of Equation (2.109) is now a joint PDF, and the likelihood function is

$$\ell(\mathbf{X}, \theta) = f_{\mathbf{X}}(\mathbf{X}, \theta) \tag{2.112}$$

which is a scalar quantity. The estimate is still obtained from solution of Equation (2.111). When the observations are independent random variables,

$$\ell(\mathbf{X}, \theta) = \prod_{i=1}^{N} f_X(X_i, \theta) \tag{2.113}$$

where the generalization of Equation (2.57) has been used.

For computational reasons, the likelihood equation of Equation (2.111) is often expressed as

$$\frac{\partial}{\partial \theta} \ln \ell = 0$$

which is permissible because the logarithm is a nondecreasing function.

If there is more than one parameter, then the above development can be repeated using a parameter vector $\boldsymbol{\theta}$ and the matrix formalism of Appendix A. Examples are developed in Chapter 7. The maximum likelihood estimate has certain desirable characteristics, in that it is a consistent estimate and is asymptotically normal; i.e., the estimate itself has a distribution which approaches a normal distribution as the number of observations grows large.

2.7.2. Estimation of a Random Variable

When the quantity to be estimated is random, then it must exhibit a joint distribution with the observed random variable. In addition, deterministic types of statements concerning the characteristics of the estimate in relation to the actual value of the quantity to be estimated cannot be made. On the other hand, the estimation process must be based upon some kind of logical reasoning, so that the normal approach is to define a reasonable loss function of the actual estimation error

$$q(\theta - \hat{\theta}). \tag{2.114}$$

This function is also a random variable, and therefore it can be used only in a probabilistic sense. Most commonly, the estimation process is chosen as that which minimizes the risk R, defined as

$$R = E[q(\theta - \hat{\theta})]. \tag{2.115}$$

Note that R is not a random quantity, but rather is a property of the distribution of q. The different estimation techniques are then derived from different ways of defining the loss function q.

2.7.2.1. Minimum Mean-Square Error Estimation. The loss functions defined above must have certain characteristics in order to be reasonable. They should be nondecreasing functions and, in most cases, should be symmetric functions. Some examples are shown in Figure 2-5. If the loss function depicted by 2-5(b) is selected, then the risk is

$$R = E[(\theta - \hat{\theta})^2] \tag{2.116}$$

and the resulting estimation process is called *minimum mean-squared error* estimation. The appropriate equations can be derived by expressing R in terms of the joint distribution of θ and X, since $\hat{\theta}$ must be a function of the observed random variable X:

$$R = E[q] = \int_{-\infty}^{\infty} \int_{-\infty}^{\infty} (\theta - \hat{\theta})^2 f_{\theta \mathrm{X}}(\theta, x)\, d\theta\, dx. \tag{2.117}$$

If $f_{\theta \mathrm{X}}(\theta, x)$ is expressed in terms of the conditional PDF,

$$f_{\theta \mathrm{X}}(\theta, x) = f_{\theta|\mathrm{X}}(\theta) f_{\mathrm{X}}(x) \tag{2.118}$$

then Equation (2.117) can be expressed as

$$R = \int_{-\infty}^{\infty} f_{\mathrm{X}}(x) \int_{\infty}^{\infty} (\theta - \hat{\theta})^2 f_{\theta|\mathrm{X}}(\theta)\, d\theta\, dx. \tag{2.119}$$

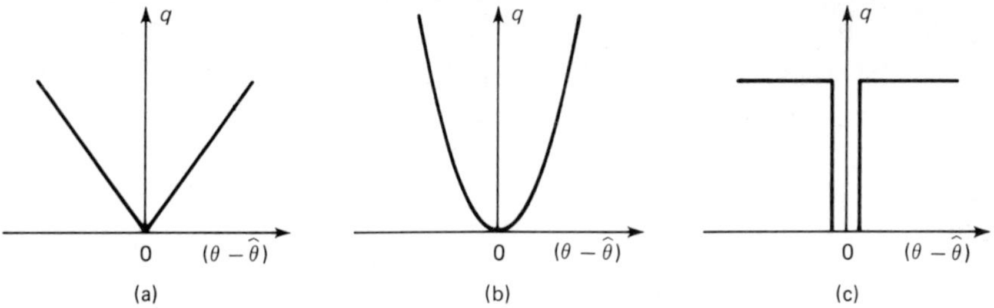

Figure 2-5. Typical loss functions.

This integral is to be minimized by proper selection of $\hat{\theta}$, which can be accomplished by minimizing only the inner integral, since $f_X(x)$, $(\theta - \hat{\theta})^2$, and $f_{\theta|X}(\theta)$ are all positive functions. The pertinent equation is

$$\frac{\partial}{\partial \hat{\theta}} \int_{-\infty}^{\infty} (\theta - \hat{\theta})^2 f_{\theta|X}(\theta)\, d\theta = 0. \tag{2.120}$$

Interchanging the operation of integration and differentiation results in

$$- \int_{\infty}^{\infty} 2(\theta - \hat{\theta}) f_{\theta|X}(\theta)\, d\theta = -2 \int_{-\infty}^{\infty} \theta f_{\theta|X}(\theta)\, d\theta + 2\hat{\theta} \int_{\infty}^{\infty} f_{\theta|X}(\theta)\, d\theta = 0. \tag{2.121}$$

Since the integral of the second term on the right-hand side represents the area under a PDF, it has a value of one, with the result

$$\hat{\theta} = \int_{-\infty}^{\infty} \theta f_{\theta|X}(\theta)\, d\theta = E_{\theta|X}[\theta] \tag{2.122}$$

stating that the minimum mean-square estimate is the mean of the conditional distribution of θ given the observation of X. Some reflection on the above process will show that a similar result would be obtained if a series of measurements $X_1, X_2, \ldots, X_N$ were obtained. The result in that case would be

$$\hat{\theta} = E_{\theta|\mathbf{X}}[\theta]. \tag{2.123}$$

Extension of these results to vector values of θ is developed in Chapter 7. This result is especially significant when θ and X have a joint distribution which is multivariate normal, since the results of Section 2.5.3 permit direct evaluation of the conditional means.

2.7.2.2. Maximum Posterior Probability Estimation. Use of the loss function shown in Figure 2-4(c) results in an estimation criterion which calls for maximization of the conditional PDF $f_{\theta|X}(\theta)$. Similarity between this method and that of maximum likelihood for parameter estimation should be obvious. For conditional distributions where the mean and the maximum of the conditional PDF coincide, as with the normal distribution, the estimate derived from this criterion is identical to the minimum mean-square error estimate.

2.8. SUBJECTIVE VERSUS OBJECTIVE PROBABILITY

Probability defined as the limiting behavior of relative frequency is usually classified as *objective probability*. Definitions of probability which reflect the degree of confidence which the observer has in the likelihood of occurrence of a particular event are classified as *subjective probability*. Subjective probability represents a personal appraisal of the nature of reality as it relates to the outcome of a single event, whereas the objective probability of a single event has no meaning. The concept of relative frequency, and hence objective probability, implies, at least in principle, that a large number of trials can be conducted. This is certainly not the case in many instances. In the example of a coin toss, however, both concepts have meaning. The objective probability of one-half relates to the percentage of tosses which would result in a head if the experiment were to be conducted a large number of times with a fair coin; the subjective probability refers to the degree of confidence an observer can put in the likelihood of obtaining a head in any given toss. The two numbers are identical; the subjective evaluation is based on past experiences in similar situations, the objective probability on the concept of a predictive judgment.

The crucial distinction between the two concepts of probability lies in the interpretation of the way in which probability applies to a single trial. In the single toss of a coin, for example, objective probability has no bearing, simply because it has meaning only in the relative frequency sense. Since subjective probabilitiy affects the confidence which the observer places in a particular outcome, it will bear directly on the outcome of a single toss. And this fact is extremely important in the consideration of decision-making when an element of uncertainty is present. If a decision is to be made a single time, then it is subjective probability which must be considered. If one were to wager on a single throw of a pair of dice, betting, say, on the occurrence of a seven, he would certainly consider the subjective probability of throwing a seven as compared to that of obtaining some other result. Common sense indicates that a probability of one-sixth of throwing a seven would require six-to-one odds in order to produce a fair situation. This rationale is based on a single throw of the dice.

From the viewpoint of objective probability the rationale is quite different, but the result is the same. If the probability of rolling a seven is one-sixth, then in a series consisting of a large number of independent throws a seven will result one-sixth of the time. Thus, if one unit were wagered on each trial, then the long-run result would be one seven for every six throws of the dice. Thus, for an even return after a large number of throws, odds of six to one would be required.

It is fortunate that objective and subjective probabilities lead to the same conclusions in those instances where they both have meaning, but it should be pointed out that there are differing opinions as to which interpretation is correct. So long as the difference is recognized, there should be no confusion as to the connotation intended.

REFERENCES

1. Anderson, T. W. *An Introduction to Multivariate Statistical Analysis.* New York: John Wiley and Sons, 1958.
2. Blum, J. R., and Rosenblatt, J. I. *Probability and Statistics.* Philadelphia: W. B. Saunders, 1972.
3. Brieman, L. *Probability and Stochastic Processes.* Boston: Houghton Mifflin, 1969.
4. Clarke, A. B., and Disney, R. L. *Probability and Random Processes for Engineers and Scientists.* New York: John Wiley and Sons, 1970.
5. Cramer, H. *The Elements of Probability Theory.* New York: John Wiley and Sons, 1955.
6. Davenport, Wilbur B., and Root, William L. *An Introduction to the Theory of Random Signals and Noise.* New York: McGraw-Hill, 1958.
7. Drake, Alvin W. *Fundamentals of Applied Probability Theory.* New York: McGraw-Hill, 1967.
8. Fagin, S. L. Recursive Linear Regression Theory, Optimal Filter Theory, and Error Analysis of Optimal Systems. *IEEE Convention Record* **12:** 216–240 (1964).
9. Feller, W. *An Introduction to Probability Theory and Its Applications.* New York: John Wiley and Sons, 1966.
10. Fry, T. C. *Probability and Its Engineering Uses.* Princeton: D. Van Nostrand, 1965.
11. Gauer, Donald P., and Thompson, Gerald L. *Programming and Probability Models in Operations Research.* Monterey, California: Brooks/Cole, 1973.
12. Morrison, Donald F. *Multivariate Statistical Methods.* New York: McGraw-Hill, 1967.
13. Naval Science Department, U.S. Naval Academy. *Naval Operations Analysis.* Annapolis: U.S. Naval Institute, 1968.
14. Papoulis, Athanasios. *Probability, Random Variables, and Stochastic Processes.* New York: McGraw-Hill, 1965.
15. Pfieffer, P. E., and Schum, P. A. *Introduction to Applied Probability.* New York: Academic Press, 1973.
16. Sage, Andrew P., and Melsa, James L. *Estimation Theory with Applications to Communications and Control.* New York: McGraw-Hill, 1971.
17. Shooman, Martin L. *Probabilistic Reliability: An Engineering Approach.* New York: McGraw-Hill, 1968.
18. Van Trees, Harry L. *Detection, Estimation, and Modulation Theory.* New York: John Wiley and Sons, 1968.

DEVELOPMENTAL EXERCISES

2.1. Derive Equation (2.4) from the basic concepts of relative frequency.

2.2. Show that the mean of the Poisson distribution is λ.

2.3. Show that the variance of the Poisson distribution is λ.

2.4. Show that the mean and variance of the uniform distribution on the interval (a, b) are $(b + a)/2$ and $(b - a)^2/12$, respectively.

2.5. Show that mean and variance of the exponential distribution with parameter λ are $1/\lambda$ and $1/\lambda^2$, respectively.

2.6. Derive Equation (2.31) from Equation (2.30).

2.7. Prove the assertions of Equation (2.52).

2.8. Prove the assertion of Equation (2.53).

2.9. Prove the assertion of Equation (2.67).

2.10. Derive Equation (2.78) from Equation (2.77).

2.11. Prove the assertion of Equation (2.73).

2.12. Prove the assertion of Equation (2.70).

2.13. Demonstrate the assertion of Section 2.5.2 using the bivariate normal distribution.

PROBLEMS

2.1. A carton contains 24 items, three of which are defective. What is the probability that, in a sample of six items chosen at random, 0, 1, 2, or 3 will be defective? Note that this computation amounts to the determination of $P_X(x)$, where X is the number of defectives chosen in six trials. (Use the hypergeometric distribution.)

2.2. Two dice are thrown 100 times. The number of fives is recorded. Find the probability function $P_X(x)$, where X is the number of fives recorded. What is the probability that at least three fives occur?

2.3. In a process that enamels electrical conductors, the average number of insulation breaks per foot is 0.02. Find the probability function $P_X(x)$, where X is a random variable representing the number of breaks in a conductor 45 feet long. The occurrence of breaks can be modeled as a Poisson distribution.

2.4. One jar contains five white marbles and fifteen red marbles. A second jar contains nine red marbles and three white marbles. A jar is selected at random and a marble is selected. It is white. What is the probability that the marble was selected from the first jar?

2.5. Two shipments of parts are received. One shipment contains 10 percent defective parts. A second shipment contains 5 percent defective parts. If each shipment contains 100 parts, and two parts are selected and determined to be good, what is the probability that the selections are made from the first shipment?

2.6. A single torpedo is to be fired at an aircraft carrier which is 1000 feet long. The angular firing errors are normally distributed with zero mean and variance of four degrees^2. If the firing range is 3000 yards, find the probability of a hit.

2.7. A jar contains two black marbles and three red marbles. A random selection is made from the jar, but its color is not revealed. The probability that it is red is $\frac{3}{5}$. Now a second selection is made and it is also red. What is the probability that the first draw resulted in a red marble?

Repeat the consideration for the second draw resulting in a black marble.

Can the result of the second draw in any physical way affect the first draw?

Does the result of the second draw affect your assessment of the probability of the first draw being red?

2.8. A tracking radar located in a radome has the following sources of random error:

Servo-noise	0.02
Receiver thermal noise	0.03
Receiver phase shift	0.015
Data unit noise	0.001

These errors are rms values (square root of the variance). Find the overall rms error, assuming that each error is independent of the others.

2.9. Three different weapons are fired at the same target. Weapons A, B, and C have probabilities of $\frac{1}{2}$, $\frac{1}{3}$, and $\frac{1}{4}$, respectively, of hitting the target. Assume independence and find the probability that:

(a) All three weapons hit the target.

(b) Exactly two weapons hit the target.

(c) The target is hit at least once.

(d) Weapon B hits the target, given that weapon A hits the target.

2.10. The performance of an inertial navigation system is evaluated by the along-track and cross-track errors. The result is the probability ellipse, a geometric figure which contains the actual position with a specified probability. When the along-track and cross-track errors are independent, identically distributed, normal random variables, the probability ellipse reduces to a circle for which the radial dimension is given by the Rayleigh distribution. The radius of the 50 percent probability circle is called the circular probable error, CEP.

(a) Find the relationship between the CEP and the error variance.

(b) Find the error variance required for a CEP of one nautical mile.

(c) Find the CEP for an error variance of 0.25 (nautical miles).^{2}

2.11. If the probability of an item being defective is 0.1, find the probability of precisely 0, 1, 2, and 3 defective items in a sample size of three.

2.12. An anti-aircraft missile has a single-shot kill probability of p. Find an expression for the kill probability of a salvo of n missiles fired at a single target, assuming that the performance of each missile is statistically independent of the performance of every other missile.

2.13. A power amplifier contains three transistors, nine resistors, and six capacitors. The percentage of initially defective transistors, resistors, and capacitors are 2, 4, and 3 percent, respectively. What is the probability that a newly manufactured amplifier has no defective components?

2.14. A production lot contains n items, k of which are defective. What is the probability that a sample of n items will include at least one defective item?

2.15. Two resistors are connected in series. Each has a resistance which is a normally distributed random variable with mean 10 and variance 0.25. If the resistances are statistically independent, what is the probability that the equivalent series resistance will exceed 21?

2.16. A hole is to be drilled in a rectangular table top. The position of the hole is determined by two rectangular coordinates x and y. The individual errors are independent, normally distributed random variables with zero mean and variance σ^2. Find the probability that the hole deviates from the desired location by more than 3σ units.

2.17. The activation point of a thermostatically controlled switch is modeled as a normally distributed random variable with unknown mean and variance. A sample is to be tested and an estimate for the mean and variance determined. How many samples are required to ensure that the sample variance will deviate from the actual variance by no more than 83 percent, with 95 percent confidence?

2.18. A measuring instrument produces questionable results for values of the measured variable lying between 1 and $\frac{4}{3}$. If a number of items is screened by this instrument, and the probability density function of the quantity to be measured is

$$f_X(x) = \begin{cases} ax^2, & 0 \leqslant x \leqslant 1 \\ \frac{2}{3}, & 1 < x \leqslant \frac{4}{3} \\ 0, & \text{otherwise} \end{cases}$$

find the average fraction of items falling in the questionable region.

2.19. The finished diameter of a precision part is modeled as a normally distributed random variable with mean 0.775 and variance 0.0001. What is the probability that the actual diameter of any given part will vary from the mean by more than 0.015?

2.20. An aircraft wing is assembled with a large number of rivets. If the number of defective rivets in any given wing assembly is approximated by a Poisson random variable X with parameter 2, find:

(a) The probability that $X \leqslant 2$.
(b) The probability that $X = 2$.
(c) The probability that $X = 0$.
(d) The probability that $X > 5$.

2.21. The probability density function of the finished diameter of a precision part is given by

$$f_X(x) = \begin{cases} (x - 0.75)/a, & 0.750 \leqslant x \leqslant 0.775 \\ (0.8 - x)/a, & 0.775 < x \leqslant 0.800, \\ 0, & \text{otherwise.} \end{cases}$$

Find the value of the probability that $X < 0.78$.

2.22. Repeat Problem 2.21 if the error in the finished problem is modeled as a normally distributed random variable with mean 0.775 and variance 0.000025.

2.23. A radioactive source is observed during five non-overlapping time intervals of six seconds each. The number of particles emitted during each period obeys a Poisson probability law with a mean rate of 0.5 particles per second. Find the probability that:

(a) In each interval three or more particles are counted.

(b) In at least one of the intervals three or more particles are counted.

2.24. If $M(N)$ is the sample mean of N independent samples of the discretely distributed random variable X, find the probability function, mean, and variance of the distribution of $M(N)$ for the following distributions of X:

(a) $P_X(x) = \mu^x e^{-\mu}/x!, \quad x = 0, 1, 2, \ldots.$

(b) $P_X(x) = p(1 - p)^{x-1}, \quad x = 1, 2, 3, \ldots.$

2.25. The random variable A has an exponential distribution with parameter λ. The value of A is to be estimated by observing the random variable X, which is the count of the number of events occurring in an experiment. If X has a Poisson distribution with parameter A, find the MMSE and MAP estimators of A.

2.26. The life of a component is modeled as a random variable with the probability density function shown in Figure P2-1. If a system contains four redundant components, find the probability that all four components are still operating at $t = 2.5$.

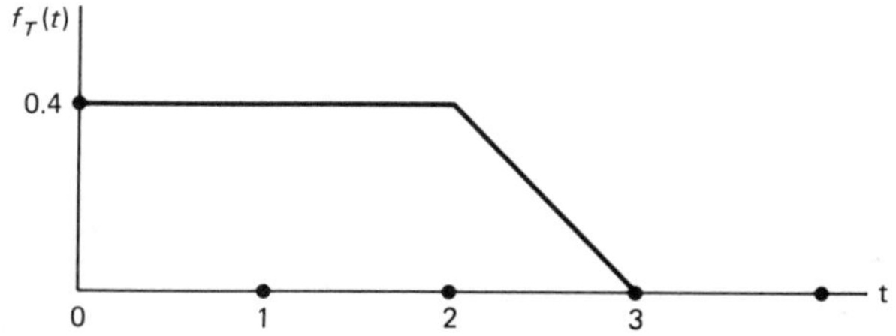

Figure P2-1.

2.27. Two communications stations are connected by parallel message channels. The transmission delay on each channel is a random variable uniformly distributed between 0 and 1 second. A message is considered received when it arrives on a channel, and verified when it arrives on the other channel.

(a) Determine the probability that a message is verified within 250 milliseconds after it is sent.

(b) Determine the probability that a message is received but not verified within 250 milliseconds after it is sent.

(c) Determine the probability that a message is verified within 250 milliseconds after it is received.

2.28. In the circuit shown in Figure P2-2 the resistance R is a random variable uniformly distributed on the interval 450 to 550 ohms. The current

$I = 0.10$ ampere and the standard resistance R is 500 ohms. Find the probability distribution associated with the calculated value of the voltage V.

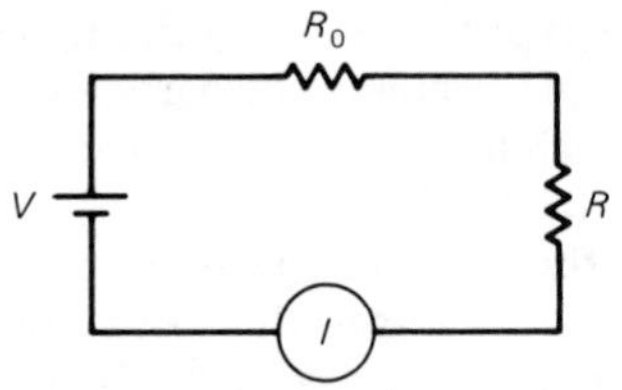

Figure P2-2.

2.29. In the circuit shown in Figure P2-3 the voltage source V is a random

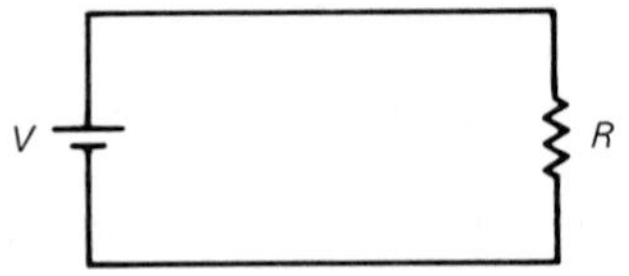

Figure P2-3.

variable uniformly distributed between 10 and 20 volts. Find the probability distributed of the power

$$P = V^2/R.$$

2.30. The noisy communication channel shown in Figure P2-4 sends a signal X

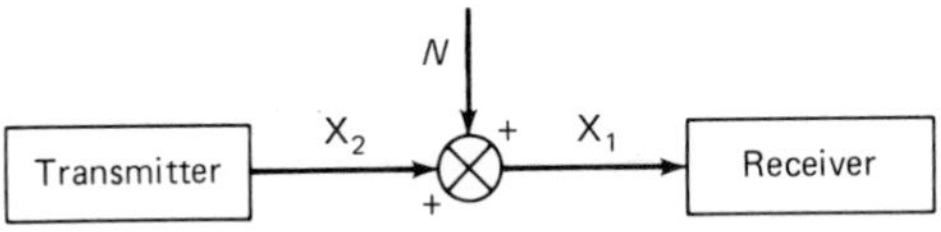

Figure P2-4.

which is a voltage uniformly distributed on the interval (0, 1). The signal is contaminated by a noise voltage N which has a normal distribution with zero mean and variance of one volt^2. Find expressions for the following probability distributions:

$$f_N(n); \quad f_{X_2}(x_2); \quad f_{X_1|X_2}(x_1); \quad f_{X_1X_2}(x_1, x_2).$$

Show that the marginal distribution of X can be expressed as

$$f_{X_1}(x_1) = \int_0^1 \frac{1}{\sqrt{2\pi}} e^{-(x_1-x_2)^2/2} \, dx_2$$

and use the table of normal probability to numerically evaluate and plot this function for the range $-2 \leqslant x \leqslant 2.5$. Show that the results indicate that $f_{X_1}(x_1)$ is a normal PDF with mean value $\frac{1}{2}$ and variance 1. By use of $f_{X_1}(x_1)$ derived above, determine the conditional probability $f_{X_1|X_2}(x_1)$. How would you suggest using this function in the receiver?

2.31. A full-wave rectifier implements the transformation

$$Y = |X|.$$

(a) If X is a random variable with probability density function $f_X(x)$, find the expression for $f_Y(y)$.
(b) If the input to a full-wave rectifier is a normally distributed random variable with zero mean and unit variance, find the probability density function of the output. Sketch this result.
(c) Give your interpretation of this result in terms of relative frequency.

2.32. A half-wave rectifier implements the transformation

$$Y = \tfrac{1}{2}(X + |X|).$$

(a) If X is a random variable with probability denisty function $f_X(x)$, find the expression for $f_Y(y)$. Dirac delta functions are permitted.
(b) If the input to a half-wave rectifier is a normally distributed random variable with zero mean and unit variance, find the probability density function of the output. Sketch the result.
(c) Give an interpretation of this result in terms of relative frequency.

2.33. X_1 and X_2 are jointly distributed normal random variables representing the amplitude of noise voltages recorded a specified time interval apart. If $\mu_1 = 1$, $\mu_2 = 2$, $\sigma_1^2 = 1$, $\sigma_2^2 = 4$, and $\rho = 0.4$, find $P[X_2 > 1|X_1 = 1]$.

2.34. A digital transmission system consists of a transmitter, a noisy channel, and a receiver. The transmitter sends a binary digit which, because of the effect of the noisy transmission channel, may not be interpreted properly at the receiver. Let the transmitted bit be the random variable X_1 and the decoded bit be the received random variable X_2. Assume the prior probability function

$$p_{X_1}(x) = \begin{cases} q, & x = 0 \\ 1 - q, & x = 1. \end{cases}$$

Also assume the conditional probabilities

$$p_{X_2|X_1}(x_2) = \begin{cases} p_0, & x = 0 \\ 1 - p_0, & x = 1 \end{cases} \quad \text{for } x_1 = 0$$

and

$$p_{X_2|X_1}(x_2) = \begin{cases} p_1, & x = 0 \\ 1 - p_1, & x = 1 \end{cases} \quad \text{for} \quad x_1 = 1.$$

(a) What is the probability that any given decoded digit is a one?
(b) Determine the probability of transmission error for both possible values of the transmitted digit.

2.35. The temperature T and the pressure P in a certain manufacturing process can be considered to be distributed according to a bivariate normal distribution $f_{TP}(t, p)$. A control system uses measurements of both T and P, but the transducers produce additive measurement noise processes which can be considered to be independent of the actual value of T and P and of one another. That is, the T and P measurements are

$$M_T = T + V_T$$

$$M_P = P + V_P$$

where V_T and V_P are the noise terms and are zero-mean, normally distributed random variables with variances σ_{VT}^2 and σ_{VP}^2. Consider the random vector

$$\mathbf{X} = \begin{bmatrix} \mathbf{X}_1 \\ \hline \mathbf{X}_2 \end{bmatrix} = \begin{bmatrix} T \\ P \\ \hline M_T \\ M_P \end{bmatrix}.$$

If

$$E[\mathbf{X}_1] = \begin{bmatrix} \mu_T \\ \mu_P \end{bmatrix}, \qquad \boldsymbol{K}_{\mathbf{X}_1} = \begin{bmatrix} \sigma_T^2 & \rho\sigma_T\sigma_P \\ \rho\sigma_T\sigma_P & \sigma_P^2 \end{bmatrix}$$

find

(a) $E[\mathbf{X}]$, the mean value vector of $\mathbf{X}$.
(b) $\boldsymbol{K}_{\mathbf{X}}$, the covariance matrix of $\mathbf{X}$.
(c) The MMSE estimate of P and T, given the measurements M_T and M_P.
(d) The limiting case when $\rho = 0$.

2.36. Consider the sums

$$Y_1 = X_1 + X_2$$

$$Y_2 = X_1 + X_2 + X_3$$

$$Y_3 = X_1 + X_2 + X_3 + X_4$$

$$Y_4 = X_1 + X_2 + X_3 + X_4 + X_5$$

where the X_i are independent random variables uniformly distributed on the interval (0, 1). Graphically represent the probability density functions

$$f_{Y_1}(y), \quad f_{Y_2}(y), \quad f_{Y_3}(y), \text{ and } f_{Y_4}(y)$$

to show how the distribution of these sums approaches a normal distribution as implied by the Central Limit Theorem. Determine the mean and variance of Y_4 and compare $f_{Y_4}(y)$ with the approximate normal distribution.

2.37. Two machines produce a certain item. The daily capacity of machine 1 is one unit and that of machine 2 is two units. Let (X_1, X_2) be the discrete random variables that represent the actual production from each machine for any given day. Each entry in Figure P2-5 represents the joint probability. That is, $P_{X_1X_2}(0, 0) = \frac{1}{8}$.

X_2 \ X_1	0	1
0	1/8	0
1	1/4	1/4
2	1/8	1/4

Figure P2-5.

Find:
(a) Marginal distributions of X_1 and X_2.
(b) The conditional distribution of X_1, given $X_2 = 2$.
(c) $E[X_1]$, $E[X_2]$, VAR$[X_1]$, VAR$[X_2]$.
(d) The distribution of $(X_1 + X_2)$.
Are X_1 and X_2 independent?

2.38. The along-range and cross-range errors in a missile guidance system are normally distributed independent random variables with zero mean. The variance of each distribution is 400 ft^2.
(a) Find the probability that the impact point lies outside a square 50 feet by 50 feet centered on the aim point.
(b) Find the probability that the impact point lies outside a rectangle 20 feet by 60 feet, centered on the aim point with the longer dimension aligned with the along-range axis.
(c) Find the probability that the impact point lies within a circle of radius 20 feet centered on the aim point.

2.39. If X_1 and X_2 have a bivariate normal distribution, show that the linear transformation

$$\mathbf{Y} = A\mathbf{X}$$

results in random variables Y_1 and Y_2 exhibiting a bivariate normal distribution. Compare the result to those of Equation (2.101).

2.40. If X_1 and X_2 are jointly distributed random variables with a probability density function $f_{\mathbf{X}}(\boldsymbol{x})$, show that the linear transformation

$$\mathbf{Y} = A\mathbf{X}$$

where

$$A = \begin{bmatrix} a_{11} & a_{12} \\ a_{21} & a_{22} \end{bmatrix}$$

produces random variables Y_1 and Y_2 with probability density function

$$f_{\mathbf{Y}}(\boldsymbol{y}) = [1/(\det A)] f_{\mathbf{X}}(a^{11}z_1 + a^{12}z_2, a^{21}z_1 + a^{22}z_2)$$

where the a^{ij} are the elements of the matrix A^{-1}.

2.41. If X_1 and X_2 are independent random variables, each exhibiting an exponential distribution with parameter λ, find the PDF of their sum.

2.42. An analysis of missile firings by sections of two aircraft equipped with two missiles each has produced the following joint probability function $p_{X_1X_2}(x_1, x_2)$, where X_1 and X_2 are the number of hits scored by the first and second aircraft, respectively:

x_2 \ x_1	0	1	2
0	$\frac{1}{9}$	$\frac{1}{18}$	$\frac{1}{6}$
1	$\frac{1}{6}$	$\frac{1}{18}$	$\frac{1}{18}$
2	$\frac{1}{18}$	$\frac{1}{6}$	$\frac{1}{6}$

$p_{X_1X_2}(x_1, x_2)$

(a) If the second aircraft scores one hit, find the probability function describing the number of hits by the first aircraft.
(b) Find the probability that $X_1 \leqslant 1$, given $X_2 = 1$.
(c) Find the probability that $X_2 = 0$, given $X_1 = 2$.
(d) Find the marginal distribution of X_1.
(e) What is the probability that $X_2 \leqslant 1$?

2.43. The random variable X is known to come from a triangular distribution with PDF as shown in Figure P2-6. The value of a is unknown but is

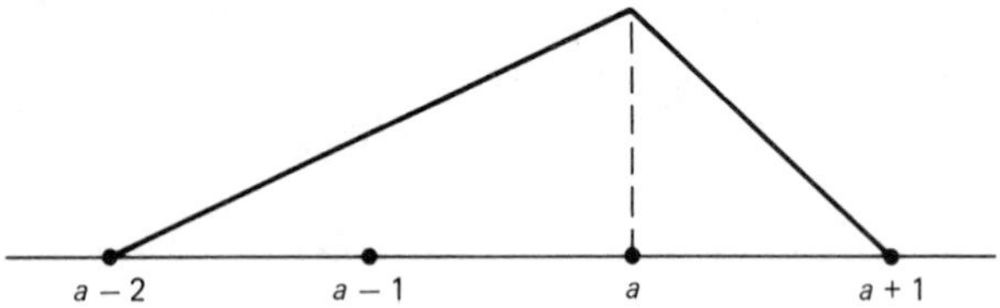

Figure P2-6.

limited to integer values. The random variable X has been observed to have the value 3.38. Find the resulting maximum likelihood estimate of a.

2.44. A discrete-time data transmission system sends a random variable A, which is uniformly distributed between 0 and 1. The transmission is contaminated by a white noise sequence $N(l)$, where the $N(l)$ are uniformly distributed between -1 and $+1$. The received sequence is

$$X(l) = A + N(l).$$

If L samples, $X(1), X(2), \ldots, X(L)$, are obtained, but the only information available is the mean

$$M = \frac{1}{L} \sum_{l=1}^{L} X(l)$$

and L is large, find:

(a) The joint distribution of the random variables M and A, $f_{MA}(m, a)$.
(b) The marginal distribution of M, $f_M(m)$.
(c) The expectation of M, $E[M]$, and the variance of M, $VAR[M]$.

Hint: Remember the Central Limit Theorem.

2.45. The time between arrivals of aircraft at an air traffic control facility is considered to be exponentially distributed with unknown parameter λ. If N samples are observed in such a manner that they are considered to be independent, find the maximum likelihood estimate of λ. That is, for the distribution

$$f_X(x) = \begin{cases} \lambda e^{-\lambda x}, & \text{for } x \geqslant 0, \\ 0, & \text{for } x < 0 \end{cases}$$

find the estimate for λ based upon the N observed values $X_1, X_2, \ldots, X_N$.

2.46. The random variable X has a normal distribution with mean μ and variance σ^2. Use the method of moments to form estimates of μ and σ^2, and show that these estimates are identical to the estimates obtained from the method of maximum likelihood.

2.47. A system contains N failure-prone subsystems, each of which has a time-to-failure modeled as an exponentially distributed random variable with parameter λ. Failure of one subsystem is independent of failure of any other subsystem.

(a) Show that the time-to-failure for the system is exponentially distributed with parameter $N\lambda$.
(b) If the time-to-failure distribution of subsystem i has parameter λ_i, find an expression for system time-to-failure.

2.48. The random variable X is uniformly distributed on the interval (a, b), where $b = a + 1$. The value of a is to be estimated from observations of N independent values of X.

(a) If 48 values of X are used, and the estimate is the sample mean, find the probability that the estimate will be within ±0.01 units of the true value.

(b) If the estimate is formed by subtracting one from the largest observed value of X, what is the probability that the estimate is within −0.02 units of the true value.

2.49. A communication system transmits a signal S over a noisy channel. The received data is given by

$$X_i = S + N_i$$

where N is a normally distributed random variable with zero mean and variance σ_N^2. If the signal is also a normal random variable, independent of N, with zero mean and variance σ_S^2, find the MMSE and MAP estimates of S, given the observed sequence $X_1, X_2, \ldots, X_L$. Find the mean-squared error.

2.50. The value of the random variable X is to be estimated from observation of a sequence of random variables Z_i, obtained as

$$Z_i = X^3 + \epsilon_i$$

where X and ϵ are independent, zero mean, normally distributed random variables with variances σ_X^2 and σ_ϵ^2. The maximum posterior probability estimate is to be used. Investigate this problem and develop an algebraic equation which can be solved for $\hat{X}$.

Chapter 3
Random Processes

3.1. THE RANDOM PROCESS

Up to this point, discussion has centered around the random event and, by a process of assigning real numbers to each possible event, the random variable. The concept of time plays no particular part in the definition of random variables, random events, or their probabilities. There are random phenomena in nature, however, which result from processes which evolve with time in accordance with some probabilistic law. Such a phenomenon, in which time plays an integral part, is termed a *random* or *stochastic process*. Examples of such random processes include Brownian motion of a particle, growth processes, various noise processes affecting radar or sonar detection, variability in the output of a manufacturing process, and radioactive decay.

To understand how a random process differs from the random variable previously discussed, consider the phenomenon of Brownian motion, in which a particle of microscopic size is immersed in a fluid and as a result is subjected to a great number of random impulses produced by collisions with the individual molecules of the fluid. The motion of the particle depends not only upon the random impulses which it receives, but also on the physical constraints imposed by Newton's Laws, a set of differential equations in which the independent variable is time. Thus, the resulting motion evolves with time, but has a definite probabilistic influence.

In inventory control, so important to a large logistics system, a problem of considerable interest is determination of when to place an order and the size of the order to be placed. Since parameters such as demand and delivery time are probabilistic in nature, the inventory at any time will be a random process depending upon the particular inventory policy and the probabilistic characteristics of demand and delivery. An extremely practical problem concerns the trade-off between the cost of maintaining a large inventory and the inability to provide the required material. Queueing theory, involving the way in which a waiting line forms as a function of the probabilistic arrivals, service time, number of servers, and serving policy, is useful in determining requirements for such things as telephone exchanges, maintenance schedules, and operating systems for large-scale computers.

In radar and sonar detection, the available observations consist of signal and noise terms, both of which may be random processes. If the characteristics of these processes are known, then certain techniques can be applied to estimate the actual signal, even though it may be masked by a large amount of noise. In target-tracking problems, the position of the target obeys Newton's Laws, but the observations are contaminated by measurement noise processes. Knowledge of the characteristics of the noise processes often permits effective estimation of the actual target track.

3.2. CHARACTERIZATION OF A RANDOM PROCESS

Taken from widely diverse fields, the above examples indicate how the random process plays a very important part in the analysis of time-dependent phenomena when probabilistic laws are involved. For this reason, a precise manner of describing a random process is required, one which will provide the necessary insight into the characteristics of such processes. Such descriptions must consider both the time dependency and the random character of the process, which in turn implies use of the fundamental concepts of random events or random variables. The idea can best be introduced by considering a particular kind of experiment, involving the concept of black boxes. Suppose that a quantity (perhaps infinite) of black boxes is involved in an experiment in the following way. Each box generates a particular time function over the interval (t_0, t_f), and the time function generated by one of the boxes will be considered to be the random process $X(t)$ as defined over this interval. The particular box, or alternatively, the time function, chosen for this distinction is determined by some sort of random selection, such as drawing lots. The situation is then as depicted in Figure 3-1, where the individual time function is labeled $X(t, s_i)$ to indicate that it is a time function generated by box number i. In this manner, the random process $X(t)$ is generated by an experiment involving a random selection, but now the outcome of the random selection is a function of time rather than a single number, as is the case when random variables are considered. This single hypothetical experiment illustrates the general concept of a random process

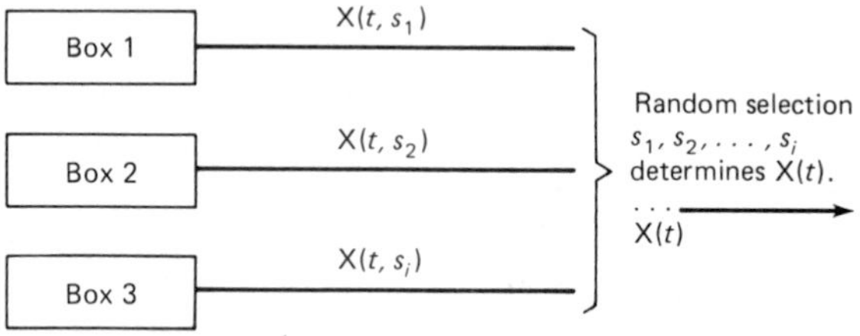

Figure 3-1. Random process experiment.

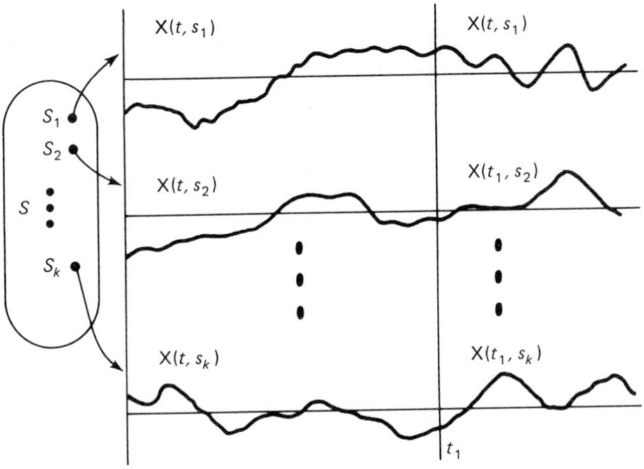

Figure 3-2. Ensemble of the random process $X(t, S)$.

quite adequately. A random process can be considered to be a time function generated by a random selection from a sample space consisting not of single points, but instead of time functions. To be very precise, the random process should be denoted as

$$X(t, S)$$

indicating that the actual time function $X(t)$ which results from the experiment is randomly selected from a sample space S of functions. The outcome of any particular experiment may be denoted as

$$X(t, s_i)$$

indicating that $X(t)$ results from selection of s_i from the sample space S. In the black box experiment discussed above, this random selection is determined by whatever method is used to pick one of the black boxes.

The more general consideration is depicted in Figure 3-2, analogous to the representation of the black box experiment of Figure 3-1. Here, the sample space S consists of the various time functions which represent possible candidates for the random process $X(t)$. Each of these time functions is termed a *sample function* or *realization* of the random process $X(t)$, and the conglomeration of these sample functions is collectively termed the *ensemble*. Each particular sample function $X(t, s_i)$ is a possible outcome of the experiment, the particular result depending upon the random selection of one of these sample functions.

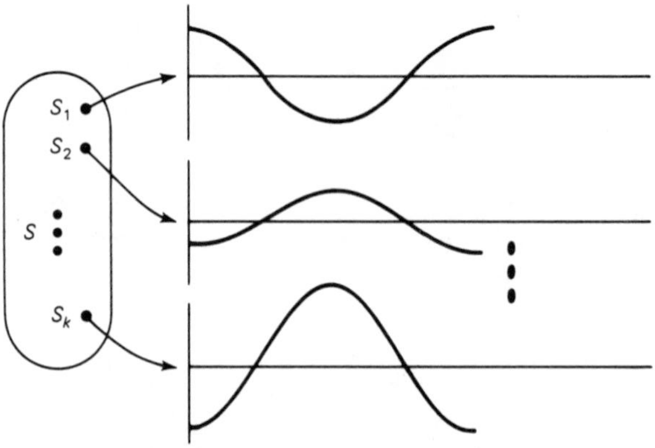

Figure 3-3. Sample space for random process $X(t) = A \cos \omega t$.

One special kind of random process for which these principles are easily envisioned is that in which the general form of the process is known. For example, consider the process

$$X(t) = A \cos \omega t \tag{3.1}$$

where A is a random variable. There is a definite time dependency in $X(t)$, but there is also a probabilistic influence represented by the random selection of A. The ensemble of sample functions consists of all functions of the form of Equation (3.1), with amplitude limited to those permitted by the distribution of A. Corresponding to Figure 3-2 is the situation shown in Figure 3-3. Here, the random selection from sample space S obviously corresponds to selection of the random variable A. Once this selection is known, the process is completely defined, so that discussion of $X(t)$ as a random process refers to what can be learned of the process prior to any knowledge of these selections.

3.2.1. Complete Characterization of a Random Process

Now that the general concept of a random process has been defined, it is necessary to consider the manner in which such a process can be characterized, using the ideas already developed in conjunction with the characterization of random variables. The procedure is rather straightforward. If the random process is observed at some particular point in time, say t_1 in Figure 3-2, the resulting value $X(t_1)$ can be construed to be a random variable since its actual value depends upon which sample function has been randomly selected to represent the process. The sample space of the random variable $X(t_1)$ consists of the values of each of the sample functions of the process, evaluated at time t_1. That is, the sample space of the random variable $X(t_1)$ is the set of values obtained by looking across the ensemble at time t_1:

$$X(t_1, s_1), \quad X(t_1, s_2), \quad X(t_1, s_3), \ldots. \tag{3.2}$$

Since $X(t_1)$ is a random variable, it is characterized by the probability density function

$$f_{X(t_1)}(x_1). \tag{3.3}$$

In a similar manner, the random variable defined by looking across the ensemble at some other time t_2 would be described by the PDF

$$f_{X(t_2)}(x_2) \tag{3.4}$$

and similarly for any other value of t.

If PDF's such as those described above could be obtained for every point in time, one might be tempted to say that the random process $X(t)$ was completely characterized, in that these PDF's represent the maximum amount of prior information which can be obtained concerning the process. Such a conclusion can easily be shown to be incorrect by recognizing that these PDF's describe marginal distributions of the individual random variables $X(t_1)$, $X(t_2)$, ... Thus, to be perfectly general in describing the distributions, one must have knowledge of the joint distribution of these random variables, which can be obtained from the marginal distributions only in the event that the random variables are independent. So, for complete characterization, it is necessary to be able to define the joint PDF

$$f_{X(t_1)X(t_2)\ldots X(t_N)}(x_1, x_2, \ldots, x_N) \tag{3.5}$$

for any set of N times, $t_1, t_2, \ldots, t_N$. The use of the vector notation introduced in Chapter 2 becomes essential here. It will sometimes be expedient to represent the joint PDF of Equation (3.5) as

$$f_{\mathbf{X}}(\mathbf{x}) = f_{X(t_1)X(t_2)\ldots X(t_N)}(x_1, x_2, \ldots, x_N). \tag{3.6}$$

In most cases, complete characterization in terms of the joint PDF is impossible to achieve, except in the event of special processes such as the Markov process to be discussed in Section 3.6. For this reason, consideration will be limited to the second-moment characterization. Although it does not contain all the information which might be obtained about a process, this characterization does permit meaningful analysis in many cases.

3.2.2. Second-Moment Characterization

If a random process is considered by looking across the ensemble at a sequence of time points, then the sequence of random variables

$$X(t_1), \quad X(t_2), \quad X(t_3), \ldots \tag{3.7}$$

is defined. If the marginal PDF's of each of these random variables is known, then the expected value of each of the random variables can be determined, resulting in the sequence of expected values

$$E[X(t_1)], \quad E[X(t_2)], \quad E[X(t_3)], \ldots. \tag{3.8}$$

Since the time points $t_1, t_2, \ldots$ are completely arbitrary, one can imagine them close enough to one another so that the continuous function

$$\mu(t) = E[X(t)] \tag{3.9}$$

is defined, representing the expected value of $X(t)$ as a function of t. This function $\mu(t)$ is termed the *mean value function* and at any point in time t_1 can be construed as

$$\mu(t_1) = \int_{-\infty}^{\infty} x f_{X(t_1)}(x)\, dx \tag{3.10}$$

where $f_{X(t_1)}(x)$ is the PDF of $X(t_1)$ as specified by Equation (3.3). The mean value function is also called the first moment of the process, and while it is useful in describing the process, it does not account for the fact that the random variable $X(t_1)$ may be–and in most cases is–dependent upon the other random variables $X(t_2)$, $X(t_3)$, etc. To consider this dependency, the second moment of the process is defined as

$$k(t_1, t_2) = \mathrm{COV}[X(t_1), X(t_2)]. \tag{3.11}$$

This function of the two arguments t_1 and t_2 is termed the *covariance kernel* of the process. It represents the covariance between the random variables $X(t_1)$–defined by looking across the ensemble at time t_1–and $X(t_2)$–defined by looking across the ensemble at time t_2. From the definition of the covariance, it is immediately determined that

$$k(t_1, t_2) = E[\{X(t_1) - \mu(t_1)\}\{X(t_2) - \mu(t_2)\}]$$
$$= E[X(t_1)X(t_2)] - \mu(t_1)\mu(t_2). \tag{3.12}$$

The first term on the right-hand side of this expression is called the *correlation function* $\rho(t_1, t_2)$ and can be obtained from the basic relationship

$$\rho(t_1, t_2) = E[X(t_1)X(t_2)]$$
$$= \int_{-\infty}^{\infty}\int_{-\infty}^{\infty} x_1 x_2 f_{X(t_1)X(t_2)}(x_1, x_2)\, dx_1\, dx_2. \tag{3.13}$$

From basic considerations, it can be seen that the covariance kernel is symmetric with respect to its argument:

$$k(t_1, t_2) = k(t_2, t_1). \tag{3.14}$$

The covariance kernel can also be shown to be a positive semidefinite function, in that

$$\int_0^T \int_0^T g(t_1)k(t_1, t_2)g(t_2)\, dt_1\, dt_2 \geqslant 0 \tag{3.15}$$

for any arbitrary function $g(t)$.

As an example of second-moment characterization, consider the simple random process defined in conjunction with Figure 3-3,

$$X(t) = A \cos \omega t$$

where ω is a constant and A is a random variable. Further assume that A is uniformly distributed on the unit interval, so that

$$f_A(a) = \begin{cases} 1, & \text{if } 0 \leqslant a \leqslant 1 \\ 0, & \text{otherwise.} \end{cases}$$

The mean value function of the random process $X(t)$ is then simply

$$\mu(t) = E[X(t)] = E[A \cos \omega t]$$
$$= E[A] \cos \omega t = \tfrac{1}{2} \cos \omega t.$$

The covariance kernel is determined as

$$\begin{aligned} k(t_1, t_2) &= E[X(t_1)X(t_2)] - \mu(t_1)\mu(t_2) \\ &= E[A^2 \cos \omega t_1 \cos \omega t_2] - (E[A]^2 \cos \omega t_1 \cos \omega t_2) \\ &= \{E[A^2] - E[A]^2\} \cos \omega t_1 \cos \omega t_2 \\ &= \text{VAR}[A] \cos \omega t_1 \cos \omega t_2 \\ &= \tfrac{1}{12} \cos \omega t_1 \cos \omega t_2. \end{aligned}$$

This combination of $\mu(t)$ and $k(t_1, t_2)$ constitutes the second-moment characterization of the random process.

As a second example, and one with a somewhat more practical character, consider the random process defined as

$$X(t) = A \cos \omega t + B \sin \omega t$$

where A and B are independent, identically distributed random variables with zero mean and variance σ^2. The mean value function of this process is

$$\mu(t) = E[X(t)] = E[A] \cos \omega t + E[B] \sin \omega t = 0.$$

The covariance kernel is

$$\begin{aligned} k(t_1, t_2) &= \text{COV}[X(t_1), X(t_2)] \\ &= E[(A \cos \omega t_1 + B \sin \omega t_1)(A \cos \omega t_2 + B \sin \omega t_2)] \\ &= \sigma^2(\cos \omega t_1 \cos \omega t_2 + \sin \omega t_1 \sin \omega t_2) \end{aligned}$$

where the fact that the random variables A and B are independent with zero means has been used to eliminate the cross product terms. By trigonometric identity, this expression can be reduced to

$$\sigma^2 \cos \omega(t_2 - t_1).$$

This particular form is an example of a wide-sense stationary process, to be discussed in the following section.

The reader should recognize that $X(t)$ could also be expressed as

$$X(t) = R \cos (\omega t + \phi)$$

where

$$R = \sqrt{A^2 + B^2}, \qquad \phi = \arctan(B/A)$$

so that this process is a sinusoid with random amplitude R and random phase ϕ. Under the stipulation that A and B are zero-mean, independent, identically distributed, normal random variables, the random variable R will exhibit a Rayleigh distribution:

$$f_R(r) = \begin{cases} \dfrac{r}{\sigma^2} e^{-(r/\sigma)^2/2} & \text{if } r \geqslant 0 \\ 0, & \text{if } r < 0 \end{cases}$$

and ϕ will be uniformly distributed on the interval $(0, 2\pi)$:

$$f_\phi(\phi) = \begin{cases} 1/2\pi, & \text{if } 0 \leqslant \phi \leqslant 2\pi \\ 0, & \text{otherwise.} \end{cases}$$

This particular random process arises often in communication theory and in the behavior of radar returns from slowly fluctuating targets.

The reader should be cautioned that these examples relate to a very special kind of random process for which an analytical expression in terms of random variables exists. The more general random process cannot be described in this way, as evidenced by the black box representation of Figure 3-1, and the functions $\mu(t)$ and $k(t_1, t_2)$ must be estimated from observation of the process, or by some other means.

3.2.3. Stationary Random Processes

The random process has both random and time-dependent characteristics. The random character of the process may in turn exhibit some kind of time dependency in that the random mechanism producing the process may itself be time-dependent. In this event, the statistical properties of the process change with time, a characteristic which must be considered a property of the general random process. In the event that such time dependency does not exist, the process is said to be *stationary*. A nonstationary process is said to be *evolutionary*.

More rigorously, a random process X(t) is said to be *strictly stationary of order* k if, for any k points $t_1, t_2, t_3, \ldots, t_k$ and any interval h, the k-dimensional random vectors

$$\begin{bmatrix} X(t_1) \\ X(t_2) \\ \vdots \\ X(t_k) \end{bmatrix} \quad \text{and} \quad \begin{bmatrix} X(t_1+h) \\ X(t_2+h) \\ \vdots \\ X(t_k+h) \end{bmatrix} \tag{3.16}$$

are identically distributed, so long as all points t_i and $t_i + h$ are included in the domain of definition of the process. This statement merely implies that a set of random variables obtained by looking across the ensemble at a given set of time points has a joint distribution dependent only upon the time intervals $(t_2 - t_1), (t_3 - t_2), \ldots, (t_k - t_{k-1})$ and not upon the values of the individual t_i themselves. If the process is strictly stationary of order k for any integer k, it is said to be *strictly stationary*.

Although the definition may seem to imply that the property of being strictly stationary is easily proven, this is not generally the case. A more useful and sometimes easily proven property is that of being *wide-sense stationary*—sometimes referred to as *weakly stationary* or *covariance stationary*. A process is said to be wide sense-stationary if its mean value function $\mu(t)$ is constant, and the covariance function $k(t_1, t_2)$ reduces to a function only of the interval $(t_2 - t_1)$. Wide-sense stationary processes are extremely important in areas such as communication theory, where much of the analysis is dependent upon the covariance function having only a single argument.

3.3. PROPERTIES OF RANDOM PROCESSES

Because of their probabilistic character, random processes must be treated differently from deterministic time functions. Operations such as differentiation and integration, for example, must be considered in a probabilistic sense, since the operations themselves produce other random processes. In this section, some of these effects will be considered, if only in a cursory manner, so that the reader will be aware of the care which must be taken in interpretation of such operations.

3.3.1. Integration and Differentiation of Random Processes

The integral of a function is normally defined as the limit of an approximating sum as the interval associated with the individual elements of the sum tends to zero. The integral of a random process is defined in the same manner, but the concept of convergence when random quantities are involved needs to be clarified. Previously, the concept of convergence in probability has been used, but for purposes of defining integration and differentiation of a random process the concept of convergence in the mean will be used. Basically, the sequence Z_n *converges in the mean* to the random variable Z if each random variable Z_n has a finite mean square and if

$$\lim_{n \to \infty} E[|Z - Z_n|^2] = 0. \tag{3.17}$$

The integral of a random process whose mean value function $\mu(t)$ and covariance function $k(t_1, t_2)$ are continuous functions of t_1 and t_2 is defined in the usual manner of the convergence of a limiting sum, where the convergence is described by Equation (3.17). The linear operations of integrating and taking expectation can then be shown to commute, with the following general results:

$$E\left[\int_a^b X(t)\, dt\right] = \int_a^b E[X(t)]\, dt = \int_a^b \mu(t)\, dt \tag{3.18}$$

$$E\left[\left|\int_a^b X(t)\, dt\right|^2\right] = E\left[\int_a^b \int_a^b X(t_1) X(t_2)\, dt_1\, dt_2\right]$$

$$= \int_a^b \int_a^b E[X(t_1) X(t_2)]\, dt_1\, dt_2 \tag{3.19}$$

$$\mathrm{VAR}\left[\int_a^b X(t)\, dt\right] = \int_a^b \int_a^b \mathrm{COV}[X(t_1), X(t_2)]\, dt_1\, dt_2$$

$$= \int_a^b \int_a^b k(t_1, t_2)\, dt_1\, dt_2 = 2 \int_a^b \int_a^{t_2} k(t_1, t_2)\, dt_1\, dt_2 \tag{3.20}$$

where the last expression results from the symmetry property of $k(t_1, t_2)$. Other useful relationships derivable from these concepts are

$$E\left[\int_a^b X(t_1)\, dt_1 \int_c^d X(t_2)\, dt_2\right] = \int_a^b \int_c^d E[X(t_1) X(t_2)]\, dt_2\, dt_1 \tag{3.21}$$

and

$$\mathrm{COV}\left[\int_a^b X(t_1)\, dt_1, \int_c^d X(t_2)\, dt_2\right] = \int_a^b \int_c^d k(t_1, t_2)\, dt_2\, dt_1. \tag{3.22}$$

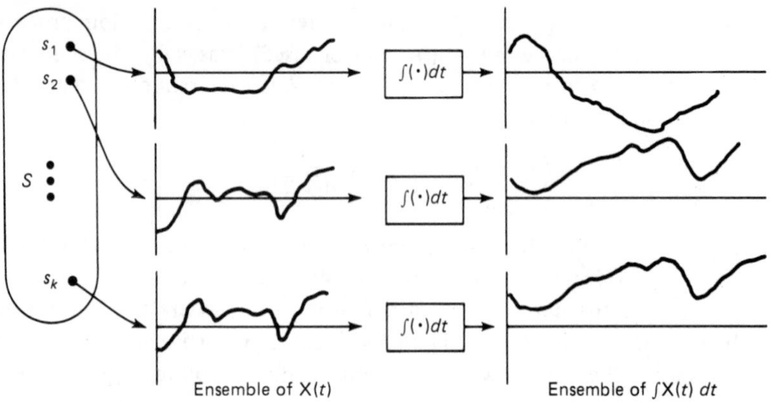

Figure 3-4. Ensemble representation of integration.

From the point of view of the ensemble description of the random process, the concept of integration can be envisioned as shown in Figure 3-4. The ensemble associated with the integral of a random process $X(t)$ is composed of the integrals of each of the sample functions of $X(t)$; thus, the random selection determining $X(t)$ also determines the value of the integral. Then Equations (3.18) through (3.22) relate the properties of the random process constituting the integral to those of $X(t)$ itself.

Differentiation is normally defined as

$$\frac{dX(t)}{dt} = \lim_{\Delta t \to 0} \frac{X(t+\Delta t) - X(t)}{\Delta t} \tag{3.23}$$

and for a random process $X(t)$ this definition holds true if convergence is again defined as in Equation (3.17). A sufficient condition for existence of the derivative with respect to convergence in the mean is that the mean value function $\mu(t)$ be differentiable, and that the mixed second derivative

$$\frac{\partial^2}{\partial t_1 \partial t_2} k(t_1, t_2) \tag{3.24}$$

exist and be continuous. Then the operations of differentiating and taking expectation can be shown to commute, with the following results:

$$E\left[\frac{dX(t)}{dt}\right] = \frac{d}{dt} E[X(t)] = \frac{d\mu(t)}{dt} \tag{3.25}$$

$$\text{COV}\left[\frac{dX(t_1)}{dt_1}, \frac{dX(t_2)}{dt_2}\right] = \frac{d}{dt_1}\frac{d}{dt_2}\text{COV}[X(t_1), X(t_2)]$$

$$= \frac{\partial^2}{\partial t_1 \partial t_2} k(t_1, t_2) \tag{3.26}$$

$$\text{COV}\left[\frac{dX(t_1)}{dt_1}, X(t_2)\right] = \frac{d}{dt_1}\text{COV}[X(t_1), X(t_2)]$$

$$= \frac{\partial}{\partial t_1} k(t_1, t_2). \tag{3.27}$$

Furthermore, if $X(t)$ is a covariance stationary process, Equation (3.26) yields

$$\text{COV}\left[\frac{dX(t_1)}{dt_1}, \frac{dX(t_2)}{dt_2}\right] = \frac{\partial^2}{\partial t_1 \partial t_2} k(t_1, t_2)$$

$$= -\frac{d^2}{d\tau^2} k(\tau) \tag{3.28}$$

where $\tau = t_2 - t_1$. Thus, the covariance kernel of the derivative process is the negative of the second derivative of the covariance kernel of the original covariance stationary process. The ensemble view of differentiation can be obtained in a manner analogous to that for integration shown in Figure 3-4.

As an example, consider the simple random process

$$X(t) = A \cos \omega t$$

where A is a random variable uniformly distributed on the unit interval. This process has been considered previously in Section 3.2.2, where it was shown that

$$\mu(t) = \tfrac{1}{2} \cos \omega t$$

$$k(t_1, t_2) = \tfrac{1}{12} \cos \omega t_1 \cos \omega t_2$$

To show that integrating and taking expectation commute, the process is integrated:

$$E\left[\int_a^b A \cos \omega t \, dt\right] = E\left[\left[\frac{A}{\omega} \sin \omega t\right]_a^b\right] = E\left[\frac{A}{\omega}(\sin \omega b - \sin \omega a)\right]$$

$$= \frac{1}{2\omega}(\sin \omega b - \sin \omega a),$$

and, from Equation (3.18),

$$E\left[\int_a^b A\cos\omega t\right] = \int_a^b \frac{1}{2}\cos\omega t\,dt = \frac{1}{2\omega}(\sin\omega b - \sin\omega a).$$

Similar relationships can be developed to illustrate the validity of Equations (3.19) through (3.22), and (3.25) through (3.28).

3.3.2. Time Averages of a Random Process

If X(t) is a wide-sense stationary random process, then from a practical standpoint one can determine by measurement the limits

$$m = \lim_{T\to\infty} \frac{1}{T}\int_0^T \mathrm{X}(t)\,dt$$

$$r(\tau) = \lim_{T\to\infty} \frac{1}{T}\int_0^T \mathrm{X}(t)\mathrm{X}(t+\tau)\,dt. \tag{3.29}$$

If X(t) were a deterministic time function, then the first equation would yield the time average, while the result of the second expression would be called the autocorrelation function. When X(t) is a random process, what do these averages really indicate? If they are obtained by observation, then what is really observed is a single sample function of the random process, and it is not at all clear that a different sample function would result in the same limiting values. The proper interpretation, of course, is to recognize that the expressions of Equation (3.29) are integrals of a random process, and as such they themselves are random variables. Only in the event that the variance of m approaches zero can the limiting value of m be considered to be a nonrandom constant. This situation occurs when virtually every sample function of X(t) produces the same limiting behavior, and if it does occur then m must surely be considered a fundamental characteristic of the process. A similar argument holds for $r(\tau)$. Just what significance such time averages have in the description of random processes, and how they can be used in a practical sense, is the subject of the following section on ergodic properties of random processes. For definitional purposes, the first expression of Equation (3.29) is termed the *time average of the random process*; the second is the *time autocorrelation function*, in contradistinction to the ensemble correlation function $\rho(t_1, t_2)$ of Equation (3.13).

3.3.3. Ergodic Theorems

In order for the theory of random processes to be useful as an analytic tool, it is necessary to determine descriptive parameters, such as the mean value function and covariance function, from observation of the random process. Normally, the physical system is such that the available observation consists of one sample function over a finite interval $(0, T)$. The mean value function and covariance kernel are ensemble properties, the determination of which would require observation of a large number of sample functions. Thus, in most practical instances, the best data available are based upon time averages rather than ensemble averages. Fortunately, in many instances the time averages formed from a record of the process observed over a finite time interval may be used as an approximation of the corresponding ensemble averages, permitting ensemble characteristics to be determined from observation of a single sample function. Such a process is said to be *ergodic*, which in essence implies that each sample function must eventually take on nearly all the modes of behavior of each of the other sample functions. To be more precise, it usually happens that one is interested only in certain characteristics of a process, such as the mean value function or the covariance kernel, and the ergodicity of the process need be considered only with respect to these characteristics. Proofs of ergodicity are beyond the scope of the present discussion, but it can be stated that the basic method is to average the appropriate time function over a finite time interval $(0, T)$, and to determine under what conditions the variance of the resulting average tends to zero as T goes to infinity. If the variance tends to zero, then the time average must tend to its expected value, the ensemble property.

For purposes of the present discussion, it need only be stressed that most processes of practical import can be considered to be ergodic, at least with respect to the characteristics of interest. Such a property is implicitly assumed for all random processes considered in what follows, except where explicitly stated to the contrary.

As an example, consider the time average m of Equation (3.29),

$$m(T) = \frac{1}{T} \int_0^T X(t)\, dt \tag{3.30}$$

written as a function of the interval length T. If the process $X(t)$ is ergodic with respect to the mean, then the variance of this random variable must approach zero as T approaches infinity. The integration must be considered in terms of convergence in the mean. For simplicity, assume the process $X(t)$ is wide-sense stationary, in which case the mean value $\mu(t)$ is constant, and the covariance function $k(t_1, t_2)$ is a function only of the time difference $(t_2 - t_1)$. Then, from the basic property concerning the commutation of the linear operations of integrating and taking expectation, it is recognized that

$$E[m(T)] = \frac{1}{T} \int_0^T \mu \, dt$$

$$= \mu \tag{3.31}$$

where μ is the constant value of $\mu(t)$. Thus, if the variance of $m(T)$ can now be shown to approach zero, then it is evident that

$$\lim_{T \to \infty} m(T) = \mu$$

and the process is ergodic with respect to the mean. It can be shown that this result occurs if and only if

$$\lim_{T \to \infty} \int_0^T k(\tau) \, d\tau = 0 \tag{3.32}$$

where

$$k(\tau) = k(t_2 - t_1)$$

is the covariance kernel of the process. Equation (3.32) then becomes the necessary and sufficient condition for ergodicity of the mean of a wide-sense stationary random process.

3.4. THE NORMAL RANDOM PROCESS

One of the most important general classes of random processes is the normal random process, arising from the concept of the multivariate normal distribution discussed in Section 2.5. It is a very specialized process, and care should be taken to realize the full impact of its definition. A random process $\mathrm{X}(t)$ is said to be a normal random process if for any set of points $t_1, t_2, \ldots, t_L$, where L is any integer, the random variables $\mathrm{X}(t_1), \mathrm{X}(t_2), \ldots, \mathrm{X}(t_L)$ defined by looking across the ensemble at the indicated points in time, have a joint multivariate normal distribution. That is, the joint probability density function is given by

$$f_{\mathbf{X}}(\mathbf{X}) = \frac{1}{(2\pi)^{L/2} \sqrt{\det \boldsymbol{K}}} \exp - \frac{1}{2} \{(\mathbf{X} - \boldsymbol{\mu})^T \boldsymbol{K}^{-1} (\mathbf{X} - \boldsymbol{\mu})\} \tag{3.33}$$

where

$$\mathbf{X} = \begin{bmatrix} X(t_1) \\ X(t_2) \\ \vdots \\ X(t_L) \end{bmatrix}, \quad \boldsymbol{x} = \begin{bmatrix} x_1 \\ x_2 \\ \vdots \\ x_L \end{bmatrix}$$

and where the mean vector $\boldsymbol{\mu}$ is obtained as

$$\boldsymbol{\mu} = \begin{bmatrix} \mu(t_1) \\ \mu(t_2) \\ \vdots \\ \mu(t_L) \end{bmatrix}.$$

The covariance matrix $\boldsymbol{K}$ is obtainable from the covariance kernel function

$$\boldsymbol{K} = \begin{bmatrix} k(t_1, t_1) & k(t_1, t_2) & \cdots & k(t_1, t_L) \\ k(t_2, t_1) & k(t_2, t_2) & \cdots & k(t_2, t_L) \\ \cdots & \cdots & \cdots & \cdots \\ k(t_L, t_1) & k(t_L, t_2) & \cdots & k(t_L, t_L) \end{bmatrix}. \tag{3.34}$$

The expressions for the mean vector and the covariance matrix result directly from the ensemble description of a random process.

From the linear transformation properties of the multivariate normal distribution, as discussed in Section 2.5.2, it is evident that linear operations on a normal random process produce other normal random processes. Since integration and differentiation are linear operations, it follows that integrals and derivatives of normal random processes are also normal random processes. The effects of nonlinear operations on normal random processes cannot be analyzed in general, but there are certain results which arise in the important area of quadratic operations. Specifically, if $X(t)$ is a normal random process with zero mean value function and covariance function $k(t_1, t_2)$, then

$$\text{VAR}[X^2(t)] = 2k^2(t, t) \tag{3.35}$$

$$\text{COV}[X^2(t_1), X^2(t_2)] = 2k^2(t_1, t_2). \tag{3.36}$$

The general result is

$$\begin{aligned}\mathrm{COV}[\mathrm{X}(t_1)\mathrm{X}(t_1+h), \mathrm{X}(t_2)\mathrm{X}(t_2+h)] &= k(t_1, t_2)k(t_1+h, t_2+h) \\ &\quad + k(t_1, t_2+h)k(t_1+h, t_2). \quad (3.37)\end{aligned}$$

The fact that quadratic operations occur often in the area of estimation and detection makes these relationships quite important, because the implications of the Central Limit Theorem make the assumption of normal processes quite reasonable in many cases.

For example, a common detection technique used in radar and sonar systems is the square-law detector, in which the received waveform $\mathrm{X}(t)$ is subjected to the nonlinear operation

$$\mathrm{Y}(t) = \mathrm{X}^2(t).$$

Because of the implications of the Central Limit Theorem, the received waveform is sometimes considered to be a zero-mean normal random process, at least when there is no signal present. Let it further be assumed that $\mathrm{X}(t)$ is wide-sense stationary. The newly defined random process $\mathrm{Y}(t)$ then has mean value function

$$\mu_{\mathrm{Y}}(t) = E[\mathrm{X}^2(t)] = k_{\mathrm{X}}(t, t) = k_{\mathrm{X}}(0)$$

where the final form results from the fact that the process is wide-sense stationary. The covariance kernel of $\mathrm{Y}(t)$ is obtained from Equation (3.36) as

$$\begin{aligned}k_{\mathrm{Y}}(t_1, t_2) &= 2k_{\mathrm{X}}^2(t_1, t_2) \\ &= 2k_{\mathrm{X}}^2(t_2 - t_1) = k_{\mathrm{Y}}(t_2 - t_1).\end{aligned}$$

Thus, $\mathrm{Y}(t)$ is also wide-sense stationary. The variance of the random variable $\mathrm{Y}(t)$ obtained by looking across the ensemble at time t is

$$\begin{aligned}\sigma_{\mathrm{Y}}^2 &= k_{\mathrm{Y}}(t, t) \\ &= 2k_{\mathrm{X}}^2(0).\end{aligned}$$

Then, at any time t, the random variable $\mathrm{Y}(t)$ has mean $k_{\mathrm{X}}(0)$ and variance $2k_{\mathrm{X}}^2(0)$, both obtainable from the statistical distribution of the received waveform. These results represent the second-moment characterization of the process $\mathrm{Y}(t)$; useful properties of the detection process can be determined by the use of the second-moment characterization.

3.5. THE RANDOM SEQUENCE

It is a fact of life that many of the continuous time processes encountered in the analysis of physical systems must be transformed to or approximated by discrete-time processes. Such a procedure is necessitated by the widespread use of digital techniques in hardware design, and by the major role that large-scale digital computer systems play in the solution of analysis problems. Just as difference equations replace differential equations to accommodate this fact, so also does the concept of a random sequence replace that of the random process. A random sequence is a collection of continuously or discretely valued random variables indexed by a discretely valued parameter. For example, the sequence

$$\mathrm{X}(0), \mathrm{X}(1), \ldots, \mathrm{X}(l), \ldots, \mathrm{X}(L) \tag{3.38}$$

represents a random sequence indexed by the discrete parameter l. To completely characterize such a random sequence it is necessary to specify the joint PDF

$$f_{\mathrm{X}(0)\mathrm{X}(1)\ldots\mathrm{X}(L)}(x_0, x_1, \ldots, x_L) \tag{3.39}$$

in a manner analogous to that considered in the study of continuous-parameter random processes. In this case, however, there are a finite (or at most a countably infinite) number of values which the index l can assume.

Partial characterizations can also be defined for random sequences, completely analogous to those defined for random processes. The most useful is the second-moment characterization using the mean-value sequence and the two-parameter covariance sequence. Thus, the mean-value sequence is defined as

$$\mu(l) = E[\mathrm{X}(l)] \tag{3.40}$$

and the covariance kernel sequence as

$$k(l, m) = \mathrm{COV}[\mathrm{X}(l), \mathrm{X}(m)]. \tag{3.41}$$

A random sequence must be considered in much the same light as a random process in terms of the ensemble representation—the mean-value sequence and covariance kernel sequence represent the mean value and covariance of random variables obtained by looking across the ensemble of the sequence at the appropriate discrete time points.

3.6. THE MARKOV PROPERTY OF RANDOM PROCESSES AND WHITE NOISE

One of the basic principles of classical physics is that of determinism, which essentially states that knowledge of the state of a system at some time t_1, and of the influences acting on the system during the interval (t_1, t_2), are sufficient to determine the state of the system at time t_2. The manner in which the system reaches the original state at time t_1 is of no consequence. An analogous principle can be stated for systems whose evolution with time depends upon probabilistic effects, in the sense that the probability that the system will be in a particular state at time t_2 is dependent upon its state at some earlier time t_1, and upon certain probabilities affecting its evolution during the interval (t_1, t_2), but not upon system behavior prior to time t_1. Random processes generated by such probabilistic systems are said to be *Markov random processes*.

3.6.1. The Continuous-Parameter Markov Processes

Continuous-parameter Markov processes have the characteristic that they are completely specified by the conditional distribution of $X(t_2)$ given $X(t_1)$ for all values of t_1 and t_2. Thus, the PDF $f_{X(t_2)|X(t_1)}(x_2)$ defined for every t_1 and t_2 in the interval of interest sufficiently describes the random process. That this statement is true can be seen from considering the Markov property implied by the above definition, which can be expressed in terms of probability density functions as

$$f_{X(t_l)|X(t_{l-1})X(t_{l-2})\ldots X(t_0)}(x_l) = f_{X(t_l)|X(t_{l-1})}(x_l). \tag{3.42}$$

This equation states that the probability distribution of the random variable $X(t_l)$, obtained by looking across the ensemble at time t_l, is dependent only on the value of $X(t_{l-1})$ and is independent of all values of $X(t)$ prior to time t_{l-1}. While this expression is presented in terms of PDF's, which might imply consideration of continuously distributed random variables only, extension to discretely distributed random variables can easily be obtained by use of the Dirac delta function. Reference to Section 2.3 will verify that the right-hand side of Equation (3.42) can be obtained from knowledge of the joint distribution $f_{X(t_l)X(t_{l-1})}(x_l, x_{l-1})$ as can the marginal distribution $f_{X(t_l)}(x_l)$. Then, by successive use of the relationship

$$\begin{aligned} &f_{X(t_l)X(t_{l-1})\ldots X(t_0)}(x_l, x_{l-1}, \ldots, x_0) \\ &\qquad = f_{X(t_l)|X(t_{l-1})\ldots X(t_0)}(x_l)\, f_{X(t_{l-1})X(t_{l-2})\ldots X(t_0)}(x_{l-1}, x_{l-2}, \ldots, x_0) \end{aligned}$$

together with Equation (3.42), it can be seen that

$$f_{X(t_l)X(t_{l-1})\dots X(t_0)}(x_l, x_{l-1}, \dots x_0)$$
$$= f_{X(t_l)|X(t_{l-1})}(x_l) f_{X(t_{l-1})|X(t_{l-2})}(x_{l-1}) \dots \times f_{X(t_1)|X(t_0)}(x_1) f_{X(t_0)}(x_0). \tag{3.43}$$

This equation states that knowledge of all the conditional PDF's and the marginal distribution of the initial value serve to provide complete characterization of the random process in the sense of Equation (3.5).

In the limiting case in which

$$f_{X(t_l)|X(t_{l-1})\dots X(t_0)}(x) = f_{X(t_l)}(x) \tag{3.44}$$

the random variable $X(t_l)$ is independent of $X(t_{l-1})$, regardless of how small the interval (t_{l-1}, t_l). Such a process is called a *white noise process*—a mathematical fiction which can never be achieved in practice because some dependency must exist when the interval (t_{l-1}, t_l) gets sufficiently small. It is a useful function, however, in the same manner as is the Dirac delta function. In fact, such a process actually implies an infinite variance, so that second-moment characterizations often use the form

$$k(t_l, t_m) = q(t_l)\,\delta(t_l - t_m) \tag{3.45}$$

which is a Dirac delta function of strength $q(t_l)$. This result indicates that the integral of the covariance kernel is finite. The value of this integral is given by $q(t_l)$, the strength of the delta function.

3.6.2. Discrete-Parameter Markov Processes

Closely analogous to the Markov random process, a *Markov random sequence* can be defined as a random sequence having the property

$$f_{X(l)|X(l-1)\dots X(0)}(x_l) = f_{X(l)|X(l-1)}(x_l) \tag{3.46}$$

where $X(l)$ is now defined only at those discrete points in time where the random sequence is defined. Such consideration also results in an expression analogous to Equation (3.43) for the complete characterization of the process. The *white noise sequence*, defined as a sequence with the characteristic

$$f_{X(l)|X(l-1)\dots X(0)}(x_l) = f_{X(l)}(x_l) \tag{3.47}$$

is a physically realizable process, as contrasted with the continuous time white noise process, because the interval between the times t_l and t_{l-1} can no longer be arbitrarily small.

Note that a random sequence generated by the first-order difference equation

$$X(l+1) = \phi(l)\,X(l) + W(l) \tag{3.48}$$

where $W(l)$ is a white noise sequence, exhibits the Markov property. For this reason, Equation (3.48) is used extensively in system modeling to generate Markov sequences. An analogous relationship holds for Markov processes, in that the differential equation

$$\frac{dX(t)}{dt} = f(t)\,X(t) + W(t) \tag{3.49}$$

generates a Markov process if $W(t)$ is a white noise process.

3.6.3. Modeling of Markov Processes

Equations (3.48) and (3.49) imply that a Markov random process can be modeled as a linear system driven by a white noise process. Because this concept is used in Chapter 7 in consideration of certain estimation techniques, it is considered further here. For the sake of clarity, the development will be limited to continuous-time processes. The model equation is

$$\dot{X}(t) = f\,X(t) + d\,W(t), \qquad f < 1 \tag{3.50}$$

where f and d are positive constants and $W(t)$ is a zero-mean white noise process with unit variance. It will be shown in Chapter 4 that the solution to Equation (3.50) can be expressed as

$$X(t) = X(t_0)\,e^{-f(t-t_0)} + \int_{t_0}^{t_f} de^{-f(t-\tau)}\,W(\tau)\,d\tau. \tag{3.51}$$

For steady-state conditions as $t_f \to \infty$, this expression reduces to

$$X(t) = \int_{t_0}^{t_f} de^{-f(t-\tau)}\,W(\tau)\,d\tau. \tag{3.52}$$

The function X(t) is obviously a random process. The mean value function is

$$E[\mathrm{X}(t)] = \int_{t_0}^{t_f} de^{-f(t-\tau)} E[\mathrm{W}(\tau)]\, d\tau = 0 \tag{3.53}$$

because W(t) is a zero-mean process. The covariance kernel of the process X(t) is

$$\begin{aligned} k(t_l, t_m) &= E[\mathrm{X}(t_l)\mathrm{X}(t_m)] \\ &= E\left[\int_{t_0}^{t_l}\int_{t_0}^{t_m} d^2 e^{-f(t_l-\tau)} e^{-f(t_m-\sigma)} \mathrm{W}(\tau)\mathrm{W}(\sigma)\, d\tau\, d\sigma\right]. \end{aligned} \tag{3.54}$$

By combining the exponential terms and interchanging the order of integration and expectation, one obtains

$$k(t_l, t_m) = \int_{t_0}^{t_l}\int_{t_0}^{t_m} d^2 e^{-f(t_l+t_m-\tau-\sigma)} E[\mathrm{W}(\tau)\mathrm{W}(\sigma)]\, d\tau\, d\sigma. \tag{3.55}$$

The expected value term within the integral is the covariance of the white noise process W(t) and, from Equation (3.45), has the value

$$E[\mathrm{W}(\tau)\mathrm{W}(\sigma)] = \delta(\tau - \sigma). \tag{3.56}$$

When the right-hand side of this expression is substituted into Equation (3.55), the integration with respect to the variable σ, because of the special properties of the delta function, yields

$$k(t_l, t_m) = \int_{0}^{t_m} d^2 e^{-f(t_l+t_m-2\tau)}\, d\tau, \qquad t_m \leqslant t_l. \tag{3.57}$$

Performing the indicated integration yields

$$k(t_l, t_m) = \frac{d^2}{2f}\,[e^{-f(t_l-t_m)} - e^{-f(t_l+t_m)}] \tag{3.58}$$

Since the assumed steady-state condition implies that t_l and t_m are large, the second term in the brackets vanishes, leaving

$$k(t_l, t_m) = \frac{d^2}{2f}\, e^{-f(t_l-t_m)}, \qquad t_m \leqslant t_l. \tag{3.59}$$

The exponential form of the covariance kernel is characteristic of Markov random processes. Also note that the covariance kernel is a function of the difference $(t_l - t_m)$, rather than a function of t_l and t_m individually, implying that the Markov process is wide-sense stationary. The variance of X(t) is given by

$$\sigma^2 = k(t, t) = d^2/2f. \tag{3.60}$$

The term $1/f$ is called the *correlation time* of the process, and it is quite common to describe a Markov process in terms of its variance and correlation time. Then, the parameters of the model of Equation (3.50) are obtained as

$$\begin{aligned} f &= 1/T \\ d &= \sqrt{2\sigma^2/T} \end{aligned} \tag{3.61}$$

where σ^2 is the variance and T the correlation time of the process.

For a discrete-time Markov sequence, the equivalent model equation is

$$X(l + 1) = \phi X(l) + \Delta W(l), \qquad \phi < 1 \tag{3.62}$$

where ϕ and Δ are positive constants and W(l) is a zero-mean white noise sequence with unit variance. The results analogous to Equations (3.59) and (3.60) are

$$\begin{aligned} k(l, m) &= \frac{\Delta^2}{1 - \phi^2} \phi^{(m-l)} \\ \sigma^2 &= k(l, l) = \Delta^2/(1 - \phi^2). \end{aligned} \tag{3.63}$$

Modeling a Markov random sequence with a specified variance σ^2 then requires

$$\Delta = \sqrt{\sigma^2(1 - \phi^2)}. \tag{3.64}$$

These models are useful in modeling noise processes associated with estimation and control of linear systems. These concepts are discussed in Chapter 7.

3.6.4. Relationships Between White Noise Processes and White Noise Sequences

Special consideration is required when relating the white noise sequence to the white noise process by the concept of vanishingly small time intervals. This procedure is used in later arguments, so it requires comment here. The continuous time, white noise process is a mathematical fiction which is useful only in describing the properties of integrals of such processes. That is, any time a white noise process is specified in analysis, that process is eventually integrated, and it is the characteristics of the integral which are of interest. The problem, then, is to relate the integral of a continuous time, white noise process $W(t)$ to the summation of a discrete time, white noise sequence $W(l)$. Without loss of generality, these processes can be assumed to have zero mean. The covariance kernels are

$$k(t, \sigma) = q_C(t)\delta(t - \sigma); \quad k(l, m) = \begin{cases} q_D(l), & \text{if } l = m \\ 0, & \text{if } l \neq m. \end{cases} \tag{3.65}$$

The integral of the continuous time process and the summation of the discrete time process are

$$Y(l) = \sum_{i=0}^{l} W(i) \quad \text{and} \quad Y(t_l) = \int_0^{t_l} W(t)\, dt \tag{3.66}$$

where $t_l = l\Delta t$, and a uniform Δt is assumed for the sequence $W(l)$.

The quantities of interest are the covariance kernels of the integral and the summation, which can be expressed as

$$k(t_l, t_m) = E[Y(t_l)Y(t_m)] = E\left[\int_0^{t_l} W(t)\, dt \int_0^{t_m} W(\sigma)\, d\sigma\right]$$

$$k(l, m) = E[Y(l)Y(m)] = E\left[\sum_{i=0}^{l} W(i) \sum_{i=0}^{m} W(j)\right]. \tag{3.67}$$

Since the expectation operation commutes with those of integration and summation, these expressions can be written as

$$k(t_l, t_m) = \int_0^{t_l}\int_0^{t_m} E[W(t)W(\sigma)]\, d\sigma\, dt = \int_0^{t_l}\int_0^{t_m} k(t, \sigma)\, d\sigma\, dt$$

$$k(l, m) = \sum_{l=0}^{l}\sum_{j=0}^{m} E[W(i)W(j)] = \sum_{i=0}^{l}\sum_{j=0}^{m} k(i, j). \tag{3.68}$$

Now, because of the white noise characteristics described by Equation (3.65), these expressions can be reduced to

$$k(t_l, t_m) = \int_0^{t_n} q_c(t)\,dt, \qquad k(l, m) = \sum_{i=0}^{n} q_D(i) \tag{3.69}$$

where t_n is the smaller of t_l and t_m, and n is the lesser of l and m. At this point, let the integral term be approximated by a summation using the time interval Δt defined in Equation (3.66),

$$k(t_l, t_m) \approx \sum_{i=0}^{n} q_c(i)\,\Delta t. \tag{3.70}$$

By comparing this expression with that of $k(l, m)$ in Equation (3.69), it is apparent that

$$q_c(l)\,\Delta t = q_D(l). \tag{3.71}$$

This relationship must be used in relating the strength of the Dirac delta function used in describing the covariance kernel of the continuous time, white noise process to the covariance kernel of the equivalent discrete time, white noise sequence. This development can also be used as a heuristic argument for the use of the Dirac delta function in describing the covariance kernel of the continuous time process, for only with such a function can the summation and integral be equated.

3.6.5. General Categories of Markov Processes

The general Markov process can be divided into four basic categories, depending upon the properties of the random variables and the parameter involved in their definition. These four categories are as follows:

I. Continuous index, continuous variable–continuous parameter Markov process.
II. Continuous index, discrete variable–continuous parameter Markov chain.
III. Discrete index, continuous variable–discrete parameter Markov process.
IV. Discrete index, discrete variable–discrete parameter Markov chain.

Markov processes, both continuous time processes and sequences, play an important role in the analysis of physical systems subject to random influence.

3.7. THE VECTOR-VALUED RANDOM PROCESS

The occasion will arise in which simultaneous consideration of more than one random process or random sequence is required. Such situations occur often in the analysis of multivariate systems, where vector-matrix notation is almost essential to the efficient management of the resulting equations. Therefore, it is convenient to define the vector random process $\mathbf{X}(t)$,

$$\mathbf{X}(t) = \begin{bmatrix} X_1(t) \\ X_2(t) \\ \vdots \\ X_N(t) \end{bmatrix}. \tag{3.72}$$

Since the various component processes comprising the vector of Equation (3.72) may not be independent or even uncorrelated, a general method of describing the joint distributions—at least for second-moment characterizations—will be required. Thus, in analogy to the scalar random process, the vector mean value function is defined as

$$\boldsymbol{\mu}(t) = E[\mathbf{X}(t)] = \begin{bmatrix} E[X_1(t)] \\ E[X_2(t)] \\ \vdots \\ E[X_N(t)] \end{bmatrix} = \begin{bmatrix} \mu_1(t) \\ \mu_2(t) \\ \vdots \\ \mu_N(t) \end{bmatrix}. \tag{3.73}$$

To describe the covariances between all the individual random variables, obtained by looking across the ensembles of the N random processes, the covariance kernel matrix function $K(t_l, t_m)$ is defined as

$$\boldsymbol{K}(t_l, t_m)$$

$$= \begin{bmatrix} \mathrm{COV}[X_1(t_l), X_1(t_m)] & \mathrm{COV}[X_1(t_l), X_2(t_m)] & \cdots & \mathrm{COV}[X_1(t_l), X_N(t_m)] \\ \mathrm{COV}[X_2(t_l), X_1(t_m)] & \mathrm{COV}[X_2(t_l), X_2(t_m)] & \cdots & \mathrm{COV}[X_2(t_l), X_N(t_m)] \\ \cdots & \cdots & \cdots & \cdots \\ \mathrm{COV}[X_N(t_l), X_1(t_m)] & \mathrm{COV}[X_N(t_l), X_2(t_m)] & \cdots & \mathrm{COV}[X_N(t_l), X_N(t_m)] \end{bmatrix}$$

$$= \begin{bmatrix} k_{11}(t_l, t_m) & k_{12}(t_l, t_m) & \cdots & k_{1N}(t_l, t_m) \\ k_{21}(t_l, t_m) & k_{22}(t_l, t_m) & \cdots & k_{2N}(t_l, t_m) \\ \cdots & \cdots & \cdots & \cdots \\ k_{N1}(t_l, t_m) & k_{N2}(t_l, t_m) & \cdots & k_{NN}(t_l, t_m) \end{bmatrix}. \tag{3.74}$$

The general term in this matrix is the covariance between $X_i(t_l)$ and $X_j(t_m)$:

$$k_{ij}(t_l, t_m) = \text{COV}[X_i(t_l), X_j(t_m)] \tag{3.75}$$

and is merely a generalization of the covariance kernel function of Equation (3.11). In fact, the diagonal terms of the matrix $\boldsymbol{K}(t_l, t_m)$, i.e., $k_{ii}(t_l, t_m)$ are the covariance kernels of the individual components of $\mathbf{X}(t)$.

Note that the covariance kernel matrix function of Equation (3.74) is not a symmetric matrix. It does, however, possess a certain kind of symmetry, in that

$$\boldsymbol{K}(t_l, t_m) = \boldsymbol{K}^T(t_m, t_l). \tag{3.76}$$

When $t_l = t_m = t$, the covariance matrix $P(t)$ is defined as

$$\boldsymbol{P}(t) = \boldsymbol{K}(t, t) \tag{3.77}$$

The general element of $\boldsymbol{P}(t)$ is $k_{ij}(t, t)$, which is the covariance between the ith and the jth element of $\mathbf{X}(t)$, both considered at the same time t. The diagonal elements of $\boldsymbol{P}(t)$ are terms like

$$k_{ii}(t, t) = \text{VAR}[X_i(t)] \tag{3.78}$$

which is the variance of the random variable obtained by looking across the ensemble of the random process $X_i(t)$ at some specified time t.

By the same reasoning, the vector-valued random sequence is defined as the sequence of vectors

$$\mathbf{X}(0), \quad \mathbf{X}(1), \quad \mathbf{X}(2), \quad \ldots, \quad \mathbf{X}(l), \ldots. \tag{3.79}$$

The vector mean-value sequence is

$$\boldsymbol{\mu}(l) = E[\mathbf{X}(l)] = \begin{bmatrix} E[X_1(l)] \\ E[X_2(l)] \\ \vdots \\ E[X_N(l)] \end{bmatrix} = \begin{bmatrix} \mu_1(l) \\ \mu_2(l) \\ \vdots \\ \mu_N(l) \end{bmatrix} \tag{3.80}$$

and the covariance matrix sequence is defined as

$$\boldsymbol{K}[l,m] = \begin{bmatrix} \mathrm{COV}[\mathrm{X}_1(l),\mathrm{X}_1(m)] & \mathrm{COV}[\mathrm{X}_1(l),\mathrm{X}_2(m)] & \cdots & \mathrm{COV}[\mathrm{X}_1(l),\mathrm{X}_N(m)] \\ \mathrm{COV}[\mathrm{X}_2(l),\mathrm{X}_1(m)] & \mathrm{COV}[\mathrm{X}_2(l),\mathrm{X}_2(m)] & \cdots & \mathrm{COV}[\mathrm{X}_2(l),\mathrm{X}_N(m)] \\ \cdots & \cdots & \cdots & \cdots \\ \mathrm{COV}[\mathrm{X}_N(l),\mathrm{X}_1(m)] & \mathrm{COV}[\mathrm{X}_N(l),\mathrm{X}_2(m)] & \cdots & \mathrm{COV}[\mathrm{X}_N(l),\mathrm{X}_N(m)] \end{bmatrix}$$

$$= \begin{bmatrix} k_{11}(l,m) & k_{12}(l,m) & \cdots & k_{1N}(l,m) \\ k_{21}(l,m) & k_{22}(l,m) & \cdots & k_{2N}(l,m) \\ \cdots & \cdots & \cdots & \cdots \\ k_{N1}(l,m) & k_{N2}(l,m) & \cdots & k_{NN}(l,m) \end{bmatrix}. \qquad (3.81)$$

Again, the covariance matrix sequence is defined as

$$\boldsymbol{P}(l) = \boldsymbol{K}[l,l] \qquad (3.82)$$

and represents the covariances among the various components of $\mathbf{X}(l)$ at a common time.

A white noise vector sequence is defined as one for which the covariance matrix sequence is

$$\boldsymbol{K}[l,m] = \begin{cases} \boldsymbol{P}(l), & \text{if } l = m \\ 0, & \text{if } l \neq m. \end{cases} \qquad (3.83)$$

That is, the random variable obtained by looking across the ensemble of X_i at time l and the random variable obtained by looking across the ensemble of X_j at time m are uncorrelated, except when $l = m$, in which case the covariances are given by the elements of the matrix $\boldsymbol{P}(l)$.

In a similar manner, a vector-valued white noise process is one for which the covariance matrix function is

$$\boldsymbol{K}(t_l, t_m) = \boldsymbol{Q}(t_l)\,\delta(t_l - t_m). \qquad (3.84)$$

where $\delta(t_l - t_m)$ is a Dirac delta function. Again, the covariance between any two random variables defined by the individual random process is zero unless $t_l = t_m$, in which case the covariances are Dirac delta functions with strengths given by the elements of the matrix $\boldsymbol{Q}(t_l)$.

Examples of the use of the vector-matrix representation of multivariate random processes must wait until the introduction of vector–matrix terminology in the analysis of multivariate dynamic systems subject to random influences. This topic is discussed in some detail in later chapters.

REFERENCES

1. Brieman, L. *Probability and Stochastic Processes.* Boston: Houghton Mifflin, 1969.
2. Clarke, A. B., and Disney, R. L. *Probability and Random Processes for Engineers and Scientists.* New York: John Wiley and Sons, 1970.
3. Davenport, Wilbur B., and Root, William R. *An Introduction to the Theory of Random Signals and Noise.* New York: McGraw-Hill, 1958.
4. Doob, J. L. *Stochastic Processes.* New York: John Wiley and Sons, 1953.
5. Larson, Harold J., and Shubert, Bruno O. *Probabilistic Methods in Engineering Sciences.* New York: John Wiley and Sons, 1979.
6. Papoulis, Athanasios. *Probability, Random Variables, and Stochastic Processes.* New York: McGraw-Hill, 1965.
7. Parzen, Emanual. *Stochastic Processes.* San Francisco: Holden-Day, 1962.
8. Shooman, Martin L. *Probabilistic Reliability: An Engineering Approach.* New York: McGraw-Hill, 1968.

DEVELOPMENTAL EXERCISES

3.1. Develop the Rayleigh PDF from the rectangular to polar coordinate transformation in the example of Section 3.2.2.

3.2. Find the necessary and sufficient condition for the sample mean of a random process to be ergodic.

3.3. Prove the assertion of Equation (3.37).

3.4. Justify the statement relating to Equation (3.48).

3.5. Consider the scalar Markov process

$$\dot{X}(t) = -f X(t) + W(t)$$

and its discrete-time representation

$$X(l) = \phi X(l-1) + W(l).$$

Use the sequence

$$X(1) = X(0) + W(0)$$
$$X(2) = X(1) + W(1)$$
$$\vdots$$
$$X(l) = X(l-1) + W(l-1)$$

to show that the general form of the covariance kernel of a stationary

Markov sequence is

$$k(l, m) = k_0 e^{-|l-m|/\tau}.$$

The value of τ is called the correlation time of the sequence.

3.6. Generalize the concept of a Markov process to define a second order Markov process, for which

$$f_{X(t_l)|X(t_{l-1})\ldots X(t_0)}(x_l) = f_{X(t_l)|X(t_{l-1})X(t_{l-2})}(x_l).$$

Show that a second-order Markov process is produced by a second-order linear system driven by white noise.

3.7. Determine the covariance kernel of a scalar stationary Markov process.

3.8. Markov processes are often used in simulation to model system noise. Determine the linear system used to generate a Markov process with mean value function $\mu(t) = \mu$ and covariance kernel $k(t, \sigma) = ke^{-|t-\sigma|/\tau}$.

3.9. A Gauss–Markov process is defined as a normal random process which exhibits the Markov property. Show that a process may exhibit both Markov and Gaussian properties.

PROBLEMS

3.1. Consider the random process generated by the expression

$$X(t) = V \sin t$$

where V is a random variable uniformly distributed between -1 and $+1$.

(a) Sketch a representation of the ensemble of sample functions.
(b) Find the mean value function $\mu(t)$.
(c) Find the covariance kernel $k(t_1, t_2)$.
(d) Find the covariance $P(t)$.

3.2. The random process $X(t)$ is defined by the differential equation

$$\dot{X}(t) = -aX(t) + W$$

where a is a positive constant and W is a zero-mean random variable which has variance σ_W^2. If $X(0) = 0$, find an expression describing the ensemble of sample functions of $X(t)$, the mean value function $\mu(t)$, the covariance kernel $k(t_1, t_2)$, and the covariance $P(t)$. Determine an approximate expression for $k(t_1, t_2)$ as t_1 and t_2 grow large.

3.3. Find the mean value function and covariance kernel of the process defined as

$$X(t) = \sum_{i=1}^{N} (A_i \cos \omega_i t + B_i \sin \omega_i t)$$

where the A_i and B_i are uncorrelated, zero-mean random variables with variances σ_i^2. Comment on the implication of this result with respect to Fourier analysis.

3.4. A particle of unit mass, constrained to rectilinear motion along the x-axis, leaves the origin at time $t = 0$, with a velocity which is a zero-mean Gaussian random variable with variance σ_0^2.

(a) Determine the means and variances of its position and velocity coordinates as functions of t. What is the PDF for the particle's position when $t = 10$? Is the two-dimensional random process (the position and velocity) a Gauss–Markov process?

(b) Suppose that, in addition to the conditions above, the particle is subject to an accelerating force which can be modeled as a zero-mean random variable which is independent of the initial velocity and has variance σ_A^2. Show that the two-dimensional process is Gauss–Markov and determine its covariance matrix as a function of time. Also determine the covariance kernel matrix.

3.5. Consider the discrete time vector stochastic sequence $X(l)$, for $l = 0, 1, 2, \ldots$, defined by

$$X(l + 1) = -\frac{1}{l + 1} X(l)$$

where $X(0)$ is a random N-vector with mean-value function $\mu_X = 0$ and covariance matrix $\boldsymbol{P}(0)$. Determine the mean value function, covariance kernel matrix, and covariance matrix for the process.

3.6. Consider the Markov sequence generated by keeping the cumulative score of the wheel-spinning game, for which

$$X(l + 1) = X(l) + W(l); \qquad X(0) = 0$$

where $X(l)$ is the cumulative score after the lth spin and $W(l)$ is a white noise sequence with uniform distribution between 0 and N.

(a) Find the mean value sequence and the covariance sequence $k(l, m)$ of $X(l)$.

(b) Show that the probability distribution of $X(l)$ approaches a normal distribution as l increases.

3.7. Consider a particle moving along one linear dimension in response to collisions with other particles. Let $N(t)$ be the number of collisions from the initial time until time t, and model $N(t)$ as a Poisson process with mean rate λ. Each time the particle suffers a collision it reverses its velocity. Thus, the velocity is either $\pm v$. If $V(t)$ denotes the velocity of the particle at time t, and the initial velocity $V(0)$ is also either $\pm v$, then

$$V(t) = V(0)(-1)^{N(t)}.$$

This form of random process is termed a *random telegraph signal.*

(a) Find the mean square displacement $E[|X(t)|^2]$, where

$$X(t) = \int_0^t V(\sigma)\, d\sigma.$$

(b) Find approximations for the mean square displacement for small t and for large t.

3.8. The discrete time process $X(l)$ is determined by the difference equation

$$X(l) = \alpha X(l-1) + \alpha^2 X(l-2) + W(l)$$

where $W(l)$ is a zero-mean normal random sequence with covariance kernel

$$k_W(l, m) = \begin{cases} \sigma_W^2, & \text{if } l = m \\ 0, & \text{if } l \neq m. \end{cases}$$

Find the mean value sequence and covariance kernel sequence of $X(l)$, assuming that $X(0) = 1$.

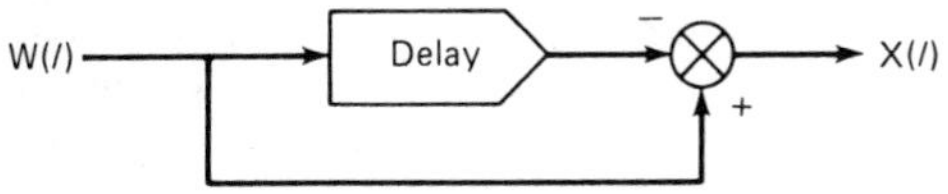

Figure P3-1.

3.9. Consider the system shown in Figure P3-1, which forms the first-order difference of the input sequence $W(l)$, that is,

$$X(l) = W(l) - W(l-1).$$

If the input sequence has zero mean value function and covariance kernel

$$k_W(l, m) = \sigma^2 e^{-\lambda(l-m)^2}$$

find the mean value function and covariance kernel of the output sequence $X(l)$.

3.10. Find the mean value function and covariance kernel of the random process described as

$$X(t) = \cos(\omega t + \phi)$$

where ϕ is a random variable uniformly distributed on the interval $(0, 2\pi)$, and ω is a random variable with PDF

$$f_\omega(\omega) = \begin{cases} \dfrac{1}{\pi(1-\omega^2)}, & \text{if } \omega > 0 \\ 0, & \text{otherwise.} \end{cases}$$

3.11. Consider the random process $X(t)$ defined as

$$X(t) = Ae^{i\omega t}$$

where A is a normally distributed random variable with mean μ and variance σ^2. Find the mean value function and covariance kernel of the process.

3.12. Find the mean value function and covariance kernel of the random process defined as

$$X(t) = A + Bt + Ct^2$$

where A, B and C are independent, normally distributed random variables with mean μ and variance σ^2.

3.13. Find the mean value function and covariance kernel of the random process defined as

$$X(t) = r \cos(\omega t + \phi)$$

where r and ω are constants and ϕ is a random variable uniformly distributed on the interval $(0, 2\pi)$.

3.14. A channel contains a series flow of items, each of fixed length l traveling at fixed velocity v. The separation between successive objects is some integral number N_l, where N is a random variable modeled by the probability function

$$P_N(n) = p(1-p)^n, \qquad n = 1, 2, \ldots$$
$$0 < p < 1.$$

(a) Find the average flow rate of the channel.
(b) At a given point, determine what fraction of gap time is occupied by gaps greater than $2l$.

3.15. Consider the scalar discrete time system described by the equation

$$X(l) = 0.9X(l-1) + 0.1Z, \qquad X(0) = 1$$

where Z is a random variable with zero mean and unit variance. Find the mean value function and covariance kernel of the random sequence $X(l)$.

3.16. A random process $X(t)$ is the increment of another random process $Y(t)$. That is,

$$X(t) = Y(t+h) - Y(t).$$

If $Y(t)$ has the covariance kernel $k_Y(t, \sigma)$, show that $X(t)$ has the covariance kernel

$$k_X(t, \sigma) = k_Y(t+h, \sigma+h) - k_Y(t+h, \sigma) - k_Y(t, \sigma+h) + k_X(t, \sigma).$$

3.17. Consider the linear oscillator system shown in Figure P3-2. The states of the system are $X_1(t)$ and $X_2(t)$, or

$$\mathbf{X}(t) = \begin{bmatrix} X_1(t) \\ X_2(t) \end{bmatrix}.$$

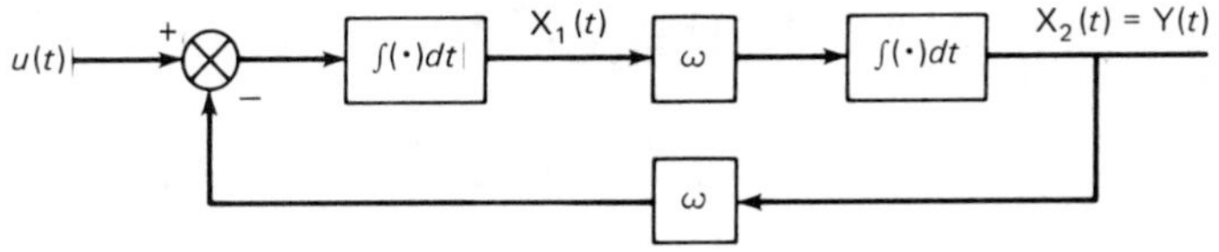

Figure P3-2.

The input $u(t)$ is zero for all t, and the initial conditions on the integrators are

$$X_1(0) = A; \qquad X_2(0) = B$$

where A and B are normally distributed random variables with zero mean and variances σ_A^2 and σ_B^2, respectively, and covariance $\rho\sigma_A\sigma_B$. Then X(t) is a vector-valued random process.

(a) If $\rho = 0$, find the mean value function $\mu(t)$.
(b) If $\rho = 0$, find the covariance matrix kernel $\boldsymbol{K}(t_1, t_2)$.
(c) If $\rho = 0$, and $\sigma_A^2 = \sigma_B^2 = \sigma^2$, find the covariance matrix kernel $\boldsymbol{K}(t_1, t_2)$.
(d) If $\rho \neq 0$, and $\sigma_A^2 = \sigma_B^2 = \sigma^2$, find the covariance matrix kernel $\boldsymbol{K}(t_2, t_2)$.
(e) Is the process X(t) wide-sense stationary in (b), (c), and (d) above?
(f) Sketch the ensembles and give the ensemble interpretation of $k_{11}(t_1, t_2)$ and $k_{12}(t_1, t_2)$.
(g) Can you completely characterize this process, or are you limited to a second-order characterization?

3.18. The random process X(t) is covariance stationary with covariance kernel $k(\tau)$. For the following covariance kernels, determine if the derivative X(t) exists and if it does, find expressions for

1. $E[X(t)\dot{X}(t)]$
2. $E[X(t)\dot{X}(t+\sigma)]$
3. $E[\dot{X}(t)X(t+\sigma)]$
4. $E[\dot{X}(t)\dot{X}(t+\sigma)]$.

(a) $k(\tau) = e^{-\lambda|\tau|}$.
(b) $k(\tau) = (\sin \omega\tau)/\tau$.
(c) $k(\tau) = 1/(\lambda^2 + \tau^2)$.

3.19. The random process X(t) is defined by the integral

$$X(t) = \frac{1}{T}\int_t^{t+T} Z(\sigma)\, d\sigma.$$

Find the covariance kernel of X(t) for the following Z(t) covariance kernels:

(a) $k_Z(t, \sigma) = \sigma^2 \min(t, \sigma)$.
(b) $k_Z(t, \sigma) = e^{-\lambda|t-\sigma|}$.

Chapter 4
Basic Tenets of System Theory

4.1. THE MATHEMATICAL MODEL EQUATIONS

Systems are described by the use of mathematical models. These models result from application of some physical law, usually applied on a macroscopic level, so that the model itself is deterministic. Random effects, if they are present to a significant degree, usually manifest themselves as random processes present as input or measurement noise. For this reason, solution of the deterministic model equations is fundamental, even in the analysis of stochastic systems. Since systems are dynamic in nature, in that they evolve with time, the model equations are differential equations and in general are nonlinear. This characteristic leads to some difficulty in obtaining solutions to the model equations, which in turn may necessitate the use of a linear approximation of the model, or to the use of linearization techniques, in solving the nonlinear equations. For this reason, the study of linear dynamic systems is of paramount importance, and the major portion of this chapter is devoted to that subject.

Mathematical models are usually generated by use of some physical law, which in most cases results in a nonlinear differential equation of the form

$$\frac{d^N x(t)}{dt^N} = f\left(x(t), \frac{dx}{dt}, \frac{d^2x}{dt^2}, \ldots, \frac{d^{N-1}x}{dt^{N-1}}, u_1(t), u_2(t), \ldots, u_P(t)\right) \quad (4.1)$$

where $x(t)$ is some physical characteristic of the system dynamical state, N is the order of the system, and $u_1, u_2, \ldots, u_P$ are the outside influences acting on the system. Before the advent of the large-scale digital computer and real time digital computation, model equations in this form were not generally solvable, but several specialized techniques were applicable. If linearization was used, the resulting linear equations were solved by frequency domain techniques such as the Laplace transform. These techniques in turn led to some specialized frequency domain procedures–notably the root-locus and Nyquist methods familiar to the control engineer–for determining the system characteristics without actually solving the model equations. More recently, an approach to the solution of system model equations using sets of first-order differential equations has come into favor, largely because it more closely matches the capabilities of the modern digital computer. There are also certain characteristics of this approach which are satisfying from a conceptual point of view, since it usually associates a variable of the solution with each physical characteristic of the model.

Since a differential equation has a unique solution for a given set of initial conditions and a given set of input or driving functions, and since the initial time is an arbitrarily chosen time reference, it is possible to define a set of variables at some time t_0 which, together with the input or driving functions for all $t > t_0$, suffices to determine the behavior of the system for all $t > t_0$. Such a set of variables constitutes what is called the *system state*, and the variables themselves are called *state variables*. The state of a system, then, is the minimum amount of information required of a system at time t_0 such that its future behavior can be determined without reference to the history of the system prior to t_0. Some reflection on this subject should convince the reader that this minimum set of variables could well be the variables involved in the equivalent set of N first-order differential equations, yielding the conclusion that an Nth order differential equation leads to N state variables, which are not unique in the sense that any set of variables linearly related by a one-to-one transformation to a set of state variables can also be considered a set of state variables.

These concepts are not new by any means, but merely constitute a different way of viewing the dynamic behavior of systems. For example, a spacecraft can be considered as a point mass with six state variables–three position coordinates and three velocity coordinates. If one knows these six values at any point in time t_1, then he can, with knowledge of the acceleration acting on the spacecraft during the interval from t_1 to t_2, determine the six state variables at time t_2. The history of the spacecraft prior to time t_1 is of no consequence; only the state at time t_1 is required. This example also suffices to illustrate the fact that the temporal ordering of t_1 and t_2 is not important, since either may be the later time with no effect on the conclusions just reached.

Then, in order to describe the system state, the system equation given by Equation (4.1) is written as N first-order differential equations

$$\dot{x}_1(t) = f_1(x_1(t), x_2(t), \ldots, x_N(t), t, u_1(t), u_2(t), \ldots, u_P(t))$$
$$\dot{x}_2(t) = f_2(x_1(t), x_2(t), \ldots, x_N(t), t, u_1(t), u_2(t), \ldots, u_P(t))$$

$$\dot{x}_N(t) = f_N(x_1(t), x_2(t), \ldots, x_N(t), t, u_1(t), u_2(t), \ldots, u_P(t))$$

or in the more concise vector form as

$$\dot{\boldsymbol{x}}(t) = \boldsymbol{f}(\boldsymbol{x}(t), t, \boldsymbol{u}(t)) \tag{4.2}$$

where $\boldsymbol{x}(t)$ is the vector comprised of the elements $x_1(t), x_2(t), \ldots, x_N(t)$, and $\boldsymbol{u}(t)$ is the vector of input functions. This matrix expression is referred to as the

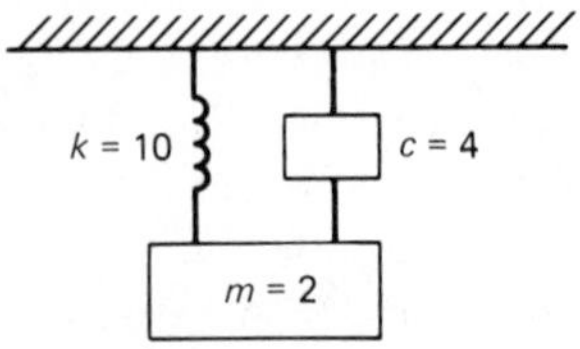

Figure 4-1. Spring–mass–damper system.

state equation and describes the time behavior of the state vector $\boldsymbol{x}(t)$. In its nonlinear form Equation (4.2) has no general solution, but in the special case in which the right-hand side is a linear form of both $\boldsymbol{x}(t)$ and $\boldsymbol{u}(t)$,

$$\dot{\boldsymbol{x}}(t) = \boldsymbol{F}(t)\boldsymbol{x}(t) + \boldsymbol{G}(t)\boldsymbol{u}(t) \tag{4.3}$$

a general solution can be obtained by the methods outlined in Appendix C.

As an example, consider the physical system composed of a spring, mass, and damper, as illustrated in Figure 4-1. The application of Newton's Laws to this system yields the equation

$$2\ddot{x} + 4\dot{x} + 10x = u$$

where $x(t)$ is the displacement of the mass m from its equilibrium position. The state equations are derived by letting

$$x_1 = x$$
$$x_2 = \dot{x}$$

with the resulting expression

$$\begin{bmatrix} \dot{x}_1 \\ \dot{x}_2 \end{bmatrix} = \begin{bmatrix} x_2 \\ -5x_1 - 2x_2 + \dfrac{u}{2} \end{bmatrix}$$

which is the state equation in the form of Equation (4.2). Note that the choice of state variables in this case yields recognizable quantities—position and velocity—as the state of the system.

It is sometimes convenient to consider the time behavior of the system as the motion of the tip of the state vector in the N-dimensional space generated by considering the state variables as coordinates. This motion is called the *trajectory* of the system in state space. In the previous example, such a trajectory would be as depicted in Figure 4-2, where the initial state at t_0 is shown

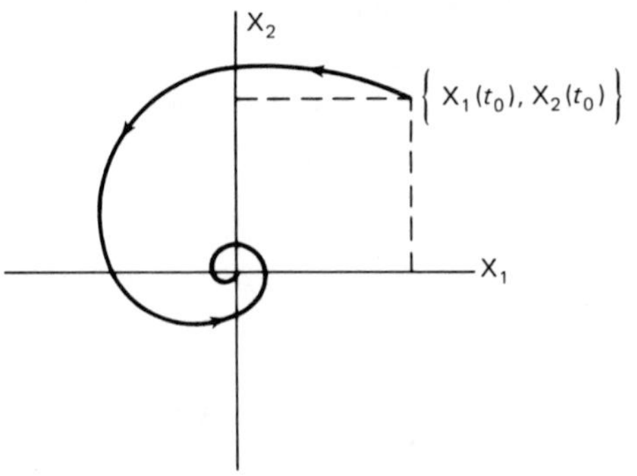

Figure 4-2. Trajectory of system of Figure 4-1 in state space.

as consisting of initial values for both position and velocity. Time, of course, is a parameter of this curve which increases in the direction of the arrows.

4.2. THE LINEAR SYSTEM

When the state equation contains the system state variables and the input function only in linear form, then the vector equation is as depicted in Equation (4.3), which is a set of linear, first order, nonhomogeneous differential equations. In their component form, these equations are

$$
\begin{aligned}
\dot{x}_1(t) &= f_{11}(t)x_1(t) + f_{12}(t)x_2(t) + \cdots + f_{1N}(t)x_N(t) \\
&\quad + g_{11}(t)u_1(t) + g_{12}(t)u_2(t) + \cdots + g_{1P}(t)u_P \\
\dot{x}_2(t) &= f_{21}(t)x_1(t) + f_{22}(t)x_2(t) + \cdots + f_{2N}(t)x_N(t) \\
&\quad + g_{21}(t)u_1(t) + g_{22}(t)u_2(t) + \cdots + g_{2P}(t)u_P \\
&\vdots \\
\dot{x}_N(t) &= f_{N1}(t)x_1(t) + f_{N2}(t)x_2(t) + \cdots + f_{NN}(t)x_N(t) \\
&\quad + g_{N1}(t)u_1(t) + g_{N2}(t)u_2(t) + \cdots + g_{NP}(t)u_P
\end{aligned}
\tag{4.4}
$$

A system described by equations of this type is termed a *linear system*, and for this type of equation a generalized solution can be obtained. The example of the previous section is seen to be a linear system, since Equation (4.4) can be written as

$$
\begin{bmatrix} \dot{x}_1 \\ \dot{x}_2 \end{bmatrix} = \begin{bmatrix} 0 & 1 \\ -5 & -2 \end{bmatrix} \begin{bmatrix} x_1 \\ x_2 \end{bmatrix} + \begin{bmatrix} 0 \\ \frac{1}{2} \end{bmatrix} u
$$

which is obviously of the form given by Equation (4.3). The generalized solution to this type of equation is obtained through the use of the transition matrix discussed in Appendix C, where it is shown that a homogeneous equation of the form

$$\dot{\boldsymbol{x}}(t) = \boldsymbol{F}(t)\boldsymbol{x}(t) \tag{4.5}$$

has a solution of the form

$$\boldsymbol{x}(t_2) = \boldsymbol{\Phi}(t_2, t_1)\boldsymbol{x}(t_1) \tag{4.6}$$

where t_1 and t_2 are any two points in time, not necessarily in any particular temporal order. The matrix $\boldsymbol{\Phi}(t_2, t_1)$ is called the *transition matrix*, or *state transition matrix* in the event that state equations are involved. This concept can be illustrated by the following simple example.

Consider a rocket sled which encounters a water brake at a point 1000 feet down its track, at a velocity of 2000 ft/sec. Assume that the drag force of the water is proportional to the velocity of the sled, and that the following characteristics apply:

mass of sled = 2000 pounds

coefficient of drag force = 124.2 pounds-sec/ft.

It is desired to determine the position and velocity two seconds after brake initiation. The state equations can easily be determined to be

$$\dot{x} = v$$

$$\dot{v} = -2v$$

or, in matrix form

$$\begin{bmatrix} \dot{x} \\ \dot{v} \end{bmatrix} = \begin{bmatrix} 0 & 1 \\ 0 & -2 \end{bmatrix} \begin{bmatrix} x \\ v \end{bmatrix}$$

Since this expression is of the form of Equation (4.5), with

$$\boldsymbol{F} = \begin{bmatrix} 0 & 1 \\ 0 & -2 \end{bmatrix}$$

it has a solution of the form given by Equation (4.6). To determine $\boldsymbol{\Phi}(t, t_0)$ the method of Appendix C is used. The eigenvalues of the $\boldsymbol{F}$ matrix above are

$$\lambda_1 = 0$$

$$\lambda_2 = -2$$

with the resulting eigenvectors

$$\begin{bmatrix} 1 \\ 0 \end{bmatrix} \text{ and } \begin{bmatrix} 1 \\ -2 \end{bmatrix}.$$

Then, the desired diagonalizing transformation matrix is formulated using the eigenvectors as the columns of the matrix:

$$\boldsymbol{P} = \begin{bmatrix} 1 & 1 \\ 0 & -2 \end{bmatrix}; \quad \boldsymbol{P}^{-1} = \begin{bmatrix} 1 & \frac{1}{2} \\ 0 & -\frac{1}{2} \end{bmatrix}$$

The state transition matrix is then

$$\boldsymbol{\Phi}(t, t_0) = \boldsymbol{P} \begin{bmatrix} e^{\lambda_1 (t-t_0)} & 0 \\ 0 & e^{\lambda_2 (t-t_0)} \end{bmatrix} \boldsymbol{P}^{-1}$$

which, in this case, yields

$$\boldsymbol{\Phi}(t, t_0) = \begin{bmatrix} 1 & \frac{1}{2} - \frac{1}{2} e^{-2t} \\ 0 & e^{-2t} \end{bmatrix}$$

when $t_0 = 0$. The desired solution is then obtained as

$$\begin{bmatrix} x(2) \\ v(2) \end{bmatrix} = \begin{bmatrix} 1 & \frac{1}{2} - \frac{1}{2} e^{-4} \\ 0 & e^{-4} \end{bmatrix} \begin{bmatrix} 1000 \\ 2000 \end{bmatrix}$$

which yields

$$x(2) = 1981.7 \text{ ft}$$

$$v(2) = 36.6 \text{ ft/sec.}$$

This same answer can be obtained by more customary methods, but this example shows how the generalized solution to the homogeneous problem can apply.

4.2.1. Solution to the Nonhomogeneous Problem

When the general nonhomogeneous problem is considered, in the form

$$\dot{\boldsymbol{x}}(t) = \boldsymbol{F}(t)\boldsymbol{x}(t) + \boldsymbol{G}(t)\boldsymbol{u}(t) \tag{4.7}$$

a generalized solution can still be obtained. Before this is developed, however, there are certain intrinsic properties of the state transition matrix which must be detailed. These are as follows:

I.

$$\boldsymbol{\Phi}(t, t) = \boldsymbol{I}. \tag{4.8}$$

II. Since $\boldsymbol{x}(t_2) = \boldsymbol{\Phi}(t_2, t_1)\boldsymbol{x}(t_1)$

and $\boldsymbol{x}(t_1) = \boldsymbol{\Phi}(t_1, t_0)\boldsymbol{x}(t_0)$

then $\boldsymbol{x}(t_2) = \boldsymbol{\Phi}(t_2, t_1)\boldsymbol{\Phi}(t_1, t_0)\boldsymbol{x}(t_0)$

with the result that

$$\boldsymbol{\Phi}(t_2, t_0) = \boldsymbol{\Phi}(t_2, t_1)\boldsymbol{\Phi}(t_1, t_0). \tag{4.9}$$

This result is valid regardless of the temporal order of t_0, t_1, and t_2.

III. Since $\boldsymbol{\Phi}(t_2, t_1)\boldsymbol{\Phi}(t_1, t_2) = \boldsymbol{\Phi}(t_2, t_2) = \boldsymbol{I}$

by virtue of I and II above, then by multiplying on the left by $\boldsymbol{\Phi}^{-1}(t_2, t_1)$, the following result is obtained:

$$\boldsymbol{\Phi}(t_1, t_2) = \boldsymbol{\Phi}^{-1}(t_2, t_1). \tag{4.10}$$

IV. A general property of any nonsingular matrix is derived as follows. Since

$$\boldsymbol{\Phi}(t, t_0)\boldsymbol{\Phi}^{-1}(t, t_0) = I$$

then the derivative with respect to t is

$$\begin{aligned}\frac{d}{dt}[\boldsymbol{\Phi}(t, t_0)\boldsymbol{\Phi}^{-1}(t, t_0)] &= \boldsymbol{\Phi}(t, t_0)\dot{\boldsymbol{\Phi}}^{-1}(t, t_0) + \dot{\boldsymbol{\Phi}}(t, t_0)\boldsymbol{\Phi}^{-1}(t, t_0)\\ &= 0.\end{aligned}$$

Multiplying on the left by $\boldsymbol{\Phi}^{-1}(t, t_0)$ and solving yields

$$\dot{\boldsymbol{\Phi}}^{-1}(t, t_0) = -\boldsymbol{\Phi}^{-1}(t, t_0)\dot{\boldsymbol{\Phi}}(t, t_0)\ \boldsymbol{\Phi}^{1}(t, t_0) \tag{4.11}$$

relating the time derivative of a matrix to that of its inverse.

Now the solution to the nonhomogeneous problem can be considered. From the developments of Appendix C, it is known that the state transition matrix must satisfy the equation

$$\dot{\boldsymbol{\Phi}}(t, t_0) = \boldsymbol{F}(t)\boldsymbol{\Phi}(t, t_0). \tag{4.12}$$

Substituting this expression into Equation (4.11) yields

$$\dot{\boldsymbol{\Phi}}^{-1}(t, t_0) = -\boldsymbol{\Phi}^{-1}(t, t_0)\boldsymbol{F}(t) \tag{4.13}$$

and multiplying on the right by $\boldsymbol{x}(t)$ finally yields

$$\dot{\boldsymbol{\Phi}}^{-1}(t, t_0)\boldsymbol{x}(t) = -\boldsymbol{\Phi}^{-1}(t, t_0)\boldsymbol{F}(t)\boldsymbol{x}(t). \tag{4.14}$$

Now, by multiplying Equation (4.7) on the left by $\boldsymbol{\Phi}^{-1}(t, t_0)$, one obtains

$$\boldsymbol{\Phi}^{-1}(t, t_0)\dot{\boldsymbol{x}}(t) = \boldsymbol{\Phi}^{-1}(t, t_0)\boldsymbol{F}(t)\boldsymbol{x}(t) + \boldsymbol{\Phi}^{-1}(t, t_0)\boldsymbol{G}(t)\boldsymbol{u}(t). \tag{4.15}$$

Summing equations (4.14) and (4.15) yields

$$\dot{\boldsymbol{\Phi}}^{-1}(t, t_0)\boldsymbol{x}(t) + \boldsymbol{\Phi}^{-1}(t, t_0)\dot{\boldsymbol{x}}(t) = \boldsymbol{\Phi}^{-1}(t, t_0)\boldsymbol{G}(t)\boldsymbol{u}(t). \tag{4.16}$$

The left-hand side of this equation constitutes a perfect derivative of the product

$$\boldsymbol{\Phi}^{-1}(t, t_0)\boldsymbol{x}(t) \tag{4.17}$$

so that the equation can be integrated over the limits t_0 to t:

$$\boldsymbol{\Phi}^{-1}(t, t_0)\boldsymbol{x}(t) - \boldsymbol{\Phi}^{-1}(t_0, t_0)\boldsymbol{x}(t_0) = \int_{t_0}^{t} \boldsymbol{\Phi}^{-1}(\sigma, t_0)\boldsymbol{G}(\sigma)\boldsymbol{u}(\sigma)\, d\sigma \tag{4.18}$$

where the dummy variable of integration σ has been used to avoid confusion between the variable of integration and the upper limit t. By recognizing the identities

$$\boldsymbol{\Phi}(t_0, t_0) = \boldsymbol{I}$$
$$\boldsymbol{\Phi}^{-1}(\sigma, t_0) = \boldsymbol{\Phi}(t_0, \sigma)$$

and solving Equation (4.18) for $\boldsymbol{x}(t)$ by multiplying on the left by $\boldsymbol{\Phi}(t, t_0)$, the following result is obtained:

$$\boldsymbol{x}(t) = \boldsymbol{\Phi}(t, t_0)\boldsymbol{x}(t_0) + \boldsymbol{\Phi}(t, t_0) \int_{t_0}^{t} \boldsymbol{\Phi}(t_0, \sigma)\boldsymbol{G}(\sigma)\boldsymbol{u}(\sigma)\, d\sigma. \tag{4.19}$$

Then, since $\boldsymbol{\Phi}(t, t_0)$ is independent of σ, it can be taken inside the integral, with the desired final result:

$$\boldsymbol{x}(t) = \boldsymbol{\Phi}(t, t_0)\boldsymbol{x}(t_0) + \int_{t_0}^{t} \boldsymbol{\Phi}(t, \sigma)\boldsymbol{G}(\sigma)\boldsymbol{u}(\sigma)\, d\sigma. \tag{4.20}$$

This equation presents a closed form solution for $\boldsymbol{x}(t)$, given the initial condition $\boldsymbol{x}(t_0)$, the input $\boldsymbol{u}(t)$ over the limits t_0 to t and the system characteristics in form of the state transition matrix $\boldsymbol{\Phi}(t_1, t_2)$. Since the matrices $\boldsymbol{F}$ and $\boldsymbol{G}$ were considered in their general context as time functions, this is a general result for linear systems.

The particular form of Equation (4.20) is significant in that it is composed of two terms, the first of which is the solution to the homogeneous equation and is not dependent upon the system input $\boldsymbol{u}(t)$. The second term is independent of the initial conditions, but does depend on the input $\boldsymbol{u}(t)$. These, of course, correspond to the complementary and particular solutions of the differential equations, as ordinarily viewed.

4.2.2. The Difference Equation Representation

In many cases, it is desirable to approximate a set of linear state equations by a set of difference equations. These can be formed directly from the general solution of Equation (4.20) in the following manner. If the discrete time values of interest are $t_1, t_2, \ldots, t_l, \ldots$, then the behavior of the system during the interval (t_{l-1}, t_l) is given by

$$\boldsymbol{x}(t_l) = \boldsymbol{\Phi}(t_l, t_{l-1})\boldsymbol{x}(t_{l-1}) + \int_{t_{l-1}}^{t_l} \boldsymbol{\Phi}(t_l, \sigma)\boldsymbol{G}(\sigma)\boldsymbol{u}(\sigma)\, d\sigma. \tag{4.21}$$

Now, if the value of $\boldsymbol{u}(t)$ is considered constant over the interval (t_{l-1}, t_l) with the value

$$\boldsymbol{u}(t) = \boldsymbol{u}(t_{l-1}), \quad \text{a constant for } t_{l-1} \leqslant t < t \tag{4.22}$$

then it can be factored out of the integral of Equation (4.21). The result is

$$\boldsymbol{x}(t_l) = \boldsymbol{\Phi}(t_l, t_{l-1})\boldsymbol{x}(t_{l-1}) + \boldsymbol{\Gamma}(t_l, t_{l-1})\boldsymbol{u}(t_{l-1}) \tag{4.23}$$

where the matrix $\boldsymbol{\Gamma}(t_l, t_{l-1})$ is defined as

$$\boldsymbol{\Gamma}(t_l, t_{l-1}) = \int_{t_{l-1}}^{t_l} \boldsymbol{\Phi}(t, \sigma)\boldsymbol{G}(\sigma)\, d\sigma. \tag{4.24}$$

Equation (4.23) is the difference equation representation of the differential equation of Equation (4.7). Such equations may arise as approximations to differential equations, or directly as a result of discrete physical systems. Solution of Equation (4.23) by recursive methods is obvious, since the set of equations

$$\begin{aligned}
\boldsymbol{x}(t_1) &= \boldsymbol{\Phi}(t_1, t_0)\boldsymbol{x}(t_0) + \boldsymbol{\Gamma}(t_1, t_0)\boldsymbol{u}(t_0)\\
\boldsymbol{x}(t_2) &= \boldsymbol{\Phi}(t_2, t_1)\boldsymbol{x}(t_1) + \boldsymbol{\Gamma}(t_2, t_1)\boldsymbol{u}(t_1)\\
&\vdots\\
\boldsymbol{x}(t_l) &= \boldsymbol{\Phi}(t_l, t_{l-1})\boldsymbol{x}(t_l) + \boldsymbol{\Gamma}(t_l, t_{l-1})\boldsymbol{u}(t_{l-1})
\end{aligned} \tag{4.25}$$

can be used to solve for $\boldsymbol{x}(t_l)$, given the initial condition $\boldsymbol{x}(t_0)$ and the input $\boldsymbol{u}(t_i)$ over the interval (t_0, t_l). There are also methods of closed form solutions for time-invariant systems, such as the multivariate generalization of the z-transform method.

The recursive solution follows directly from substitution of terms from Equation (4.25). That is, the equation for $\boldsymbol{x}(t_1)$, when substituted into the $\boldsymbol{x}(t_2)$ equation yields

$$\boldsymbol{x}(t_2) = \boldsymbol{\Phi}(t_2, t_1)[\boldsymbol{\Phi}(t_1, t_0)\boldsymbol{x}(t_0) + \boldsymbol{\Gamma}(t_1, t_0)\boldsymbol{u}(t_0)] + \boldsymbol{\Gamma}(t_2, t_1)\boldsymbol{u}(t_1). \tag{4.26}$$

By virtue of the properties of the transition matrix this expression reduces to

$$\boldsymbol{x}(t_2) = \boldsymbol{\Phi}(t_2, t_0)\boldsymbol{x}(t_0) + \boldsymbol{\Phi}(t_2, t_1)\,\boldsymbol{\Gamma}(t_1, t_0)\boldsymbol{u}(t_0) + \boldsymbol{\Gamma}(t_2, t_1)\boldsymbol{u}(t_1). \tag{4.27}$$

This equation can be substituted into the $\boldsymbol{x}(t_3)$ equation, and so on, until the general expression is obtained:

$$\boldsymbol{x}(t_l) = \boldsymbol{\Phi}(t_l, t_0)\boldsymbol{x}(t_0) + \sum_{m=1}^{l} \boldsymbol{\Phi}(t_l, t_m)\boldsymbol{\Gamma}(t_m, t_{m-1})\boldsymbol{\mu}(t_{m-1}). \tag{4.28}$$

That this result is the difference equation analog to Equation (4.20) is obvious.

In order to partially relieve the burden of cumbersome notation, the discrete time equations will be expressed with the general subscripts l and m as arguments. Thus, Equation (4.23) would be represented as

$$\boldsymbol{x}(l) = \boldsymbol{\Phi}(l, l-1)\boldsymbol{x}(l-1) + \boldsymbol{\Gamma}(l, l-1)\boldsymbol{u}(l-1) \tag{4.29}$$

where $\boldsymbol{x}(l)$ implies the discrete time function $\boldsymbol{x}(t_l)$.

4.3. PROPERTIES OF SYSTEMS

There are three general properties of systems which may affect their dynamic behavior and which are amenable to explicit analysis in certain cases. These are *stability*, *observability*, and *controllability*. These properties are defined in the following manner, for the system

$$\dot{\boldsymbol{x}}(t) = \boldsymbol{f}(\boldsymbol{x}(t), \boldsymbol{u}(t), t). \tag{4.30}$$

I. If, for $\boldsymbol{u}(t) = 0$ and $\boldsymbol{f}(0, t) = 0$, the scalar $\boldsymbol{x}^T(t)\boldsymbol{x}(t)$ approaches zero as t approaches ∞, independently of t_0 and $\boldsymbol{x}(t_0)$, then the system is said to exhibit uniform, asymptotic, global stability. Uniform refers to the independence of t_0; global refers to independence of $\boldsymbol{x}(t_0)$; and asymptotic refers to the behavior as t approaches ∞.

II. A system is said to be controllable if the available input $\boldsymbol{u}(t)$ is sufficient to bring the system from any initial state $\boldsymbol{x}(t_0)$ to any desired final state $\boldsymbol{x}(t)$.

III. A system is said to be observable if the system output, which in its most general form is given by

$$\boldsymbol{y}(t) = \boldsymbol{h}(\boldsymbol{x}(t), t) \tag{4.31}$$

yields sufficient information to determine the state at any time t. System output is discussed further in Section 4.3.3.

While these concepts can apply equally well to both linear and nonlinear systems, in the latter case there are no standardized tests which may be applied.

Also, in the case of nonlinear systems, the global characteristic may not apply, while the system may be locally stable, controllable, or observable. For this reason, concentration is again placed on linear systems.

4.3.1. Stability of Systems

There is a general criterion for uniform, asymptotic, global stability which is valid for either linear or nonlinear systems. This criterion is due to Lyapunov, and can be stated as follows. If, for the homogeneous form of Equation (4.30)

$$\dot{\boldsymbol{x}}(t) = \boldsymbol{f}(\boldsymbol{x}(t), t) \tag{4.32}$$

there exists a scalar function $V(\boldsymbol{x}(t), t)$ such that:

I. The partial derivatives $\partial V/\partial \boldsymbol{x}$ and $\partial V/\partial t$ are continuous
II. $V(\boldsymbol{x}(t), t) > 0$ for all $\boldsymbol{x} \neq 0$ and for all t
III. $dV/dt = \dot{V} < 0$ for all $\boldsymbol{x}(t) \neq 0$ and for all t
IV. $V(\boldsymbol{x}(t), t) \rightarrow \infty$ as $\boldsymbol{x}^T(t)\boldsymbol{x}(t) \rightarrow \infty$

then the system of Equation (4.32) has uniform, asymptotic, global stability.

This result reduces the determination of stability to a problem of finding a Lyapunov function $V(\boldsymbol{x}(t), t)$ with the desired properties. There is no general method for doing this for nonlinear systems, although several special techniques for particular kinds of equations have been developed. For linear systems, quite general methods can be obtained.

As an example involving a nonlinear system, consider the two-state system,

$$\dot{x}_1 = -\frac{6x_1}{(1 + x_1^2)^2} + 2x_2$$

$$\dot{x}_2 = -2x_1 - \frac{2x_2}{(1 + x_1^2)^2}.$$

The problem is to determine the stability of this system. As a candidate Lyapunov function, consider

$$V(\boldsymbol{x}(t), t) = V(\boldsymbol{x}(t)) = x_1^2(t) + x_2^2(t).$$

This function obviously satisfies conditions I, II, and IV, and it now must be determined if it also satisfies condition III. If $\boldsymbol{V}(x(t))$ is differentiated with respect to time, and then $\dot{x}_1$ and $\dot{x}_2$ eliminated by means of the system equations, the result is

$$\dot{V}(\boldsymbol{x}(t)) = -\frac{12x_2^2(t) + 4x_2^2(t)}{(1 + x_1^2(t))^2}$$

which is negative for all $\boldsymbol{x} \neq 0$. Thus, $V(\boldsymbol{x}(t))$ as defined above is a valid Lyapunov function, and the system of equations has uniform global asymptotic stability.

For a time-invariant, linear system, the homogeneous equation is

$$\dot{\boldsymbol{x}}(t) = \boldsymbol{F}\boldsymbol{x}(t) \tag{4.33}$$

where $\boldsymbol{F}$ is a constant matrix. In this case, if the quadratic form of the symmetric, positive definite matrix $\boldsymbol{A}$

$$V(\boldsymbol{x}(t)) = \boldsymbol{x}^T(t)\boldsymbol{A}\boldsymbol{x}(t) \tag{4.34}$$

is taken as a candidate Lyapunov function, then

$$\dot{V}(\boldsymbol{x}(t)) = \boldsymbol{x}^T(t)\boldsymbol{A}\dot{\boldsymbol{x}}(t) + \dot{\boldsymbol{x}}^T(t)\boldsymbol{A}\boldsymbol{x}(t) \tag{4.35}$$

or, upon substitution from Equation (4.33),

$$\dot{V}(\boldsymbol{x}(t)) = \boldsymbol{x}^T(t)[\boldsymbol{F}^T\boldsymbol{A} + \boldsymbol{A}\boldsymbol{F}]\boldsymbol{x}(t). \tag{4.36}$$

This expression is a quadratic form of the constant matrix $[\boldsymbol{F}^T\boldsymbol{A} + \boldsymbol{A}\boldsymbol{F}]$, and if it is negative for all values of $\boldsymbol{x}(t)$ then the system is stable. This condition is met if the matrix itself is negative definite; or, as it is more commonly stated, the matrix

$$-\boldsymbol{F}^T\boldsymbol{A} - \boldsymbol{A}\boldsymbol{F} = \boldsymbol{Q} \tag{4.37}$$

must be positive definite. This result is equivalent to the statement that, for any positive definite, symmetric matrix $\boldsymbol{Q}$, there is a positive definite, symmetric matrix $\boldsymbol{A}$ such that

$$-[\boldsymbol{F}^T\boldsymbol{A} + \boldsymbol{A}\boldsymbol{F}] = \boldsymbol{Q} \tag{4.38}$$

or, specifically for $\boldsymbol{Q} = \boldsymbol{I}$,

$$-[\boldsymbol{F}^T\boldsymbol{A} + \boldsymbol{A}\boldsymbol{F}] = \boldsymbol{I}. \tag{4.39}$$

Thus, a necessary and sufficient condition for the stability of the system of

Equation (4.33) is the existence of a positive definite, symmetric matrix $\boldsymbol{A}$ which satisfies this equation. Note also that, for linear, time-invariant systems, this condition implies uniform, global, asymptotic stability.

As an example, consider again the spring-mass-damper system of Figure 4-1. For this system

$$\boldsymbol{F} = \begin{bmatrix} 0 & 1 \\ -5 & -2 \end{bmatrix}$$

and application of Equation (4.38) yields

$$\begin{bmatrix} -10a_1 & a_1 - 2a_2 - 5a_3 \\ a_1 - 2a_2 - 5a_3 & 2a_1 - 4a_3 \end{bmatrix} = \begin{bmatrix} 1 & 0 \\ 0 & 1 \end{bmatrix}$$

where

$$\boldsymbol{A} = \begin{bmatrix} a_1 & a_2 \\ a_2 & a_3 \end{bmatrix}$$

is a symmetric matrix. If these equations can be solved for a_1, a_2, and a_3, and the resulting symmetric matrix $\boldsymbol{A}$ is positive definite, then the system is stable. Solution yields

$$\begin{bmatrix} 17/10 & 1/10 \\ 1/10 & 3/10 \end{bmatrix}$$

which is readily seen to be positive definte by application of Sylvester's theorem as described in Appendix A. Thus, the system is proven to be stable, as expected from physical considerations.

A physical realization of the meaning of stability can be obtained from examining the modal response of the spring-mass-damper system following the method outlined in Appedix C. The eigenvalues of the $\boldsymbol{F}$ matrix can be determined to be

$$\lambda_1 = -1 + 2i$$

$$\lambda_2 = -1 - 2i$$

which indicates a solution of the form

$$\boldsymbol{x}(t) = \boldsymbol{P} \begin{bmatrix} e^{(-1+2i)(t-t_0)} & 0 \\ 0 & e^{(-1-2i)(t-t_0)} \end{bmatrix} \boldsymbol{P}^{-1}\boldsymbol{x}(t_0)$$

where the matrix $\boldsymbol{P}$ is a transformation matrix composed of eigenvectors of $\boldsymbol{F}$, resulting in expressions for $x_1(t)$ and $x_2(t)$. For example,

$$x_1(t) = C_1 e^{(-1+2i)(t-t_0)} + C_2 e^{(-1-2i)(t-t_0)}$$

for which the exponential terms can be factored,

$$x_1(t) = e^{-(t-t_0)}[C_1 e^{2i(t-t_0)} + C_2 e^{-2i(t-t_0)}].$$

The imaginary terms lead to sinusoidal responses, as outlined in Appendix C, and the $e^{-(t-t_0)}$ term approaches zero as $t \to \infty$. A little thought will indicate that the real exponent is dependent upon the real part of the eigenvalues of $\boldsymbol{F}$. Since the time response for large t is determined by this term, it serves as an indicator of the stability of the system. That this is a general property is easily ascertained from the general form of the solution to the homogeneous equation

$$\boldsymbol{x}(t) = \boldsymbol{P} \begin{bmatrix} e^{\lambda_1(t-t_0)} & & & \\ & e^{\lambda_2(t-t_0)} & & \\ & & \ddots & \\ & & & e^{\lambda_N(t-t_0)} \end{bmatrix} \boldsymbol{P}^{-1}\boldsymbol{x}(t_0) \tag{4.40}$$

as derived in Appendix C. This conclusion indicates that eigenvalues with positive real parts lead to modal responses which grow without bound with increasing t, and therefore lead to unstable systems. Negative real parts, on the other hand, indicate stable systems.

Since the eigenvalues of the $\boldsymbol{F}$ matrix are identical to the roots of the characteristic equation obtained when the system is expressed as an Nth-order differential equation, the stability is dependent upon the roots of the characteristic equation having negative real parts. The methods of the classical approach to stability analysis all lead to ways of determining the characteristic roots of the equation and, as such, are equivalent to determining the real parts of the eigenvalues of $\boldsymbol{F}$. The Lyapunov method as expressed by Equation (4.39) does precisely this and in this form can be shown to be equivalent to Routh's criterion, so familiar to the control engineer.

For an unstable system, Equation (4.38) will lead to one of two results. Either the resulting equations in a_1, a_2, and a_3 will be inconsistent , or the solution will lead to matrix $\boldsymbol{A}$ which is not positive definite. For example, the system

$$\boldsymbol{F} = \begin{bmatrix} -1 & 1 \\ -3 & 2 \end{bmatrix}$$

leads to the equation

$$\begin{bmatrix} 2a_1 + 6a_2 & -a_1 - a_2 + 3a_3 \\ -a_1 - a_2 + 3a_3 & -2a_2 - 4a_3 \end{bmatrix} = \begin{bmatrix} 1 & 0 \\ 0 & 1 \end{bmatrix}$$

with the result, upon solution,

$$\boldsymbol{A} = \begin{bmatrix} -7 & 15/6 \\ 15/6 & -3/2 \end{bmatrix}.$$

The negative leading term indicates that this matrix is not positive definite; therefore the system is unstable. Determination of the eigenvalues yields

$$\lambda_1 = \frac{1}{2} + \frac{i\sqrt{3}}{2}$$

$$\lambda_2 = \frac{1}{2} - \frac{i\sqrt{3}}{2}$$

and the positive real parts of the eigenvalues indicate the source of the instability.

For discrete time systems described by the difference equation given in Equation (4.23), Lyapunov stability criteria can also be developed. For the time-invariant case, the homogeneous equation is

$$\boldsymbol{x}(t_l) = \boldsymbol{\Phi}(t_l, t_{l-1})\boldsymbol{x}(t_{l-1}) \tag{4.41}$$

where now $\boldsymbol{\Phi}$ is a function only of the time difference $(t_l - t_{l-1})$ because the system is time invariant, as outlined in Appendix C. Furthermore, if a constant interval is specified, as is often the case, this expression further reduces to

$$\boldsymbol{x}(l) = \boldsymbol{\Phi}\boldsymbol{x}(l-1) \tag{4.42}$$

where $\boldsymbol{\Phi}$ is now constant, and the notation of Equation (4.29) has been employed. The discrete time analog of the Lyapunov function of Equation (4.34) is

$$V(\boldsymbol{x}(l)) = \boldsymbol{x}^T(l)\boldsymbol{A}\boldsymbol{x}(l) \tag{4.43}$$

where $\boldsymbol{A}$ is a positive definite, symmetric matrix. Then the difference

$$\Delta V(\boldsymbol{x}(l)) = V(\boldsymbol{x}(l)) - V(\boldsymbol{x}(l-1)) \tag{4.44}$$

must be less than zero for all nonzero values of $\boldsymbol{x}(l)$. Expanding this expression by substituting from Equations (4.42) and (4.43) yields

$$\begin{aligned}\Delta V(x(l)) &= \boldsymbol{x}^T(l)\boldsymbol{A}\boldsymbol{x}(l) - \boldsymbol{x}^T(l-1)\boldsymbol{A}\boldsymbol{x}(l-1)\\ &= \boldsymbol{x}^T(l-1)\boldsymbol{\Phi}^T\boldsymbol{A}\boldsymbol{\Phi}\boldsymbol{x}(l-1) - \boldsymbol{x}^T(l-1)\boldsymbol{A}\boldsymbol{x}(l-1)\\ &= \boldsymbol{x}^T(l-1)[\boldsymbol{\Phi}^T\boldsymbol{A}\boldsymbol{\Phi} - \boldsymbol{A}]\boldsymbol{x}(l-1).\end{aligned} \tag{4.45}$$

This quantity is less than zero for all $\boldsymbol{x}(l-1)$ if the matrix

$$\boldsymbol{\Phi}^T\boldsymbol{A}\boldsymbol{\Phi} - \boldsymbol{A}$$

is negative definite, or, equivalently, the symmetric matrix

$$\boldsymbol{Q} = \boldsymbol{A} - \boldsymbol{\Phi}^T\boldsymbol{A}\,\boldsymbol{\Phi} \tag{4.46}$$

must be positive definite. Specifically, for $\boldsymbol{Q} = \boldsymbol{I}$,

$$\boldsymbol{I} = \boldsymbol{A} - \boldsymbol{\Phi}^T\boldsymbol{A}\,\boldsymbol{\Phi} \tag{4.47}$$

and a necessary and sufficient condition for stability of the system described by Equation (4.42) is the existence of a positive definite symmetric matrix $\boldsymbol{A}$ which satisfies this equation.

4.3.2. Controllability of Systems

The second system property which is of interest is controllability. Strictly speaking, this property is formally defined in the following manner. A system is said to be controllable over the interval (t_1, t_2) if a finite control function $\boldsymbol{u}(t)$ for $t_1 \leqslant t \leqslant t_2$ can be found which will transfer the system from any given state $\boldsymbol{x}(t_1)$ to any other state $\boldsymbol{x}(t_2)$. In making such a formal definition, no intuitive feel for just what constitutes the property of controllability is conveyed. For this reason, controllability is introduced here in its special application to linear, time-invariant systems of the form

$$\dot{\boldsymbol{x}}(t) = \boldsymbol{F}\boldsymbol{x}(t) + \boldsymbol{G}\boldsymbol{u}(t). \tag{4.48}$$

For illustrative purposes, a particular system will be considered:

$$\dot{x}_1 = -x_1 + u$$

$$\dot{x}_2 = x_1 - 2x_2 + u$$

for which it can be seen that

$$\boldsymbol{F} = \begin{bmatrix} -1 & 0 \\ 1 & -2 \end{bmatrix}; \quad G = \begin{bmatrix} 1 \\ 1 \end{bmatrix}.$$

Now the transformation

$$\boldsymbol{x} = \boldsymbol{P}\boldsymbol{z}$$

will be considered, where $\boldsymbol{P}$ is a transformation matrix composed of eigenvectors of $\boldsymbol{F}$. Solution for eigenvectors and eigenvalues yields

$$\begin{matrix} \lambda_1 = -1; \\ \lambda_2 = -2; \end{matrix} \quad \boldsymbol{P} = \begin{bmatrix} 1 & 0 \\ 1 & 1 \end{bmatrix}; \quad \boldsymbol{P}^{-1} = \begin{bmatrix} 1 & 0 \\ -1 & 1 \end{bmatrix}.$$

Then, substitution for $\boldsymbol{x}$ into the original equation yields

$$\dot{\boldsymbol{z}} = \boldsymbol{P}^{-1}\boldsymbol{F}\boldsymbol{P}\boldsymbol{z} + \boldsymbol{P}^{-1}\boldsymbol{G}\boldsymbol{u}$$

$$= \begin{bmatrix} \lambda_1 & 0 \\ 0 & \lambda_2 \end{bmatrix} \begin{bmatrix} z_1 \\ z_2 \end{bmatrix} + \begin{bmatrix} 1 \\ 0 \end{bmatrix} u.$$

In component form, these equations are

$$\dot{z}_1 = \lambda_1 z_1 + u = -z_1 + u$$

$$\dot{z}_2 = \lambda_2 z_2 = -2z_2$$

The equations are uncoupled and, in their homogeneous forms, represent the modes of response of the system. But note that the second mode is not affected by the input u. That is, the dynamic behavior of the system is such that the mode z_2 responds to a set of initial conditions only, regardless of the system input u. Since states x_1 and x_2 are composed of linear combinations of the modes, the states themselves cannot be completely controlled by the input u. This system, then, does not comply with the requirements of controllability as outlined above and therefore must be classed as a noncontrollable system. In this special case of a linear, time-invariant system, the existence of a mode of response which cannot be affected by the system input leads directly to a noncontrollable system. For time-varying linear and nonlinear systems this modal concept does not apply, but the basic concept of controllability as defined above should present no difficulty.

For a linear, time-invariant, discrete time system of the general form

$$\begin{aligned} \boldsymbol{x}(l) &= \boldsymbol{\Phi}(l, l-1)\boldsymbol{x}(l-1) + \boldsymbol{\Gamma}(l, l-1)\boldsymbol{u}(l-1) \\ &= \boldsymbol{\Phi}\boldsymbol{x}(l-1) + \boldsymbol{\Gamma}\boldsymbol{u}(l-1) \end{aligned} \tag{4.49}$$

where the second form results from constant $(t_l - t_{l-1})$ and constant values for $\boldsymbol{\Phi}$ and $\boldsymbol{\Gamma}$, the concept of controllability can be easily demonstrated. From Equation (4.49), indexed down by one,

$$\boldsymbol{x}(l-1) = \boldsymbol{\Phi}\boldsymbol{x}(l-2) + \boldsymbol{\Gamma}\boldsymbol{u}(l-2) \tag{4.50}$$

and substitution of this expression into Equation (4.49) yields

$$\boldsymbol{x}(l) = \boldsymbol{\Phi}[\boldsymbol{x}(l-2) + \boldsymbol{\Gamma}\boldsymbol{u}(l-2)] + \boldsymbol{\Gamma}\boldsymbol{u}(l-1). \tag{4.51}$$

Similarly, for $\boldsymbol{x}(l-2)$, the expression

$$\boldsymbol{x}(l-2) = \boldsymbol{\Phi}\boldsymbol{x}(l-3) + \boldsymbol{\Gamma}\boldsymbol{u}(l-3) \tag{4.52}$$

is substituted, yielding

$$\boldsymbol{x}(l) = \boldsymbol{\Phi}\{\boldsymbol{\Phi}[\boldsymbol{\Phi}\boldsymbol{x}(l-3) + \boldsymbol{\Gamma}\boldsymbol{u}(l-3)] + \boldsymbol{\Gamma}\boldsymbol{u}(l-2)\} + \boldsymbol{u}(l-1). \tag{4.53}$$

If this procedure is continued to its logical conclusion, the result is

$$\boldsymbol{x}(l) = \boldsymbol{\Phi}^l\boldsymbol{x}(0) + \sum_{i=0}^{l-1} \boldsymbol{\Phi}^{l-i-1}\boldsymbol{\Gamma}\boldsymbol{u}(i) \tag{4.54}$$

relating $\boldsymbol{x}(l)$ to the initial condition $\boldsymbol{x}(0)$, and the input $\boldsymbol{u}(i)$ over the interval (t_0, t_{l-1}). This equation can be expressed as

$$\boldsymbol{x}(l) - \boldsymbol{\Phi}^l\boldsymbol{x}(0) = [\boldsymbol{\Gamma} \mid \boldsymbol{\Phi}\boldsymbol{\Gamma} \mid \cdots \mid \boldsymbol{\Phi}^{l-2}\boldsymbol{\Gamma} \mid \boldsymbol{\Phi}^{l-1}\boldsymbol{\Gamma}] \begin{bmatrix} \boldsymbol{u}(0) \\ \vdots \\ \boldsymbol{u}(l-2) \\ \boldsymbol{u}(l-1) \end{bmatrix}. \tag{4.55}$$

By expressing the relationship in this form, the following simple interpretation of controllability of discrete systems is obtained. The left-hand side of this equation is a vector dependent upon two quantities—the initial condition and the final condition. The question of controllability involves the existence of a control sequence which will bring the system from some arbitrary initial condition to some arbitrary final condition in l stages, which in turn is determined by the existence or nonexistence of a solution for the vector

$$\begin{bmatrix} \boldsymbol{u}(0) \\ \vdots \\ \boldsymbol{u}(l-2) \\ \boldsymbol{u}(l-1) \end{bmatrix}$$

of Equation (4.55). Since this equation is linear, a necessary and sufficient condition that it have a solution is that the matrix

$$[\boldsymbol{\Gamma} \mid \boldsymbol{\Phi\Gamma} \mid \cdots \mid \boldsymbol{\Phi}^{l-2}\boldsymbol{\Gamma} \mid \boldsymbol{\Phi}^{l-1}\boldsymbol{\Gamma}] \tag{4.56}$$

have maximum rank. Maximum rank, of course, is the same as the dimension of the state vector $\boldsymbol{x}$, which implies that the maximum value of l in Equation (4.56) is N, the dimension of the state vector. Here, then, is a simply applied and easily interpretable test for controllability of discrete linear systems in l stages—the matrix of Equation (4.56) must have maximum rank.

A general statement for controllability of discrete, linear, time-invariant systems follows from the l-stage controllability conditions above. A discrete, linear, time-invariant system is controllable if and only if the matrix

$$[\boldsymbol{\Gamma} \mid \boldsymbol{\Phi\Gamma} \mid \boldsymbol{\Phi}^2\boldsymbol{\Gamma} \mid \cdots \mid \boldsymbol{\Phi}^{N-P-1}\boldsymbol{\Gamma} \mid \boldsymbol{\Phi}^{N-P}\boldsymbol{\Gamma}] \tag{4.57}$$

has maximum rank. Here, N is the dimension of the state vector and P is the dimension of the control sequence $\boldsymbol{u}(l)$.

It can be shown that a similar test for controllability of continuous, time-invariant, linear systems exists, in that a necessary and sufficient condition for controllability is that the matrix

$$[\boldsymbol{G} \mid \boldsymbol{FG} \mid \cdots \mid \boldsymbol{F}^{N-P-1}\boldsymbol{G} \mid \boldsymbol{F}^{N-P}\boldsymbol{G}] \tag{4.58}$$

have maximum rank, where N is the dimension of the state vector and P is the dimension of the control function $\boldsymbol{u}(t)$. In the previous example, $N = 2$, $P = 1$ and the matrix is

$$[\boldsymbol{G} \mid \boldsymbol{FG}] = \begin{bmatrix} 1 & -1 \\ 1 & -1 \end{bmatrix}$$

which has rank 1, thus indicating a noncontrollable system. The approach through the modal analysis is informative in that it demonstrates the actual mechanism causing the effect, but it is impractical as a test when the simple test of Equation (4.58) is available.

4.3.3. Observability of Systems

The third property of a system is that of observability. Before the concept of observability can be considered, distinction must be made between the state vector of a system and the output of a system. Discussion to this point has centered on the state of a system, which for all practical purposes consists of a number of variables in which coupled first-order differential equations can be written which describe the system dynamics. The output of a system is the set of externally measurable functions available for estimation or control purposes. The definition of system output is usually more a function of what one can reasonably measure than it is of intrinsic dynamic properties. For example, in the spring-mass-damper system of Section 4.2, the position of the mass would probably be easily measured, whereas measurement of velocity would be more difficult.

In the most general case, the output function is some function of the state vector and perhaps time, expressed as

$$\boldsymbol{y}(t) = \boldsymbol{h}(\boldsymbol{x}, t) \tag{4.59}$$

and in the case of a linear system, this function has linear form,

$$\boldsymbol{y}(t) = \boldsymbol{H}(t)\boldsymbol{x}(t). \tag{4.60}$$

Note here that a linear system must, in addition to exhibiting a linear state equation as implied by Equation (4.3), also exhibit a linear output equation. Observability, then, concerns the usefulness of available measurement data in the form of the output function in determining values for the state variables of a system.

A system is said to be observable over the interval (t_1, t_2) if the system state at time t_2 can be determined from observations of the output during the interval (t_1, t_2). This concept can best be illustrated by the examples of the previous section, with the output equation specified as

$$\boldsymbol{y}(t) = \boldsymbol{H}\boldsymbol{x}(t)$$

where

$$\boldsymbol{H} = [-1 \quad 1].$$

Under the change of variable previously specified, the output equation is

$$y = HPz = [0 \quad 1] \begin{bmatrix} z_1 \\ z_2 \end{bmatrix}$$

or,

$$y = z_2.$$

Thus, mode z_1 does not affect the output y. By reasoning similar to that used in examining the concept of controllability, it can be seen that the output y does not yield sufficient information to determine the values of the states x_1 and x_2, no matter what length of time is available for the observation. This system is then nonobservable, in the sense of the above definition.

Tests for observability can be derived for both continuous and discrete systems and can take the following forms. A necessary and sufficient condition for a continuous, linear, time-invariant system to be observable is that the matrix

$$[H^T \,\vdots\, F^T H^T \,\vdots\, \cdots \,\vdots\, (F^T)^{N-M-1} H^T \,\vdots\, (F^T)^{N-M} H^T] \tag{4.61}$$

has maximum rank, where N is the dimension of the state vector and M is the dimension of the measurement vector y. For a discrete time system, the corresponding matrix is

$$[H^T \,\vdots\, (\Phi^T)^{-1} H^T \,\vdots\, \cdots \,\vdots\, (\Phi^T)^{-(N-M-1)} H^T \,\vdots\, (\Phi^T)^{-(N-M)} H^T] \tag{4.62}$$

where $(\Phi^T)^{-(N-M)}$ symbolically represents the product of the inverse of Φ^T with itself $(N - M)$ times. Analysis similar to that leading to Equation (4.56) will verify the appropriateness of these tests.

In the previous example, the test of Equation (4.61) yields

$$[H^T \,\vdots\, F^T H^T] = \begin{bmatrix} -1 & 2 \\ 1 & -2 \end{bmatrix}$$

which is of rank 1, thus verifying that the system is nonobservable, as already shown by the modal analysis.

To illustrate the application of the controllability and observability tests to discrete time systems, the difference equation equivalent of the example system described in this and the previous section will be considered. From the developments of Section 4.2.2, the difference equation is

$$x(t_l) = \Phi(t_l, t_{l-1}) x(t_{l-1}) + \Gamma(t_l, t_{l-1}) u(t_{l-1})$$

where

$$\Gamma(t_l, t_{l-1}) = \int_{t_{l-1}}^{t_l} \Phi(t_l, \sigma) G(\sigma)\, d\sigma. \tag{4.63}$$

In the present situation, the transition matrix $\Phi(t_l, t_{l-1})$ can be readily obtained from the eigenvalues and eigenvectors previously determined in Section 4.3.2. Thus,

$$\Phi(t_l, t_{l-1}) = P \begin{bmatrix} e^{\lambda_1(t_l - t_{l-1})} & 0 \\ 0 & e^{\lambda_2(t_l - t_{l-1})} \end{bmatrix} P^{-1}$$

$$= \begin{bmatrix} e^{-(t_l - t_{l-1})} & 0 \\ e^{-(t_l - t_{l-1})} - e^{-2(t_l - t_{l-1})} & e^{-2(t_l - t_{l-1})} \end{bmatrix}.$$

For numerical convenience, the time interval $(t_l - t_{l-1})$ will be considered to be constant at the value ln 2, yielding

$$\Phi(t_l, t_{l-1}) = \begin{bmatrix} \frac{1}{2} & 0 \\ \frac{1}{4} & \frac{1}{4} \end{bmatrix}.$$

The expression for $\Gamma(t_l, t_{l-1})$ is

$$\int_{t_{l-1}}^{t_l} \begin{bmatrix} e^{-(t_l - \sigma)} & 0 \\ e^{-(t_l - \sigma)} - e^{-2(t_l - \sigma)} & e^{-2(t_l - \sigma)} \end{bmatrix} \begin{bmatrix} 1 \\ 1 \end{bmatrix} d\sigma$$

$$= \int \begin{bmatrix} e^{-(t_l - \sigma)} \\ e^{-(t_l - \sigma)} \end{bmatrix} d\sigma = \begin{bmatrix} e^{-(t_l - \sigma)} \\ e^{-(t_l - \sigma)} \end{bmatrix}_{t_{l-1}}^{t_l}.$$

Evaluating the limits and using $(t_l - t_{l-1}) = \ln 2$, yield

$$\Gamma(t_{l-1}, t_l) = \Gamma = \begin{bmatrix} \frac{1}{2} \\ \frac{1}{2} \end{bmatrix}.$$

The difference equation can now be written as

$$x(l) = \Phi x(l - 1) + \Gamma u(l - 1)$$

where Φ and Γ are given by the expressions developed above. Now the test for controllability can be constructed from Equation (4.56) as

$$[\Gamma \,\vdots\, \Phi\Gamma] = \begin{bmatrix} \frac{1}{2} & \frac{1}{4} \\ \frac{1}{2} & \frac{1}{4} \end{bmatrix}$$

which has rank 1, indicating that the system is noncontrollable. This is a test for two stages, $l = 2$. The reader can verify that similar results are obtained for all $l > 2$. Similarly, the observability test from Equation (4.62) is

$$[\boldsymbol{H}^T \,\vdots\, (\boldsymbol{\Phi}^T)^{-1}\boldsymbol{H}^T] = \begin{bmatrix} -1 & -4 \\ 1 & 4 \end{bmatrix}$$

which also has rank 1, indicating that the system is nonobservable.

4.3.4. General Results for Time-Varying Systems

There are equivalent tests for time-varying systems, and these are mentioned here for the purpose of completeness. For controllability of continuous, time-varying, linear systems, the matrix

$$\int_{t_0}^{t} \boldsymbol{\Phi}(t, \sigma)\boldsymbol{G}(\sigma)\boldsymbol{G}^T(\sigma)\boldsymbol{\Phi}^T(t, \sigma)\, d\sigma \tag{4.64}$$

must be positive definite. For a discrete system, the analogous matrix is

$$\sum_{i=1}^{l-1} \boldsymbol{\Phi}(t_{l-1}, t_i)\boldsymbol{\Gamma}(t_i)\boldsymbol{\Gamma}^T(t_i)\boldsymbol{\Phi}^T(t_{l-1}, t). \tag{4.65}$$

The tests for observability are the positive definiteness of the matrices

$$\int_{t_0}^{t} \boldsymbol{\Phi}^T(\sigma, t)\boldsymbol{H}^T(\sigma)\boldsymbol{H}(\sigma)\boldsymbol{\Phi}(\sigma, t)\, d\sigma \tag{4.66}$$

for continuous systems, and

$$\sum_{i=1}^{l-1} \boldsymbol{\Phi}^T(t_i, t_l)\boldsymbol{H}^T(t_i)\boldsymbol{H}(t_i)\boldsymbol{\Phi}(t_i, t_l) \tag{4.67}$$

for discrete systems.

4.4. UNCERTAINTIES IN LINEAR SYSTEMS

Linear system characteristics have now been described in considerable detail, and it has been shown that the solution to the continuous system can be expressed in quite general terms for both the homogeneous and nonhomogeneous cases. These considerations have been deterministic in nature, in that the initial conditions and the system input have been considered to be precisely definable quantities. In many cases of practical import, either or both of these quantities may be of a random nature, in that $\boldsymbol{x}(t_0)$ may be a random vector and $\boldsymbol{u}(t)$ a vector random process. In this instance, these quantities cannot be precisely defined, and one must resort to some kind of probabilistic description of the system state $\mathbf{X}(t)$, now a random process. The most common approach is to use the second-moment characterization as discussed in Chapter 3. This approach leads to the determination of the mean value function and the covariance matrix function for the case of continuous time systems, and to the analogous quantities for discrete time systems. Since slightly different approaches are used for the continuous and discrete time systems, each will be treated separately.

4.4.1. Uncertainties in Continuous Systems

For the continuous time system, it has been previously shown that the homogeneous equation

$$\dot{\boldsymbol{x}}(t) = \boldsymbol{F}(t)\boldsymbol{x}(t)$$

has a solution of the form

$$\boldsymbol{x}(t) = \boldsymbol{\Phi}(t, t_0)\boldsymbol{x}(t_0).$$

In many applications, such a system will be responding to a set of initial conditions which are not completely known but instead constitute a set of random variables. The state vector $\boldsymbol{x}(t)$ is then a vector-valued random process as described in Chapter 3, and denoted by $\mathbf{X}(t)$. As such, it can be partially described by the covariance matrix kernel

$$K(t_1, t_2) = E[\{\mathbf{X}(t_1) - \mu(t_1)\}\{\mathbf{X}(t_2) - \mu(t_2)\}^T] \tag{4.68}$$

and the mean value function

$$\mu(t) = E[\mathbf{X}(t)]. \tag{4.69}$$

In most instances, the information of interest is contained in the covariance matrix

$$P(t) = K(t, t) \tag{4.70}$$

and very concise equations for this matrix in terms of the initial uncertainties can be developed.

The initial value of the covariance matrix is

$$P(t_0) = E[\{\mathbf{X}(t_0) - \mu(t_0)\}\{\mathbf{X}(t_0 - \mu(t_0)\}^T] \tag{4.71}$$

and, from basic considerations

$$\mathbf{X}(t_1) = \Phi(t_1, t_0)\mathbf{X}(t_0)$$

$$\mathbf{X}(t_2) = \Phi(t_2, t_0)\mathbf{X}(t_0). \tag{4.72}$$

The mean value function at times t_1 and t_2 is

$$\mu(t_1) = \Phi(t_1, t_0)\mu(t_0)$$

$$\mu(t_2) = \Phi(t_2, t_0)\mu(t_0). \tag{4.73}$$

Now, substituting Equations (4.72) and (4.73) into the expression for $K(t_1, t_2)$ as given by Equation (4.68) yields

$$\begin{aligned} K(t_1, t_2) &= E[\{\Phi(t_1, t_0)[\mathbf{X}(t_0) - \mu(t_0)]\}\{\Phi(t_2, t_0)[\mathbf{X}(t_0) - \mu(t_0)]\}^T] \\ &= \Phi(t_1, t_0)E[\{\mathbf{X}(t_0) - \mu(t_0)\}\{\mathbf{X}(t_0) - \mu(t_0)\}^T]\Phi^T(t_2, t_0) \\ &= \Phi(t_1, t_0)P(t_0)\Phi^T(t_2, t_0) \end{aligned} \tag{4.74}$$

Thus, the covariance matrix kernel $K(t_1, t_2)$ can be obtained from $P(t_0)$, which describes the initial uncertainties of the system. When $t_1 = t_2 = t$,

$$K(t, t) = P(t) = \Phi(t, t_0)P(t_0)\Phi^T(t, t_0) \tag{4.75}$$

giving the covariance matrix $P(t)$ as a function of the initial uncertainties.

For the particular case of a homogeneous linear dynamic system, a relationship between $P(t)$ and $K(t_1, t_2)$ can be shown. If the identity

$$P(t_0) = \Phi(t_0, t_1)P(t_1)\Phi^T(t_0, t_1) \tag{4.76}$$

is substituted into Equation (4.74), the result is

$$K(t_1, t_2) = \Phi(t_1, t_0)\Phi(t_0, t_1)P(t_1)\Phi^T(t_0, t_1)\Phi^T(t_2, t_0). \tag{4.77}$$

By use of the special characteristics the transition matrix outlined in Section 4.2.1, this equation reduces to

$$K(t_1, t_2) = P(t_1)\Phi^T(t_2, t_1). \tag{4.78}$$

In an analogous fashion, it can be shown that

$$K(t_2, t_1) = \Phi(t_2, t_1)P(t_1). \tag{4.79}$$

These expressions indicate that knowledge of $P(t)$ is sufficient to describe the random vector function $\mathbf{X}(t)$ through second-moment considerations, since $K(t_1, t_2)$ can be inferred from knowledge of $P(t)$. This useful result permits the relationship of Equation (4.75) to be used in the analysis of linear stochastic systems.

As an example of these principles, consider the water brake problem introduced in Section 4.2. There it was shown that the state transition matrix is

$$\Phi(t, 0) = \begin{bmatrix} 1 & \frac{1}{2} - \frac{1}{2}e^{-2t} \\ 0 & e^{-2t} \end{bmatrix}.$$

Now the initial position and velocity are considered to be independent random variables with the following characteristics:

Position $X(t_0)$: mean value = 1,000 ft; variance = 10,000 ft^2.
Velocity $V(t_0)$: mean value = 2,000 ft/sec; variance = 20,000 ft^2/sec^2.

Such a situation can arise when only estimates of these initial conditions are available, where the estimate is the mean of the particular distribution. The initial covariance matrix $P(0)$ is

$$P(0) = \begin{bmatrix} 10{,}000 & 0 \\ 0 & 20{,}000 \end{bmatrix}.$$

At some later time t, the covariance matrix is, from Equation (4.75),

$$\boldsymbol{P}(t) = \begin{bmatrix} 1 & \frac{1}{2} - \frac{1}{2}e^{-2t} \\ 0 & e^{-2t} \end{bmatrix} \begin{bmatrix} 10{,}000 & 0 \\ 0 & 20{,}000 \end{bmatrix} \begin{bmatrix} 1 & 0 \\ \frac{1}{2} - \frac{1}{2}e^{-2t} & e^{-2t} \end{bmatrix}$$

$$= \begin{bmatrix} \frac{3}{2} - e^{-2t} - \frac{1}{4}e^{-4t} & e^{-2t} - e^{-4t} \\ e^{-2t} - e^{-4t} & 2e^{-4t} \end{bmatrix} \times 10^4.$$

For $t = 2$ seconds, this matrix yields

$$\boldsymbol{P}(2) = \begin{bmatrix} 14820 & 180 \\ 180 & 6 \end{bmatrix}.$$

This result is interpreted as follows:

$$\text{VAR}[\text{X}(2)] = 14{,}820 \text{ ft}^2$$

$$\text{VAR}[\text{V}(2)] = 6 \text{ ft}^2/\text{sec}^2$$

$$\text{COV}[\text{X}(2), \text{V}(2)] = 180 \text{ ft}^2/\text{sec}.$$

Thus, the variance of position has increased, that of velocity has decreased, and these two random variables are no longer independent, since the covariance has a nonzero value. The mean values are obtained from Equation (4.73):

$$E[\text{X}(2)] = 1981.2 \text{ ft}$$

$$E[\text{V}(2)] = 36.6 \text{ ft/sec}.$$

If the mean values constitute an estimate, then the covariance matrix indicates the quality of the estimate.

In the event that the system has a deterministic input during the interval (t_1, t_2), then the nonhomogeneous equation applies:

$$\dot{\mathbf{X}}(t) = \boldsymbol{F}\mathbf{X}(t) + \boldsymbol{G}\boldsymbol{u}(t). \tag{4.80}$$

As previously derived, the solution for this equation is

$$\mathbf{X}(t) = \boldsymbol{\Phi}(t, t_0)\mathbf{X}(t_0) + \int_{t_0}^{t} \boldsymbol{\Phi}(t, \sigma)\boldsymbol{G}(\sigma)\boldsymbol{u}(\sigma)\, d\sigma. \tag{4.81}$$

The integral term is not random, since $\boldsymbol{u}(t)$ is not random. Let the integral be denoted by $\boldsymbol{z}(t, t_0)$,

$$\boldsymbol{z}(t, t_0) = \int_{t_0}^{t} \boldsymbol{\Phi}(t, \sigma)\boldsymbol{G}(\sigma)\boldsymbol{u}(\sigma)\, d\sigma \tag{4.82}$$

so that

$$\mathbf{X}(t) = \boldsymbol{\Phi}(t, t_0)\mathbf{X}(t_0) + \boldsymbol{z}(t, t_0). \tag{4.83}$$

Also, the mean value of $\mathbf{X}(t)$ is

$$\begin{aligned}\boldsymbol{\mu}(t) &= E[\boldsymbol{\Phi}(t, t_0)\mathbf{X}(t_0) + \boldsymbol{z}(t, t_0)]\\ &= \boldsymbol{\Phi}(t, t_0)\boldsymbol{\mu}(t_0) + \boldsymbol{z}(t, t_0)\end{aligned} \tag{4.84}$$

where the latter form results because $\boldsymbol{z}(t, t_0)$ is not random. The covariance matrix kernel is now

$$\begin{aligned}\boldsymbol{K}(t_1, t_2) &= E[\{\mathbf{X}(t_1) - \boldsymbol{\mu}(t_1)\}\{\mathbf{X}(t_2) - \boldsymbol{\mu}(t_2)\}^T]\\ &= E[\{\boldsymbol{\Phi}(t_1, t_0)[\mathbf{X}(t_0) - \boldsymbol{\mu}(t_0)]\}\{\boldsymbol{\Phi}(t_2, t_0)[\mathbf{X}(t_0) - \boldsymbol{\mu}(t_0)]\}^T]\\ &= \boldsymbol{\Phi}(t_1, t_0)\boldsymbol{P}(t_0)\boldsymbol{\Phi}^T(t_2, t_0).\end{aligned} \tag{4.85}$$

This expression is identical to the previous result of Equation (4.74), indicating that a deterministic input does not affect the uncertainties in a linear system. The covariance matrix $\boldsymbol{P}(t)$ is still given by Equation (4.75).

On the other hand, if the input is itself a random process designated as $\mathbf{U}(t)$, then Equation (4.82) yields the random quantity $\mathbf{Z}(t, t_0)$. The counterpart of Equation (4.84) is then

$$\boldsymbol{\mu}(t) = \boldsymbol{\Phi}(t, t_0)\boldsymbol{\mu}(t_0) + E[\mathbf{Z}(t, t_0)] \tag{4.86}$$

where

$$E[\mathbf{Z}(t, t_0)] = \int_{t_0}^{t} \boldsymbol{\Phi}(t, \sigma)\boldsymbol{G}(\sigma)E[\mathbf{U}(\sigma)]\, d\sigma \tag{4.87}$$

and $E[\mathbf{U}(t)]$ is the mean value function of the random process $\mathbf{U}(t)$. For simplicity in the present derivation, assume that both $\mathbf{X}(t_0)$ and $\mathbf{U}(t)$ have zero means. From the previous derivations, it can be seen that mean values affect the results in a very predictable way, and as long as only central moments such as variances are considered, no loss of generality is produced by the assumption of a zero mean. Thus, for those considerations which follow it will be assumed that

$$E[\mathbf{X}(t_0)] = \boldsymbol{\mu}(t_0) = 0$$
$$E[\mathbf{U}(t)] = 0, \quad \text{for all } t \tag{4.88}$$

and, as a result,

$$\boldsymbol{\mu}(t) = 0$$
$$E[\mathbf{Z}(t, t_0)] = 0.$$

Now, since $\mathbf{U}(t)$ is a random process, it is described by a second moment characterization consisting of the mean value function, assumed to be zero above, and a covariance matrix kernel

$$K_{\mathbf{U}}(t_1, t_2) = E[\mathbf{U}(t_1)\mathbf{U}^T(t_2)] \tag{4.89}$$

where the particular form of this expression results from the zero-mean assumption. With these assumptions and results, the covariance matrix kernel of the process $\mathbf{X}(t)$ can be constructed as

$$\begin{aligned} K(t_1, t_2) &= E[\mathbf{X}(t_1)\mathbf{X}^T(t_2)] \\ &= E[\{\boldsymbol{\Phi}(t_1, t_0)\mathbf{X}(t_0) + \mathbf{Z}(t_1, t_0)\}\{\boldsymbol{\Phi}(t_2, t_0)\mathbf{X} + \mathbf{Z}(t_2, t_0)\}^T] \\ &= E[\boldsymbol{\Phi}(t_1, t_0)\mathbf{X}(t_0)\mathbf{X}^T(t_0)\boldsymbol{\Phi}^T(t_2, t_0) + \boldsymbol{\Phi}(t_1, t_0)\mathbf{X}(t_0)\mathbf{Z}(t_2, t_0) \\ &\quad + \mathbf{Z}(t_1, t_0)\mathbf{X}^T(t_0)\boldsymbol{\Phi}^T(t_2, t_0) + \mathbf{Z}(t_1, t_0)\mathbf{Z}^T(t_2, t_0)]. \end{aligned} \tag{4.90}$$

When the expectation of this expression is taken, the second and third terms lead to components involving the forms

$$E[\mathbf{X}(t_0)\mathbf{Z}^T(t_2, t_0)] \quad \text{and} \quad E[\mathbf{Z}(t_1, t_0)\mathbf{X}^T(t_0)] \tag{4.91}$$

which are composed of covariances between $\mathbf{X}(t_0)$ and the term resulting from the input process subsequent to t_0. Normally these two random terms are independent, with the result that these covariances are zero, and therefore the expressions of Equation (4.91) are zero. This result yields

$$K(t_1, t_2) = \boldsymbol{\Phi}(t_1, t_0)P(t_0)\boldsymbol{\Phi}^T(t_2, t_0) + E[\mathbf{Z}(t_1, t_0)\mathbf{Z}^T(t_2, t_0)]. \tag{4.92}$$

The first term is identical to that obtained from the homogeneous case or the nonhomogeneous case with deterministic input. The second term represents the uncertainty produced by the random input process $\mathbf{U}(t)$, and is given by

$$E[\mathbf{Z}(t_1, t_0)\mathbf{Z}^T(t_2, t_0)]$$
$$= E\left[\left[\int_{t_0}^{t_1} \mathbf{\Phi}(t_1, \sigma)\boldsymbol{G}(\sigma)\mathbf{U}(\sigma)\, d\sigma\right]\left[\int_{t_0}^{t_2} \mathbf{\Phi}(t_2, \nu)\boldsymbol{G}(\nu)\mathbf{U}(\nu)\, d\nu\right]^T\right]$$
$$= E\left[\int_{t_0}^{t_1}\int_{t_0}^{t_2} \mathbf{\Phi}(t_1, \sigma)\boldsymbol{G}(\sigma)\mathbf{U}(\sigma)\mathbf{U}^T(\nu)\boldsymbol{G}^T(\nu)\mathbf{\Phi}^T(t_2, \nu)\, d\nu\, d\sigma\right]. \quad (4.93)$$

The variable of integration ν has been introduced into the second integral to avoid confusion with σ when the product of integrals is written as a double integral. Since $\mathbf{U}(\sigma)$ and $\mathbf{U}(\nu)$ are the only random quantities in this integral, the operations of integration and taking expectation can be interchanged, with the result

$$E[\mathbf{Z}(t_1, t_0)\mathbf{Z}^T(t_2, t_0)]$$
$$= \int_{t_0}^{t_1}\int_{t_0}^{t_2} \mathbf{\Phi}(t_1, \sigma)\boldsymbol{G}(\sigma)\boldsymbol{K}_{\mathbf{U}}(\sigma, \nu)\,\boldsymbol{G}^T(\nu)\mathbf{\Phi}^T(t_2, \nu)\, d\nu\, d\sigma. \quad (4.94)$$

This integral then represents the contribution to $\boldsymbol{K}(t_1, t_2)$, the covariance matrix kernel of $\mathbf{X}(t)$, due to the randomness of the input $\mathbf{U}(t)$. As might be expected, such contribution depends on the covariance matrix kernel of $\mathbf{U}(t)$, $\boldsymbol{K}_{\mathbf{U}}(t_1, t_2)$.

If a further assumption regarding the characteristics of $\mathbf{U}(t)$ is made, namely that it is a white noise process as described in Section 3.6, then

$$\boldsymbol{K}_{\mathbf{U}}(t_1, t_2) = \boldsymbol{Q}(t_1)\delta(t_1 - t_2) \quad (4.95)$$

and integration of Equation (4.94) with respect to ν can be accomplished using the sampling property of the delta function, with the result

$$E[\mathbf{Z}(t_1, t_0)\mathbf{Z}^T(t_2, t_0] = \int_{t_0}^{t_1} \mathbf{\Phi}(t_1, \sigma)\boldsymbol{G}(\sigma)\boldsymbol{Q}(\sigma)\boldsymbol{G}^T(\sigma)\mathbf{\Phi}^T(t_2, \sigma)\, d\sigma. \quad (4.96)$$

In this case, the final form for the covariance matrix kernel of $\mathbf{X}(t)$ is

$$\boldsymbol{K}(t_1, t_2) = \boldsymbol{\Phi}(t_1, t_0)\boldsymbol{P}(t_0)\boldsymbol{\Phi}^T(t_2, t_0) + \int_{t_0}^{t_1} \boldsymbol{\Phi}(t_1, \sigma)\boldsymbol{G}(\sigma)\boldsymbol{Q}(\sigma)\boldsymbol{G}^T(\sigma)\boldsymbol{\Phi}^T(t_2, \sigma)\, d\sigma. \quad (4.97)$$

When $t_1 = t_2 = t$, the covariance matrix $\boldsymbol{P}(t)$ results,

$$\boldsymbol{K}(t, t) = \boldsymbol{P}(t) = \boldsymbol{\Phi}(t, t_0)\boldsymbol{P}(t_0)\boldsymbol{\Phi}^T(t, t_0) + \int_{t_0}^{t} \boldsymbol{\Phi}(t, \sigma)\boldsymbol{G}(\sigma)\boldsymbol{Q}(\sigma)\boldsymbol{G}^T(\sigma)\boldsymbol{\Phi}^T(t, \sigma)\, d\sigma. \quad (4.98)$$

The contributions to $\boldsymbol{P}(t)$ resulting from initial uncertainties and from the random input are readily identified as the first and second terms, respectively, of this expression.

As an example of these principles the water brake problem previously considered will be modified. Now, in addition to the initial uncertainties, assume that the rocket thrust is not zero at brake initiation, but rather is a white noise random process with covariance kernel

$$K_{\mathrm{U}}(t_1, t_2) = 3.86 \times 10^7\, \delta(t_1 - t_2) \text{ pound}^2$$

and constant mean value function

$$\boldsymbol{\mu}_{\mathrm{U}}(t) = 62{,}100 \text{ pounds.}$$

The initial conditions are random variables with the following properties:

Position $\mathrm{X}(t_0)$: mean value = 1,000 ft; variance = 10,000 ft^2
Velocity $\mathrm{V}(t_0)$: mean value = 2,000 ft/sec; variance = 20,000 $\mathrm{ft}^2/\mathrm{sec}^2$.

It is desired to find the mean value function and covariance matrix two seconds after brake initiation. The transition matrix has been previously determined as

$$\boldsymbol{\Phi}(t) = \begin{bmatrix} 1 & \frac{1}{2} - \frac{1}{2}e^{-2t} \\ 0 & e^{-2t} \end{bmatrix}$$

for $t_0 = 0$. To determine the mean value function at two seconds, the expression is

$$E[\mathbf{X}(t)] = E\ \mathbf{\Phi}(t, t_0)\mathbf{X}(t_0) + \int_{t_0}^{t} \mathbf{\Phi}(t, \sigma)\mathbf{G}(\sigma)\mathbf{U}(\sigma)\, d\sigma$$

$$= \mathbf{\Phi}(t, t_0)E[\mathbf{X}(t_0)] + \int_{t_0}^{t} \mathbf{\Phi}(t, \sigma)\mathbf{G}(\sigma)E[\mathbf{U}(\sigma)]\, d\sigma.$$

In the present case, $E[\mathbf{U}]$ in the proper units is

$$62{,}100\ (32.2/2000) = 10^3\ \text{ft/sec}^2$$

with the result

$$E[\mathbf{X}(2)] = \begin{bmatrix} 1 & \frac{1}{2} - \frac{1}{2}e^{-4} \\ 0 & e^{-4} \end{bmatrix}\begin{bmatrix} 1000 \\ 2000 \end{bmatrix} + \int_0^2 \begin{bmatrix} 1 & \frac{1}{2} - \frac{1}{2}e^{-2(2-\sigma)} \\ 0 & e^{-2(2-\sigma)} \end{bmatrix}\begin{bmatrix} 0 \\ 1 \end{bmatrix} \times 10^3\, d\sigma.$$

Performing the multiplication and integration yields

$$E[\mathbf{X}(2)] = \begin{bmatrix} 1981.2 \\ 36.6 \end{bmatrix} + 10^3 \int_0^2 \begin{bmatrix} \frac{1}{2} - \frac{1}{2}e^{-2(2-\sigma)} \\ e^{-2(2-\sigma)} \end{bmatrix} d\sigma$$

$$= \begin{bmatrix} 1981.2 \\ 36.6 \end{bmatrix} + 10^3 \begin{bmatrix} \frac{1}{2}\sigma - \frac{1}{4}e^{-2(2-\sigma)} \\ \frac{1}{2}e^{-2(2-\sigma)} \end{bmatrix}_0^2.$$

Finally, from evaluation of the limits,

$$E[\mathbf{X}(2)] = \begin{bmatrix} 1981.2 \\ 36.6 \end{bmatrix} + \begin{bmatrix} 755 \\ 491 \end{bmatrix} = \begin{bmatrix} 2736 \\ 528 \end{bmatrix}.$$

Thus, the contribution due to the initial conditions and that due to the input $\mathbf{U}(t)$ are clearly identified.

To determine the covariance matrix at two seconds, the equation is

$$\boldsymbol{P}(t) = \mathbf{\Phi}(t, t_0)\boldsymbol{P}(t_0)\mathbf{\Phi}^T(t, t_0) + \int_{t_0}^{t} \mathbf{\Phi}(t, \sigma)\boldsymbol{G}(\sigma)\boldsymbol{Q}(\sigma)\boldsymbol{G}^T(\sigma)\mathbf{\Phi}^T(\sigma, t)\, d\sigma$$

where $\boldsymbol{Q}(t)$ in the proper units is

$$3.86 \times 10^7 (32.2/2000)^2 = 10^4 \text{ ft}^2/\text{sec}^4.$$

Making the proper substitutions yields, for $t = 2$, $t_0 = 0$,

$$\boldsymbol{P}(2) = \begin{bmatrix} 1 & \frac{1}{2} - \frac{1}{2}e^{-4} \\ 0 & e^{-4} \end{bmatrix} \begin{bmatrix} 10^4 & 0 \\ 0 & 2 \times 10^4 \end{bmatrix} \begin{bmatrix} 1 & 0 \\ \frac{1}{2} - \frac{1}{2}e^{-4} & e^{-4} \end{bmatrix}$$

$$+ 10^4 \int_0^2 \begin{bmatrix} 1 & \frac{1}{2} - \frac{1}{2}e^{-2(2-\sigma)} \\ 0 & e^{-2(2-\sigma)} \end{bmatrix} \begin{bmatrix} 0 \\ 1 \end{bmatrix} [0 \quad 1] \begin{bmatrix} 1 & 0 \\ \frac{1}{2} - \frac{1}{2}e^{-2(2-\sigma)} & e^{-2(2-\sigma)} \end{bmatrix} d\sigma.$$

Performing the indicated operations finally yields

$$\boldsymbol{P}(2) = \begin{bmatrix} 14280 & 180 \\ 180 & 6 \end{bmatrix} + \begin{bmatrix} 3170 & 1205 \\ 1205 & 2500 \end{bmatrix} = \begin{bmatrix} 17450 & 1385 \\ 1385 & 2506 \end{bmatrix}.$$

Here again, from the first term one can determine the contribution to the uncertainty at two seconds due to the initial uncertainty, and from the second the uncertainty due to the random input.

4.4.2. Uncertainties in Discrete Systems

The discrete linear system is described by the difference equation

$$\mathbf{X}(l) = \boldsymbol{\Phi}(l, l-1)\mathbf{X}(l-1) + \boldsymbol{\Gamma}(l, l-1)\mathbf{U}(l-1) \tag{4.99}$$

where $\mathbf{X}(l)$ and $\mathbf{U}(l)$ are now random sequences. Again, for mathematical convenience and without loss of generality, $\mathbf{X}(0)$ and $\mathbf{U}(0)$ will be assumed to have zero means. Note that

$$\mathbf{X}(1) = \boldsymbol{\Phi}(1,0)\mathbf{X}(0) + \boldsymbol{\Gamma}(1,0)\mathbf{U}(0).$$

also has a zero mean, as will all $\mathbf{X}(l)$. Thus, the above assumptions imply that

$$E[\mathbf{X}(l)] = \boldsymbol{\mu}(l) = 0. \tag{4.100}$$

In this case, the covariance matrix kernel of $\mathbf{X}(l)$ is

$$\begin{aligned} \boldsymbol{K}(l, m) &= E[\{\mathbf{X}(l) - \boldsymbol{\mu}(l)\}\{\mathbf{X}(m) - \boldsymbol{\mu}(m)\}^T] \\ &= E[\mathbf{X}(l)\mathbf{X}^T(m)]. \end{aligned} \tag{4.101}$$

Substituting from Equation (4.99) into this expression yields

$$\boldsymbol{K}(l,m) = E[\{\boldsymbol{\Phi}(l,l-1)\mathbf{X}(l-1) + \boldsymbol{\Gamma}(l,l-1)\mathbf{U}(l-1)\}\{\boldsymbol{\Phi}(m,m-1)\mathbf{X}(m-1) + \boldsymbol{\Gamma}(m,m-1)\mathbf{U}(m-1)\}^T]. \quad (4.102)$$

When this product is taken, some cross product terms will result, which will be independent if $\mathbf{U}(l)$ is a white noise sequence; that is, if

$$\boldsymbol{K}_{\mathbf{U}}(l,m) = \begin{cases} \boldsymbol{Q}(l), & l = m \\ 0, & l \neq m. \end{cases}$$

Then the cross product terms vanish and Equation (4.102) reduces to

$$\boldsymbol{K}(l,m) = \boldsymbol{\Phi}(l,l-1)\boldsymbol{K}(l-1,m-1)\boldsymbol{\Phi}^T(m,m-1) + \boldsymbol{\Gamma}(l,l-1)\boldsymbol{Q}(l)\boldsymbol{\Gamma}^T(m,m-1). \quad (4.103)$$

This matrix difference equation for $\boldsymbol{K}(l, m)$ can be solved by the usual recursive methods.

Ordinarily, the covariance matrix $\boldsymbol{P}(l)$ is the quantity of interest, in which case the equation reduces to

$$\boldsymbol{P}(l) = \boldsymbol{K}(l,l) = \boldsymbol{\Phi}(l,l-1)\boldsymbol{P}(l-1)\boldsymbol{\Phi}^T(l,l-1) + \boldsymbol{\Gamma}(l,l-1)\boldsymbol{Q}(l-1)\boldsymbol{\Gamma}^T(l,l-1) \quad (4.104)$$

This difference equation in the single independent variable l describes the behavior of the covariance matrix $\boldsymbol{P}(l)$.

4.5. THE NONLINEAR SYSTEM

The linear system has been investigated in great detail because its characteristics permit such detailed analysis. Such is not the case when nonlinear systems are considered, for there are no general approaches to the solution of nonlinear differential equations. Without such a general approach, the practice of solving nonlinear equations is reduced to the act of recognizing a particular method as it applies in certain cases, or to the use of approximation methods. The approximation techniques will be considered briefly here since they are generally applicable regardless of the specific form of the equation.

4.5.1. The Perturbation Method Approach

Although there are several approximation techniques for the solution of nonlinear equations, the perturbation technique is probably the most widely used approach because it leads to a set of linear equations which describe small changes or perturbation around a reference solution. Because linear systems can be analyzed in detail by quite general methods, this approach quite often provides interpretation of system characteristics not available with other methods. The basic premise of the perturbation method is that a reference solution of the nonlinear equation

$$\dot{\boldsymbol{x}}(t) = \boldsymbol{f}(\boldsymbol{x}(t), \boldsymbol{u}(t), t) \tag{4.105}$$

can be obtained. This is usually accomplished by some numerical technique of sufficient accuracy. Starting with some specified initial condition $\boldsymbol{x}^*(t_0)$ and utilizing some specified input $\boldsymbol{u}^*(t)$, the system equation is solved, yielding the reference solution $\boldsymbol{x}^*(t)$ over the desired time interval. Now, in the event that the initial condition is not $\boldsymbol{x}^*(t_0)$, but rather is changed by some small amount $\delta\boldsymbol{x}(t_0)$, and in a similar manner the input function is changed to $\boldsymbol{u}^*(t) + \delta\boldsymbol{u}(t)$, modification by the perturbations $\delta\boldsymbol{x}(t_0)$ and $\delta\boldsymbol{u}(t)$ will change the system equation to

$$\dot{\boldsymbol{x}}^* + \delta\dot{\boldsymbol{x}} = \boldsymbol{f}(\boldsymbol{x}^* + \delta\boldsymbol{x}, \boldsymbol{u}^* + \delta\boldsymbol{u}, t) \tag{4.106}$$

where $\delta\dot{\boldsymbol{x}}$ and $\delta\boldsymbol{x}$ are the perturbations in $\boldsymbol{x}$ and $\dot{\boldsymbol{x}}$ produced by the perturbations in $\boldsymbol{x}(t_0)$ and $\boldsymbol{u}(t)$. The close similarity between this concept and that of the variations of Appendix B should be evident. Expansion of Equation (4.106) in a multidimensional Taylor series yields

$$\boldsymbol{x}^* + \delta\dot{\boldsymbol{x}} = f(\boldsymbol{x}^*, \boldsymbol{u}^*, t) + \left[\frac{\partial f}{\partial \boldsymbol{x}}\right]_{\boldsymbol{x}^*, \boldsymbol{u}^*} \delta\boldsymbol{x} + \left[\frac{\partial f}{\partial \boldsymbol{u}}\right]_{\boldsymbol{x}^*, \boldsymbol{u}^*} \delta\boldsymbol{u} + \text{h.o.t.} \tag{4.107}$$

where h.o.t. indicates terms of quadratic and higher orders. So long as $\delta\boldsymbol{x}(t)$ and $\delta\boldsymbol{u}(t)$ are small, the higher-order terms are vanishingly small, and a useful approximation is

$$\dot{\boldsymbol{x}}^* + \delta\dot{\boldsymbol{x}} = f(\boldsymbol{x}^*, \boldsymbol{u}^*, t) + \left[\frac{\partial f}{\partial \boldsymbol{x}}\right]_{\boldsymbol{x}^*, \boldsymbol{u}^*} \delta\boldsymbol{x} + \left[\frac{\partial f}{\partial \boldsymbol{u}}\right]_{\boldsymbol{x}^*, \boldsymbol{u}^*} \delta\boldsymbol{u} \tag{4.108}$$

where the specific evaluation of the partial derivatives along the solution $\boldsymbol{x}^*(t)$, $\boldsymbol{u}^*(t)$ is implied. From Equation (4.105) the $\boldsymbol{x}^*$ and $\boldsymbol{f}$ terms above can be seen to cancel, yielding finally

$$\delta\dot{\boldsymbol{x}} = \left[\frac{\partial \boldsymbol{f}}{\partial \boldsymbol{x}}\right]_{\boldsymbol{x}^*,\, \boldsymbol{u}^*} \delta\boldsymbol{x} + \left[\frac{\partial \boldsymbol{f}}{\partial \boldsymbol{u}}\right]_{\boldsymbol{x}^*,\, \boldsymbol{u}^*} \delta\boldsymbol{u} \tag{4.109}$$

as a set of linear equations describing the time behavior of the perturbations $\delta\boldsymbol{x}$ around the reference solution $\boldsymbol{x}^*(t)$. As such, these equations are equivalent to Equation (4.3) and can be analyzed using any of the techniques discussed in Section 4.2. One must realize, however, that this set of equations is a time-varying set because the partial derivatives

$$\left[\frac{\partial \boldsymbol{f}}{\partial \boldsymbol{x}}\right] \quad \text{and} \quad \left[\frac{\partial \boldsymbol{f}}{\partial \boldsymbol{u}}\right] \tag{4.110}$$

are evaluated for each instant of time using the appropriate values of $\boldsymbol{x}^*(t)$ and $\boldsymbol{u}^*(t)$ as obtained from the reference solution. Once the solution for $\delta\boldsymbol{x}(t)$ is obtained, then the expression

$$\boldsymbol{x}(t) = \boldsymbol{x}^*(t) + \delta\boldsymbol{x}(t) \tag{4.111}$$

can be used to determine the system state. The main limitation to this method is the requirement for maintaining $\delta\boldsymbol{x}(t)$ and $\delta\boldsymbol{u}(t)$ small enough so that the linear approximation is valid.

As an example, consider the simple one-dimensional problem of a particle in the gravitational field of a much heavier object, in which a drag force proportional to the square of the velocity is produced. If x_1 is the position and x_2 the velocity coordinate, the system homogeneous equations are

$$\dot{x}_1 = x_2$$

$$\dot{x}_2 = \frac{k_1}{m x_1^2} - \frac{k_2}{m} x_2^2$$

where k_1 and k_2 are constants associated with the gravitational and drag forces, respectively. In the standard form, these equations are

$$\begin{bmatrix} \dot{x}_1 \\ \dot{x}_2 \end{bmatrix} = \begin{bmatrix} x_2 \\ \dfrac{k_1}{m x_1^2} - \dfrac{k_2}{m} x_2^2 \end{bmatrix}$$

or

$$\dot{\boldsymbol{x}} = \boldsymbol{f}(\boldsymbol{x}).$$

Obviously, this equation cannot be put into linear form. The perturbation equation, however, is obtained using the matrix

$$\frac{\partial \boldsymbol{f}}{\partial \boldsymbol{x}} = \begin{bmatrix} \dfrac{\partial f_1}{\partial x_1} & \dfrac{\partial f_1}{\partial x_2} \\ \dfrac{\partial f_2}{\partial x_1} & \dfrac{\partial f_2}{\partial x_2} \end{bmatrix} = \begin{bmatrix} 0 & 1 \\ -\dfrac{2k_1}{mx_1^3} & -\dfrac{2k_1 x_2}{m} \end{bmatrix}$$

as obtained using the matrix formalism discussed in Appendix A. Then the linear perturbation equation is

$$\begin{bmatrix} \delta\dot{x}_1(t) \\ \delta\dot{x}_2(t) \end{bmatrix} = \begin{bmatrix} 0 & 1 \\ -\dfrac{2k_1}{mx_1^2(t)} & -\dfrac{2k_2 x_2(t)}{m} \end{bmatrix} \begin{bmatrix} \delta x_1(t) \\ \delta x_2(t) \end{bmatrix}$$

which is the linearized equation describing the time behavior of δx_1 and δx_2, small perturbations from the reference solution $\boldsymbol{x}^*(t)$. This equation is in the standard homogeneous linear form

$$\dot{\boldsymbol{x}}(t) = \boldsymbol{F}(t)\boldsymbol{x}(t)$$

where the time-varying characteristic of $\boldsymbol{F}$ is emphasized. From this equation, the transition matrix $\boldsymbol{\Phi}(t, t_0)$ can be obtained, such that

$$\delta\dot{\boldsymbol{x}}(t) = \boldsymbol{\Phi}(t, t_0)\delta\boldsymbol{x}(t_0)$$

which gives the perturbation $\delta\boldsymbol{x}(t)$ at any time. In turn, the state of the system is obtained as

$$\boldsymbol{x}(t) = \boldsymbol{x}^*(t) + \delta\boldsymbol{x}(t).$$

The use of these equations is as follows. A reference solution $\boldsymbol{x}^*(t)$ is obtained, starting from some particular initial condition $\boldsymbol{x}^*(t_0)$. This solution is obtained by some suitable means, usually numerical integration of the equations. Then, if one desires the effect of small changes in $\boldsymbol{x}(t_0)$, he need not solve the nonlinear equations again, but rather he can determine the transition matrix $\boldsymbol{\Phi}(t, t_0)$ describing the behavior of the small perturbations $\delta\boldsymbol{x}(t)$ and can obtain the desired information by the use of the linear equations.

In the event that the nonlinear equations are not homogeneous—that is, there is an input function $\boldsymbol{u}(t)$—then the perturbation equations take the form of Equation (4.109). In the present example, if an input function $\boldsymbol{u}(t)$ is applied, in the form of some kind of drag-producing speed brake, for which the resulting acceleration is

$$a = -\frac{k_3 x_2^2}{m} \sin u$$

then the input function is the speed brake angle u. Thus,

$$\boldsymbol{f}(\boldsymbol{x}(t), \boldsymbol{u}(t)) = \begin{bmatrix} x_2 \\ \dfrac{k_1}{m x_1^2} - \dfrac{k_2 x_2^2}{m} - \dfrac{k_3 x_2^2}{m} \sin u \end{bmatrix}$$

and the pertinent matrices are

$$\left[\frac{\partial \boldsymbol{f}}{\partial \boldsymbol{x}}\right] = \begin{bmatrix} 0 & 1 \\ -\dfrac{2k_1}{m x_1^3} & -\dfrac{2k_2 x_2}{m} - \dfrac{2k_3 x_2}{m} \sin u \end{bmatrix}$$

$$\left[\frac{\partial \boldsymbol{f}}{\partial \boldsymbol{u}}\right] = \begin{bmatrix} 0 \\ -\dfrac{k_3 x_2^2}{m} \cos u \end{bmatrix}.$$

The nonhomogeneous linear differential equation for the perturbation $\delta \boldsymbol{x}(t)$ is

$$\delta \dot{\boldsymbol{x}}(t) = \boldsymbol{F}(t) \delta \boldsymbol{x}(t) + \boldsymbol{G}(t) \delta \boldsymbol{u}(t)$$

which is of the standard form. This equation permits evaluation of the perturbations $\delta \boldsymbol{x}(t)$ due to changes in the initial condition $\delta \boldsymbol{x}(t_0)$ and due to perturbations $\delta \boldsymbol{u}(t)$ in the input. Note that determination of the reference solution requires a reference input function $\boldsymbol{u}^*(t)$, and that $\delta \boldsymbol{u}(t)$ is the perturbation around this reference input.

4.5.2. Discrete Time Systems

Linearization techniques for discrete time systems follow in direct analogy to those for continuous time systems. The general discrete time difference equation is

$$\boldsymbol{x}(l) = \boldsymbol{f}(\boldsymbol{x}(l-1), \boldsymbol{u}(l-1), l). \tag{4.112}$$

If a reference solution $\boldsymbol{x}^*(l)$ is determined with input function $\boldsymbol{u}^*(l)$, then the perturbation sequences

$$\delta\boldsymbol{x}(l) = \boldsymbol{x}(l) - \boldsymbol{x}^*(l)$$
$$\delta\boldsymbol{u}(l) = \boldsymbol{u}(l) - \boldsymbol{u}^*(l) \tag{4.113}$$

are defined. The perturbation equations are then

$$\delta\boldsymbol{x}(l) = \left[\frac{\partial f}{\partial \boldsymbol{x}}\right]_{\boldsymbol{x}^*,\,\boldsymbol{u}^*} \delta\boldsymbol{x}(l-1) + \left[\frac{\partial f}{\partial \boldsymbol{u}}\right]_{\boldsymbol{x}^*,\,\boldsymbol{u}^*} \delta\boldsymbol{u}(l-1). \tag{4.114}$$

If the partial derivative matrices have the associations

$$\left[\frac{\partial f}{\partial \boldsymbol{x}}\right]_{\boldsymbol{x}^*,\,\boldsymbol{u}^*} \rightarrow \boldsymbol{\Phi}(l, l-1)$$
$$\left[\frac{\partial f}{\partial \boldsymbol{u}}\right]_{\boldsymbol{x}^*,\,\boldsymbol{u}^*} \rightarrow \boldsymbol{\Gamma}(l, l-1) \tag{4.115}$$

then Equation (4.114) is of the form of Equation (4.29), and any of the linear techniques of Section 4.2 can be applied.

4.5.3. Uncertainties in Nonlinear Systems

The application of perturbation techniques to the analysis of uncertainty effects in nonlinear systems should be apparent. The perturbation analysis applies directly to the uncertainties themselves, in that random deviations from a deterministic reference solution can be considered to be perturbations, in which case the treatment of uncertainties in linear systems developed in Section 4.2.4 applies directly.

In the example of the previous section $\delta\mathbf{X}^*(t_0)$ can be considered a random variation in the initial condition $\boldsymbol{x}(t_0)$, and $\mathbf{U}(t)$ a random process describing the deviation of $\boldsymbol{u}(t)$ from its deterministic reference value $\boldsymbol{u}^*(t)$. Then, if the covariance matrix

$$\boldsymbol{P}(t_0) = \begin{bmatrix} \mathrm{VAR}[\mathrm{X}_1(t_0)] & \mathrm{COV}[\mathrm{X}_1(t_0), \mathrm{X}_2(t_0)] \\ \mathrm{COV}[\mathrm{X}_1(t_0), \mathrm{X}_2(t_0)] & \mathrm{VAR}[\mathrm{X}_2(t_0)] \end{bmatrix}$$

and covariance function

$$\boldsymbol{K}_{\mathrm{U}}(t_1, t_2) = \mathrm{COV}[\mathbf{U}(t_1), \mathbf{U}(t_2)]$$

are known, the covariance matrix at any time t can be determined using Equations (4.92) and (4.94):

$$\boldsymbol{P}(t) = \boldsymbol{\Phi}(t, t_0)\boldsymbol{P}(t_0)\boldsymbol{\Phi}^T(t, t_0) + \int_{t_0}^{t}\int_{t_0}^{t} \boldsymbol{\Phi}(t, \sigma)\boldsymbol{G}(\sigma)\boldsymbol{K}_{\mathbf{U}}(\sigma, \nu)\boldsymbol{G}^T(\nu)\boldsymbol{\Phi}^T(t, \nu)\, d\nu\, d\sigma \quad (4.116)$$

where $\boldsymbol{\Phi}(t, t_0)$ and $\boldsymbol{G}(t)$ are derived from the linearized perturbation equation. The propagation of errors in nonlinear systems is often investigated by a procedure similar to this one. As will be shown in Chapter 7, techniques of nonlinear filtering also use the general concepts developed here.

REFERENCES

1. Brockett, Roger W. Poles, Zeros, and Feedback: State Space Interpretation. *IEEE Transactions on Automatic Control*, **AC-10:** 129–135 (April 1965).
2. Bryson, Arthur E., and Ho, Yu-Chi. *Applied Optimal Control.* Waltham, Massachusetts: Blaisdell, 1969.
3. Derusso, Paul M., Roy, Rob J., and Close, Charles M. *State Variables for Engineers.* New York, John Wiley & Sons, 1965.
4. Hahn, W. *Theory and Application of Lyapunov's Direct Method.* Englewood Cliffs, New Jersey: Prentice-Hall, 1963.
5. Jacquot, Raymond G. *Modern Digital Control Systems.* New York: Marcel Dekker, 1981.
6. Kuo, Benjiman C. *Analysis and Synthesis of Sampled-Data Control Systems.* Englewood Cliffs, New Jersey: Prentice-Hall, 1963.
7. ——. *Automatic Control Systems.* Englewood Cliffs, New Jersey: Prentice Hall, 1975.
8. Ogata, Katsuhiko. *Modern Control Engineering.* Englewood Cliffs, New Jersey: Prentice-Hall, 1970.
9. ——. *State Space Analysis of Control Systems.* Englewood Cliffs, New Jersey: Prentice Hall, 1967.
10. Schultz, Donald G., and Melsa, James L. *State Functions and Linear Control Systems.* New York: McGraw-Hill, 1967.
11. Van Loan, Charles F. Computing Integrals Involving the Matrix Exponential. *IEEE Transactions on Automatic Control*, **AC-23:** 129–135 (June 1978).

DEVELOPMENTAL EXERCISES

4.1. Show that a canonical form of the state space representation of the Nth order single input–single output system described by the equation

$$y^{(N)}(t) + a_{N-1}y^{(N-1)}(t) + a_{N-2}y^{(N-2)}(t) + \cdots + a_1\dot{y}(t) + a_0 y(t) = gu(t)$$

where $y^{(N)}(t) = d^N y/dt^N$, is given by

$$\dot{\mathbf{x}}(t) = \boldsymbol{F}(t)\mathbf{x}(t) + \boldsymbol{G}u(t)$$

$$y(t) = \boldsymbol{H}\mathbf{x}(t)$$

where

$$\mathbf{x}(t) = \begin{bmatrix} y(t) \\ \dot{y}(t) \\ \vdots \\ y^{(N-2)}(t) \\ y^{(N-1)}(t) \end{bmatrix}; \quad \boldsymbol{F} = \begin{bmatrix} 0 & 1 & 0 & \cdots & 0 & 0 \\ 0 & 0 & 1 & \cdots & 0 & 0 \\ \cdots & \cdots & \cdots & \cdots & \cdots & \cdots \\ 0 & 0 & 0 & \cdots & 0 & 1 \\ -a_0 & -a_1 & -a_2 & \cdots & -a_{N-2} & -a_{N-1} \end{bmatrix}$$

$$\boldsymbol{G} = \begin{bmatrix} 0 \\ 0 \\ \vdots \\ 0 \\ -g \end{bmatrix}; \quad \boldsymbol{H} = [1 \quad 0 \quad \cdots \quad 0 \quad 0].$$

State variables formulated in the this manner are known as *phase variables*. They do not, in general, represent real physical variables of the system.

4.2. Generalize the results of Exercise 4.1 to account for systems of the form

$$y^{(N)}(t) + a_{N-1}y^{(N-1)}(t) + \cdots + a_1\dot{y}(t) + a_0 y(t)$$
$$= b_P u^{(P)}(t) + b_{P-1}u^{(P-1)}(t) + \cdots + b_1\dot{u}(t) + b_0 u(t).$$

4.3. Use the results of this chapter to develop the impulse-response solution to a single input-single output linear system,

$$y(t) = \int_{t_0}^{t} h(t, \tau)\, u(\tau)\, d\tau.$$

4.4. Find the response of a linear system with transition matrix $\boldsymbol{\Phi}(t, \tau)$ to an initial condition consisting of all zeros, except for the ith component, which is unity (1).

4.5. Generalize the concept of Exercise 4.4 to a nonlinear system for which numerical solutions are available. How could the state transition matrix of the linearized perturbation variables be obtained?

4.6. Relate the concept of controllability of single input–single output systems to pole cancellation in the transfer function.
4.7. Apply the pole cancellation concept of Exercise 4.6 to the observability of single input–single output systems.
4.8. Give some ideas for the development of Lyapunpov functions for nonlinear systems.

PROBLEMS

4.1. Find a set of state variables and the resulting vector–matrix differential (state) equations for systems described by the following differential equations:
(a) $\ddot{x}(t) + 3\dot{x}(t) + 2x(t) = u(t)$
(b) $\dddot{x}(t) + 4\ddot{x}(t) + 6\dot{x}(t) + 4x(t) = u(t)$
(c) $\dddot{x}(t) + 4\ddot{x}(t) + 5\dot{x}(t) + 2x(t) = u(t)$
(d) $\ddot{x}_1(t) - 10\dot{x}_2(t) + x_1(t) = u_1(t)$
$\ddot{x}_2(t) + 6x_2(t) = u_2(t)$.

4.2. Determine the state transition matrix corresponding to the following system matrices:

(a) $\begin{bmatrix} 1 & 1 \\ 1 & 1 \end{bmatrix}$ (b) $\begin{bmatrix} 3 & 2 \\ -1 & 6 \end{bmatrix}$ (c) $\begin{bmatrix} -2 & 5 \\ -4 & -3 \end{bmatrix}$

(d) $\begin{bmatrix} 0 & 1 & 0 \\ 0 & 0 & 1 \\ 0 & 0 & 0 \end{bmatrix}$ (e) $\begin{bmatrix} 0 & 1 & 0 \\ 0 & 0 & 1 \\ 0 & 0 & -1 \end{bmatrix}$ (f) $\begin{bmatrix} 0 & 1 & 0 \\ 0 & 0 & 1 \\ -6 & -11 & -6 \end{bmatrix}$

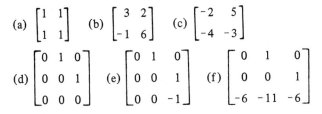

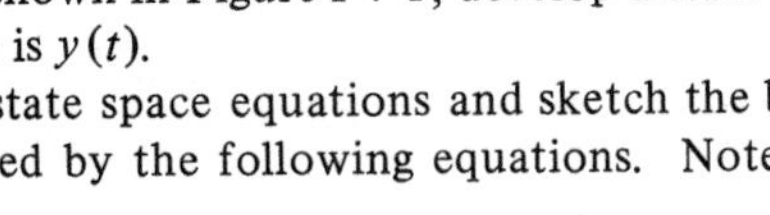

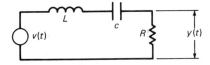

Figure P4-1.

4.3. For the circuit shown in Figure P4-1, develop a state space representation. The observation is $y(t)$.
4.4. Determine the state space equations and sketch the block diagram for the systems described by the following equations. Note that $y(t)$ is the system output.
(a) $\ddot{y}(t) + 2\dot{y}(t) + 5y(t) = u(t)$
(b) $\ddot{y}_1(t) + 3\dot{y}_1(t) + 2y_2(t) = u_1(t)$
$\ddot{y}_2(t) + 3\dot{y}_1(t) + y_2(t) = u_2(t)$
(c) $\ddot{y}_1(t) + 3\dot{y}_1(t) + 2y_2(t) = u_1(t) + 2\dot{u}_1(t) + u_2(t)$
$\ddot{y}_2(t) + 3\dot{y}_1(t) + y_2(t) = u_2(t) + 2\dot{u}_2(t) + u_1(t)$

4.5. Determine the state space equations for a system described by the following differential equations:

(a) $\ddot{x}(t) + 2\dot{x}(t) + x(t) = 0$

(b) $\dddot{x}(t) + 3\ddot{x}(t) + 2\dot{x}(t) + 2x(t) = 0$

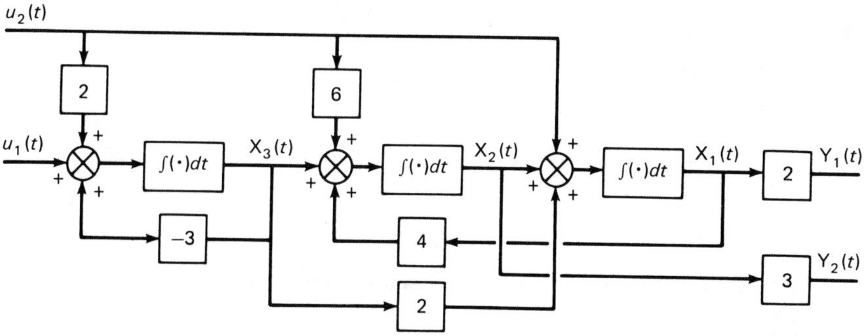

Figure P4-2.

4.6. Determine the state space equations for the system with block diagram shown in Figure P4-2.

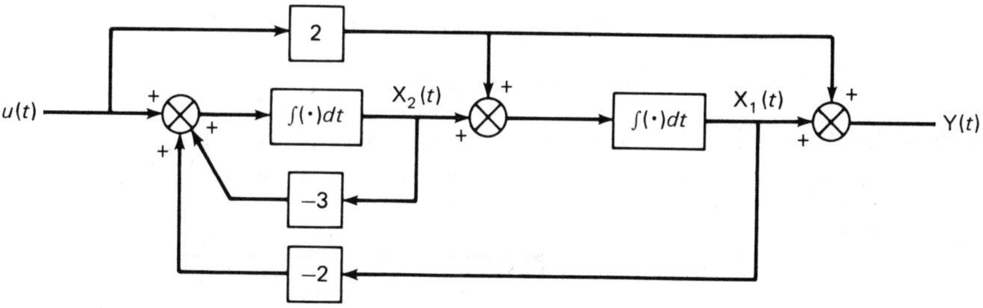

Figure P4-3.

4.7. Determine the state space equations for the system with block diagram shown in Figure P4-3. Find the state transition matrix for this system.

4.8. The yaw motion of an aircraft for small values of the yaw angle θ is given by the equation

$$J\ddot{\theta}(t) = -k_1\phi(t) - k_2\theta(t) - k_3\omega(t)$$

where J is the moment of inertia, $\omega(t)$ is the yaw angular velocity, and $\phi(t)$ is the rudder deflection angle. The constants k_1, k_2, and k_3 reflect the various aerodynamic characteristics.

(a) Determine the state variable form of the system equation, using θ and ω as state variables.

(b) If measurements are made with both angle and rate measuring devices, determine the observation equation.

(c) Suppose that k_1 and k_2 are very small and can be neglected. Determine the transition matrix for the system.

(d) Using the result of (c), find the discrete time difference equation in state space form.

4.9. The motion of a statellite of mass m around a planet of mass M, written in polar coordinates r and θ, is

$$\ddot{r} = r(t)\dot{\theta}^2(t) - \frac{GM}{r(t)^2} + \frac{u_r(t)}{m}$$

$$\ddot{\theta} = \frac{-2\dot{r}(t)\dot{\theta}(t)}{r(t)} + \frac{u_\theta(t)}{m}$$

where u_r and u_θ are the radial and tangential thrust forces. The equations constitute a fourth-order system.

(a) Write the state space equations for this system using state variables r, a, θ, and ω, where $a = \dot{r}$ and $\omega = \dot{\theta}$.

(b) If measurement of the range variable r is made, determine the measurement equation.

(c) Linearize these equations around a reference solution $x^*(t)$ and write the complete state space equation in the form

$$\delta\dot{\boldsymbol{x}}(t) = \boldsymbol{F}(t)\delta\boldsymbol{x}(t) + \boldsymbol{G}(t)\delta\boldsymbol{u}(t)$$

$$\boldsymbol{y}(t) = \boldsymbol{H}(t)\boldsymbol{x}(t).$$

4.10. Consider the continuous-time system shown in Figure P4-4. Use a Δt of 1 second and determine the difference equation for the system and the value of the state vector for $t = 1$ and $t = 2$, if the initial values are zero.

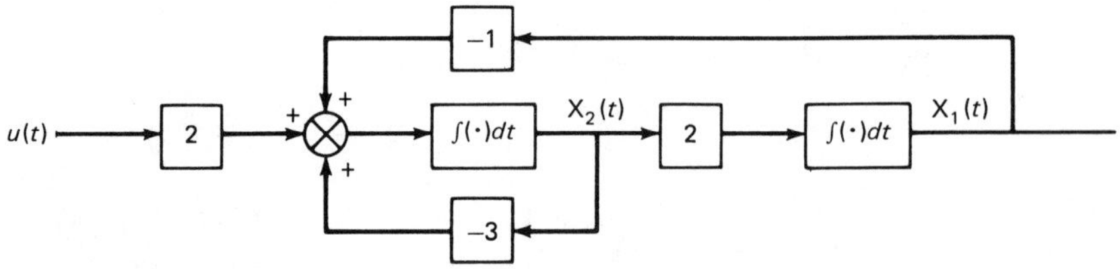

Figure P4-4.

4.11. A DC motor is modeled by the following equations:

$$J\ddot{\theta} + \beta\dot{\theta} = K_t i(t)$$

$$Li(t) + Ri(t) = V - K_v\dot{\theta}$$

where θ is the angular position
i is the armature current
J is the motor-load moment of inertia
β is the viscous damping constant
K_t is the torque constant of the motor
L is the armature inductance
R is the armature resistance
V is the applied voltage
K_v is the back emf constant of the motor.

Represent this model in state space form and sketch the block diagram.

4.12. Consider the time-dependent system described by the system matrix

$$\boldsymbol{F}(t) = \begin{bmatrix} 0 & 1/(t+1)^2 \\ 0 & 0 \end{bmatrix}$$

(a) Use the matrix differential equation

$$\dot{\boldsymbol{\Phi}}(t, \tau) = \boldsymbol{F}(t)\boldsymbol{\Phi}(t, \tau)$$

to derive the transition matrix for this system.
(b) For $\tau = 0$ and $x(0) = \left[\begin{smallmatrix}1\\1\end{smallmatrix}\right]$, find an expression for $x(t)$.
(c) For $\tau = 1$ and $x(1) = \left[\begin{smallmatrix}1\\1\end{smallmatrix}\right]$, find an expression for $x(t)$.
(d) For part (b), determine $x(1)$, and for part (c) determine $x(2)$. Both of these values result from the same initial conditions and $\Delta t = (t - t_0)$. Why are they different?

4.13. Determine the state space equations for the system shown in Figure P4-5.

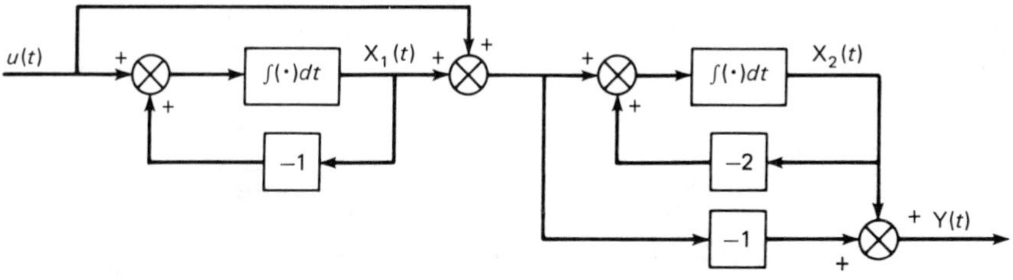

Figure P4-5.

(a) What special form of the $\boldsymbol{F}$ matrix is produced?
(b) Determine the observability and controllability of this system.
(c) Determine the transfer function $y(s)/u(s)$ for this system and comment.

4.14. Consider the system depicted in Figure P4-6. Determine the state space

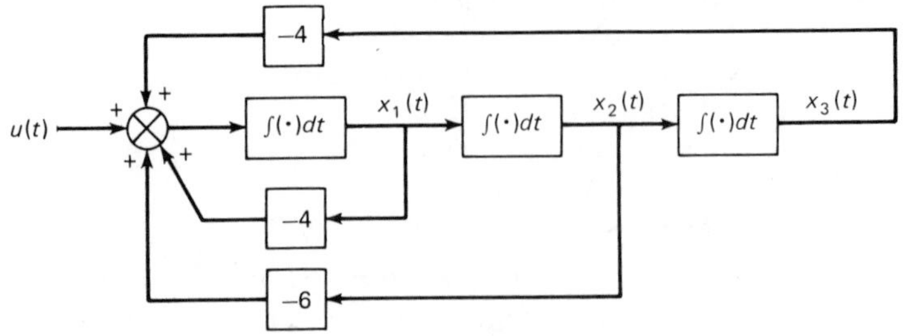

Figure P4-6.

representation and find the state transition matrix.

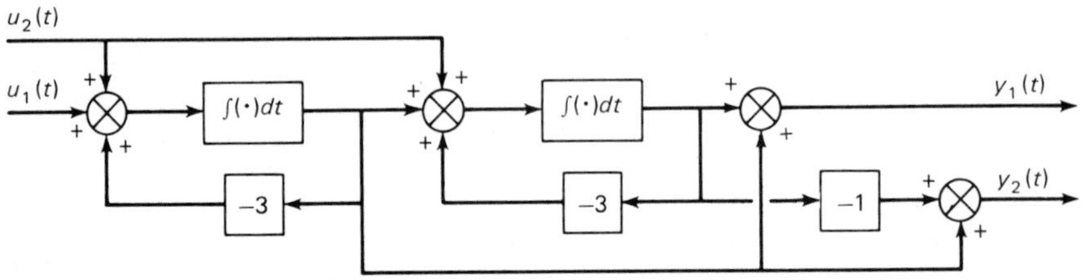

Figure P4-7.

4.15. A linear system is depicted in Figure P4-7. Determine the state space representation of this system, the modes of response, and the controllability and observability of the system.

4.16. Describe the electrical circuit shown in Figure P4-8 in state space form, and find the state transition matrix.

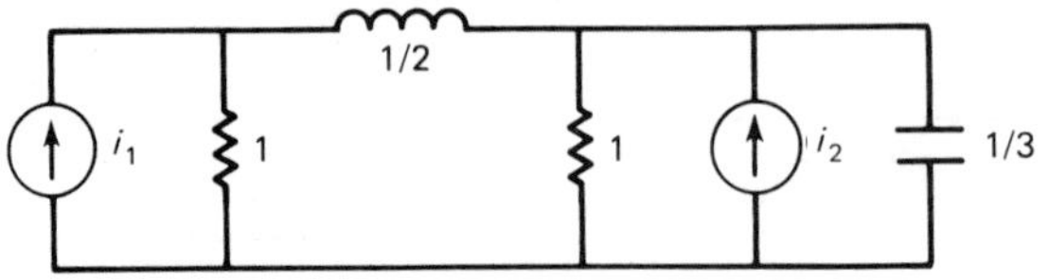

Figure P4-8.

4.17. A third-order system is described by the equations

$$\dot{x}_1(t) = 3x_1(t) + 5x_2(t) + x_3(t) + u_1(t)$$

$$\dot{x}_2(t) = x_3(t)$$

$$\dot{x}_3(t) = -2x_2(t) + 3x_3(t) + u_1(t) + u_3(t).$$

Find the response modes of this system by evaluating the eigenvalues of the $\boldsymbol{F}$ matrix.

4.18. Consider the system described by the equations

$$\dot{\boldsymbol{x}}(t) = \boldsymbol{F}\boldsymbol{x}(t) + \boldsymbol{G}u(t)$$

where

$$\boldsymbol{x}(t) = \begin{bmatrix} x_1(t) \\ x_2(t) \end{bmatrix}; \quad \boldsymbol{F} = \begin{bmatrix} 0 & 1 \\ -1 & -2 \end{bmatrix}; \quad \boldsymbol{G} = \begin{bmatrix} 1 \\ -1 \end{bmatrix}$$

and $u(t)$ is a scalar control function. The measurement equation is

$$\boldsymbol{y}(t) = \boldsymbol{H}\boldsymbol{x}(t)$$

where

$$\boldsymbol{y}(t) = \begin{bmatrix} y_1(t) \\ y_2(t) \end{bmatrix}; \quad \boldsymbol{H} = \begin{bmatrix} 1 & 0 \\ 0 & 1 \end{bmatrix}.$$

(a) Determine the controllability of the system.
(b) Determine the observability of the system.
(c) Find the transformation which diagonalizes $\boldsymbol{F}$ and interpret the result in terms of (a) and (b).

4.19. Consider the system described by the equations

$$\begin{aligned} \dot{x}_1(t) &= -2x_1(t) + x_2(t) \\ \dot{x}_2(t) &= -2x_1(t) + x_3(t) \\ \dot{x}_3(t) &= x_2(t) + x_3(t) + u_1(t) + u_2(t) \\ \dot{x}_4(t) &= x_4(t) + u_2(t) \end{aligned}$$

and the measurement equation

$$y(t) = x_1(t).$$

Determine the controllability and observability of this system.

4.20. Consider the pendulum system shown in Figure P4-9. Equal and opposite

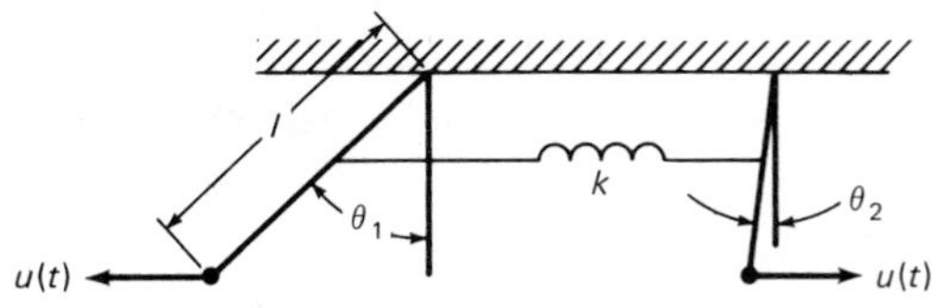

Figure P4-9.

forces u act on the masses m. The equations of motion are

$$ml^2\ddot{\theta}_1(t) + ka^2(\theta_1(t) - \theta_2(t)) + mgl\theta_1(t) + u(t) = 0$$
$$ml^2\ddot{\theta}_2(t) + ka^2(\theta_2(t) - \theta_1(t)) + mgl\theta_2(t) + u(t) = 0.$$

Show that this system is not controllable. Further show that if one of the forces is eliminated, the system is then controllable.

4.21. Consider the linear system described by the diagram of Figure P4-10.

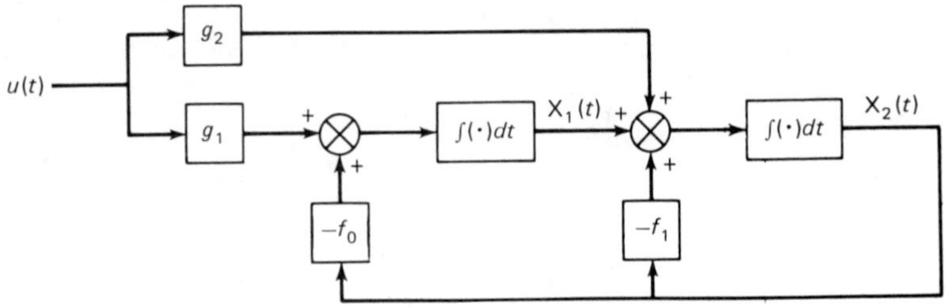

Figure P4-10.

(a) Determine the state space representation of this system.
(b) Is the system observable?
(c) Is the system controllable?

4.22. Consider the system described by the equations

$$\dot{x}_1(t) = -x_1(t) + u_1(t)$$
$$\dot{x}_2(t) = x_1(t) - 2x_2(t) + u_2(t)$$

and the measurement equation

$$y(t) = -x_1(t) + x_2(t).$$

(a) Sketch the block diagram of this system.
(b) Determine if the system is controllable and observable.
(c) Verify the results of (b) by a modal analysis.

4.23. Consider the system described by the equation

$$\dot{x}_1(t) = x_2(t) + u_2(t)$$
$$\dot{x}_2(t) = -2x_1(t) - 3x_2(t) + u_1(t) - 2u_2(t).$$

The observation equation is

$$y(t) = x_1(t).$$

(a) Find the modes of response by diagonalizing the system matrix.
(b) From results of (a), determine if the system is observable and controllable.
(c) Verify the results of (b) by use of the controllability and observability tests.

4.24. Consider the second order system with system matrix

$$\boldsymbol{F} = \begin{bmatrix} 0 & 1 \\ -2 & -2 \end{bmatrix}$$

(a) Determine the modes of response of this system.
(b) Find the state transition matrix, and the response of the system to initial conditions $x_1(0)$, $x_2(0)$.

4.25. Consider the nonlinear system described by the equation

$$\dot{x}_1(t) = x_1(t) + x_1^2(t) - 2x_1(t)x_2(t) + u(t)$$
$$\dot{x}_2(t) = x_1(t) - x_2(t).$$

Determine the equilibrium points ($\dot{x} = 0$) for this system. Linearize the equations around these points and determine the stability of the linearized equations.

4.26. Consider the nonlinear system described by the equations

$$\dot{x}_1(t) = x_1(t) - 2x_2(t)$$
$$\dot{x}_2(t) = -2x_1(t) - x_2(t) + x_2^2(t).$$

(a) Find the equilibrium points for this system and determine the stability of the linearized equations around these points.
(b) Use the function

$$V(\boldsymbol{x}) = x_1^2 + x_2^2$$

as a Lyapunov function to test the system stability.

4.27. Consider the first-order nonlinear equation

$$\dot{x}(t) = -x(t) + ax^3(t), \qquad a > 0.$$

Use Lyapunov functions

$$V(x) = \frac{1}{2}x^2 \quad \text{and} \quad V(x) = a\left(\frac{x^2}{2} - \frac{1}{2a}\right)^2$$

to investigate the stability of the equation in the vicinity of the points where $\dot{x}(t) = 0$.

4.28. Consider the nonlinear equation

$$\dot{x}(t) = -x(t)[x^2(t) + 1]$$

Prove that the system is globally stable.

4.29. Consider the second-order discrete time system

$$\mathbf{X}(l) = \mathbf{\Phi X}(l-1) + \mathbf{\Delta} \mathrm{W}(l-1)$$

where

$$\mathbf{X}(l) = \begin{bmatrix} X_1(l) \\ X_2(l) \end{bmatrix}; \quad \mathbf{\Phi} = \begin{bmatrix} -\frac{1}{2} & -\frac{1}{2} \\ 1 & 0 \end{bmatrix}; \quad \mathbf{\Delta} = \begin{bmatrix} 1 \\ 0 \end{bmatrix}.$$

If $\mathrm{W}(l)$ is a zero-mean white noise sequence with covariance kernel

$$k_{\mathrm{W}}(l, m) = \begin{cases} 1, & \text{if } l = m \\ 0, & \text{if } l \neq m \end{cases}$$

and the initial value of the vector $\mathbf{X}(0)$ is zero, find the mean value function and covariance kernel of the sequence $\mathbf{X}(l)$.

4.30. Consider the system described by the equations

$$\dot{\mathbf{X}}(t) = \boldsymbol{F}(t)\mathbf{X}(t) + \boldsymbol{D}(t)\mathrm{W}(t)$$

where

$$\mathbf{X}(t) = \begin{bmatrix} X_1(t) \\ X_2(t) \end{bmatrix}; \quad \boldsymbol{F}(t) = \begin{bmatrix} 0 & t \\ t & 0 \end{bmatrix}; \quad \boldsymbol{D}(t) = \begin{bmatrix} 1 \\ 0 \end{bmatrix}$$

and $\mathrm{W}(t)$ is scalar white noise process with covariance kernel $q\delta(t - \sigma)$.

(a) Find the transition matrix $\mathbf{\Phi}(t, \tau)$ for this system.

(b) If the output process is

$$\mathrm{Y}(t) = \boldsymbol{H}\mathbf{X}(t), \quad \text{where } \boldsymbol{H} = [0 \quad 1]$$

find an expression for the mean function and covariance kernel of the output process $\mathrm{Y}(t)$.

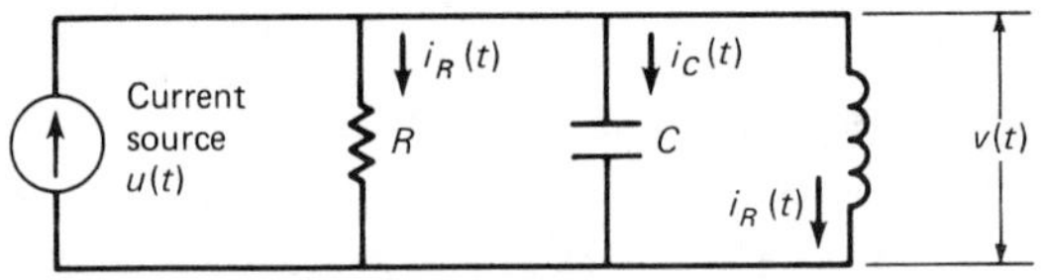

Figure P4-11.

4.31. Consider the electrical network shown in Figure P4-11. Define the state vector of this network to be

$$\mathbf{x}(t) = \begin{bmatrix} i_L(t) \\ i_R(t) \end{bmatrix}.$$

(a) Determine the state space representation of the system. (Assume subcritical damping. $2R < L/C$.)
(b) Find the response of the system to an initial condition

$$\mathbf{x}(t_0) = \begin{bmatrix} 1 \\ 0 \end{bmatrix}$$

with $u(t) = 0$.
(c) Repeat (b) if the input is $u(t) = 1$.
(d) Determine the steady state solutions for (b) and (c).

4.32. In the network of Problem 4.31, let the current source produce a zero-mean white noise process with covariance kernel $q\delta(t - \sigma)$.
(a) Find expressions for the mean value function and covariance kernel of the state vector $\mathbf{X}(t)$.
(b) Consider the initial condition to be a random variable with mean and covariance matrix

$$\begin{bmatrix} 0 \\ 1 \end{bmatrix} \quad \text{and} \quad \begin{bmatrix} \sigma_1^2 & 0 \\ 0 & \sigma_2^2 \end{bmatrix}$$

and repeat (a).
(c) Determine the steady state values for (a) and (b).
(d) Determine the covariance matrix $\boldsymbol{P}(t)$ under steady state conditions. Interpret these results in terms of those obtained in (c).

4.33. Consider the system

$$\begin{bmatrix} \dot{x}_1(t) \\ \dot{x}_2(t) \end{bmatrix} = \begin{bmatrix} 0 & 1 \\ -\omega^2 & 0 \end{bmatrix} \begin{bmatrix} x_1(t) \\ x_2(t) \end{bmatrix}$$

where $\mathbf{X}(t)$ is a two-vector and ω = constant. Assume that $\mathbf{X}(0)$ is normal with zero mean and covariance matrix equal to the 2×2 identity matrix.
(a) Determine the covariance matrix $\boldsymbol{P}_{\mathbf{X}}(t)$.
(b) Determine the covariance matrix kernel $\boldsymbol{K}_{\mathbf{X}}(t_1, t_2)$.

4.34. The equation of motion of a pendulum in a viscous fluid is

$$\ddot{\mathrm{X}}(t) + 2c\dot{\mathrm{X}}(t) + (\omega_0^2 + c^2) = \mathrm{W}(t)$$

in which $X(t)$ is the displacement of the pendulum from its rest position, c is the damping coefficieint, $2\pi/\omega_0$ is the damped period, and $W(t)$ is the driving force per unit mass.

(a) Show that, after the effect of initial conditions, the solution to this equation can be expressed as

$$X(t) = \int_{-\infty}^{t} e^{-\alpha(t-\sigma)} \left(\frac{\sin \omega_0 (t - \sigma)}{\omega_0} \right) W(\sigma)\, d\sigma.$$

(b) Find an expression for the covariance kernel of $X(t)$ in terms of the covariance kernel of $W(t)$.

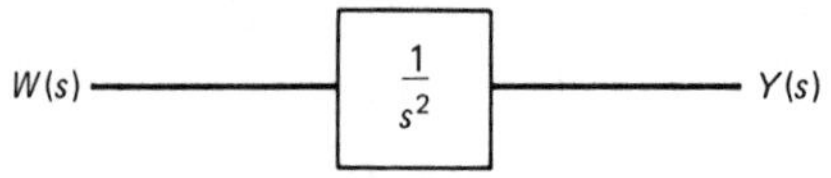

Figure P4-12.

4.35. Consider the second-order system shown in Figure P4-12. If $W(t)$ is a white noise process with covariance kernel $q\delta(t - \sigma)$, and the state variables are defined as

$$X_1(t) = Y(t)$$
$$X_2(t) = \dot{Y}(t)$$

find the covariance matrix kernel $\boldsymbol{K}_X(t, \sigma)$.

4.36. A system is modeled by the following equations:

$$\dot{X}_1(t) = X_1(t) + X_2(t) + U(t)$$
$$\dot{X}_2(t) = 2X_2(t)$$

(a) Find the state transition matrix for this system.

(b) If the state at $t = 0$ is estimated to be

$$\hat{X}_1 = \tfrac{1}{3}$$
$$\hat{X}_2 = 1$$

find the estimated state at $t = 1$, if $E[U(t)] = 0$ during the interval and no additional measurements are available.

(c) If the estimate at the state $t = 0$ has error variances of 1 and 2, respectively, for the estimates $\hat{X}_1(t)$ and $\hat{X}_2(t)$, find the variance of the error in the estimate of $X_2(t)$ at $t = 1$.

(d) Find the covariance of the errors in the estimates of $X_1(t)$ and $X_2(t)$ at $t = 1$, under the conditions of (c) above. How is this to be interpreted?

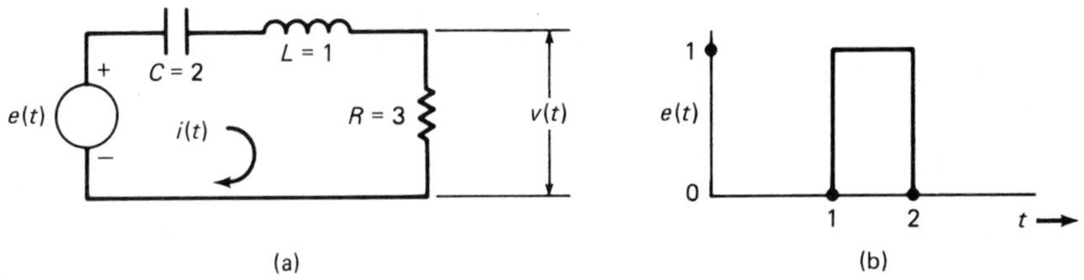

Figure P4-13.

4.37. Consider the electrical circuit shown in Figure P4-13(a). Find the state space representation of this system. Use $i(t)$ and $\dot{i}(t)$ as the state variables, and $v(t)$ as the output. For the voltage pulse shown in Figure P4-13(b) as the input, find an expression for the output voltage $v(t)$ for any general set of initial conditions.

4.38. For Problem 4.37, let the initial conditions be a random vector with zero mean and covariance matrix

$$\begin{bmatrix} 4 & 0 \\ 0 & 9 \end{bmatrix}.$$

If the input voltage is a white noise sequence with covariance kernel $2\delta(t - \sigma)$, find expressions for the mean value function and covariance kernel of the output voltage $V(t)$.

4.39. The approximate equations of motion of a rocket in vertical flight are

$$\dot{M}(t) = -F(t)/c; \qquad M(t)\dot{V}(t) = F(t) - M(t)g; \qquad \dot{H} = V(t)$$

where

$$M = \text{mass}, \qquad F = \text{thrust}, \qquad H = \text{altitude} \qquad V = \text{velocity},$$

c = specific impulse of rocket (pound of thrust per pound/sec of propellant expended).

Suppose that $H(0)$ and $V(0)$ are random variables with zero mean and variances σ_H^2 and σ_V^2, and the initial mass has expected value m_0 and variance σ_M^2. The thrust $F(t)$ is a white noise process with constant mean value function f_0 and covariance kernel $q\delta(t - \sigma)$. Further assume that the initial values of H, V, and M are uncorrelated. Find the mean values and covariances of the random processes $H(t)$, $V(t)$, and $M(t)$.

4.40. Consider the system described by the time-dependent differential equation

$$\ddot{x}(t) + f_1(t)\,\dot{x}(t) + f_2(t)\,x(t) = g_1(t)\,\dot{u}(t) + g_2 u(t).$$

The measurement is $y(t) = x(t)$. Develop a state space representation of this system. Identify the $\boldsymbol{F}$, $\boldsymbol{G}$, and $\boldsymbol{H}$ matrices.

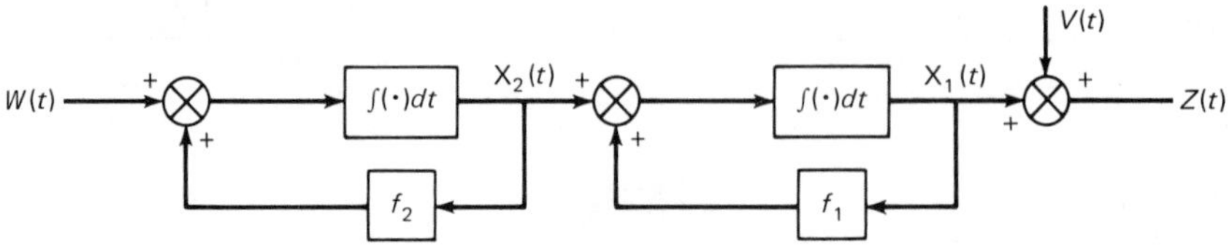

Figure P4-14.

4.41. A second-order system is shown in Figure P4-14. W(t) and V(t) are independent white noise processes with covariance kernels $q\delta(t - \sigma)$ and $r\delta(t - \sigma)$, respectively.

(a) Find the state equation and measurement equation for this system.

(b) Determine the steady state covariance matrix kernel $\boldsymbol{K}_{\mathbf{X}}(t, \sigma)$ and covariance matrix $\boldsymbol{P}(t)$.

(c) Write the expression for the output covariance matrix in terms of the input properties.

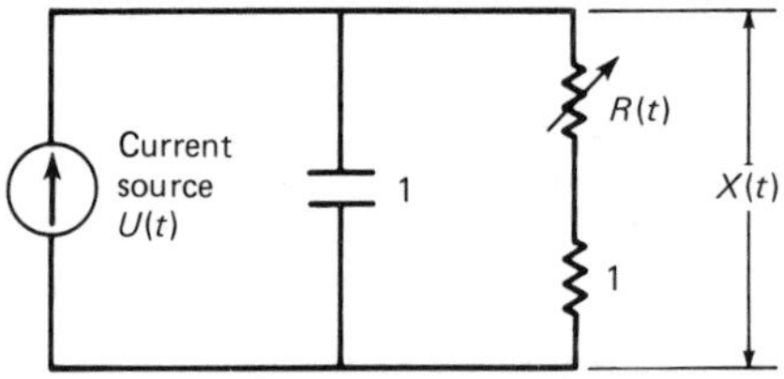

Figure P4-15.

4.42. Consider the circuit shown in Figure P4-15. The variable resistance produces the function

$$R(t) = t \text{ ohms}, \qquad t > 0.$$

(a) The state vector X(t) for this system is a scalar. Find the state transition matrix (in this case a scalar function). Note that this system is time dependent.

(b) If the input current U(t) is a random process with mean value function $\sin t$ ($t > 0$) and covariance kernel

$$k(t, \sigma) = e^{-|t-\sigma|}$$

find the mean value function and covariance kernel of the output voltage X(t).

(c) Repeat part (b) if the initial voltage is a random variable with mean 1 and variance 4.

4.43. A charged particle moving in an electromagnetic field with the electric intensity vector $\boldsymbol{E}$ and the initial velocity vector both perpendicular to a constant magnetic induction of magnitude B, has the motion equations

$$m\dot{v}_1(t) = qE_1(t) + qBv_2(t)$$
$$m\dot{v}_2(t) = qE_2(t) - qBv_1(t)$$

where v_1, v_2 and E_1, E_2 are the components of the velocity and electric intensity, respectively, and q is the charge on the particle.

(a) Express these motion equations in state space form, using the electrical intensities E_1 and E_2 as the input functions. Find the state transition matrix.

(b) If the vector $\boldsymbol{E}$ has magnitude B/m and rotates in a clockwise direction at an angular rate of $\omega = qB/m$, then the input vector is

$$\boldsymbol{u}(t) = \begin{bmatrix} (B/m)\cos\omega t \\ -(B/m)\sin\omega t \end{bmatrix}.$$

Find an expression for the state vector, if the initial displacements are both zero, $v_1(0) = 0$, and $v_2(0) = 1$.

(c) Sketch the trajectory of the particle in the (x_1, x_2) coordinate system.

(d) Determine the kinetic energy of the particle as a function of time at the points where it crosses the coordinate axes.

4.44. Consider the system of Problem 4.43, but now the initial velocity is a random vector with mean value and covariance matrix

$$E[\mathbf{X}] = \begin{bmatrix} 0 \\ 0 \\ 0 \\ 1 \end{bmatrix}, \quad \boldsymbol{K} = \begin{bmatrix} 0 & 0 & 0 & 0 \\ 0 & 0 & 0 & 0 \\ 0 & 0 & \sigma_1^2 & 0 \\ 0 & 0 & 0 & \sigma_2^2 \end{bmatrix}.$$

That is, the mean value of V_1 is 0, but it has variance σ_1^2; the mean value of V_2 is 1, and it has variance σ_2^2. Suppose also that the magnitude of the electric field is a white noise process with mean value function B/m and covariance kernel $r\delta(t - \sigma)$. Find the mean value function and covariance kernel of the system state vector.

4.45. A discrete-time system is described by the difference equation

$$\mathbf{X}(l) = \mathbf{\Phi}\mathbf{X}(l-1) + \mathbf{W}(l-1)$$

where

$$\mathbf{X}(l) = \begin{bmatrix} X_1(l) \\ X_2(l) \end{bmatrix}; \quad \mathbf{\Phi} = \begin{bmatrix} 0 & -0.3 \\ 0.2 & 0.5 \end{bmatrix}; \quad \mathbf{W}(l) = \begin{bmatrix} W_1(l) \\ W_2(l) \end{bmatrix}.$$

The measurement equation is

$$\mathbf{Y}(l) = \boldsymbol{H}\mathbf{X}(l)$$

where

$$\mathbf{Y}(l) = \begin{bmatrix} Y_1(l) \\ Y_2(l) \end{bmatrix}; \quad \boldsymbol{H} = \begin{bmatrix} -0.24 & 0.6 \\ -1 & 0 \end{bmatrix}.$$

If the initial state $\mathbf{X}(0)$ is a random vector with zero mean and covariance

$$\boldsymbol{P}(0) = \begin{bmatrix} 1 & 0 \\ 0 & 0 \end{bmatrix}$$

and the input $\mathbf{W}(l)$ is a zero-mean white noise sequence with covariance kernel

$$\boldsymbol{K}(l, m) = \begin{cases} \begin{bmatrix} 0.8 & 0.7 \\ 0.7 & 1.2 \end{bmatrix}, & \text{if } l = m \\ 0, & \text{if } l \neq m. \end{cases}$$

(a) Find the mean value function and covariance matrix of the system state vector $\mathbf{X}(l)$ for $l = 1, 2, 3$, and 4.
(b) Find the covariance kernel of the system state vector, $\boldsymbol{K}(l, m)$, for $l = 2$, $m = 4$.
(c) Find the covariance matrix of the output vector $\mathbf{Y}(l)$ for $l = 1, 2, 3, 4$.

4.46. Consider the discrete time system described by the equation

$$\boldsymbol{x}(l) = \mathbf{\Phi}\boldsymbol{x}(l-1) + \mathbf{\Gamma}u(l-1)$$

where

$$\mathbf{\Phi} = \begin{bmatrix} 0.2 & -0.1 \\ -0.1 & 0.2 \end{bmatrix}; \quad \mathbf{\Gamma} = \begin{bmatrix} 0.5 \\ 0.5 \end{bmatrix}$$

Determine the observability and controllability of this system.

4.47. Consider the discrete time system described by the equation

$$x_1(l) = x_1(l-1) - 0.2x_2(l-1) + u(l-1)$$
$$x_2(l) = 0.4x(l-1) + 0.4x_2(l-1) + u(l-1).$$

The measurement equation is

$$y(l) = x_1(l) + 0.2x_2(l).$$

Determine the observability and controllability of this system.

4.48. Write a state space equation in terms of state variables x_1, x_2, x_3, and x_4 for the system described by the equations

$$\ddot{y}_1(t) + 3\dot{y}_1(t) + 2y_2(t) = u_1(t)$$
$$\ddot{y}_2(t) + \dot{y}_1(t) + y_2(t) = u_2(t)$$

and sketch the block diagram. The functions $y_1(t)$ and $y_2(t)$ constitute the output of the system.

Chapter 5
Modal Control of Linear Dynamic Systems

5.1. STATE SPACE APPROACH TO LINEAR SYSTEM CONTROL

The classical treatment of control system analysis and design has been by frequency domain techniques utilizing the Laplace transform. There are several reasons for this approach, not the least of which was the nonavailability of the computational power of the modern digital computer. Several elegant techniques, such as the Nyquist, Bode, and root-locus methods for evaluating the characteristics of the response of the linear system by frequency domain methods, have been used successfully for many years. With the advent of the high-speed digital computer and the subsequent reduction in the cost of computation as the industry matured, the state space technique as a time domain approach to linear control system analysis gained wide usage. Particularly in the area of optimal control and estimation theory, the systematic approach afforded by state space techniques provided a useful impetus to some of the more recent developments. The purpose of this chapter is not to present a thorough treatment of state space analysis of control systems, but rather to convey the fundamental characteristics of a feedback system when viewed in a state space context. No argument is made for or against frequency domain techniques, for they remain today a very useful tool in feedback system analysis and as such should be considered to complement the state space approach.

5.1.1. Linear State Variable Feedback

There are two general types of control which can be exercised over a system. These are known as *open-loop* and *closed-loop* control. Open-loop control entails the application of a control function which is independent of the actual system state and therefore is a function of time only. For the general linear system given by the state equation

$$\dot{\boldsymbol{x}}(t) = \boldsymbol{F}(t)\boldsymbol{x}(t) + \boldsymbol{G}(t)\boldsymbol{u}(t) \tag{5.1}$$

open-loop control would consist of determining a control function $\boldsymbol{u}(t)$ over some interval (t_0, t_f) which would cause the system state $\boldsymbol{x}(t)$ to evolve in some prescribed way. The concepts of stability and controllability discussed in Chapter 4 obviously apply in this instance. In Chapter 6, certain optimal control techniques leading to control functions of this nature are considered.

Closed-loop control, on the other hand, uses the actual values of the states themselves to generate a control function and is related to the common concept of feedback control so aptly treated by the frequency domain techniques. The closed-loop control function is an explicit function of the states, and perhaps of time, so that the system equation can be written as

$$\dot{\boldsymbol{x}}(t) = \boldsymbol{F}(t)\boldsymbol{x}(t) + \boldsymbol{G}(t)\boldsymbol{u}((\boldsymbol{x}(t), t) \tag{5.2}$$

and the system block diagram would be as shown in Figure 5-1. Note that $\boldsymbol{u}((\boldsymbol{x}(t), t)$ has no dynamics associated with it, but rather is an algebraic function of the state vector $\boldsymbol{x}(t)$. This feature is characteristic of state variable feedback systems, since any lead or lag compensation can effectively be produced by feedback of different states. That is, if

$$\boldsymbol{x}(t) = \begin{bmatrix} x_1(t) \\ \dot{x}_1(t) \\ \ddot{x}_1(t) \end{bmatrix} = \begin{bmatrix} x_1(t) \\ x_2(t) \\ x_3(t) \end{bmatrix}$$

then lead compensation with respect to $x_1(t)$ is obtained by feedback of $x_2(t)$, and so on. The particular form of the feedback controller will depend upon the desired closed-loop characteristics of the system—that is, by the response of the entire system including the feedback controller. Discussion here will be limited to linear feedback control. It will be shown in Chapter 6 that the optimum con-

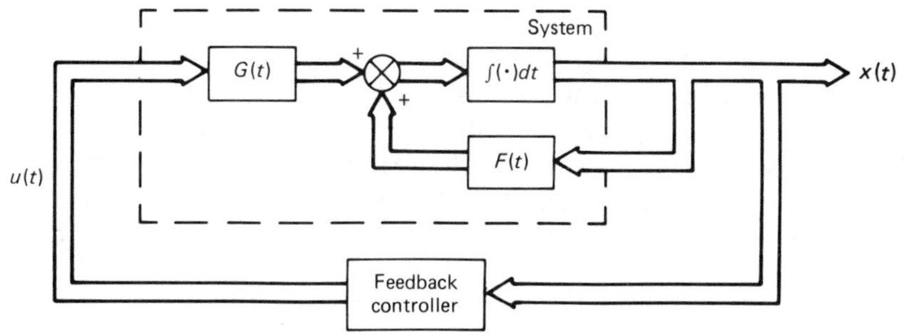

Figure 5-1. State variable feedback control.

trol function for a very general class of optimization criteria is a linear state variable feedback system.

Under the assumption of a linear controller, the control function can be expressed as

$$\boldsymbol{u}(\boldsymbol{x}(t), t) = \boldsymbol{C}(t)\boldsymbol{x}(t) \tag{5.3}$$

where $\boldsymbol{C}(t)$ is an $N \times N$ matrix of feedback gains. Substitution of this expression into Equation (5.2) yields

$$\dot{\boldsymbol{x}}(t) = [\boldsymbol{F}(t) + \boldsymbol{G}(t)\boldsymbol{C}(t)]\boldsymbol{x}(t) \tag{5.4}$$

—a linear homogeneous equation. The solution to this equation can be described by the transition matrix method as

$$\boldsymbol{x}(t) = \tilde{\boldsymbol{\Phi}}(t, t_0)\boldsymbol{x}(t_0) \tag{5.5}$$

where $\tilde{\boldsymbol{\Phi}}(t, t_0)$ is determined by use of the closed-loop system matrix

$$\tilde{\boldsymbol{F}}(t) = [\boldsymbol{F}(t) + \boldsymbol{G}(t)\boldsymbol{C}(t)]. \tag{5.6}$$

The solution of Equation (5.4) is obtained by the techniques discussed in Chapter 4 and Appendix C. When $\boldsymbol{F}(t)$, $\boldsymbol{G}(t)$, and $\boldsymbol{C}(t)$ are time independent, Equation (5.4) is

$$\begin{aligned} \dot{\boldsymbol{x}}(t) &= [\boldsymbol{F} + \boldsymbol{G}\boldsymbol{C}]\boldsymbol{x}(t) \\ &= \tilde{\boldsymbol{F}}\boldsymbol{x}(t) \end{aligned} \tag{5.7}$$

and the response characteristics of the system are determined by the eigenvalues of the matrix $\tilde{\boldsymbol{F}}$, as discussed in Section 4.3.1. This concept is important in feedback system design, since the synthesis process consists of determining the matrix $\boldsymbol{C}$ so that the eigenvalues of the $\tilde{\boldsymbol{F}}$ matrix produce the desired system response characteristics. This process is analogous to the technique of pole placement in frequency domain analysis, since the roots of the closed-loop characteristic equation are identical to the eigenvalues of the matrix $\tilde{\boldsymbol{F}}$.

As a simple example, consider the spring-mass-damper system of Figure 4-1. This second-order time-invariant system is described by the equation

$$\dot{\boldsymbol{x}}(t) = \boldsymbol{F}\boldsymbol{x}(t) + \boldsymbol{G}\boldsymbol{u}(t)$$

$$\boldsymbol{x}(t) = \begin{bmatrix} s(t) \\ v(t) \end{bmatrix}; \quad \boldsymbol{F} = \begin{bmatrix} 0 & 1 \\ -5 & -2 \end{bmatrix}; \quad \boldsymbol{G} = \begin{bmatrix} 1 & 0 \\ 0 & 1 \end{bmatrix}; \quad \boldsymbol{u}(t) = \begin{bmatrix} u_1(t) \\ u_2(t) \end{bmatrix}$$

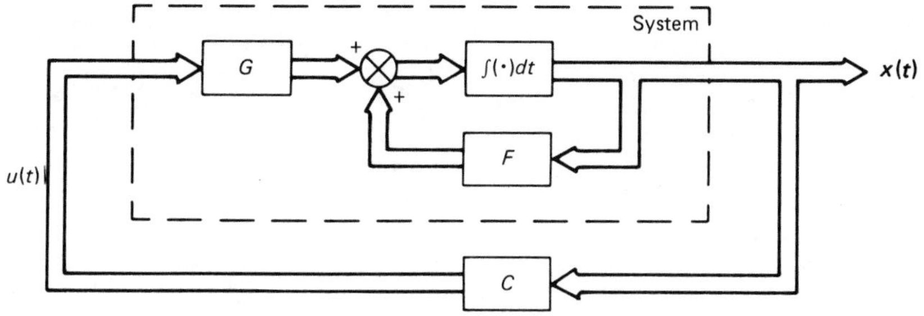

Figure 5-2. Linear state variable feedback control.

in which state variable feedback control is to be accomplished. The system block diagram is shown in Figure 5-2, which describes the general case of time-invariant, linear state variable feedback control. Then, from Equation (5.7), the closed-loop system equation is

$$\dot{x}(t) = \tilde{F}x(t)$$

where

$$\tilde{F} = \begin{bmatrix} 0 & 1 \\ -5 & -2 \end{bmatrix} + \begin{bmatrix} c_{11} & c_{12} \\ c_{21} & c_{22} \end{bmatrix}$$

and the c_{ij} are the elements of the matrix C. The system closed-loop response is then determined by the eigenvalues of the matrix

$$\tilde{F} = \begin{bmatrix} c_{11} & 1 + c_{12} \\ -5 + c_{21} & -2 + c_{22} \end{bmatrix}.$$

Suppose that C is the identity matrix, in which case

$$\tilde{F} = \begin{bmatrix} 1 & 1 \\ -5 & -1 \end{bmatrix}$$

and the eigenvalues of $\tilde{F}$ are determined to be

$$\lambda_1 = 2i; \quad \lambda_2 = -2i.$$

The closed-loop system response is seen to be an undamped sinusoid, as implied by the conjugate pair of imaginary eigenvalues. On the other hand, if

$$C = \begin{bmatrix} -3 & 1 \\ 4 & 2 \end{bmatrix}$$

then

$$\tilde{\boldsymbol{F}} = \begin{bmatrix} -3 & 2 \\ -1 & 0 \end{bmatrix}$$

which has the eigenvalues

$$\lambda_1 = -1; \qquad \lambda_2 = -2$$

and the closed-loop system is overdamped.

The closed-loop state transition matrix can be determined using the method outlined in Appendix C. The diagonalizing matrix of eigenvectors is

$$\boldsymbol{P} = \begin{bmatrix} 1 & 2 \\ 1 & 1 \end{bmatrix}; \qquad \boldsymbol{P}^{-1} = \begin{bmatrix} -1 & 2 \\ 1 & -1 \end{bmatrix}$$

from which the closed-loop state transition matrix is determined to be

$$\tilde{\boldsymbol{\Phi}}(t, t) = \boldsymbol{P} \begin{bmatrix} e^{-(t-t_0)} & 0 \\ 0 & e^{-2(t-t_0)} \end{bmatrix} \boldsymbol{P}^{-1}$$

$$= \begin{bmatrix} -e^{-(t-t_0)} + 2e^{-2(t-t_0)} & 2e^{-(t-t_0)} - 2^{-2(t-t_0)} \\ -e^{-(t-t_0)} + e^{-2(t-t_0)} & 2e^{-(t-t_0)} - e^{-2(t-t_0)} \end{bmatrix}.$$

This transition matrix describes the time domain response of the closed-loop system to an initial condition vector $\boldsymbol{x}(t_0)$; that is,

$$\boldsymbol{x}(t) = \tilde{\boldsymbol{\Phi}}(t, t_0)\boldsymbol{x}(t_0).$$

The reader should compare the transition matrix above to that of the example in Appendix C. Even though the eigenvalues are identical in these two examples, the system matrices and the resulting transition matrices are different. This situation merely reflects the fact that different matrices may have identical eigenvalues. This concept can be developed further in the specific context of the current example. Let the control matrix $\boldsymbol{C}$ be

$$\boldsymbol{C} = \begin{bmatrix} -\frac{1}{2} & -2 \\ \frac{23}{4} & -\frac{9}{2} \end{bmatrix}$$

in which case the closed-loop system matrix is

$$\tilde{F} = \begin{bmatrix} -\frac{1}{2} & -1 \\ \frac{3}{4} & -\frac{5}{2} \end{bmatrix}.$$

The eigenvalues of this matrix are again

$$\lambda_1 = -1; \qquad \lambda_2 = -2$$

but the diagonalizing matrix of eigenvectors is

$$P = \begin{bmatrix} 2 & 2 \\ 1 & 2 \end{bmatrix}; \qquad P^{-1} = \begin{bmatrix} \frac{3}{4} & -\frac{1}{2} \\ -\frac{1}{4} & \frac{1}{2} \end{bmatrix}.$$

The closed-loop transition matrix is determined to be

$$\tilde{\Phi}(t, t_0) = \begin{bmatrix} \frac{3}{2}e^{-(t-t_0)} - \frac{1}{2}e^{-2(t-t_0)} & -e^{-(t-t_0)} + e^{-2(t-t_0)} \\ \frac{3}{4}e^{-(t-t_0)} - \frac{3}{4}e^{-2(t-t_0)} & -\frac{1}{2}e^{-(t-t_0)} + \frac{3}{2}e^{-2(t-t_0)} \end{bmatrix}.$$

This result is yet another state transition matrix associated with system matrices with eigenvalues -1 and -2. The particular point to be noted is that two different control matrices have been used in the example problem, each producing the same eigenvalues but different time responses. This concept is an example of complete eigenstructure assignment, which is developed in the following section.

5.1.2. The Eigenstructure Characteristics of Linear State Variable Feedback Systems

Virtually any closed-loop response characteristic possible with a time-invariant system of order N can be obtained by use of the linear, time-invariant, state variable feedback controller described above. The general procedure would be to solve for the eigenvalues of the closed-loop system by use of the equation

$$\det[F + GC - I\lambda] = 0 \tag{5.8}$$

as illustrated in the example of the previous section. This procedure will yield an Nth order polynomial equation in λ, with the elements of the gain matrix C appearing in the coefficients. If the system is controllable, then the elements of C can be selected to produce any desired eigenvalues, so long as complex eigenvalues occur in conjugate pairs.

The example of the previous section has illustrated an important aspect of the kind of modal analysis being considered here, and that is the idea of the shape of a dynamical mode of response. In each of the closed-loop transition matrices the two modes

$$e^{-t} \quad \text{and} \quad e^{-2t}$$

are present, but in different linear combinations. That the modes are identical results from the fact that the closed-loop eigenvalues are identical in the two cases considered. That the linear combinations are different results from the distinctly different diagonalizing matrices, which in turn result from the use of the two different gain matrices. In a practical sense, the particular manner in which dynamic modes of response appear in the closed-loop transition matrix is termed the *shape* of the mode, and is determined by the eigenvector corresponding to that mode of response in the diagonalizing matrix of eigenvectors Thus, the general response of the mode is determined by the eigenvalue, and its shape by the eigenvector associated with that particular eigenvalue. An analogous situation exists in classical control theory with poles and zeros of the closed-loop transfer function.

These concepts can perhaps be seen more clearly when the practical problem of system synthesis is considered. The general linear system is described by Equations (5.2) and (5.3), and the closed-loop response is given by Equation (5.5). For linear, time-invariant systems with distinct eigenvalues, the closed-loop transition matrix is given by the expression

$$\tilde{\boldsymbol{\Phi}}(t, t_0) = \boldsymbol{PDP}^{-1} \tag{5.9}$$

where $\boldsymbol{D}$ is the diagonal matrix

$$\boldsymbol{D} = \begin{bmatrix} e^{\lambda_1(t-t_0)} & & & \\ & e^{\lambda_2(t-t_0)} & & \\ & & \ddots & \\ & & & e^{\lambda_N(t-t_0)} \end{bmatrix}.$$

The synthesis problem consists of determining the feedback matrix $\boldsymbol{C}$ in Equation (5.6) so that the desired closed-loop system response is produced. The modes of the closed-loop response are produced by the eigenvalues of the closed-loop system matrix $\tilde{\boldsymbol{F}}$, while the shapes of the modes are determined by the associated eigenvectors selected by the designer. Both of these characteristics result from the selection of the $\boldsymbol{C}$ matrix.

To formalize the considerations, assume that a particular set of closed-loop eigenvalues and associated eigenvectors has been selected. Note that this selection is tantamount to specification of the closed-loop state transition matrix. At this point, the known quantities are the system matrix $\boldsymbol{F}$ and the closed-loop state transition matrix in the form of Equation (5.9). That is, the diagonalizing matrix $\boldsymbol{P}$ is specified in the statement of the design objectives. The closed-loop system matrix can be obtained from the relationship

$$\tilde{\boldsymbol{F}} = \boldsymbol{P}\boldsymbol{\Lambda}\boldsymbol{P}^{-1} \tag{5.10}$$

where $\boldsymbol{\Lambda}$ is the diagonal matrix of eigenvalues,

$$\boldsymbol{\Lambda} = \begin{bmatrix} \lambda_1 & & & \\ & \lambda_2 & & \\ & & \ddots & \\ & & & \lambda_N \end{bmatrix}. \tag{5.11}$$

Combining these expressions with Equation (5.7) results in an expression for the matrix product $\boldsymbol{GC}$,

$$\begin{aligned} \boldsymbol{GC} &= \tilde{\boldsymbol{F}} - \boldsymbol{F} \\ &= \boldsymbol{P}\boldsymbol{\Lambda}\boldsymbol{P}^{-1} - \boldsymbol{F}. \end{aligned} \tag{5.12}$$

Thus, specification of the eigenvalues and eigenvectors results in an expression for the product $\boldsymbol{GC}$. In the special case in which the input distribution matrix $\boldsymbol{G}$ is nonsingular, a unique value of $\boldsymbol{C}$ results,

$$\boldsymbol{C} = \boldsymbol{G}^{-1}[\boldsymbol{P}\boldsymbol{\Lambda}\boldsymbol{P}^{-1} - \boldsymbol{F}]. \tag{5.13}$$

Nonsingular input distribution matrices are a rarity in practical systems, since they imply a number of inputs equal to the number of states. Such was the situation in the examples of the previous section. In fact, if the first example there is considered as a design problem, and the $\boldsymbol{F}$, $\boldsymbol{G}$, $\boldsymbol{P}$ and $\boldsymbol{\Lambda}$ matrices are given as

$$\boldsymbol{F} = \begin{bmatrix} 0 & 1 \\ -5 & -2 \end{bmatrix}; \quad \boldsymbol{G} = \begin{bmatrix} 1 & 0 \\ 0 & 1 \end{bmatrix}; \quad \boldsymbol{P} = \begin{bmatrix} 1 & 2 \\ 1 & 1 \end{bmatrix}; \quad \boldsymbol{\Lambda} = \begin{bmatrix} -1 & 0 \\ 0 & -2 \end{bmatrix}$$

then Equation (5.13) yields

$$C = \begin{bmatrix} 1 & 0 \\ 0 & 1 \end{bmatrix} \left[\begin{bmatrix} 1 & 2 \\ 1 & 1 \end{bmatrix} \begin{bmatrix} -1 & 0 \\ 0 & -2 \end{bmatrix} \begin{bmatrix} -1 & 2 \\ 1 & -1 \end{bmatrix} - \begin{bmatrix} 0 & 1 \\ -5 & -2 \end{bmatrix} \right]$$

$$= \begin{bmatrix} -3 & 1 \\ 4 & 2 \end{bmatrix}$$

which is the specified control matrix in the example.

In most practical instances, the input distribution matrix $\boldsymbol{G}$ is not square, in which event there may be no solutions for $\boldsymbol{G}$, there may be a unique solution, or there may be many solutions.

As an example of the use of eigenstructure characteristics, consider the time-invariant system equations

$$\dot{x}_1 = x_2 + u_1$$
$$\dot{x}_2 = x_3$$
$$\dot{x}_3 = 4x_1 + 4x_2 - x_3 + u_2.$$

The system matrix $\boldsymbol{F}$ and input distribution matrix $\boldsymbol{G}$ are seen to be

$$\boldsymbol{F} = \begin{bmatrix} 0 & 1 & 0 \\ 0 & 0 & 1 \\ 4 & 4 & -1 \end{bmatrix}; \qquad \boldsymbol{G} = \begin{bmatrix} 1 & 0 \\ 0 & 0 \\ 0 & 1 \end{bmatrix}.$$

The eigenvalues of $\boldsymbol{F}$ can be determined to be

$$\lambda_1 = -1; \quad \lambda_2 = -2; \quad \lambda_3 = 2$$

and the open-loop system is seen to be unstable, as evidenced by the positive eigenvalue. Linear state variable feedback is to be used to produce a closed-loop response characterized by eigenvalues

$$\lambda_1 = -2; \quad \lambda_2 = -3; \quad \lambda_3 = -4.$$

If the shape of the modes corresponding to these eigenvalues is of no interest, the problem is one of determining a feedback gain matrix $\boldsymbol{C}$ which produces the desired eigenvalues of the closed-loop system matrix $\tilde{\boldsymbol{F}}$ as implied by Equation

(5.7). In this case,

$$\tilde{F} = F + GC = \begin{bmatrix} 0 & 1 & 0 \\ 0 & 0 & 1 \\ 4 & 4 & -1 \end{bmatrix} + \begin{bmatrix} 1 & 0 \\ 0 & 0 \\ 0 & 1 \end{bmatrix} \begin{bmatrix} c_{11} & c_{12} & c_{13} \\ c_{21} & c_{22} & c_{23} \end{bmatrix}$$

$$= \begin{bmatrix} c_{11} & 1 + c_{12} & c_{13} \\ 0 & 0 & 1 \\ 4 + c_{21} & 4 + c_{22} & -1 + c_{23} \end{bmatrix}$$

and the desire is to select the c_{ij} so that this matrix has the desired eigenvalues. The selected eigenvalues produce the characteristic equation

$$(\lambda + 2)(\lambda + 3)(\lambda + 4) = \lambda^3 + 9\lambda^2 + 26\lambda + 24 = 0.$$

This polynomial in λ must be identical to the determinantal equation

$$\det \begin{bmatrix} c_{11} - \lambda & 1 + c_{12} & c_{13} \\ 0 & -\lambda & 1 \\ 4 + c_{21} & 4 + c_{22} & -1 + c_{23} - \lambda \end{bmatrix} = 0$$

Thus, the coefficients in these two polynomials can be equated and the resulting three equations in the six unknown c_{ij} constitute the requirements on the selection on the c_{ij}. Obviously, there are an infinite number of combinations which satisfy these equations. One solution is

$$C = \begin{bmatrix} -3 & -1 & 0 \\ -4 & -12 & -5 \end{bmatrix}$$

which yields

$$\tilde{F} = \begin{bmatrix} -3 & 0 & 0 \\ 0 & 0 & 1 \\ 0 & -8 & -6 \end{bmatrix}.$$

That this closed-loop matrix has the desired characteristic equation is readily determined.

The closed-loop transition matrix resulting from this particular choice of feedback matrix can be determined to be

$$\tilde{\Phi}(t, t_0) = \begin{bmatrix} e^{-3(t-t_0)} & 0 & 0 \\ 0 & 2e^{-2(t-t_0)} - e^{-4(t-t_0)} & \frac{1}{2}e^{-(t-t_0)} - \frac{1}{2}e^{-2(t-t_0)} \\ 0 & -4e^{-2(t-t_0)} + 4e^{-2(t-t_0)} & -e^{-(t-t_0)} + 2e^{-4(t-t_0)} \end{bmatrix}.$$

The shapes of the three modes are implied by the specific form of this transition matrix. Other feedback gain matrices producing the desired eigenvalues will produce different mode shapes.

The process followed in the example above is to select a feedback gain matrix by some number of arbitrary decisions on values of the elements of the $\boldsymbol{C}$ matrix. The mode shapes are then determined, possibly by determining the diagonalizing matrix of eigenvectors for the resulting closed-loop system matrix $\tilde{\boldsymbol{F}}$.

If the eigenstructure concept is applied, both the eigenvalues and the mode shapes are simultaneously selected using the concepts developed in this section. Specification of the mode shapes is tantamount to the specification of a diagonalizing matrix of eigenvectors. In the present example, let the desired matrix of eigenvectors be

$$\boldsymbol{P} = \begin{bmatrix} 0 & 2 & 9 \\ -1 & -1 & 0 \\ 2 & 3 & 0 \end{bmatrix}; \quad \boldsymbol{P}^{-1} = \begin{bmatrix} 0 & -3 & -1 \\ 0 & 2 & 1 \\ \frac{1}{9} & -\frac{4}{9} & -\frac{2}{9} \end{bmatrix}.$$

That this matrix of eigenvectors specifies the mode shape is obvious, since the closed-loop transition matrix is immediately determined from Equation (5.9):

$$\tilde{\Phi}(t, t_0) = \begin{bmatrix} 0 & 2 & 9 \\ -1 & -1 & 0 \\ 2 & 3 & 0 \end{bmatrix} \begin{bmatrix} e^{-2(t-t_0)} & & \\ & e^{-3(t-t_0)} & \\ & & e^{-4(t-t_0)} \end{bmatrix} \begin{bmatrix} 0 & -3 & -1 \\ 0 & 2 & 1 \\ \frac{1}{9} & -\frac{4}{9} & -\frac{2}{9} \end{bmatrix}$$

$$= \begin{bmatrix} e^{-4(t-t_0)} & 4e^{-3(t-t_0)} - 4e^{-4(t-t_0)} & 2e^{-3(t-t_0)} - 2e^{-4(t-t_0)} \\ 0 & 3e^{-2(t-t_0)} - 2e^{-3(t-t_0)} & e^{-2(t-t_0)} - e^{-3(t-t_0)} \\ 0 & -6e^{2(t-t_0)} + 6e^{-3(t-t_0)} & -2e^{2(t-t_0)} + 3e^{-3(t-t_0)} \end{bmatrix}.$$

The reader should note that, because of the diagonal matrix in the expression for $\tilde{\Phi}(t, t_0)$, zeros occurring in the rows of the $\boldsymbol{P}$ matrix serve to ban the corresponding mode from the response of the associated state variable. That is, the zero in the (1, 1) position excludes the e^{-2t} mode from the response of the state variable $x_1(t)$. The remaining two zeros serve to exclude the mode e^{-4t} from the $x_2(t)$ and $x_3(t)$ responses, respectively. These characteristics are evident in the expression for the closed-loop transition matrix $\tilde{\boldsymbol{\Phi}}(t, t_0)$.

So far, the closed-loop state transition matrix has been specified. It is now desired to determine a feedback gain matrix, if one or more exist, which produces this result. From Equation (5.10), the closed-loop system matrix $\tilde{\boldsymbol{F}}$ corresponding to the desired closed-loop system matrix can be determined:

$$\tilde{\boldsymbol{F}} = \begin{bmatrix} 0 & 2 & 9 \\ -1 & -1 & 0 \\ 2 & 3 & 0 \end{bmatrix} \begin{bmatrix} -2 & & \\ & -3 & \\ & & -4 \end{bmatrix} \begin{bmatrix} 0 & -3 & -1 \\ 0 & 2 & 1 \\ \frac{1}{9} & -\frac{4}{9} & -\frac{2}{9} \end{bmatrix}$$

$$= \begin{bmatrix} 4 & 4 & 2 \\ 0 & 0 & 1 \\ 0 & -6 & -5 \end{bmatrix}.$$

Then, using Equation (5.12), the matrix produced $\boldsymbol{GC}$ is determined to be

$$\boldsymbol{GC} = \tilde{\boldsymbol{F}} - \boldsymbol{F} = \begin{bmatrix} -4 & 4 & 2 \\ 0 & 0 & 1 \\ 0 & -6 & -5 \end{bmatrix} - \begin{bmatrix} 0 & 1 & 0 \\ 0 & 0 & 1 \\ 4 & 4 & -1 \end{bmatrix}$$

$$= \begin{bmatrix} -4 & 3 & 2 \\ 0 & 0 & 0 \\ -4 & -10 & -4 \end{bmatrix}.$$

Since the input distribution matrix $\boldsymbol{G}$ is known, this relationship suffices to determine the feedback matrix $\boldsymbol{C}$, if there indeed exists a $\boldsymbol{C}$ which satisfies this relationship. In this case,

$$\boldsymbol{GC} = \begin{bmatrix} 1 & 0 \\ 0 & 0 \\ 0 & 1 \end{bmatrix} \begin{bmatrix} c_{11} & c_{12} & c_{13} \\ c_{21} & c_{22} & c_{23} \end{bmatrix} = \begin{bmatrix} c_{11} & c_{12} & c_{13} \\ 0 & 0 & 0 \\ c_{21} & c_{22} & c_{23} \end{bmatrix}$$

and there is a unique solution,

$$\boldsymbol{C} = \begin{bmatrix} -4 & 3 & 2 \\ -4 & -10 & -4 \end{bmatrix}.$$

The reader can easily verify that this value of $\boldsymbol{C}$ produces the desired $\tilde{\boldsymbol{F}}$ and $\tilde{\boldsymbol{\Phi}}$ matrices. It should also be noted that fortuitous circumstance produced a unique solution for $\boldsymbol{C}$. For example, the row of zeros in the matrix $[\tilde{\boldsymbol{F}} - \boldsymbol{F}]$ is necessary if there is to be any value of $\boldsymbol{C}$ which satisfies the equations above. This observation in turn implies that the choice of the diagonalizing matrix of eigenvectors, and thus the specification of mode shapes, is not completely arbitrary. The following section considers these concepts in more detail, and in a manner amenable to digital computer analysis.

5.1.3. System Synthesis by State Variable Feedback: Eigenvalue Assignment

One of the features of modern control theory is that it permits the use of many of the elegant techniques of matrix theory which, in turn, are well adapted to digital computer implementation. Nowhere is this feature more evident than in the area of system synthesis, where transformation of the system equations to certain very useful canonical forms permits rather systematic procedures to be developed. When eigenvalues are to be assigned, without regard to the resulting mode shapes, a most useful transformation is that to a form known as the *generalized control canonical form*. This transformation is applied to time-invariant systems of the form

$$\dot{\boldsymbol{x}}(t) = \boldsymbol{F}\boldsymbol{x}(t) + \boldsymbol{G}\boldsymbol{u}(t) \tag{5.14}$$

and the problem is to solve for the value of the control matrix $\boldsymbol{C}$ which will produce the specified eigenvalues of the matrix

$$\tilde{\boldsymbol{F}} = \boldsymbol{F} + \boldsymbol{G}\boldsymbol{C}. \tag{5.15}$$

The generalized control canonical form is obtained by a systematic procedure which transforms Equation (5.14) to the form

$$\dot{\boldsymbol{z}}(t) = \boldsymbol{A}\boldsymbol{z}(t) + \boldsymbol{B}\boldsymbol{v}(t) \tag{5.16}$$

where the matrix $\boldsymbol{A}$ has the specified form

$$\boldsymbol{A} = \left[\begin{array}{c|c|c|c} \boldsymbol{A}_1 & 0 & \cdots & 0 \\ \hline 0 & \boldsymbol{A}_2 & \cdots & 0 \\ \hline \cdots & \cdots & \cdots & \cdots \\ \hline 0 & 0 & \cdots & \boldsymbol{A}_p \end{array}\right]. \tag{5.17}$$

The submatrices $\boldsymbol{A}_1$ through $\boldsymbol{A}_P$, where P is the dimension of the control input $\boldsymbol{u}$, in turn have the special form

$$\boldsymbol{A}_i = \begin{bmatrix} 0 & 1 & 0 & \cdots & 0 \\ 0 & 0 & 1 & \cdots & 0 \\ \cdots & \cdots & \cdots & \cdots & \cdots \\ 0 & 0 & 0 & \cdots & 1 \\ 0 & 0 & 0 & \cdots & 0 \end{bmatrix}. \tag{5.18}$$

These matrices are of dimension $N_i \times N_i$.

Furthermore, the matrix $\boldsymbol{B}$ has the form

$$\boldsymbol{B} = \left[\begin{array}{c|c|c|c} \boldsymbol{b}_1 & & \cdots & 0 \\ \hline 0 & \boldsymbol{b}_2 & \cdots & 0 \\ \hline \cdots & \cdots & \cdots & \cdots \\ \hline 0 & 0 & \cdots & \boldsymbol{b}_P \end{array}\right] \tag{5.19}$$

where the vectors $\boldsymbol{b}_i$ in turn have the form

$$\boldsymbol{b}_i = \begin{bmatrix} 0 \\ 0 \\ \vdots \\ 1 \end{bmatrix} \tag{5.20}$$

and are of dimension N_i, corresponding to that of the $\boldsymbol{A}_i$ matrix of Equation (5.18).

The dimension of $\boldsymbol{A}_i$ is the same as the row dimension of the vector $\boldsymbol{b}_i$. A sixth-order system, for example, might have a generalized control canonical form given by

$$\boldsymbol{A} = \left[\begin{array}{ccc|cc|c} 0 & 1 & 0 & 0 & 0 & 0 \\ 0 & 0 & 1 & 0 & 0 & 0 \\ 0 & 0 & 0 & 0 & 0 & 0 \\ \hline 0 & 0 & 0 & 0 & 1 & 0 \\ 0 & 0 & 0 & 0 & 0 & 0 \\ \hline 0 & 0 & 0 & 0 & 0 & 0 \end{array}\right]; \quad \boldsymbol{B} = \left[\begin{array}{c|c|c} 0 & 0 & 0 \\ 0 & 0 & 0 \\ 1 & 0 & 0 \\ \hline 0 & 0 & 0 \\ 0 & 1 & 0 \\ \hline 0 & 0 & 1 \end{array}\right].$$

The order of the submatrices of $\boldsymbol{A}$ and $\boldsymbol{B}$, in this case 3, 2, and 1, are called the *control invariants* of the system and are uniquely determined by the system matrix $\boldsymbol{F}$ and input distribution matrix $\boldsymbol{G}$. The number of control invariants is equal to P, the dimension of the control input.

The reader should note that the eigenvalues of $\boldsymbol{A}$ are all zero, and that the special form of the matrix makes the selection of feedback gains to produce a specified set of eigenvalues rather easy. Such assignment, followed by a similarity transformation back to the original state space, completes the synthesis process.

The transformation to the generalized control canonical form is accomplished by three steps, as follows:

I. A similarity transformation of the state variable $\boldsymbol{x}(t)$ using a nonsingular transformation matrix $\boldsymbol{T}$:

$$\boldsymbol{x}(t) = \boldsymbol{T}\boldsymbol{z}(t). \tag{5.21}$$

The resulting system matrix $\boldsymbol{T}^{-1}\boldsymbol{F}\boldsymbol{T}$, has the same eigenvalues as $\boldsymbol{F}$, since eigenvalues are invariant under a similarity transformation.

II. A transformation of the control variable $\boldsymbol{u}(t)$ using a nonsingular matrix transformation $\boldsymbol{R}$,

$$\boldsymbol{u}(t) = \boldsymbol{R}\boldsymbol{w}(t). \tag{5.22}$$

The result is a system described by the equation

$$\dot{\boldsymbol{z}}(t) = \boldsymbol{T}^{-1}\boldsymbol{F}\boldsymbol{T}\boldsymbol{z}(t) + \boldsymbol{T}^{-1}\boldsymbol{G}\boldsymbol{R}\boldsymbol{w}(t) \tag{5.23}$$

which retains the eigenvalues of the original system.

III. State variable feedback of the form

$$\begin{aligned}\boldsymbol{w}(t) &= \boldsymbol{v}(t) - \hat{\boldsymbol{C}}\boldsymbol{q}(t)\\ &= \boldsymbol{v}(t) - \hat{\boldsymbol{C}}\boldsymbol{T}^{-1}\boldsymbol{x}(t)\end{aligned} \tag{5.24}$$

is introduced. The closed-loop system equations are

$$\dot{\boldsymbol{z}}(t) = [\boldsymbol{T}^{-1}\boldsymbol{F}\boldsymbol{T} - \boldsymbol{T}^{-1}\boldsymbol{G}\boldsymbol{R}\hat{\boldsymbol{C}}]\boldsymbol{z}(t) + [\boldsymbol{T}^{-1}\boldsymbol{G}\boldsymbol{R}]\boldsymbol{v}(t). \tag{5.25}$$

This set of equations has all zero eigenvalues.

There is enough flexibility in the selection of the matrices $\boldsymbol{T}$, $\boldsymbol{R}$, and $\hat{\boldsymbol{C}}$ that Equation (5.25) can be put into the form of Equation (5.16). There are several algorithms for determining these transformation matrices. One of them is presented here as a step-by-step procedure, without justification.

The specific procedure requires knowledge of the control invariants, which are the dimensions of the submatrices of Equation (5.17). Since the control invariants are uniquely determined by the system matrix $\boldsymbol{F}$ and control distribution matrix $\boldsymbol{G}$, they should be determinable prior to the calculation of the matrices $\boldsymbol{A}$ and $\boldsymbol{B}$. Such is in fact the case, and the control invariants can be determined from the controllability matrix, Equation (4.58),

$$[\boldsymbol{G} \mid \boldsymbol{F}\boldsymbol{G} \mid \boldsymbol{F}^2\boldsymbol{G} \mid \cdots \mid \boldsymbol{F}^{N-P}\boldsymbol{G}]. \tag{5.26}$$

Developments in Chapter 4 indicate that, if this composite matrix has full rank, then the system is controllable—meaning, in the time-invariant case, that all modes of system response are affected by the available input. If the input distribution matrix is partitioned into column vectors,

$$\boldsymbol{G} = [\boldsymbol{g}_1 \mid \boldsymbol{g}_2 \mid \cdots \mid \boldsymbol{g}_P]$$

then the controllability matrix can be expressed as a partitioned matrix of column vectors,

$$[\boldsymbol{g}_1 \mid \boldsymbol{g}_2 \mid \cdots \mid \boldsymbol{g}_P \mid \boldsymbol{F}\boldsymbol{g}_1 \mid \boldsymbol{F}\boldsymbol{g}_2 \mid \cdots \mid \boldsymbol{F}\boldsymbol{g}_P \mid \cdots \mid \boldsymbol{F}^{N-P}\boldsymbol{g}_1 \mid \boldsymbol{F}^{N-P}\boldsymbol{g}_2 \mid \cdots \mid \boldsymbol{F}^{N-P}\boldsymbol{g}_P]. \tag{5.27}$$

If the system is controllable, then this matrix has full rank, equal to the dimension of $\boldsymbol{F}$. If the system is not controllable, then the eigenvalues corresponding to the noncontrollable modes cannot be altered. For controllable systems, N of the vectors constituting the controllability matrix of Equation (5.27) are linearly independent. In order to determine the control invariants, these column vectors are rearranged so that each of the first N vectors is linearly independent of the preceeding vectors. The vectors are taken in the order

$$\boldsymbol{g}_1, \boldsymbol{F}\boldsymbol{g}_1, \boldsymbol{F}^2\boldsymbol{g}_1, \ldots, \boldsymbol{F}^{N_1-1}\boldsymbol{g}_1 \tag{5.28}$$

until the first dependent vector is reached. This procedure then determines the value of the first control invariant, N_1. Then the vectors

$$\boldsymbol{g}_2, \boldsymbol{F}\boldsymbol{g}_2, \boldsymbol{F}^2\boldsymbol{g}_2, \ldots, \boldsymbol{F}^{N_2-1}\boldsymbol{g}_2 \tag{5.29}$$

are selected, until the first dependent vector is reached. This process continues until N linearly independent vectors have been selected. The resulting $N \times N$ matrix

$$\boldsymbol{M} = [\boldsymbol{g}_1 \mid \boldsymbol{F}\boldsymbol{g}_1 \mid \cdots \mid \boldsymbol{F}^{N_1-1}\boldsymbol{g}_1 \mid \boldsymbol{g}_2 \mid \boldsymbol{F}\boldsymbol{g}_2 \mid \cdots \mid \boldsymbol{F}^{N_2-1}\boldsymbol{g}_2 \mid \cdots \mid \boldsymbol{g}_P \mid \boldsymbol{F}\boldsymbol{g}_P \mid \cdots \mid \boldsymbol{F}^{N_P-1}\boldsymbol{g}_p] \quad (5.30)$$

implies the values of control invariants for the system, which are

$$N_1, N_2, N_3, \ldots, N_P. \quad (5.31)$$

There are P control invariants, where P is the dimension of the control vector $\boldsymbol{u}$, and they sum to N, the dimension of the system. Once the control invariants are determined, they are put in decreasing order. The columns of the input distribution matrix $\boldsymbol{G}$ are then rearranged, together with the elements of the control vector $\boldsymbol{u}$, to conform with this new order of control invariants. Such rearrangement does not change the system equations.

The system equations are still in the form of Equation (5.14), but now the $\boldsymbol{G}$ matrix and $\boldsymbol{u}$ vector have been rearranged to conform with the ordering of the control invariants. The resulting matrix $\boldsymbol{M}$ of Equation (5.30) can now be partitioned into P submatrices of dimension $N \times N_i$, in order of descending N_i:

$$\boldsymbol{M} = [\boldsymbol{M}_1 \mid \boldsymbol{M}_2 \mid \cdots \mid \boldsymbol{M}_P]. \quad (5.32)$$

Since $\boldsymbol{M}$ has full rank, it has an inverse

$$\boldsymbol{M}^{-1} = \begin{bmatrix} \boldsymbol{M}_1^{-1} \\ \hline \boldsymbol{M}_2^{-1} \\ \hline \vdots \\ \hline \boldsymbol{M}_P^{-1} \end{bmatrix}. \quad (5.33)$$

The $\boldsymbol{M}_i^{-1}$ are not inverse matrices—they are not even square. Rather, the nomenclature implies that $\boldsymbol{M}_i^{-1}$ is a submatrix of the inverse of the nonsingular matrix $\boldsymbol{M}$ of Equation (5.32). If $\boldsymbol{m}_i^T$ is the row vector which is the last row of the inverse matrix partition $\boldsymbol{M}_i^{-1}$, then the inverse of the transformation matrix is determined in the following manner:

$$T^{-1} = \begin{bmatrix} m_1^T \\ \hdashline m_1^T F \\ \hdashline \vdots \\ \hdashline m_1^T F^{N_1 - 1} \\ \hdashline \vdots \\ \hdashline m_P^T \\ \hdashline m_P^T F \\ \hdashline \vdots \\ \hdashline m_P^T F^{N_P - 1} \end{bmatrix}. \tag{5.34}$$

This matrix is used in the similarity transformation of Step I. The transformation matrix $\boldsymbol{R}$ of Equation (5.22) can be obtained from the fact that Equation (5.25) requires that the matrix

$$T^{-1}GR \tag{5.35}$$

must be of the special form given by Equation (5.19). Finally, the feedback matrix $\hat{\boldsymbol{C}}$ is determined from the requirement that the matrix

$$[T^{-1}FT - T^{-1}GR\hat{C}] \tag{5.36}$$

of Equation (5.25) must have the special form given by Equation (5.17). These procedures, although tedious by hand, can be accomplished quite efficiently by a digital computer. In fact, there are even more efficient methods for use in computer implementation.

The utility of the generalized control canonical form in system synthesis is immediately evident when state variable feedback is used to produce selected eigenvalues in the system described by the canonical form. That is, if state variable feedback is used in the form

$$v(t) = Nz(t) \tag{5.37}$$

then the closed-loop system matrix in the canonical variables is

$$\dot{z}(t) = [\boldsymbol{A} + \boldsymbol{BN}]\,z(t). \tag{5.38}$$

It is now desired to select the matrix $\boldsymbol{N}$ so that the desired closed-loop eigenvalues result. Because of the special form of the $\boldsymbol{A}$ and $\boldsymbol{B}$ matrices, it is possible to specify that the matrix $[\boldsymbol{A} + \boldsymbol{BN}]$ be partitioned in the same manner as matrix $\boldsymbol{A}$, as indicated by Equation (5.17), and that each of the $N_i \times N_i$ diagonal submatrices have the form

$$\begin{bmatrix} 0 & 1 & 0 & \cdots & 0 \\ 0 & 0 & 1 & \cdots & 0 \\ \cdots & \cdots & \cdots & \cdots & \cdots \\ 0 & 0 & 0 & \cdots & 1 \\ -a_0 & -a_1 & -a_2 & \cdots & -a_{N_i-1} \end{bmatrix}. \tag{5.39}$$

In this matrix, the a_i are the coefficients of a factor of the desired characteristic equation containing those eigenvalues to be assigned by this submatrix. For example, if $N_i = 3$, then three eigenvalues can be assigned by this submatrix, and the factor of the characteristic equation is

$$\begin{aligned} Q_i(\lambda) &= (\lambda - \lambda_j)(\lambda - \lambda_k)(\lambda - \lambda_l) \\ &= \lambda^3 + a_2\lambda^2 + a_1\lambda + a_0 \end{aligned} \tag{5.40}$$

where λ_j, λ_k, and λ_l are the three eigenvalues assigned by this submatrix. Complex eigenvalues must be assigned in complex pairs. The characteristic equation itself is the product of all the factors. The number of factors is $\boldsymbol{P}$, the dimension of the input vector $\boldsymbol{u}$.

If the matrix comprised of the diagonal submatrices of Equation (5.39) is designated by $\tilde{\boldsymbol{A}}$, then

$$\tilde{\boldsymbol{A}} = \boldsymbol{A} + \boldsymbol{BN} \tag{5.41}$$

and the problem reduces to one of solving for $\boldsymbol{N}$. This task is made easy by the fact that the $\boldsymbol{B}$ matrix, as given by Equations (5.19) and (5.20), has the special property that

$$\boldsymbol{B}^T\boldsymbol{B} = \boldsymbol{I}. \tag{5.42}$$

Thus, by solving Equation (5.41) for $\boldsymbol{BN}$ and then multiplying on the left by $\boldsymbol{B}^T$, one obtains

$$N = B^T [\tilde{A} - A] \tag{5.43}$$

which represents a closed-form expression for N.

At this point in the development, the system described by the generalized control canonical form has been controlled in a manner which produces the desired closed-loop eigenvalues. Transformation back to the original state variables is provided by Equation (5.21), and since eigenvalues are invariant under a similarity transformation, the original system will also exhibit the desired-closed loop eigenvalues. The required control matrix C can be determined from Equations (5.22), (5.24), and (5.37):

$$u = R[N - \hat{C}]T^{-1}x \tag{5.44}$$

or

$$C = R[N - \hat{C}]T^{-1}. \tag{5.45}$$

This result completes the development. The matrix C determined by Equation (5.45) is, in general, not unique–a characteristic observed in developments of the previous section. Different values of C yielding this same set of closed-loop eigenvalues can be obtained by replacing zeros in some of the off-diagonal elements of A–comprised of diagonal submatrices described by Equation (5.39)–by arbitrary values in such a manner that the eigenvalues of A are not affected.

To illustrate use of the generalized control canonical form, the example of the previous section will be considered. The system is described by the matrices

$$F = \begin{bmatrix} 0 & 1 & 0 \\ 0 & 0 & 1 \\ 4 & 4 & -1 \end{bmatrix}; \quad G = \begin{bmatrix} 1 & 0 \\ 0 & 0 \\ 0 & 1 \end{bmatrix}.$$

Here, $N = 3, P = 2$, and the open-loop eigenvalues are

$$\lambda_1 = 1; \quad \lambda_2 = -2; \quad \lambda_3 = 2.$$

It is desired to determine feedback matrix C so that the closed-loop system has eigenvalues

$$\lambda_1 = -2; \quad \lambda_2 = -3; \quad \lambda_3 = -4.$$

Previously, the problem was approached by solving for $\boldsymbol{C}$ directly from the requirement that the closed-loop system matrix have the desired eigenvalues. Here, the system will be transformed to the generalized control canonical form and the procedure of this section applied. The matrix $\boldsymbol{M}$ of Equation (5.30), with $N = 3, P = 2$, is

$$\boldsymbol{M} = \left[\begin{array}{c|cc} 1 & 0 & 0 \\ 0 & 0 & 1 \\ 0 & 1 & -1 \end{array}\right].$$

The control invariants are therefore $N_1 = 2$ and $N_2 = 1$, and rearranging the input distribution vector $\boldsymbol{G}$ to reflect this descending order yields the new value of $\boldsymbol{G}$,

$$\boldsymbol{G} = \begin{bmatrix} 0 & 1 \\ 0 & 0 \\ 1 & 0 \end{bmatrix}.$$

The elements of the input vector $\boldsymbol{u}$ are revised to accommodate this interchange. The revised $\boldsymbol{M}$ matrix is

$$\boldsymbol{M} = \left[\begin{array}{cc|c} 0 & 0 & 1 \\ 0 & 1 & 0 \\ 1 & -1 & 0 \end{array}\right]; \qquad \boldsymbol{M}^{-1} = \left[\begin{array}{ccc} 0 & 1 & 1 \\ 0 & 1 & 0 \\ \hline 1 & 0 & 0 \end{array}\right]$$

where the inverse matrix corresponds to Equations (5.32) and (5.33) in the following form:

$$\boldsymbol{M}^{-1} = \left[\begin{array}{c} \boldsymbol{M}_1^{-1} \\ \hline \boldsymbol{M}_2^{-1} \end{array}\right]; \qquad N_1 = 2, \quad N_2 = 1.$$

The transformation matrix $\boldsymbol{T}$, from Equation (5.34), is determined from

$$\boldsymbol{T}^{-1} = \left[\begin{array}{c} \boldsymbol{m}_1^T \\ \hline \boldsymbol{m}_1^T\boldsymbol{F} \\ \hline \boldsymbol{m}_2^T \end{array}\right] = \begin{bmatrix} 0 & 1 & 0 \\ 0 & 0 & 1 \\ 1 & 0 & 0 \end{bmatrix}$$

from which

$$\boldsymbol{T} = \begin{bmatrix} 0 & 0 & 1 \\ 1 & 0 & 0 \\ 0 & 1 & 0 \end{bmatrix}$$

The matrix $\boldsymbol{R}$ can now be determined from the requirement that the matrix of Equation (5.35) must have the special form of Equation (5.19), which in this case is

$$\boldsymbol{T}^{-1}\boldsymbol{G}\boldsymbol{R} = \left[\begin{array}{c|c} 0 & 0 \\ 1 & 0 \\ \hline 0 & 1 \end{array}\right] = \boldsymbol{B}$$

or

$$\begin{bmatrix} 0 & 1 & 0 \\ 0 & 0 & 1 \\ 1 & 0 & 0 \end{bmatrix}\begin{bmatrix} 0 & 1 \\ 0 & 0 \\ 1 & 0 \end{bmatrix}\begin{bmatrix} r_{11} & r_{12} \\ r_{21} & r_{22} \end{bmatrix} = \begin{bmatrix} 0 & 0 \\ 1 & 0 \\ 0 & 1 \end{bmatrix}.$$

This equation can be solved for the components of the matrix $\boldsymbol{R}$,

$$\begin{bmatrix} 0 & 0 \\ r_{11} & r_{12} \\ r_{21} & r_{22} \end{bmatrix} = \begin{bmatrix} 0 & 0 \\ 1 & 0 \\ 0 & 1 \end{bmatrix}; \qquad \boldsymbol{R} = \begin{bmatrix} 1 & 0 \\ 0 & 1 \end{bmatrix}.$$

Finally, the matrix $\hat{\boldsymbol{C}}$ can be determined from the fact that the matrix of Equation (5.36) must have the special form of Equation (5.17). In this case

$$\boldsymbol{T}^{-1}\boldsymbol{F}\boldsymbol{T} = \begin{bmatrix} 0 & 1 & 0 \\ 4 & -1 & 4 \\ 1 & 0 & 0 \end{bmatrix}$$

and

$$\boldsymbol{T}^{-1}\boldsymbol{G}\boldsymbol{R}\hat{\boldsymbol{C}} = \begin{bmatrix} 0 & 0 \\ 1 & 0 \\ 0 & 1 \end{bmatrix}\begin{bmatrix} \hat{c}_{11} & \hat{c}_{12} & \hat{c}_{13} \\ \hat{c}_{21} & \hat{c}_{22} & \hat{c}_{23} \end{bmatrix}$$

$$= \begin{bmatrix} 0 & 0 & 0 \\ \hat{c}_{11} & \hat{c}_{12} & \hat{c}_{13} \\ \hat{c}_{21} & \hat{c}_{22} & \hat{c}_{23} \end{bmatrix}.$$

Then

$$T^{-1}FT - T^{-1}GR\hat{C} = \begin{bmatrix} 0 & 1 & 0 \\ 4-\hat{c}_{11} & -1-\hat{c}_{12} & 4-\hat{c}_{13} \\ 1-\hat{c}_{21} & -\hat{c}_{22} & -\hat{c}_{23} \end{bmatrix}.$$

This matrix is equated to the form of Equation (5.17), in this case

$$\left[\begin{array}{cc|c} 0 & 1 & 0 \\ 0 & 0 & 0 \\ \hline 0 & 0 & 0 \end{array}\right] = A.$$

Equating coefficients yields, for matrix $\hat{C}$,

$$\hat{C} = \begin{bmatrix} 4 & -1 & 4 \\ 1 & 0 & 0 \end{bmatrix}.$$

The reader should verify that these values of T, R, and $\hat{C}$ do actually produce the generalized control canonical form of Equation (5.16).

With the canonical form now established, the eigenvalues can be assigned to the closed-loop matrix $\tilde{A}$ of Equation (5.41). The two eigenvalues -2 and -3 will be assigned by the 2×2 submatrix, and the eigenvalue -4 by the 1×1 submatrix. The characteristic equation factors are

$$Q_2 = (\lambda + 2)(\lambda + 3) = \lambda^2 + 5\lambda + 6$$

with the 2×2 submatrix corresponding to Equation (5.39) being

$$\begin{bmatrix} 0 & 1 \\ -6 & -5 \end{bmatrix}$$

and

$$Q_1 = \lambda + 4$$

with the corresponding 1×1 submatrix $[-4]$. The desired canonical form closed-loop system matrix then follows as

$$\tilde{A} = \left[\begin{array}{cc|c} 0 & 1 & 0 \\ -6 & -5 & 0 \\ \hline 0 & 0 & -4 \end{array}\right]$$

which does indeed exhibit the desired eigenvalues. The feedback matrix N can now be determined, using Equation (5.43):

$$N = \begin{bmatrix} 0 & 1 & 0 \\ 0 & 0 & 1 \end{bmatrix} \begin{bmatrix} 0 & 0 & 0 \\ -6 & -5 & 0 \\ 0 & 0 & -4 \end{bmatrix}$$

$$= \begin{bmatrix} -6 & -5 & 0 \\ 0 & 0 & -4 \end{bmatrix}.$$

Finally, the feedback matrix C in the original system can be determined from Equation (5.45),

$$C = \begin{bmatrix} 1 & 0 \\ 0 & 1 \end{bmatrix} \begin{bmatrix} -10 & -4 & -4 \\ -1 & 0 & -4 \end{bmatrix} \begin{bmatrix} 0 & 1 & 0 \\ 0 & 0 & 1 \\ 1 & 0 & 0 \end{bmatrix}$$

$$= \begin{bmatrix} -4 & -10 & -4 \\ -4 & -1 & 0 \end{bmatrix}.$$

This value of C yields the closed-loop system matrix

$$\tilde{F} = [F + GC] = \begin{bmatrix} -4 & 0 & 0 \\ 0 & 0 & 1 \\ 0 & -6 & -5 \end{bmatrix}$$

which does indeed exhibit the desired eigenvalues. This solution represents yet another choice of the feedback gain matrix C that produces the desired eigenvalues. The fact that A is not unique as a matrix exhibiting the desired eigenvalues is the reason for multiple solutions for C.

The synthesis of systems with complex eigenvalues presents no additional problems. It is only necessary to ensure that the factors of the characteristic equation, such as that illustrated by Equation (5.40), contain the complex eigenvalues in conjugate pairs. This, in turn, ensures real values for the coefficients $a_0, a_1, \ldots, a_{N-1}$, and the synthesis proceeds as outlined above.

While this procedure may seem complex, it is easily implemented on a digital computer, and it represents a systematic approach which is valid for systems of any order.

5.1.4. System Synthesis by State Variable Feedback: Eigenstructure Assignment

When the system synthesis problem consists of assigning both eigenvalues and their associated eigenvectors–thus specifying the closed-loop response of the system, including both modes and mode shapes–systematic procedures based upon generalized matrix theory can be developed. The example problem of Section 5.1.2 illustrates the general concept involved, but it stops short of considering whether or not a solution exists, or how to determine what mode shapes can be obtained. This section addresses these questions.

The general system equations are given by Equation (5.14), and the closed-loop system matrix by Equation (5.15). The purpose is to select the gain matrix $\boldsymbol{C}$ so that the closed-loop eigenvalues have certain specified values, and the mode shapes associated with these eigenvalues are in accordance with some specific design criteria. The latter requirement is tantamount to specifying the diagonalizing matrix of eigenvectors $\boldsymbol{P}$ in the expression of Equation (5.10),

$$\tilde{\boldsymbol{F}} = \boldsymbol{P}\boldsymbol{\Lambda}\boldsymbol{P}^{-1}.$$

The columns of $\boldsymbol{P}$ are the eigenvectors associated with the respective eigenvalues, and as such each column–considered as the vector $\boldsymbol{p}_i$–must satisfy the defining equation for an eigenvector of the matrix $\tilde{\boldsymbol{F}}$,

$$\tilde{\boldsymbol{F}}\boldsymbol{p}_i = \lambda_i \boldsymbol{p}_i \tag{5.46}$$

or, by substitution from Equation (5.15),

$$[\boldsymbol{F} + \boldsymbol{G}\boldsymbol{C}]\,\boldsymbol{p}_i = \lambda_i \boldsymbol{p}_i. \tag{5.47}$$

This equation represents the restrictions placed on the eigenvector $\boldsymbol{p}_i$. Multiplying the left-hand side and factoring the $\boldsymbol{p}_i$ yields

$$[\boldsymbol{F} - \boldsymbol{I}\lambda_i]\,\boldsymbol{p}_i + \boldsymbol{G}\boldsymbol{C}\boldsymbol{p}_i = 0. \tag{5.48}$$

Since $\boldsymbol{C}\boldsymbol{p}_i$ is a column vector, this expression can be represented in partitioned form as

$$[\boldsymbol{F} - \boldsymbol{I}\lambda_i \,\vdots\, \boldsymbol{G}]\begin{bmatrix} \boldsymbol{p}_i \\ \hdashline \boldsymbol{C}\boldsymbol{p}_i \end{bmatrix} = 0 \tag{5.49}$$

This equation determines the restriction on the eigenvector $\boldsymbol{p}_i$–it must have a value that satisfies Equation (5.49). This restriction is met if the $(N + P) \times 1$

partitioned vector

$$\begin{bmatrix} p_i \\ \hdashline Cp_i \end{bmatrix} \tag{5.50}$$

lies in the null space of the $N \times (N + P)$ matrix

$$[F - I\lambda_i \,\vdots\, G]. \tag{5.51}$$

The synthesis procedure, then, is to determine the null space of the matrix of Equation (5.51), and then to select a vector of the form of Equation (5.50) which gives a satisfactory mode shape for the mode corresponding to eigenvalue λ_i. When all the eigenvectors have been selected in this manner, then the lower partitions of the selected partitioned vectors are written as a $P \times N$ matrix

$$R = [Cp_1 \,\vdots\, Cp_2 \,\vdots\, \cdots \,\vdots\, Cp_N]. \tag{5.52}$$

The upper partitions are the selected eigenvectors, while the lower partitions, combined to form the $P \times N$ matrix of Equation (5.52), serve to specify the feedback matrix C, since Equation (5.52) is equivalent to the matrix equation

$$R = CP. \tag{5.53}$$

Then, since P is nònsingular, this expression can be solved for C,

$$C = RP^{-1}. \tag{5.54}$$

Thus, it is seen that C is unique, once a permissible set of eigenvectors has been specified. The normal limitations apply in this procedure, in that the selected eigenvectors must be linearly independent—so that P is nonsingular—and that complex eigenvalues must occur in conjugate pairs.

The determination of the null space of a matrix is a well defined problem of matrix theory. With respect to the matrix of Equation (5.51), the following procedure applies:

I. Form the $(N + 2P) \times (N + P)$ partitioned matrix

$$\begin{bmatrix} F - I\lambda_i & G \\ \hdashline I & 0 \\ \hdashline 0 & I \end{bmatrix}. \tag{5.55}$$

II. Use elementary column operations on this matrix to obtain P all-zero columns in the submatrix $[F - I\lambda_i]$.

III. The remainder of these colums determine the null space of the matrix of Equation (5.51). Any linear combination of the $(N+P) \times 1$ vectors defined by the remainder of these columns lies in the null space of the matrix of Equation (5.51).

The synthesis process consists of first specifying the desired eigenvalues, and then determining the null space of the matrix of Equation (5.51) for each of these eigenvalues. From the null space is selected a vector of the form of Equation (5.50), in which the upper partitions are the selected eigenvectors. The lower partitions are then used to form the matrix of Equation (5.52), which is then used to determine the feedback matrix C in accordance with Equation (5.54). The computation involved—even for low-order systems—is considerable, but the matrix techniques used lend themselves quite well to digital computer implementation.

The method outlined here is valid for both real and complex eigenvalues. In the event that complex eigenvalues are selected, they must occur in conjugate pairs. Digital computer programs intended to handle complex eigenvalues must provide for the complex arithmetic involved.

To illustrate these techniques, the second example of Section 5.1.2 will be considered using the systematic approach developed here. The system is defined by the matrices

$$F = \begin{bmatrix} 0 & 1 & 0 \\ 0 & 0 & 1 \\ 4 & 4 & -1 \end{bmatrix}; \qquad G = \begin{bmatrix} 1 & 0 \\ 0 & 0 \\ 0 & 1 \end{bmatrix}$$

and the desired eigenvalues are

$$\lambda_1 = -2; \qquad \lambda_2 = -3; \qquad \lambda_3 = -4.$$

The matrix of Equation (5.51) is

$$[F - I\lambda \mid G] = \left[\begin{array}{ccc|cc} -\lambda & 1 & 0 & 1 & 0 \\ 0 & -\lambda & 1 & 0 & 0 \\ 4 & 4 & -1-\lambda & 0 & 1 \end{array}\right].$$

To determine the null space, the matrix of Equation (5.55) is formed:

$$\left[\begin{array}{ccc|cc} -\lambda & 1 & 0 & 1 & 0 \\ 0 & -\lambda & 1 & 0 & 0 \\ 4 & 4 & -1-\lambda & 0 & 1 \\ \hline 1 & 0 & 0 & 0 & 0 \\ 0 & 1 & 0 & 0 & 0 \\ 0 & 0 & 1 & 0 & 0 \\ \hline 0 & 0 & 0 & 1 & 0 \\ 0 & 0 & 0 & 0 & 1 \end{array}\right].$$

Now, elementary column operations are used to produce two columns with all zeros in the upper partition. Substituting $\lambda = \lambda_1 = -2$ and performing the elementary operations yields, as one possible result,

$$\left[\begin{array}{ccccc} 0 & 0 & 0 & 1 & 0 \\ 0 & 0 & 1 & 0 & 0 \\ 0 & 0 & 1 & 0 & 1 \\ \hline 1 & 0 & 0 & 0 & 0 \\ 0 & 1 & 0 & 0 & 0 \\ 0 & -2 & 1 & 0 & 0 \\ -2 & -1 & 0 & 1 & 0 \\ -4 & -2 & 0 & 0 & 1 \end{array}\right].$$

Any linear combination of the two vectors

$$\left[\begin{array}{c} 1 \\ 0 \\ 0 \\ \hline -2 \\ -4 \end{array}\right] \quad \text{and} \quad \left[\begin{array}{c} 0 \\ 1 \\ -2 \\ \hline -1 \\ -2 \end{array}\right]$$

lies in the desired null space. For $\lambda = -3$, these procedures yield the vectors

$$\left[\begin{array}{c} 1 \\ 0 \\ 0 \\ \hline -3 \\ -4 \end{array}\right] \quad \text{and} \quad \left[\begin{array}{c} 0 \\ 1 \\ -3 \\ \hline -1 \\ 2 \end{array}\right].$$

For $\lambda = -4$,

$$\left[\begin{array}{c} 1 \\ 0 \\ 0 \\ \hline -4 \\ -4 \end{array}\right] \quad \text{and} \quad \left[\begin{array}{c} 0 \\ 1 \\ -4 \\ \hline -1 \\ -8 \end{array}\right].$$

Now the eigenvectors can be selected from linear combinations of the upper partitions of these vectors. Suppose, for $\lambda = -2$, the linear combination

$$(0)\left[\begin{array}{c} 1 \\ 0 \\ 0 \\ \hline -2 \\ -4 \end{array}\right] - (1)\left[\begin{array}{c} 0 \\ 1 \\ -2 \\ \hline -1 \\ -2 \end{array}\right] = \left[\begin{array}{c} 0 \\ -1 \\ 2 \\ \hline 1 \\ 2 \end{array}\right]$$

is selected, to yield the eigenvalue given by the upper partition. Similarly, for $\lambda = -3$, let the selected combination be

$$(2)\left[\begin{array}{c} 1 \\ 0 \\ 0 \\ \hline -3 \\ -4 \end{array}\right] - (1)\left[\begin{array}{c} 0 \\ 1 \\ -3 \\ \hline -1 \\ 2 \end{array}\right] = \left[\begin{array}{c} 2 \\ -1 \\ 3 \\ \hline -5 \\ -10 \end{array}\right]$$

and, for $\lambda = -4$,

$$(9)\begin{bmatrix} 1 \\ 0 \\ 0 \\ \hdashline -4 \\ -4 \end{bmatrix} + (0)\begin{bmatrix} 0 \\ 1 \\ 4 \\ \hdashline -1 \\ -8 \end{bmatrix} = \begin{bmatrix} 9 \\ 0 \\ 0 \\ \hdashline -36 \\ -36 \end{bmatrix}.$$

The matrix of diagonalizing eigenvectors has thus been chosen to be

$$\boldsymbol{P} = \begin{bmatrix} 0 & 2 & 9 \\ -1 & -1 & 0 \\ 2 & 3 & 0 \end{bmatrix}; \quad \boldsymbol{P}^{-1} = \begin{bmatrix} 0 & -3 & -1 \\ 0 & 2 & 1 \\ \frac{1}{9} & -\frac{4}{9} & -\frac{2}{9} \end{bmatrix}$$

which is the same as the example of Section 5.1.2. The matrix $\boldsymbol{R}$ of Equation (5.52) is formed from the lower partitions of the linear combinations,

$$\begin{bmatrix} 1 & -5 & -36 \\ 2 & -10 & -36 \end{bmatrix}.$$

Finally, the feedback matrix $\boldsymbol{C}$ is determined from Equation (5.54),

$$\boldsymbol{C} = \begin{bmatrix} 1 & -5 & -36 \\ -2 & -10 & -36 \end{bmatrix} \begin{bmatrix} 0 & -3 & -1 \\ 0 & 2 & 1 \\ \frac{1}{9} & -\frac{4}{9} & -\frac{2}{9} \end{bmatrix}$$

$$= \begin{bmatrix} -4 & 3 & 2 \\ -4 & -10 & -4 \end{bmatrix}.$$

This result is identical to that obtained previously in Section 5.1.2, since the matrix $\boldsymbol{C}$ which yields a specific set of eigenvalues and associated eigenvectors is unique.

The general system depicted in Figure 5-1 can easily be modified to include the effect of a reference signal $\boldsymbol{r}(t)$, as shown in Figure 5-3, where a linear state variable feedback controller has again been assumed. The general closed-loop system equation is then obtained by inclusion of a forcing function in Equation

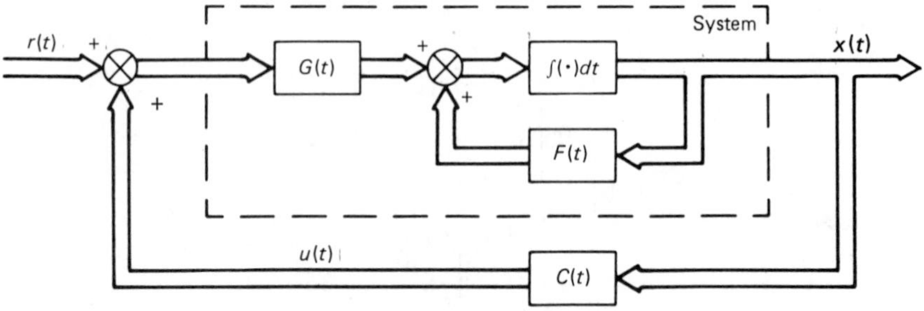

Figure 5-3. Linear state variable feedback control with reference function $\boldsymbol{r}(t)$.

(5.4), yielding

$$\dot{\boldsymbol{x}}(t) = [\boldsymbol{F}(t) + \boldsymbol{G}(t)\boldsymbol{C}(t)]\boldsymbol{x}(t) + \boldsymbol{G}(t)\boldsymbol{r}(t). \tag{5.56}$$

The stability and system response characteristics determined by the homogeneous closed-loop equation remain unchanged.

5.1.5. Linear State Variable Feedback with a Scalar Control Function

In many situations of practical import, the control function is a scalar, in which case Equation (5.1) is

$$\dot{\boldsymbol{x}}(t) = \boldsymbol{F}(t)\boldsymbol{x}(t) + \boldsymbol{g}(t)u(t) \tag{5.57}$$

where vector $\boldsymbol{g}(t)$ now determines the distribution of the scalar input function $u(t)$ to the various state derivatives. The closed-loop system matrix is then

$$\tilde{\boldsymbol{F}}(t) = \boldsymbol{F}(t) + \boldsymbol{g}(t)\boldsymbol{c}^T(t) \tag{5.58}$$

where $\boldsymbol{c}$ is a vector of feedback gains. When $\boldsymbol{F}$, $\boldsymbol{g}$, and $\boldsymbol{c}$ are time independent, the eigenvalues of

$$\tilde{\boldsymbol{F}} = \boldsymbol{F} + \boldsymbol{g}\boldsymbol{c}^T \tag{5.59}$$

determine the closed-loop system characteristics. The derivation of the closed-loop transfer function of the frequency domain representation of this problem is presented in Appendix D.

If the example of Section 5.1.1 is limited to a control function which affects only the state variable $v(t)$, then the closed-loop system equation is

$$\dot{\boldsymbol{x}}(t) = \boldsymbol{F}\boldsymbol{x}(t) + \boldsymbol{g}\boldsymbol{c}^T\boldsymbol{x}(t)$$

where

$$\boldsymbol{F} = \begin{bmatrix} 0 & 1 \\ -5 & 2 \end{bmatrix}; \quad \boldsymbol{g} = \begin{bmatrix} 0 \\ 1 \end{bmatrix}; \quad \boldsymbol{c} = \begin{bmatrix} c_1 \\ c_2 \end{bmatrix}.$$

Then, the closed-loop system matrix is

$$\tilde{\boldsymbol{F}} = \boldsymbol{F} + \boldsymbol{g}\boldsymbol{c}^T = \begin{bmatrix} 0 & 1 \\ -5 & -2 \end{bmatrix} + \begin{bmatrix} 0 & 0 \\ c_1 & c_2 \end{bmatrix} = \begin{bmatrix} 0 & 1 \\ c_1 - 5 & c_2 - 2 \end{bmatrix}.$$

The eigenvalues of this matrix are determined as solutions of the equations

$$\det \begin{bmatrix} -\lambda & 1 \\ c_1 - 5 & c_2 - 2 - \lambda \end{bmatrix} = \lambda^2 + (2 - c_2)\lambda + 5 - c_1 = 0.$$

The gain terms c_1 and c_2 are then selected to produce desired values of λ_1 and λ_2. In general, the solution is

$$\lambda = \frac{c_2}{2} - 1 \pm \sqrt{\frac{c_2^2}{4} - c_2 + c_1 - 4}.$$

For stability,

$$\frac{c_2}{2} - 1 < 0$$

in order for the real part of λ to be negative, or, equivalently,

$$c_2 > 2.$$

Suppose that c_2 is chosen to be 4, and critical damping is desired. Critical damping requires the term under the radical to be zero, from which the value of c_1 can be determined as

$$c_1 = 4.$$

Substitution into the eigenvalue equation readily verifies that

$$\lambda_1 = \lambda_2 = 1$$

so that the system is critically damped with time constant equal to 1. The physical aspects of this example are more reasonable than those of the example in Section 5.1.1, since control of a spring-mass-damper system can only be physically accomplished by an acceleration input.

If the generalized control canonical form is applied to systems of the form of Equation (5.57), the results are considerably simpler than in the general multiple input case. The number of control invariants is equal to the dimension of the input vector, in this case one. The matrices $\boldsymbol{A}$ and $\boldsymbol{B}$ of the canonical form of Equation (5.16) are then

$$\boldsymbol{A} = \begin{bmatrix} 0 & 1 & 0 & \cdots & 0 \\ 0 & 0 & 1 & \cdots & 0 \\ \cdots & \cdots & \cdots & \cdots & \cdots \\ 0 & 0 & 0 & \cdots & 1 \\ 0 & 0 & 0 & \cdots & 0 \end{bmatrix}; \quad \boldsymbol{B} = \begin{bmatrix} 0 \\ 0 \\ \vdots \\ 0 \\ 1 \end{bmatrix}. \tag{5.60}$$

The transformation matrix $\boldsymbol{R}$ of Equation (5.22) is a scalar r, and the $\hat{\boldsymbol{C}}$ matrix of Equation (5.24) is a vector $\hat{\boldsymbol{c}}$. The $\boldsymbol{G}$ matrix of Equation (5.26) has only one column, $\boldsymbol{g}$, and therefore the $\boldsymbol{M}$ matrix of Equation (5.30) must have the form

$$\boldsymbol{M} = [\boldsymbol{g} \mid \boldsymbol{F}\boldsymbol{g} \mid \cdots \mid \boldsymbol{F}^{N-1}\boldsymbol{g}] \tag{5.61}$$

and the expression for $\boldsymbol{M}^{-1}$ given by Equation (5.33) has only a single partition. Thus, i in Equation (5.33) has only the single value 1, and only the last row of $\boldsymbol{M}^{-1}$ is used in Equation (5.34) to form the inverse of the transformation matrix $\boldsymbol{T}$:

$$\boldsymbol{T}^{-1} = \begin{bmatrix} \boldsymbol{m}^T \\ \hline \boldsymbol{m}^T\boldsymbol{F} \\ \hline \vdots \\ \hline \boldsymbol{m}^T\boldsymbol{F}^{N-1} \end{bmatrix} \tag{5.62}$$

where the vector $\boldsymbol{m}^T$ is the last row of $\boldsymbol{M}^{-1}$. Because of the single control invariant, the matrix $\boldsymbol{N}$ of Equation (5.43) is simply

$$\boldsymbol{N} = [-a_0 \quad -a_1 \quad -a_2 \quad \cdots \quad -a_{N-1}] \tag{5.63}$$

where the elements are the coefficients of the desired characteristic equation

$$Q(\lambda) = (\lambda - \lambda_1)(\lambda - \lambda_2) \ldots (\lambda - \lambda_N). \tag{5.64}$$

The desired feedback control matrix follows directly from Equation (5.45).

To illustrate the application of these techniques to a single-input system, consider the system described by the following matrices:

$$\boldsymbol{F} = \begin{bmatrix} -\frac{3}{2} & \frac{7}{2} & \frac{1}{2} \\ 0 & 0 & 1 \\ -\frac{1}{2} & \frac{5}{2} & \frac{1}{2} \end{bmatrix}; \quad \boldsymbol{G} = \begin{bmatrix} 1 \\ 0 \\ 1 \end{bmatrix}.$$

The matrix $\boldsymbol{M}$ of Equation (5.61) is

$$\boldsymbol{M} = \begin{bmatrix} 1 & -1 & 5 \\ 0 & 1 & 0 \\ 1 & 0 & 3 \end{bmatrix}; \quad \boldsymbol{M}^{-1} = \begin{bmatrix} -\frac{3}{2} & -\frac{3}{2} & \frac{5}{2} \\ 0 & 1 & 0 \\ \frac{1}{2} & \frac{1}{2} & -\frac{1}{2} \end{bmatrix}.$$

The inverse of the transformation matrix $\boldsymbol{T}$ follows from Equation (5.62), with $\boldsymbol{m}$ equal to the bottom row of $\boldsymbol{M}^{-1}$ above:

$$\boldsymbol{T}^{-1} = \begin{bmatrix} \frac{1}{2} & \frac{1}{2} & -\frac{1}{2} \\ -\frac{1}{2} & \frac{1}{2} & \frac{1}{2} \\ \frac{1}{2} & -\frac{1}{2} & \frac{1}{2} \end{bmatrix}; \quad \boldsymbol{T} = \begin{bmatrix} 1 & 0 & 1 \\ 1 & 1 & 0 \\ 0 & 1 & 1 \end{bmatrix}.$$

The value of $\boldsymbol{R}$, now the scalar r, is determined from Equation (5.35) to be 1. The matrix $\boldsymbol{C}$ can be determined by equating the matrix of Equation (5.36) to the canonical form of $\boldsymbol{A}$, which yields

$$\begin{bmatrix} 0 & 1 & 0 \\ 0 & 0 & 1 \\ 2 & 3 & -1 \end{bmatrix} - \begin{bmatrix} 0 \\ 0 \\ 1 \end{bmatrix} [\hat{c}_1 \;\; \hat{c}_2 \;\; \hat{c}_3] = \begin{bmatrix} 0 & 1 & 0 \\ 0 & 0 & 1 \\ 0 & 0 & 0 \end{bmatrix}.$$

Solution for $\hat{c}_1, \hat{c}_2$, and $\hat{c}_3$ yields

$$\hat{C} = [2 \;\; 3 \;\; -1].$$

If the desired eigenvalues are

$$\lambda_1 = -2; \quad \lambda_2 = -3; \quad \lambda_3 = -4$$

the characteristic equation is

$$Q(\lambda) = \lambda^3 + 9\lambda^2 + 26\lambda + 24 = 0.$$

Thus, from Equation (5.63), the matrix $\boldsymbol{N}$ is

$$\boldsymbol{N} = [-24 \quad -26 \quad -9]$$

and the feedback matrix $\boldsymbol{C}$, now the row vector $\boldsymbol{c}^T$, is determined directly from Equation (5.45),

$$\boldsymbol{c}^T = [-26 \quad -29 \quad -8] \begin{bmatrix} \frac{1}{2} & \frac{1}{2} & -\frac{1}{2} \\ -\frac{1}{2} & \frac{1}{2} & \frac{1}{2} \\ \frac{1}{2} & -\frac{1}{2} & \frac{1}{2} \end{bmatrix}$$

$$= [-\tfrac{5}{2} \quad -\tfrac{47}{2} \quad -\tfrac{11}{2}].$$

This feedback matrix produces the desired eigenvalues. From this development, it would seem that the vector $\boldsymbol{c}$ is unique, since no arbitrary assignments were made. Such is, in fact, the case for single-input systems.

Further insight can be gained by applying the technique of eigenstructure assignment to the general single-input system. The null space of the matrix of Equation (5.51) must be determined, but now $\boldsymbol{G}$ is a column vector $\boldsymbol{g}$, yielding

$$[\boldsymbol{F} - \boldsymbol{I}\lambda_i \,|\, \boldsymbol{g}]. \tag{5.65}$$

This matrix has dimension $N \times (N + 1)$, and the null space consists of scalar multiples of a single vector, implying that the eigenvector associated with λ_i is unique, and therefore that the mode shape is determined by the eigenvalue assignment. This conclusion is in agreement with previous results.

Another general approach is made possible by the special form of a single-input system. A transformation matrix $\boldsymbol{T}$ is found which transforms the system directly into a form

$$\dot{\boldsymbol{z}}(t) = \boldsymbol{A}\boldsymbol{z}(t) + \boldsymbol{b}u(t) \tag{5.66}$$

where $\boldsymbol{A}$ is of the form of Equation (5.39), in which the a_i are the coefficients in the open-loop characteristic equation, and $\boldsymbol{b}$ has the special form of Equation (5.20); that is,

$$\boldsymbol{T}^{-1}\boldsymbol{F}\boldsymbol{T} = \boldsymbol{A}; \qquad \boldsymbol{b} = \boldsymbol{T}^{-1}\boldsymbol{g} = \begin{bmatrix} 0 \\ 0 \\ \vdots \\ 1 \end{bmatrix}. \tag{5.67}$$

The particular form of Equation (5.66) is known as the control canonical form for single-input systems. The matrix $\boldsymbol{T}$ can be determined from the expression

$$\boldsymbol{T} = \boldsymbol{M}\hat{\boldsymbol{M}}^{-1} \tag{5.68}$$

where the matrix $\boldsymbol{M}$ is given by Equation (5.61) and $\hat{\boldsymbol{M}}$ by the expression

$$\hat{\boldsymbol{M}} = [\boldsymbol{b} \,\vdots\, \boldsymbol{A}\boldsymbol{b} \,\vdots\, \boldsymbol{A}^2\boldsymbol{b} \,\vdots\, \cdots \,\vdots\, \boldsymbol{A}^{N-1}\boldsymbol{b}]. \tag{5.69}$$

The relationship of Equation (5.68) follows from a general result involving equivalent forms of a linear system—that is, forms derivable from one another by similarity transformations. If the quantities from Equation (5.67) are substituted into Equation (5.69), the result is

$$\begin{aligned} \hat{\boldsymbol{M}} &= [\boldsymbol{T}^{-1}\boldsymbol{g} \,\vdots\, \boldsymbol{T}^{-1}\boldsymbol{F}\boldsymbol{g} \,\vdots\, \boldsymbol{T}^{-1}\boldsymbol{F}^2\boldsymbol{g} \,\vdots\, \cdots \,\vdots\, \boldsymbol{T}^{-1}\boldsymbol{F}^{N-1}\boldsymbol{g}] \\ &= \boldsymbol{T}^{-1}[\boldsymbol{g} \,\vdots\, \boldsymbol{F}\boldsymbol{g} \,\vdots\, \boldsymbol{F}^2\boldsymbol{g} \,\vdots\, \cdots \,\vdots\, \boldsymbol{F}^{N-1}\boldsymbol{g}]. \end{aligned} \tag{5.70}$$

Substitution from Equation (5.61) then yields

$$\hat{\boldsymbol{M}} = \boldsymbol{T}^{-1}\boldsymbol{M}. \tag{5.71}$$

Equation (5.68) then follows directly. The reader should note that this result does not depend upon the particular value of $\boldsymbol{g}$ in Equation (5.67) and is therefore valid for any two equivalent systems related by a similarity transformation. The matrices $\boldsymbol{M}$ and $\hat{\boldsymbol{M}}$ in Equation (5.68) are the controllability matrices of the two equivalent systems, implying a requirement for controllability. An equivalent relationship involving the observability matrix can be derived,

$$\boldsymbol{T} = [\hat{\boldsymbol{N}}\boldsymbol{N}^{-1}]^T \tag{5.72}$$

where $\boldsymbol{N}$ and $\hat{\boldsymbol{N}}$ are the observability matrices of the original system and the transformed system, respectively. This relationship is used in the development of canonical forms for observers in Section 5.2.5.

If the system eigenvalues are known—or at least the system characteristic equation is known—the transformed system matrix $\boldsymbol{A}$ in the canonical form can be determined. The vector $\boldsymbol{b}$ is specified in Equation (5.67), and therefore the controllability matrix $\hat{\boldsymbol{M}}$ can be formed. Since the controllability matrix of the original system, $\boldsymbol{M}$, is also known, the necessary transformation matrix $\boldsymbol{T}$ can be determined from Equation (5.68). Then, working with the canonical form of Equation (5.66), state variable feedback of the form

$$u(t) = \hat{\boldsymbol{c}}^T \boldsymbol{z}(t) \tag{5.73}$$

is used to create a closed-loop system with the desired eigenvalues. That is, the closed-loop system matrix in the canonical form is

$$\tilde{\boldsymbol{A}} = [\boldsymbol{A} + \boldsymbol{b}\hat{\boldsymbol{c}}^T] = \begin{bmatrix} 0 & 1 & 0 & \cdots & 0 \\ 0 & 0 & 0 & \cdots & 0 \\ \cdots & \cdots & \cdots & \cdots & \cdots \\ 0 & 0 & 0 & \cdots & 1 \\ (-a_0 + \hat{c}_1) & (-a_1 + \hat{c}_2) & (-a_2 - \hat{c}_3) & \cdots & (-a_{N-1} - \hat{c}_N) \end{bmatrix} \tag{5.74}$$

and this matrix is to be equated to that of the canonical form of the desired closed-loop system, which has as its bottom row the coefficients of the desired closed-loop characteristic equation. Equating the bottom rows of these matrices yields the value of the control vector $\hat{\boldsymbol{c}}$. Transformation back to the original state variables produces the desired feedback vector $\boldsymbol{c}^T$,

$$\boldsymbol{c}^T = \hat{\boldsymbol{c}}^T \boldsymbol{T}^{-1}. \tag{5.75}$$

This technique is less complex than that using the generalized control canonical form, but it is applicable only to single-input systems.

An illustration of this technique is provided by the example previously worked using the generalized approach. The system was specified by the matrices

$$\boldsymbol{F} = \begin{bmatrix} -\frac{3}{2} & \frac{7}{2} & \frac{1}{2} \\ 0 & 0 & 1 \\ -\frac{1}{2} & \frac{5}{2} & \frac{1}{2} \end{bmatrix}; \quad \boldsymbol{g} = \begin{bmatrix} 1 \\ 0 \\ 1 \end{bmatrix}.$$

The desired eigenvalues are

$$\lambda_1 = -2; \qquad \lambda_2 = -3; \qquad \lambda_3 = -4.$$

The open-loop characteristic equation is

$$Q(\lambda) = \lambda^3 + \lambda^2 - 3\lambda - 2 = 0.$$

The control canonical form is thus described by

$$\boldsymbol{A} = \begin{bmatrix} 0 & 1 & 0 \\ 0 & 0 & 1 \\ 2 & 3 & -1 \end{bmatrix}; \qquad \boldsymbol{b} = \begin{bmatrix} 0 \\ 0 \\ 1 \end{bmatrix}.$$

To find the matrix transformation of this system to the control canonical form for single-input systems, Equation (5.68) is used. The matrix $\boldsymbol{M}$ has been previously determined to be

$$\boldsymbol{M} = \begin{bmatrix} 1 & -1 & 5 \\ 0 & 1 & 0 \\ 1 & 0 & 3 \end{bmatrix}.$$

The matrix $\hat{\boldsymbol{M}}$ is given by Equation (5.69), which yields

$$\hat{\boldsymbol{M}} = \begin{bmatrix} 0 & 0 & 1 \\ 0 & 1 & -1 \\ 1 & -1 & 4 \end{bmatrix}; \qquad \hat{\boldsymbol{M}}^{-1} = \begin{bmatrix} -3 & 1 & 1 \\ 1 & 1 & 0 \\ 1 & 0 & 0 \end{bmatrix}.$$

Then, from Equation (5.68)

$$\boldsymbol{T} = \begin{bmatrix} 1 & -1 & 5 \\ 0 & 1 & 0 \\ 1 & 0 & 3 \end{bmatrix} \begin{bmatrix} -3 & 1 & 1 \\ 1 & 1 & 0 \\ 1 & 0 & 0 \end{bmatrix} = \begin{bmatrix} 1 & 0 & 1 \\ 1 & 1 & 0 \\ 0 & 1 & 1 \end{bmatrix}; \qquad \boldsymbol{T}^{-1} = \begin{bmatrix} \frac{1}{2} & \frac{1}{2} & -\frac{1}{2} \\ -\frac{1}{2} & \frac{1}{2} & \frac{1}{2} \\ \frac{1}{2} & -\frac{1}{2} & \frac{1}{2} \end{bmatrix}$$

which agrees with a previous result. The closed-loop canonical form system matrix is of the form of Equation (5.39), in which the a_i are the coefficients of the closed-loop characteristic equation. In this case,

$$\tilde{A} = \begin{bmatrix} 0 & 1 & 0 \\ 0 & 0 & 1 \\ -24 & -26 & -9 \end{bmatrix}.$$

Equation (5.74) can now be used to determine the matrix $\hat{C}$, or, in this case, the row vector $\hat{c}^T$.

$$\begin{bmatrix} 0 & 1 & 0 \\ 0 & 0 & 1 \\ -24 & -26 & -9 \end{bmatrix} = \begin{bmatrix} 0 & 1 & 0 \\ 0 & 0 & 1 \\ 2 & 3 & -1 \end{bmatrix} + \begin{bmatrix} 0 \\ 0 \\ 1 \end{bmatrix} [\hat{c}_1 \quad \hat{c}_2 \quad \hat{c}_3]$$

The solution follows as

$$\hat{c}^T = [-26 \quad -29 \quad -8].$$

Finally, the feedback vector c^T is determined from Equation (5.75):

$$c^T = [-26 \quad -29 \quad -8] \begin{bmatrix} \frac{1}{2} & \frac{1}{2} & -\frac{1}{2} \\ -\frac{1}{2} & \frac{1}{2} & \frac{1}{2} \\ \frac{1}{2} & -\frac{1}{2} & \frac{1}{2} \end{bmatrix}$$

$$= [-\tfrac{5}{2} \quad -\tfrac{47}{2} \quad -\tfrac{11}{2}]$$

which agrees with a previous result.

The approach to the synthesis of single-input systems using the control canonical form is a specialization of that using the generalized control canonical form. The transformation matrix T, for example, is identical in either approach. The reader should note other analogies.

5.1.6. Controllability Considerations

When applying the concepts of state variable feedback, one makes an implicit assumption of controllability of the system being considered. The concept of controllability is discussed in Section 4.3.2 and relates to the capability of transferring a system from one arbitrary state to another. If a system is not controllable, then it seems fruitless to attempt design of a feedback controller to accomplish arbitrary state transitions. It is, in fact, relatively easy to show that the closed-loop system of Figure 5-2 is controllable if and only if the open-loop system is controllable. This fact does not preclude the use of a feedback controller to effect control of the controllable modes of a system.

When a system is uncontrollable, the controllability matrix

$$M = [G \,\vdots\, FG \,\vdots\, F^2G \,\vdots\, \cdots \,\vdots\, F^{N-P}G] \tag{5.76}$$

has rank less than N, in accordance with the developments of Chapter 4. In an uncontrollable system, the rank deficiency in the controllability matrix $\boldsymbol{M}$ is equal to the number of uncontrollable modes—those modes whose corresponding eigenvalues cannot be changed by state variable feedback. The eigenvalues corresponding to uncontrollable modes can be readily identified using the fact that the matrix of Equation (5.55), when formed using one of these eigenvalues, is not of full rank. This property is limited to distinct eigenvalues.

While the eigenvalue corresponding to an uncontrollable mode cannot be changed by state variable feedback, its associated eigenvector can be assigned by the designer, subject to the general limitations described in Section 5.1.4. In general, there will be more flexibility in assigning eigenvectors associated with uncontrollable modes than with assignment of eigenvectors associated with controllable modes. This added flexibility results from the reduced rank of the matrix of Equation (5.51), which essentially expands the null space of that matrix. Eigenvector assignment for uncontrollable modes is accomplished simultaneously with—and in the same manner as—assignment for controllable modes, using the general procedures of Section 5.1.4.

It is an important design consideration that mode shaping of uncontrollable modes by use of eigenvector assignment can be used to determine, to some extent, the influence of these modes on each of the state variables.

5.2. LINEAR OBSERVERS

A physical system typically will not provide all state variables for use in generating the feedback control function $\boldsymbol{u}(t)$. More realistically, an output function $\boldsymbol{y}(t)$ is available, which is some function of the system state. The general equations, as discussed in Chapter 4, are

$$\begin{aligned} \dot{\boldsymbol{x}}(t) &= \boldsymbol{f}(\boldsymbol{x}(t), \boldsymbol{u}(t), t) \\ \boldsymbol{y}(t) &= \boldsymbol{h}(\boldsymbol{x}(t), t) \end{aligned} \tag{5.77}$$

and, for the linear system,

$$\begin{aligned} \dot{\boldsymbol{x}}(t) &= \boldsymbol{F}(t)\boldsymbol{x}(t) + \boldsymbol{G}(t)\boldsymbol{u}(t) \\ \boldsymbol{y}(t) &= \boldsymbol{H}(t)\boldsymbol{x}(t). \end{aligned} \tag{5.78}$$

Note that the linearity of the output process is also implied by the term *linear*

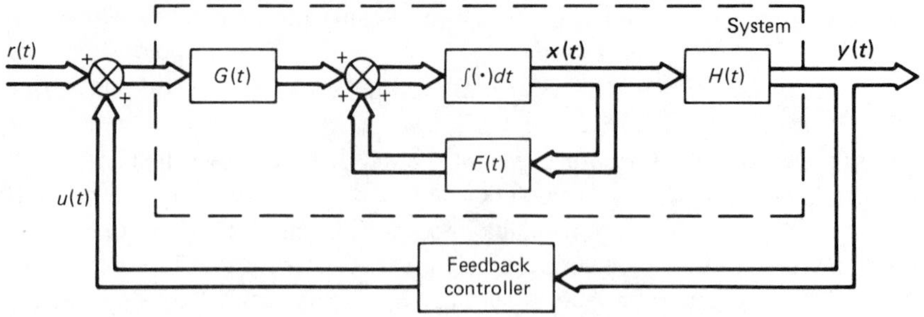

Figure 5-4. Linear output function feedback control.

system. The control problem is then one of finding the feedback controller which accepts as input the observation $\boldsymbol{y}(t)$ and produces the control function $\boldsymbol{u}(t)$. This situation is depicted in Figure 5-4, in which the general case with provision for reference signal $\boldsymbol{r}(t)$ is shown. The feedback controller itself will be considered to be linear, so that the closed-loop equations can be developed as

$$\begin{aligned}\dot{\boldsymbol{x}}(t) &= \boldsymbol{F}(t)\boldsymbol{x}(t) + \boldsymbol{G}(t)\boldsymbol{C}(t)\boldsymbol{y}(t) + \boldsymbol{G}(t)\boldsymbol{r}(t)\\ &= \boldsymbol{F}(t)\boldsymbol{x}(t) + \boldsymbol{G}(t)\boldsymbol{C}(t)\boldsymbol{H}(t)\boldsymbol{x}(t) + \boldsymbol{G}(t)\boldsymbol{r}(t)\\ &= [\boldsymbol{F}(t) + \boldsymbol{G}(t)\boldsymbol{C}(t)\boldsymbol{H}(t)]\boldsymbol{x}(t) + \boldsymbol{G}(t)\boldsymbol{r}(t).\end{aligned} \tag{5.79}$$

Thus, the closed-loop system matrix is

$$\tilde{\boldsymbol{F}}(t) = \boldsymbol{F}(t) + \boldsymbol{G}(t)\boldsymbol{C}(t)\boldsymbol{H}(t) \tag{5.80}$$

and, in the time-invariant case,

$$\tilde{\boldsymbol{F}} = \boldsymbol{F} + \boldsymbol{GCH}. \tag{5.81}$$

Closed-loop system response is then determined by the eigenvalues of this matrix. Selection of the control matrix $\boldsymbol{C}$ in this case is not so clearly defined as it is when all the states are available for observation. For one thing, some loss of flexibility in eigenvalue placement is incurred by the effect of the $\boldsymbol{H}$ matrix. This fact can be illustrated by considering the limiting case of a single input-single output system, in which the $\boldsymbol{G}$ and $\boldsymbol{H}$ matrices are of dimension $N \times 1$ and $1 \times N$, respectively, and $\boldsymbol{C}$ is a scalar. In classical control theory, it is a well known fact that a system of this kind generally does not have the required capability for arbitrary root assignment. Instead, some compensation in the form of feedback dynamics is required. On the other hand, if all state variables can be observed, then Equation (5.59) can be used to effect the maximum possible control over closed-loop response available with a single-input system.

There are also other techniques for determining the feedback matrix $\boldsymbol{C}$ to accomplish certain objectives with regard to closed-loop system response, most of which require feedback of all state variables. For these reasons, the concept of system synthesis by use of techniques designed to compensate for the inclusion of the observation matrix $\boldsymbol{H}$ in the system equations will be developed.

5.2.1. Observers for Time-Invariant Linear Systems

The requirement for feedback of all system state variables–characteristic of many design techniques–presents a problem when a system such as that shown in Figure 5-4 is encountered, because the available observation is not the state vector $\boldsymbol{x}(t)$ but is instead the measurement vector $\boldsymbol{y}(t)$. Rather than lose the potential of these design techniques, one may use the concept of an observer –an auxiliary dynamic system which estimates the state vector from observation of the measurement vector $\boldsymbol{y}(t)$. Note that this is not estimation in the statistical sense, since there is no random effect present here. Instead, it may be considered as a process in which sequential observations of the measurement vector provide information on the values of all the state variables. A very special case would arise when $\boldsymbol{H}$ is a nonsingular matrix, in which case the observer equation would simply be

$$\hat{\boldsymbol{x}}(t) = \boldsymbol{H}^{-1}\boldsymbol{y}(t) \tag{5.82}$$

where $\hat{\boldsymbol{x}}(t)$ is the estimate of $\boldsymbol{x}(t)$ produced by the observer. In most instances of practical import, however, $\boldsymbol{H}$ is singular and the transformation of Equation (5.82) does not exist.

The general time-invariant linear system is represented by Equation (5.78), where $\boldsymbol{H}$ is normally a singular matrix:

$$\begin{aligned}\dot{\boldsymbol{x}}(t) &= \boldsymbol{F}\boldsymbol{x}(t) + \boldsymbol{G}\boldsymbol{u}(t)\\ \boldsymbol{y}(t) &= \boldsymbol{H}\boldsymbol{x}(t).\end{aligned} \tag{5.83}$$

The objective is to determine observer equations so that $\hat{\boldsymbol{x}}(t)$ can be formed from the measurement of $\boldsymbol{y}(t)$. One method would be simply to build a model of the system, so that with knowledge of $\boldsymbol{x}(t_0)$ and $\boldsymbol{u}(t)$, the value of $\hat{\boldsymbol{x}}(t)$ could be determined from the system equations

$$\dot{\hat{\boldsymbol{x}}}(t) = \boldsymbol{F}\hat{\boldsymbol{x}}(t) + \boldsymbol{G}\boldsymbol{u}(t). \tag{5.84}$$

Then, the measurements $\boldsymbol{y}(t)$ are not even needed. The problem with this method, of course, is that the system model and input function are not known precisely. To compensate for this lack of precision, a correction term of the form

$$\boldsymbol{K}[\boldsymbol{y}(t) - \boldsymbol{H}\hat{\boldsymbol{x}}(t)]$$

can be added to the right-hand side of Equation (5.84) to yield

$$\dot{\hat{\boldsymbol{x}}}(t) = \boldsymbol{F}\hat{\boldsymbol{x}}(t) + \boldsymbol{G}\boldsymbol{u}(t) + \boldsymbol{K}[\boldsymbol{y}(t) - \boldsymbol{H}\hat{\boldsymbol{x}}(t)] \tag{5.85}$$

where the matrix $\boldsymbol{K}$ has been incorporated to permit selective weighting of the elements of the correction vector. The observer defined by this equation is linear, which permits use of linear theory in its analysis. The error in the observer is defined as

$$\boldsymbol{\epsilon}(t) = \hat{\boldsymbol{x}}(t) - \boldsymbol{x}(t) \tag{5.86}$$

which, by differentiating, yields

$$\dot{\boldsymbol{\epsilon}}(t) = \dot{\hat{\boldsymbol{x}}}(t) - \dot{\boldsymbol{x}}(t). \tag{5.87}$$

Then substituting from Equations (5.83) and (5.85) yields the differential equation for $\boldsymbol{\epsilon}(t)$,

$$\dot{\boldsymbol{\epsilon}}(t) = \boldsymbol{F}\boldsymbol{\epsilon}(t) + \boldsymbol{K}[\boldsymbol{y}(t) - \boldsymbol{H}\hat{\boldsymbol{x}}(t)]. \tag{5.88}$$

Further substitution for $\boldsymbol{y}(t)$ from Equation (5.83) finally yields

$$\dot{\boldsymbol{\epsilon}}(t) = [\boldsymbol{F} - \boldsymbol{K}\boldsymbol{H}]\,\boldsymbol{\epsilon}(t) \tag{5.89}$$

which is a time-invariant, homogeneous, linear differential equation. If this system of equations is stable, then for sufficiently large t

$$\boldsymbol{\epsilon}(t) \rightarrow 0$$

and the observer will function as desired. The response of the system of Equation (5.89) is determined by the eigenvalues of the matrix

$$[\boldsymbol{F} - \boldsymbol{K}\boldsymbol{H}] \tag{5.90}$$

and therefore can be manipulated by selection of the observer gain matrix $\boldsymbol{K}$. In fact, if the system described by Equation (5.83) is observable, the eigenvalues of the matrix of Equation (5.90) can be chosen to have any desired values, so long as complex values occur in conjugate pairs.

The expression of Equation (5.85) can be written as

$$\dot{\hat{\boldsymbol{x}}}(t) = [\boldsymbol{F} - \boldsymbol{K}\boldsymbol{H}]\hat{\boldsymbol{x}}(t) + \boldsymbol{G}\boldsymbol{u}(t) + \boldsymbol{K}\boldsymbol{y}(t) \tag{5.91}$$

which clearly identifies the observer as a linear system with forcing function

$$\boldsymbol{G}\boldsymbol{u}(t) + \boldsymbol{K}\boldsymbol{y}(t) \tag{5.92}$$

and also relates observer response to the matrix of Equation (5.90). The observer of Equation (5.85) or (5.91) will track the system state $\boldsymbol{x}(t)$, with dynamic characteristics determined by the eigenvalues of the matrix of Equation (5.90).

5.2.2. Use of Observers in System Synthesis

The reasons for considering observers are varied. The general concept of estimating the state of a linear system, when only functions of some of the state variables are available for measurement, may arise in a variety of situations. In most cases, the information obtained from an observer is used in some kind of decision process which, in turn, may affect the system states themselves. Thus, a feedback system analogy is appropriate in many instances.

A situation in which certain of the system state variables are not directly measureable can arise in a variety of ways. In some instances physical characteristics dictate the availability of state variables for direct measurement. In others, economic factors may preclude direct measurement of one or more of the state variables, but it may be economically feasible to estimate a state variable for which direct measurement is prohibitively costly.

Given the general feedback system analogy in the use of observers, it is appropriate to consider them in terms of their potential use in feedback control systems in which the feedback of all state variables is desired, but in which the state variables themselves are not available for measurement. The implied procedure, then, is to design the feedback controller as though the entire state vector were available for use, and then design the observer to produce an estimate of the state vector for use by the controller. If such a procedure is applied, then there must be some concern over the effect that the observer has on closed-loop system response. To illustrate, consider the general time invariant system described by the equation

$$\begin{aligned} \dot{\boldsymbol{x}}(t) &= \boldsymbol{F}\boldsymbol{x}(t) + \boldsymbol{G}\boldsymbol{u}(t) \\ \boldsymbol{y}(t) &= \boldsymbol{x}(t). \end{aligned} \tag{5.93}$$

Here, the observation process actually consists of measurement of the state vector itself. No observer is needed, so that the principles of Section 5.1 apply directly, and the feedback system is as depicted in Figure 5-2.

Suppose that a control matrix $\boldsymbol{C}$ has been selected to produce the desired closed-loop system response. Now consider the same system, but with the observation process

$$\boldsymbol{y}(t) = \boldsymbol{H}\boldsymbol{x}(t) \tag{5.94}$$

where $\boldsymbol{H}$ is a singular matrix. By use of the observer of Equation (5.85), the estimate $\hat{\boldsymbol{x}}(t)$ can be formed from the observation process $\boldsymbol{y}(t)$, and then the control function $\boldsymbol{u}(t)$ determined as

$$\boldsymbol{u}(t) = \boldsymbol{C}\hat{\boldsymbol{x}}(t). \tag{5.95}$$

This procedure is depicted in Figure 5-5, where the general case with reference signal $\boldsymbol{r}(t)$ is shown. The easiest way to analyze this system is to use state variables $\hat{\boldsymbol{x}}(t)$ and $\boldsymbol{\epsilon}(t)$, recognizing the closed-loop system as a system of order $2N$. Recall that the choice of the particular set of state variables used in describing a system is arbitrary so long as they completely describe the state of the system. The vector $\boldsymbol{\epsilon}(t)$, as described by Equation (5.86), is a linear combination of $\boldsymbol{x}(t)$ and $\hat{\boldsymbol{x}}(t)$ and is therefore a permissible choice. The state equation can then be obtained by substituting from Equations (5.86) and (5.94) into Equation (5.85). The result is

$$\dot{\hat{\boldsymbol{x}}}(t) = [\boldsymbol{F} + \boldsymbol{G}\boldsymbol{C}]\hat{\boldsymbol{x}}(t) - [\boldsymbol{K}\boldsymbol{H}]\boldsymbol{\epsilon}(t). \tag{5.96}$$

By combining Equation (5.96) with Equation (5.89), a system of order $2N$ is produced:

$$\begin{bmatrix} \dot{\boldsymbol{\epsilon}}(t) \\ \dot{\hat{\boldsymbol{x}}}(t) \end{bmatrix} = \begin{bmatrix} \boldsymbol{F} - \boldsymbol{K}\boldsymbol{H} & 0 \\ -\boldsymbol{K}\boldsymbol{H} & \boldsymbol{F} + \boldsymbol{G}\boldsymbol{C} \end{bmatrix} \begin{bmatrix} \boldsymbol{\epsilon}(t) \\ \hat{\boldsymbol{x}}(t) \end{bmatrix}. \tag{5.97}$$

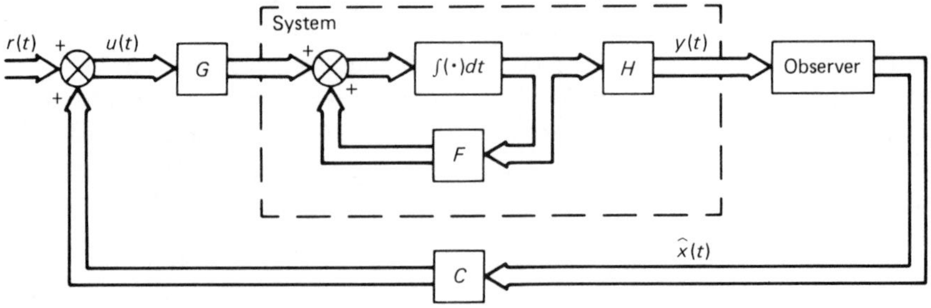

Figure 5-5. Use of an observer in linear feedback control.

This result is a time-invariant, homogeneous, linear differential equation, but it is of a special form which implies that there is no coupling from $\hat{\boldsymbol{x}}(t)$ to $\boldsymbol{\epsilon}(t)$, in turn implying that the eigenvalues of the system consist of the eigenvalues of $[\boldsymbol{F} - \boldsymbol{KH}]$ and the eigenvalues of $[\boldsymbol{F} + \boldsymbol{GC}]$. Since the observer gain matrix $\boldsymbol{K}$ affects only the first set of eigenvalues—while the feedback controller gain matrix $\boldsymbol{C}$ affects only the second—separate and independent design of the observer and feedback controller is possible. Note that the eigenvalues are just those of the observer, as indicated by Equation (5.90), and those of the closed-loop system with state variable feedback, as implied by Equation (5.7).

The major conclusion to be drawn here is that the closed-loop response of a system designed using a state variable feedback controller and the observer of Equation (5.85) consists of the individual responses of the observer and of the closed-loop system with state variable feedback, ensuring the user that coupling of the two systems will not produce modes of response which are unexpected.

To illustrate these principles, consider the fourth-order single-input system described by the equation

$$\begin{bmatrix} \dot{x}_1 \\ \dot{x}_2 \\ \dot{x}_3 \\ \dot{x}_4 \end{bmatrix} = \begin{bmatrix} 0 & 1 & 0 & 0 \\ 1 & 0 & 0 & 0 \\ 0 & 0 & 0 & 1 \\ f & 0 & 0 & 0 \end{bmatrix} \begin{bmatrix} x_1 \\ x_2 \\ x_3 \\ x_4 \end{bmatrix} + \begin{bmatrix} 0 \\ 1 \\ 1 \\ -1 \end{bmatrix} u.$$

The eigenvalues of the system matrix can be determined to be

$$\lambda_1 = 1; \quad \lambda_2 = 0; \quad \lambda_3 = -1; \quad \lambda_4 = 0$$

which implies one unstable mode of system response caused by the positive eigenvalue. Using the linear state variable feedback control discussed in Section 5.1.1 results in the control function

$$u = \boldsymbol{c}^T \boldsymbol{x} = c_1 x_1 + c_2 x_2 + c_3 x_3 + c_4 x_4.$$

The closed-loop system response is obtained from the eigenvalues of the matrix

$$\begin{bmatrix} 0 & 1 & 0 & 0 \\ 1 + c_1 & c_2 & c_3 & c_4 \\ 0 & 0 & 0 & 1 \\ f - c_1 & -c_2 & -c_3 & -c_4 \end{bmatrix}$$

as required by Equation (5.7). The eigenvalue equation is

$$\lambda^4 + (c_4 - c_2)\lambda^3 + (c_3 - c_1 - 1)\lambda^2 - (f + 1)c_4\lambda - (f + 1)c_3 = 0.$$

Now, if it is desired to produce a closed-loop system response corresponding to the eigenvalues

$$\lambda_1 = -1; \quad \lambda_2 = -1; \quad \lambda_3 = -1 + i; \quad \lambda_4 = -1 - i$$

the eigenvalue equation would be

$$(\lambda + 1)(\lambda + 1)(\lambda + 1 - i)(\lambda + 1 + i) = \lambda^4 + 4\lambda^3 + 7\lambda^2 + 6\lambda + 2 = 0.$$

By equating coefficients in these two equations, one can determine the values of c_1, c_2, c_3, and c_4 to be:

$$c_1 = -8 - \frac{2}{(1+f)}; \quad c_2 = \frac{-2}{(1+f)}; \quad c_3 = -4 - \frac{6}{(1+f)}; \quad c_4 = \frac{-6}{(1+f)}.$$

This technique requires feedback of all state variables. The reader should note that f can have any value, so that the closed-loop eigenvalues are independent of f even though the control vector $\boldsymbol{c}$ depends on the specific value of f.

Now suppose that the observation process is such that only the state variable x_3 can be measured. If the feedback of this single state variable is attempted, the resulting closed-loop system response is determined, as indicated by Equation (5.81), by the eigenvalues of the matrix

$$[\boldsymbol{F} + \boldsymbol{GCH}]$$

where now $\boldsymbol{C}$ is the scalar c, and

$$\boldsymbol{H} = [0 \quad 0 \quad 1 \quad 0]$$

yielding

$$[\boldsymbol{F} + \boldsymbol{GCH}] = \begin{bmatrix} 0 & 1 & 0 & 0 \\ 1 & 0 & c & 0 \\ 0 & 0 & 0 & 1 \\ f & 0 & -c & 0 \end{bmatrix}.$$

The eigenvalue equation for this matrix is

$$\lambda^4 + (c - 1)\lambda^2 - c(1 + f) = 0.$$

Fundamental considerations from basic feedback control theory imply that some of the solutions for λ must have positive real parts, because all powers of λ are not present in this equation. Thus, the system cannot be stabilized. This result illustrates the loss of flexibility in manipulating the eigenvalues caused by the fact that all state variables are not available for use in the feedback controller. Classical control theory would introduce compensation in the form of dynamic elements in the feedback path. Here, the use of an observer as described by Equation (5.85) will be considered. The specific observer equation is

$$\begin{bmatrix} \dot{\hat{x}}_1 \\ \dot{\hat{x}}_2 \\ \dot{\hat{x}}_3 \\ \dot{\hat{x}}_4 \end{bmatrix} = \begin{bmatrix} 0 & 1 & 0 & 0 \\ 1 & 0 & 0 & 0 \\ 0 & 0 & 0 & 1 \\ f & 0 & 0 & 0 \end{bmatrix} \begin{bmatrix} \hat{x}_1 \\ \hat{x}_2 \\ \hat{x}_3 \\ \hat{x}_4 \end{bmatrix} + \begin{bmatrix} k_1 \\ k_2 \\ k_3 \\ k_4 \end{bmatrix} [x_3 - \hat{x}_3] + \begin{bmatrix} 0 \\ 1 \\ 0 \\ -1 \end{bmatrix} u.$$

The eigenvalues of the observer are obtained from Equation (5.90), which yields

$$[\boldsymbol{F} - \boldsymbol{KH}] = \begin{bmatrix} 0 & 1 & -k_1 & 0 \\ 1 & 0 & -k_2 & 0 \\ 0 & 0 & -k_3 & 1 \\ 0 & 0 & -k_4 & 0 \end{bmatrix}$$

The eigenvalue equation is

$$\lambda^4 + k_3\lambda^3 + (k_4 - 1)\lambda^2 + (fk_1 - k_3)\lambda + fk_2 - k_4 = 0.$$

The values of the k_i may be selected to yield any desired observer response. For example, the eigenvalues

$$\lambda_1 = -3; \quad \lambda_2 = -3; \quad \lambda_3 = -3 + i3; \quad \lambda_4 = -3 - i3$$

can be obtained by equating coefficients in the desired eigenvalue equation

$$(\lambda + 3)(\lambda + 3)(\lambda + 3 - i3)(\lambda + 3 + i3) = \lambda^4 + 12\lambda^3 + 63\lambda^2 + 162\lambda + 162 = 0$$

with those of the equation in terms of the k_i. This procedure results in the following value for the k_i:

$$k_1 = 174/f; \quad k_2 = 266/f; \quad k_3 = 12; \quad k_4 = 64.$$

If this observer is then used with the feedback controller developed above for state variable feedback control in the manner depicted by Figure 5-5, the resulting system will have a response described by the eight eigenvalues

$$\lambda_1 = -1; \quad \lambda_2 = -2; \quad \lambda_3 = -1 + i; \quad \lambda_4 = -1 - i;$$

$$\lambda_5 = -3; \quad \lambda_6 = -3; \quad \lambda_7 = -3 + i3; \quad \lambda_8 = -3 - i3$$

which are the eigenvalues of the closed-loop response of the system with state variable feedback, and the eigenvalues of the observer.

5.2.3. Observability Considerations

The use of an observer in the manner described above to generate estimates of the state variables of the system implies certain system characteristics which have not been discussed to this point. These characteristics relate to system observability. In Chapter 4, the concept of observability was introduced, and it was shown there that certain combinations of system matrix $\boldsymbol{F}$ and observation matrix $\boldsymbol{H}$ can produce systems in which particular modes of response are not observable. That is, the contribution to the system state variables due to these modes cannot be observed by measurement of the available output vector $\boldsymbol{y}(t)$. One would expect, then, that the ability of an observer of the type described by Equation (5.85) to estimate the state vector would be dependent upon system observability. Such is, in fact, the case, and use of observers in the sense described above is limited to observable systems. In those instances in which the state variables themselves can be divided into an observable and a nonobservable group, an observer can still be used to estimate the observable states. The developments that follow assume system observability.

These concepts can be illustrated by the example of the previous section. The combination of

$$\boldsymbol{F} = \begin{bmatrix} 0 & 1 & 0 & 0 \\ 1 & 0 & 0 & 0 \\ 0 & 0 & 0 & 1 \\ f & 0 & 0 & 0 \end{bmatrix}; \qquad \boldsymbol{H} = [0 \quad 0 \quad 1 \quad 0]$$

produces an observable system. This characteristic can readily be shown by application of Equation (4.61). On the other hand, if the observation process is described by

$$H = [1 \quad 0 \quad 0 \quad 0]$$

the system is not observable, which is also easily shown by application of Equation (4.61). The effect of the lack of observability can be seen when the observer equation is formed for this system. Equation (5.90) implies that the observer eigenvalues are obtained from

$$[F - KH] = \begin{bmatrix} -k_1 & 1 & 0 & 0 \\ 1 - k_2 & 0 & 0 & 0 \\ -k_3 & 0 & 0 & 1 \\ f - k_4 & 0 & 0 & 0 \end{bmatrix}.$$

The eigenvalue equation for this matrix is

$$\lambda^2(\lambda^2 + k_1\lambda + 1 - k_2) = 0$$

from which

$$\lambda_1 = 0; \quad \lambda_2 = 0; \quad \lambda_3 = -\frac{k_1}{2} + \sqrt{\frac{k_1^2}{4} + k_2 - 1}; \quad \lambda_4 = -\frac{k_1}{2} - \sqrt{\frac{k_1^2}{4} + k_2 - 1}.$$

The two zero eigenvalues imply nonconvergence of two modes of response of the estimation error equations as described by Equation (5.88), in turn implying that the observer cannot be used to estimate the state vector of this system.

5.2.4. Observer Design

The systematic approach to state variable feedback provided by the developments of Section 5.1.3 and 5.1.4 can be applied equally well to observer design. This fact stems directly from the developments of Section 5.2.3. Specifically, as implied by (5.97) and the associated discussion, observer design consists of the assignment of eigenvalues and possibly the associated eigenvectors of the matrix of Equation (5.90)

$$[F - KH] \tag{5.98}$$

by suitable choice of the matrix K. This situation is analogous to that encountered in Section 5.1.3, where eigenvalues of the matrix of Equation (5.15),

$$[F + GC] \tag{5.99}$$

were selected by suitable choice of the matrix $\boldsymbol{C}$. The only differences lie in the algebraic sign and in the order of the matrix products $\boldsymbol{KH}$ and $\boldsymbol{GC}$. The two situations can be made equivalent by transposing Equation (5.98) to produce the matrix

$$[\boldsymbol{F} - \boldsymbol{KH}]^T = \boldsymbol{F}^T - \boldsymbol{H}^T\boldsymbol{K}^T \tag{5.100}$$

Since the eigenvalues of a matrix are invariant under the transpose operation, the techniques of Sections 5.1.3 and 5.1.4 are directly applicable. In the specific case of eigenvector assignment, the eigenvectors selected are those of the matrix of Equation (5.100). That is, the diagonalizing matrix of eigenvalues, $\boldsymbol{P}$, is selected so that

$$\boldsymbol{P}\boldsymbol{\Lambda}\boldsymbol{P}^{-1} = [\boldsymbol{F} - \boldsymbol{KH}]^T \tag{5.101}$$

where $\boldsymbol{\Lambda}$ is the diagonal matrix of eigenvalues. Transposing this equation yields

$$(\boldsymbol{P}^{-1})^T\boldsymbol{\Lambda}\,\boldsymbol{P}^T = [\boldsymbol{F} - \boldsymbol{KH}]. \tag{5.102}$$

By comparing this expression to the results of Section 5.1.4, it is seen that the assigned eigenvectors of the matrix $[\boldsymbol{F} - \boldsymbol{KH}]$ are the rows of the matrix $\boldsymbol{P}^{-1}$. That is, the columns of $\boldsymbol{P}$ are assigned using the techniques of Section 5.1.4, but it is the rows of $\boldsymbol{P}^{-1}$ which are the eigenvectors associated with the eigenvalues of the matrix $[\boldsymbol{F} - \boldsymbol{KH}]$.

It should be recognized that the matrices $\boldsymbol{F}$ and $\boldsymbol{H}$ must satisfy a condition analogous to that specified by Equation (5.26), in that the partitioned matrix

$$[\boldsymbol{H}^T \,\vdots\, \boldsymbol{F}^T\boldsymbol{H}^T \,\vdots\, (\boldsymbol{F}^T)^2\boldsymbol{H}^T \,\vdots\, \cdots \,\vdots\, (\boldsymbol{F}^T)^{(N-M}\boldsymbol{H}^T] \tag{5.103}$$

have maximum rank, where M is the dimension of the measurement vector $\boldsymbol{y}$. This requirement is identical to that specified for observability in Equation (4.61), implying that a requirement for eigenvalue assignment in the observer is an observable system, which is an untuitively satisfying conclusion.

5.2.5. Observer Design for Systems with a Scalar Output Function

For the specific application to systems with a scalar output function, the system can be transformed to an observer canonical form

$$\begin{aligned} \dot{\boldsymbol{z}}(t) &= \boldsymbol{A}\boldsymbol{z}(t) + \boldsymbol{T}^{-1}\boldsymbol{G}\boldsymbol{u}(t) \\ y(t) &= \boldsymbol{b}^T\boldsymbol{z}(t) \end{aligned} \tag{5.104}$$

where

$$A = T^{-1}FT = \begin{bmatrix} 0 & 0 & \cdots & 0 & -a_0 \\ 1 & 0 & \cdots & 0 & -a_1 \\ 0 & 1 & \cdots & 0 & -a_2 \\ \cdots & & & & \\ 0 & 0 & \cdots & 0 & -a_{N-1} \end{bmatrix}; \qquad b = T^{-1}h = \begin{bmatrix} 0 \\ 0 \\ 0 \\ \vdots \\ 1 \end{bmatrix} \tag{5.105}$$

and the a_i are the coefficients of the open-loop characteristic equation. This canonical form is known once the eigenvalues—or at least the coefficients of the characteristic equation—of the system are known. The transformation matrix T can be obtained by a development analogous to that leading to Equation (5.68). Since the canonical form and the original system are equivalent, the transformation matrix can be obtained from the two observability matrices

$$\begin{aligned} N &= [h \mid F^T h \mid (F^T)^2 h \mid \cdots \mid (F^T)^{(N-1)} h] \\ \hat{N} &= [b \mid A^T b \mid (A^T)^2 b \mid \cdots \mid (A^T)^{(N-1)} b]. \end{aligned} \tag{5.106}$$

That is, by substituting the expression for b in Equation (5.105) into Equation (5.106), one obtains

$$\hat{N} = T^T N \quad \text{or} \quad T = [\hat{N}N^{-1}]^T \tag{5.107}$$

which is the relationship of Equation (5.72).

Once the canonical form is obtained, the observer gains, in the form of a gain vector $\hat{k}$, can be determined so that the eigenvalues of the matrix of Equation (5.90)—as applied to the system of Equation (5.104)—are the desired values. In this case, Equation (5.90) yields

$$\tilde{A} = [A - \hat{k}b^T] = \begin{bmatrix} 0 & 0 & \cdots & 0 & (-a_0 - k_1) \\ 1 & 0 & \cdots & 0 & (-a_1 + k_2) \\ 0 & 1 & \cdots & 0 & (-a_2 + k_3) \\ \cdots & & & & \\ 0 & 0 & \cdots & 1 & (-a_{N-1} + k_N) \end{bmatrix} \tag{5.108}$$

and equating this matrix to the desired form of $\tilde{A}$, with the coefficients of the desired characteristic equation constituting the right-hand column, yields the observer gain vector $\hat{k}$. Transformation back to the original system,

$$k = T\hat{k} \tag{5.109}$$

yields the desired observer gain vector $\boldsymbol{k}$. As with feedback techniques related to the single-input control canonical form, observer design using the observer canonical form is less complex than the general approach, but it is limited to single-output systems.

To illustrate the use of the observer canonical form, consider the system described by the equations

$$\boldsymbol{F} = \begin{bmatrix} -\frac{3}{2} & 0 & -\frac{1}{2} \\ \frac{7}{2} & 0 & \frac{5}{2} \\ \frac{1}{2} & 1 & \frac{1}{2} \end{bmatrix}; \quad \boldsymbol{G} = \begin{bmatrix} 1 & 0 \\ 0 & 1 \\ 1 & 0 \end{bmatrix}; \quad \boldsymbol{h} = \begin{bmatrix} 1 \\ 0 \\ 1 \end{bmatrix}.$$

The characteristic equation is

$$\lambda^3 + \lambda^2 - 3\lambda - 2 = 0$$

from which the observer canonical form is

$$\boldsymbol{A} = \begin{bmatrix} 0 & 0 & 2 \\ 1 & 0 & 3 \\ 0 & 1 & -1 \end{bmatrix}; \quad \boldsymbol{b} = \begin{bmatrix} 0 \\ 0 \\ 1 \end{bmatrix}.$$

If the desired observer eigenvalues are

$$\lambda_1 = -3; \qquad \lambda_2 = -4; \qquad \lambda_3 = -5$$

the desired characteristic equation is

$$\lambda^3 + 12\lambda^2 + 47\lambda + 60 = 0.$$

Then, Equation (5.108) yields

$$\tilde{\boldsymbol{A}} = \begin{bmatrix} 0 & 0 & -60 \\ 1 & 0 & -47 \\ 0 & 1 & -12 \end{bmatrix} = [\boldsymbol{A} - \hat{\boldsymbol{k}}\boldsymbol{b}^T] = \begin{bmatrix} 0 & 0 & (2 - \hat{k}_1) \\ 1 & 0 & (3 - \hat{k}_2) \\ 0 & 1 & (-1 - \hat{k}_3) \end{bmatrix}.$$

Equating the last rows of these matrices yields

$$\hat{\boldsymbol{k}} = \begin{bmatrix} \hat{k}_1 \\ \hat{k}_2 \\ \hat{k}_3 \end{bmatrix} = \begin{bmatrix} 62 \\ 50 \\ 11 \end{bmatrix}.$$

The transformation matrix T is obtained from Equation (5.107), with

$$N = \begin{bmatrix} 1 & -1 & 5 \\ 0 & 1 & 0 \\ 1 & 0 & 3 \end{bmatrix}; \quad N^{-1} = \begin{bmatrix} -\frac{3}{2} & -\frac{3}{2} & \frac{5}{2} \\ 0 & 1 & 0 \\ \frac{1}{2} & \frac{1}{2} & -\frac{1}{2} \end{bmatrix}; \quad \hat{N} = \begin{bmatrix} 0 & 0 & 1 \\ 0 & 1 & -1 \\ 1 & -1 & 4 \end{bmatrix}$$

as given by Equation (5.106). Thus

$$T = [\hat{N}N^{-1}]^T = \left[\begin{bmatrix} 0 & 0 & 1 \\ 0 & 1 & -1 \\ 1 & -1 & 4 \end{bmatrix}\begin{bmatrix} -\frac{3}{2} & -\frac{3}{2} & \frac{5}{2} \\ 0 & 1 & 0 \\ \frac{1}{2} & \frac{1}{2} & -\frac{1}{2} \end{bmatrix}\right]^T = \begin{bmatrix} \frac{1}{2} & -\frac{1}{2} & \frac{1}{2} \\ \frac{1}{2} & \frac{1}{2} & -\frac{1}{2} \\ -\frac{1}{2} & \frac{1}{2} & \frac{1}{2} \end{bmatrix}$$

that is, the transformation $T^{-1}FT$ yields the desired matrix A, which is easily verified. This transformation matrix is needed to transform back to the original system using Equation (5.109),

$$k = T\hat{k} = \begin{bmatrix} \frac{1}{2} & -\frac{1}{2} & \frac{1}{2} \\ \frac{1}{2} & \frac{1}{2} & -\frac{1}{2} \\ -\frac{1}{2} & \frac{1}{2} & \frac{1}{2} \end{bmatrix}\begin{bmatrix} 62 \\ 50 \\ 11 \end{bmatrix} = \begin{bmatrix} \frac{23}{2} \\ \frac{101}{2} \\ \frac{1}{2} \end{bmatrix}$$

to yield the desired observer gain matrix. This solution can be verified by evaluating Equation (5.90):

$$[F - KH] = [F - kh^T] = \begin{bmatrix} -\frac{3}{2} & 0 & -\frac{1}{2} \\ \frac{7}{2} & 0 & \frac{5}{2} \\ \frac{1}{2} & 1 & \frac{1}{2} \end{bmatrix} - \begin{bmatrix} \frac{23}{2} \\ \frac{101}{2} \\ \frac{1}{2} \end{bmatrix}[1 \quad 0 \quad 1]$$

$$= \begin{bmatrix} -13 & 0 & -12 \\ -47 & 0 & -48 \\ 1 & 1 & 1 \end{bmatrix}.$$

It is easily verified that this matrix has the specified eigenvalues -3, -4, and -5. The observer equation is, from Equation (5.91),

$$\begin{bmatrix} \dot{\hat{x}}_1(t) \\ \dot{\hat{x}}_2(t) \\ \dot{\hat{x}}_3(t) \end{bmatrix} = \begin{bmatrix} -13 & 0 & -12 \\ -47 & 0 & -48 \\ 1 & 1 & 1 \end{bmatrix}\begin{bmatrix} \hat{x}_1(t) \\ \hat{x}_2(t) \\ \hat{x}_3(t) \end{bmatrix} + \begin{bmatrix} 1 & 0 \\ 0 & 1 \\ 1 & 0 \end{bmatrix}\begin{bmatrix} u_1(t) \\ u_2(t) \end{bmatrix} + \begin{bmatrix} \frac{23}{2} \\ \frac{101}{2} \\ \frac{1}{2} \end{bmatrix} y(t)$$

and a block diagram of the system and observer is shown in Figure 5-6.

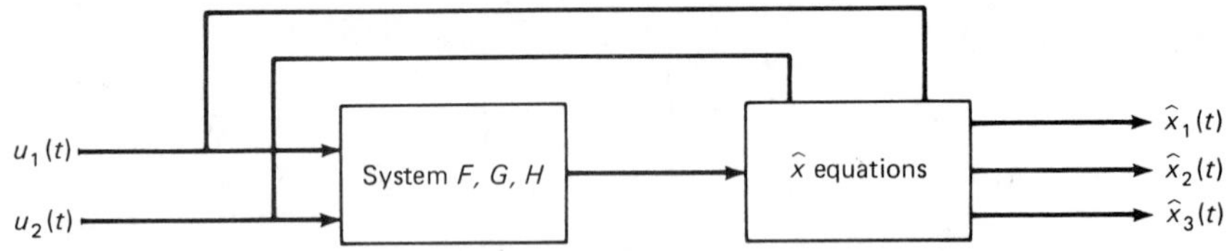

Figure 5-6. Block diagram of observer example.

5.2.6. Reduced-Order Observers

The observer described by Equation (5.85) estimates all components of the state vector, even those which may be measured directly. That is, for some very important types of systems, the observation matrix $\boldsymbol{H}$ is such that the measurement vector $\boldsymbol{y}(t)$ consists of selected components of the state vector $\boldsymbol{x}(t)$. The examples of the previous section are of this type. It seems a reasonable idea to estimate only those states not directly measurable, and a general treatment of this problem can be developed. Observers designed for this purpose are termed *reduced-order observers* and, as implied, are of lower order than those observers which estimate the entire state vector.

The system model for use in this development is

$$\begin{aligned} \dot{\boldsymbol{x}}(t) &= \boldsymbol{F}\boldsymbol{x}(t) + \boldsymbol{G}\boldsymbol{u}(t) \\ \boldsymbol{y}(t) &= \boldsymbol{H}\boldsymbol{x}(t) \end{aligned} \tag{5.110}$$

where $\boldsymbol{y}(t)$ has dimension M, a more general problem than that discussed above, since the particular form of the observation matrix $\boldsymbol{H}$ is not specified. The concept developed here is that, if $\boldsymbol{x}(t)$ has dimension N and $\boldsymbol{y}(t)$ has dimension $M < N$, then the observer should have to be of an order of no more than $N - M$. To proceed with the development, the state vector is changed from $\boldsymbol{x}(t)$ to

$$\begin{bmatrix} \boldsymbol{q}(t) \\ \hline \boldsymbol{y}(t) \end{bmatrix} \tag{5.111}$$

where $\boldsymbol{q}(t)$ is an $N - M$-dimensioned vector chosen in a manner to be described below. The only requirement for such a change in state vectors is that the vector of Equation (5.111) be formed from a nonsingular transformation on $\boldsymbol{x}(t)$, as discussed in Chapter 4. This transformation is represented by

$$\begin{bmatrix} \boldsymbol{q}(t) \\ \hline \boldsymbol{y}(t) \end{bmatrix} = \begin{bmatrix} \boldsymbol{P} \\ \hline \boldsymbol{H} \end{bmatrix} \boldsymbol{x}(t) \tag{5.112}$$

where the matrix $\boldsymbol{P}$ is selected so that the partitioned transformation matrix is nonsingular. Then, the inverse transformation is

$$x(t) = \begin{bmatrix} P \\ \hline H \end{bmatrix}^{-1} \begin{bmatrix} q(t) \\ \hline y(t) \end{bmatrix}. \tag{5.113}$$

By use of basic transformation operations, the system equations in terms of this new state vector can be shown to be

$$\begin{bmatrix} \dot{q}(t) \\ \hline \dot{y}(t) \end{bmatrix} = \begin{bmatrix} P \\ \hline H \end{bmatrix} [F] \begin{bmatrix} P \\ \hline H \end{bmatrix}^{-1} \begin{bmatrix} q(t) \\ \hline y(t) \end{bmatrix} + \begin{bmatrix} P \\ \hline H \end{bmatrix} [G] u(t) \tag{5.114}$$

which is easily obtained from the principles outlined in Appendix A. The two matrices in this equation can be partitioned appropriately to produce the alternate form

$$\begin{bmatrix} \dot{q}(t) \\ \hline \dot{y}(t) \end{bmatrix} = \begin{bmatrix} F_{qq} & F_{qy} \\ \hline F_{yq} & F_{yy} \end{bmatrix} \begin{bmatrix} q(t) \\ \hline y(t) \end{bmatrix} + \begin{bmatrix} G_q \\ \hline G_y \end{bmatrix} u(t) \tag{5.115}$$

which is in the standard form of a time-invariant linear system. Now an open-loop observer can be designed for this system, as given by Equation (5.84):

$$\begin{bmatrix} \dot{\hat{q}}(t) \\ \hline \dot{\hat{y}}(t) \end{bmatrix} = \begin{bmatrix} F_{qq} & F_{qy} \\ \hline F_{yq} & F_{yy} \end{bmatrix} \begin{bmatrix} \hat{q}(t) \\ \hline \hat{y}(t) \end{bmatrix} + \begin{bmatrix} G_q \\ \hline G_y \end{bmatrix} u(t) \tag{5.116}$$

with

$$\begin{bmatrix} \hat{q}(t_0) \\ \hline \hat{y}(t_0) \end{bmatrix} = 0.$$

Acutally, only the $\hat{q}(t)$ equation is needed, so that the observer equation can be reduced to

$$\dot{\hat{q}}(t) = F_{qq}\hat{q}(t) + F_{qy}y(t) + G_q u(t) \tag{5.117}$$

where the measured quantity $y(t)$ has replaced the estimated value $\hat{y}(t)$. From this estimate and the measured value $y(t)$, the state estimate can be obtained as

$$\hat{x}(t) = \begin{bmatrix} P \\ \hline H \end{bmatrix}^{-1} \begin{bmatrix} \hat{q}(t) \\ \hline y(t) \end{bmatrix}. \tag{5.118}$$

The order of the observer is obviously $N - M$, since only the $\hat{\boldsymbol{q}}(t)$ equations are implemented.

Equation (5.117) is a linear system driven by two input functions, $\boldsymbol{y}(t)$ and $\boldsymbol{u}(t)$. The system matrix is $\boldsymbol{F}_{qq}$; therefore the response of the observer is determined by the eigenvalues of $\boldsymbol{F}_{qq}$. This feature can be shown by determining the equation for the error in estimating $\boldsymbol{q}(t)$,

$$\dot{\boldsymbol{\epsilon}}(t) = \boldsymbol{F}_{qq}\,\boldsymbol{\epsilon}(t) \tag{5.119}$$

obtained by subtracting Equation (5.117) from the appropriate part of Equation (5.115). The net result is that, if the matrix $\boldsymbol{P}$ is selected so that the eigenvalues of the submatrix $\boldsymbol{F}_{qq}$ have negative real parts, then $\boldsymbol{\epsilon}(t)$ converges to zero as t increases, and the observer is functional.

An expression relating $\boldsymbol{F}_{qq}$, $\boldsymbol{F}_{qy}$, and $\boldsymbol{P}$ can be obtained from the implied relationship of Equation (5.114) to (5.115):

$$\begin{bmatrix} \boldsymbol{P} \\ \hline \boldsymbol{H} \end{bmatrix} [\boldsymbol{F}] \begin{bmatrix} \boldsymbol{P} \\ \hline \boldsymbol{H} \end{bmatrix}^{-1} = \left[\begin{array}{c|c} \boldsymbol{F}_{qq} & \boldsymbol{F}_{qy} \\ \hline \boldsymbol{F}_{yq} & \boldsymbol{F}_{yy} \end{array}\right]. \tag{5.120}$$

Multiplying on the right by the matrix

$$\begin{bmatrix} \boldsymbol{P} \\ \hline \boldsymbol{H} \end{bmatrix}$$

yields

$$\begin{bmatrix} \boldsymbol{P} \\ \hline \boldsymbol{H} \end{bmatrix} [\boldsymbol{F}] = \left[\begin{array}{c|c} \boldsymbol{F}_{qq} & \boldsymbol{F}_{qy} \\ \hline \boldsymbol{F}_{yq} & \boldsymbol{F}_{yy} \end{array}\right] \begin{bmatrix} \boldsymbol{P} \\ \hline \boldsymbol{H} \end{bmatrix} \tag{5.121}$$

which upon expansion produces

$$\begin{bmatrix} \boldsymbol{PF} \\ \hline \boldsymbol{HF} \end{bmatrix} = \begin{bmatrix} \boldsymbol{F}_{qq}\boldsymbol{P} + \boldsymbol{F}_{qy}\boldsymbol{H} \\ \hline \boldsymbol{F}_{yq}\boldsymbol{P} + \boldsymbol{F}_{yy}\boldsymbol{H} \end{bmatrix}. \tag{5.122}$$

The upper partition of this equation yields

$$\boldsymbol{PF} = \boldsymbol{F}_{qq}\boldsymbol{P} + \boldsymbol{F}_{qy}\boldsymbol{H}. \tag{5.123}$$

This expression relates $\boldsymbol{F}_{qq}$, $\boldsymbol{F}_{qy}$, and $\boldsymbol{P}$, and is required for the transformation of Equation (5.112). Thus, in a practical sense, the $\boldsymbol{F}_{qq}$ and $\boldsymbol{F}_{qy}$ matrices can be selected, and a satisfactory $\boldsymbol{P}$ matrix determined from this relationship.

The particular form of Equation (5.118) indicates that the reduced-order ob-

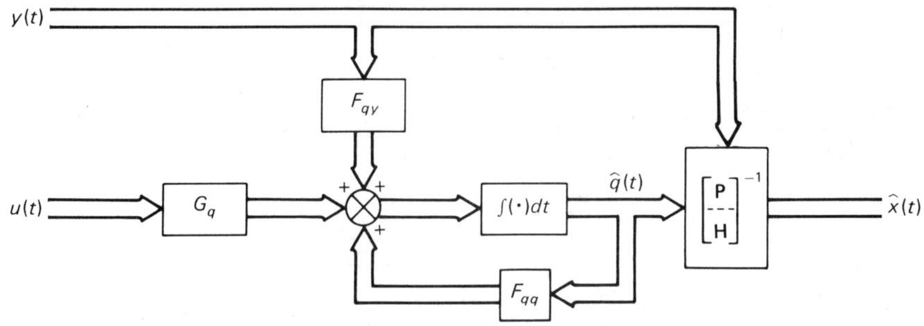

Figure 5-7. The reduced-order observer.

server has a feedforward loop, in that the input function $\boldsymbol{u}(t)$ appears in the observer output directly, with no intervening dynamics. This characteristic is found in any reduced-order observer and has the potential disadvantage that there is no smoothing between input and output, so that any noise on the input signal is transferred directly to the output. The reduced-order observer is illustrated in Figure 5-7, where the feedforward nature of the observer is easily seen.

By use of a procedure analogous to that leading to Equation (5.97), it can be shown that when the reduced-order observer is used in a state variable feedback system with feedback gain matrix $\boldsymbol{C}$, as illustrated in Figure 5-7, the $2N - M$ system composed of $\boldsymbol{\epsilon}(t)$ and $\hat{\boldsymbol{x}}(t)$ is described by the equation

$$\begin{bmatrix} \dot{\boldsymbol{\epsilon}}(t) \\ \hline \dot{\hat{\boldsymbol{x}}}(t) \end{bmatrix} = \left[\begin{array}{c|c} \boldsymbol{F}_{qq} & \boldsymbol{0} \\ \hline \begin{bmatrix} \boldsymbol{P} \\ \hline \boldsymbol{H} \end{bmatrix}^{-1} \begin{bmatrix} \boldsymbol{0} \\ \hline \boldsymbol{F}_{yq} \end{bmatrix} & \boldsymbol{F} + \boldsymbol{GC} \end{array}\right] \begin{bmatrix} \boldsymbol{\epsilon}(t) \\ \hline \boldsymbol{x}(t) \end{bmatrix}. \tag{5.124}$$

The conclusion here is that eigenvalues of $\boldsymbol{F}_{qq}$, as functions of the matrix $\boldsymbol{P}$, can be selected completely independently of the feedback gain matrix $\boldsymbol{C}$, allowing independent design of the state variable feedback controller and the reduced-order observer, as was the case with the full-state observer. Furthermore, as was the case previously, the eigenvalues of the $2N - M$-dimensioned, closed-loop system depicted in Figure 5-8 has N eigenvalues equal to those of the closed-loop state variable feedback system, and has $N - M$ eigenvalues which are equal to those of the reduced-order observer as determined from the matrix $\boldsymbol{F}_{qq}$.

The developments of this section are based upon an implicit assumption that the matrix $\boldsymbol{H}$ has rank M, the dimension of the measurement vector $\boldsymbol{y}(t)$. If such is not the case, then the nonsingular matrix of Equation (5.112) cannot be obtained. This situation presents no problem, however, since a new measurement vector $\boldsymbol{y}(t)$ of dimension equal to the rank of the matrix $\boldsymbol{H}$ can be obtained by a nonsquare linear transformation on $\boldsymbol{y}(t)$. The analysis then proceeds as outlined above.

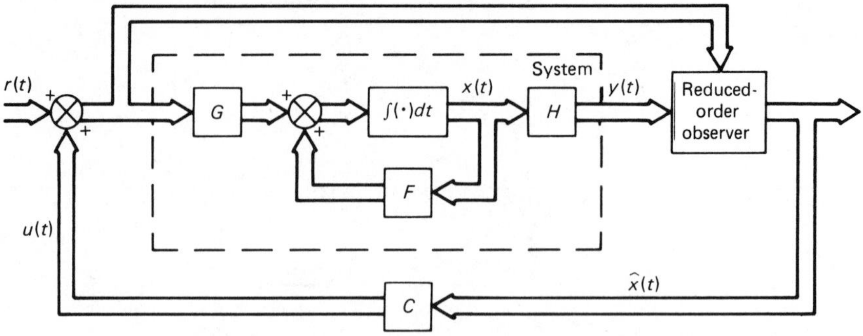

Figure 5-8. The reduced-order observer used with state variable feedback.

5.2.7. Reduced-Order Observer Design for Systems with a Scalar Output Function

The principles outlined in the previous section imply that a reduced-order observer of order $N - 1$ can be obtained for the single-output system described by the equations

$$\dot{\boldsymbol{x}}(t) = \boldsymbol{F}\boldsymbol{x}(t) + \boldsymbol{G}\boldsymbol{u}(t)$$
$$y(t) = \boldsymbol{h}^T\boldsymbol{x}(t). \tag{5.125}$$

The process for determining the reduced-order observer can be systematized by the introduction of a second observer canonical form analogous to that considered in Section 5.25. Any system of the form of Equation (5.125) can be transformed by a similarity transformation to the canonical form

$$\dot{\boldsymbol{z}}(t) = \boldsymbol{A}\boldsymbol{z}(t) + \boldsymbol{T}^{-1}\boldsymbol{G}\boldsymbol{u}(t)$$
$$y(t) = \boldsymbol{b}^T\boldsymbol{x}(t) \tag{5.126}$$

where the row vector $\boldsymbol{b}^T$ has the special form

$$\boldsymbol{b}^T = [1 \quad 0 \quad 0 \quad \cdots \quad 0] \tag{5.127}$$

and the canonical form system matrix $\boldsymbol{A}$ is

$$\boldsymbol{A} = \begin{bmatrix} -a_{N-1} & 1 & 0 & \cdots & 0 \\ -a_{N-2} & 0 & 1 & \cdots & 0 \\ \cdots & \cdots & \cdots & \cdots & \cdots \\ -a_1 & 0 & 0 & \cdots & 1 \\ -a_0 & 0 & 0 & \cdots & 0 \end{bmatrix} \tag{5.128}$$

in which the a_i are the coefficients in the characteristic equation

$$\lambda^N + a_{N-1}\lambda^{N-1} + \cdots + a_1\lambda + a_0 = 0 \tag{5.129}$$

and thus determine the eigenvalues of the system.

The particular transformation matrix $\boldsymbol{T}$ producing this second observer canonical form can be obtained from Equation (5.107), with

$$\begin{aligned} N &= [\boldsymbol{h} \mid \boldsymbol{F}^T\boldsymbol{h} \mid (\boldsymbol{F}^T)^2\boldsymbol{h} \mid \cdots \mid (\boldsymbol{F}^T)^{(N-1)}\boldsymbol{h}] \\ \hat{N} &= [\boldsymbol{b} \mid \boldsymbol{A}^T\boldsymbol{b} \mid (\boldsymbol{A}^T)^2\boldsymbol{b} \mid \cdots \mid (\boldsymbol{A}^T)^{(N-1)}\boldsymbol{b}] \end{aligned} \tag{5.130}$$

where $\boldsymbol{b}$ has the special form implied by Equation (5.127). The similarity transformation producing the canonical form of Equation (5.126) is

$$\begin{aligned} \boldsymbol{A} &= \boldsymbol{T}^{-1}\boldsymbol{F}\boldsymbol{T} \\ \boldsymbol{b} &= \boldsymbol{h}^T\boldsymbol{T}. \end{aligned} \tag{5.131}$$

This canonical form will now be used in reduced-order observer design, using the concepts developed in Section 5.2.6.

The procedures of Section 5.2.6 relating to the determination of reduced-order observers do not, in general, result in a unique observer. This conclusion still holds when the procedures are applied to the system in the canonical form of Equation (5.126). However, with a system in the canonical form, certain arbitrary selections can be made which result in a systematic development of a reduced-order observer. In particular, the matrix $\boldsymbol{F}_{qq}$ of Equation (5.117), whose eigenvalues determine the observer modal responses, is selected to be the second observer canonical form

$$\boldsymbol{F}_{qq} = \begin{bmatrix} -b_{N-2} & 1 & 0 & \cdots & 0 \\ -b_{N-3} & 0 & 1 & \cdots & 0 \\ \cdots & \cdots & \cdots & \cdots & \cdots \\ -b_1 & 0 & 0 & \cdots & 1 \\ -b_0 & 0 & 0 & \cdots & 0 \end{bmatrix} \tag{5.132}$$

The b_i are the coefficients in the characteristic equations for the desired observer eigenvalues,

$$\lambda^{N-1} + b_{N-2}\lambda^{N-2} + \cdots + b_1\lambda + b_0 = 0. \tag{5.133}$$

The relationship of Equation (5.123), applied to this situation, is

$$\boldsymbol{PA} = \boldsymbol{F}_{qq}\boldsymbol{P} + \boldsymbol{f}_{qy}\boldsymbol{b}^T \tag{5.134}$$

where the vector $\boldsymbol{f}_{qy}$ is the $(N-1) \times 1$ matrix $\boldsymbol{F}_{qy}$. Now, if the transformation matrix $\boldsymbol{P}$ is taken to be the $(N-1) \times N$ matrix

$$\boldsymbol{P} = [\boldsymbol{F}_{qq} \mid \boldsymbol{e}] \tag{5.135}$$

where $\boldsymbol{e}$ is the elementary vector

$$\boldsymbol{e} = \begin{bmatrix} 0 \\ 0 \\ \vdots \\ 1 \end{bmatrix} \tag{5.136}$$

then the relationship of Equation (5.134) is

$$[\boldsymbol{F}_{qq} \mid \boldsymbol{e}]\boldsymbol{A} - \boldsymbol{F}_{qq}[\boldsymbol{F}_{qq} \mid \boldsymbol{e}] = \boldsymbol{f}_{qy}\boldsymbol{b}^T. \tag{5.137}$$

Performing the indicated multiplication and reducing the result produce some fortuitous cancellations, producing finally

$$\begin{bmatrix} b_{N-2}(a_{N-1} - b_{N-2}) - a_{N-2} + b_{N-3} & 0 & \cdots & 0 \\ b_{N-2}(a_{N-1} - b_{N-2}) - a_{N-3} + b_{N-4} & 0 & \cdots & 0 \\ \cdots & \cdots & \cdots & \cdots \\ b_1(a_{N-1} - b_{N-2}) - a_1 + b_0 & 0 & \cdots & 0 \\ b_0(a_{N-1} - b_{N-2}) - a_0 & 0 & \cdots & 0 \end{bmatrix} = \begin{bmatrix} f_1 & 0 & \cdots & 0 \\ f_2 & 0 & \cdots & 0 \\ \cdots & \cdots & \cdots & \cdots \\ f_{N-2} & 0 & \cdots & 0 \\ f_{N-1} & 0 & \cdots & 0 \end{bmatrix} \tag{5.138}$$

where the f_i are the elements of $\boldsymbol{f}_{qy}$. Equating elements yields expressions for the f_i,

$$\begin{aligned} f_1 &= b_{N-2}(a_{N-1} - b_{N-2}) - a_{N-2} + b_{N-3} \\ f_2 &= b_{N-3}(a_{N-1} - b_{N-2}) - a_{N-3} + b_{N-4} \\ &\vdots \\ f_{N-2} &= b_1(a_{N-1} - b_{N-2}) - a_1 + b_0 \\ f_{N-1} &= b_0(a_{N-1} - b_{N-2}) - a_0. \end{aligned} \tag{5.139}$$

Thus, the f_i are determined by the coefficients of the system characteristic equation (a_i) and those of the observer characteristic equation (b_i). This equation provides the $(N-1) \times 1$ matrix $\boldsymbol{F}_{qy}$ of Equation (5.117)—in the form of the vector $\boldsymbol{f}_{qy}$—which, together with the selected value of $\boldsymbol{F}_{qq}$ and $\boldsymbol{P}$ as given by Equations (5.132) and (5.135), satisfy the conditions required by Equation (5.123). Thus, the reduced-order observer of Equation (5.117) can be formed as

$$\dot{\hat{\boldsymbol{q}}}(t) = \boldsymbol{F}_{qq}\hat{\boldsymbol{q}}(t) + \boldsymbol{f}_{qy}y(t) + \boldsymbol{G}_q\boldsymbol{u}(t). \tag{5.140}$$

The input distribution matrix of Equation (5.140) is obtained from Equations (5.114), (5.115), and (5.126):

$$\boldsymbol{G}_q = \boldsymbol{P}\boldsymbol{T}^{-1}\boldsymbol{G} \tag{5.141}$$

where the matrix $\boldsymbol{P}$ is given by Equation (5.135), and $\boldsymbol{T}$ is the transformation matrix producing the second observer canonical form. The solution for $\hat{\boldsymbol{q}}(t)$, together with the measurement $\boldsymbol{y}(t)$, form the estimate of the state vector of Equation (5.111):

$$\begin{bmatrix} \hat{\boldsymbol{q}}(t) \\ \hdashline \boldsymbol{y}(t) \end{bmatrix}$$

which can then be transformed to estimates of the canonical variables of Equation (5.126), and then to estimates of the original variables of Equation (5.125),

$$\hat{\boldsymbol{x}}(t) = \boldsymbol{T}\begin{bmatrix} \boldsymbol{P} \\ \hdashline \boldsymbol{H} \end{bmatrix}^{-1}\begin{bmatrix} \hat{\boldsymbol{q}}(t) \\ \hdashline \boldsymbol{y}(t) \end{bmatrix}. \tag{5.142}$$

The systematic approach presented here is readily adaptable to digital computer implementation.

The reader may have noted that the preceding development could well have been carried out using the first observer canonical form of Equation (5.104). The approach used here illustrates the flexibility of use of the canonical forms.

To illustrate the use of these techniques, consider the system described by the equations

$$\boldsymbol{F} = \begin{bmatrix} -\frac{3}{2} & 0 & -\frac{1}{2} \\ \frac{7}{2} & 0 & \frac{5}{2} \\ \frac{1}{2} & 1 & \frac{1}{2} \end{bmatrix}; \quad \boldsymbol{G} = \begin{bmatrix} 0 \\ 0 \\ 1 \end{bmatrix}; \quad \boldsymbol{H} = [1 \quad 0 \quad 1]$$

with characteristic equation

$$\lambda^3 + \lambda^2 - 3\lambda - 2 = 0.$$

A reduced-order ($N - M = 2$) observer is desired for this system. From the development of this example in the previous section, the observability matrix is

$$\boldsymbol{N} = \begin{bmatrix} 1 & -1 & 5 \\ 0 & 1 & 0 \\ 1 & 0 & 3 \end{bmatrix}; \quad \boldsymbol{N}^{-1} = \begin{bmatrix} -\frac{3}{2} & -\frac{3}{2} & \frac{5}{2} \\ 0 & 1 & 0 \\ \frac{1}{2} & \frac{1}{2} & -\frac{1}{2} \end{bmatrix}.$$

The second observer canonical form of Equations (5.126), (5.127), and (5.128) is defined by

$$\boldsymbol{A} = \begin{bmatrix} 1 & 1 & 0 \\ 3 & 0 & 1 \\ 2 & 0 & 0 \end{bmatrix}; \quad \boldsymbol{b} = \begin{bmatrix} 1 \\ 0 \\ 0 \end{bmatrix}$$

From these expressions, the observability matrix of the canonical form can be determined in accordance with Equation (5.130):

$$\hat{\boldsymbol{N}} = \begin{bmatrix} 1 & -1 & 4 \\ 0 & 1 & -1 \\ 0 & 0 & 1 \end{bmatrix}.$$

Now the transformation matrix $\boldsymbol{T}$, which transforms the original system to the second observer canonical form, can be obtained from Equation (5.107):

$$\boldsymbol{T} = [\hat{\boldsymbol{N}}\boldsymbol{N}^{-1}]^T = (\boldsymbol{N}^{-1})^T\hat{\boldsymbol{N}}^T = \begin{bmatrix} -\frac{3}{2} & 0 & \frac{1}{2} \\ -\frac{3}{2} & 1 & \frac{1}{2} \\ \frac{5}{2} & 0 & -\frac{1}{2} \end{bmatrix}\begin{bmatrix} 1 & 0 & 0 \\ -1 & 1 & 0 \\ 4 & -1 & 1 \end{bmatrix}$$

$$= \begin{bmatrix} \frac{1}{2} & -\frac{1}{2} & \frac{1}{2} \\ -\frac{1}{2} & \frac{1}{2} & \frac{1}{2} \\ \frac{1}{2} & \frac{1}{2} & -\frac{1}{2} \end{bmatrix}$$

$$\boldsymbol{T}^{-1} = \begin{bmatrix} 1 & 0 & 1 \\ 0 & 1 & 1 \\ 1 & 1 & 0 \end{bmatrix}.$$

If the desired eigenvalues are

$$\lambda_1 = -3; \quad \lambda_2 = -4$$

the characteristic equation is

$$\lambda^2 + 7\lambda + 12 = 0.$$

The b_i of Equation (5.132) are then

$$b_0 = 12; \quad b_1 = 7$$

and the a_i of Equation (5.128) are, from the second observer canonical form,

$$a_0 = -2; \quad a_1 = -3; \quad a_2 = 1.$$

The vector $\boldsymbol{f}_{qy}$ can now be determined directly from Equation (5.139):

$$\begin{aligned} f_1 &= b_1(a_2 - b_1) - a_1 + b_0 = -27 \\ f_2 &= b_0(a_2 - b_1) - a_0 \qquad = -70. \end{aligned}$$

The observer equation in the estimate $\hat{\boldsymbol{q}}$ is given by Equation (5.140). The only term not defined so far is the input distribution matrix $\boldsymbol{G}_q$, which is obtained from Equation (5.141):

$$\boldsymbol{G}_q = \boldsymbol{P}\boldsymbol{T}^{-1}\boldsymbol{G} = \begin{bmatrix} -7 & 1 & 0 \\ -12 & 0 & 1 \end{bmatrix} \begin{bmatrix} 1 & 0 & 1 \\ 0 & 1 & 1 \\ 1 & 1 & 0 \end{bmatrix} \begin{bmatrix} 0 \\ 1 \\ 1 \end{bmatrix}$$

$$= \begin{bmatrix} -5 \\ -11 \end{bmatrix}$$

where the matrix $\boldsymbol{P}$ is obtained from Equation (5.135). The equation for the estimate $\hat{\boldsymbol{q}}$ is then

$$\begin{bmatrix} \dot{\hat{q}}_1(t) \\ \dot{\hat{q}}_2(t) \end{bmatrix} = \begin{bmatrix} -7 & 1 \\ -12 & 0 \end{bmatrix} \begin{bmatrix} \hat{q}_1(t) \\ \hat{q}_2(t) \end{bmatrix} + \begin{bmatrix} -27 \\ -70 \end{bmatrix} y(t) + \begin{bmatrix} -5 \\ -11 \end{bmatrix} u(t).$$

From the estimate $\hat{\boldsymbol{q}}(t)$ and the measurement $\boldsymbol{y}(t)$, the estimate of the original state vector is given by Equation (5.142), in which

$$\boldsymbol{P} = [\boldsymbol{F}_{qq} \,\vdots\, \boldsymbol{e}] = \begin{bmatrix} -7 & 1 & 0 \\ -12 & 0 & 1 \end{bmatrix}$$

$$\boldsymbol{H} = \boldsymbol{b} = [1 \quad 0 \quad 0].$$

Then

$$\begin{bmatrix} P \\ \hline H \end{bmatrix} = \begin{bmatrix} -7 & 1 & 0 \\ -12 & 0 & 1 \\ 1 & 0 & 0 \end{bmatrix}; \quad \begin{bmatrix} P \\ \hline H \end{bmatrix}^{-1} = \begin{bmatrix} 0 & 0 & 1 \\ 1 & 0 & 7 \\ 0 & 1 & 12 \end{bmatrix}$$

and

$$T\begin{bmatrix} P \\ \hline H \end{bmatrix}^{-1} = \begin{bmatrix} \frac{1}{2} & -\frac{1}{2} & \frac{1}{2} \\ -\frac{1}{2} & \frac{1}{2} & \frac{1}{2} \\ \frac{1}{2} & \frac{1}{2} & -\frac{1}{2} \end{bmatrix}\begin{bmatrix} 0 & 0 & 1 \\ 1 & 0 & 7 \\ 0 & 1 & 12 \end{bmatrix}$$

$$= \begin{bmatrix} -1 & 1 & -4 \\ 1 & 1 & 8 \\ 1 & -1 & 6 \end{bmatrix}.$$

Thus, the estimate of the original state vector is obtained as

$$\begin{bmatrix} \hat{x}_1(t) \\ \hat{x}_2(t) \\ \hat{x}_3(t) \end{bmatrix} = \begin{bmatrix} -1 & 1 & -4 \\ 1 & 1 & 8 \\ 1 & -1 & 6 \end{bmatrix}\begin{bmatrix} \hat{q}_1(t) \\ \hat{q}_2(t) \\ y(t) \end{bmatrix}.$$

The system and observer are shown in Figure 5-9.

The reader should note that if there were two output functions, a first-order observer would result; if there were three outputs, the observer would be implemented simply by Equation (5.82), if the $\boldsymbol{H}$ matrix were nonsingular.

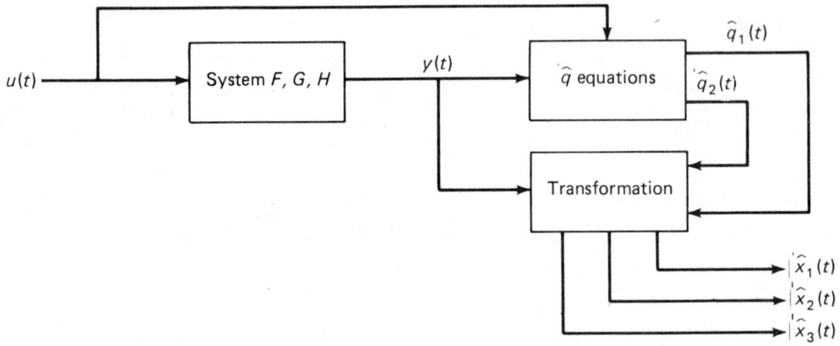

Figure 5-9. Block diagram of reduced-order observer example.

5.2.8. General Reduced-Order Observer Design Concepts

A general approach to reduced-order observer design can be obtained by forming a general observer canonical form somewhat along the lines of the generalized control canonical form of Section 5.1.3. The general time invariant system equations are

$$\dot{\boldsymbol{x}}(t) = \boldsymbol{F}\boldsymbol{x}(t) + \boldsymbol{G}\boldsymbol{u}(t)$$
$$\boldsymbol{y}(t) = \boldsymbol{H}\boldsymbol{x}(t).$$

It is desired to transform this system to an equivalent system of the form

$$\begin{aligned} \dot{\boldsymbol{z}}(t) &= \boldsymbol{A}\boldsymbol{z}(t) + \boldsymbol{T}^{-1}\boldsymbol{G}\boldsymbol{u}(t) \\ \boldsymbol{v}(t) &= \boldsymbol{B}\boldsymbol{z}(t) \end{aligned} \tag{5.143}$$

in which the matrices $\boldsymbol{A}$ and $\boldsymbol{B}$ have the special form

$$\boldsymbol{A} = \left[\begin{array}{c|c|c|c} \boldsymbol{A}_{11} & \boldsymbol{A}_{12} & \cdots & \boldsymbol{A}_{1M} \\ \hline \boldsymbol{A}_{21} & \boldsymbol{A}_{22} & \cdots & \boldsymbol{A}_{2M} \\ \hline \cdots & \cdots & \cdots & \cdots \\ \hline \boldsymbol{A}_{M1} & \boldsymbol{A}_{M2} & \cdots & \boldsymbol{A}_{MM} \end{array}\right]$$

$$\boldsymbol{B} = \left[\begin{array}{cccc|cccc|c|cccc} 1 & 0 & \cdots & 0 & 0 & 0 & \cdots & 0 & \cdots & 0 & 0 & \cdots & 0 \\ 0 & 0 & \cdots & 0 & 1 & 0 & \cdots & 0 & \cdots & 0 & 0 & \cdots & 0 \\ \cdots & & & & & & & & & & & & \cdots \\ 0 & 0 & \cdots & 0 & 0 & 0 & \cdots & 0 & \cdots & 1 & 0 & \cdots & 0 \end{array}\right]. \tag{5.144}$$

The diagonal submatrices of $\boldsymbol{A}$ are in the second observer canonical form

$$\boldsymbol{A}_{ii} = \begin{bmatrix} -a_{N-1} & 1 & \cdots & 0 \\ -a_{N-2} & 0 & \cdots & 0 \\ \cdots & \cdots & \cdots & \cdots \\ -a_1 & 0 & \cdots & 1 \\ -a_0 & 0 & \cdots & 0 \end{bmatrix} \tag{5.145}$$

where the a_i are, in general, different for each submatrix. The off-diagonal matrices have the form

$$A_{ij} = \begin{bmatrix} X & 0 & \cdots & 0 \\ X & 0 & \cdots & 0 \\ \cdots & \cdots & \cdots & \cdots \\ X & 0 & \cdots & 0 \end{bmatrix} \tag{5.146}$$

where the X's indicate elements which can be nonzero. The dimensions of A_{ii} are $N_i \times N_i$, where the N_i can be termed the measurement invariants of the system. In direct analogy to the control invariants of Section 5.1.3, the measurement invariants are M in number—the dimension of the measurement vector—and sum to N, the dimension of the state vector. The dimensions of the submatrices of B are $M \times N_i$.

For example, a sixth-order system with a measurement vector of dimension three might have the canonical form

$$A = \left[\begin{array}{c|cc|ccc} 1 & 3 & 0 & 2 & 0 & 0 \\ \hline 0 & -4 & 1 & 3 & 0 & 0 \\ 1 & -7 & 0 & 2 & 0 & 0 \\ \hline 0 & 2 & 0 & 3 & 1 & 0 \\ 0 & -2 & 0 & 2 & 0 & 1 \\ 0 & 1 & 0 & -1 & 0 & 0 \end{array}\right]; \quad B = \left[\begin{array}{c|cc|ccc} 1 & 0 & 0 & 0 & 0 & 0 \\ 0 & 1 & 0 & 0 & 0 & 0 \\ 0 & 0 & 0 & 1 & 0 & 0 \end{array}\right].$$

Here, the measurement invariants are $N_1 = 1$, $N_2 = 2$, and $N_3 = 3$. The characteristic equation is the product of those equations implied by the diagonal submatrices. In this case,

$$(\lambda - 1)(\lambda^2 + 4\lambda + 7)(\lambda^3 - 3\lambda^2 - 2\lambda + 1) = 0$$

in accordance with the form of Equation (5.145).

The transformation to the generalized observer canonical form is accomplished in two steps, as follows:

I. A similarity transformation using a nonsingular transformation matrix T,

$$x(t) = Tz(t) \tag{5.147}$$

is determined, so that the matrix A of Equation (5.144) is obtained as

$$A = T^{-1}FT. \tag{5.148}$$

II. A transformation of the measurement vector using a nonsingular transformation matrix $\boldsymbol{R}$,

$$\boldsymbol{v}(t) = \boldsymbol{R}\boldsymbol{y}(t) \tag{5.149}$$

is made, so that the resulting system

$$\begin{aligned} \dot{\boldsymbol{z}}(t) &= \boldsymbol{T}^{-1}\boldsymbol{F}\boldsymbol{T}\boldsymbol{z}(t) + \boldsymbol{T}^{-1}\boldsymbol{G}\boldsymbol{u}(t) \\ \boldsymbol{v}(t) &= \boldsymbol{R}\boldsymbol{H}\boldsymbol{T}\boldsymbol{z}(t) \end{aligned} \tag{5.150}$$

is of the desired form of Equation (5.143).

The first step is accomplished by forming the observability matrix of Equation (4.61)

$$[\boldsymbol{H}^T \mid \boldsymbol{F}^T\boldsymbol{H}^T \mid (\boldsymbol{F}^T)^2\boldsymbol{H}^T \mid \cdots \mid (\boldsymbol{F}^T)^{(N-M)}\boldsymbol{H}^T] \tag{5.151}$$

which, by partitioning $\boldsymbol{H}$ into column vectors, can be expressed as

$$\begin{aligned} [\boldsymbol{h}_1 \mid \boldsymbol{h}_2 \mid \cdots \mid \boldsymbol{h}_M \mid \boldsymbol{F}^T\boldsymbol{h}_1 \mid \boldsymbol{F}^T\boldsymbol{h}_2 \mid \cdots \mid \boldsymbol{F}^T\boldsymbol{h}_M \mid \\ \cdots \mid (\boldsymbol{F}^T)^{(N-M)}\boldsymbol{h}_1^T \mid (\boldsymbol{F}^T)^{(N-M)}\boldsymbol{h}_2^T \mid \cdots \mid (\boldsymbol{F}^T)^{(N-M)}\boldsymbol{h}_M^T]. \end{aligned} \tag{5.152}$$

If the system is observable, this matrix has full rank, equal to the dimension of $\boldsymbol{F}$. To determine the measurement invariants, these column vectors are rearranged so that each of the first N vectors is linearly independent of the preceding vectors. The vectors are taken in the sequence

$$\boldsymbol{h}_1, \boldsymbol{F}^T\boldsymbol{h}_1, (\boldsymbol{F}^T)^2\boldsymbol{h}_1, \ldots, (\boldsymbol{F}^T)^{(N_M-1)}\boldsymbol{h}_1 \tag{5.153}$$

until the first dependent vector is reached. The integer N_M, obtained in this manner, is the Mth measurement invariant. Then the vectors

$$\boldsymbol{h}_2, \boldsymbol{F}^T\boldsymbol{h}_2, (\boldsymbol{F}^T)^2\boldsymbol{h}_2, \ldots, (\boldsymbol{F}^T)^{(N_{M-1}-1)}\boldsymbol{h}_2 \tag{5.154}$$

are selected, determining the measurement invariant N_{M-1}, and so on until N linearly independent vectors have been obtained, and the value of the N_1 measurement invariant is determined. The resulting $N \times N$ matrix

$$\begin{aligned} N = [\boldsymbol{h}_1 \mid \boldsymbol{F}^T\boldsymbol{h}_1 \mid \cdots \mid (\boldsymbol{F}^T)^{(N_M-1)}\boldsymbol{h}_1 \mid \boldsymbol{h}_2 \mid \cdots \mid (\boldsymbol{F}^T)^{(N_{M-1}-1)}\boldsymbol{h}_2 \mid \\ \cdots \mid \boldsymbol{h}_M \mid \boldsymbol{F}^T\boldsymbol{h}_M \mid \cdots \mid (\boldsymbol{F}^T)^{(N_1-1)}\boldsymbol{h}_M] \end{aligned} \tag{5.155}$$

implies the values of the measurement invariants of the system, $N_1, N_2, \ldots, N_M$. Once the measurement invariants are determined, they should be rearranged into increasing order. The rows of the measurement matrix H, together with the elements of the measurement vector $\boldsymbol{y}$, are permuted to reflect this reordering of the measurement invariants. This permutation does not change the system equations, but merely changes the location of the scalar measurements within the measurement vector $\boldsymbol{y}$. The resulting matrix $\boldsymbol{N}$ of Equation (5.155) can now be partitioned into $\boldsymbol{M}$ submatrices of dimension $N \times N_i$, in decreasing order of N:

$$\boldsymbol{N} = [\boldsymbol{N}_1 \,\vdots\, \boldsymbol{N}_2 \,\vdots\, \cdots \,\vdots\, \boldsymbol{N}_M]. \tag{5.156}$$

Since $\boldsymbol{N}$ has full rank, it has an inverse:

$$\boldsymbol{N}^{-1} = \begin{bmatrix} \boldsymbol{N}_1^{-1} \\ \hline \boldsymbol{N}_2^{-1} \\ \hline \vdots \\ \hline \boldsymbol{N}_M^{-1} \end{bmatrix}. \tag{5.157}$$

The submatrix $\boldsymbol{N}_i^{-1}$ is meant to imply the ith submatrix of $\boldsymbol{N}^{-1}$. The submatrix $\boldsymbol{N}_i^{-1}$ is not square, and therefore is singular. If the last row of the submatrix $\boldsymbol{N}_i^{-1}$ is the row vector $\boldsymbol{n}_i^T$, then the desired transformation matrix $\boldsymbol{T}$ can be obtained as

$$\boldsymbol{T} = [\boldsymbol{n}_M \,\vdots\, \boldsymbol{F}\boldsymbol{n}_M \,\vdots\, \cdots \,\vdots\, \boldsymbol{F}^{(N_M-1)}\boldsymbol{n}_M \,\vdots\, \boldsymbol{n}_{M-1} \,\vdots\, \cdots \quad \boldsymbol{F}^{(N_{M-1}-1)}\boldsymbol{n}_{M-1} \,\vdots\, \cdots \,\vdots\, \boldsymbol{n}_1 \,\vdots\, \boldsymbol{F}\boldsymbol{n}_1 \,\vdots\, \cdots \quad \boldsymbol{F}^{(N_1-1)}\boldsymbol{n}_1]. \tag{5.159}$$

Once this transformation matrix is determined, then the canonical form system matrix $\boldsymbol{A}$ can be found from Equation (5.148). The required transformation matrix $\boldsymbol{R}$, of Equation (5.149), can then also be determined from the relationship implied by Equations (5.143) and (5.150),

$$\boldsymbol{RHT} = \boldsymbol{B} \tag{5.160}$$

in which $\boldsymbol{B}$ has the special form of Equation (5.144).

The general characteristics of the canonical form equations are clear. The overall system is divided into M subsystems, each associated with a measurement invariant N_i and of dimension equal to N_i, and each with a single output which is the first state variable of the subsystem. Furthermore, each subsystem is coupled

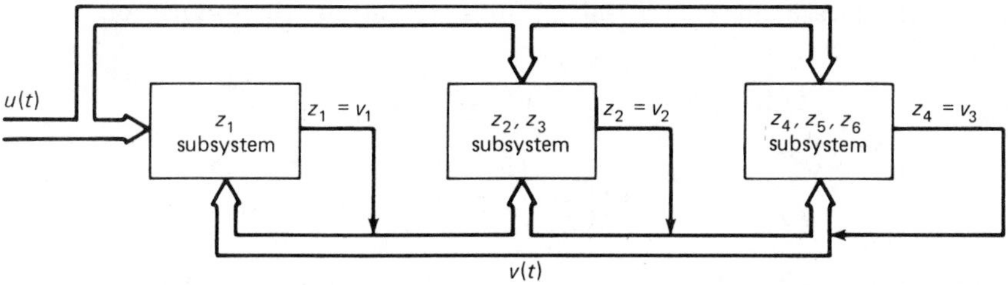

Figure 5-10. Subsystem representation of canonical form.

to another subsystem only through the state variable which is the output of that subsystem. This feature, for the sixth-order example considered above, is represented in Figure 5-10. The coupling between the systems is determined by the non-zero columns of the off-diagonal matrices $\boldsymbol{A}_{ij}$ of Equation (5.146), but is provided entirely through the output of vector $\boldsymbol{v}(t)$.

Since the output of each of the subsystems is directly observable, a reduced-order observer of dimension $N - 1$ can be formulated for each subsystem, in accordance with the procedures developed in Section 5.2.7. The outputs of other subsystems, since they are known, can be treated as inputs to the subsystem. Once the reduced-order observers have been designed, transformation back to the original system can be accomplished using the matrices $\boldsymbol{T}^{-1}$ and $\boldsymbol{R}^{-1}$. The resulting observer will be of order

$$(N_1 - 1) + (N_2 - 1) + \cdots + (N_M - 1). \tag{5.161}$$

But, since there are M measurement invariants N_i, and they must sum to N–the dimension of state vector–this expression reduces to

$$N - M \tag{5.162}$$

which agrees with the general results of Section 5.2.6. For the sixth-order example considered previously and represented by Figure 5-8, an observer of order three should be possible. From Figure 5-10 it is easily seen that the first subsystem requires no observer, the second requires a first-order observer, and the third requires a second-order observer.

To illustrate the application of this general approach to a relatively simple problem, consider the system described by the equations

$$\boldsymbol{F} = \begin{bmatrix} 3 & 0 & 0 \\ 5 & 0 & -2 \\ 1 & 1 & 3 \end{bmatrix}; \quad \boldsymbol{G} = \begin{bmatrix} 1 \\ 0 \\ 1 \end{bmatrix}; \quad \boldsymbol{H} = \begin{bmatrix} 0 & 0 & 1 \\ 1 & 0 & 3 \end{bmatrix}.$$

To find the measurement invariants, the observability matrix of Equation (5.151) is formed:

$$\begin{bmatrix} 0 & 1 & 1 & 6 \\ 0 & 0 & 1 & 3 \\ 1 & 3 & 3 & 9 \end{bmatrix}.$$

Selecting the columns of this matrix in the sequence implied by Equations (5.153) and (5.154) yields, for the matrix $\boldsymbol{N}$ of Equation (5.155),

$$\boldsymbol{N} = \begin{bmatrix} 0 & 1 & 1 \\ 0 & 1 & 0 \\ 1 & 3 & 3 \end{bmatrix}; \qquad \boldsymbol{N}^{-1} = \begin{bmatrix} -3 & 0 & 1 \\ 0 & 1 & 0 \\ \hline 1 & -1 & 0 \end{bmatrix}.$$

Then, from Equation (5.159), the transformation matrix $\boldsymbol{T}$ can be obtained with

$$\boldsymbol{n}_2 = \begin{bmatrix} 1 \\ -1 \\ 0 \end{bmatrix}; \qquad \boldsymbol{n}_1 = \begin{bmatrix} 0 \\ 1 \\ 0 \end{bmatrix}$$

$$\boldsymbol{T} = [\boldsymbol{n}_2 \,\vdots\, \boldsymbol{F}\boldsymbol{n}_1 \,\vdots\, \boldsymbol{n}_1] = \begin{bmatrix} 1 & 0 & 0 \\ -1 & 0 & 1 \\ 0 & 1 & 0 \end{bmatrix}$$

$$\boldsymbol{T}^{-1} = \begin{bmatrix} 1 & 0 & 0 \\ 0 & 0 & 1 \\ 1 & 1 & 0 \end{bmatrix}.$$

The transformation implied by Equation (5.148) then yields

$$\boldsymbol{A} = \begin{bmatrix} 1 & 0 & 0 \\ 0 & 0 & 1 \\ 1 & 1 & 0 \end{bmatrix} \begin{bmatrix} 3 & 0 & 0 \\ 5 & 0 & 2 \\ 1 & 1 & 3 \end{bmatrix} \begin{bmatrix} 1 & 0 & 0 \\ -1 & 0 & 1 \\ 0 & 1 & 0 \end{bmatrix} = \left[\begin{array}{c|cc} 3 & 0 & 0 \\ \hline 0 & 3 & 1 \\ 8 & -2 & 0 \end{array}\right]$$

which is of the general form of Equation (5.144). The two subsystem equations are immediately obvious, and the characteristic equation is seen to be

$$(\lambda - 3)(\lambda^2 - 3\lambda + 2) = 0 = (\lambda - 3)(\lambda - 2)(\lambda - 1)$$

which agrees with that obtained from the system matrix $\boldsymbol{F}$.

The second part of the transformation is to find the matrix $\boldsymbol{R}$ of Equation (5.149), which is determined from the relationship of Equation (5.160):

$$\begin{bmatrix} r_{11} & r_{12} \\ r_{21} & r_{22} \end{bmatrix} \begin{bmatrix} 0 & 0 & 1 \\ 1 & 0 & 3 \end{bmatrix} \begin{bmatrix} 1 & 0 & 0 \\ -1 & 0 & 1 \\ 0 & 1 & 0 \end{bmatrix} = \left[\begin{array}{c|cc} 1 & 0 & 0 \\ 0 & 1 & 0 \end{array}\right].$$

Solution yields

$$\boldsymbol{R} = \begin{bmatrix} -3 & 1 \\ 1 & 0 \end{bmatrix}.$$

The final computation in the determination of the canonical form is the input distribution matrix from Equation (5.150)

$$\boldsymbol{T}^{-1}\boldsymbol{G} = \begin{bmatrix} 1 & 0 & 0 \\ 0 & 0 & 1 \\ 1 & 1 & 0 \end{bmatrix} \begin{bmatrix} 1 \\ 0 \\ 1 \end{bmatrix} = \begin{bmatrix} 1 \\ 1 \\ 1 \end{bmatrix}.$$

The canonical form equations are now known;

$$\begin{bmatrix} \dot{z}_1(t) \\ \dot{z}_2(t) \\ \dot{z}_3(t) \end{bmatrix} = \begin{bmatrix} 3 & 0 & 0 \\ 0 & 3 & 1 \\ 8 & -2 & 0 \end{bmatrix} \begin{bmatrix} z_1(t) \\ z_2(t) \\ z_3(t) \end{bmatrix} + \begin{bmatrix} 1 \\ 1 \\ 1 \end{bmatrix} u(t)$$

$$\begin{bmatrix} v_1(t) \\ v_2(t) \end{bmatrix} = \begin{bmatrix} 1 & 0 & 0 \\ 0 & 1 & 0 \end{bmatrix} \begin{bmatrix} z_1(t) \\ z_2(t) \\ z_3(t) \end{bmatrix}.$$

The first subsystem is

$$\dot{z}_1(t) = 3z_1(t) + u(t)$$

$$v_1(t) = z_1(t)$$

and the second subsystem is

$$\begin{bmatrix} \dot{z}_2(t) \\ \dot{z}_3(t) \end{bmatrix} = \begin{bmatrix} 3 & 1 \\ -2 & 0 \end{bmatrix} \begin{bmatrix} z_2(t) \\ z_3(t) \end{bmatrix} + \begin{bmatrix} 1 \\ 1 \end{bmatrix} u(t) + 8v_1(t)$$

$$v_2(t) = z_2(t).$$

The first subsystem needs no observer, while the second requires only a first-order observer. Since the subsystem equation is in the second observer canonical form, the developments of Section 5.2.7 apply directly without need for determining the transformation matrix. Suppose the desired observer eigenvalue is -4, in which case the matrix $\boldsymbol{F}_{qq}$ of Equation (5.132) is the scalar 4. The system matrix $\boldsymbol{A}$ of Equation (5.128) is

$$\boldsymbol{A} = \begin{bmatrix} 3 & 1 \\ -2 & 0 \end{bmatrix}.$$

Then, in terms of the elements of these matrices,

$$a_1 = -3; \qquad a_0 = 2; \qquad b_0 = -4$$

in which case Equation (5.139) yields

$$f_1 = -4(-3+4) - 2 = -6.$$

This result is the value of the vector–in this case a scalar–$\boldsymbol{f}_{qy}$ of Equatior (5.140). Finally, the matrix $\boldsymbol{G}_q$ of the observer equation is obtained from Equation (5.141)

$$\boldsymbol{G}_q = [-6 \quad 1]\begin{bmatrix} 1 \\ 1 \end{bmatrix} = -5.$$

The transformation matrix $\boldsymbol{T}$ in Equation (5.141) is the identity matrix here, because the second-order subsystem to which the technique is being applied is already in canonical form.

All coefficients in Equation (5.140) have now been determined, so that the observer equation can now be written as

$$\dot{\hat{q}}(t) = -4\hat{q}(t) - 6v_2(t) - 5[u(t) + 8v_1(t)].$$

The estimate of the state of the second subsystem is, from Equation (5.142) with $\boldsymbol{T}$ the identity matrix,

$$\begin{bmatrix} \dot{\hat{z}}_2(t) \\ \dot{\hat{z}}_3(t) \end{bmatrix} = \begin{bmatrix} -6 & 1 \\ 1 & 0 \end{bmatrix}^{-1} \begin{bmatrix} \hat{q}(t) \\ v_2(t) \end{bmatrix} = \begin{bmatrix} 0 & 1 \\ 1 & 6 \end{bmatrix} \begin{bmatrix} \hat{q}(t) \\ v_2(t) \end{bmatrix} = \begin{bmatrix} v_2(t) \\ \hat{q}(t) + 6v_2(t) \end{bmatrix}$$

while the estimate of the state of the first-order subsystem is merely the measurement $v_1(t)$. Thus, for the total system in canonical form, the state estimate is

$$\hat{\boldsymbol{z}}(t) = \begin{bmatrix} v_1(t) \\ v_2(t) \\ \hat{q}(t) + 6v_2(t) \end{bmatrix}.$$

Now the transformation back to the original system must be made. From Equation (5.149)

$$\begin{bmatrix} v_1(t) \\ v_2(t) \end{bmatrix} = \begin{bmatrix} -3 & 1 \\ 1 & 0 \end{bmatrix} \begin{bmatrix} y_1(t) \\ y_2(t) \end{bmatrix} = \begin{bmatrix} -3y_1(t) + y_2(t) \\ y_1(t) \end{bmatrix}$$

and, from Equation (5.147)

$$\hat{\boldsymbol{x}}(t) = \begin{bmatrix} 1 & 0 & 0 \\ -1 & 0 & 1 \\ 0 & 1 & 0 \end{bmatrix} \begin{bmatrix} -3y_1(t) + y_2(t) \\ y_1(t) \\ 6y_1(t) + \hat{q}(t) \end{bmatrix}$$

or, finally,

$$\begin{bmatrix} \hat{x}_1(t) \\ \hat{x}_2(t) \\ \hat{x}_3(t) \end{bmatrix} = \begin{bmatrix} -3y_1(t) + y_2(t) \\ -3y_1(t) - y_2(t) + \hat{q}(t) \\ y_1(t) \end{bmatrix}.$$

The $\hat{q}(t)$ equation can also be simplified to

$$\hat{q}(t) = -4\hat{q}(t) + 112y_1(t) - 40y_2(t) - 5u(t)$$

which now completes the observer design. Note that, in hindsight, the measurement matrix

$$\boldsymbol{H} = \begin{bmatrix} 0 & 0 & 1 \\ 1 & 0 & 3 \end{bmatrix}$$

would immediately yield the observer equations for $\hat{x}_1(t)$ and $\hat{x}_3(t)$. Such is the case any time a $\hat{q}$ term does not explicitly appear in the expression for the estimate.

5.2.9. Observers for Systems with Scalar Control Functions

For the important class of systems with scalar control functions, as described by the state equation

$$\dot{\boldsymbol{x}}(t) = \boldsymbol{F}\boldsymbol{x}(t) + \boldsymbol{g}u(t)$$
$$\boldsymbol{y}(t) = \boldsymbol{H}\boldsymbol{x}(t) \tag{5.163}$$

the state variable feedback control function is given by

$$u(t) = \boldsymbol{c}^T\boldsymbol{x}(t) \tag{5.164}$$

as discussed in Section 5.1.5. In this case, it is possible that a reduced order observer of order less than $N - M$ can be obtained. To consider this case, it is necessary to extend the discussion of observability of Section 4.3.3 to include the concept of the observability index of a system. Briefly stated, the observability index of a time invariant, linear system as described by Equation (5.163) is the smallest positive integer ν for which the partitioned matrix

$$[\boldsymbol{H}^T \,\vdots\, \boldsymbol{F}^T\boldsymbol{H}^T \,\vdots\, (\boldsymbol{F}^T)^2\boldsymbol{H}^T \,\vdots\, \cdots \,\vdots\, (\boldsymbol{F}^T)^{(\nu-1)}\boldsymbol{H}^T] \tag{5.165}$$

has rank N, where N is the dimension of $\boldsymbol{x}(t)$. For an observable system,

$$N/M < \nu < N - M + 1 \tag{5.166}$$

which follows from comparison of Equations (5.165) and (4.61). The possibility of a reduced-order observer can now be expressed in form of the following statement. For a system with scalar input function, the state of the system can be estimated by an observer with $\nu - 1$ eigenvalues, where ν is the observability index of the system. The systematic approach to the derivation of the minimum order observer consists of transforming the system to the canonical form of the previous section, then recognizing that it is the linear function of Equation (5.164) which is to be estimated, rather than the state variables themselves. This concept will not be discussed further here, except to note that if the output function $y(t)$ is also a scalar, then $M = 1$ and Equation (5.166) then requires that $\nu = N$. This situation is treated in the example problem of Section 5.2.5.

5.2.10. Use of Observers to Estimate Bias Errors

The systems discussed in this chapter are considered to be noise free. That is, all quantities relating to system evolution are deterministic. When noisy input or measurement functions are encountered, then the techniques described in

Chapter 7 for estimation and control relating to stochastic systems should be used. However, the observers discussed here can be useful in estimating one particular kind of error term known as the *fixed bias*. It often occurs in measurement systems that an error term appears that is constant with time, or at the very least changes slowly. This occurrence is represented by an observation process described by

$$\boldsymbol{y}(t) = \boldsymbol{H}\boldsymbol{x}(t) + \boldsymbol{b} \tag{5.167}$$

where $\boldsymbol{b}$ is an M-dimensional constant vector whose value is unknown. If the probability distribution of $\boldsymbol{b}$ is known, then the statistical estimation procedures of Chapter 7 should be used. If $\boldsymbol{b}$ is merely a completely unknown parameter, then it can be modeled by the differential equation

$$\dot{\boldsymbol{b}}(t) = 0. \tag{5.168}$$

Then, the equations can be augmented to produce a system of order $N + M$, in which case the system equations are

$$\begin{bmatrix} \dot{\boldsymbol{x}}(t) \\ \hline \dot{\boldsymbol{b}}(t) \end{bmatrix} = \left[\begin{array}{c|c} \boldsymbol{F} & \boldsymbol{0} \\ \hline \boldsymbol{0} & \boldsymbol{0} \end{array}\right] \begin{bmatrix} \boldsymbol{x}(t) \\ \hline \boldsymbol{b}(t) \end{bmatrix} + \begin{bmatrix} \boldsymbol{G} \\ \hline \boldsymbol{0} \end{bmatrix} \boldsymbol{u}(t)$$

$$\boldsymbol{y}(t) = [\boldsymbol{H} \mid \boldsymbol{I}] \begin{bmatrix} \boldsymbol{x}(t) \\ \hline \boldsymbol{b}(t) \end{bmatrix}. \tag{5.169}$$

Since this equation set is of the form of Equation (5.83), an observer can therefore be designed for this system by the techniques outlined previously. The output of the observer will consist of the vector

$$\begin{bmatrix} \hat{\boldsymbol{x}}(t) \\ \hline \hat{\boldsymbol{b}}(t) \end{bmatrix} \tag{5.170}$$

so that the bias error is estimated along with the state vector $\boldsymbol{x}(t)$. Furthermore, since the system of Equation (5.169) has no bias error itself, the state vector estimate is automatically compensated for the presence of the measurement bias. This feature can easily be seen from the form of the observer, which is

$$\begin{bmatrix} \dot{\hat{\boldsymbol{x}}}(t) \\ \hline \dot{\hat{\boldsymbol{b}}}(t) \end{bmatrix} = \left[\begin{array}{c|c} \boldsymbol{F} & \boldsymbol{0} \\ \hline \boldsymbol{0} & \boldsymbol{0} \end{array}\right] \begin{bmatrix} \hat{\boldsymbol{x}}(t) \\ \hline \hat{\boldsymbol{b}}(t) \end{bmatrix} + \begin{bmatrix} \boldsymbol{G} \\ \hline \boldsymbol{0} \end{bmatrix} \boldsymbol{u}(t) + \boldsymbol{K}[\boldsymbol{y}(t) - \boldsymbol{H}\hat{\boldsymbol{x}}(t) - \hat{\boldsymbol{b}}(t)]. \tag{5.171}$$

To illustrate these concepts, a second-order system with scalar input and measurement processes is considered. The system equations are

$$\dot{\boldsymbol{x}}(t) = \boldsymbol{F}\boldsymbol{x}(t) + \boldsymbol{g}u(t)$$

$$y(t) = \boldsymbol{H}\boldsymbol{x}(t)$$

where

$$\boldsymbol{F} = \begin{bmatrix} 0 & 1 \\ f & 0 \end{bmatrix}; \quad \boldsymbol{g} = \begin{bmatrix} 0 \\ 1 \end{bmatrix}; \quad \boldsymbol{H} = [0 \quad 1].$$

The observer of Equation (5.85) for this system can be found immediately by determining the eigenvalues of the matrix

$$[\boldsymbol{F} - \boldsymbol{K}\boldsymbol{H}] = \begin{bmatrix} 0 & 1 - k_1 \\ f & -k_2 \end{bmatrix}$$

as prescribed by Equation (5.90). The eigenvalue equation is

$$\lambda^2 + k_2\lambda + f(1 - k_1) = 0.$$

The observer gain terms k_1 and k_2 can then be selected to produce the desired observer response characteristics. Now suppose that the scalar fixed bias term b is present, such that the measurement process is

$$y(t) = \boldsymbol{H}\boldsymbol{x}(t) + b.$$

The system equations are then augmented so that the revised equations are

$$\begin{bmatrix} \dot{x}_1(t) \\ \dot{x}_2(t) \\ \dot{b}(t) \end{bmatrix} = \begin{bmatrix} 0 & 1 & 0 \\ f & 0 & 0 \\ 0 & 0 & 0 \end{bmatrix} \begin{bmatrix} x_1(t) \\ x_2(t) \\ b(t) \end{bmatrix} + \begin{bmatrix} 0 \\ 1 \\ 0 \end{bmatrix} u(t); \quad y(t) = [0 \quad 1 \quad 1] \begin{bmatrix} x_1(t) \\ x_2(t) \\ b(t) \end{bmatrix}.$$

The observer design is obtained from the eigenvalues of the matrix

$$[\boldsymbol{F} - \boldsymbol{K}\boldsymbol{H}] = \begin{bmatrix} 0 & 1 - k_1 & -k_1 \\ f & -k_2 & -k_2 \\ 0 & -k_3 & -k_3 \end{bmatrix}.$$

The eigenvalue equation is

$$\lambda^3 + (k_2 + k_3)\lambda^2 + f(k_1 + 1)\lambda + k_3 f = 0$$

from which the eigenvalues can be specified by selection of the k_i. The observer equation is

$$\begin{bmatrix} \dot{\hat{x}}_1(t) \\ \dot{\hat{x}}_2(t) \\ \dot{\hat{b}}(t) \end{bmatrix} = \begin{bmatrix} 0 & 1 & 0 \\ f & 0 & 0 \\ 0 & 0 & 0 \end{bmatrix} \begin{bmatrix} \hat{x}_1(t) \\ \hat{x}_2(t) \\ \hat{b}(t) \end{bmatrix} + \begin{bmatrix} 0 \\ 1 \\ 0 \end{bmatrix} u(t) + \begin{bmatrix} k_1 \\ k_2 \\ k_3 \end{bmatrix} [y(t) - \hat{x}_2(t) - \hat{b}(t)].$$

The compensation for the bias term b is clearly evident. Note also that the equation for the estimate $\hat{b}$ is

$$\dot{\hat{b}}(t) = k_3 [y(t) - \hat{x}_2(t) - \hat{b}(t)]$$

indicating that steady state is reached when

$$\hat{x}_2(t) = y(t) - \hat{b}$$

which further illustrates the compensation present.

5.3. DISCRETE TIME SYSTEMS

Virtually all the concepts in the preceding sections can be applied to discrete time systems of the form

$$\boldsymbol{x}(l) = \boldsymbol{\Phi}(l, l-1)\boldsymbol{x}(l-1) + \boldsymbol{\Gamma}(l, l-1)\boldsymbol{u}(l-1) \tag{5.172}$$

which constitutes an important application because of the modern tendency toward digital control logic and sampled data systems. The system of Equation (5.172) is depicted in Figure 5-11. The relationship of this difference equation to the system differential equation of Equation (5.1) is discussed in Section 4.2.2.

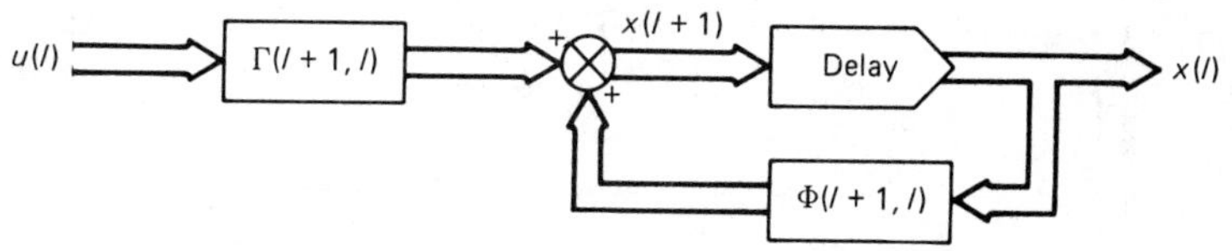

Figure 5-11. The discrete time linear system.

5.3.1. Discrete Time Linear State Variable Feedback

A linear, feedback controller utilizing all the state variables has the form

$$\boldsymbol{u}(l) = \boldsymbol{C}(l)\boldsymbol{x}(l) \tag{5.173}$$

and the resulting closed-loop system is shown in Figure 5-12. The closed-loop difference equation is then obtained by substituting into Equation (5.172), which yields the homogeneous equation

$$\boldsymbol{x}(l) = [\boldsymbol{\Phi}(l, l-1) + \boldsymbol{\Gamma}(l, l-1)\boldsymbol{C}(l)]\boldsymbol{x}(l-1). \tag{5.174}$$

In analogy to the continuous time case, the closed-loop response of the system is determined by the matrix function

$$\tilde{\boldsymbol{\Phi}}(l, l-1) = \boldsymbol{\Phi}(l, l-1) + \boldsymbol{\Gamma}(l, l-1)\boldsymbol{C}(l). \tag{5.175}$$

When the system is time invariant and the time interval $(t_l - t_{l-1})$ is constant, Equation (5.174) reduces to

$$\begin{aligned}\boldsymbol{x}(l) &= [\boldsymbol{\Phi} + \boldsymbol{\Gamma C}]\boldsymbol{x}(l-1)\\ &= \tilde{\boldsymbol{\Phi}}\boldsymbol{x}(l-1).\end{aligned} \tag{5.176}$$

The response of this system of homogeneous difference equations is determined by the eigenvalues of the matrix $\tilde{\boldsymbol{\Phi}}$, as is evident from the treatment in Appendix C. For stability, the eigenvalues must lie within the unit circle. Complex eigenvalues imply an oscillatory solution, while real eigenvalues imply a non-oscillating solution. The relationship of the z-transformation to these conclusions is obvious. The eigenvalues, of course, are obtained from the determinantal equation

$$\det[\tilde{\boldsymbol{\Phi}} - \boldsymbol{I}\lambda] = \det[\boldsymbol{\Phi} + \boldsymbol{\Gamma C} - \boldsymbol{I}\lambda] = 0. \tag{5.177}$$

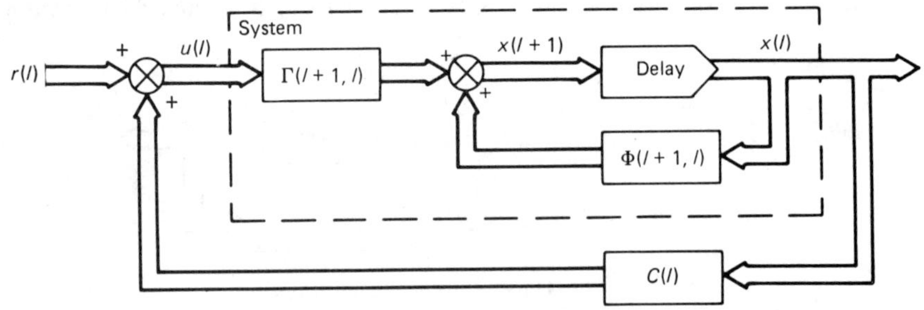

Figure 5-12. Discrete time state variable feedback.

This expression yields an Nth-order polynomial equation in λ, with the elements of the gain matrix $\boldsymbol{C}$ appearing in the coefficients. If the system is controllable, then the elements of $\boldsymbol{C}$ can be selected to produce any desired eigenvalues, so long as complex eigenvalues occur in conjugate pairs. Procedures generally analogous to those of Sections 5.1.3-5.1.5 apply to eigenvalue and eigenstructure assignment. When a reference signal $\boldsymbol{r}(l)$ is considered, as is depicted in Figure 5-12, the general closed-loop system equation is then modified to

$$\boldsymbol{x}(l) = [\boldsymbol{\Phi}(l, l-1) + \boldsymbol{\Gamma}(l, l-1)\boldsymbol{C}(l)]\boldsymbol{x}(l-1) + \boldsymbol{\Gamma}(l, l-1)\boldsymbol{r}(l-1). \quad (5.178)$$

The stability and system response characteristics, determined by the homogeneous equations, remain unchanged. When the control function is a scalar, Equation (5.172) can be written as

$$\boldsymbol{x}(l) = \boldsymbol{\Phi}(l, l-1)\boldsymbol{x}(l-1) + \boldsymbol{\gamma}(l, l-1)u(l-1) \quad (5.179)$$

where $\boldsymbol{\gamma}$ is the vector form of the input matrix $\boldsymbol{\Gamma}$. In a state variable feedback control system, the scalar control function $u(l)$ is obtained as

$$u(l) = \boldsymbol{c}^T(l)\boldsymbol{x}(l) \quad (5.180)$$

where $\boldsymbol{c}$ is a vector of feedback gains. In the time-invariant case, the eigenvalue equation is

$$\det[\boldsymbol{\Phi} + \boldsymbol{c}^T \boldsymbol{\gamma} - \boldsymbol{I}\lambda] = 0. \quad (5.181)$$

This particular case is treated further in Appendix D.

5.3.2. Observers for Discrete Systems

For time-invariant, discrete-time systems described by the equations

$$\begin{aligned} \boldsymbol{x}(l) &= \boldsymbol{\Phi}\boldsymbol{x}(l-1) + \boldsymbol{\Gamma}\boldsymbol{u}(l-1) \\ \boldsymbol{y}(l) &= \boldsymbol{H}\boldsymbol{x}(l) \end{aligned} \quad (5.182)$$

where $\boldsymbol{H}$ is a singular matrix, the concept of an observer analogous to that of Equation (5.85) for continuous time systems can be applied. In this case, the observer difference equation is

$$\hat{\boldsymbol{x}}(l) = \boldsymbol{\Phi}\hat{\boldsymbol{x}}(l-1) + \boldsymbol{\Gamma}\boldsymbol{u}(l-1) + \boldsymbol{K}[\boldsymbol{y}(l) - \boldsymbol{H}\hat{\boldsymbol{x}}(l)] \quad (5.183)$$

where $\boldsymbol{K}$ is a matrix of observer gains and $\hat{\boldsymbol{x}}(l)$ is the estimated state. The estimation error is given by

$$\boldsymbol{\epsilon}(l) = \hat{\boldsymbol{x}}(l) - \boldsymbol{x}(l) \tag{5.184}$$

and substitution from Equations (5.182) and (5.183) yields

$$\boldsymbol{\epsilon}(l) = \boldsymbol{\Phi}\boldsymbol{\epsilon}(l-1) + \boldsymbol{K}[\boldsymbol{y}(l) - \boldsymbol{H}\hat{\boldsymbol{x}}(l)]. \tag{5.185}$$

Further substitution for $\boldsymbol{y}(l)$ from the second expression of Equation (5.182) produces the equation for $\boldsymbol{\epsilon}(l)$

$$\boldsymbol{\epsilon}(l) = \boldsymbol{\Phi}\boldsymbol{\epsilon}(l-1) - \boldsymbol{K}\boldsymbol{H}\boldsymbol{\epsilon}(l). \tag{5.186}$$

Solving this equation for $\boldsymbol{\epsilon}(l)$ yields

$$\boldsymbol{\epsilon}(l) = [\boldsymbol{I} + \boldsymbol{K}\boldsymbol{H}]^{-1}\boldsymbol{\Phi}\boldsymbol{\epsilon}(l-1). \tag{5.187}$$

This is a homogeneous difference equation for the error $\boldsymbol{\epsilon}(l)$. If this equation is stable, then

$$\boldsymbol{\epsilon}(l) \rightarrow 0$$

as l increases, and the observer will function as desired. The response of the system of Equation (5.187) is determined by the eigenvalues of the matrix

$$[\boldsymbol{I} + \boldsymbol{K}\boldsymbol{H}]^{-1}\boldsymbol{\Phi} \tag{5.188}$$

and can therefore be manipulated by selection of the observer gain matrix $\boldsymbol{K}$.

When the observer of Equation (5.183) is used in a discrete time state variable feedback control system, as depicted in Figure 5-13, the resulting system is of order $2N$, with an obvious choice of state variables being represented by the partitioned vector

$$\begin{bmatrix} \boldsymbol{x}(l) \\ \hline \hat{\boldsymbol{x}}(l) \end{bmatrix}. \tag{5.189}$$

Since the error term $\boldsymbol{\epsilon}(l)$ is a linear function of $\boldsymbol{x}(l)$ and $\hat{\boldsymbol{x}}(l)$, as indicated by Equation (5.184), a second choice of state variables is represented by

$$\begin{bmatrix} \boldsymbol{\epsilon}(l) \\ \hline \hat{\boldsymbol{x}}(l) \end{bmatrix} \tag{5.190}$$

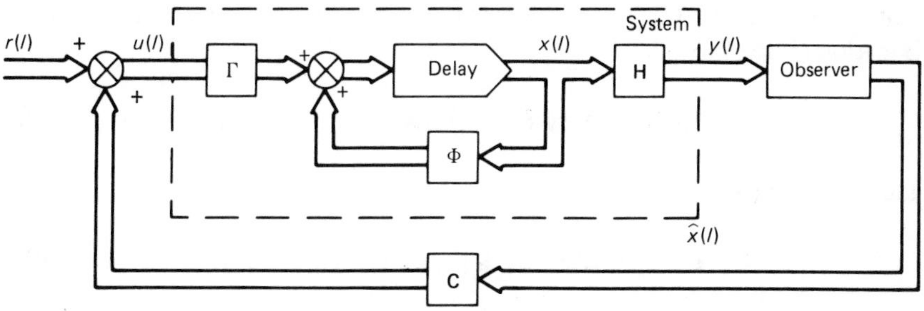

Figure 5-13. Use of an observer in discrete time state variable feedback.

in a manner analogous to the treatment of the continuous time case in Section 5.2.2. Then, the state equation is obtained by substituting from Equations (5.173), (5.182), and (5.184) into Equation (5.183). The result is

$$\hat{\boldsymbol{x}}(l) = [\boldsymbol{\Phi} + \boldsymbol{\Gamma}\boldsymbol{C}]\hat{\boldsymbol{x}}(l-1) - \boldsymbol{KH}\boldsymbol{\epsilon}(l) \tag{5.191}$$

which, upon further substitution from Equation (5.187), yields

$$\hat{\boldsymbol{x}}(l) = [\boldsymbol{\Phi} + \boldsymbol{\Gamma}\boldsymbol{C}]\hat{\boldsymbol{x}}(l-1) - \boldsymbol{KH}[\boldsymbol{I} + \boldsymbol{KH}]^{-1}\boldsymbol{\Phi}\boldsymbol{\epsilon}(l-1). \tag{5.192}$$

Then the complete state equation is

$$\left[\begin{array}{c} \boldsymbol{\epsilon}(l) \\ \hline \hat{\boldsymbol{x}}(l) \end{array}\right] = \left[\begin{array}{c|c} [\boldsymbol{I} + \boldsymbol{KH}]^{-1}\boldsymbol{\Phi} & \boldsymbol{0} \\ \hline -\boldsymbol{KH}[\boldsymbol{I} + \boldsymbol{KH}]^{-1}\boldsymbol{\Phi} & \boldsymbol{\Phi} + \boldsymbol{\Gamma}\boldsymbol{C} \end{array}\right] \left[\begin{array}{c} \boldsymbol{\epsilon}(l-1) \\ \hline \hat{\boldsymbol{x}}(l-1) \end{array}\right]. \tag{5.193}$$

As was the conclusion for continuous time systems in an analogous situation, this equation implies that the observer gain matrix $\boldsymbol{K}$ can be chosen to produce the desired eigenvalues of the matrix given by Equation (5.188) independently of the choice of the eigenvalues produced by the feedback gain matrix $\boldsymbol{C}$. The elements of $\boldsymbol{C}$ are chosen to produce the desired closed-loop, state variable feedback response in accordance with Equation (5.177).

The matrix inversion of Equation (5.193) may present a problem in eigenvalue selection and may be reduced in complexity by use of a matrix identity from Appendix A,

$$[\boldsymbol{I} + \boldsymbol{KH}]^{-1} = \boldsymbol{I} - \boldsymbol{K}[\boldsymbol{I} + \boldsymbol{HK}]^{-1}\boldsymbol{H}. \tag{5.194}$$

For single-output systems, the matrix inversion on the right-hand side reduces to a scalar operation.

The computational problems introduced by the matrix inversion of Equation (5.194) can be circumvented by using an observer which does not use the current measurement $\boldsymbol{y}(l)$, but rather uses the previous value $\boldsymbol{y}(l-1)$, modifying Equation (5.183) to

$$\hat{\boldsymbol{x}}(l) = \boldsymbol{\Phi}\hat{\boldsymbol{x}}(l-1) + \boldsymbol{\Gamma u}(l-1) + \boldsymbol{K}[\boldsymbol{y}(l-1) - \boldsymbol{H}\hat{\boldsymbol{x}}(l-1)] \tag{5.195}$$

and Equation (5.185) to

$$\boldsymbol{\epsilon}(l) = \boldsymbol{\Phi\epsilon}(l-1) + \boldsymbol{K}[\boldsymbol{y}(l-1) - \boldsymbol{H}\hat{\boldsymbol{x}}(l-1)]. \tag{5.196}$$

By continuing in accordance with the previous development, one can determine the homogeneous equation for $\boldsymbol{\epsilon}(l)$ by substituting from Equations (5.182) and (5.184),

$$\boldsymbol{\epsilon}(l) = [\boldsymbol{\Phi} - \boldsymbol{KH}]\,\boldsymbol{\epsilon}(l-1). \tag{5.197}$$

This result is analogous to Equation (5.89) obtained for continuous time systems, and the response of the observer can be seen to depend upon the eigenvalues of the matrix

$$[\boldsymbol{\Phi} - \boldsymbol{KH}]. \tag{5.198}$$

Use of this observer in the feedback system shown in Figure 5-12 yields the state equation, corresponding to Equation (5.193):

$$\begin{bmatrix} \boldsymbol{\epsilon}(l) \\ \hline \hat{\boldsymbol{x}}(l) \end{bmatrix} = \left[\begin{array}{c|c} \boldsymbol{\Phi} - \boldsymbol{KH} & \boldsymbol{0} \\ \hline -\boldsymbol{KH} & \boldsymbol{\Phi} + \boldsymbol{\Gamma C} \end{array}\right] \begin{bmatrix} \boldsymbol{\epsilon}(l-1) \\ \hline \hat{\boldsymbol{x}}(l-1) \end{bmatrix}. \tag{5.199}$$

Here again, the selections of $\boldsymbol{K}$ and $\boldsymbol{C}$ are independent of one another, and in this case the matrix inversion problem is not encountered. The price to be paid is that the estimate $\hat{\boldsymbol{x}}(l)$ does not include the effect of measurement $\boldsymbol{y}(l)$, as implied by Equation (5.195). Analogously to previous results, the determinant of the matrix $[\boldsymbol{\Phi} - \boldsymbol{KH}]$ is used to select eigenvalues of the observer, while the determinant of the matrix $[\boldsymbol{\Phi} + \boldsymbol{\Gamma C}]$ is used to select eigenvalues of the closed-loop state variable feedback response. Again, procedures generally analogous to those developed for continuous time systems in Sections 5.2.4 and 5.2.5 apply to the design of discrete time observers.

5.3.3. Reduced-Order Observers for Discrete Time Systems

The concept of a reduced-order observer considered in Section 5.2.6 for continuous time systems can be applied to discrete time systems as well. The approach is analogous to that of Section 5.2.6. The system of order N, with an

M-dimensional measurement vector $\boldsymbol{y}(l)$, is described in terms of the state vector

$$\begin{bmatrix} \boldsymbol{q}(l) \\ \hdashline \boldsymbol{y}(l) \end{bmatrix} = \begin{bmatrix} \boldsymbol{P} \\ \hdashline \boldsymbol{H} \end{bmatrix} \boldsymbol{x}(l) \tag{5.200}$$

where $\boldsymbol{q}(l)$ is a vector of dimension $N - M$. This approach is permissible because the newly designed state vector is linear function of $\boldsymbol{x}(l)$. The matrix $\boldsymbol{P}$ must be selected so that the transformation is nonsingular, so that

$$\boldsymbol{x}(l) = \begin{bmatrix} \boldsymbol{P} \\ \hdashline \boldsymbol{H} \end{bmatrix}^{-1} \begin{bmatrix} \boldsymbol{q}(l) \\ \hdashline \boldsymbol{y}(l) \end{bmatrix}. \tag{5.201}$$

The new system equations are, from Equation (5.182),

$$\begin{bmatrix} \boldsymbol{q}(l) \\ \hdashline \boldsymbol{y}(l) \end{bmatrix} = \begin{bmatrix} \boldsymbol{P} \\ \hdashline \boldsymbol{H} \end{bmatrix} \boldsymbol{\Phi} \begin{bmatrix} \boldsymbol{P} \\ \hdashline \boldsymbol{H} \end{bmatrix}^{-1} \begin{bmatrix} \boldsymbol{q}(l-1) \\ \hdashline \boldsymbol{y}(l-1) \end{bmatrix} + \begin{bmatrix} \boldsymbol{P} \\ \hdashline \boldsymbol{H} \end{bmatrix} \boldsymbol{\Gamma} \boldsymbol{u}(l-1). \tag{5.202}$$

By partitioning the two matrices in this equation, one obtains

$$\begin{bmatrix} \boldsymbol{q}(l) \\ \hdashline \boldsymbol{y}(l) \end{bmatrix} = \begin{bmatrix} \boldsymbol{\Phi}_{qq} & \boldsymbol{\Phi}_{qy} \\ \hdashline \boldsymbol{\Phi}_{yq} & \boldsymbol{\Phi}_{yy} \end{bmatrix} \begin{bmatrix} \boldsymbol{q}(l-1) \\ \hdashline \boldsymbol{y}(l-1) \end{bmatrix} + \begin{bmatrix} \boldsymbol{\Gamma}_q \\ \hdashline \boldsymbol{\Gamma}_y \end{bmatrix} \boldsymbol{u}(l-1) \tag{5.203}$$

An open-loop observer can be designed for this system with the equation

$$\begin{bmatrix} \hat{\boldsymbol{q}}(l) \\ \hdashline \hat{\boldsymbol{y}}(l) \end{bmatrix} = \begin{bmatrix} \boldsymbol{\Phi}_{qq} & \boldsymbol{\Phi}_{qy} \\ \hdashline \boldsymbol{\Phi}_{yq} & \boldsymbol{\Phi}_{yy} \end{bmatrix} \begin{bmatrix} \hat{\boldsymbol{q}}(l-1) \\ \hdashline \hat{\boldsymbol{y}}(l-1) \end{bmatrix} + \begin{bmatrix} \boldsymbol{\Gamma}_q \\ \hdashline \boldsymbol{\Gamma}_y \end{bmatrix} \boldsymbol{u}(l-1) \tag{5.204}$$

with

$$\begin{bmatrix} \hat{\boldsymbol{q}}(0) \\ \hat{\boldsymbol{y}}(0) \end{bmatrix} = 0.$$

Since the $\boldsymbol{y}(l)$ estimate is not to be used, only the $\hat{\boldsymbol{q}}(l)$ equation is needed:

$$\hat{\boldsymbol{q}}(l) = \boldsymbol{\Phi}_{qq} \hat{\boldsymbol{q}}(l-1) + \boldsymbol{\Phi}_{qy} \boldsymbol{y}(l-1) + \boldsymbol{\Gamma}_q \boldsymbol{u}(l-1) \tag{5.205}$$

where the measured quantity $\boldsymbol{y}(l)$ has replaced $\hat{\boldsymbol{y}}(l)$. With this estimate and the measured $\boldsymbol{y}(l)$ values, the state estimate can be obtained as

$$\hat{\boldsymbol{x}}(l) = \begin{bmatrix} \boldsymbol{P} \\ \hdashline \boldsymbol{H} \end{bmatrix}^{-1} \begin{bmatrix} \hat{\boldsymbol{q}}(l) \\ \hdashline \boldsymbol{y}(l) \end{bmatrix}. \tag{5.206}$$

The order of the observer is obviously $N - M$ since only the $\hat{\boldsymbol{q}}(l)$ equations are implemented.

By subtracting Equation (5.205) from the appropriate part of Equation (5.203), one obtains the equation for the error in estimating $\boldsymbol{q}(l)$:

$$\boldsymbol{\epsilon}(l) = \hat{\boldsymbol{q}}(l) - \boldsymbol{q}(l) = \boldsymbol{\Phi}_{qq}\,\boldsymbol{\epsilon}(l-1). \tag{5.207}$$

Thus, the eigenvalues of the matrix $\boldsymbol{\Phi}_{qq}$ determine the response of the reduced-order observer. They must lie within the unit circle for the observer to be stable. As in the continuous time problem, the eigenvalues of $\boldsymbol{\Phi}_{qq}$, as determined by the selection of the matrix $\boldsymbol{P}$, can be chosen independently of the eigenvalues of the closed-loop state variable feedback system, which are determined by the selection of the $\boldsymbol{C}$ matrix.

5.3.4. Deadbeat Response in Discrete Time Linear Systems

While discrete time and continuous time systems permit analogous treatment in many areas of practical importance, there are certain areas in which their different natures manifest themselves. One of these areas relates to a particular closed-loop eigenvalue assignment in discrete-time systems. The system described by Equation (5.172), when subjected to state variable feedback control, has closed-loop characteristics determined by the eigenvalues and eigenvectors of the closed-loop transition matrix of Equation (5.176). The determinantal equation of Equation (5.177) yields the eigenvalues, and the selection of the gain matrix $\boldsymbol{C}$ is made in a manner that produces the desired response. Suppose that $\boldsymbol{C}$ is selected so that

$$\det[\tilde{\boldsymbol{\Phi}} - \boldsymbol{I}\lambda] = \det[\boldsymbol{\Phi} + \boldsymbol{\Gamma}\boldsymbol{C} - \boldsymbol{I}\lambda] = \lambda^N \tag{5.208}$$

which assigns all eigenvalues the value zero. Since, according to the Cayley–Hamilton theorem of matrix theory, every matrix must satisfy its own characteristic equation, the closed-loop transition matrix must satisfy

$$\tilde{\boldsymbol{\Phi}}^N = \boldsymbol{0}.$$

This feature has direct implication on the closed-loop system characteristic, in that it implies that

$$\boldsymbol{x}(l+N) = \tilde{\boldsymbol{\Phi}}^N \boldsymbol{x}(l) = \boldsymbol{0}. \tag{5.209}$$

This result implies that the feedback matrix $\boldsymbol{C}$ so chosen brings the state of the system to zero in at most N steps, where N is the dimension of the system. A system with this characteristic is said to exhibit a *state deadbeat response*, and the controller is said to be a *state deadbeat controller.*

The design of a deadbeat controller follows directly from the general developments of Sections 5.1.3 and 5.1.4. For example, for a system with a scalar control function, the control canonical form of Equation (5.66)—if deadbeat control is desired—has the canonical system matrix

$$\boldsymbol{A} = \begin{bmatrix} 0 & 1 & 0 & \cdots & 0 \\ 0 & 0 & 1 & \cdots & 0 \\ \cdots & \cdots & \cdots & \cdots & \cdots \\ 0 & 0 & 0 & \cdots & 1 \\ 0 & 0 & 0 & \cdots & 0 \end{bmatrix} \tag{5.210}$$

which implies that the control vector $\hat{\boldsymbol{c}}$ in the canonical form is obtained as

$$\hat{c}_i = a_{i-1}, \qquad i = 1, 2, \ldots, N \tag{5.211}$$

where the a_i are the coefficients in the open-loop characteristic equation. The original system feedback gains are then determined from Equation (5.75).

For the general multivariable case, it is clear that the canonical form of Equation (5.16) describes a system with all zero eigenvalues. In fact, the state variable feedback of Equation (5.24) in Step III of the development of the canonical form corresponds precisely to the selection of the gain matrix $\boldsymbol{C}$ to produce all zero eigenvalues. Thus, the developments of Section 5.1.3 apply directly to the design of deadbeat controllers for discrete time systems.

To illustrate the design of a deadbeat controller, consider the system with a scalar input described by the equations

$$\boldsymbol{\Phi} = \begin{bmatrix} -\frac{3}{2} & \frac{7}{2} & \frac{1}{2} \\ 0 & 0 & 1 \\ -\frac{1}{2} & \frac{5}{2} & \frac{1}{2} \end{bmatrix}; \qquad \boldsymbol{\Gamma} = \boldsymbol{\gamma} = \begin{bmatrix} 1 \\ 0 \\ 1 \end{bmatrix}.$$

The open-loop characteristic equation is

$$\lambda^3 + \lambda^2 - 3\lambda - 2 = 0$$

which immediately identifies the gain vector in the canonical form as

$$\hat{\boldsymbol{c}} = \begin{bmatrix} -2 \\ -3 \\ 1 \end{bmatrix}.$$

All that remains is to determine the transformation matrix which, from Equation (5.68) and previous development of this example, is

$$\boldsymbol{T} = \begin{bmatrix} 1 & 0 & 1 \\ 1 & 1 & 0 \\ 0 & 1 & 1 \end{bmatrix}; \quad \boldsymbol{T}^{-1} = \begin{bmatrix} \frac{1}{2} & \frac{1}{2} & -\frac{1}{2} \\ -\frac{1}{2} & \frac{1}{2} & \frac{1}{2} \\ \frac{1}{2} & -\frac{1}{2} & \frac{1}{2} \end{bmatrix}.$$

Then, from Equation (5.75), the feedback gains in the original system are

$$\boldsymbol{c}^T = \hat{\boldsymbol{c}}^T \boldsymbol{T}^{-1} = [-2 \quad -3 \quad 1] \begin{bmatrix} \frac{1}{2} & \frac{1}{2} & -\frac{1}{2} \\ -\frac{1}{2} & \frac{1}{2} & \frac{1}{2} \\ \frac{1}{2} & -\frac{1}{2} & \frac{1}{2} \end{bmatrix}$$

$$= [1 \quad -3 \quad 0].$$

To verify that this gain vector produces the desired deadbeat response, the closed-loop transition matrix is determined:

$$\tilde{\boldsymbol{\Phi}} = \boldsymbol{\Phi} + \gamma \boldsymbol{c}^T$$

$$= \begin{bmatrix} -\frac{3}{2} & \frac{7}{2} & \frac{1}{2} \\ 0 & 0 & 1 \\ -\frac{1}{2} & \frac{5}{2} & \frac{1}{2} \end{bmatrix} + \begin{bmatrix} 1 \\ 0 \\ 1 \end{bmatrix} [1 \quad -3 \quad 0]$$

$$= \begin{bmatrix} -\frac{1}{2} & \frac{1}{2} & \frac{1}{2} \\ 0 & 0 & 1 \\ \frac{1}{2} & -\frac{1}{2} & \frac{1}{2} \end{bmatrix}.$$

This matrix has characteristic equation

$$\lambda^3 = 0$$

implying all zero eigenvalues, as desired. The reader may verify that $\tilde{\boldsymbol{\Phi}}^3 = \boldsymbol{0}$, as required for deadbeat control.

An observer can also be designed with deadbeat response, implying that the estimation error will die out in a number of steps no greater than the order of the observer. If the observer is used to estimate the system state for feedback control purposes, and a state deadbeat controller is used, then the resulting system will take the system to the zero state in a number of steps equal to the sum of the orders of the system and the estimator.

While deadbeat controllers offer attractive performance features, like other theoretically optimal systems they may require control effort which exceeds that available. Transient behavior of the system may also present a problem with the use of this type of controller.

5.4. INPUT-OUTPUT REPRESENTATION

In many instances, more than just the closed-loop response characteristics of a system are required. The relationship between the input function and the output function may be of specific interest. To permit maximum flexibility in the design of closed-loop input-output systems, an additional input distribution matrix $\boldsymbol{D}$ is defined, which modifies Figure 5-3 in a manner shown in Figure 5-14. Here, the input function is $\boldsymbol{r}(t)$, the available measurement is $\boldsymbol{y}(t)$, and the relationship between these two functions is to be examined. For time-invariant systems, the matrices $\boldsymbol{F}, \boldsymbol{G}, \boldsymbol{H}, \boldsymbol{C}$, and $\boldsymbol{D}$ are constant matrices, and the closed-loop response characteristics of the system depend on the eigenvalues of the matrix

$$[\boldsymbol{F} + \boldsymbol{GC}] \tag{5.212}$$

as discussed in Section 5.1.1. The closed-loop system equations are obtained by a slight modification of Equation (5.79) to account for the distribution matrix $\boldsymbol{D}$:

$$\dot{\boldsymbol{x}}(t) = [\boldsymbol{F}(t) + \boldsymbol{G}(t)\boldsymbol{C}(t)\boldsymbol{H}(t)]\boldsymbol{x}(t) + \boldsymbol{G}(t)\boldsymbol{D}(t)\boldsymbol{r}(t). \tag{5.213}$$

The time-invariant equation is

$$\dot{\boldsymbol{x}}(t) = [\boldsymbol{F} + \boldsymbol{GCH}]\boldsymbol{x}(t) + \boldsymbol{GDr}(t) \tag{5.214}$$

and it is this expression which is of interest here.

The general input-output relationship can be obtained by multiplying Equa-

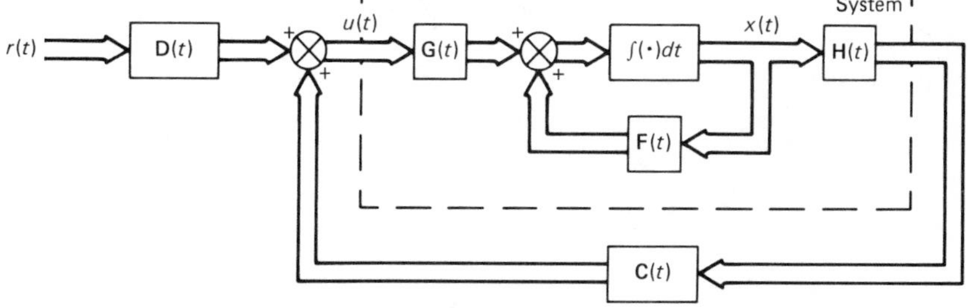

Figure 5-14. Linear state variable feedback control with distribution matrix $\boldsymbol{D}$.

tion (5.214) on the left by $\boldsymbol{H}$:

$$\dot{\boldsymbol{y}}(t) = \boldsymbol{H}\dot{\boldsymbol{x}}(t) = \boldsymbol{H}[\boldsymbol{F} + \boldsymbol{GCH}]\boldsymbol{x}(t) + \boldsymbol{HGDr}(t). \tag{5.215}$$

This equation relates components of $\boldsymbol{y}(t)$ to components of $\boldsymbol{r}(t)$, and indicates that coupling occurs both through the matrix $\boldsymbol{HGD}$ and through the state equation of the system. Various special cases may arise in which this equation is useful in selecting the matrices $\boldsymbol{C}$ and $\boldsymbol{D}$ to produce certain desirable characteristics. Often, one such characteristic is input-output decoupling, in which one component of the input vector controls only one or more components of the output vector, and these components are controlled only by that component of the input vector. In multiple input-multiple output systems, such a characteristic greatly simplifies the control of the system.

5.4.1. Systems with State Variable Feedback

If all the state variables are available for feedback, then $\boldsymbol{H} = \boldsymbol{I}$, and Equation (5.215) reduces to

$$\dot{\boldsymbol{y}}(t) = \dot{\boldsymbol{x}}(t) = [\boldsymbol{F} + \boldsymbol{GC}]\boldsymbol{x}(t) + \boldsymbol{GDr}(t). \tag{5.216}$$

For the case in which $\boldsymbol{r}(t)$ is of the same dimension as $\boldsymbol{x}(t)$, input-output decoupling would require that each component of $\boldsymbol{x}(t)$ be controlled by one, and only one, element of $\boldsymbol{r}(t)$. This feature would obviously require that the matrices

$$[\boldsymbol{F} + \boldsymbol{GC}] \quad \text{and} \quad [\boldsymbol{GD}] \tag{5.217}$$

be diagonal. Since $\boldsymbol{F}$ and $\boldsymbol{G}$ are system matrices and therefore fixed, this diagonalization must be accomplished by selection of the matrices $\boldsymbol{C}$ and $\boldsymbol{D}$. This is easily obtained if $\boldsymbol{G}$ is a nonsingular matrix, for then

$$\boldsymbol{GD} = \boldsymbol{I} \quad \text{implies} \quad \boldsymbol{D} = \boldsymbol{G}^{-1}. \tag{5.218}$$

Furthermore,

$$[\boldsymbol{F} + \boldsymbol{GC}] = -\boldsymbol{I}\alpha \tag{5.219}$$

implies

$$\boldsymbol{C} = \boldsymbol{G}^{-1}[-\boldsymbol{I}\alpha - \boldsymbol{F}]$$

where α is a positive scalar constant. The resulting closed-loop equations are then

$$\dot{\boldsymbol{x}}(t) = -\alpha \boldsymbol{x}(t) + \boldsymbol{r}(t) \tag{5.220}$$

and the utmost of simplification is obtained. Note that the selection of $-\boldsymbol{I}$ in Equation (5.219) ensures that the resulting system is stable.

As an example, consider the system described by the equations

$$\dot{\boldsymbol{x}}(t) = \boldsymbol{F}\boldsymbol{x}(t) + \boldsymbol{G}\boldsymbol{u}(t)$$
$$\boldsymbol{y}(t) = \boldsymbol{x}(t)$$

where

$$\boldsymbol{F} = \begin{bmatrix} 0 & 1 \\ -5 & -2 \end{bmatrix}; \quad \boldsymbol{G} = \begin{bmatrix} 1 & -1 \\ 1 & 2 \end{bmatrix}; \quad \boldsymbol{H} = \begin{bmatrix} 1 & 0 \\ 0 & 1 \end{bmatrix}.$$

It is desired to select matrices $\boldsymbol{C}$ and $\boldsymbol{D}$ such that there is input-output decoupling. The system block diagram is shown in Figure 5-15. From Equation (5.218) the value of $\boldsymbol{D}$ is obtained as

$$\boldsymbol{D} = \boldsymbol{G}^{-1} = \frac{1}{3}\begin{bmatrix} 2 & 1 \\ -1 & 1 \end{bmatrix}$$

and from Equation (5.219) with $\alpha = 1$,

$$\boldsymbol{C} = \frac{1}{3}\begin{bmatrix} 2 & 1 \\ -1 & 1 \end{bmatrix}\begin{bmatrix} -1 & -1 \\ 5 & 1 \end{bmatrix} = \frac{1}{3}\begin{bmatrix} 3 & -1 \\ 6 & 2 \end{bmatrix}.$$

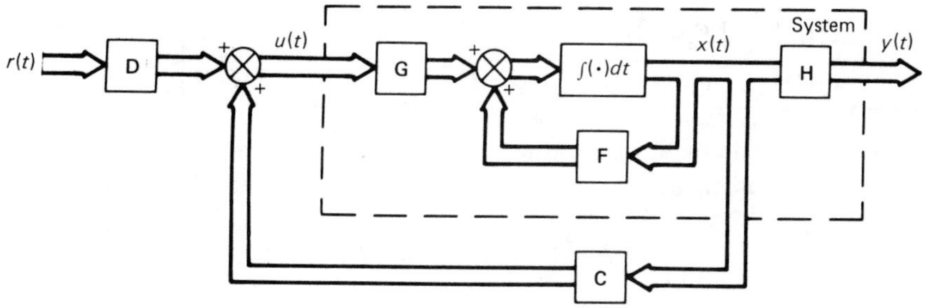

Figure 5-15. State variable feedback for input–output decoupling.

Substituting these values into Equation (5.216) yields the expected result

$$\dot{\boldsymbol{x}}(t) = -\boldsymbol{x}(t) + \boldsymbol{r}(t).$$

The time constant here is 1, but any value could have been selected by proper choice of α in Equation (5.219).

The specification of Equation (5.219) could be modified to

$$[\boldsymbol{F} + \boldsymbol{G}\boldsymbol{C}] = -\begin{bmatrix} -\alpha_1 & & \\ & \ddots & \\ & & -\alpha_N \end{bmatrix} = \boldsymbol{A} \tag{5.221}$$

which implies

$$\boldsymbol{C} = \boldsymbol{G}^{-1}[-\boldsymbol{A} - \boldsymbol{F}].$$

Here, each output responds with its own characteristic time constant.

If the matrix $\boldsymbol{G}$ is singular, then complete input-output decoupling is not obtainable by this direct method, since Equation (5.218) cannot be used to generate an N-dimensional identity matrix.

5.4.2. Systems with Number of Inputs Less than Number of States

When all state variables are available for feedback, but the dimension of the input vector is less than that of the state vector, then certain of the states may be identified with each component of the input vector, and noninteraction established in that context. Specifically, the matrix $\boldsymbol{H}$ in Equation (5.215) is the identity matrix, the matrix $\boldsymbol{C}$ has dimension $N \times M$, where M is the dimension of the control input $\boldsymbol{r}(t)$. The idea is to establish decoupling between the components of $\boldsymbol{r}(t)$ and the first M components of the state vector $\boldsymbol{x}(t)$. By partitioning Equation (5.215) in accordance with these dimensions, one obtains

$$\begin{bmatrix} \dot{\boldsymbol{x}}_1(t) \\ \hline \dot{\boldsymbol{x}}_2(t) \end{bmatrix} = \left[\begin{array}{c|c} \boldsymbol{F}_{11} + \boldsymbol{G}_1\boldsymbol{C}_1 & \boldsymbol{F}_{12} + \boldsymbol{G}_1\boldsymbol{C}_2 \\ \hline \boldsymbol{F}_{21} + \boldsymbol{G}_2\boldsymbol{C}_1 & \boldsymbol{F}_{22} + \boldsymbol{G}_2\boldsymbol{C}_2 \end{array}\right] \begin{bmatrix} \boldsymbol{x}_1(t) \\ \hline \boldsymbol{x}_2(t) \end{bmatrix} + \begin{bmatrix} \boldsymbol{G}_1\boldsymbol{D} \\ \hline \boldsymbol{G}_2\boldsymbol{D} \end{bmatrix} r(t). \tag{5.222}$$

The block diagram of this system would be as shown in Figure 5-15, with $\boldsymbol{H} = \boldsymbol{I}$. To obtain decoupling between $\boldsymbol{r}(t)$ and $\boldsymbol{x}(t)$, it is necessary that the matrices

$$[\boldsymbol{F}_{11} + \boldsymbol{G}_1\boldsymbol{C}_1] \quad \text{and} \quad \boldsymbol{G}_1\boldsymbol{D} \tag{5.223}$$

be diagonal, and that

$$\boldsymbol{F}_{12} + \boldsymbol{G}_1\boldsymbol{C}_2 = \boldsymbol{0}. \tag{5.224}$$

If the matrix G_1 is nonsingular, then this requirement is met by selecting C_1, C_2, and D such that

$$G_1 D = I; \qquad F_{11} + G_1 C_1 = -I; \qquad F_{12} - G_1 C_2 = 0. \tag{5.225}$$

The solution is

$$D = G_1^{-1}; \qquad C_1 = G_1^{-1}[-I - F_{11}]; \qquad C_2 = -G_1^{-1} F_{12} \tag{5.226}$$

This result can easily be verified by substitution into Equation (5.222). Diagonal matrices other than the identity can be used in Equation (5.225) with only minor modification of these results.

The closed-loop response of the system with the matrices D and G selected as in Equation (5.226) is determined by the eigenvalues of the matrix

$$[F + GC] = \left[\begin{array}{c|c} -I & 0 \\ \hline F_{21} + G_2 G_1^{-1}[-I - F_{11}] & F_{22} - G_2 G_1^{-1} F_{12} \end{array}\right]. \tag{5.227}$$

Because of the special form of this matrix, the eigenvalues are the eigenvalues of

$$-I \quad \text{and} \quad [F_{22} - G_2 G_1^{-1} F_{12}]. \tag{5.228}$$

The eigenvalues of $-I$ are all -1, but those of the second matrix are determined by the system defined matrices F_{22}, F_{12}, G_1, and G_2. The particular submatrices of F and G appearing here are dependent upon the specification of decoupling on the part of $x(t)$ and $r(t)$, and are not under control of the designer. Hence, for certain systems, the specification of decoupling may lead to closed-loop instability.

5.4.3. Input-Output Decoupling Using State Variable Feedback: The General Case

The general input-output relationship for a time-invariant linear system is given by Equation (5.215). The two previous sections have dealt with decoupling for a special case in which the measurement matrix H in Equation (5.215) is the identity matrix, implying that the output of the system is the state vector itself. A straightforward extension of these developments considers the case involving other nonsingular H matrices by use of the general transformation

$$y(t) = Hx(t). \tag{5.229}$$

Rarely in practice is the H matrix nonsingular, or the number of inputs equal to the number of states. More commonly, a system will be established with an

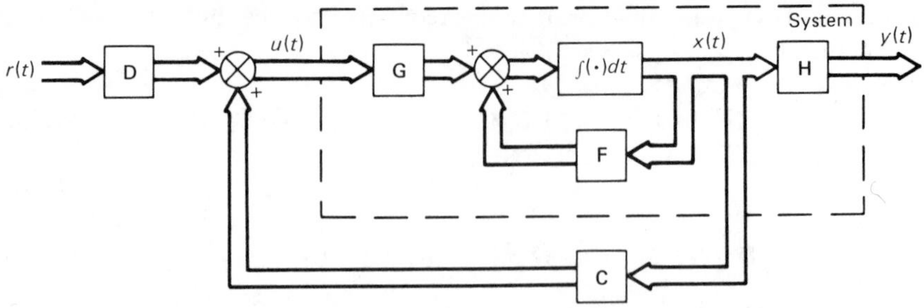

Figure 5-16. Decoupling by state variable feedback.

equal number of inputs and outputs, generally less than the number of states, and the desire is to have complete input-output decoupling in which one input function controls one, and only one, output function. The general system to be considered here—depicted in Figure 5-16—uses state variable feedback as discussed in Section 5.1.4 and is described by the equations

$$\dot{\boldsymbol{x}}(t) = [\boldsymbol{F} + \boldsymbol{GC}]\boldsymbol{x}(t) + \boldsymbol{GDr}(t)$$
$$\boldsymbol{y}(t) = \boldsymbol{Hx}(t). \tag{5.230}$$

Both the input vector $\boldsymbol{r}(t)$ and the output vector $\boldsymbol{y}(t)$ are of dimension M. The problem, then, is to select the $P \times N$ gain matrix $\boldsymbol{C}$ and the $P \times M$ matrix $\boldsymbol{D}$ so that input-output decoupling is accomplished. It seems intuitive that such decoupling may not be possible for all systems of this form, and such is indeed the case. In fact, the determination of necessary and sufficient conditions for decoupling is an important part of the theory.

The development of the theory of input-output decoupling by state variable feedback is rather involved, and for this reason only the pertinent results are given here. The idea, of course, is to modify the modal structure of the system so that the desired decoupling is effected. Whether or not the required modal restructuring is possible can be determined by application of the following test. A succession of vectors is formed, for fixed i, as

$$\hat{\boldsymbol{g}}_i^T = \boldsymbol{h}_i^T \boldsymbol{F}^k \boldsymbol{G}, \qquad k = 0, 1, 2, \ldots, N$$
$$i = 1, 2, \ldots, M \tag{5.231}$$

in which $\boldsymbol{h}_i^T$ is the ith row of the measurement matrix $\boldsymbol{H}$. For each $\boldsymbol{h}_i$, this succession of row vectors is formed in order to determine the smallest value of k for which $\hat{\boldsymbol{g}}_i$ is not null. For each i, the smallest value of k for which the row vector of Equation (5.231) is not null is given the symbol d_i. If null values are obtained for all values of k, then

$$d_i = (N - 1).$$

Then, a matrix $\hat{G}$ is formed by the row vectors $\hat{g}_i^T$, using $k = d_i$ in each case

$$\hat{G} = \begin{bmatrix} h_1^T F^{d_1} G \\ \hdashline h_2^T F^{d_2} G \\ \hdashline \vdots \\ \hdashline h_M^T F^{d_M} G \end{bmatrix}. \tag{5.232}$$

If $\hat{G}$ formed in this manner is nonsingular, then input–output decoupling is possible with state variable feedback. This condition is both necessary and sufficient.

To determine the actual C and D matrices, the following procedure can be used. Form the matrix $\hat{F}$, in which the ith row is

$$\hat{f}_i^T = h_i^T F^{d_i+1} \tag{5.233}$$

where the d_i are obtained as outlined above and h_i^T is the ith row of the measurement matrix H. Then form the matrix sum

$$\hat{M} = \sum_{k=0}^{\max d_i} M_k H F^k \tag{5.234}$$

where the M_k are arbitrary diagonal matrices which permit the designer some freedom in assigning the eigenvalues of the decoupled system, as discussed below. The expression yielding the feedback gain matrix C is

$$C = \hat{G}^{-1} [\hat{M} - \hat{F}]. \tag{5.235}$$

The value of the matrix D is given by

$$D = N\hat{G}^{-1} \tag{5.236}$$

where N is an arbitrary diagonal matrix which permits the designer some flexibility in establishing input–output sensitivities.

Even though the closed-loop system resulting from application of these techniques will exhibit the desired coupling, the transient response of the system may well be undesirable—even unstable—depending upon the resulting closed-loop eigenvalues. The designer has some control over these eigenvalues. In fact, at least

$$M + \sum_{i=1}^{M} d_i \tag{5.237}$$

of the closed-loop eigenvalues can be arbitrarily varied by selection of the $\boldsymbol{M}_k$ diagonal matrices of Equation (5.234).

To illustrate these procedures, consider the system

$$\boldsymbol{F} = \begin{bmatrix} 3 & 0 & 0 \\ 5 & 0 & -2 \\ 1 & 1 & 3 \end{bmatrix}; \quad \boldsymbol{G} = \begin{bmatrix} 1 & 0 \\ 0 & 1 \\ 1 & 1 \end{bmatrix}; \quad \boldsymbol{H} = \begin{bmatrix} 0 & 0 & 1 \\ 1 & 0 & 3 \end{bmatrix}.$$

This system has open-loop eigenvalues

$$\lambda_1 = 1; \quad \lambda_2 = 2; \quad \lambda_3 = 3$$

and is therefore unstable. If decoupling procedures are applied, the system must be simultaneously stabilized. To determine if this system can, in fact, be decoupled, the matrix $\hat{\boldsymbol{G}}$ of Equation (5.232) is formed. Since

$$\boldsymbol{h}_1^T\boldsymbol{G} = [0 \quad 0 \quad 1] \begin{bmatrix} 1 & 0 \\ 0 & 1 \\ 1 & 1 \end{bmatrix} = [1 \quad 1]$$

and

$$\boldsymbol{h}_2^T\boldsymbol{G} = [1 \quad 0 \quad 3] \begin{bmatrix} 1 & 0 \\ 0 & 1 \\ 1 & 1 \end{bmatrix} = [4 \quad 3]$$

then

$$d_1 = 0; \quad d_2 = 0; \quad \max d_i = 0$$

and

$$\hat{\boldsymbol{G}} = \begin{bmatrix} 1 & 1 \\ 4 & 3 \end{bmatrix}; \quad \hat{\boldsymbol{G}}^{-1} = \begin{bmatrix} -3 & 1 \\ 4 & -1 \end{bmatrix}.$$

Since $\hat{G}$ is nonsingular, it is possible to decouple the system. To determine the gain matrix $\boldsymbol{C}$, the matrix $\hat{\boldsymbol{F}}$ is determined from Equation (5.233),

$$\hat{f}_1^T = [0 \quad 0 \quad 1]\begin{bmatrix} 3 & 0 & 0 \\ 5 & 0 & -2 \\ 1 & 1 & 3 \end{bmatrix}; \qquad \hat{f}_2^T = [1 \quad 0 \quad 3]\begin{bmatrix} 3 & 0 & 0 \\ 5 & 0 & -2 \\ 1 & 1 & 3 \end{bmatrix}$$

$$\hat{\boldsymbol{F}} = \begin{bmatrix} 1 & 1 & 3 \\ 6 & 3 & 9 \end{bmatrix}.$$

From Equation (5.234) the matrix $\hat{\boldsymbol{M}}$ is

$$\hat{\boldsymbol{M}} = \boldsymbol{MH}.$$

Since $\boldsymbol{M}$ is an arbitrary nonsingular matrix which affects the closed-loop eigenvalues, let it have diagonal elements m_1 and m_2. Then

$$\hat{\boldsymbol{M}} = \begin{bmatrix} m_1 & 0 \\ 0 & m_2 \end{bmatrix}\begin{bmatrix} 0 & 0 & 1 \\ 1 & 0 & 3 \end{bmatrix} = \begin{bmatrix} 0 & 0 & m_1 \\ m_2 & 0 & 3m_2 \end{bmatrix}.$$

Now, the gain matrix $\boldsymbol{C}$ can be determined from Equation (5.235)

$$\boldsymbol{C} = \begin{bmatrix} -3 & 1 \\ 4 & -1 \end{bmatrix}\left[\begin{bmatrix} 0 & 0 & m_1 \\ m_2 & 0 & 3m_2 \end{bmatrix} - \begin{bmatrix} 1 & 1 & 3 \\ 6 & 3 & 9 \end{bmatrix}\right]$$

$$= \begin{bmatrix} -3 + m_2 & 0 & 3m_2 - 3m_1 \\ 2 - m_2 & -1 & -3 + 4m_1 - 3m_2 \end{bmatrix}.$$

The matrix $\boldsymbol{D}$ is then determined from Equation (5.236), and the arbitrary diagonal matrix $\boldsymbol{N}$ is here taken to be the identity matrix,

$$\boldsymbol{D} = \begin{bmatrix} -3 & 1 \\ 4 & -1 \end{bmatrix}.$$

Then, the resulting closed-loop system of Equation (5.230), with the $\boldsymbol{F}$, $\boldsymbol{G}$, $\boldsymbol{C}$, $\boldsymbol{D}$, and $\boldsymbol{H}$ matrices as specified here, exhibits input–output decoupling. The closed-loop system matrix and input distribution matrices are

$$[\boldsymbol{F} + \boldsymbol{GC}] = \begin{bmatrix} m_2 & 0 & 3m_2 - 3m_1 \\ 7 - m_2 & -1 & -5 + 4m_1 - 3m_2 \\ 0 & 0 & m_1 \end{bmatrix}; \qquad \boldsymbol{GD} = \begin{bmatrix} -3 & 1 \\ 4 & -1 \\ 1 & 0 \end{bmatrix}.$$

The transient response of the decoupled system is determined by the eigenvalues of $[\boldsymbol{F}+\boldsymbol{GC}]$. The determinantal or characteristic equation can be found to be

$$(\lambda - m_1)(\lambda^2 + \lambda(1 - m_2) - m_3) = 0.$$

The roots of this equation, and therefore the eigenvalues of the system, are

$$\lambda_1 = m_1; \qquad \lambda_2 = m_2; \qquad \lambda_3 = -1.$$

Let $m_1 = -3$ and $m_2 = -2$, yielding

$$\lambda_1 = -3; \qquad \lambda_2 = -2; \qquad \lambda_3 = -1$$

in which case the closed-loop system matrix is

$$[\boldsymbol{F}+\boldsymbol{GC}] = \begin{bmatrix} -2 & 0 & 3 \\ 9 & -1 & -11 \\ 0 & 0 & -3 \end{bmatrix}.$$

In order to gain some intuitive appreciation of the decoupling process, one might be tempted to transform this closed-loop system to an observer canonical form, but it can easily be determined by use of Equation (4.61) that the closed-loop system is nonobservable and therefore cannot be transformed to canonical form. As an alternative approach, the system can be transformed to the modal form described in Appendix C. The diagonalizing matrix of eigenvectors is determined to be

$$\boldsymbol{P} = \begin{bmatrix} 0 & 1 & 1 \\ 1 & -9 & -\frac{19}{3} \\ 0 & 0 & -\frac{1}{3} \end{bmatrix}; \qquad \boldsymbol{P}^{-1} = \begin{bmatrix} 9 & 1 & 8 \\ 1 & 0 & 3 \\ 0 & 0 & -3 \end{bmatrix}$$

and the similarity transformation

$$\boldsymbol{x}(t) = \boldsymbol{P}\boldsymbol{z}(t)$$

yields

$$\boldsymbol{P}^{-1}[\boldsymbol{F}+\boldsymbol{GC}]\boldsymbol{P} = \begin{bmatrix} -1 & 0 & 0 \\ 0 & -2 & 0 \\ 0 & 0 & -3 \end{bmatrix}$$

as expected. The transformed input distribution matrix and measurement matrix are

$$P^{-1}GD = \begin{bmatrix} -15 & 8 \\ 0 & 1 \\ -3 & 0 \end{bmatrix}; \qquad HP = \begin{bmatrix} 0 & 0 & -\frac{1}{3} \\ 0 & 1 & 0 \end{bmatrix}.$$

Then, the $z(t)$ equations are

$$\dot{z}_1(t) = -z_1(t) - 15r_1(t) + 8r_2(t)$$
$$\dot{z}_2(t) = -2z_2(t) + r_2(t)$$
$$\dot{z}_3(t) = -3z_3(t) - 3r_1(t)$$

with measurement equations

$$y_1(t) = -z_3(t)/3$$
$$y_2(t) = z_2(t).$$

The manner of decoupling is clear. Only input $r_1(t)$ affects the z_3 mode, and output $y_1(t)$ is affected only by $z_3(t)$. Similarly, only input $r_2(t)$ affects the z_2 mode, and output $y_2(t)$ is affected only by $z_2(t)$. The z_1 mode is affected by either input, but it is a nonobservable mode–it does not affect the output functions. This fact, of course, is why the decoupled system gives indication of nonobservability.

5.4.4. Input-Output Decoupling Using Output Feedback

The previous section considered the use of complete state variable feedback in order to achieve input-output decoupling, as indicated by Figure 5-16. If the states themselves are not available for feedback, then it may still be possible to decouple the system by use of output feedback, in which case the closed-loop equations corresponding to Equation (5.230) is obtained by using as the input function

$$\boldsymbol{u}(t) = \hat{\boldsymbol{C}}\boldsymbol{H}\boldsymbol{x}(t) + \boldsymbol{D}\boldsymbol{r}(t). \qquad (5.237)$$

Thus, the closed-loop equation is

$$\dot{\boldsymbol{x}}(t) = [\boldsymbol{F} + \boldsymbol{G}\hat{\boldsymbol{C}}\boldsymbol{H}]\boldsymbol{x}(t) + \boldsymbol{G}\boldsymbol{D}\boldsymbol{r}(t) \qquad (5.238)$$

and the system corresponds to that of Figure 5-17. Now, since the system with

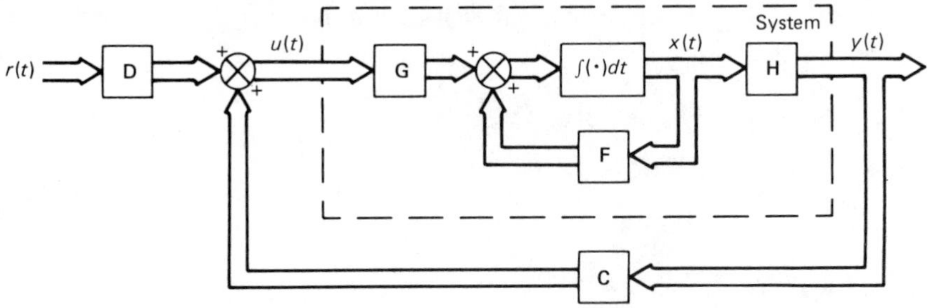

Figure 5-17. Decoupling by output feedback.

state variable feedback, as described by Equation (5.230), can be decoupled if and only if the matrix $\hat{G}$ of Equation (5.232) is nonsingular, then the system of Equation (5.238) can be decoupled if and only if the system can be decoupled with state variable feedback gain matrix $\boldsymbol{C}$, and an output feedback matrix $\hat{\boldsymbol{C}}$ can be found such that

$$\boldsymbol{C} = \hat{\boldsymbol{C}}\boldsymbol{H}. \tag{5.239}$$

Since $\boldsymbol{H}$ is singular, solution for $\hat{\boldsymbol{C}}$ is not straightforward. In fact, there may be no solution for $\hat{\boldsymbol{C}}$, or the solution may not be unique. One approach to the solution of this equation is to form the partitioned matrices

$$[\boldsymbol{C}_1 \mid \boldsymbol{C}_2] = \hat{\boldsymbol{C}}[\boldsymbol{H}_1 \mid \boldsymbol{H}_2] \tag{5.240}$$

in which the columns on both sides of Equation (5.239) have been rearranged to produce the nonsingular $P \times P$ submatrix $\boldsymbol{H}_1$, where P is the dimension of the input vector $\boldsymbol{u}(t)$. Then there exists two relationships,

$$\begin{aligned} \boldsymbol{C}_1 &= \hat{\boldsymbol{C}}\boldsymbol{H}_1 \\ \boldsymbol{C}_2 &= \hat{\boldsymbol{C}}\boldsymbol{H}_2 \end{aligned} \tag{5.241}$$

of which the former can be solved for $\hat{\boldsymbol{C}}$ by inverting the submatrix $\boldsymbol{H}_1$,

$$\hat{\boldsymbol{C}} = \boldsymbol{C}_1\boldsymbol{H}_1^{-1}. \tag{5.242}$$

Substitution of this expression into the latter relationship of Equation (5.241) yields

$$\boldsymbol{C}_2 = \boldsymbol{C}_1\boldsymbol{H}_1^{-1}\boldsymbol{H}_2. \tag{5.243}$$

If this equation is consistent, then so also is Equation (5.239), in which case Equation (5.242) yields a value for $\hat{\boldsymbol{C}}$. As a result, the necessary and sufficient conditions for output feedback decoupling are:

I. The nonsingularity of the matrix of Equation (5.232), implying the possibility of state variable feedback decoupling.
II. The consistency of Equation (5.243), implying that the state feedback decoupling requirements can be met by use of output feedback.

The matrix of Equation (5.242) is not unique, since its columns were selected in arbitrary order to produce the nonsingular submatrix $\boldsymbol{H}_1$.

If condition I is met, but condition II is not, indicating that state variable feedback decoupling is possible—but that the requirements cannot be met by output feedback—then an observer can be employed to provide estimates of the state variables. By virtue of the developments of Section 5.2, the observer eigenvalues can be selected independently of the selection of system eigenvalues, so that a well defined observer will produce only short-lived transient deviations from the desired decoupling properties.

To illustrate these concepts, consider again the example problem of the previous section,

$$\boldsymbol{F} = \begin{bmatrix} 3 & 0 & 0 \\ 5 & 0 & -2 \\ 1 & 1 & 3 \end{bmatrix}; \quad \boldsymbol{G} = \begin{bmatrix} 1 & 0 \\ 0 & 1 \\ 1 & 1 \end{bmatrix}; \quad \boldsymbol{H} = \begin{bmatrix} 0 & 0 & 1 \\ 1 & 0 & 3 \end{bmatrix}.$$

It has already been shown that this system can be decoupled by state variable feedback. If, in addition, Equation (5.239) applied to this system has a solution for $\hat{\boldsymbol{C}}$, the system can also be decoupled by output feedback. The question of the existence of a matrix $\hat{\boldsymbol{C}}$ satisfying this requirement can be addressed directly here because of the small number of dimensions involved. That is, the equations

$$\begin{bmatrix} -5 & 0 & 3 \\ 4 & -1 & 9 \end{bmatrix} = \begin{bmatrix} c_{11} & c_{12} \\ c_{21} & c_{22} \end{bmatrix} \begin{bmatrix} 0 & 0 & 1 \\ 1 & 0 & 3 \end{bmatrix}$$

where the c_{ij} are elements of $\hat{\boldsymbol{C}}$, must be consistent. Completing the multiplication on the right-hand side yields

$$\begin{bmatrix} -5 & 0 & 3 \\ 4 & -1 & 9 \end{bmatrix} = \begin{bmatrix} c_{12} & 0 & c_{11} + 3c_{12} \\ c_{22} & 0 & c_{21} + 3c_{22} \end{bmatrix}$$

which has no solution. Thus, although the system can be decoupled by state variable feedback, it cannot be decoupled by output feedback. In this case, an observer may be used to estimate the state, and this estimate then used with the state variable feedback matrix developed in the previous section.

On the other hand, suppose the measurement matrix $\boldsymbol{H}$ is

$$\boldsymbol{H} = \begin{bmatrix} -1 & 0 & \frac{3}{5} \\ \frac{4}{9} & -\frac{1}{9} & -1 \end{bmatrix}$$

Then, Equation (5.239) yields

$$\begin{bmatrix} -5 & 0 & 3 \\ 4 & -1 & 9 \end{bmatrix} = \begin{bmatrix} c_{11} & c_{12} \\ c_{21} & c_{22} \end{bmatrix} \begin{bmatrix} -1 & 0 & -\frac{3}{5} \\ \frac{4}{9} & -\frac{1}{9} & 1 \end{bmatrix}$$

$$= \begin{bmatrix} -c_{11} + \dfrac{4c_{12}}{9} & \dfrac{-c_{12}}{9} & \dfrac{-3c_{11}}{5} + c_{12} \\ -c_{21} + \dfrac{4c_{22}}{9} & \dfrac{-c_{22}}{9} & \dfrac{-c_{22}}{9} - \dfrac{3c_{12}}{5} + c_{22} \end{bmatrix}$$

which has the solution

$$\hat{\boldsymbol{C}} = \begin{bmatrix} c_{11} & c_{12} \\ c_{21} & c_{22} \end{bmatrix} = \begin{bmatrix} 5 & 0 \\ 0 & 9 \end{bmatrix}.$$

Thus, this system can be decoupled by output feedback. The general procedures of Equations (5.240) through (5.243) could be used here, but the direct approach used above more clearly illustrates the principles involved. For problems of larger dimension, or for systematic computer-based procedures, this direct approach is not suitable.

Finally, the closed-loop system matrix of the system with output feedback is

$$[\boldsymbol{F} + \boldsymbol{GCH}] = \begin{bmatrix} -5 & 0 & 3 \\ 4 & -1 & 9 \\ -1 & -1 & 6 \end{bmatrix} - \begin{bmatrix} 1 & 0 \\ 0 & 1 \\ 1 & 1 \end{bmatrix} \begin{bmatrix} 5 & 0 \\ 0 & 9 \end{bmatrix} \begin{bmatrix} -1 & 0 & \frac{3}{5} \\ \frac{4}{9} & -\frac{1}{9} & -1 \end{bmatrix}$$

$$= \begin{bmatrix} -2 & 0 & 3 \\ 9 & -1 & -11 \\ 0 & 0 & -3 \end{bmatrix}.$$

This is the same closed-loop system matrix as that produced by the state variable feedback approach. This is not a surprising result, but it further elucidates the relationship between the two methods.

5.5. FURTHER CONSIDERATIONS

One of the great advantages of the use of modern control theory and the state space analysis of systems is that it makes available to the analyst the powerful techniques of matrix theory. Not only does the use of these techniques permit more lucid description of system structure, but it also blends well with the practical use of the modern digital computer. Throughout this chapter, several systematic techniques have been presented, some without development, for achieving desired results through the use of matrix theory. Many times, in dealing with matrices and matrix operations, the technique best suited for the development of ideas and concepts is not that best suited for digital computer implementation. In fact, efficient computer-based algorithms seldom lend themselves well to conceptual understanding of the principles involved. For this reason, the algorithms presented here as systematic procedures for achieving desired goals have been superseded by more computationaly efficient algorithms better suited to digital computer implementation. Some of the references at the end of this chapter reflect the latest developments in this very important area.

A second concept receiving a great deal of attention lately is the "robustness" of controllers and observers designed using the concepts and techniques presented in this chapter. Robustness refers to the sensitivity of the performance of a controller or observer to small changes in system parameters. Most of the underlying principles of controller and observer design depend upon rather precise interaction between the various modal responses of a linear system. These interactions may be severely degraded by small variations of system parameters from their design values. Recent developments in the sensitivity analysis of controller and observer designs have provided some insight into this critical area, and alternative designs of robust controllers and observers are beginning to appear.

Once a controller has been synthesized, there remains the task of implementation. Today, the trend is toward digital implementation of control logic, principally through the use of microprocessors in one form or the other. The range of microprocessor capability is so wide that it can be said that there is a microprocessor for every application. From small, single chip, four and eight-bit computers to the very capable sixteen and thirty-two bit processors capable of addressing multi-megabyte memories, the designer has a wide range of computational power from which to select. These processors make possible today the implementation of complex control logic virtually impossible a decade ago. But the use of all-digital control logic introduces new concerns regarding precision, computational capability, interface speed, and memory requirements, all of which must be established in the design process. Some of the references at the end of this chapter relate the latest developments in this area.

5.5.1. State Variable Feedback

While the contents of Section 5.1 introduce the reader to state variable feedback concepts, many further developments have been described in the literature. More efficient algorithms–better suited to digital computer computation–for development of the control canonical forms are presented in References 10 and 26. The use of output feedback for eigenvalue assignment is developed in References 29, 47, and 48. The eigenstructure assignment problem is treated in Reference 3, in which a computationaly efficient method for obtaining the null space of a matrix is developed, and in Reference 46, in which the design of a state variable feedback controller with specified response is accomplished by converting the design problem to a more easily solved estimation problem. In Reference 41, a technique which decomposes the feedback design problem into two reduced-order subsystem design problems is developed. The relationship between the time and frequency domain approach to controller design is used in Reference 33, which presents an iterative design algorithm for eigenvalue assignment in linear compensators by combining state space and root locus methods. Special techniques for single input-single output systems are developed in Reference 22. In Reference 16, a numerical procedure for solution of the eigenstructure problem for multivariable systems is presented.

The concept of robustness of controllers is considered in Reference 11, with respect to closed-loop stability, asymptotic regulation, and other performance characteristics. A computer-based algorithm for determining bounds on parameter variation for robustness of closed-loop stability is developed in Reference 15. In References 1, 20, and 30, procedures are developed for assigning closed-loop eigenvalues and minimizing their sensitivity with respect to system parameters, while in Reference 39 a robustness analysis with respect to stability considers simultaneous linear and nonlinear perturbations in the feedback path.

The actual implementation of digital control logic is considered in Reference 25, where the implementation of state feedback controllers, observers, optimal controllers, regulators, and Kalman filters by digital logic is developed. This reference also has an excellent discussion of sampling and quantization errors, multiplication errors, the effect of finite word length, and other concerns. In Reference 27, a microprocessor is used to implement several digital controller designs, and the resulting programs are presented. This reference also contains very useful information on the effects of the use of digital hardware. The performance of digital state regulators using analog-to-digital converters is considered in Reference 42, where the effects of finite word length in floating point computation on regulator performance are investigated. In Reference 18, the use of an LSI-11 processor is considered for implementation of the optimal controller problem to be developed in Chapter 6. Requirements for random access and read only memory are determined, and the effects of the processor and interface speed and precision are investigated.

Finally, some additional transformation properties relating the control canonical form to the Jordan canonical form are presented in Reference 37, and generalization of the concept of eigenvalues and eigenvectors for time-varying systems is developed in Reference 52.

5.5.2. Linear Observers

The developments of Section 5.2 deal exclusively with observers for linear, time-invariant systems. In Reference 53, the concept of an observer for time-varying linear systems is considered. The extension of observer theory to the situation in which the measurements are contaminated by noise is considered in Reference 21 for the continuous-time observer, and in Reference 50 for the discrete time observer. The question of robustness of observers is addressed in References 6 and 36. In Reference 14, the transfer function properties of observer-based controllers are discussed. This reference points out how the use of an observer can make a nonrobust controller from what would be a robust state variable feedback controller. It also presents a design adjustment procedure to improve the robustness of observer-based controllers.

Further developments in the design of reduced-order observers are presented in Reference 5, while frequency domain procedures for minimal order observers are developed in Reference 43. In Reference 30, reduced-order observer design for linear time-invariant systems with unknown input functions is considered. For large-scale systems—systems with many state variables—the observer design process is hampered by the large number of equations which must be solved simultaneously. Techniques presented in Reference 4 specifically for large-scale systems address this problem.

In some observer applications, the state variables themselves are not really required, but instead the important quantity is some scalar linear function of the state variables. Certain optimal control results are of this form. For such a system, it is possible to design an observer of order less than $(N - M)$. This concept is developed in References 7 and 34.

5.5.3. Discrete Time Observers

The concept of deadbeat control was briefly discussed in Section 5.3.4. This topic is considered further in Reference 2, in which the performance of a state variable feedback controller is compared to that of an observer-based controller.

The use of deadbeat control is also discussed in Reference 40. Because deadbeat control is closely related to the discrete time version of the minimum time problem, this topic is also considered in some of the Chapter 6 references.

The discrete time observer is further developed in Reference 50, in which a stochastic observer is considered and compared to the Kalman filter equations developed in Chapter 7.

5.5.4. Input-Output Decoupling

The concepts of input-output decoupling were first developed in Reference 17. More recent developments lie in the area of output feedback decoupling (see Reference 28), and state variable decoupling in which a subset of the output is to be decoupled (Reference 44). In Reference 23, a frequency domain approach to input-output decoupling into single input-multiple output subsystems is considered.

REFERENCES

1. Ackermann, Jurgen. Parameter Space Design of Robust Control Systems. *IEEE Transactions on Automatic Control* **AC-25:** 1058–1072 (December 1980).
2. Akashi, Hajime, and Imai, Hiroyuki. A Basis for the Controllable Canonical Form of Linear Time-Invariant Multi-Input Systems. *IEEE Transactions or Automatic Control,* **AC-23:** 742–745 (August 1978).
3. Apelvich, J. D. A Simple Method For Finding a Basis for the Null Space of a Matrix. *IEEE Transactions on Automatic Control,* **AC-21:** 402–403 (June 1976).
4. Arbel, Ami and Tse. Observer Design for Large-Scale Linear Systems. *IEEE Transactions on Automatic Control,* **AC-24:** 469–475 (June, 1979).
5. Balestrino, A. and Celentano, G. Pole Assignment in Linear Multivariable Systems Using Observers of Reduced Order. *IEEE Transactions on Automatic Control,* **AC-24:** 144–146 (February 1979).
6. Brattacharyya, S. P. The Structure of Robust Observers. *IEEE Transactions on Automatic Control,* **AC-21:** 581–588 (August 1976).
7. ——. Observer Design for Linear Systems with Unknown Inputs. *IEEE Transactions on Automatic Control,* **AC-23:** 483–485 (June 1978).
8. Brown, Frank Markham. An Observer-Canonical State-Representation for a System of Linear Variable Difference Equations. *IEEE Transactions on Automatic Control,* **AC-21:** 541–544 (August 1976).
9. Bryson, Arthur E., and Luenberger, David G. The Synthesis of Regulator Logic Using State-Variable Concepts. *Proceedings of the IEEE* **58:** 1803–1811 (November 1970).
10. Chan, W. S., and Wang, Y. T. A Basis for the Controllable Canonical Form of Linear Time-Invariant Multiinput Systems. *IEEE Transactions on Automatic Control,* **AC-23:** 742–745 (August 1978).
11. Davidson, Edward J., and Ferguson, Ian J. The Design of Controllers for the Multivariable Robust Servomechanism Problem Using Parameter Optimization Methods. *IEEE Transactions on Automatic Control,* **AC-26:** 93–110 (February 1981).

12. D'Azzo, John D., and Houpis, Constantine H. *Linear Control System Analysis and Design*. New York: McGraw-Hill, 1981.
13. Derusso, Paul M., Roy, Rob J., and Close, Charles M. *State Variables for Engineers*. New York: John Wiley and Sons, 1965.
14. Doyle, J. C. and Stein, G. Robustness with Observers. *IEEE Transactions on Automatic Control*, **AC-24:** 607–611 (August 1979).
15. Eslami, Mansour, and Russell, David L. On Stability with Large Parameter Variations: Stemming From the Direct Method of Lyapunov. *IEEE Transactions on Automatic Control*, **AC-25:** 1231–1234 (December 1980).
16. Fahmy, M. M., and O'Reilly, J. On Eigenstructure Assignment in Linear Multivariable Systems. *IEEE Transactions on Automatic Control*, **AC-27:** 690–693 (June 1982).
17. Falb, Peter, and Wolovich, William. Decoupling in the Design and Synthesis of Multivariable Control Systems. *IEEE Transactions on Automatic Control*, **AC-12:** 651–659 (December 1967).
18. Fanor, Florence A., and Edens, Richard S. Microprocessor Requirements for Implementing Modern Control Logic. *IEEE Transactions on Automatic Control*, **AC-25:** 461–468 (June 1980).
19. Foss, Alan S. Critique of Chemical Process Control Theory. *IEEE Transactions on Automatic Control*, **AC-18:** 646–652 (December 1973).
20. Gomath, K., Prabhu, S. S., and Pai, M. A. A Suboptimal Controller for Minimum Sensitivity of Closed-Loop Eigenvalues to Parameter Variations. *IEEE Transactions on Automatic Control*, **AC-25:** 587–588 (June 1980).
21. Hang Chang-Chieh. A New Form of Stable Adaptive Observer. *IEEE Transactions on Automatic Control*, **AC-21:** 544–547 (August 1976).
22. Hickin, J. Pole Assignment in Single-Input Linear Systems. *IEEE Transactions on Automatic Control*, **AC-25:** 282–284 (April 1980).
23. Hisamura, T., and Nakao, H. Output Feedback Decoupling of Multivariable Systems. *IEEE Transactions on Automatic Control*, **AC-23:** 74–76 (February 1978).
24. Howze, J. W. Necessary and Sufficient Conditions for Decoupling Using Output Feedback. *IEEE Transactions on Automatic Control*, **AC-18:** 44–46 (February 1973).
25. Jacquot, Raymond G. *Modern Digital Control Systems*. New York: Marcel Dekker, 1981.
26. Jordan, D., and Sridor, B. An Efficient Algorithm for Calculation of the Luenberger Canonical Form. *IEEE Transactions on Automatic Control*, **AC-18:** 292–295 (June 1973).
27. Katz, Paul. *Digital Control Using Microprocessors*. Englewood Cliffs, New Jersey: Prentice Hall International, 1981.
28. Kim, H. Y., and Shapiro, E. Y. On Output Feedback Decoupling. *IEEE Transactions on Automatic Control*, **AC-26:** 782–784 (June 1981).
29. Kimura, Hidenori. A Further Result on the Pole Assignment by Output Feedback. *IEEE Transactions on Automatic Control*, **AC-22:** 458–463 (June 1977).
30. Klien, G., and Moore, B. C. Eigenvalue–Generalized Eigenvector Assignment with State Feedback. *IEEE Transaction on Automatic Control*, **AC-22** (February 1977).
31. Kudva, P., Viswanadham, N., and Ramakrishna, A. Observers for Linear Systems with Unknown Inputs. *IEEE Transactions on Automatic Control*, **AC-25:** 113–115 (February 1980).
32. Kuo, Benjiman C. *Automatic Control Systems*. Englewood Cliffs, New Jersey: Prentice Hall, 1975.
33. Lee, Gordan, Jordan, David, and Sohrwardy, Munir. A Pole Assignment Algorithm for Multivariable Control Systems. *IEEE Transactions on Automatic Control*, **AC-24:** 357–362 (April 1979).

34. Luenberger, D. G. An Introduction to Observers. *IEEE Transactions on Automatic Control*, **AC-16:** 596–602 (December 1971).
35. ——. Observers for Multivariate Systems. *IEEE Transactions on Automatic Control*, **AC-11:** 190 (April 1966).
36. Missaghie, M. M., and Fairman, F. W. Sensitivity Reducing Observers for Optimal Feedback Control. *IEEE Transactions on Automatic Control*, **AC-22:** 952–956 (December 1977).
37. Mita, Tsutomu, and Arakawa, Hiro. On Eigenvectors of the Canonical Matrix for Multiple-Input Controllable Systems. *IEEE Transactions on Automatic Control*, **AC-22:** 262–263 (April 1977).
38. Molander, Per, and Willems, Jan C. Synthesis of State Feedback Control Laws with a Specified Gain and Phase Margin. *IEEE Transactions on Automatic Control*, **AC-25:** 928–931 (October 1980).
39. Owens, D. H., and Chotai, A. Robust Stability of Multivariable Feedback Systems with Respect to Linear and Non-Linear Feedback Perturbations. *IEEE Transactions on Automatic Control*, **AC-27:** 254–256 (February 1982).
40. Pachter, M. An Explicit Pole-Assigning Feedback Formula with Application to Dead-Beat Feedback Construction in Discrete Linear Systems. *IEEE Transactions on Automatic Control*, **AC-22:** 263–265 (April 1977).
41. Phillips, R. G. A Two-Stage Design of Linear Feedback Controls. *IEEE Transactions on Automatic Control*, **AC-25:** 1220–1223 (December 1980).
42. Rink, Raymond, and Chong, Hoi. Performances of State Regulator Systems with Floating-Point Computation. *IEEE Transactions on Automatic Control*, **AC-24:** 411–421 (June 1979).
43. Russel, D. W., and Bullock, T. E. A Frequency Domain Approach to Minimal-Order Observer Design for Several Linear Functions of the State. *IEEE Transactions on Automatic Control*, **AC-22:** 600–604 (August 1977).
44. Sato, Stephen M., and Loprestic, Philip V. On the Generalization of State Feedback Decoupling Theory. *IEEE Transactions on Automatic Control*, **AC-16:** 133–139 (April 1971).
45. Schultz, Donald G., and Melsa, James L. *State Functions and Linear Control Systems.* New York: McGraw-Hill, 1967.
46. Sinha, A. K., and Mahalanabis, A. K. On the Design of Multivariable Control Systems. *IEEE Transactions on Automatic Control*, **AC-21:** 403–405 (June 1976).
47. ——. A New Condition for Output Feedback Decoupling of Multivariable Systems. *IEEE Transactions on Automatic Control*, **AC-24:** 476–478 (June 1979).
48. Srinathkumar, S. Eigenvalue/Eigenvector Assignment Using Output Feedback. *IEEE Transactions on Automatic Control*, **AC-23:** 79–81 (February 1978).
49. Van Dooren, Paul M. The Generalized Eigenstructure Problem in Linear System Theory. *IEEE Transactions on Automatic Control*, **AC-26:** 111–129 (February 1981).
50. Weiss, Haim. On the Structure of the Luenberger Observer in Discrete-Time Linear Stochastic. *IEEE Transactions on Automatic Control*, **AC-22:** 871–873 (October 1977).
51. Wonham, W. M. *Linear Multivariate Control–A Geometric Approach.* New York: Springer-Verlag, 1974.
52. Wu Min-Yen. A New Concept of Eigenvalues and Eigenvectors and Its Applications. *IEEE Transactions on Automatic Control*, **AC-25:** 824–826 (August 1980).
53. Yuskel, Y. Onder, and Bongiorno, Joseph J. Observers for Linear Multivariable Systems with Application. *IEEE Transaction on Automatic Control*, **AC-16:** 603–613 (December 1971).

DEVELOPMENTAL EXERCISES

5.1. Write the general set of $2N$ equations by augmenting the system equations with the observer equations. Draw general conclusions from this representation.

5.2. Develop the frequency domain (Laplace transform) representation of state variable feedback.

5.3. Develop the PID (proportional-integral-derivative) controller in state variable form. Limit consideration to a single input–single output system, for which the control function is

$$u(t) = k_1 y(t) + k_2 \dot{y}(t) + \int_0^t k_3 y(\sigma)\, d\sigma.$$

PROBLEMS

5.1. A third-order system is described by the system of equations

$$\dot{x}_1(t) = 3x_1(t) + 5x_2(t) + x_3(t) + u_1(t)$$
$$\dot{x}_2(t) = x_3(t)$$
$$\dot{x}_3(t) = -2x_2(t) + 3x_3(t) + u_1(t) + u_2(t).$$

Find the state variable feedback gains which will produce a closed-loop response characterized by eigenvalues $-1 \pm i1$ and -4.

5.2. For the system described by the set of equations

$$\dot{x}_1(t) = x_1(t) + 6x_2(t) - 3x_3(t) + u_1(t)$$
$$\dot{x}_2(t) = -x_1(t) - x_2(t) + x_3(t) + u_2(t)$$
$$\dot{x}_3(t) = -2x_1(t) + 2x_2(t) + u_3(t)$$

find the state variable feedback gains which will produce a closed loop system with eigenvalues $-1 \pm i1$, and -4. Sketch the block diagram of the closed-loop system.

5.3. Consider the system described by the equations

$$\dot{x}_1(t) = -2x_1(t) + x_2(t)$$
$$\dot{x}_2(t) = -2x_1(t) + x_3(t)$$
$$\dot{x}_3(t) = x_2(t) + x_3(t) + u_1(t) + u_2(t)$$
$$\dot{x}_4(t) = x_4(t) + u_2(t).$$

Find the state variable feedback gains to yield closed loop eigenvalues $-5 \pm i2$, -6, and -10.

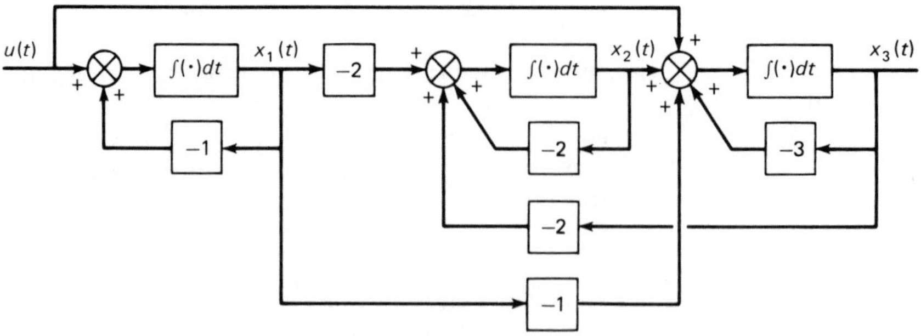

Figure P5-1.

5.4. A third-order system is shown in Figure P5-1.

(a) Find the eigenvalues of the system.

(b) Determine the state variable feedback gains required to produce closed-loop eigenvalues -4, -5, and -6.

5.5. A second order system is described by the equations

$$\dot{x}_1(t) = -x_1(t) + u_1(t)$$
$$\dot{x}_2(t) = x_1(t) - 2x_2(t) + u_2(t)$$

and the measurement equation is

$$y(t) = -x_1(t) + x_2(t).$$

(a) Design an observer for this system with response characterized by critical damping with time constant 0.1. Explain any problems you might have.

(b) Determine the state variable feedback gains required to produce closed-loop response characterized by critical damping with time constant = 1. Explain any problem you might have.

(c) Test the system for observability and controllability.

5.6. A second-order system is described by the equations

$$\dot{x}_1(t) = x_2(t)$$
$$\dot{x}_2(t) = -10x_2(t) + 10u(t).$$

(a) Use time interval $(t_l - t_{l-1}) = \frac{1}{10}$ second and determine the discrete time equations for this system.

(b) Find the state variable feedback gains required to produce discrete time closed-loop eigenvalues of $0.8 \pm i0.25$.

5.7. In the discrete time system of Problem 5.6, the measurement is $y(l) = x_1(l)$. Design a second-order observer for this system, with observer response characterized by eigenvectors $0.3 \pm i0.5$. Argue for the appropriateness of these observer response characteristics.

5.8. A third-order system is described by the following equations:

$$\dot{x}_1(t) = x_2(t) + u_1(t)$$
$$\dot{x}_2(t) = x_3(t)$$
$$\dot{x}_3(t) = 4x_1(t) + 4x_2(t) - x_3(t) + u_2(t).$$

The measurement equation is

$$y_1(t) = x_1(t)$$
$$y_2(t) = x_2(t) + x_3(t).$$

(a) Show that the uncontrolled system is unstable.
(b) Use state variable feedback to produce a closed-loop response with eigenvalues -2, -3, and -4.
(c) Design a third-order observer for this system. Produce observer eigenvalues of -10, -11, and -12.

5.9. Consider the linear system

$$\dot{\boldsymbol{x}}(t) = \boldsymbol{F}\boldsymbol{x}(t) + \boldsymbol{G}u(t)$$

where

$$\boldsymbol{x}(t) = \begin{bmatrix} x_1(t) \\ x_2(t) \end{bmatrix}; \quad \boldsymbol{F} = \begin{bmatrix} 0 & 1 \\ 0 & 0 \end{bmatrix}; \quad \boldsymbol{G} = \begin{bmatrix} 0 \\ 1 \end{bmatrix}$$

and $u(t)$ is a scalar input. The measurement equation is

$$z(t) = x_2(t).$$

(a) Design an observer for this system. Select observer parameters so that the observer exhibits critical damping.
(b) Sketch the system and observer block diagram.
(c) Relate this example to a physical system of some kind.

5.10. For the system of Problem 5.9, design the reduced-order observer.

5.11. Design a reduced-order observer for the system

$$\boldsymbol{x}(t) = \boldsymbol{F}\boldsymbol{x}(t) + \boldsymbol{G}u(t)$$

where

$$\boldsymbol{x}(t) = \begin{bmatrix} x_1(t) \\ x_2(t) \end{bmatrix}; \quad \boldsymbol{F} = \begin{bmatrix} -2 & 1 \\ 0 & -1 \end{bmatrix}; \quad \boldsymbol{G} = \begin{bmatrix} 0 \\ 1 \end{bmatrix}$$

and the measurement equation is

$$z(t) = x_1(t).$$

5.12. Design a second-order observer for the system of Problem 5.11.

5.13. Design a reduced-order observer for the system shown in Figure P5-2.

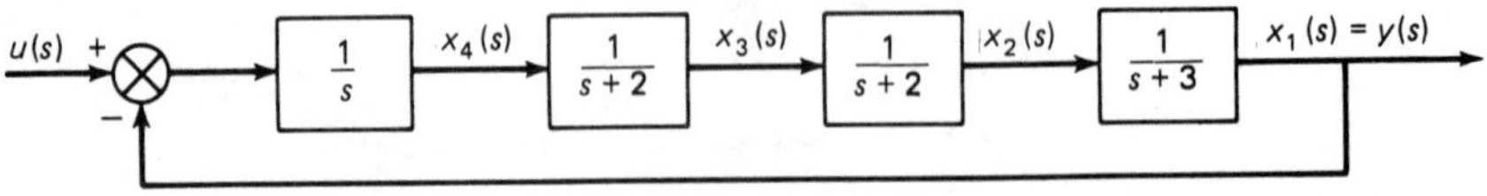

Figure P5-2.

5.14. Consider the second-order system described by the equations

$$\dot{x}_1(t) = x_1(t) + u_1(t)$$

$$\dot{x}_2(t) = u_2(t)$$

and measurement equation

$$y(t) = 2x_1(t) - x_2(t).$$

Design a second-order observer for this system. Select the observer time constants so that the observer response time is one-tenth that of the system.

5.15. For the system of Problem 5.14, design a reduced-order observer.

5.16. A second-order system is described by the equations

$$\dot{x}_1(t) = x_2(t)$$

$$\dot{x}_2(t) = -5x_2(t) + 10u(t).$$

Find the state variable feedback coefficients which will cause the closed loop system response to be governed by eigenvalues

$$\frac{10}{\sqrt{12}} \pm i\frac{10}{\sqrt{12}}.$$

5.17. For the system of Problem 5.16, consider the case in which the available output is $y(t) = x_1(t)$. Design a second-order observer to be used for implementation of the state variable feedback control developed in Problem 5.16. The observer is to exhibit critically damped response with time constant = 0.02. Sketch the resulting controller, including both state variable feedback and observer.

5.18. Repeat Problem 5.17, only now use a reduced-order observer.

5.19. A second-order system is described by the equations

$$\dot{x}_1(t) = -x_1(t) + 3x_2(t) + u(t)$$
$$\dot{x}_2(t) = -2x_2(t) + u(t).$$

The measurement is $y(t) = x_1(t)$. Design a second-order observer which exhibits real eigenvalues -9 and -10. Are these values proper for this system?

5.20. For the system of Problem 5.19, design a reduced-order observer.

5.21. For the system of Problem 5.20, determine the state variable feedback gains necessary to produce closed-loop eigenvalues of -5 and -6. Find the response of the closed-loop system to a unit step input in $x_2(t)$.

5.22. For the system of Problem 5.19 and the state variable control of Problem 5.21, sketch the closed-loop system including both observer and controller. Find the response of this system to a unit step input in $x_2(t)$.

5.23. Consider the second-order system described by the block diagram of Figure P5-3. Design a second-order observer for this system. The observer response is to be characterized by critical damping with time constant 0.01.

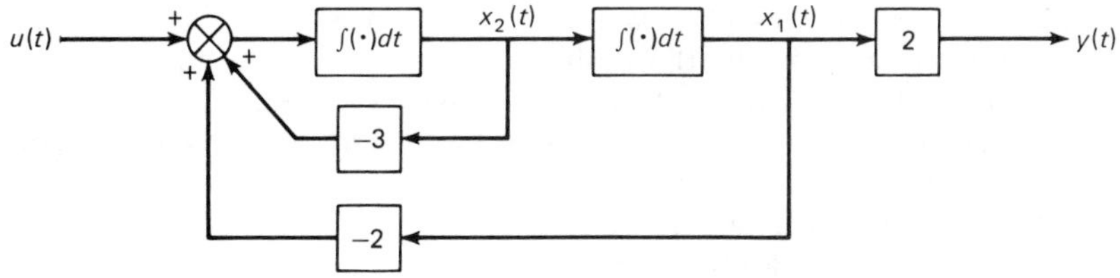

Figure P5-3.

5.24. Design a reduced-order observer for the system of Problem 5.23 and sketch the block diagram.

5.25. For the system of Problem 5.24, determine the frequency domain representation for the closed-loop system and sketch the block diagram.

5.26. A third-order system is described by the block diagram of Figure P5-4.

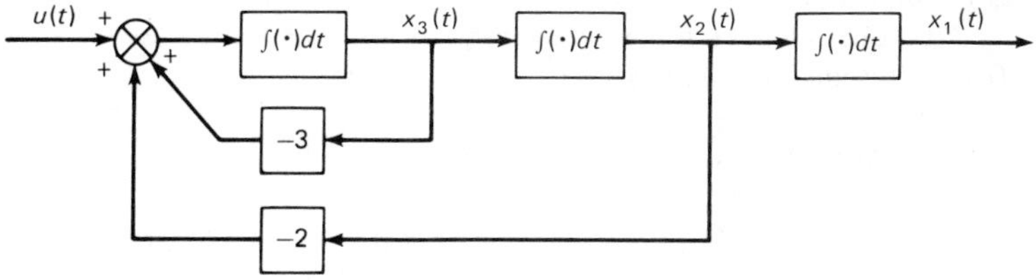

Figure P5-4.

(a) Find the eigenvalues of this system and sketch the response of the system to a step input.

(b) Determine the state variable feedback gains necessary to have a closed-loop response characterized by eigenvalues $-1 \pm i$, and -2. Sketch the response of the closed loop system to a step input.

5.27. Design a third-order observer for the system of Problem 5.26 for the case in which the measurement is $x_1(t)$.

5.28. Consider the second-order system described by the block diagram of Figure P5-5.

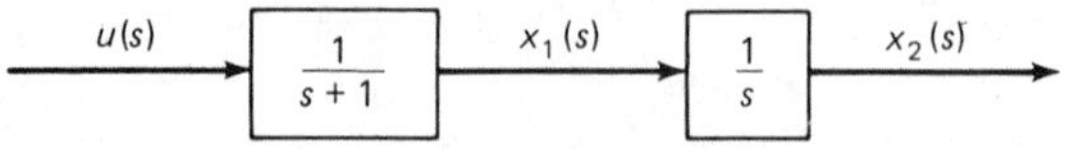

Figure P5-5.

(a) Determine the discrete time representation of this system for time interval $(t_l - t_{l-1}) = 1$ sec.

(b) Design a two-state observer for this system. Sketch the block diagram of the system and observer.

5.29. For the discrete-time system of Problem 5.28, design the reduced-order observer. Sketch the block diagram of the system and observer.

5.30. A radar system tracks targets in north and east coordinates. From radar position measurements, the position and velocity of the target are to be determined. North and east channels are independent.

(a) Design a continuous time observer with critically damped response and time constant of 5 seconds.

(b) For a time interval $(t_l - t_{l-1})$ of $\frac{1}{2}$ second, design a discrete time observer with critically damped response.

5.31. A spring–mass–damper system has the following characteristics:

spring constant: 8; mass: 2; damping constant: 800.

(a) Find the eigenvalues of the system matrix of this system. What time constant is implied?

(b) Determine the discrete time equations for time interval $(t_l - t_{l-1}) =$ 0.05 sec.

(c) Determine the discrete time state variable feedback gains to produce response characterized by eigenvalues $0.5 \pm i0.2$.

(d) If the measurement is $y(l)$ = displacement, design a second-order observer for this system.

5.32. An aircraft attitude control system is represented by the equations

$$\dot{x}_1(t) = x_2(t)$$

$$\dot{x}_2(t) = u(t).$$

(a) Determine the discrete-time representation of this system with interval $(t_l - t_{l-1}) = 0.1$ sec.
(b) Find the state variable feedback gains to produce closed-loop response characterized by multiple eigenvalues with value 0.8.
(c) If the measurement is $y(l) = x_1(l)$, design a second-order observer for this system. Sketch the block diagram of system, controller, and observer.

5.33. For the discrete time system of Problem 5.32, design a reduced-order observer. Sketch the system, controller, and observer.

5.34. A second-order system is described by the equations

$$\dot{x}_1(t) = -2x_1(t) + 2x_2(t)$$

$$\dot{x}_2(t) = 0.5x_1(t) - 0.75x_2(t) + 0.5u(t).$$

(a) Find the discrete time version of these equations for time interval $(t_l - t_{l-1}) = \frac{1}{4}$ sec.
(b) Find the eigenvalues of the discrete time system and discuss their implications.
(c) Determine the state variable feedback gains necessary for the closed-loop discrete time system to exhibit response characterized by eigenvalues $\frac{1}{2} \pm i\frac{1}{10}$.
(d) If the measurement is $y(l) = x(l)$, design a second-order observer for this system. Sketch the block diagram of the system, controller, and observer.

5.35. For the discrete time system of Problem 5.34, design a reduced-order observer. Sketch the system, controller, and observer.

Chapter 6
Optimal Control of Dynamic Systems

6.1. APPLICATION OF THE CALCULUS OF VARIATIONS

In Chapter 4, the concept of the state of a dynamic system was introduced, and characteristics of the solution of the resulting first-order, coupled differential equations were discussed. In Chapter 5, these concepts were applied to the specification of feedback control systems with certain desired performance characteristics. Here, optimization of system performance characteristics is investigated. The approach to be used will be general, in that it will be applicable to any system in which the dynamic behavior is governed by a set of state equations in the standard form

$$\dot{\boldsymbol{x}}(t) = \boldsymbol{f}(\boldsymbol{x}(t), \boldsymbol{u}(t), t). \tag{6.1}$$

in which $\boldsymbol{f}$ is, in general, a nonlinear function of the system state vector $\boldsymbol{x}(t)$, the control function $\boldsymbol{u}(t)$, and perhaps explicitly a function of time. The dimension if this equation is N, the number of state variables, and the dimension of $\boldsymbol{u}(t)$ is P, the number of control inputs. Specifically, the optimization techniques afforded by the calculus of variations, as reviewed in Appendix B, will be applied to the general system described by Equation (6.1). The purpose is to solve a very general optimization problem in very general terms, with the intent that any specific problem will fall within this general category. When one considers the state of a dynamic system and the general kind of optimization problems which he may be required to solve, a functional of the general form

$$J = \phi[\boldsymbol{x}(t_f), t_f] + \int_{t_0}^{t_f} L(\boldsymbol{x}(t), \boldsymbol{u}(t), t)\, dt \tag{6.2}$$

would seem appropriate to include most practical situations. The scalar functional J, termed the *performance index*, is a function of the final state $\boldsymbol{x}(t_f)$, of the final time t_f, and of the integral of some function of the state $\boldsymbol{x}(t)$, the control $\boldsymbol{u}(t)$, and, perhaps, time. It can be seen that this form is quite adequate for trajectory problems, where ϕ would represent some function of the terminal conditions and L would perhaps represent the fuel flow rate, or for some social science problem where ϕ may represent the total number of low-income housing units built and L the rate of displacement of families from homes caused by the construction schedules. Proper selection of parameters in ϕ and L give proper weighting to the two terms.

The objective is to maximize or minimize J by the proper selection of the control function $\boldsymbol{u}(t)$. Definition of $\boldsymbol{u}(t)$, together with a proper set of boundary conditions for $\boldsymbol{x}(t)$, suffices to define $\boldsymbol{x}(t)$ for all time. There are several ramifications of the problem, however, depending upon how the boundary conditions are specified, whether the terminal time t_f is fixed or can itself be varied in the optimization process, and what constraints are placed on the control function $\boldsymbol{u}(t)$. This problem falls within the general framework of the calculus of variations as discussed in Appendix B. Since this is a general problem with practical application, it has developed its own terminology and special forms. For this reason, the pertinent equations will be derived in their entirety here, and the development will not rely on the results of the constrained optimization problem of the calculus of variations treated in Appendix B.

6.1.1. Free Terminal Conditions with Fixed Terminal Time

This case is the simplest, since it involves a fixed time interval and the only boundary conditions on $\boldsymbol{x}(t)$ are the initial conditions. The objective is to determine $\boldsymbol{u}(t)$ such that the functional

$$J = \phi[\boldsymbol{x}(t_f)] + \int_{t_0}^{t_f} L(\boldsymbol{x}(t), \boldsymbol{u}(t), t)\, dt \tag{6.3}$$

is optimized, subject to the constraint

$$\dot{\boldsymbol{x}}(t) = \boldsymbol{f}(\boldsymbol{x}(t), \boldsymbol{u}(t), t) \tag{6.4}$$

where $\boldsymbol{x}(t_0)$ is specified and $\boldsymbol{x}(t_f)$ is completely free. The basic approach is to adjoin the constraint with Lagrange multiplier functions $\boldsymbol{\lambda}(t)$, to yield

$$\bar{J} = \phi[\boldsymbol{x}(t_f)] + \int_{t_0}^{t_f} \{L(\boldsymbol{x}, \boldsymbol{u}, t) + \boldsymbol{\lambda}^T(t)[\boldsymbol{f}(\boldsymbol{x}, \boldsymbol{u}, t) - \dot{\boldsymbol{x}}]\}\, dt. \tag{6.5}$$

Before proceeding, it is necessary to define a new function

$$H(\boldsymbol{x}(t), \boldsymbol{u}(t), \boldsymbol{\lambda}(t), t) = L(\boldsymbol{x}, \boldsymbol{u}, t) + \boldsymbol{\lambda}^T(t)\boldsymbol{f}(\boldsymbol{x}, \boldsymbol{u}, t) \tag{6.6}$$

which, when substituted into Equation (6.5), yields

$$\bar{J} = \phi[\boldsymbol{x}(t_f)] + \int_{t_0}^{t_f} [H(\boldsymbol{x}, \boldsymbol{u}, \boldsymbol{\lambda}, t) - \boldsymbol{\lambda}^T \dot{\boldsymbol{x}}] \, dt. \tag{6.7}$$

The function $H(\boldsymbol{x}, \boldsymbol{u}, \boldsymbol{\lambda}, t)$ is called the *Hamiltonian function* and will play an important role in the developments to follow. In accordance with the general procedures of the calculus of variations, the change in $\bar{J}$ due to variations $\delta\boldsymbol{\lambda}(t)$ and $\delta\boldsymbol{u}(t)$, and to the resulting variations $\delta\dot{\boldsymbol{x}}(t)$ and $\delta\boldsymbol{x}(t)$, should be determined and equated to zero. Since the $\delta\boldsymbol{\lambda}(t)$ and $\delta\boldsymbol{u}(t)$ variations are independent, and since the variation $\delta\boldsymbol{\lambda}(t)$ obviously results in the term

$$\int_{t_0}^{t_f} \delta\boldsymbol{\lambda}^T [\boldsymbol{f}(\boldsymbol{x}, \boldsymbol{u}, t) - \dot{\boldsymbol{x}}] \, dt \tag{6.8}$$

which must be equated to zero, this variation results in a restatement of the constraint of Equation (6.4). Then, the change in $\bar{J}$ due to the variation $\delta\boldsymbol{u}(t)$ and the resulting $\delta\dot{\boldsymbol{x}}(t)$ and $\delta\boldsymbol{x}(t)$ variations can be considered independently. Thus, the change in $\bar{J}$ due to these variations is

$$\Delta\bar{J} = \frac{\partial\phi}{\partial\boldsymbol{x}} \delta\boldsymbol{x}\bigg|_{t=t_f} + \int_{t_0}^{t_f} \left[\frac{\partial H}{\partial\boldsymbol{x}} \delta\boldsymbol{x} + \frac{\partial H}{\partial\boldsymbol{u}} \delta\boldsymbol{u} - \boldsymbol{\lambda}^T \delta\dot{\boldsymbol{x}}\right] dt. \tag{6.9}$$

where the partial derivative terms are defined in terms of the formalism developed in Appendix A. The last term can be integrated by parts by making a change of variable

$$\int_{t_0}^{t_f} \boldsymbol{\lambda}^T \delta\dot{\boldsymbol{x}} \, dt = \int_{t_0}^{t_f} \boldsymbol{\lambda}^T \frac{d(\delta\boldsymbol{x})}{dt} \, dt = \int_{t=t_0}^{t=t_f} \boldsymbol{\lambda}^T \, d(\delta\boldsymbol{x})$$

$$= \boldsymbol{\lambda}^T \delta\boldsymbol{x}\bigg|_{t=t_0}^{t=t_f} - \int_{t=t_0}^{t=t_f} d\boldsymbol{\lambda}^T \, \delta\boldsymbol{x}. \tag{6.10}$$

The integral part of this expression can be changed back to an integral over t by multiplying by dt/dt, with the result

$$\int_{t_0}^{t_f} \boldsymbol{\lambda}^T \delta\dot{\boldsymbol{x}} \, dt = \boldsymbol{\lambda}^T \delta\boldsymbol{x}\bigg|_{t=t_0}^{t=t_f} - \int_{t_0}^{t_f} \dot{\boldsymbol{\lambda}}^T \delta\boldsymbol{x} \, dt. \tag{6.11}$$

Now, substituting this result into Equation (6.9) and grouping terms yield

$$\Delta \bar{J} = \left[\frac{\partial \phi}{\partial \boldsymbol{x}} - \boldsymbol{\lambda}^T\right]\delta \boldsymbol{x}\Big|_{t=t_f} + \int_{t_0}^{t_f}\left\{\left[\frac{\partial H}{\partial \boldsymbol{x}} + \dot{\boldsymbol{\lambda}}^T\right]\delta \boldsymbol{x} + \frac{\partial H}{\partial \boldsymbol{u}}\,\delta \boldsymbol{u}\right\}dt + \boldsymbol{\lambda}^T\delta \boldsymbol{x}\Big|_{t=t_0}. \tag{6.12}$$

The necessary condition for a minimum or maximum is that $\Delta \bar{J}$ vanish for arbitrary choices of $\delta \boldsymbol{x}(t)$ and $\delta \boldsymbol{u}(t)$, requiring in turn that the coefficients of these terms in the above equation vanish. This requirement yields the following set of equations.

$$\text{I.}\quad \dot{\boldsymbol{\lambda}}^T = -\frac{\partial H}{\partial \boldsymbol{x}}, \qquad \text{II.}\quad \frac{\partial H}{\partial \boldsymbol{u}} = 0,$$

$$\text{III.}\quad \boldsymbol{\lambda}^T\delta \boldsymbol{x}\big|_{t=t_0} = 0, \qquad \text{IV.}\quad \boldsymbol{\lambda}^T(t_f) = \frac{\partial \phi}{\partial \boldsymbol{x}}. \tag{6.13}$$

The additional necessary condition which results from the expression of Equation (6.8) is the original constraint equation

$$\text{V.}\quad \dot{\boldsymbol{x}} = \boldsymbol{f}(\boldsymbol{x}, \boldsymbol{u}, t).$$

These results constitute a set of $2N$ differential equations in $\boldsymbol{\lambda}(t)$ and $\boldsymbol{x}(t)$ (I and V), a set of M algebraic equations which can be used to eliminate $\boldsymbol{u}(t)$ from the differential equations (II), and the required set of boundary conditions (III and IV). In a typical problem, the initial condition $\boldsymbol{x}(t_0)$ will be specified, in which case III above is automatically satisfied and yields no information, and the remaining N boundary conditions are obtained from IV, which yields the terminal values $\boldsymbol{\lambda}(t_f)$. The result is a two-point boundary value problem, which can be extremely difficult to solve, especially when the differential equations are nonlinear. In fact, in most cases iterative numerical solutions are required.

As an example, consider the scalar problem treated in Appendix B, that of minimizing the performance index

$$J = \int_0^1 (u^2 + x)\,dt = \int_0^1 L(x, u)\,dt$$

subject to the constraint

$$\dot{x} = u = f(x, u); \qquad x(0) = 2.$$

The terminal constraint term $\phi[x(t_f)]$ is zero, and the Hamiltonian function is

$$H(x, u, \lambda) = L(x, u) + \lambda f(x, u)$$
$$= u^2 + x + \lambda u.$$

The necessary conditions are

$$\text{I.} \quad \dot{\lambda} = -\frac{\partial H}{\partial x} = -1; \qquad \text{II.} \quad \frac{\partial H}{\partial u} = 2u + \lambda = 0$$

$$\text{III. satisfied;} \qquad \text{IV.} \quad \lambda(1) = \frac{\partial \phi}{\partial x} = 0.$$

The two first-order differential equations are

$$\dot{x} = -\lambda/2$$
$$\dot{\lambda} = -1$$

with boundary conditions

$$x(0) = 2$$
$$\lambda(1) = 0.$$

In this simple case, the differential equations are linear and coupled in one direction only. Solution yields

$$\lambda = 1 - t$$
$$x = \frac{t^2}{4} - \frac{t}{2} + 2.$$

That this approach through the Hamiltonian function is no different from the straightforward calculus of variations approach can be seen by comparing this development with that in Appendix B. The introduction of the Hamiltonian function provides more concise treatment of the necessary conditions, and, as will be shown later, this function has an even more fundamental role in the optimization process.

6.1.2. Terminal Constraints with Fixed Terminal Time

The previous problem is of limited practical importance because the terminal conditions are completely free. A somewhat more interesting case arises when the terminal conditions are constrained in some way. Specifically, consider the problem of minimizing or maximizing the functional

$$J = \phi[\boldsymbol{x}(t_f)] + \int_{t_0}^{t_f} L(\boldsymbol{x}(t), \boldsymbol{u}(t), t)\, dt$$

subject to (6.14)

$$\dot{\boldsymbol{x}}(t) = \boldsymbol{f}(\boldsymbol{x}(t), \boldsymbol{u}(t), t)$$

and the additional terminal constraint

$$\boldsymbol{\psi}[\boldsymbol{x}(t_f)] = 0.$$

Here, $\boldsymbol{\psi}$ is a q-dimensional vector function representing q functions of the terminal state $\boldsymbol{x}(t_f)$, which are constrained to be zero. Note that this general case includes the situation in which the states themselves are constrained at the terminal time. The method used to account for this additional constraint is to adjoin the constraint to the performance index with a set of Lagrange multipliers represented by the vector $\boldsymbol{\nu}$:

$$\bar{J} = \phi[\boldsymbol{x}(t_f)] + \boldsymbol{\nu}^T \boldsymbol{\psi}[\boldsymbol{x}(t_f)] + \int_{t_0}^{t_f} [H(\boldsymbol{x}, \boldsymbol{u}, \boldsymbol{\lambda}, t) - \boldsymbol{\lambda}^T \dot{\boldsymbol{x}}]\, dt \qquad (6.15)$$

where the Hamiltonian function $H(\boldsymbol{x}, \boldsymbol{u}, \boldsymbol{\lambda}, t)$ has been introduced as in previous developments. Since the constraints of Equation (6.14) are to be met only at the single point $t = t_f$, they do not appear within the integral. The change in $\bar{J}$ due to variations $\delta\boldsymbol{u}(t)$ and $\delta\boldsymbol{x}(t)$ is obtained as in the previous section, with the additional term

$$\left.\frac{\partial \boldsymbol{\psi}}{\partial \boldsymbol{x}}\, \delta\boldsymbol{x}\right|_{t_f}$$

produced because of the added constraint. Again grouping terms yields

$$\Delta\bar{J} = \left[\frac{\partial \phi}{\partial \boldsymbol{x}} - \boldsymbol{\lambda}^T + \boldsymbol{\nu}^T \frac{\partial \boldsymbol{\Psi}}{\partial \boldsymbol{x}}\right]_{t=t_f} \delta\boldsymbol{x}(t_f)$$
$$+ \int_{t_0}^{t_f} \left\{\left[\frac{\partial H}{\partial x} + \dot{\lambda}^T\right] \delta\boldsymbol{x} + \frac{\partial H}{\partial \boldsymbol{u}}\, \delta\boldsymbol{u}\right\} dt - \left.\lambda^T \delta\boldsymbol{x}\right|_{t=t_0}. \qquad (6.16)$$

The necessary conditions derived by equating the various coefficients to zero are identical to those of Equation (6.13), with the exception

$$\text{IV.}\quad \boldsymbol{\lambda}^T(t_f) = \left[\frac{\partial \phi}{\partial \boldsymbol{x}} + \boldsymbol{\nu}^T \frac{\partial \boldsymbol{\psi}}{\partial \boldsymbol{x}}\right]_{t=t_f}. \tag{6.17}$$

It can be seen that the terminal constraints affect only the boundary conditions for the Lagrange multiplier function. The remaining equations are unchanged.

In many cases of interest, the terminal constraints will be in the form of specification of some of the states themselves, rather than in the more general form of Equation (6.14). In this event, those states which are specified at the terminal time will not appear in the $\phi[\boldsymbol{x}(t_f)]$ function, and those states which do not appear in this function will not be specified at the terminal time. For simplicity, let the states be ordered such that

$$\left.\begin{matrix} x_1(t_f) \\ x_2(t_f) \\ \vdots \\ x_q(t_f) \end{matrix}\right\}\text{are specified}$$

$$\left.\begin{matrix} x_{q+1}(t_f) \\ x_{q+2}(t_f) \\ \vdots \\ x_N(t_f) \end{matrix}\right\}\text{are free.}$$

Then, the $\boldsymbol{\psi}[\boldsymbol{x}(t_f)]$ vector-valued function is

$$\boldsymbol{\psi}[\boldsymbol{x}(t_f)] = \begin{bmatrix} x_1(t_f) - a_1 \\ x_2(t_f) - a_2 \\ \vdots \\ x_q(t_f) - a_q \end{bmatrix} = 0 \tag{6.18}$$

where $a_1, a_2, \ldots, a_q$ are the specified values of $x_1(t_f), x_2(t_f), \ldots, x_q(t_f)$. Then, the matrix

$$\frac{\partial \boldsymbol{\psi}}{\partial \boldsymbol{x}} = \begin{bmatrix} \frac{\partial \psi_1}{\partial x_1} & \frac{\partial \psi_1}{\partial x_2} & \cdots & \frac{\partial \psi_1}{\partial x_N} \\ \frac{\partial \psi_2}{\partial x_1} & \frac{\partial \psi_2}{\partial x_2} & \cdots & \frac{\partial \psi_2}{\partial x_N} \\ \cdots & \cdots & \cdots & \cdots \\ \frac{\partial \psi_q}{\partial x_1} & \frac{\partial \psi_q}{\partial x_2} & \cdots & \frac{\partial \psi_q}{\partial x_N} \end{bmatrix} = \begin{bmatrix} 1 & 0 & 0 & \cdots & 0 & \cdots & 0 \\ 0 & 1 & 0 & \cdots & 0 & \cdots & 0 \\ \cdots & \cdots & \cdots & \cdots & \cdots & \cdots & \cdots \\ 0 & 0 & 0 & \cdots & 1 & \cdots & 0 \end{bmatrix}. \tag{6.19}$$

The product term of Equation (6.17) is then determined to be

$$\boldsymbol{\nu}^T \frac{\partial \boldsymbol{\psi}}{\partial \boldsymbol{x}} = [\nu_1 \quad \nu_2 \quad \cdots \quad \nu_q \quad 0 \quad 0 \quad \cdots \quad 0] \tag{6.20}$$

and the remaining term is

$$\frac{\partial \phi}{\partial \boldsymbol{x}} = \left[0 \quad 0 \quad \cdots \quad 0 \quad \frac{\partial \phi}{\partial x_{q+1}} \quad \frac{\partial \phi}{\partial x_{q+2}} \quad \cdots \quad \frac{\partial \phi}{\partial x_N}\right] \tag{6.21}$$

since those states which appear in $\boldsymbol{\psi}[\boldsymbol{x}(t_f)]$ do not appear in $\phi[\boldsymbol{x}(t_f)]$. Combining Equations (6.20) and (6.21) to generate the boundary conditions $\boldsymbol{\lambda}(t_f)$ yields

$$\boldsymbol{\lambda}^T(t_f) = \left[\nu_1 \quad \nu_2 \quad \cdots \quad \nu_q \quad \frac{\partial \phi}{\partial x_{q+1}} \quad \frac{\partial \phi}{\partial x_{q+2}} \quad \cdots \quad \frac{\partial \phi}{\partial x_N}\right]. \tag{6.22}$$

Some reflection on this development will show that the general result is not dependent on the order in which the states are numbered, with the general conclusion that the boundary conditions for the Lagrange multiplier functions are given by

$$\begin{aligned} \lambda_i(t_f) &= \nu_i, && \text{if } x_i(t_f) \text{ is specified} \\ \lambda_i(t_f) &= \frac{\partial \phi}{\partial x_i}, && \text{if } x_i(t_f) \text{ is free.} \end{aligned} \tag{6.23}$$

It should be emphasized that this result is a special case of the more general situation for which Equation (6.17) applies.

As an example of these principles, consider the idealized orbital injection problem in two dimensions as described in Figure 6-1. The vehicle of mass m is

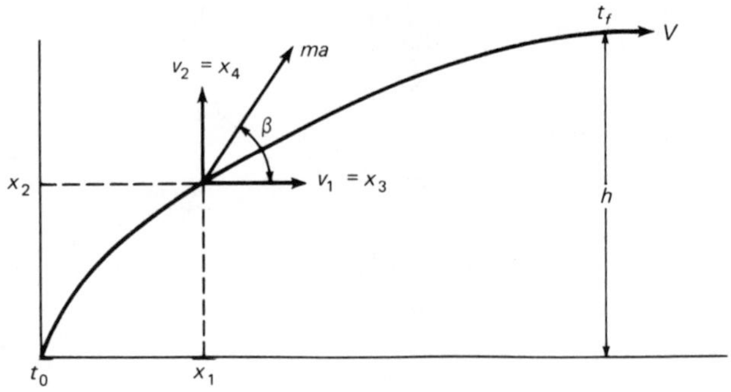

Figure 6-1. Idealized orbit injection.

acted upon by thrust ma at angle β as shown. The states are the two position coordinates x_1 and x_2, and two velocity coordinates x_3 and x_4. The control $u(t)$, in this case, is the scalar $\beta(t)$. The problem is to determine the control law equation for $\beta(t)$ required to bring the vehicle from rest at the origin to a path parallel to the x_1 axis, at a specified height h, and with maximum final velocity.

This problem is amenable to direct solution using the general principles of the calculus of variations. The general problem, however, has already been solved, and if this particular problem can be formulated in such a way that it takes the general form, the necessary conditions for an optimum solution can be immediately stated. The general problem was outlined by Equations (6.3) and (6.4), and the analogous equations here are

$$J = x_3(t_f)$$

and

$$\dot{x}_1 = x_3$$

$$\dot{x}_2 = x_4$$

$$\dot{x}_3 = a \cos \beta$$

$$\dot{x}_4 = a \sin \beta.$$

Then, the functional J is defined by the quantities

$$\phi[x(t_f)] = x_3(t_f)$$

$$L(x, u, t) = 0.$$

The boundary conditions for the problem are

$$x_1(t_0) = x_2(t_0) = x_3(t_0) = x_4(t_0) = 0$$

$$x_1(t_f) \text{ free}; \quad x_2(t_f) = h; \quad x_3(t_f) \text{ free}; \quad x_4(t_f) = 0.$$

Since this problem has been put into the standard form, the general results of Equation (6.13), as amended by Equation (6.17), can be used. The Hamiltonian function is

$$H = L + \lambda^T f = 0 + [\lambda_1 \lambda_2 \lambda_3 \lambda_4] \begin{bmatrix} x_3 \\ x_4 \\ a \cos \beta \\ a \sin \beta \end{bmatrix}$$

$$= \lambda_1 x_3 + \lambda_2 x_4 + \lambda_3 (a \cos \beta) + \lambda_4 (a \sin \beta)$$

and the necessary conditions are

$$\dot{\lambda}^T = -\frac{\partial H}{\partial \boldsymbol{x}}: \quad \dot{\lambda}_1 = -\frac{\partial H}{\partial x_1}; \quad \dot{\lambda}_2 = -\frac{\partial H}{\partial x_2}; \quad \dot{\lambda}_3 = -\frac{\partial H}{\partial x_3}; \quad \dot{\lambda}_4 = -\frac{\partial H}{\partial x_4}$$

or

$$\dot{\lambda}_1 = 0; \quad \dot{\lambda}_2 = 0; \quad \dot{\lambda}_3 = -\lambda_1; \quad \dot{\lambda}_4 = -\lambda_2.$$

This set of first-order differential equations must be solved simultaneously with the four state equations. The remaining necessary condition is

$$\partial H / \partial \boldsymbol{u} = 0: \quad -\lambda_3 (a \sin \beta) + \lambda_4 (a \cos \beta) = 0$$

or

$$\tan \beta = \lambda_3 / \lambda_4.$$

This algebraic equation can be used to eliminate β from the differential equations. Boundary conditions are given by

$$x_1(t_0) = 0, \quad \lambda_1(t_f) = \frac{\partial \phi}{\partial x_1} = 0$$

$$x_2(t_0) = 0, \quad \lambda_2(t_f) = \nu_2$$

$$x_3(t_0) = 0, \qquad \lambda_3(t_f) = \frac{\partial \phi}{\partial x_3} = 1$$

$$x_4(t_0) = 0, \qquad \lambda_4(t_f) = \nu_4$$

where the particular form results from the fact that x_2 and x_4 are specified at the final time, while x_1 and x_3 are free.

In this particular case, solution of the $\boldsymbol{\lambda}$ equations is simple because they are not coupled with the $\boldsymbol{x}$ equations. The solution of the $\boldsymbol{\lambda}$ equations, with application of the boundary conditions above, yields

$$\lambda_1(t) = 0; \quad \lambda_2(t) = \nu_2; \quad \lambda_3(t) = 1; \quad \lambda_4(t) = \nu_2(t_f - t) + \nu_4.$$

Then substitution of these results into the control equation yields the control law

$$\tan \beta = \nu_2(t_f - t) + \nu_4.$$

Note that this result still contains two unknown constants ν_2 and ν_4. To see how these constants are evaluated, it is necessary to recall that the $\boldsymbol{x}$ and $\boldsymbol{\lambda}$ equations must be solved simultaneously, and that only fortuitous circumstances have permitted the independent solution of the λ equations. To solve the $\boldsymbol{x}$ equations, the control law must be substituted into the $\boldsymbol{x}$ equations, and this result then solved for the terminal state $\boldsymbol{x}(t_f)$, still in terms of the constants ν_2 and ν_4. Proper matching in the terminal values $x_2(t_f)$ and $x_4(t_f)$ will yield two equations which can then be solved for ν_2 and ν_4. In particular, the set of equations

$$\dot{x}_1 = x_3$$

$$\dot{x}_2 = x_4$$

$$\dot{x}_3 = a \cos\left[\arctan\{\nu_2(t_f - t) + \nu_4\}\right]$$

$$\dot{x}_4 = a \sin\left[\arctan\{\nu_2(t_f - t) + \nu_4\}\right]$$

must be solved for $\boldsymbol{x}(t_f)$, in terms of ν_2 and ν_4. Even in this simple case the equations are nonlinear in the control function $\beta(t)$. Proper matching of the boundary conditions requires iteration techniques characteristic of the two-point boundary value problem.

6.1.3. Terminal Constraints with Free Terminal Time: The General Case

To this point, only those optimization problems with the final time t_f specified have been considered. In many applications, the extra degree of freedom afforded by a variable final time can result in a more favorable value of the performance index J. To maintain the generality of the result the following problem will be investigated. It is desired to maximize or minimize the performance index

$$J = \phi[\boldsymbol{x}(t_f), t_f] + \int_{t_0}^{t_f} L(\boldsymbol{x}(t), \boldsymbol{u}(t), t)\, dt \tag{6.24}$$

subject to the differential constraint equation

$$\dot{\boldsymbol{x}}(t) = \boldsymbol{f}(\boldsymbol{x}(t), \boldsymbol{u}(t), t)$$

and the terminal constraint

$$\boldsymbol{\psi}[\boldsymbol{x}(t_f)] = 0$$

where now the terminal time t_f is no longer specified. Note that t_f now appears explicitly in the ϕ term.

Before considering the general solution to this problem, it must be recognized that the total change in a given state x_i at the terminal time t_f results from two effects—one due to the variation $\delta x_i(t)$ and the other due to the change in the terminal time dt_f, as shown in Figure 6-2. It can be seen that this total change $dx_i(t_f)$ is given by

$$dx_i(t_f) = \delta x_i(t_f) + \dot{x}_i(t_f)\, dt_f \tag{6.25}$$

where the second term is the change due to the change in the final time. With this result, the general optimization problem can be examined. As before, the

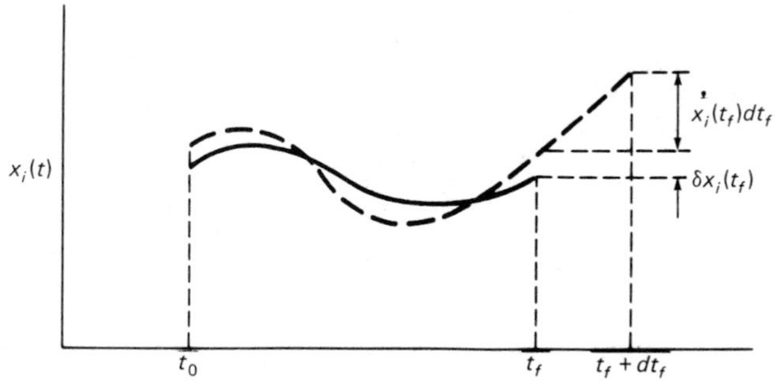

Figure 6-2. Total change in $x_i(t_f)$.

constraints are adjoined to the performance index to yield

$$\bar{J} = \phi[\boldsymbol{x}(t_f), t_f] + \boldsymbol{\nu}^T \boldsymbol{\psi}[\boldsymbol{x}(t_f)] + \int_{t_0}^{t_f} [H(\boldsymbol{x}(t), \boldsymbol{u}(t), t) - \boldsymbol{\lambda}^T(t)\dot{\boldsymbol{x}}]\, dt \quad (6.26)$$

where the Hamiltonian function H has been used in the manner prescribed by Equation (6.6). Now, the change in $\bar{J}$ must be determined for arbitrary variations $\delta\boldsymbol{u}(t)$, the resulting variations $\delta\boldsymbol{x}(t)$ and $\delta\dot{\boldsymbol{x}}(t)$, and an arbitrary change dt_f in the terminal time. Following the general procedures of the previous section yields

$$\Delta\bar{J} = \left[\frac{\partial\phi}{\partial\boldsymbol{x}} + \boldsymbol{\nu}^T \frac{\partial\boldsymbol{\psi}}{\partial\boldsymbol{x}}\right] d\boldsymbol{x}(t_f) - \boldsymbol{\lambda}^T\delta\boldsymbol{x}\Big|_{t=t_f} + \int_{t_0}^{t_f} \left\{\left[\frac{\partial H}{\partial\boldsymbol{x}} + \dot{\boldsymbol{\lambda}}^T\right]\delta\boldsymbol{x} + \frac{\partial H}{\partial\boldsymbol{u}}\delta\boldsymbol{u}\right\} dt$$

$$+ \boldsymbol{\lambda}^T\delta\boldsymbol{x}\Big|_{t=t_0} + \frac{d}{dt_f}\left[\int_{t_0}^{t_f} (H - \boldsymbol{\lambda}^T\dot{\boldsymbol{x}})\, dt\right] dt_f \quad (6.27)$$

The first four terms are identical to the previous result of Equation (6.16), with the exception of $d\boldsymbol{x}(t_f)$ replacing $\delta\boldsymbol{x}(t_f)$ in the first term. The last term is the change in $\bar{J}$ due to the change in the terminal time t_f and can be evaluated by Leibniz' rule to yield

$$[H - \boldsymbol{\lambda}^T\dot{\boldsymbol{x}}]_{t=t_f}\, dt_f.$$

Now, by replacing $\delta\boldsymbol{x}(t_f)$ in the second term with the appropriate expression from Equation (6.25), and then combining this result with the expression above, the $\dot{\boldsymbol{x}}$ term can be eliminated. The result, after combining terms, is

$$\Delta\bar{J} = \left[\frac{\partial\phi}{\partial\boldsymbol{x}} + \boldsymbol{\nu}^T \frac{\partial\boldsymbol{\psi}}{\partial\boldsymbol{x}} - \boldsymbol{\lambda}^T\right]_{t=t_f} d\boldsymbol{x}(t_f) + \int_{t_0}^{t_f} \left\{\left[\frac{\partial H}{\partial\boldsymbol{x}} + \dot{\boldsymbol{\lambda}}^T\right]\delta\boldsymbol{x} + \frac{\partial H}{\partial\boldsymbol{u}}\delta\boldsymbol{u}\right\} dt$$

$$+ \left[H + \frac{\partial\phi}{\partial t}\right]_{t=t_f} dt_f + \boldsymbol{\lambda}^T\delta\boldsymbol{x}\Big|_{t=t_0}. \quad (6.28)$$

By equating this expression to zero, the necessary conditions for minimization or maximization are obtained. It can be seen that the result is identical to that obtained previously, with one additional condition given by VI below:

$$\text{I.}\quad \dot{\boldsymbol{\lambda}}^T = -\frac{\partial H}{\partial \boldsymbol{x}} \qquad \text{II.}\quad \frac{\partial H}{\partial \boldsymbol{u}} = 0$$

$$\text{III.}\quad \boldsymbol{\lambda}^T \delta \boldsymbol{x}\big|_{t=t_0} = 0 \qquad \text{IV.}\quad \boldsymbol{\lambda}^T(t_f) = \left[\frac{\partial \phi}{\partial \boldsymbol{x}} + \boldsymbol{\nu}^T \frac{\partial \boldsymbol{\psi}}{\partial \boldsymbol{x}}\right]_{t=t_f}$$

$$\text{V.}\quad \dot{\boldsymbol{x}} = \boldsymbol{f}(\boldsymbol{x}, \boldsymbol{u}, t) \qquad \text{VI.}\quad \left[H + \frac{\partial \phi}{\partial t}\right]_{t=t_f} dt_f = 0. \tag{6.29}$$

These results represent the most general case. The additional condition represented by VI above is called the *transversality condition* and yields a scalar equation which is used to solve for the one additional unknown, t_f. If t_f is fixed, then dt_f is identically zero, and condition VI yields no information.

These results can be used to solve the classical minimum time problem, as represented by the following example. Suppose that, in the situation described by Figure 6-1, it is desired to bring the vehicle from a state of rest at the origin to a specified final altitude h with a specified final velocity V_f, and it is further desired to accomplish this task in minimum time $(t_f - t_0)$. Again, this specific problem can be expressed in terms of the general optimization problem described by Equation (6.24) and the subsequent discussion. This can be done in one of two ways,

$$\phi = 0, \qquad L = 1$$

or

$$\phi = t_f - t_0, \qquad L = 0$$

either of which results in

$$J = t_f - t_0$$

as required. Application of the necessary conditions results in the same set of equations as those obtained in the previous example,

$$\dot{\lambda}_1 = 0; \quad \dot{\lambda}_2 = 0; \quad \dot{\lambda}_3 = -\lambda_1; \quad \dot{\lambda}_4 = -\lambda_2; \quad \tan\beta = \lambda_4/\lambda_3.$$

Now, however, the boundary conditions are

$$x_1(t_0) = 0; \quad \lambda_1(t_f) = \frac{\partial \phi}{\partial x_1} = 0$$

$$x_2(t_0) = 0; \quad \lambda_2(t_f) = \nu_2$$

$$x_3(t_0) = 0; \quad \lambda_3(t_f) = \nu_3$$

$$x_4(t_0) = 0; \quad \lambda_4(t_f) = \nu_4$$

since the $x_2(t_f)$, $x_3(t_f)$, and $x_4(t_f)$ are specified. Solution of these equations with the specified boundary conditions yields

$$\tan \beta = \frac{\lambda_4}{\lambda_3} = \frac{\nu_2(t_f - t) + \nu_4}{\nu_3}.$$

Again, substitution of this expression into the state equations and subsequent solution (by numerical methods) will yield three equations in three unknowns, ν_2, ν_3, ν_4, when the boundary conditions are evaluated. The remaining unknown is t_f, the terminal time, and the required additional equation is obtained from the transversality condition VI of Equation (6.29):

$$\left[\frac{\partial \phi}{\partial t} + H\right]_{t=t_f} = 0 = H\big|_{t=t_f}$$

or, upon substitution,

$$1 + \nu_3 a(t_f) \cos \beta(t_f) + \nu_4 a(t_f) \sin \beta(t_f) = 0.$$

By use of the expression

$$\tan \beta(t_f) = \nu_3/\nu_4$$

this equation can be reduced to

$$\cos \beta(t_f) + \nu_3 a(t_f) = 0.$$

This equation and the remaining boundary conditions

$$x_2(t_f) = h$$

$$x_3(t_f) = V$$

$$x_4(t_f) = 0$$

constitute the four simultaneous equations which must be solved for ν_2, ν_3, ν_4, and t_f.

As a final note, it should be mentioned that, if the Hamiltonian function H is not an explicit function of time—that is

$$\partial H/\partial t = 0 \tag{6.30}$$

—then it can be shown that H must be constant everywhere along the optimal path. To prove this contention, consider the total time derivative

$$\frac{dH}{dt} = \frac{\partial H}{\partial \boldsymbol{x}}\dot{\boldsymbol{x}} + \frac{\partial H}{\partial \boldsymbol{u}}\dot{\boldsymbol{u}} + \frac{\partial H}{\partial \boldsymbol{\lambda}}\dot{\boldsymbol{\lambda}} + \frac{\partial H}{\partial t} \tag{6.31}$$

which is obtained from the chain rule of differentiation. Along the optimal path,

$$\partial H/\partial \boldsymbol{u} = 0 \tag{6.32}$$

as one of the necessary conditions. In addition, since

$$H = L + \boldsymbol{\lambda}^T \boldsymbol{f} = L + \boldsymbol{f}^T \boldsymbol{\lambda}$$

then

$$\partial H/\partial \boldsymbol{\lambda} = \boldsymbol{f}^T = \dot{\boldsymbol{x}}. \tag{6.33}$$

Substituting these expressions into Equation (6.31) yields

$$\begin{aligned} \frac{dH}{dt} &= \frac{\partial H}{\partial \boldsymbol{x}}\dot{\boldsymbol{x}} + \dot{\boldsymbol{x}}^T \dot{\boldsymbol{\lambda}} + \frac{\partial H}{\partial t} \\ &= \frac{\partial H}{\partial \boldsymbol{x}}\dot{\boldsymbol{x}} + \dot{\boldsymbol{\lambda}}^T \dot{\boldsymbol{x}} + \frac{\partial H}{\partial t} \\ &= \left[\frac{\partial H}{\partial \boldsymbol{x}} + \dot{\boldsymbol{\lambda}}^T\right]\dot{\boldsymbol{x}} + \frac{\partial H}{\partial t}. \end{aligned} \tag{6.34}$$

The term within the brackets is also zero along the optimal path, since condition I of Equation (6.29) must be satisfied, with the final result

$$dH/dt = \partial H/\partial t. \tag{6.35}$$

Thus, the explicit and implicit dependency of H on t are identical, and if Equation (6.30) holds, then

$$dH/dt = 0 \tag{6.36}$$

along the optimal path, and H must be constant. This result can sometimes be quite useful in the solution of optimal control problems. It contains nothing new but merely combines the previously derived necessary conditions into a different form. Note also that if the terminal time is not specified, then the transversality condition yields

$$H\bigg|_{t=t_f} = -\frac{\partial \phi}{\partial t}\bigg|_{t=t_f} \tag{6.37}$$

which, together with Equation (6.34), specifies H at every point along the optimal path if H is not an explicit function of time.

6.2. OPTIMIZATION WITH CONTROL VARIABLE CONSTRAINTS

In the situations considered to this point, an implicit assumption has been made that the elements of the control vector $\boldsymbol{u}(t)$ have not been constrained in any way. Regardless of the magnitude of the elements of $\boldsymbol{u}(t)$ required by the control equations, the system was considered capable of providing such control. In practice, this situation is seldom encountered because saturation effects are almost always present to some extent, since all practical controllers can deliver only a finite amount of energy. Such effects may be the result of some physical limitation inherent in the basic system, or they may be imposed arbitrarily by the application of operating practices. In either case, the imposition of these constraints can drastically affect the optimal control problem because it violates the basic assumption of an unconstrained control vector. In order to consider this important aspect of optimal control, the problem is reformulated as follows. It is desired to select the control vector $\boldsymbol{u}(t)$ which minimizes the performance index

$$J = \phi[\boldsymbol{x}(t_f), t_f] + \int_{t_0}^{t_f} L(\boldsymbol{x}(t), \boldsymbol{u}(t), t)\, dt \tag{6.38}$$

subject to the constraint equation

$$\dot{\boldsymbol{x}}(t) = f(\boldsymbol{x}(t), \boldsymbol{u}(t), t) \tag{6.39}$$

and the terminal constraint

$$\boldsymbol{\psi}[\boldsymbol{x}(t_f)] = 0 \tag{6.40}$$

with the additional constraint that

$$\boldsymbol{c}(u, t) \leqslant 0. \tag{6.41}$$

This last constraint requires that some set of time varying functions of the control vector can never be positive. The control variable constraints, such as

$$u_p(t) \leqslant C_p, \quad p = 1, 2, \ldots, P. \tag{6.42}$$

can easily be expressed in this way. The details of the solution to this important problem are beyond the scope of the present discussion, but it is very concisely stated by Pontryagin's Maximum Principle, discussed in the next section. Formal development of the Maximum Principle by use of dynamic programming is presented in Chapter 8.

6.2.1. The Maximum Principle

The solution of the problem stated above is given in terms of the Hamiltonian function previously defined

$$H(\boldsymbol{x}, \boldsymbol{u}, \boldsymbol{\lambda}, t) = L(\boldsymbol{x}(t), \boldsymbol{u}(t), t) + \boldsymbol{\lambda}^T(t)\boldsymbol{f}(\boldsymbol{x}(t), \boldsymbol{u}(t), t) \tag{6.43}$$

and is stated as follows. The optimal solution is obtained by selecting $\boldsymbol{u}(t)$ so that two necessary conditions are met at every point along the optimal path.

I. $\dot{\boldsymbol{\lambda}}^T = -\partial H/\partial \boldsymbol{x}$

II. $\boldsymbol{u}(t)$ is chosen as that value which satisfies Equation (6.41) and yields the minimum value of H. (6.44)

Thus, the Hamiltonian function plays a much more important role in the theory of optimal control than merely reducing the notational complexity. It is fundamental to the basic theory, a fact not obvious in the previous derivations. When the constraint of Equation (6.41) is removed, then the minimum value of H is determined by use of the equation

$$\partial H/\partial \boldsymbol{u} = 0 \tag{6.45}$$

and the previously derived results are obtained. The result expressed in equation (6.44) is known as the *Pontryagin Maximum Principle*. The term *maximum* is used because the original development dealt with a function which was the negative of the Hamiltonian function of Equation (6.43). In the modern literature, Equation (6.44) is sometimes referred to as the *Minimum Principle*, which is perhaps more descriptive. The development of the Maximum Principle was a substantial generalization of the early work in the calculus of variations, but did not occur until the mid-1950s. It can be seen that the Maximum Principle could be used to develop all the previously obtained results and represents, in this context, the most general development of optimal control theory. It should be noted, however, that II of Equation (6.44) is a condition rather than an equation, so that logical decisions play an important role in specification of the optimal control $\boldsymbol{u}(t)$, whereas in use of the equations resulting from the unconstrained control problem, only substitution into these equations is required. The reader may compare this condition and its use to the Kuhn–Tucker conditions generated by the point optimization problem with inequality constraints, as discussed in Appendix B.

If the inequality constraint of Equation (6.41) is in the form of limitations on the individual elements of $\boldsymbol{u}(t)$, that is,

$$u_p(t) \leqslant C_p, \qquad p = 1, 2, \ldots, P. \tag{6.46}$$

then application of II of Equation (6.44) leads to the result

$$\begin{aligned} \frac{\partial H}{\partial u_p} &= 0, \qquad \text{if } u_p(t) \leqslant C_p \\ \frac{\partial H}{\partial u_p} &\leqslant 0, \qquad \text{if } u_p(t) = C_p. \end{aligned} \tag{6.47}$$

This result must be true because the first expression yields the minimum value of H for the unconstrained problem, which is also the minimum for the constrained problem if the resulting $u_p(t)$ does not exceed the constraint value. On the other hand, if the second condition exists, then a value of $u_p(t)$ resulting from $\partial H/\partial u_p = 0$ which exceeds C_p must necessarily result in a lower value of H than that produced by the constraint value. That is, a lower value of H can only be obtained by exceeding the constraint. Such logical conclusions are required in the solution of problems of this type and are analogous to the procedures required in the use of the Kuhn–Tucker conditions for problem solution. In this sense, the Maximum Principle may be more useful as a tool for understanding the characteristics of optimal solutions than for generating the solutions themselves.

As an example of these principles, consider the simple scalar problem of minimizing the performance index

$$J = ax(t_f) + \tfrac{1}{2} \int_{t_0}^{t_f} u^2(t)\, dt$$

subject to the differential constraint equation

$$\dot{x}(t) = x(t) + u(t)$$

and the control variable constraint

$$|u(t)| \leqslant 1$$

where $x(t)$ and $u(t)$ are scalars. The constraint is really two constraints:

$$u \leqslant 1$$
$$-u \leqslant 1.$$

The Hamiltonian function for this system is

$$H = L + \lambda^T f = \tfrac{1}{2}u^2 + \lambda(x + u).$$

The necessary conditions are

I. $\dot{\lambda} = -\lambda$

II. $u(t)$ selected as that value lying between +1 and −1 which minimizes H.

The boundary condition for the differential equation is obtained from Equation (6.23), with $x(t_f)$ free,

$$\lambda(t_f) = \left.\frac{\partial \phi}{\partial x}\right|_{t=t_f} = a.$$

Again, the λ equation is not coupled to the x equation, so it can be solved independently to yield

$$\lambda(t) = ae^{t_f - t}.$$

Now the second condition must be applied, to the effect that $u(t)$ must be selected to minimize H. So long as $u(t)$ as computed from $\partial H/\partial u = 0$ does not violate the control variable constraints, it is the optimum $u(t)$. That is,

$$\frac{\partial H}{\partial u} = 0 = u + \lambda$$

or

$$u(t) = -\lambda(t) = -ae^{t_f - t}.$$

When $u(t)$ as computed from this expression lies outside the permissible region, the optimum value is chosen as that which minimizes H. In this case, $u(t)$ as computed from the above equation is always negative, so that the only constraint violation which can occur is

$$u(t) < -1$$

in which case the value of u which minimizes H is -1. This conclusion results from the fact that u^2 is always positive, as is λ previously calculated for positive values of a. In order to minimize H, u must therefore be negative and have the largest possible magnitude, which yields -1. An analogous result is obtained for negative a. Finally, the optimum control $u(t)$ can then be defined as

$$u(t) = \begin{cases} -ae^{t_f - t}, & \text{if } -1 \leqslant ae^{t_f - t} \leqslant 1 \\ -1, & \text{if } -ae^{t_f - t} < -1 \\ +1, & \text{if } -ae^{t_f - t} > 1. \end{cases}$$

In Figure 6-3, a block diagram of the system described by this example is shown,

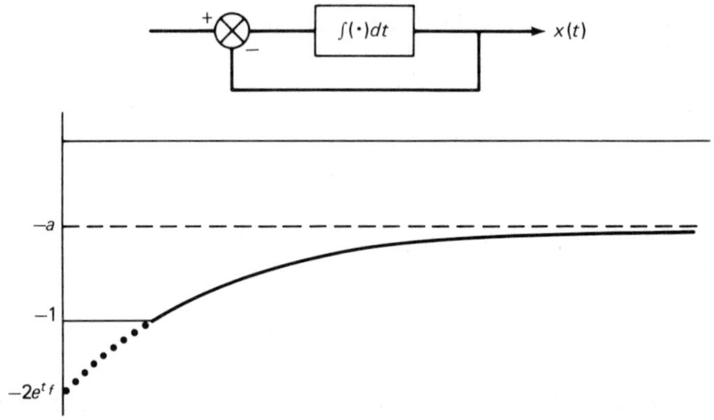

Figure 6-3. Optimum control function for example problem.

together with a time history of the optimal control function $u(t)$. One can easily see that the result obtained produces a saturation effect, which may have been suspected from intuitive reasoning. Application of the maximum principle has given theoretical justification of the optimality of this approach. The reader should examine the results obtained when a and t_f are varied, especially since these parameters determine the relative effects of $x(t_f)$ and the integral of u^2 on the performance index.

A second example will serve to further elucidate the meaning behind the maximum principle. The system equations here are extremely simple:

$$\dot{x} = u; \qquad x(0) = 2; \qquad x(1) = 3$$

and the performance index to be minimized is

$$J = \int_0^1 [u^2(t) + x(t)]\, dt.$$

The Hamiltonian function follows immediately as

$$H = u^2 + x + \lambda u.$$

The maximum principle states that H is to be minimized with respect to u and, since there are no constraints on $u(t)$, this minimization is accomplished by solving the equation

$$\frac{\partial H}{\partial u} = 0 = 2u + \lambda$$

from which

$$\lambda = -2u.$$

The remaining condition is

$$\dot{\lambda} = -\frac{\partial H}{\partial x} = -1$$

with boundary conditions $\lambda(1) = \nu$, where ν must be determined from the boundary conditions in the problem statement. Solution follows easily as

$$\lambda(t) = -t + (\nu + 1); \qquad u(t) = \tfrac{1}{2}t - \tfrac{1}{2}(\nu + 1).$$

Use of the boundary conditions to evaluate the constant ν yields the value $-5/2$, from which

$$\lambda(t) = -t - \tfrac{3}{2}; \qquad u(t) = \tfrac{1}{2}t + \tfrac{3}{4}.$$

and the resulting expression for $x(t)$ is

$$x(t) = \tfrac{1}{4}t^2 + \tfrac{3}{4}t + 2.$$

This result could have been obtained without the use of the maximum principle, the only application of which was to produce the relation $\partial H/\partial u = 0$. The value of $H(x, u, \lambda, t)$ along the optimal path is obtained by substituting into the defining expression of Equation (6.43),

$$H = L + \boldsymbol{\lambda}^T \boldsymbol{f} = u^2 + x + \lambda u$$

which in turn yields

$$H(x, u, \lambda, t) = x - u^2 = \tfrac{1}{4}t^2 + \tfrac{3}{4}t + 2 - (\tfrac{1}{2}t + \tfrac{3}{4})^2 = \tfrac{23}{16}.$$

As required by discussion leading to Equation (6.36), this value is independent of any of the variables and of t.

Suppose that the control variable $u(t)$ is constrained in the following way:

$$-\tfrac{1}{4} \leqslant u(t) \leqslant \tfrac{9}{8}, \qquad \text{for } 0 \leqslant t \leqslant 1.$$

Examination of the previous results, as presented in Figure 6-4, shows that the solution obtained there does indeed violate these constraints, and furthermore that the violation occurs at $t = \frac{3}{4}$.

To consider the constraint, one could merely cause the control function to saturate at the value $\frac{9}{8}$ for $t > \frac{3}{4}$, but there is no assurance that the resulting $u(t)$ is truly optimum. The maximum principle permits consideration of this constraint in an optimum manner. Application of the maximum principle requires $u(t)$ to be the value that minimizes H while satisfying the constraints. In this case, $u(t)$ must be determined from $\partial H/\partial u = 0$, so long as the result does not violate the constraint. Otherwise the minimizing value meeting the constraint must be used, implying that $u(t)$ is obtained from

$$u(t) = \begin{cases} \tfrac{1}{2}t - \tfrac{1}{2}c, & \text{for } 0 \leqslant t \leqslant \tfrac{9}{4} + c \\ \text{minimizing value}, & \text{for } \tfrac{9}{4} + c < t \leqslant 1 \end{cases}$$

where $c = \nu + 1$. The range on t merely reflects the constraint $u(t) \leqslant \frac{9}{8}$. The

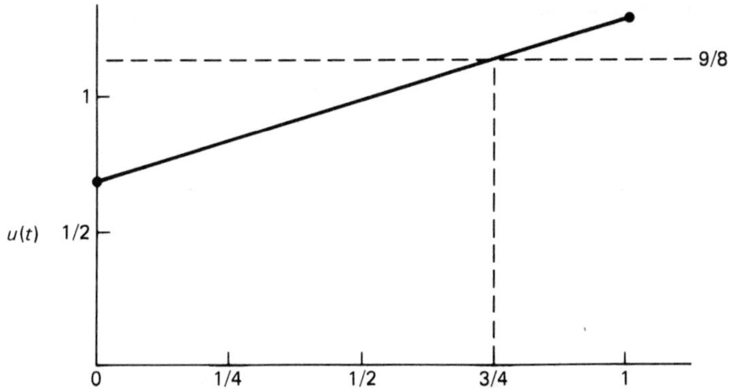

Figure 6-4. Solution of unconstrained problem.

$x(t)$ equation follows as

$$x(t) = \tfrac{1}{2}t^2 - \tfrac{1}{2}ct + 2$$

and, at $t = \frac{9}{4} + c$, yields

$$x(\tfrac{9}{4} + c) = -\tfrac{1}{4}c^2 + \tfrac{209}{64}.$$

For $t > \frac{9}{4} + c$, the optimizing value of $u(t)$ must be selected from those values which satisfy the constraints, and the value to be selected is that which minimizes H. Since

$$H = u^2 + x + \lambda u = u^2 + x + (c - t)u$$

and the x term can be ignored when selecting u to minimize H, the function to be minimized is

$$u^2 + (c - t)u \leqslant u^2 - \tfrac{9}{4}u.$$

The selection of $u = \frac{9}{8}$ is obvious. Then c can be evaluated from the $x(t)$ equation for $t > \frac{9}{4} + c$:

$$x(t) = -\tfrac{1}{4}c^2 + \tfrac{209}{64} + \tfrac{9}{8}(t - \tfrac{9}{4} - c).$$

At $t = 1, x(1) = 3$, and solving the resulting equation for c yields

$$c = -\tfrac{9}{4} + \tfrac{1}{2}\sqrt{2}.$$

The time of saturation is

$$t_s = \tfrac{1}{2}\sqrt{2} = 0.707$$

and the optimal control $u(t)$ for $t < 0.707$ is

$$u(t) = \tfrac{1}{2}t + (\tfrac{9}{8} - \tfrac{1}{4}\sqrt{2}).$$

The resulting $x(t)$ is then

$$x(t) = \tfrac{1}{4}t^2 + (\tfrac{9}{8} - \tfrac{1}{4}\sqrt{2})t + 2$$

in order to meet the boundary condition at $t = 0$. The final conclusion with regard to $u(t)$ is

$$u(t) = \begin{cases} \tfrac{1}{2}t + (\tfrac{9}{8} - \tfrac{1}{4}\sqrt{2}), & \text{for } 0 \leqslant t \leqslant \tfrac{1}{2}\sqrt{2} \\ \tfrac{9}{8}, & \text{for } \tfrac{1}{2}\sqrt{2} < t \leqslant 1 \end{cases}$$

as depicted in Figure 6-5. Comparison of the two solutions is given below, where t_s is the saturation time:

Unconstrained	Constrained
$u(t) = 0.5t + 0.75$	$u(t) = \begin{cases} 0.5t + 0.77144, & t \leqslant 0.707 \\ 1.125, & t > 0.707 \end{cases}$
$x(t) = 0.25t^2 + 0.75t + 2$	$x(t) = \begin{cases} 0.25t^2 + 0.77144t + 2, & t \leqslant 0.707 \\ 1.125t + 1.875, & t > 0.707 \end{cases}$
$t_s = 0.75$	$t_s = 0.707$

When one considers saturation, it must be realized that, if saturation actually occurs when the unconstrained optimal control function is being used, the terminal boundary condition will not be satisfied. The difference in the control functions prior to saturation represents the anticipation of saturation effects by the solution to the constrained problem.

This example can be further exploited by determining the Hamiltonian function H over the time interval (0, 1) for the two situations described above. In the first instance, the value of H is

$$\begin{aligned} H &= u^2 + x + \lambda u \\ &= (0.5t + 0.75)^2 + 0.25t^2 + 0.75t + 2 - 2(0.5t + 0.75)^2 \\ &= 1.438 \end{aligned}$$

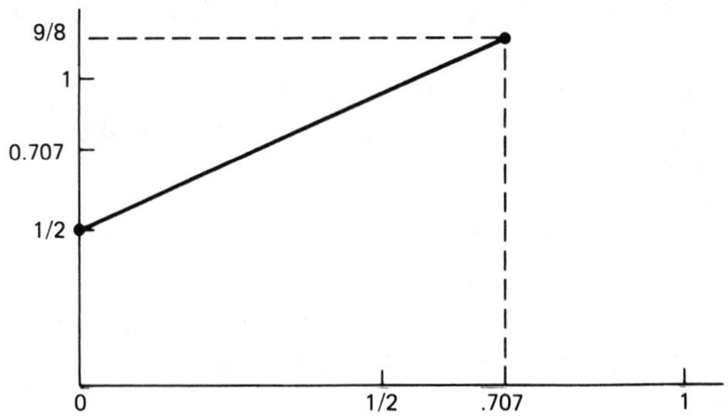

Figure 6-5. Solution of constrained problem.

which is independent of t, as required by the fact that H does not contain t explicitly. In the second instance, before saturation

$$H = 0.25t^2 + 0.77144t + 2 - (0.5t + 0.77144)^2 = 1.405$$

again independent of t. After saturation at $t_s = 0.707$,

$$H = 1.125^2 + 1.875 + 1.125t - (t + 0.77144)(1.125) = 1.405$$

as required by the theory. Further insight can be gained by comparing the value of the performance index J in these two situations. In the unconstrained problem,

$$J = \int_0^1 [u^2(t) + x(t)]\, dt$$

$$= \int_0^1 [(0.5t + 0.75)^2 + 0.25t^2 + 0.75t + 2]\, dt = 3.479.$$

For the constrained problem,

$$J = \int_0^{0.707} [(0.5t + 0.75)^2 + 0.25t^2 + 0.75t + 2]\, dt$$

$$+ \int_{0.707}^1 [1.125^2 + 2.703 + 1.125(t - 0.75)]\, dt = 3.482.$$

These results verify the expected increase in the minimum value of the performance index.

To further illustrate the concepts presented previously, the boundary condition is now modified so that $x(1)$ is free, still maintaining the control function constraints. As before, the necessary conditions yield

$$\lambda(t) = -t + c; \qquad u(t) = -2\lambda(t)$$

so long as $u(t)$ does not violate the constraint. The value of c is now determined from the boundary condition $\lambda(1) = 0$, which is obtained from IV of Equation (6.13). Then,

$$\lambda(t) = 1 - t.$$

So long as $u(t)$ is not at the saturation limit, the maximum principle yields $u = \frac{1}{2}t - \frac{1}{2}$ which is true so long as $t \geqslant \frac{1}{2}$. For $t < \frac{1}{2}$, the value of u produced by this expression violates the lower control function constraint. In this region, then, $u(t)$ must be selected as that permissible value which minimizes H. Since

$$H = u^2 + x + \lambda u \quad \text{and} \quad \lambda > \tfrac{1}{2}$$

$u(t)$ should be selected as its most negative permissible value, or $-\frac{1}{4}$. The desired control function $u(t)$ is then

$$u(t) = \begin{cases} -\frac{1}{4}, & \text{for } 0 \leqslant t < \frac{1}{2} \\ \frac{1}{2}t - \frac{1}{2}, & \text{for } \frac{1}{2} \leqslant t \leqslant 1 \end{cases}$$

yielding for $x(t)$

$$x(t) = \begin{cases} 2 - \frac{1}{4}t, & \text{for } 0 \leqslant t < \frac{1}{2} \\ \frac{1}{4}t^2 - \frac{1}{2}t + \frac{33}{16}, & \text{for } \frac{1}{2} \leqslant t \leqslant 1. \end{cases}$$

The final value $x(1)$ is

$$x(1) = \tfrac{1}{4} - \tfrac{1}{2} + \tfrac{33}{16} = \tfrac{29}{16}.$$

The value of H for both intervals–$(0, \frac{1}{2})$ and $(\frac{1}{2}, 1)$–is 1.8125. The value of J produced by this control function is 1.927 and reflects the additional freedom which the optimization process has when the terminal state is free. The unconstrained problem with free terminal state yields

$$u(t) = \tfrac{1}{2}t - \tfrac{1}{2}$$
$$x(t) = \tfrac{1}{4}t^2 - \tfrac{1}{2}t + 2.$$

The terminal value is $x(1) = 1.75$, and the value of J is 1.916.

Before leaving this example, one additional point can be made by considering the situation in which the constraint on $u(t)$ is $u(t) \leqslant 1$ in the fixed terminal state problem. The obvious solution here is $u(t) = 1$ for all t, for this function is the only possible one which can meet the required terminal boundary condition. Any constraint less than this value will result in a terminal state which cannot be met. Such terminal states are termed *unreachable*, since the control function constraints are such that no choice of $\boldsymbol{u}(t)$ can attain them.

6.2.2. Minimum Time Problems

The minimum time problem was briefly mentioned in conjunction with the example given in Section 6.1.3. Stated specifically, the general minimum time problem entails transferring the system

$$\dot{\boldsymbol{x}}(t) = \boldsymbol{f}(\boldsymbol{x}(t), \boldsymbol{u}(t), t) \tag{6.48}$$

from an initial state $\boldsymbol{x}(t_0)$ to a prescribed final state $\boldsymbol{x}(t_f)$ in minimum time. This situation is produced by the general optimal control problem considered in Section 6.1.3, with performance index

$$J = t_f - t_0 \tag{6.49}$$

which can be obtained by one of two associations,

$$\begin{aligned} &\phi = 0; \qquad L = 1 \\ \text{or} \quad & \\ &\phi = t_f - t_0; \quad L = 0 \end{aligned} \tag{6.50}$$

both of which yield identical results. The Hamiltonian H is then

$$H(\boldsymbol{x}(t), \boldsymbol{\lambda}(t), \boldsymbol{u}(t), t) = \boldsymbol{\lambda}^T(t)\boldsymbol{f}(\boldsymbol{x}(t), \boldsymbol{u}(t), t). \tag{6.51}$$

Even with this simplification, the general equations are still nonlinear with split boundary conditions and are therefore not solvable in the general case.

In most instances, minimum time problems will entail some kind of control variable constraint, in which case the Maximum Principle replaces II of Equation (6.29). This case has important consequences with respect to an important subset of minimum time problems—those linear in the control function $\boldsymbol{u}(t)$. If the system equations are of the form

$$\dot{\boldsymbol{x}}(t) = \boldsymbol{f}(\boldsymbol{x}(t), t) + \boldsymbol{G}(\boldsymbol{x}(t), t)\boldsymbol{u}(t) \tag{6.52}$$

then the nonlinearity of the system is restricted to the system function $\boldsymbol{f}$, and the system equations are linear in the control function $\boldsymbol{u}(t)$. The input matrix $\boldsymbol{G}$ may be explicitly dependent upon the state $\boldsymbol{x}(t)$, as implied above. If, in addition, the control function $\boldsymbol{u}(t)$ is subject to a set of constraints

$$C_p^l \leqslant u_p(t) \leqslant C_p^u \tag{6.53}$$

where the superscripts represent lower and upper constraint limits, then some general conclusions can be reached. The Hamiltonian H is then

$$H = \boldsymbol{\lambda}^T [\boldsymbol{f} + \boldsymbol{G}\boldsymbol{u}] = \boldsymbol{\lambda}^T \boldsymbol{f} + \boldsymbol{\lambda}^T \boldsymbol{G}\boldsymbol{u} \tag{6.54}$$

and $\boldsymbol{u}(t)$ must be selected to minimize H. The linear term $\boldsymbol{\lambda}^T \boldsymbol{G}\boldsymbol{u}$, if expanded, yields

$$\begin{aligned} \boldsymbol{\lambda}^T \boldsymbol{G}\boldsymbol{u} &= (\lambda_1 g_{11} + \lambda_2 g_{21} + \cdots + \lambda_N g_{N1}) u_1 \\ &\quad + (\lambda_1 g_{12} + \lambda_2 g_{22} + \cdots + \lambda_N g_{N2}) u_2 + \cdots \\ &\quad + (\lambda_1 g_{1P} + \lambda_2 g_{2P} + \cdots + \lambda_N g_{NP}) u_P \end{aligned} \tag{6.55}$$

in which g_{ij} is the ijth element of the matrix $\boldsymbol{G}$. This result clearly expresses the $\boldsymbol{u}(t)$ dependence of the Hamiltonian H as being linear. If the components $u_1, u_2, \ldots, u_P$ are independent of one another, then the minimum value of H is obtained when each component is in its maximum or minimum saturation state, depending upon whether its coefficient in Equation (6.55) is negative or positive. Equation (6.55) can be written in more concise form as

$$\boldsymbol{\lambda}^T \boldsymbol{G}\boldsymbol{u} = \sum_{p=1}^{P} \boldsymbol{\lambda}^T \boldsymbol{g}_p u_p \tag{6.56}$$

where the vector $\boldsymbol{g}_p$ is the pth column of matrix $\boldsymbol{G}$. The term

$$\boldsymbol{\lambda}^T(t) \boldsymbol{g}_p(\boldsymbol{x}(t), t) \tag{6.57}$$

is called the *switching function* for the pth component $u_p(t)$, since its algebraic sign determines whether $u_p(t)$ should be saturated at its largest or smallest value. The interesting thing to note is that, if the system is described by Equation (6.51), then there must be control variable constraints, for if there were not the control law would call for infinite values of the control variables. Furthermore, even though $\boldsymbol{u}(t)$ may be variable continuously over its permissible range, only the saturation values are ever used.

Such systems, for which maximum control effort is required throughout the entire interval (t_0, t_f), are called *bang-bang control systems*, after early full deflection systems developed principally for reasons of reduced complexity. The reader may wish to compare these results with the analogous results of linear programming. In both cases, the objective functions and equations of constraint are linear, and in both cases the optimum solution occurs on the boundary of the permissible region.

In summary, the control function $\boldsymbol{u}(t)$ for the system given by Equation (6.52) is obtained, component by component, from the equation

$$u_p(t) = \begin{cases} C_p^u, & \text{if } \boldsymbol{\lambda}^T(t)\boldsymbol{g}_p(\boldsymbol{x}(t), t) < 0 \\ C_p^l, & \text{if } \boldsymbol{\lambda}^T(t)\boldsymbol{g}_p(\boldsymbol{x}(t), t) > 0 \\ \text{undetermined}, & \text{if } \boldsymbol{\lambda}^T(t)\boldsymbol{g}_p(\boldsymbol{x}(t), t) = 0. \end{cases} \tag{6.58}$$

The last consideration, which occurs when the switching function is zero for some finite time interval, constitutes what is called a *singular condition*. Such instances are sometimes encountered and one must be aware of the possibility that they may occur. Singular conditions are briefly discussed in Section 6.3.9.

6.2.3. Minimum Time Problems Involving Time Invariant Linear Systems

When the system is linear and time invariant, that is, when the system equations are of the form

$$\dot{\boldsymbol{x}}(t) = \boldsymbol{F}\boldsymbol{x}(t) + \boldsymbol{G}\boldsymbol{u}(t) \tag{6.59}$$

where $\boldsymbol{F}$ and $\boldsymbol{G}$ are constant matrices, further conclusions can be reached regarding the minimum time problem. Since this type of problem is included in the general category of problems described by Equation (6.52), the optimal control $\boldsymbol{u}(t)$ is a bang-bang system with switching of the elements of $\boldsymbol{u}(t)$ as described by Equation (6.58). Two additional characteristics of this system are stated here without proof:

I. If all the eigenvalues of $\boldsymbol{F}$ have negative real parts, then there is an optimal control $\boldsymbol{u}(t)$ that will transfer any initial state $\boldsymbol{x}(t_0)$ to the origin.

II. If all the eigenvalues of $\boldsymbol{F}$ are real, then each component $u_p(t)$ can switch at most $(N-1)$ times, where N is the order of the system.

Note that the switching functions are now of the form

$$\boldsymbol{\lambda}^T(t)\boldsymbol{g}_p \tag{6.60}$$

where the $\boldsymbol{g}_p$ vectors are now constant.

As an example of these principles, consider the one-dimensional problem of transferring a particle from some initial position $-s$ and velocity v to a state of rest at the origin, as illustrated in Figure 6-6. The objective is to accomplish this transfer in minimum time. Here, the initial state of the particle is given by the position and velocity coordinates shown and the state equation is

$$\begin{bmatrix} \dot{x}_1 \\ \dot{x}_2 \end{bmatrix} = \begin{bmatrix} 0 & 1 \\ 0 & 0 \end{bmatrix} \begin{bmatrix} x_1 \\ x_2 \end{bmatrix} + \begin{bmatrix} 0 \\ 1 \end{bmatrix} u(t)$$

where the scalar control $u(t)$ is the acceleration acting on the particle. The available acceleration is continuously variable between the two values $\pm C$. This problem obviously fits the general description of Equations (6.53) and (6.59), with the following associations:

$$-C \leqslant u(t) \leqslant C;$$

$$\boldsymbol{F} = \begin{bmatrix} 0 & 1 \\ 0 & 0 \end{bmatrix}; \quad \boldsymbol{G} = \begin{bmatrix} 0 \\ 1 \end{bmatrix}.$$

The switching function in this case is

$$\boldsymbol{\lambda}^T\boldsymbol{G} = \lambda_2$$

and the Hamiltonian function is

$$H = \lambda_1 x_2 + \lambda_2 u$$

from which the $\boldsymbol{\lambda}$ equation can be obtained:

$$\dot{\boldsymbol{\lambda}}^T = -\partial H/\partial \boldsymbol{x}.$$

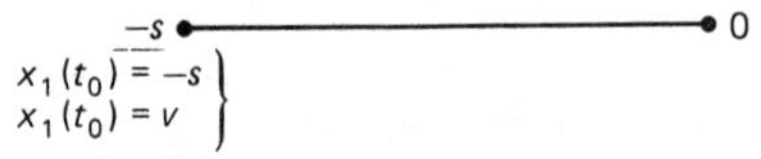

Figure 6-6. One-dimensional translation problem.

In component form, this equation is

$$\dot{\lambda}_1 = 0$$
$$\dot{\lambda}_2 = -\lambda_1$$

with the boundary conditions

$$\lambda_1(t_f) = \nu_1$$
$$\lambda_2(t_f) = \nu_2$$

resulting from the fact that $x_1(t_f)$ and $x_2(t_f)$ are specified. Again in this case, the $\boldsymbol{\lambda}$ equations are not coupled with the state equations and can therefore be solved directly to yield

$$\lambda_1 = \nu_1$$
$$\lambda_2 = \nu_2 + \nu_1(t_f - t).$$

If $\lambda_2(t)$ is known, then $u(t)$ can be determined as follows

$$u(t) = -C, \quad \text{if } \lambda_2(t) > 0$$
$$u(t) = +C, \quad \text{if } \lambda_2(t) < 0.$$

Furthermore, since λ_2 is a linear function of t, it can change sign at most once, implying that $u(t)$ also can change sign at most one time. Also, since the final time is not fixed, the transversality condition of Equation (6.29) must apply, which yields

$$\left[\frac{\partial \phi}{\partial t} + H\right]_{t=t_f} = 0 = 1 + \lambda_2(t_f)u(t_f).$$

From this expression one obtains,

$$\lambda_2(t_f)u(t_f) = \nu_2 u(t_f) = -1$$

or, upon solving for ν_2,

$$\nu_2 = -1/u(t_f).$$

Now $u(t_f)$ is either $+C$ or $-C$, because of the nature of the particular problem.

Suppose it is $-C$. Then

$$\lambda_2 = 1/C$$

and

$$\lambda_2(t) = (1/C) + \nu_1(t_f - t)$$

which changes sign at time

$$t_s = t_f + (1/\nu_1 C)$$

if this expression yields

$$t_0 < t_s < t_f.$$

On the other hand, if $u(t_f)$ has the value $+C$, then similar considerations yield

$$\lambda_2(t) = -(1/C) + \nu_1(t_f - t)$$

and

$$t_s = -(1/\nu_1 C) + t_f.$$

These relationships are still in terms of the unknowns ν_1 and t_f. To evaluate these parameters, it is necessary to substitute the control $u(t)$ in terms of these unknowns into the state equations and to solve for the states in terms of the unknowns. Matching of the boundary conditions then provides the two required equations. In the present case, suppose again that $u(t_f) = -C$. This supposition makes sense from an intuitive viewpoint because of the requirement for zero velocity at the terminal time, if the initial position coordinate is negative as implied by Figure 6-6. Then $\nu_2 = 1/C$ and

$$\lambda_2(t) = 1/C + \nu_1(t_f - t).$$

For $t > t_s$,

$$\dot{x}_1 = x_2$$
$$\dot{x}_2 = -C$$

which, when solved with the boundary conditions

$$x_1(t_f) = x_2(t_f) = 0$$

yields

$$x_1(t) = -\tfrac{1}{2}C(t_f - t)^2$$
$$x_2(t) = C(t_f - t).$$

For $t < t_s$,

$$\dot{x}_1 = x_2$$
$$\dot{x}_2 = C.$$

Solution of these equations, with boundary conditions obtained from above at $t = t_s$, yields

$$x_1(t) = \left[\frac{(t_f - t_s)^2}{2} + (t_f - t_s)(t - t_s) + \frac{(t - t_s)^2}{2}\right]C$$
$$x_2(t) = [(t_f - t_s) + (t - t_s)]\,C$$

These are two equations in two unknowns t_f and t_s, or equivalently, t_f and ν_1. By evaluating these equations at the initial point

$$x_1(t_0) = -s$$
$$x_2(t_0) = v$$

the two unknowns ν_1 and t_f can be determined. A similar procedure applies when $u(t_f)$ has a positive value; just which situation applies will depend on the initial conditions. In fact, a question of reachable states arises here because there are certain initial conditions for which the desired terminal conditions cannot be obtained simply because the available acceleration is not sufficient.

As a particularly simple and easily interpreted special case of this problem, consider the situation in which

$$t_0 = 0; \quad x_1(0) = -s; \quad x_2(0) = 0.$$

Then, from evaluation of the $x_1(t)$ and $x_2(t)$ equations at $t_0 = 0$,

$$t_s = t_f/2$$
$$t_f = \sqrt{s/C}$$

and the optimum control $u(t)$ is

$$u(t) = \begin{cases} +C, & \text{for } t \leqslant \sqrt{s/C} \\ -C, & \text{for } t > \sqrt{s/C}. \end{cases}$$

Substitution of these values into the state equations will readily verify that they produce the required terminal conditions.

6.2.4. Minimum Energy Problems

A second general type of optimal control problem is that in which some function of the control expenditure is to be minimized. Again, consider the special situation in which the system equation is given by Equation (6.52),

$$\dot{\boldsymbol{x}}(t) = \boldsymbol{f}(\boldsymbol{x}(t), t) + G(\boldsymbol{x}(t), t)\boldsymbol{u}(t)$$

and the performance index is

$$J = \phi(\boldsymbol{x}(t_f), t_f) + \tfrac{1}{2} \int_{t_0}^{t_f} [\boldsymbol{u}^T(t)\boldsymbol{B}\boldsymbol{u}(t)]\, dt \tag{6.61}$$

where $\boldsymbol{B}$ is a diagonal matrix. Such a problem is termed a minimum energy problem, because the objective function is quadratic in $\boldsymbol{u}(t)$. The Hamiltonian H for this system is

$$H = \tfrac{1}{2}\boldsymbol{u}^T\boldsymbol{B}\boldsymbol{u} + \boldsymbol{\lambda}^T\boldsymbol{f} + \boldsymbol{\lambda}^T\boldsymbol{G}\boldsymbol{u} \tag{6.62}$$

and the necessary conditions are

$$\dot{\boldsymbol{\lambda}}^T = -\partial H/\partial \boldsymbol{x} \tag{6.63}$$

and $\boldsymbol{u}(t)$ selected to minimize H. The first of these conditions yields the differential equation for $\boldsymbol{\lambda}(t)$, while the second requires that the elements of $\boldsymbol{u}(t)$ be selected such that $\frac{1}{2}\boldsymbol{u}^T\boldsymbol{B}\boldsymbol{u} + \boldsymbol{\lambda}^T\boldsymbol{G}\boldsymbol{u}$ is minimized. This requirement specifies that $\boldsymbol{u}(t)$ satisfy the equation

$$\frac{\partial H}{\partial \boldsymbol{u}} = 0 = \boldsymbol{u}^T\boldsymbol{B} + \boldsymbol{\lambda}^T\boldsymbol{G} \tag{6.64}$$

which yields

$$\boldsymbol{u}(t) = -\boldsymbol{B}^{-1}(t)\boldsymbol{G}^T(\boldsymbol{x}(t), t)\boldsymbol{\lambda}(t) \tag{6.65}$$

so that the solution for $\boldsymbol{\lambda}(t)$ in turn yields $\boldsymbol{u}(t)$. However, solving for $\boldsymbol{\lambda}(t)$ may involve solution of a nonlinear, two-point boundary value problem.

If a control function constraint is applied, then the value of $\boldsymbol{u}(t)$ as given by

Equation (6.64) may violate the constraint, in which case $\boldsymbol{u}(t)$ must be selected from permissible values to minimize H in accordance with the maximum principle. If each element $u_p(t)$ is independent of the others and C_p^u and C_p^l are the upper and lower constraint values, then $\boldsymbol{u}(t)$ is selected to satisfy the requirement

$$u_p(t) = \begin{cases} C_p^u, & \text{if } b_p(C_p^u)^2 + \boldsymbol{\lambda}^T \boldsymbol{g}_p C_p^u \leqslant b_p(C_p^l)^2 + \boldsymbol{\lambda}^T C_p^l \\ C_p^l, & \text{if } b_p(C_p^l)^2 + \boldsymbol{\lambda}^T \boldsymbol{g}_p C_p^l > b_p(C_p^u)^2 + \boldsymbol{\lambda}^T C_p^u \end{cases} \tag{6.66}$$

where b_p is the pth diagonal element of the matrix $\boldsymbol{B}$ and the vector $\boldsymbol{g}_p$ is the pth column of the matrix $\boldsymbol{G}$. These results are merely a statement of the maximum principle as it relates to Equation (6.62), requiring that u_p be selected as the upper limit C_p^u or the lower limit C_p^l, depending on which yields the smaller value of H. This specification is more complex than that of the minimum time problem because of the presence of the quadratic term $\boldsymbol{u}^T \boldsymbol{B} \boldsymbol{u}$ in the performance index.

6.2.5. Minimum Energy Problems Involving Time Invariant Linear Systems

When the system is linear and time invariant, then the system equations are

$$\dot{\boldsymbol{x}}(t) = \boldsymbol{F}\boldsymbol{x}(t) + \boldsymbol{G}\boldsymbol{u}(t) \tag{6.67}$$

where $\boldsymbol{F}$ and $\boldsymbol{G}$ are constant matrices. The minimum energy problem involving a system such as this results in a special case of the equations presented in the previous section,

$$\boldsymbol{u}(t) = -\boldsymbol{B}^{-1}\boldsymbol{G}^T\boldsymbol{\lambda}(t) \tag{6.68}$$

where $\boldsymbol{G}$ is now a constant matrix, and

$$\dot{\boldsymbol{\lambda}} = -\left[\frac{\partial H}{\partial \boldsymbol{x}}\right]^T = -\boldsymbol{F}^T\boldsymbol{\lambda}. \tag{6.69}$$

Here, the $\boldsymbol{\lambda}(t)$ equations are linear, homogeneous, and uncoupled from the $\boldsymbol{x}(t)$ equations, responding only to the boundary conditions specified by Equation (6.13) or Equation (6.29). Upon solution for $\boldsymbol{\lambda}(t)$, the control function $\boldsymbol{u}(t)$ follows immediately from Equation (6.68). If there is a control variable constraint, then the elements of $\boldsymbol{u}(t)$ must be selected in accordance with Equation (6.66), with the vectors $\boldsymbol{g}_p$ now constant.

An example of a minimum energy problem involving a linear, time invariant system was developed in Section 6.2.1 as an application of the Maximum Principle.

It should be noted that the result of this section can also be applied to linear, time dependent systems, in that the $\boldsymbol{F}$ and $\boldsymbol{G}$ matrices can be made functions of time with no change in the overall result, other than the fact that Equation (6.69) for determining $\boldsymbol{\lambda}(t)$ will be a time dependent, homogeneous differential equation.

6.2.6. Minimum Control Effort Problems

Yet another kind of optimal control problem involves a performance index which is linear in the absolute value of the control function $\boldsymbol{u}(t)$,

$$J = \phi(\boldsymbol{x}(t_f), t_f) + \int_{t_0}^{t_f} \boldsymbol{b}^T |\boldsymbol{u}(t)| \, dt \tag{6.70}$$

where $|\boldsymbol{u}(t)|$ signifies a vector of absolute values. Such a form would result, for example, from a minimum fuel specification for a thrust control problem. As before, the special situation in which the system equations are linear in the control function $\boldsymbol{u}(t)$,

$$\dot{\boldsymbol{x}}(t) = \boldsymbol{f}(\boldsymbol{x}(t), \boldsymbol{u}(t), t) + \boldsymbol{G}((\boldsymbol{x}(t), t))\boldsymbol{u}(t) \tag{6.71}$$

is considered. The Hamiltonian H for this system is

$$H = \boldsymbol{b}^T |\boldsymbol{u}(t)| + \boldsymbol{\lambda}^T \boldsymbol{f} + \boldsymbol{\lambda}^T \boldsymbol{G}\boldsymbol{u}(t) \tag{6.72}$$

which is also linear in $|\boldsymbol{u}(t)|$. The maximum principle then requires that $\boldsymbol{u}(t)$ be selected to minimize H, and will obviously result in some sort of maximum effort control system, since H is piecewise linear in $\boldsymbol{u}(t)$. Thus, some form of control variable constraint is required, represented by the inequalities

$$C_p^l \leqslant u_p(t) \leqslant C_p^u, \quad p = 1, 2, \ldots, P \tag{6.73}$$

where the superscripts refer to the upper and lower constraints. If the components of the control function $\boldsymbol{u}(t)$ are independently selected, then the Hamiltonian is minimized by minimizing the contribution due to each component individually. Thus, for the pth component, it is desired to minimize

$$b_p |u_p(t)| + \boldsymbol{\lambda}^T(t) \boldsymbol{g}_p(\boldsymbol{x}(t), t) u_p(t) \tag{6.74}$$

where $\boldsymbol{g}_p$ is the pth column of the matrix $\boldsymbol{G}$, and b_p is the pth element of vector $\boldsymbol{b}$. Equation (6.74) is equivalent to

$$\begin{aligned} &(b_p + \boldsymbol{\lambda}^T \boldsymbol{g})u_p, \quad &\text{if } u_p(t) \geqslant 0 \\ &(-b_p + \boldsymbol{\lambda}^T \boldsymbol{g})u_p, \quad &\text{if } u_p(t) < 0 \end{aligned} \tag{6.75}$$

which merely reflects the definition of the absolute value function. For convenience, the elements of $\boldsymbol{b}$ will be considered positive, corresponding to most practical applications. Then, if

$$\boldsymbol{\lambda}^T \boldsymbol{g}_p > b_p \tag{6.76}$$

the minimum value of the first expression is zero, obtained with $u_p(t) = 0$. The minimum value of the second expression is obtained with $u_p(t) = C_p^l$, since the quantity within the parentheses is positive. On the other hand, if

$$\boldsymbol{\lambda}^T \boldsymbol{g}_p < -b_p \tag{6.77}$$

then the minimum of the first expression is obtained with $u_p(t) = C_p^u$ since the quantity within the parentheses is negative. The second expression also attains its minimum with $u_p(t) = C_p^u$. For

$$-b_p < \boldsymbol{\lambda}^T \boldsymbol{g}_p < b_p \tag{6.78}$$

the minimum value of both expressions is zero, and is attained with $u_p(t) = 0$. Finally, for

$$\boldsymbol{\lambda}^T \boldsymbol{g}_p = b_p \tag{6.79}$$

the first expression can be made equal to zero by selecting $u_p(t) = 0$, while the second expression yields zero for all nonpositive values of $u_p(t)$. A similar argument holds for

$$\boldsymbol{\lambda}^T \boldsymbol{g}_p = -b_p. \tag{6.80}$$

In summary, the optimal control component $u_p(t)$ is given by

$$u_p(t) = \begin{cases} C_p^u, & \text{if } \boldsymbol{\lambda}^T \boldsymbol{g}_p < -b_p \\ C_p^l, & \text{if } \boldsymbol{\lambda}^T \boldsymbol{g}_p > b_p \\ 0, & \text{if } -b_p < \boldsymbol{\lambda}^T \boldsymbol{g}_p < b_p \\ \text{any nonnegative value,} & \text{if } \boldsymbol{\lambda}^T \boldsymbol{g}_p = -b_p \\ \text{any nonpositive value,} & \text{if } \boldsymbol{\lambda}^T \boldsymbol{g}_p = b_p. \end{cases} \tag{6.81}$$

This result implies that the optimal control law is similar to a bang-bang system, but with the addition of coast intervals when the control function is zero. Recall that Equation (6.81) must be applied to each component $u_p(t)$ of $\boldsymbol{u}(t)$.

6.2.7. Minimum Control Effort Problems for Time Invariant Systems

For a linear, constant coefficient system, in which the system equation is

$$\dot{\boldsymbol{x}}(t) = \boldsymbol{F}\boldsymbol{x}(t) + \boldsymbol{G}\boldsymbol{u}(t) \tag{6.82}$$

where $\boldsymbol{F}$ and $\boldsymbol{G}$ are constant matrices, application of the principles of the previous section yields the result of Equation (6.81), except that now the vectors $\boldsymbol{g}_p$ are constant vectors. In addition, the $\boldsymbol{\lambda}(t)$ equation, as obtained from the necessary condition

$$\dot{\boldsymbol{\lambda}}^T(t) = -\partial H/\partial \boldsymbol{x} \tag{6.83}$$

is a linear homogeneous equation, uncoupled from the $\boldsymbol{x}(t)$ equations, with constant coefficients,

$$\dot{\boldsymbol{\lambda}}(t) = -\boldsymbol{F}^T\boldsymbol{\lambda}(t). \tag{6.84}$$

As with the minimum energy problem, these results hold when $\boldsymbol{F}$ and $\boldsymbol{G}$ are functions of time, but the $\boldsymbol{\lambda}(t)$ equations then have time varying coefficients.

To exemplify a minimum control effort problem, consider the example problem of Section 6.2.1, with the scalar system equation

$$\dot{x}(t) = x(t) + u(t)$$
$$|u(t)| \leqslant 1$$

but now with performance index

$$J = ax(t_f) + \int_{t_0}^{t_f} |u(t)|\, dt.$$

The Hamiltonian H is

$$H = |u(t)| + \lambda x + \lambda u$$

from which the $\lambda(t)$ equation is

$$\dot{\lambda} = -\lambda$$

with solution

$$\lambda(t) = ae^{t_f - t}.$$

Here,

$$\boldsymbol{b} = 1; \quad C^l = -1; \quad C^u = 1; \quad G = 1$$

and Equation (6.81) yields

$$u(t) = \begin{cases} 1, & \text{if } ae^{t_f - t} < -1 \\ 0, & \text{if } -1 < ae^{t_f - t} < 1 \\ -1, & \text{if } ae^{t_f - t} > 1. \end{cases}$$

Compare this result with that of the minimum energy example in Section 6.2.1:

$$u(t) = \begin{cases} 1, & \text{if } ae^{t_f - t} < -1 \\ -ae^{t_f - t}, & \text{if } -1 \leqslant ae^{t_f - t} \leqslant 1 \\ -1, & \text{if } ae^{t_f - t} > 1. \end{cases}$$

The occurrence of a coasting interval in the minimum control effort problem is evident. Actual solutions will depend upon the initial condition $x(t_0)$; for some values of $x(t_0)$, the final desired value $x(t_f)$ may be impossible to attain.

6.2.8. The Constrained Linear Problem

When both the performance index and the system equations are linear,

$$\dot{\boldsymbol{x}}(t) = \boldsymbol{F}(t)\boldsymbol{x}(t) + \boldsymbol{G}(t)\boldsymbol{u}(t)$$

$$J = \phi[\boldsymbol{x}(t_f), t_f] + \int_{t_0}^{t_f} [\boldsymbol{a}^T\boldsymbol{x}(t) + \boldsymbol{b}^T\boldsymbol{u}(t)]\, dt \tag{6.85}$$

where $\boldsymbol{a}$ and $\boldsymbol{b}$ are vectors, the Hamiltonian H is

$$\begin{aligned} H &= L + \boldsymbol{\lambda}^T\boldsymbol{f} = \boldsymbol{a}^T\boldsymbol{x} + \boldsymbol{b}^T\boldsymbol{u} + \boldsymbol{\lambda}^T\boldsymbol{F}\boldsymbol{x} + \boldsymbol{\lambda}^T\boldsymbol{G}\boldsymbol{u} \\ &= (\boldsymbol{a}^T + \boldsymbol{\lambda}^T\boldsymbol{F})\boldsymbol{x} + (\boldsymbol{b}^T + \boldsymbol{\lambda}^T\boldsymbol{G})\boldsymbol{u}. \end{aligned} \tag{6.86}$$

The maximum principle requires that $\boldsymbol{u}(t)$ be selected to minimize H, which in turn requires that each element of $\boldsymbol{u}(t)$ be as large as possible or as small as possible, depending on the algebraic sign of the corresponding component of the switching function

$$(\boldsymbol{b}^T + \boldsymbol{\lambda}^T \boldsymbol{G}) \tag{6.87}$$

further requiring control function constraints

$$C_p^l \leqslant u_p(t) \leqslant C_p^u, \quad p = 1, 2, \ldots, P \tag{6.88}$$

as in previous cases. This development produces a bang-bang optimal control law, but with a switching function different from those encountered previously.

6.2.9. Singular Intervals in Optimal Control

In several instances resulting from the use of the maximum principle, intervals were encountered in which the necessary conditions provide no information concerning the relationship between the control function $\boldsymbol{u}(t)$ and remaining variables of the problem. Such conditions were termed *singular conditions*, and imply that the control function $\boldsymbol{u}(t)$ has no effect during the interval over which the singular condition exists. This consideration immediately relates the concept of a singular condition to the characteristics of the Hamiltonian function, in that a singular condition will exist whenever the Hamiltonian is of such a form that the Maximum Principle does not yield sufficient information to define $u(t)$ in terms of the remaining variables $\boldsymbol{x}(t)$ and $\boldsymbol{\lambda}(t)$. In the case of the minimum time problem of Section 6.2.2, this condition obviously exists for $\boldsymbol{\lambda}^T \boldsymbol{g} = 0$, as indicated by Equation (6.55), and is explicitly described by the control function specification of Equation (6.58).

The idea that the control function can be undefined implies that singular conditions might be related to the concept of controllability of the system, which indeed can be shown for certain types of problems. In the linear, minimum time problem, the system must be noncontrollable for a singular condition to exist. Conversely, if the system is controllable, then a singular condition cannot exist. For the linear, minimum energy problem, the necessary condition for existence of a singular interval is either noncontrollability of the system or singularity of the matrix $\boldsymbol{F}$, or both.

Singular conditions apparently occur most commonly when the Hamiltonian function is linear in one or more of the control variables, yet nonlinear in one or more of the state variables. Both of the problems considered above are of this kind.

6.2.10. Corner Conditions in Optimal Control

One implicit assumption present in the application of the variational approach to the optimal control problem is that the functions $\boldsymbol{x}(t)$ and $\dot{\boldsymbol{x}}(t)$ be continuous; i.e., the set of admissible functions $\boldsymbol{x}(t)$ is restricted to those which are continuous and have continuous first derivatives. When the maximum principle is applied, this restriction can be lifted and discontinuous control functions $\boldsymbol{u}(t)$ result in some cases, producing bang-bang type control laws. In such an instance, the control function $\boldsymbol{u}(t)$ is said to be piecewise continuous, resulting in an $\boldsymbol{x}(t)$ which is piecewise smooth. The point at which the discontinuity in $\boldsymbol{u}(t)$ occurs is called a *corner*. To determine the conditions which must hold at a corner—that is, at a point where $\boldsymbol{u}(t)$ exhibits a discontinuity—it is necessary to define the time at which the discontinuity occurs as t_c and then to consider the two incrementally adjacent times t_{c^-} and t_{c^+}. By definition, at a corner

$$\boldsymbol{u}(t_{c^-}) \neq \boldsymbol{u}(t_{c^+}). \tag{6.89}$$

When the maximum principle is applied, however, it can be seen that the Lagrange multiplier function $\boldsymbol{\lambda}(t)$ and the Hamiltonian H are continuous across a corner. In fact, as shown in Section 6.1.3, H is a constant along the optimal path unless it is explicitly dependent on t. Furthermore, the derivatives

$$\partial H/\partial \boldsymbol{x} \tag{6.90}$$

are also continuous over the optimal path, so that the conditions which must be met at a corner are

$$\begin{aligned} \boldsymbol{\lambda}(t_{c^-}) &= \boldsymbol{\lambda}(t_{c^+}) \\ H\Big|_{t_{c^-}} &= H\Big|_{t_{c^+}} \\ \left.\frac{\partial H}{\partial \boldsymbol{u}}\right|_{t_{c^-}} &= \left.\frac{\partial H}{\partial \boldsymbol{u}}\right|_{t_{c^+}} \end{aligned} \tag{6.91}$$

These conditions are satisfied by application of the maximum principle but must be explicitly considered if a variational approach is applied separately for the intervals (t_0, t_{c^-}) and (t_{c^+}, t_f).

When the performance index is

$$J = \int_{t_0}^{t_f} L\,[\boldsymbol{x}(t), \dot{\boldsymbol{x}}(t), t]\; dt \tag{6.92}$$

these requirements can be expressed as

$$\left.\frac{\partial L}{\partial \dot{\boldsymbol{x}}}\right|_{t_{c^-}} = \left.\frac{\partial L}{\partial \dot{\boldsymbol{x}}}\right|_{t_{c^+}} \tag{6.93}$$

and

$$L - \frac{\partial L}{\partial \dot{x}} \dot{x} \bigg|_{t_{c-}} = L - \frac{\partial L}{\partial \dot{x}} \dot{x} \bigg|_{t_{c+}} . \tag{6.94}$$

These conditions are known as the *Weirstrauss-Erdman corner conditions* and are a classical result in the calculus of variations for determining the necessary conditions when a discontinuity in $\boldsymbol{x}(t)$ is permitted.

6.3. UNCONSTRAINED LINEAR PROBLEMS

The fact that the two-point boundary value problem can be solved for linear systems leads to the expression of a very general type of linear problem—that with a quadratic performance index. In other words, the performance index is composed of quadratic forms, in the following manner:

$$J = \tfrac{1}{2}[\boldsymbol{x}^T(t_f)\boldsymbol{S}_f\boldsymbol{x}(t_f)] + \tfrac{1}{2}\int_{t_0}^{t_f} [\boldsymbol{x}^T(t)\boldsymbol{A}(t)\boldsymbol{x}(t) + \boldsymbol{u}^T(t)\boldsymbol{B}(t)\boldsymbol{u}(t)]\, dt \tag{6.95}$$

where $\boldsymbol{A}$ and $\boldsymbol{S}_f$ are symmetric, positive semidefinite matrices, an $\boldsymbol{B}$ is a symmetric, positive definite matrix. The system to be considered is linear; therefore, the state equations have the general form

$$\dot{\boldsymbol{x}}(t) = \boldsymbol{F}(t)\boldsymbol{x}(t) + \boldsymbol{G}(t)\boldsymbol{u}(t). \tag{6.96}$$

Also, the system will be unconstrained in the sense that the terminal values of the states are considered to be free and that there are no limitations on the control vector $\boldsymbol{u}(t)$. The latter consideration may be somewhat of a limitation for reasons discussed previously, but the former is purposely formulated in this way in order to permit a tradeoff consideration between the manner in which desired terminal conditions are met and the manner in which the functional component of J is optimized. In most practical problems involving terminal constraints, the approximate meeting of these constraints is usually satisfactory and has the advantage of permitting more leeway in selecting other parameters of the problem. In an intercept problem involving a missile and a target, for example, a near miss is an acceptable terminal condition and may be preferred if a considerable saving in fuel can be achieved.

The particular form of the performance index of Equation (6.95) permits the individual weighting of all components of the states, control vector, and terminal states, as well as cross product terms involving the elements of each of these vectors, by proper selection of the elements of the three matrices $\boldsymbol{A}$, $\boldsymbol{B}$, and $\boldsymbol{S}_f$. If the terminal states are considered to be of prime importance, then the diagonal elements of $\boldsymbol{S}_f$ should be large compared to the elements of $\boldsymbol{A}$ and $\boldsymbol{B}$, assuming that J is to be minimized. If the control vector is of prime importance, then the elements of $\boldsymbol{B}$ should be large in comparison with those of the other two matrices. This general form, then, provides the designer with systematic procedures for specifying system performance.

As was the case in previous developments, the very general problem specified by Equations (6.95) and (6.96) will be considered, with the result that any specific problem which can be put into this form will already have been solved. All that remains is to correlate the terms of the general and specific problems. The two types of linear control problems to be considered here are the linear feedback controller and the linear regulator. These terms have very broad definitions, as follows:

I. A *linear controller* is feedback control law which brings a system from some initial state $\boldsymbol{x}(t_0)$ to a state close to some desired terminal condition, using reasonable values of the control vector and producing reasonable values of the state vector in the interval (t_0, t_f).

II. A *regulator* is a feedback control law which maintains the state of a system close to some desired reference state during the interval (t_0, t_f), using reasonable values of the control vector. Usually, t_f is considered to be located far in the future.

Naturally, the specific meaning of close and reasonable in these definitions is subject to interpretation, but it is in precisely this situation that the particular form of the performance index is useful. It should be noted that the reference values for $\boldsymbol{x}(t_f)$, $\boldsymbol{x}(t)$, and $\boldsymbol{u}(t)$–with respect to which the terms *close* and *reasonable* are defined–are assumed to be zero in the performance index of Equation (6.95), but inclusion of nonzero values is straightforward. Also, since this approach has usefulness primarily in linear problems resulting from linearization of nonlinear systems, zero reference values are encountered quite often. General solutions for both of these problems will now be developed.

6.3.1. The Linear Controller and the Riccati Equation

For the general linear system with quadratic performance index, the Hamiltonian function is

$$\begin{aligned} H(\boldsymbol{x}(t), \boldsymbol{\lambda}(t), \boldsymbol{u}(t), t) &= L(\boldsymbol{x}(t), \boldsymbol{u}(t), t) + \boldsymbol{\lambda}^T(t)\boldsymbol{f}(\boldsymbol{x}(t), \boldsymbol{u}(t), t) \\ &= \tfrac{1}{2}[\boldsymbol{x}^T(t)\boldsymbol{A}(t)\boldsymbol{x}(t) + \boldsymbol{u}^T(t)\boldsymbol{B}(t)\boldsymbol{u}(t)] \\ &\quad + \boldsymbol{\lambda}^T(t)\boldsymbol{F}(t)\boldsymbol{x}(t) + \boldsymbol{\lambda}^T(t)\boldsymbol{G}(t)\boldsymbol{u}(t). \end{aligned} \tag{6.97}$$

Following the general procedures for optimization produces the necessary conditions

$$\dot{\boldsymbol{\lambda}}^T = -\frac{\partial H}{\partial \boldsymbol{x}} = -\boldsymbol{x}^T\boldsymbol{A} - \boldsymbol{\lambda}^T\boldsymbol{F}$$

$$\frac{\partial H}{\partial \boldsymbol{u}} = 0 = \boldsymbol{u}^T\boldsymbol{B} + \boldsymbol{\lambda}^T\boldsymbol{G}. \tag{6.98}$$

Transposing the first equation and solving the second yield

$$\dot{\boldsymbol{\lambda}} = -\boldsymbol{A}\boldsymbol{x} - \boldsymbol{F}^T\boldsymbol{\lambda}$$

$$\boldsymbol{u} = -\boldsymbol{B}^{-1}\boldsymbol{G}^T\boldsymbol{\lambda} \tag{6.99}$$

where the symmetry of matrices $\boldsymbol{A}$ and $\boldsymbol{B}$ has been used. The boundary condition for $\boldsymbol{\lambda}(t)$, as obtained from Equation (6.13), is

$$\boldsymbol{\lambda}^T(t_f) = \left.\frac{\partial \phi}{\partial \boldsymbol{x}}\right|_{t=t_f} = \boldsymbol{x}^T(t_f)\boldsymbol{S}_f$$

or

$$\boldsymbol{\lambda}(t_f) = \boldsymbol{S}_f\boldsymbol{x}(t_f). \tag{6.100}$$

Substitution of $\boldsymbol{u}$ from Equation (6.99) into the state equation results in

$$\dot{\boldsymbol{x}} = \boldsymbol{F}\boldsymbol{x} - \boldsymbol{G}\boldsymbol{B}^{-1}\boldsymbol{G}^T\boldsymbol{\lambda} \tag{6.101}$$

which, together with the $\boldsymbol{\lambda}$ equation, can be written in matrix from as

$$\begin{bmatrix} \dot{\boldsymbol{x}} \\ \hline \dot{\boldsymbol{\lambda}} \end{bmatrix} = \left[\begin{array}{c|c} \boldsymbol{F} & -\boldsymbol{G}\boldsymbol{B}^{-1}\boldsymbol{G}^T \\ \hline -\boldsymbol{A} & -\boldsymbol{F}^T \end{array}\right] \begin{bmatrix} \boldsymbol{x} \\ \hline \boldsymbol{\lambda} \end{bmatrix}. \tag{6.102}$$

This equation is the standard form for homogeneous linear equations and therefore has a solution in terms of a state transition matrix,

$$\begin{bmatrix} \boldsymbol{x}(t_1) \\ \hline \boldsymbol{\lambda}(t_1) \end{bmatrix} = \left[\begin{array}{c|c} \boldsymbol{\Phi}_{11}(t_1, t_2) & \boldsymbol{\Phi}_{12}(t_1, t_2) \\ \hline \boldsymbol{\Phi}_{21}(t_1, t_2) & \boldsymbol{\Phi}_{22}(t_1, t_2) \end{array}\right] \begin{bmatrix} \boldsymbol{x}(t_2) \\ \hline \boldsymbol{\lambda}(t_2) \end{bmatrix} \tag{6.103}$$

where the state transition matrix has been partitioned as indicated by the partitioning of the vector. Now, if

$$t_1 = t \quad \text{and} \quad t_2 = t_f$$

solving for $\boldsymbol{x}(t)$ and $\boldsymbol{\lambda}(t)$ yields

$$\begin{aligned} \boldsymbol{x}(t) &= [\boldsymbol{\Phi}_{11}(t, t_f) + \boldsymbol{\Phi}_{12}(t, t_f)\boldsymbol{S}_f]\boldsymbol{x}(t_f) \\ \boldsymbol{\lambda}(t) &= [\boldsymbol{\Phi}_{21}(t, t_f) + \boldsymbol{\Phi}_{22}(t, t_f)\boldsymbol{S}_f]\boldsymbol{x}(t_f) \end{aligned} \tag{6.104}$$

where Equation (6.100) has been used to eliminate $\boldsymbol{\lambda}(t_f)$. Furthermore, if $t = t_0$ in the $\boldsymbol{x}(t)$ equation above, solving for $\boldsymbol{x}(t_f)$ yields

$$\boldsymbol{x}(t_f) = [\boldsymbol{\Phi}_{11}(t_0, t_f) + \boldsymbol{\Phi}_{12}(t_0, t_f)\boldsymbol{S}_f]^{-1}\boldsymbol{x}(t_0). \tag{6.105}$$

Then, substitution of this expression into the $\boldsymbol{\lambda}(t)$ expression of Equation (6.104) yields

$$\begin{aligned} \boldsymbol{\lambda}(t) &= [\boldsymbol{\Phi}_{21}(t, t_f) + \boldsymbol{\Phi}_{22}(t, t_f)\boldsymbol{S}_f]\,[\boldsymbol{\Phi}_{11}(t_0, t_f) + \boldsymbol{\Phi}_{12}(t_0, t_f)\boldsymbol{S}_f]^{-1}\boldsymbol{x}(t_0) \\ &= \boldsymbol{S}(t, t_0)\boldsymbol{x}(t_0) \end{aligned} \tag{6.106}$$

where the definition of the matrix $\boldsymbol{S}(t, t_0)$ is implied. If $\boldsymbol{\lambda}(t)$ is known, then the control function $\boldsymbol{u}(t)$ from Equation (6.99) is

$$\boldsymbol{u}(t) = -\boldsymbol{B}^{-1}(t)\boldsymbol{G}^T(t)\boldsymbol{S}(t, t_0)\boldsymbol{x}(t_0) \tag{6.107}$$

implying that the initial condition $\boldsymbol{x}(t_0)$ serves to specify the control vector $\boldsymbol{u}(t)$ for the interval (t_0, t_f), for fixed t_f. This development is in essence a general solution to the two-point boundary value problem for the particular case of linear systems and quadratic performance indices. The matrix $\boldsymbol{S}(t, t_0)$ can, at least in principle, be determined by use of the system characteristics as contained in the matrix $\boldsymbol{F}(t)$. All other terms of Equation (6.107) are known from the problem specification.

The fact that the control vector $\boldsymbol{u}(t)$ is completely determined from knowledge of the initial state $\boldsymbol{x}(t_0)$ implies an open loop type of control, in which no check is made on the development of the system state with time to insure that the optimum path is actually being followed. That is, if $\boldsymbol{u}(t)$ is determined for the interval (t_0, t_f), then it in turn prescribes an optimum path $\boldsymbol{x}(t)$ during the interval. In most cases of practical interest this optimum path is not actually achieved, for several reasons. The two primary causes of this divergence of the desired path and that actually produced are improper modeling and random effects acting on the system. Improper modeling is the result of applying linear analysis to nonlinear systems or of imprecise determination of the matrices $\boldsymbol{F}$ and $\boldsymbol{G}$. The random effects which can act on a system include those factors about which the analyst or designer may have information, but which he cannot predict or control. As will be shown in the next chapter, these effects can be accounted for in a probabilistic way, but for the present purposes they represent a disturbance which produces a nonoptimal path $\boldsymbol{x}(t)$ even though the control vector has been properly computed using Equation (6.107).

To account for these effects, the designer usually provides for some kind of feedback control, in which the system state $\boldsymbol{x}(t)$ is monitored at specific points during the interval (t_0, t_f) to determine whether or not the optimal path is actually being followed. Suppose that the first such observation is made at time t_1, and it is found that the state $\boldsymbol{x}(t_1)$ does not fall on the previously computed optimal path $\boldsymbol{x}(t)$. Given the fact that this deviation from the optimal path has occurred, what is the best procedure to follow from time t_1? This question can be answered by considering t_1 as a new initial condition and by recomputing the optimum path from the expression

$$\boldsymbol{u}(t) = -\boldsymbol{B}^{-1}(t)\boldsymbol{G}^T(t)\boldsymbol{S}(t, t_1)\boldsymbol{x}(t_1). \tag{6.108}$$

The result of such a procedure is shown symbolically in Figure 6-7, in which the time history of one of the states $x_i(t)$ is represented. The dashed line indicates the originally determined optimal path computed from $\boldsymbol{x}(t_0)$. Since $\boldsymbol{u}(t)$ is different depending upon whether it is derived from $\boldsymbol{x}(t_0)$ or $\boldsymbol{x}(t_1)$, a new control function is obtained after the observation at time t_1. If this process is repeated at regular intervals $t_1, t_2, t_3, \ldots$, then at each time point a new optimal path is defined from that time to t_f, and is superseded when the next observation is made. Such a procedure results in a sampled-data feedback control, represented symbolically in Figure 6-8. Note that, at the time of each observation t_l, a new

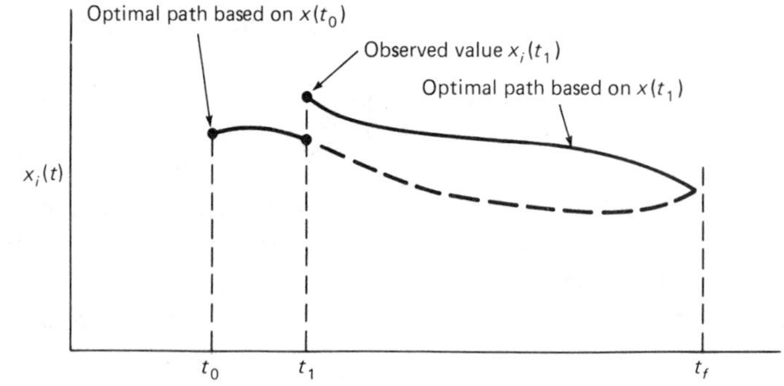

Figure 6-7. Optimal path of state $x_i(t)$.

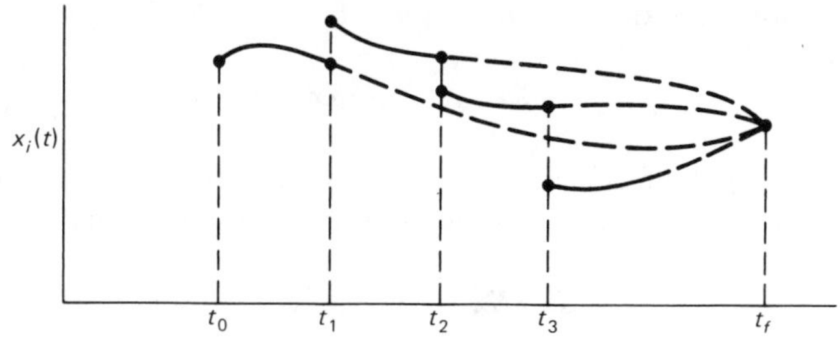

Figure 6-8. A sampled-data feedback control.

control vector $\boldsymbol{u}(t)$ is defined over the interval (t_l, t_f), but it is used only until the time of the next observation, at which point a new computation is made.

If the time between sampling instants becomes very small in the above procedure, then in the limit as

$$\Delta t = t_{l+1} - t_l \to 0 \tag{6.109}$$

the general expression for the sampled-data feedback control,

$$\boldsymbol{u}(t) = -\boldsymbol{B}^{-1}(t)\boldsymbol{G}^T(t)\boldsymbol{S}(t, t_l)\boldsymbol{x}(t_l) \tag{6.110}$$

which is valid only over the interval (t_l, t_{l+1}), becomes

$$\boldsymbol{u}(t) = -\boldsymbol{B}^{-1}(t)\boldsymbol{G}^T(t)\boldsymbol{S}(t)\boldsymbol{x}(t) \tag{6.111}$$

and a continuous time feedback control law is defined. The matrix $\boldsymbol{S}(t)$, from previous consideration, is given by the continuous time form of Equation (6.106)

$$\boldsymbol{S}(t) = [\boldsymbol{\Phi}_{21}(t, t_f) + \boldsymbol{\Phi}_{22}(t, t_f)\boldsymbol{S}_f]\,[\boldsymbol{\Phi}_{11}(t, t_f) + \boldsymbol{\Phi}_{12}(t, t_f)\boldsymbol{S}_f]^{-1}. \tag{6.112}$$

Although the matrix $\boldsymbol{S}(t)$ is derivable as a function of time from knowledge of the system matrix $\boldsymbol{F}(t)$, using Equation (6.112), a preferable method can be derived from the fact that, along the optimal path, $\boldsymbol{\lambda}(t)$ and $\boldsymbol{x}(t)$ are related by the expression

$$\boldsymbol{\lambda}(t) = \boldsymbol{S}(t)\boldsymbol{x}(t). \tag{6.113}$$

This conclusion follows directly from Equation (6.106) and the limiting process defined above. If this expression is substituted into Equation (6.101), the result is

$$\dot{\boldsymbol{x}} = \boldsymbol{F}\boldsymbol{x} - \boldsymbol{G}\boldsymbol{B}^{-1}\boldsymbol{G}^T\boldsymbol{S}\boldsymbol{x}. \tag{6.114}$$

Also, substitution into the first expression of Equation (6.99) yields

$$\dot{\boldsymbol{\lambda}} = -\boldsymbol{A}\boldsymbol{x} - \boldsymbol{F}^T\boldsymbol{S}\boldsymbol{x}. \tag{6.115}$$

By differentiating $\boldsymbol{\lambda}$ of Equation (6.113) with respect to time and substituting the result into the left-hand side of the above expression, one obtains

$$\dot{\boldsymbol{S}}\boldsymbol{x} + \boldsymbol{S}\dot{\boldsymbol{x}} = -\boldsymbol{A}\boldsymbol{x} - \boldsymbol{F}^T\boldsymbol{S}\boldsymbol{x} \tag{6.116}$$

which, upon substitution from Equation (6.114), yields, in factored form,

$$[\dot{\boldsymbol{S}} + \boldsymbol{S}\boldsymbol{F} + \boldsymbol{F}^T\boldsymbol{S} - \boldsymbol{S}\boldsymbol{G}\boldsymbol{B}^{-1}\boldsymbol{G}^T\boldsymbol{S} + \boldsymbol{A}]\boldsymbol{x} = 0. \tag{6.117}$$

This equation holds for general $\boldsymbol{x}(t)$ only if the expression within the brackets is zero, which in turn implies satisfaction of the matrix differential equation

$$\dot{\boldsymbol{S}}(t) = -\boldsymbol{S}(t)\boldsymbol{F}(t) - \boldsymbol{F}^T(t)\boldsymbol{S}(t) + \boldsymbol{S}(t)\boldsymbol{G}(t)\boldsymbol{B}^{-1}(t)\boldsymbol{G}^T(t)\boldsymbol{S}(t) - \boldsymbol{A}(t). \tag{6.118}$$

The boundary condition for this equation is obtained from comparison of Equation (6.100) with Equation (6.113) evaluated at $t = t_f$:

$$\boldsymbol{S}(t_f) = \boldsymbol{S}_f. \tag{6.119}$$

Equation (6.118) is known as the *matrix Riccati equation* and is solved backward in time from the known terminal condition $\boldsymbol{S}_f$. Once $\boldsymbol{S}(t)$ is determined, the optimum feedback control law is obtained directly as

$$\boldsymbol{u}(t) = -\boldsymbol{B}^{-1}(t)\boldsymbol{G}^T(t)\boldsymbol{S}(t)\boldsymbol{x}(t). \tag{6.120}$$

Because of the third term on the right-hand side of Equation (6.118) the Riccati equation is nonlinear, but it can be quite readily integrated by use of numerical techniques on a digital computer. The attractive result is that iterative procedures are not required to solve the two-point boundary value problem. Also, several special techniques for solution of equations of this form have recently been developed. Using the fact that $\boldsymbol{S}_f$ is positive definite, it can easily be shown that $\boldsymbol{S}(t)$ is positive definite for all t.

In the special case in which the matrix $\boldsymbol{A}$ is zero, then the nonlinear Riccati equation can be reduced to a linear equation in $\boldsymbol{S}^{-1}(t)$ by means of the property derived in Chapter 4,

$$\dot{\boldsymbol{S}}^{-1} = \boldsymbol{S}^{-1}\dot{\boldsymbol{S}}\boldsymbol{S}^{-1}. \tag{6.121}$$

Specifically, if Equation (6.118) is multiplied on the left and right by $\boldsymbol{S}^{-1}$, the result is, when $\boldsymbol{A}$ is zero,

$$\boldsymbol{S}^{-1}\dot{\boldsymbol{S}}\boldsymbol{S}^{-1} = -\boldsymbol{F}\boldsymbol{S}^{-1} - \boldsymbol{S}^{-1}\boldsymbol{F}^T + \boldsymbol{G}\boldsymbol{B}^{-1}\boldsymbol{G}^T \tag{6.122}$$

which, in view of Equation (6.121), yields

$$\dot{\boldsymbol{S}}^{-1}(t) = \boldsymbol{F}(t)\boldsymbol{S}^{-1}(t) + \boldsymbol{S}^{-1}(t)\boldsymbol{F}^T(t) - \boldsymbol{G}(t)\boldsymbol{B}^{-1}(t)\boldsymbol{G}^T(t). \tag{6.123}$$

This equation is linear in $\boldsymbol{S}^{-1}$, with the boundary condition

$$\boldsymbol{S}^{-1}(t_f) = \boldsymbol{S}_f^{-1}. \tag{6.124}$$

This expression is useful, especially in hand computation, since it replaces the nonlinear Riccati equation with the linear result given in Equation (6.123). The disadvantage, of course, is the requirement for inversion of the $\boldsymbol{S}^{-1}$ matrix once it has been obtained.

As an example of these techniques, consider the second-order system described by the equations

$$\dot{x}_1 = x_2$$
$$\dot{x}_2 = u$$

which is representative of linear motion in one dimension. The performance index is specified to be

$$J = \tfrac{1}{2}[c_1 x_1^2(t_f) + c_2 x_2^2(t_f)] + \tfrac{1}{2}\int_{t_0}^{t_f} u^2(t)\,dt.$$

This problem is similar to that considered in Section 6.2.2, but now the performance index is quadratic. To correlate this problem with the general problem already solved, the following associations are made:

$$\boldsymbol{F} = \begin{bmatrix} 0 & 1 \\ 0 & 0 \end{bmatrix}; \quad \boldsymbol{G} = \begin{bmatrix} 0 \\ 1 \end{bmatrix}; \quad \boldsymbol{S}_f = \begin{bmatrix} c_1 & 0 \\ 0 & c_2 \end{bmatrix}; \quad \boldsymbol{A} = 0; \quad \boldsymbol{B} = 1.$$

Since $\boldsymbol{A}$ is zero, the linear expression of Equation (6.123) may be used. Let the matrices $\boldsymbol{S}$ and $\boldsymbol{S}^{-1}$ be represented in component form as

$$\boldsymbol{S} = \begin{bmatrix} s_1 & s_2 \\ s_2 & s_3 \end{bmatrix} \quad \text{and} \quad \boldsymbol{S}^{-1} = \begin{bmatrix} s^1 & s^2 \\ s^2 & s^3 \end{bmatrix}.$$

The symmetric form of $\boldsymbol{S}$, as required by its defining equation, has been assumed. Then, from Equation (6.123)

$$\begin{bmatrix} \dot{s}^1 & \dot{s}^2 \\ \dot{s}^2 & \dot{s}^3 \end{bmatrix} = \begin{bmatrix} 0 & 1 \\ 0 & 0 \end{bmatrix}\begin{bmatrix} s^1 & s^2 \\ s^2 & s^3 \end{bmatrix} + \begin{bmatrix} s^1 & s^2 \\ s^2 & s^3 \end{bmatrix}\begin{bmatrix} 0 & 0 \\ 1 & 0 \end{bmatrix} - \begin{bmatrix} 0 \\ 1 \end{bmatrix}[1][0 \quad 1]$$

$$= \begin{bmatrix} s^2 & s^3 \\ 0 & 0 \end{bmatrix} + \begin{bmatrix} s^2 & 0 \\ s^3 & 0 \end{bmatrix} - \begin{bmatrix} 0 & 0 \\ 0 & 1 \end{bmatrix}.$$

This expression represents three ordinary, coupled, differential equations, which are

$$\dot{s}^1 = 2s^2$$
$$\dot{s}^2 = s^3$$
$$\dot{s}^3 = -1$$

with boundary condition

$$\boldsymbol{S}^{-1}(t_f) = \boldsymbol{S}_f^{-1} = \begin{bmatrix} 1/c_1 & 0 \\ 0 & 1/c_2 \end{bmatrix}.$$

The solution of these equations is

$$s^3 = (t_f - t) + (1/c_2)$$
$$s^2 = -[(t_f - t)^2/2] - [(t_f - t)/c_2]$$
$$s^1 = [(t_f - t)^3/3] + [(t_f - t)^2/c_2] + (1/c_1)$$

which yields

$$\boldsymbol{S}^{-1}(t) = \begin{bmatrix} [(t_f - t)^3/3] + [(t_f - t)^2/c_2] + (1/c_1) & -[(t_f - t)^2/2] - [(t_f - t)/c_2] \\ -[(t_f - t)^2/2] - [(t_f - t)/c_2] & (t_f - t) + (1/c_2) \end{bmatrix}.$$

Inversion yields the matrix $\boldsymbol{S}$:

$$\boldsymbol{S} = \begin{bmatrix} s_1 & s_2 \\ s_2 & s_3 \end{bmatrix}$$

$$= \frac{1}{D(t)}\begin{bmatrix} (t_f - t) + (1/c_2) & [(t_f - t)^2/2] + [(t_f - t)/c_2] \\ [(t_f - t)^2/2] + [(t_f - t)/c_2] & [(t_f - t)^3/3] + [(t_f - t)^2/c_2] + (1/c_1) \end{bmatrix}$$

where $D(t)$ is the determinant and has the value

$$D(t) = [(1/c_2) + (t_f - t)]\{(t_f - t)^3/3 + (1/c_1)\} - [(t_f - t)^4/4]$$

Now that $S(t)$ has been obtained, the optimal feedback control law follows from Equation (6.120):

$$\begin{aligned} u(t) &= -B^{-1}G^T S(t)x(t) \\ &= -[1][0 \quad 1]\begin{bmatrix} s_1 & s_2 \\ s_2 & s_3 \end{bmatrix}\begin{bmatrix} x_1 \\ x_2 \end{bmatrix} \\ &= -s_2 x_1(t) - s_3 x_2(t). \end{aligned}$$

Then, combining these results produces the control law

$$u(t) = -\left[\frac{\dfrac{(t_f - t)^2}{2} + \dfrac{(t_f - t)}{c_2}}{D(t)}\right] x_1(t) - \left[\frac{\dfrac{(t_f - t)^3}{3} + \dfrac{(t_f - t)^2}{c_2} + \dfrac{1}{c_1}}{D(t)}\right] x_2(t).$$

This equation gives the feedback control law which will minimize the performance index J as prescribed in the problem formulation. Just what this minimization means is dependent upon the relative values of c_1, c_2, and unity (the coefficient of the u^2 term). When c_1 is large, $x_1(t_f)$ will tend to be near zero, possibly at the expense of large values of $x_2(t_f)$ and u. On the other hand, if both c_1 and c_2 are small compared to unity, then $u(t)$ will tend to be small, at the expense of large terminal values of x_1 an x_2. This kind of tradeoff relationship often exists in most systems of practical import; the type of analysis presented here provides a systematic approach to the design of feedback controllers in such situations.

For example, suppose that $x_2(t_f)$ is of no concern, in which case $c_2 \to 0$ and the resulting feedback control law is

$$u(t) = -\left[\frac{(t_f - t)}{\dfrac{(t_f - t)^3}{3} + \dfrac{1}{c_1}}\right] x_1(t) - \left[\frac{(t_f - t)^2}{\dfrac{(t_f - t)^3}{3} + \dfrac{1}{c_1}}\right] x_2(t).$$

This expression is obtained by multiplying the previous equation in the numerator and the denominator by c_2, and then taking the limit as c_2 gets very small. It gives the optimum control function when the tradeoff is between the value of terminal value of x_1 and the integral of $u^2(t)$. Note that in the limiting case $c_1 \to 0$ this expression yields $u(t) = 0$, which gives the minimum value of $u(t)$ if the terminal conditions are not considered. On the other hand, when $c_1 \to \infty$

then

$$u(t) = -[(t_f - t)^2/3]x_1(t) - [3/(t_f - t)]x_2(t)$$

which is the result of classical analysis with no regard for the value of $u(t)$.

6.3.2. The Linear Regulator

One of the basic stipulations of the regulator problem is that the final time is very far in the future. That is, the regulator attempts to maintain the state of the system close to some reference value, with use of a reasonable amount of control, and with no regard for the terminal states. The terms *close* and *reasonable* are defined by the designer and serve to determine the matrices $\boldsymbol{A}$ and $\boldsymbol{B}$ in the performance index

$$J = \frac{1}{2} \int_{t_0}^{t_f} [\boldsymbol{x}^T(t)\boldsymbol{A}(t)\boldsymbol{x}(t) + \boldsymbol{u}^T(t)\boldsymbol{B}(t)\boldsymbol{u}(t)]\, dt. \tag{6.125}$$

For reasons discussed previously, the reference value for $\boldsymbol{x}(t)$ in this situation is assumed to be zero, but generalization to nonzero reference values is possible. Now consider the controller problem of the last section, with the general performance index given by Equation (6.95), and assume that the terminal time is far in the future. The approach to this problem involves solution of the matrix Riccati equation backward in time from the terminal time t_f, using the boundary condition

$$\boldsymbol{S}(t_f) = \boldsymbol{S}_f. \tag{6.126}$$

If the Riccati equation is time invariant–the matrices $\boldsymbol{F}$, $\boldsymbol{G}$, $\boldsymbol{A}$, and $\boldsymbol{B}$ are not functions of time–then the solution for $\boldsymbol{S}(t)$ may reach steady state conditions for t far removed from t_f. But this relationship between t and t_f is a basic characteristic of the regulator problem and permits use of steady state values in determining the control vector. This concept is shown symbolically in Figure 6-9, where the *ij*th element of the $\boldsymbol{S}$ matrix is shown reaching steady state conditions while responding to different boundary conditions at the terminal time. It is important to realize that steady state responses are independent of the particular boundary condition, a fact that permits deletion of the term $\boldsymbol{x}^T\boldsymbol{S}_f\boldsymbol{x}$ from the performance index of Equation (6.125). The steady state solution of the Riccati equation can be determined from solution of the algebraic equation

$$\dot{\boldsymbol{S}} = 0 = -\boldsymbol{S}\boldsymbol{F} - \boldsymbol{F}^T\boldsymbol{S} + \boldsymbol{S}\boldsymbol{G}\boldsymbol{B}^{-1}\boldsymbol{G}^T\boldsymbol{S} - \boldsymbol{A}. \tag{6.127}$$

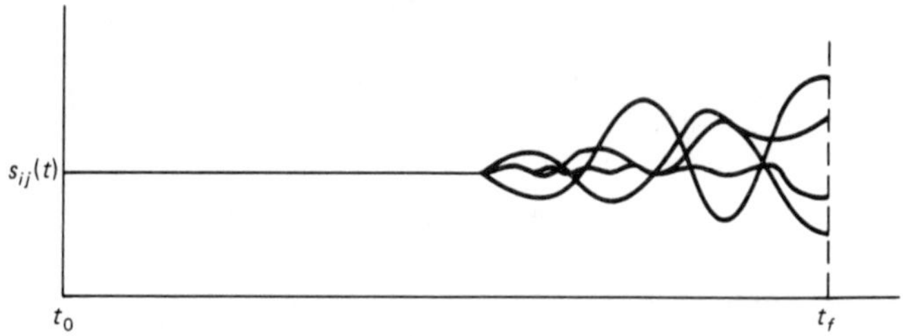

Figure 6-9. Asymptotic behavior of Riccati equations solved backward in time.

This set of simultaneous equations can be solved for $\boldsymbol{S}$, the steady state value of $\boldsymbol{S}(t)$. The optimal control vector is then simply

$$\boldsymbol{u}(t) = -\boldsymbol{B}^{-1}\boldsymbol{G}^T\boldsymbol{S}\boldsymbol{x}(t) \tag{6.128}$$

which is a time invariant linear function of the state.

As an example, consider the terminal controller of the previous section, except that now it is desired to maintain $x_1(t)$ and $x_2(t)$ near zero over a long period of time, while simultaneously using an acceptable amount of control. The performance index is

$$J = \frac{1}{2}\int_{t_0}^{t_f} (a_1 x_1^2 + a_2 x_2^2 + u^2)\, dt$$

and the relative weighting of $x_1(t)$, $x_2(t)$, and $u(t)$ in performing the minimization is determined by the values of a_1 and a_2 compared to unity. Here, correspondence to the general problem is obtained by defining the matrices

$$\boldsymbol{F} = \begin{bmatrix} 0 & 1 \\ 0 & 0 \end{bmatrix}; \quad \boldsymbol{G} = \begin{bmatrix} 0 \\ 1 \end{bmatrix}; \quad \boldsymbol{A} = \begin{bmatrix} a_1 & 0 \\ 0 & a_2 \end{bmatrix}; \quad \boldsymbol{B} = [1].$$

The algebraic Riccati equation used to determine the steady state solution is

$$-\begin{bmatrix} s_1 & s_2 \\ s_2 & s_3 \end{bmatrix}\begin{bmatrix} 0 & 1 \\ 0 & 0 \end{bmatrix} - \begin{bmatrix} 0 & 0 \\ 1 & 0 \end{bmatrix}\begin{bmatrix} s_1 & s_2 \\ s_2 & s_3 \end{bmatrix} + \begin{bmatrix} s_1 & s_2 \\ s_2 & s_3 \end{bmatrix}\begin{bmatrix} 0 \\ 1 \end{bmatrix}[1][0 \quad 1]\begin{bmatrix} s_1 & s_2 \\ s_2 & s_3 \end{bmatrix} - \begin{bmatrix} a_1 & 0 \\ 0 & a_2 \end{bmatrix} = 0$$

Performing the necessary multiplications and equating terms yield the set of algebraic equations

$$s_2^2 = a_1$$
$$-s_1 + s_2 s_3^2 = 0$$
$$-2s_2 + s_3^2 = a_2$$

which have the solution

$$s_1 = \sqrt{a_1} + (a_2 + 2\sqrt{a_1})$$
$$s_2 = \sqrt{a_1}$$
$$s_3 = \sqrt{a_2 + 2\sqrt{a_1}}.$$

The optimal feedback control is obtained from Equation (6.128):

$$u(t) = -\boldsymbol{B}^{-1}\boldsymbol{G}^T\boldsymbol{S}\boldsymbol{x}(t) = -[1][0 \quad 1]\begin{bmatrix} s_1 & s_2 \\ s_2 & s_3 \end{bmatrix}\begin{bmatrix} x_1(t) \\ x_2(t) \end{bmatrix}$$
$$= -s_2 x_1(t) - s_3 x_2(t)$$

or, upon substitution for s_2 and s_3,

$$u(t) = -\sqrt{a_1}x_1(t) - [\sqrt{a_2 + 2\sqrt{a_1}}]x_2(t).$$

This expression gives the feedback control law which minimizes J, so long as t_f is far in the future. In block diagram form, this control law is shown in Figure 6-10.

The regulator tends to maintain the values of $x_1(t)$ and $x_2(t)$ near zero, in the face of disturbance inputs to the system. Note that if the value of $x_2(t)$ is not important, then a_2 is selected to be zero, and the control law is

$$u(t) = -\sqrt{a_1}x_1(t) - [\sqrt{2\sqrt{a_1}}]x_2(t)$$

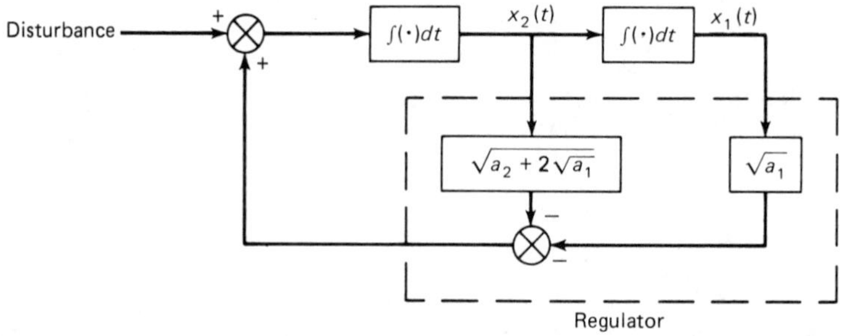

Figure 6-10. Block diagram of linear regulator.

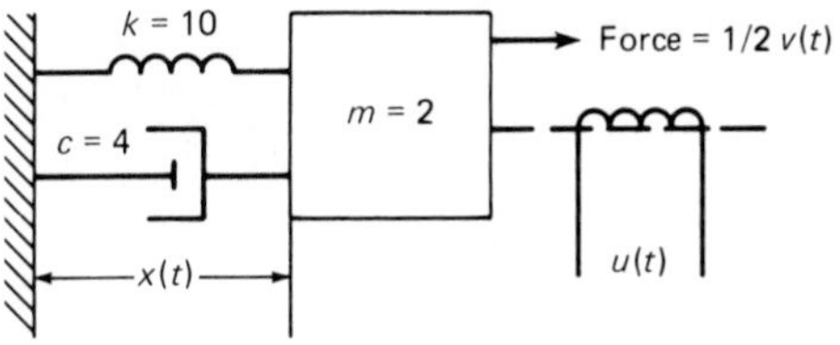

Figure 6-11. Spring-mass-damper model.

which still depends on both states. This result is in accord with the general characteristic that the optimum control is a linear function of the state, regardless of the form of the particular performance index.

As a second example, consider the spring-mass-damper problem of Section 4.2, and depicted in Figure 6-11 in a slightly modified form. This model is a practical one, since many real-world systems can be adequately modeled by a second-order system. The system is described by the matrices

$$\boldsymbol{F} = \begin{bmatrix} 0 & 1 \\ -5 & -2 \end{bmatrix}; \quad \boldsymbol{G} = \begin{bmatrix} 0 \\ \frac{1}{2} \end{bmatrix}.$$

Suppose that it is desired to design an optimal regulator to maintain the state of this system near its equilibrium value using the performance index

$$J = \int_{t_0}^{t_f} [56x_1^2 + 24x_2^2 + (u^2/4)]\, dt.$$

The $\boldsymbol{A}$ and $\boldsymbol{B}$ matrices of Equation (6.125) corresponding to this performance index are

$$\boldsymbol{A} = \begin{bmatrix} 56 & 0 \\ 0 & 24 \end{bmatrix}; \quad \boldsymbol{B} = [\tfrac{1}{4}].$$

Since this is a regulator problem, the algebraic Riccati equation can be used. From Equation (6.127), then,

$$-\begin{bmatrix} s_1 & s_2 \\ s_2 & s_3 \end{bmatrix}\begin{bmatrix} 0 & 1 \\ -5 & -2 \end{bmatrix} - \begin{bmatrix} 0 & -5 \\ 1 & -2 \end{bmatrix}\begin{bmatrix} s_1 & s_2 \\ s_2 & s_3 \end{bmatrix}$$

$$+ \begin{bmatrix} s_1 & s_2 \\ s_2 & s_3 \end{bmatrix}\begin{bmatrix} 0 \\ \frac{1}{2} \end{bmatrix}[4][0 \quad \tfrac{1}{2}]\begin{bmatrix} s_1 & s_2 \\ s_2 & s_3 \end{bmatrix} - \begin{bmatrix} 56 & 0 \\ 0 & 24 \end{bmatrix} = 0.$$

There are three independent equations:

$$10s_2 + s_2^2 - 56 = 0$$
$$-s_1 + 2s_2 + 5s_3 + s_2 s_3 = 0$$
$$-2s_2 + 4s_3 + s_3^2 - 24 = 0$$

with solution

$$s_1 = 44; \quad s_2 = 4; \quad s_3 = 4.$$

Although there are other solutions, that specified above is the only solution resulting in a positive definite matrix solution

$$\boldsymbol{S} = \begin{bmatrix} 44 & 4 \\ 4 & 4 \end{bmatrix}.$$

The optimal feedback control function then follows from Equation (6.128):

$$u(t) = -[4][0 \quad \tfrac{1}{2}] \begin{bmatrix} 44 & 4 \\ 4 & 4 \end{bmatrix} \begin{bmatrix} x_1(t) \\ x_2(t) \end{bmatrix}$$
$$= [-8 \quad -8] \begin{bmatrix} x_1(t) \\ x_2(t) \end{bmatrix}.$$

Thus, the optimal feedback gain matrix is

$$\boldsymbol{C} = [-8 \quad -8].$$

The response characteristics of the optimum closed-loop system can be inferred from the closed-loop system matrix of Equation (5.6):

$$\tilde{\boldsymbol{F}} = \boldsymbol{F} + \boldsymbol{G}\boldsymbol{C} = \begin{bmatrix} 0 & 1 \\ -5 & -2 \end{bmatrix} + \begin{bmatrix} 0 \\ \tfrac{1}{2} \end{bmatrix} [-8 \quad -8]$$
$$= \begin{bmatrix} 0 & 1 \\ -9 & -6 \end{bmatrix}$$

which has the eigenvalues

$$\lambda_1 = -3; \quad \lambda_2 = -3.$$

Thus, the optimum system is critically damped with time constant equal to $\frac{1}{3}$. The open-loop system has eigenvalues

$$\lambda_1 = -1 + 2i; \qquad \lambda_2 = -1 - 2i$$

indicating an underdamped oscillatory response. The critically damped optimal response is a result of the particular performance index selected. Different performance indices, of course, produce different transient characteristics.

The reader should note that the control system as specified in this problem implements state variable feedback. This feature implies that both state variables—position and velocity—are capable of being measured and used for feedback purposes. In a practical system of this kind, velocity measurements would be difficult or impossible to obtain, leaving the designer with only position measurements available for feedback control purposes. A logical choice in such a situation is to use a linear observer to estimate the velocity, and then to use this estimate in the optimal state variable feedback system. This concept is discussed further in Section 6.5.

6.3.3. The Linear Follower

A generalization of the linear regulator is obtained when the system is required to follow, or track, a time varying reference signal $\boldsymbol{r}(t)$, which is known. Here, the performance index is

$$J = \tfrac{1}{2}[\boldsymbol{x}(t_f) - \boldsymbol{r}(t_f)]^T \boldsymbol{S}_f [\boldsymbol{x}(t_f) - \boldsymbol{r}(t_f)] + \tfrac{1}{2}\int_{t_0}^{t_f} \{[\boldsymbol{x}(t) - \boldsymbol{r}(t)]^T \boldsymbol{A}(t)[\boldsymbol{x}(t) - \boldsymbol{r}(t)] + [\boldsymbol{u}^T(t)\boldsymbol{B}(t)\boldsymbol{u}(t)]\}\, dt$$

and the purpose is to maintain $\boldsymbol{x}(t)$ close to the reference value $\boldsymbol{r}(t)$. Proceeding as in the case of the linear controller results in

$$H = L + {}^T f = \tfrac{1}{2}[(\boldsymbol{x} - \boldsymbol{r})^T \boldsymbol{A}(\boldsymbol{x} - \boldsymbol{r}) + \boldsymbol{u}^T \boldsymbol{B}\boldsymbol{u}] + \boldsymbol{\lambda}^T \boldsymbol{F}\boldsymbol{x} + \boldsymbol{\lambda}^T \boldsymbol{G}\boldsymbol{u} \tag{6.129}$$

and

$$\dot{\boldsymbol{\lambda}}^T = -\frac{\partial H}{\partial \boldsymbol{x}} = -(\boldsymbol{x} - \boldsymbol{r})^T \boldsymbol{A} - \boldsymbol{\lambda}^T \boldsymbol{F}. \tag{6.130}$$

Analogously to the derivation of Equation (6.102), there results for this case

$$\begin{bmatrix} \dot{\boldsymbol{x}} \\ \hline \dot{\boldsymbol{\lambda}} \end{bmatrix} = \left[\begin{array}{c|c} \boldsymbol{F} & -\boldsymbol{G}\boldsymbol{B}^{-1}\boldsymbol{G}^T \\ \hline -\boldsymbol{A} & -\boldsymbol{F}^T \end{array}\right] \begin{bmatrix} \boldsymbol{x} \\ \hline \boldsymbol{\lambda} \end{bmatrix} + \begin{bmatrix} \boldsymbol{0} \\ \hline \boldsymbol{A}\boldsymbol{r} \end{bmatrix}. \tag{6.131}$$

The homogeneous part of this equation is identical to Equation (6.102), but the boundary condition is now

$$\boldsymbol{\lambda}(t_f) = \boldsymbol{S}_f \boldsymbol{x}(t_f) - \boldsymbol{S}_f \boldsymbol{r}(t_f) \tag{6.132}$$

as required by Equation (6.13). Furthermore, these boundary conditions must be applied to the general solution rather than the homogeneous solution. The homogeneous solution will be of the form of Equation (6.103) and the particular integral will be

$$\int_{t_1}^{t_2} \left[\begin{array}{c|c} \boldsymbol{\Phi}_{11}(t_2, \sigma) & \boldsymbol{\Phi}_{12}(t_2, \sigma) \\ \hline \boldsymbol{\Phi}_{21}(t_2, \sigma) & \boldsymbol{\Phi}_{22}(t_2, \sigma) \end{array}\right] \left[\begin{array}{c} \boldsymbol{0} \\ A(\sigma) r(\sigma) \end{array}\right] d\sigma = \left[\begin{array}{c} \boldsymbol{s}_1(t_1, t_2) \\ \hline \boldsymbol{s}_2(t_1, t_2) \end{array}\right]. \tag{6.133}$$

Combining this result with Equation (6.103) then yields the general solution

$$\begin{aligned} \boldsymbol{x}(t_1) &= \boldsymbol{\Phi}_{11}(t_1, t_2)\boldsymbol{x}(t_2) + \boldsymbol{\Phi}_{12}(t_1, t_2)\boldsymbol{\lambda}(t_2) + \boldsymbol{s}_1(t_1, t_2) \\ \boldsymbol{\lambda}(t_1) &= \boldsymbol{\Phi}_{21}(t_1, t_2)\boldsymbol{x}(t_2) + \boldsymbol{\Phi}_{22}(t_1, t_2)\boldsymbol{\lambda}(t_2) + \boldsymbol{s}_2(t_1, t_2). \end{aligned} \tag{6.134}$$

Letting

$$t_1 = t_f \quad \text{and} \quad t_2 = t$$

in this equation and substituting into Equation (6.132) yields, upon solving for $\boldsymbol{\lambda}(t)$,

$$\begin{aligned} \boldsymbol{\lambda}(t) = {} & [\boldsymbol{\Phi}_{22}(t_f, t) - \boldsymbol{S}_f \boldsymbol{\Phi}_{12}(t_f, t)]^{-1} [\boldsymbol{S}_f \boldsymbol{\Phi}_{11}(t_f, t) - \boldsymbol{\Phi}_{21}(t_f, t)] x(t) \\ & + [\boldsymbol{\Phi}_{22}(t_f, t) - \boldsymbol{S}_f \boldsymbol{\Phi}_{12}(t_f, t)]^{-1} [\boldsymbol{S}_f \boldsymbol{s}_1(t_f, t) - \boldsymbol{s}_2(t_f, t) - \boldsymbol{S}_f \boldsymbol{x}(t_f)]. \end{aligned} \tag{6.135}$$

By defining the matrix $\boldsymbol{S}(t)$ and the vector $\boldsymbol{s}(t)$ appropriately, Equation (6.135) can be written as

$$\boldsymbol{\lambda}(t) = \boldsymbol{S}(t)\boldsymbol{x}(t) + \boldsymbol{s}(t). \tag{6.136}$$

Substituting this expression into Equation (6.131) and performing the matrix arithmetic then yield

$$\left[\begin{array}{c} \dot{\boldsymbol{x}} \\ \hline \dot{\boldsymbol{\lambda}} \end{array}\right] = \left[\begin{array}{c} \boldsymbol{Fx} - \boldsymbol{GB}^{-1}\boldsymbol{G}^T\boldsymbol{Sx} - \boldsymbol{GB}^{-1}\boldsymbol{G}^T\boldsymbol{s} \\ \hline -\boldsymbol{Ax} - \boldsymbol{F}^T\boldsymbol{Sx} - \boldsymbol{F}^T\boldsymbol{s} + \boldsymbol{Ar} \end{array}\right]. \tag{6.137}$$

By differentiating $\boldsymbol{\lambda}(t)$ as given by Equation (6.135) and substituting the result into the $\boldsymbol{\lambda}(t)$ equation above, one obtains

$$\dot{\boldsymbol{S}}\boldsymbol{x} + \boldsymbol{S}\dot{\boldsymbol{x}} + \dot{\boldsymbol{s}} = -\boldsymbol{A}\boldsymbol{x} - \boldsymbol{F}^T\boldsymbol{S}\boldsymbol{x} - \boldsymbol{F}^T\boldsymbol{s} + \boldsymbol{A}\boldsymbol{r} \tag{6.138}$$

which, upon substitution for $\dot{\boldsymbol{x}}$ from Equation (6.137) yields

$$[\dot{\boldsymbol{S}} + \boldsymbol{S}\boldsymbol{F} + \boldsymbol{F}^T\boldsymbol{S} - \boldsymbol{S}\boldsymbol{G}\boldsymbol{B}^{-1}\boldsymbol{G}^T\boldsymbol{S} + \boldsymbol{A}]\boldsymbol{x} + \dot{\boldsymbol{s}} - [\boldsymbol{S}\boldsymbol{G}\boldsymbol{B}^{-1}\boldsymbol{G}^T - \boldsymbol{F}^T]\boldsymbol{s} + \boldsymbol{A}\boldsymbol{r} = 0. \tag{6.139}$$

Since $\boldsymbol{x}(t)$ and $\boldsymbol{s}(t)$ vary independently, satisfaction of this equation requires that

$$\dot{\boldsymbol{S}}(t) = -\boldsymbol{S}(t)\boldsymbol{F}(t) - \boldsymbol{F}^T(t)\boldsymbol{S}(t) + \boldsymbol{S}(t)\boldsymbol{G}(t)\boldsymbol{B}^{-1}(t)\boldsymbol{G}^T(t)\boldsymbol{S}(t) - \boldsymbol{A}(t) \tag{6.140}$$

and

$$\dot{\boldsymbol{s}}(t) = [\boldsymbol{S}(t)\boldsymbol{G}(t)\boldsymbol{B}^{-1}(t)\boldsymbol{G}^T(t) - \boldsymbol{F}^T(t)]\boldsymbol{s}(t) - \boldsymbol{A}(t)\,\boldsymbol{r}(t). \tag{6.141}$$

The first of these equations is the matrix Riccati equation derived in the solution to the linear controller problem. The second is a linear vector differential equation driven by the function $\boldsymbol{A}(t)\boldsymbol{r}(t)$. It is an equation with time varying coefficients, however, since the coefficient matrix contains $\boldsymbol{S}(t)$. The boundary conditions for these two equations are obtained from Equations (6.132) and (6.136), which yield for $t = t_f$

$$\boldsymbol{\lambda}(t_f) = \boldsymbol{S}(t_f)\boldsymbol{x}(t_f) + \boldsymbol{s}(t_f) = \boldsymbol{S}_f\boldsymbol{x}(t_f) - \boldsymbol{S}_f\boldsymbol{r}(t_f). \tag{6.142}$$

Again, since $\boldsymbol{x}(t_f)$ and $\boldsymbol{r}(t_f)$ may vary independently, this equation further requires that

$$\boldsymbol{S}(t_f) = \boldsymbol{S}_f, \quad \boldsymbol{s}(t_f) = -\boldsymbol{S}_f\boldsymbol{r}(t_f). \tag{6.143}$$

Thus, the solution to this problem is similar to that for the linear regulator, except for the additional complexity of solving the linear vector differential equation given in Equation (6.141). This equation is also solved backward in time from the terminal boundary condition. The optimal control function is obtained from the second necessary condition

$$\partial H/\partial \boldsymbol{u} = 0$$

which, from Equations (6.99) and (6.136), yields

$$\begin{aligned} \boldsymbol{u}(t) &= -\boldsymbol{B}^{-1}(t)\boldsymbol{G}^T(t)\boldsymbol{\lambda}(t) \\ &= -\boldsymbol{B}^{-1}(t)\boldsymbol{G}^T(t)\boldsymbol{S}(t)\boldsymbol{x}(t) - \boldsymbol{B}^{-1}(t)\boldsymbol{G}^T(t)\boldsymbol{s}(t). \end{aligned} \tag{6.144}$$

When $\boldsymbol{r}(t) = 0$, this result reduces to the linear regulator solution, since $\boldsymbol{s}(t)$ is then identically zero.

The derivation of the linear controller solution in Section 6.3.1 could well have proceeded in a similar manner, yielding Equation (6.118) directly without resort to the limiting process. The limiting concept helps to bridge the gap between the open-loop and feedback structure of optimal controllers, and for this reason it was used in the earlier derivation.

A special case arises in the instance in which a linear controller is required to maintain the state vector near some constant value $\boldsymbol{r}$ over the interval (t_0, t_f), as opposed to a zero value as implied in the linear controller solution of Section 6.3.1. In this instance, $\boldsymbol{r}(t)$ has the constant value $\boldsymbol{r}$; the differential equation for $\boldsymbol{s}(t)$ then has a constant driving function.

6.3.3.1. The Linear Regulator with Nonzero Set Point. A second special case arises in the event that a linear regulator has a reference value other than zero, as specified in the treatment in Section 6.3.2. The same steady state argument which results in the algebraic form of the Riccati equation can now be made with regard to the differential equation for $\boldsymbol{s}(t)$. That is, when the matrices $\boldsymbol{F}$, $\boldsymbol{G}$, $\boldsymbol{B}$, and $\boldsymbol{A}$, and the vector $\boldsymbol{r}$, are constant, Equation (6.141) may reach a steady state condition $\boldsymbol{s}(t) = \boldsymbol{s}$ in which

$$\dot{\boldsymbol{s}}(t) = 0 = [\boldsymbol{S}\boldsymbol{G}\boldsymbol{B}^{-1}\boldsymbol{G}^T - \boldsymbol{F}^T]\boldsymbol{s} - \boldsymbol{A}\boldsymbol{r}. \tag{6.145}$$

This algebraic equation has the solution

$$\boldsymbol{s} = [\boldsymbol{S}\boldsymbol{G}\boldsymbol{B}^{-1}\boldsymbol{G}^T - \boldsymbol{F}^T]^{-1}\boldsymbol{A}\boldsymbol{r} \tag{6.146}$$

which, when substituted into Equation (6.144) for the control function, yields

$$\boldsymbol{u}(t) = -\boldsymbol{B}^{-1}\boldsymbol{G}^T\boldsymbol{S}\boldsymbol{x}(t) - \boldsymbol{B}^{-1}\boldsymbol{G}^T[\boldsymbol{S}\boldsymbol{G}\boldsymbol{B}^{-1}\boldsymbol{G}^T - \boldsymbol{F}^T]^{-1}\boldsymbol{A}\boldsymbol{r}. \tag{6.147}$$

The second term is a constant; the time varying character of $\boldsymbol{u}(t)$ is generated solely by feedback of the state vector $\boldsymbol{x}(t)$.

6.3.3.2. The Linear Regulator with Time Varying Reference Function. It should be apparent that a time invariant system—a system with $\boldsymbol{F}, \boldsymbol{G}, \boldsymbol{B}$, and $\boldsymbol{A}$ matrices constant—does not imply a constant value for the reference signal $\boldsymbol{r}(t)$. Indeed, the more general form of Equation (6.141) for the time invariant system is

$$\dot{s}(t) = [SGB^{-1}G^T - F^T]s(t) - Ar(t). \tag{6.148}$$

The steady state characteristics then apply to the Riccati equation and perhaps to the transient part of the solution to Equation (6.148). The term *transient* here applies to those terms which tend to vanish for small values of t as the equation is solved backward in time from the terminal condition $s(t_f)$ given by Equation (6.143). In any event, Equation (6.148) must be solved as part of the system design process.

6.3.3.3. The Output Follower. It is a simple process to modify the linear follower so that the output function $y(t)$, given by the expression

$$y(t) = H(t)x(t) \tag{6.149}$$

will follow some input function $r(t)$. Of course, $r(t)$ is now of the dimension of $y(t)$. The system is optimized using the performance index

$$J = \tfrac{1}{2}[y(t_f) - r(t_f)]^T S_f [y(t_f) - r(t_f)]$$
$$+ \tfrac{1}{2}\int_{t_0}^{t_f} [[y(t) - r(t)]^T A(t)[y(t) - r(t)] + u^T(t)B(t)u(t)]\,dt. \tag{6.150}$$

Here, the purpose is to maintain $y(t)$ close to the reference value $r(t)$. Analogously to the formation of Equation (6.129), the Hamiltonian function is

$$H = L + \lambda^T f = \tfrac{1}{2}[(Hx - r)^T A(Hx - r) + u^T Bu] + \lambda^T Fx + \lambda^T Gu \tag{6.151}$$

where Equation (6.149) has been used to eliminate $y(t)$. Corresponding to Equation (6.130), the λ equation is

$$\dot{\lambda}^T = -\frac{\partial H}{\partial x} = -(Hx - r)^T AH - \lambda^T F. \tag{6.152}$$

Then, using the necessary condition

$$\partial H/\partial u = 0 \tag{6.153}$$

produces the revised version of Equation (6.131):

$$\left[\begin{array}{c} \dot{x} \\ \hline \dot{\lambda} \end{array}\right] = \left[\begin{array}{c|c} F & -GB^{-1}G^T \\ \hline -H^TAH & -F^T \end{array}\right]\left[\begin{array}{c} x \\ \hline \lambda \end{array}\right] + \left[\begin{array}{c} 0 \\ \hline H^TAr \end{array}\right]. \tag{6.154}$$

The boundary condition for $\boldsymbol{\lambda}$ is obtained from Equation (6.13), substituting for $\boldsymbol{y}(t)$ from Equation (6.149):

$$\boldsymbol{\lambda}(t_f) = \boldsymbol{H}(t_f)^T \boldsymbol{S}_f \boldsymbol{H}(t_f)\boldsymbol{x}(t_f) - \boldsymbol{H}^T(t_f)\boldsymbol{S}_f \boldsymbol{r}(t_f). \tag{6.155}$$

Then, following the general developments leading to Equation (6.137) and letting

$$\boldsymbol{\lambda}(t) = \boldsymbol{S}(t)\boldsymbol{x}(t) + \boldsymbol{s}(t) \tag{6.156}$$

lead to the expression

$$\begin{bmatrix} \dot{\boldsymbol{x}} \\ \hline \dot{\boldsymbol{\lambda}} \end{bmatrix} = \begin{bmatrix} \boldsymbol{Fx} - \boldsymbol{GB}^{-1}\boldsymbol{G}^T\boldsymbol{Sx} - \boldsymbol{GB}^{-1}\boldsymbol{G}^T\boldsymbol{s} \\ \hline -\boldsymbol{H}^T\boldsymbol{AHx} - \boldsymbol{F}^T\boldsymbol{Sx} - \boldsymbol{F}^T\boldsymbol{Sx} + \boldsymbol{H}^T\boldsymbol{Ar} \end{bmatrix}. \tag{6.157}$$

As before, $\boldsymbol{\lambda}(t)$ from Equation (6.156) is differentiated and the result substituted for $\dot{\boldsymbol{\lambda}}$ in the equation above to yield

$$\dot{\boldsymbol{S}}\boldsymbol{x} + \boldsymbol{S}\dot{\boldsymbol{x}} + \dot{\boldsymbol{s}} = -\boldsymbol{H}^T\boldsymbol{AHx} - \boldsymbol{F}^T\boldsymbol{Sx} - \boldsymbol{F}^T\boldsymbol{s} + \boldsymbol{H}^T\boldsymbol{Ar}. \tag{6.158}$$

Then, substituting for $\dot{\boldsymbol{x}}$ from Equation (6.157) and combining terms yield

$$[\dot{\boldsymbol{S}} + \boldsymbol{SF} + \boldsymbol{F}^T\boldsymbol{S} - \boldsymbol{SGB}^{-1}\boldsymbol{G}^T\boldsymbol{S} + \boldsymbol{H}^T\boldsymbol{AH}]\boldsymbol{x} + \dot{\boldsymbol{s}} = [\boldsymbol{SGB}^{-1}\boldsymbol{G}^T - \boldsymbol{F}^T]\boldsymbol{s} + \boldsymbol{H}^T\boldsymbol{Ar}. \tag{6.159}$$

Since $\boldsymbol{x}(t)$ and $\boldsymbol{s}(t)$ vary independently, satisfaction of this equation requires that

$$\dot{\boldsymbol{S}}(t) = -\boldsymbol{S}(t)\boldsymbol{F}(t) - \boldsymbol{F}^T(t)\boldsymbol{S}(t) + \boldsymbol{S}(t)\boldsymbol{G}(t)\boldsymbol{B}^{-1}(t)\boldsymbol{G}^T(t)\boldsymbol{S}(t) - \boldsymbol{H}^T(t)\boldsymbol{A}(t)\boldsymbol{H}(t) \tag{6.160}$$

and

$$\dot{\boldsymbol{s}}(t) = [\boldsymbol{S}(t)\boldsymbol{G}(t)\boldsymbol{B}^{-1}(t)\boldsymbol{G}^T(t) - \boldsymbol{F}^T(t)]\boldsymbol{s}(t) - \boldsymbol{H}^T(t)\boldsymbol{A}(t)\boldsymbol{r}(t). \tag{6.161}$$

The first equation is the matrix Riccati equation, slightly modified to include the effect of the matrix $\boldsymbol{H}(t)$. The second is identical to Equation (6.141), except for the inclusion of the $\boldsymbol{H}(t)$ matrix in the driving term. These equations must be solved backward in time from boundary conditions obtained by evaluation of Equation (6.156) at t_f, and substituting from Equation (6.155), yielding

$$\boldsymbol{\lambda}(t_f) = \boldsymbol{S}(t_f)\boldsymbol{x}(t_f) + \boldsymbol{s}(t_f) = \boldsymbol{H}^T(t_f)\boldsymbol{S}_f\boldsymbol{H}(t_f)\boldsymbol{x}(t_f) - \boldsymbol{H}^T(t_f)\boldsymbol{S}_f\boldsymbol{r}(t_f). \tag{6.162}$$

Since $\boldsymbol{x}(t_f)$ and $\boldsymbol{r}(t_f)$ may vary independently, this equation requires that

$$\boldsymbol{S}(t_f) = \boldsymbol{H}^T(t_f)\boldsymbol{S}_f\boldsymbol{H}(t_f)$$
$$\boldsymbol{s}(t_f) = -\boldsymbol{H}^T(t_f)\boldsymbol{S}_f\boldsymbol{r}(t_f). \tag{6.163}$$

These are the terminal boundary conditions from which the $\boldsymbol{S}(t)$ and $\boldsymbol{s}(t)$ equations must be solved. Extension of these results to the regulator case is accomplished by only a slight variation of the developments of the two previous sections.

6.4. DISCRETE TIME SYSTEMS

The all important application of optimal control techniques to discrete time systems can be obtained directly from the discrete time versions of Equations (6.1) and (6.2). That is, the system is described by the general difference equation

$$\boldsymbol{x}(l) = \boldsymbol{f}(\boldsymbol{x}(l-1), \boldsymbol{u}(l-1), l) \tag{6.164}$$

and the performance index by the summation of terms

$$J = \phi(\boldsymbol{x}(L)) + \sum_{l=1}^{L} j[\boldsymbol{x}(l-1), \boldsymbol{u}(l-1), l] \tag{6.165}$$

where j is some function of $\boldsymbol{x}$, $\boldsymbol{u}$, and perhaps l. It is now desired to select the sequence $\boldsymbol{u}(l)$ so that J is minimized, and a terminal constraint given by the general expression

$$\boldsymbol{\psi}(\boldsymbol{x}(l)) = 0 \tag{6.166}$$

is met. In a manner analogous to the developments of Section 6.1.2, the constraints are adjoined to the performance index to yield

$$\bar{J} = \phi(\boldsymbol{x}(L)) + \boldsymbol{\nu}^T\boldsymbol{\psi}(\boldsymbol{x}(L)) + \sum_{l=1}^{L} [j(\boldsymbol{x}(l-1), \boldsymbol{u}(l-1), l)$$
$$+ \boldsymbol{\lambda}^T(l)[\boldsymbol{f}(\boldsymbol{x}(l-1), \boldsymbol{u}(l-1), l) - \boldsymbol{x}(l)]. \tag{6.167}$$

The discrete Hamiltonian function H is then defined as

$$H(\boldsymbol{x}(l), \boldsymbol{\lambda}(l), \boldsymbol{u}(l), l) = j(\boldsymbol{x}(l), \boldsymbol{u}(l), l+1) + \boldsymbol{\lambda}^T(l+1)\boldsymbol{f}(\boldsymbol{x}(l), \boldsymbol{u}(l), l+1). \tag{6.168}$$

The process then proceeds by considering the differential change in $\bar{J}$ produced by differential changes in the sequence $\boldsymbol{u}(l)$, with the result that the necessary condition for $\boldsymbol{\lambda}(l)$ is

$$\boldsymbol{\lambda}^T(l) = -\frac{\partial H(\boldsymbol{x}(l), \boldsymbol{\lambda}(l), \boldsymbol{u}(l), l)}{\partial \boldsymbol{x}(l)} \tag{6.169}$$

or,

$$\boldsymbol{\lambda}^T(l) = \boldsymbol{\lambda}^T(l+1)\left[\frac{\partial \boldsymbol{f}(\boldsymbol{x}(l), \boldsymbol{u}(l), l+1)}{\partial \boldsymbol{x}(l)}\right] + \frac{\partial j(\boldsymbol{x}(l), \boldsymbol{u}(l), l+1)}{\partial \boldsymbol{x}(l)}. \tag{6.170}$$

This expression is a difference equation for the sequence $\boldsymbol{\lambda}^T(l)$, which must be solved backward from the terminal boundary condition

$$\boldsymbol{\lambda}^T(L) = \left[\frac{\partial \phi}{\partial \boldsymbol{x}(l)} + \boldsymbol{\nu}^T \frac{\partial \boldsymbol{\psi}}{\partial \boldsymbol{x}(l)}\right]. \tag{6.171}$$

The second necessary condition is

$$\left[\frac{\partial H(\boldsymbol{x}(l), \boldsymbol{\lambda}(l), \boldsymbol{u}(l), l)}{\partial \boldsymbol{u}(l)}\right] = 0$$

or

$$\left[\frac{\partial j(\boldsymbol{x}(l), \boldsymbol{u}(l), l+1)}{\partial \boldsymbol{u}(l)}\right] + \boldsymbol{\lambda}^T(l+1)\left[\frac{\boldsymbol{f}(\boldsymbol{x}(l), \boldsymbol{u}(l), l+1)}{\partial \boldsymbol{u}(l)}\right] = 0. \tag{6.172}$$

In summary then, Equations (6.164) and (6.170) must be solved simultaneously for $\boldsymbol{x}(l)$ and $\boldsymbol{\lambda}(l)$, using Equation (6.172) to determine the $\boldsymbol{u}(l)$ sequence. In general, the boundary conditions on $\boldsymbol{\lambda}(l)$ are terminal conditions specified by Equation (6.171). The general result is a nonlinear, two-point boundary value difference equation.

6.4.1. Linear Controllers and Regulators

If the difference equation describing the system is linear:

$$\boldsymbol{x}(l) = \boldsymbol{\Phi}(l, l-1)\boldsymbol{x}(l-1) + \boldsymbol{\Gamma}(l, l-1)\boldsymbol{u}(l-1) \tag{6.173}$$

and the performance index is quadratic:

$$J = \tfrac{1}{2}[\boldsymbol{x}^T(L)\boldsymbol{S}_L\boldsymbol{x}(L)] + \tfrac{1}{2}\sum_{l=0}^{L-1}[\boldsymbol{x}^T(l)\boldsymbol{A}(l)\boldsymbol{x}(l) + \boldsymbol{u}^T(l)\boldsymbol{B}(l)\boldsymbol{u}(l)] \tag{6.174}$$

then the generalization of the continuous time results to the discrete time case yields the following. The optimal control sequence is a linear function of the state:

$$\boldsymbol{u}(l) = \boldsymbol{C}(l)\boldsymbol{x}(l) \tag{6.175}$$

where the feedback gain matrix $\boldsymbol{C}(l)$ is given by the expression

$$\boldsymbol{C}(l) = -[\boldsymbol{B}(l) + \boldsymbol{\Gamma}^T(l+1,l)\boldsymbol{S}(l+1)\boldsymbol{\Gamma}(l+1,l)]^{-1}\boldsymbol{\Gamma}^T(l+1,l)\boldsymbol{S}(l+1)\boldsymbol{\Phi}(l+1,l) \tag{6.176}$$

and the matrix $\boldsymbol{S}(l)$ by

$$\boldsymbol{S}(l) = \boldsymbol{A}(l) + \boldsymbol{\Phi}^T(l+1,l)\boldsymbol{S}(l+1)[\boldsymbol{\Phi}(l+1,l) + \boldsymbol{\Gamma}(l+1,l)\boldsymbol{C}(l)]. \tag{6.177}$$

This pair of difference equations must be solved recursively from the final condition

$$\boldsymbol{S}(L) = \boldsymbol{S}_L \tag{6.178}$$

and represents the discrete time form of the matrix Riccati equation. These equations are derived in Section 8.4.1 using dynamic programming, but they also follow directly from application of the discrete Maximum Principle. Thus, from Equations (6.170), (6.173), and (6.174),

$$\boldsymbol{\lambda}(l) = \boldsymbol{\Phi}^T(l+1,l)\boldsymbol{\lambda}(l+1) + \boldsymbol{A}(l)\boldsymbol{x}(l) \tag{6.179}$$

and from Equations (6.172) and (6.174),

$$\begin{aligned} \boldsymbol{B}(l)\boldsymbol{u}(l) + \boldsymbol{\Gamma}^T(l+1,l)\boldsymbol{\lambda}(l+1) &= 0 \\ \boldsymbol{u}(l) &= -\boldsymbol{B}^{-1}(l)\boldsymbol{\Gamma}^T(l+1,l)\boldsymbol{\lambda}(l+1). \end{aligned} \tag{6.180}$$

Substitution of $\boldsymbol{u}(l)$ from Equation (6.180) into Equation (6.173), indexed up by one, yields

$$\boldsymbol{x}(l+1) = \boldsymbol{\Phi}(l+1,l)\boldsymbol{x}(l) - \boldsymbol{\Gamma}(l+1,l)\boldsymbol{B}^{-1}(l)\boldsymbol{\Gamma}^T(l+1,l)\boldsymbol{\lambda}(l+1). \tag{6.181}$$

This expression and Equation (6.179) constitute a set of coupled difference equations which must be solved, using the split boundary conditions $\boldsymbol{x}(0)$ and $\boldsymbol{\lambda}(L)$. The terminal condition on $\boldsymbol{\lambda}$ is obtained from Equation (6.171)

$$\boldsymbol{\lambda}(L) = \boldsymbol{S}_L\boldsymbol{x}(L). \tag{6.182}$$

Because of the analogy between the current development and that of Section 6.3.1, it is reasonable to assume a solution of the form

$$\boldsymbol{\lambda}(l) = \boldsymbol{S}(l)\boldsymbol{x}(l) \tag{6.183}$$

in which case Equations (6.179) and (6.181) constitute the following set of recursive equations:

$$\begin{aligned} \boldsymbol{x}(l+1) &= \boldsymbol{\Phi}(l+1,l)\boldsymbol{x}(l) - \boldsymbol{\Gamma}(l+1,l)\boldsymbol{B}^{-1}(l)\boldsymbol{\Gamma}^T(l+1,l)\boldsymbol{S}(l+1)\boldsymbol{x}(l+1) \\ \boldsymbol{S}(l)\boldsymbol{x}(l) &= \boldsymbol{A}(l)\boldsymbol{x}(l) + \boldsymbol{\Phi}^T(l+1,l)\boldsymbol{S}(l+1)\boldsymbol{x}(l+1). \end{aligned} \tag{6.184}$$

Solving the first of these expressions for $\boldsymbol{x}(l+1)$ yields

$$\boldsymbol{x}(l+1) = [\boldsymbol{I} + \boldsymbol{\Gamma}(l+1,l)\boldsymbol{B}^{-1}(l)\boldsymbol{\Gamma}^T(l+1,l)\boldsymbol{S}(l+1)]^{-1}\boldsymbol{\Phi}(l+1,l)\boldsymbol{x}(l) \tag{6.185}$$

which, when substituted into the second expression, yields

$$\begin{aligned} \boldsymbol{S}(l)\boldsymbol{x}(l) = \boldsymbol{A}(l)\boldsymbol{x}(l) &+ \boldsymbol{\Phi}^T(l+1,l)\boldsymbol{S}(l+1) \\ &\cdot [\boldsymbol{I} + \boldsymbol{\Gamma}(l+1,l)\boldsymbol{B}^{-1}(l)\boldsymbol{\Gamma}^T(l+1,l)]^{-1}\boldsymbol{\Phi}(l+1,l)\boldsymbol{x}(l). \end{aligned} \tag{6.186}$$

This expression must be valid for arbitrary values of $\boldsymbol{x}(l)$, which is true only if the following equation is satisfied:

$$\begin{aligned} \boldsymbol{S}(l) = \boldsymbol{A}(l) &+ \boldsymbol{\Phi}^T(l+1,l)\boldsymbol{S}(l+1) \\ &\cdot [\boldsymbol{I} + \boldsymbol{\Gamma}(l+1,l)\boldsymbol{B}^{-1}(l)\boldsymbol{\Gamma}^T(l+1,l)\boldsymbol{S}(l+1)]^{-1}\boldsymbol{\Phi}(l+1,l). \end{aligned} \tag{6.187}$$

The $\boldsymbol{S}(l+1)$ term may be taken inside the brackets, yielding

$$\begin{aligned} \boldsymbol{S}(l) = \boldsymbol{A}(l) &+ \boldsymbol{\Phi}^T(l+1,l) \\ &\cdot [\boldsymbol{S}^{-1}(l+1) + \boldsymbol{\Gamma}(l+1,l)\boldsymbol{B}^{-1}(l)\boldsymbol{\Gamma}^T(l+1,l)]^{-1}\boldsymbol{\Phi}(l+1,l). \end{aligned} \tag{6.188}$$

Equations (6.187) and (6.188) are discrete time versions of the matrix Riccati equation. They must be solved backward in time from the terminal condition

$$\boldsymbol{S}(L) = \boldsymbol{S}_L. \tag{6.189}$$

The particular form of the equation may be modified by use of the matrix identity of Equation (A.61), to yield

$$\begin{aligned} \boldsymbol{S}(l) = \boldsymbol{A}(l) &+ \boldsymbol{\Phi}^T(l+1,l)\boldsymbol{S}(l+1)\boldsymbol{\Phi}(l+1,l) - \boldsymbol{\Phi}^T(l+1,l)\boldsymbol{S}(l+1)\boldsymbol{\Gamma}(l+1,l) \\ &\cdot [\boldsymbol{B}(l) + \boldsymbol{\Gamma}^T(l+1,l)\boldsymbol{S}(l+1)\boldsymbol{\Gamma}(l+1,l)]^{-1}\boldsymbol{\Gamma}^T(l+1,l)\boldsymbol{S}(l+1)\boldsymbol{\Phi}(l+1,l). \end{aligned} \tag{6.190}$$

Let the term $\boldsymbol{C}(l)$ be defined as

$$\boldsymbol{C}(l) = -[\boldsymbol{B}(l) + \boldsymbol{\Gamma}^T(l+1,l)\boldsymbol{S}(l+1)\boldsymbol{\Gamma}(l+1,l)]^{-1}\boldsymbol{\Gamma}^T(l+1,l)\boldsymbol{S}(l+1)\boldsymbol{\Phi}(l+1,l) \tag{6.191}$$

in which case Equation (6.190) is written as

$$\begin{aligned}\boldsymbol{S}(l) &= \boldsymbol{A}(l) + \boldsymbol{\Phi}^T(l+1,l)\boldsymbol{S}(l+1)\boldsymbol{\Phi}(l+1,l)\\ &\quad + \boldsymbol{\Phi}^T(l+1,l)\boldsymbol{S}(l+1)\boldsymbol{\Gamma}(l+1,l)\boldsymbol{C}(l)\\ &= \boldsymbol{A}(l) + \boldsymbol{\Phi}^T(l+1,l)\boldsymbol{S}(l+1)[\boldsymbol{\Phi}(l+1,l) + \boldsymbol{\Gamma}(l+1,l)\boldsymbol{C}(l)].\end{aligned} \tag{6.192}$$

Equations (6.191) and (6.192) are identical to Equations (6.176) and (6.177), thus completing the derivation of these recursive equations by use of the Maximum Principle. All that remains is to show that the optimal feedback control is actually given by Equation (6.175). This fact follows from solving for $\boldsymbol{u}(l)$ from Equations (6.180) and (6.183):

$$\boldsymbol{u}(l) = -\boldsymbol{B}^{-1}(l)\boldsymbol{\Gamma}^T(l+1,l)\boldsymbol{S}(l+1)\boldsymbol{x}(l+1). \tag{6.193}$$

What is needed is a feedback equation in terms of $\boldsymbol{x}(l)$, which can be obtained by solving for the product $\boldsymbol{S}(l+1)\boldsymbol{x}(l+1)$ from the second expression of Equation (6.184),

$$\boldsymbol{S}(l+1)\boldsymbol{x}(l+1) = (\boldsymbol{\Phi}^{-1}(l+1,l))^T[\boldsymbol{S}(l) - \boldsymbol{A}(l)]\boldsymbol{x}(l) \tag{6.194}$$

and substituting into Equation (6.193) to produce the expression

$$\boldsymbol{u}(l) = \boldsymbol{B}^{-1}(l)\boldsymbol{\Gamma}^T(l+1,l)(\boldsymbol{\Phi}^{-1}(l+1,l))^T[\boldsymbol{S}(l) - \boldsymbol{A}(l)]\boldsymbol{x}(l). \tag{6.195}$$

Upon substitution for $\boldsymbol{S}(l)$ from Equation (6.188), this expression yields

$$\begin{aligned}\boldsymbol{u}(l) = -\boldsymbol{B}^{-1}(l)\boldsymbol{\Gamma}^T(l+1,l)[&\boldsymbol{S}^{-1}(l+1)\\ &+ \boldsymbol{\Gamma}(l+1,l)\boldsymbol{B}^{-1}(l)\boldsymbol{\Gamma}^T(l+1,l)]^{-1}\boldsymbol{\Phi}(l+1,l)\boldsymbol{x}(l).\end{aligned} \tag{6.196}$$

By successive application of the matrix identity of Equation (A.61), this expression can be reduced to

$$\begin{aligned}\boldsymbol{u}(l) &= -[\boldsymbol{B}(l)\\ &\quad + \boldsymbol{\Gamma}^T(l+1,l)\boldsymbol{S}(l+1)\boldsymbol{\Gamma}(l+1,l)]^{-1}\boldsymbol{\Gamma}^T(l+1,l)\boldsymbol{S}(l+1)\boldsymbol{\Phi}(l+1,l)\boldsymbol{x}(l)\\ &= \boldsymbol{C}(l)x(l)\end{aligned} \tag{6.197}$$

and the development is complete.

In summary, the discrete time, optimal, linear controller is a state variable feedback controller with the gain matrix given by any of Equations (6.195), (6.196), or (6.197). The corresponding discrete time matrix Riccati equation is given by any of Equations (6.187), (6.188), or (6.190), or by the combination of Equations (6.176) and (6.177). The most commonly encountered form of the discrete-time, optimal, linear controller is that represented by Equations (6.175) through (6.177).

In the event that the system is time invariant, and the quadratic weights do not vary with time, the discrete time, optimal controller equations become

$$\begin{aligned} \boldsymbol{u}(l) &= \boldsymbol{C}(l)\boldsymbol{x}(l) \\ \boldsymbol{C}(l) &= -[\boldsymbol{B} + \boldsymbol{\Gamma}^T\boldsymbol{S}(l+1)\boldsymbol{\Gamma}]^{-1}\boldsymbol{\Gamma}^T\boldsymbol{S}(l+1)\boldsymbol{\Phi} \\ \boldsymbol{S}(l) &= \boldsymbol{A} + \boldsymbol{\Phi}^T\boldsymbol{S}(l+1)[\boldsymbol{\Phi} + \boldsymbol{\Gamma}\boldsymbol{C}(l)] \end{aligned} \tag{6.198}$$

The discrete time regulator equations are obtained by recognizing that $\boldsymbol{C}(l)$ and $\boldsymbol{S}(l)$ in Equation (6.198) are constant for the regulator problem, so that

$$\begin{aligned} \boldsymbol{u}(l) &= \boldsymbol{C}\boldsymbol{x}(l) \\ \boldsymbol{C} &= -[\boldsymbol{B} + \boldsymbol{\Gamma}^T\boldsymbol{S}\boldsymbol{\Gamma}]^{-1}\boldsymbol{\Gamma}^T\boldsymbol{S}\boldsymbol{\Phi} \end{aligned} \tag{6.199}$$

and $\boldsymbol{S}$ is given by the solution to the discrete time algebraic Riccati equation derived from Equation (6.188) or from Equation (6.190):

$$\begin{aligned} \boldsymbol{S} &= \boldsymbol{A} + \boldsymbol{\Phi}^T[\boldsymbol{S}^{-1} + \boldsymbol{\Gamma}\boldsymbol{B}^{-1}\boldsymbol{\Gamma}^T]^{-1}\boldsymbol{\Phi} \\ &= \boldsymbol{A} + \boldsymbol{\Phi}^T\boldsymbol{S}\boldsymbol{\Phi} - \boldsymbol{\Phi}^T\boldsymbol{S}\boldsymbol{\Gamma}[\boldsymbol{B} + \boldsymbol{\Gamma}^T\boldsymbol{S}\boldsymbol{\Gamma}]^{-1}\boldsymbol{\Gamma}^T\boldsymbol{S}\boldsymbol{\Phi}. \end{aligned} \tag{6.200}$$

Solution of this algebraic equation for the matrix $\boldsymbol{S}$ is tantamount to solving the discrete-time, optimal regulator problem.

6.4.2. Linear Followers

The development of the discrete time linear follower parallels closely that of the continuous time linear follower. If the reference input function is $\boldsymbol{r}(l)$, and it is desired to have the system state $\boldsymbol{x}(l)$ follow $\boldsymbol{r}(l)$, the resulting performance index is

$$\begin{aligned} J = \tfrac{1}{2}[\boldsymbol{x}(L) - \boldsymbol{r}(L)]^T\boldsymbol{S}_L[\boldsymbol{x}(L) - \boldsymbol{r}(L)] \\ + \sum_{l=0}^{L-1} \{[\boldsymbol{x}(l) - \boldsymbol{r}(l)]^T\boldsymbol{A}(l)[\boldsymbol{x}(l) - \boldsymbol{r}(l)] + \boldsymbol{u}^T(l)\boldsymbol{B}(l)\boldsymbol{u}(l)\}. \end{aligned} \tag{6.201}$$

With this modification, developments paralleling those of the previous section yield

$$\begin{aligned} \boldsymbol{s}(l) = -\boldsymbol{\Phi}^T(l+1, l)[\boldsymbol{S}^{-1}(l+1) \\ + \boldsymbol{\Gamma}(l+1, l)\boldsymbol{B}^{-1}(l)\boldsymbol{\Gamma}^T(l+1, l)]^{-1}\boldsymbol{\Gamma}(l+1, l)\boldsymbol{B}^{-1}(l)\boldsymbol{\Gamma}^T(l+1, l)\boldsymbol{s}(l+1) \\ + \boldsymbol{A}(l)\boldsymbol{r}(l) \end{aligned} \tag{6.202}$$

or, alternatively,

$$\boldsymbol{s}(l) = \boldsymbol{C}^T(l)\boldsymbol{\Gamma}^T(l+1, l)\boldsymbol{s}(l+1) + \boldsymbol{A}(l)\boldsymbol{r}(l) \tag{6.203}$$

which is obtained by substituting the value of $\boldsymbol{C}^T(l)$ implied by Equation (6.196). The value of $\boldsymbol{S}(l)$ is obtained from one of the discrete matrix Riccati equation forms given in Equations (6.187), (6.188), or (6.190). Equation (6.202) must be solved backward in time from the terminal condition

$$\boldsymbol{s}(L) = -\boldsymbol{S}_L\boldsymbol{r}(L) \tag{6.204}$$

in direct analogy to the continuous time case.

For the time invariant discrete time system, the matrices $\boldsymbol{\Phi}$, $\boldsymbol{\Gamma}$, $\boldsymbol{A}$, and $\boldsymbol{B}$ are constant, yielding for $\boldsymbol{s}(l)$ the time invariant difference equation

$$\boldsymbol{s}(l) = -\boldsymbol{\Phi}^T[\boldsymbol{S}^{-1} + \boldsymbol{\Gamma}\boldsymbol{B}^{-1}\boldsymbol{\Gamma}^T]^{-1}\boldsymbol{\Gamma}\boldsymbol{B}^{-1}\boldsymbol{\Gamma}^T\boldsymbol{s}(l+1) + \boldsymbol{A}\boldsymbol{r}(l) \tag{6.205}$$

where $\boldsymbol{S}$ is the steady state solution of the Riccati equation. This expression applies when a regulator has a time varying reference value $\boldsymbol{r}(l)$. If $\boldsymbol{r}(l) = \boldsymbol{r}$ is a constant set point, then Equation (6.205) is an algebraic equation with the solution

$$\boldsymbol{s} = [\boldsymbol{I} + \boldsymbol{\Phi}^T[\boldsymbol{S}^{-1} + \boldsymbol{\Gamma}\boldsymbol{B}^{-1}\boldsymbol{\Gamma}^T]^{-1}\boldsymbol{\Gamma}\boldsymbol{B}^{-1}\boldsymbol{\Gamma}^T]^{-1}\boldsymbol{A}\boldsymbol{r}. \tag{6.206}$$

In all of these situations, the feedback control function is given by the expression

$$\boldsymbol{u}(l) = \boldsymbol{C}(l)\boldsymbol{x}(l) + \boldsymbol{B}^{-1}(l)\boldsymbol{\Gamma}^T(l)\boldsymbol{s}(l) \tag{6.207}$$

where $\boldsymbol{C}(l)$ is the optimum gain given in Equations (6.176) and (6.191).

The discrete time output follower equations are obtained using the performance index

$$\begin{aligned} J = \tfrac{1}{2}[\boldsymbol{y}^T(L) - \boldsymbol{r}(l)]^T\boldsymbol{S}_L[\boldsymbol{y}(L) - \boldsymbol{r}(L)] \\ + \tfrac{1}{2}\sum_{L=0}^{L-1}\{[\boldsymbol{y}(l) - \boldsymbol{r}(l)]^T\boldsymbol{A}(l)[\boldsymbol{y}(l) - \boldsymbol{r}(l)] + \boldsymbol{u}^T(l)\boldsymbol{B}(l)\boldsymbol{u}(l)\} \end{aligned} \tag{6.208}$$

where, in accordance with the output equation,

$$\boldsymbol{y}(l) = \boldsymbol{H}(l)\boldsymbol{x}(l). \tag{6.209}$$

Substituting this expression into the performance index of Equation (6.208) and repeating the derivation of the optimal control equations produce the equations

$$\boldsymbol{S}(l) = \boldsymbol{H}^T(l)\boldsymbol{A}(l)\boldsymbol{H}(l) + \boldsymbol{\Phi}^T(l+1,l)[\boldsymbol{S}^{-1}(l+1) + \boldsymbol{\Gamma}(l+1,l)\boldsymbol{B}^{-1}(l)\boldsymbol{\Gamma}^T(l+1,l)]^{-1}\boldsymbol{\Phi}(l+1,l) \tag{6.210}$$

and

$$\boldsymbol{s}(l) = -\boldsymbol{\Phi}^T(l+1,l)[\boldsymbol{S}^{-1}(l+1) + \boldsymbol{\Gamma}(l+1,l)\boldsymbol{B}^{-1}(l)\boldsymbol{\Gamma}^T(l+1,l)]^{-1}\boldsymbol{\Gamma}(l+1,l)\boldsymbol{B}^{-1}(l)\boldsymbol{\Gamma}^T(l+1,l)\boldsymbol{s}(l+1) + \boldsymbol{H}^T(l)\boldsymbol{A}(l)\boldsymbol{r}(l). \tag{6.211}$$

Specialization to the time invariant case is a straightforward extension of these results.

6.4.3. The Time-Optimal Problem and Deadbeat Control

In Section 5.3.4, the concept of state deadbeat response of discrete time linear systems was introduced. Deadbeat response is obtained when all closed-loop eigenvalues are assigned to the origin. The system is then taken to zero in at most N steps, where N is the dimension of the system. From the general discussion of eigenvalue assignment in Sections 5.1 and 5.3, it is known that eigenvectors, and thus the shape of the modal responses, of a system are not uniquely defined by the eigenvalue assignment—except for the case of the single input-single output system. Because of this fact, and because deadbeat response is closely related to the time-optimal control problem, it is not surprising that optimal control theory has been applied to the design of deadbeat controllers.

The minimum time problem was discussed in Section 6.2.2 for continuous time systems, where it was shown that, for any system linear in the control function $\boldsymbol{u}(t)$, the optimum control signal is of the bang-bang or full deflection type. This result is possible because the switching functions are continuous time functions with no restrictions on the times at which switching occurs. The discrete time problem does not have this feature, in that the control functions are discrete and have constant values over the sample period. However, it has been shown that deadbeat control can be obtained from the discrete-time, linear optimal control problem with a quadratic performance index. The procedure is to formulate the problem with only the terminal state penalized. That is, the performance index is

$$J = \boldsymbol{x}^T(L)\boldsymbol{x}(L) \tag{6.212}$$

implying that

$$\boldsymbol{A} = \boldsymbol{B} = 0$$
$$\boldsymbol{S}(L) = \boldsymbol{I}.$$

The feedback gain and Riccati equation are then, from Equation (6.198),

$$\begin{aligned} \boldsymbol{C}(l) &= -[\boldsymbol{\Gamma}^T\boldsymbol{S}(l+1)\boldsymbol{\Gamma}]^{-1}\boldsymbol{\Gamma}\boldsymbol{S}(l+1)\boldsymbol{\Phi} \\ \boldsymbol{S}(l) &= \boldsymbol{\Phi}^T\boldsymbol{S}(l+1)[\boldsymbol{\Phi} + \boldsymbol{\Gamma}\boldsymbol{C}] \end{aligned} \tag{6.213}$$

with the initial condition

$$\boldsymbol{S}(L) = \boldsymbol{I}.$$

Under certain not too restrictive conditions, the feedback gain matrix $\boldsymbol{C}(l)$ produces a deadbeat characteristic.

A less restrictive development uses the performance index

$$J = \tfrac{1}{2} \sum_{l=1}^{L} [\boldsymbol{x}^T(l)\boldsymbol{A}\boldsymbol{x}(l) + \boldsymbol{u}^T(l)\boldsymbol{B}\boldsymbol{u}(l)]$$

and the solution of the algebraic Riccati equation given by the second expression of Equation (6.200). The solution is obtained in the case for $\boldsymbol{B} = 0$, which implies no penalty on the control function. The quadratic weight matrix $\boldsymbol{A}$ must be selected in a special manner which causes the closed-loop poles to approach the origin. Details can be found in the references cited in Section 6.6.4.

6.5. INACCESSIBLE STATE VARIABLES

The techniques of this chapter provide a very powerful approach to the design and synthesis of control systems with various optimality criteria. State variable feedback is a characteristic of all solutions obtained, but it is not generally characteristic of real-world systems that all state variables are accessible for control purposes. This fact is normally attributable to one or both of the following system characteristics:

I. Rather than the state variables themselves, some linear combination of state variables constitutes the available observation, including as a special case the condition in which some but not all of the state variables are accessible. The reasons for inaccessible state variables are varied; they may be physical limitations or even economic considerations.

II. Those observations that are available are contaminated by measurement noise.

In the former instance, one approach is to form an estimate of the state vector from operations on the available measurements. This concept is treated in some detail in Chapter 5, where the general idea of system design by state variable feedback is considered. It is shown there that a linear observer can be designed to provide the required estimation. This concept is developed in the next section. The latter situation is discussed in detail in Chapter 7.

6.5.1. Optimal Control of Linear, Time Invariant Systems with Inaccessible State Variables

The linear optimal control problems considered in this chapter have all resulted in some form of state variable feedback as the optimal control law. In the event that one or more of the state variables are not accessible as a system output, then some other approach to the control problem must be found. Rather than discard the state variable feedback concept with its very powerful and systematic design and synthesis procedures, it seems reasonable to use the linear observer theory developed in Section 5.2 to provide estimates for the inaccessible state variables, and then use these estimates in the optimal feedback controller. Either a full-order or a reduced-order observer may be used, each with its own particular characteristics as discussed in Section 5.2. Such an arrangement is illustrated in Figure 6-12, in which a full-order observer is shown. The reader should compare Figure 6-12 to Figure 5-5, depicting the use of an observer in the general state variable feedback control problem. That these two figures are equivalent results from the fact that the optimal linear feedback control problem results in state variable feedback. Therefore, any conclusions reached in Section 5.2.2

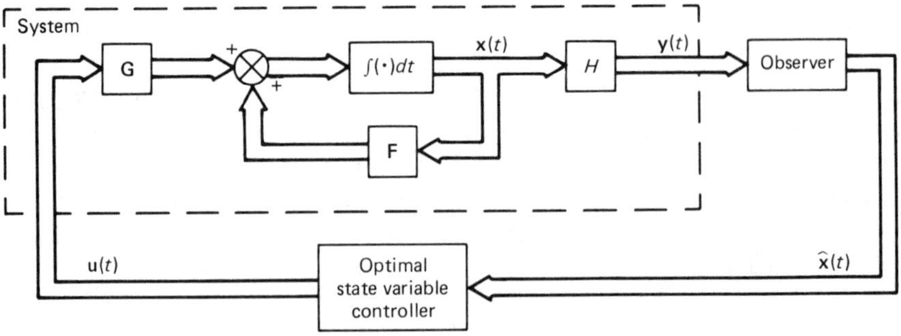

Figure 6-12. Optimal linear control using an observer.

regarding the use of observers in system synthesis also apply to their use in optimal state variable feedback systems.

To illustrate the combined use of an observer and a regulator, the spring-mass-damper example of Section 6.3.2 will be considered. The matrices describing the system are

$$\boldsymbol{F} = \begin{bmatrix} 0 & 1 \\ -5 & -2 \end{bmatrix}; \quad \boldsymbol{G} = \begin{bmatrix} 0 \\ \frac{1}{2} \end{bmatrix}.$$

In the development of Section 6.3.2, the optimal regulator which minimized the performance index

$$J = \int_{t_0}^{t_f} [56x_1 + 24x_2 + (u^2/4)]\, dt$$

was shown to be state variable feedback with gain matrix

$$\boldsymbol{C} = [-8 \quad -8].$$

The resulting system was critically damped with eigenvalues of -3. The assumed configuration of the regulator provides observation of both state variables for feedback purposes. A more realistic problem involves measurement of only $x_1(t)$, the position variable, yielding the measurement matrix

$$\boldsymbol{H} = [1 \quad 0].$$

An observer will be designed to provide an estimate of the nonaccessible state variable $x_2(t)$. A full-order observer–one providing estimates for both $x_1(t)$ and $x_2(t)$–can be used, or a reduced-order observer can be used to estimate only $x_2(t)$. Both procedures will be considered here.

The full-order observer is to be critically damped with eigenvalues of -5. The full-order observer follows easily from Equations (5.106) through (5.109). The canonical form system matrix and the observability matrices of Equation (5.106) are

$$\boldsymbol{A} = \begin{bmatrix} 0 & -5 \\ 1 & -2 \end{bmatrix}; \quad \boldsymbol{N} = \begin{bmatrix} 1 & 0 \\ 0 & 1 \end{bmatrix}; \quad \hat{\boldsymbol{N}} = \begin{bmatrix} 0 & 1 \\ 1 & -2 \end{bmatrix}$$

and the transformation matrix $\boldsymbol{T}$ of Equation (5.107) is

$$\boldsymbol{T} = \begin{bmatrix} 0 & 1 \\ 1 & -2 \end{bmatrix}.$$

The desired eigenvalues are both −5, so the desired characteristic equation is

$$\lambda^2 + 10\lambda + 25 = 0.$$

Then, from Equation (5.108),

$$\begin{bmatrix} 0 & -25 \\ 1 & -10 \end{bmatrix} = \begin{bmatrix} 0 & -5 \\ 1 & -2 \end{bmatrix} - \begin{bmatrix} \hat{k}_1 \\ \hat{k}_2 \end{bmatrix} [0 \quad 1]$$

which yields the solution

$$\hat{k}_1 = 20; \qquad \hat{k}_2 = 8.$$

Transformation to the original system is, from Equation (5.109),

$$\begin{bmatrix} k_1 \\ k_2 \end{bmatrix} = \begin{bmatrix} 0 & 1 \\ 1 & -2 \end{bmatrix} \begin{bmatrix} 20 \\ 8 \end{bmatrix} = \begin{bmatrix} 8 \\ 4 \end{bmatrix}.$$

The observer equation is obtained from Equation (5.91),

$$\begin{bmatrix} \dot{\hat{x}}_1(t) \\ \dot{\hat{x}}_2(t) \end{bmatrix} = \begin{bmatrix} -8 & 1 \\ -9 & -2 \end{bmatrix} \begin{bmatrix} \hat{x}_1(t) \\ \hat{x}_2(t) \end{bmatrix} + \begin{bmatrix} 0 \\ \frac{1}{2} \end{bmatrix} u(t) + \begin{bmatrix} 8 \\ 4 \end{bmatrix} x_1(t).$$

The closed-loop control system is then as shown in Figure 6-13. The reader should note that the observer is a linear system driven by two input functions, $x_1(t)$ and $u(t)$.

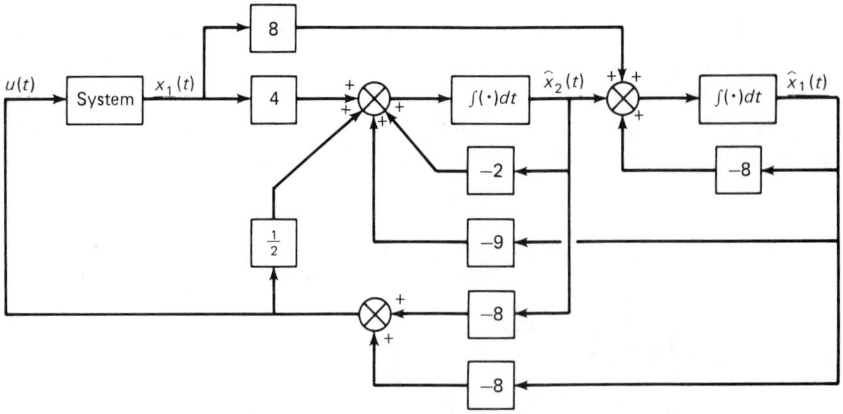

Figure 6-13. Optimal regulator using a full-order observer.

If a reduced-order observer is to be used, the procedures of Section 5.2.7 can be applied, although the computation is almost trivial. The desired eigenvalue is -5, and the matrix of Equation (5.132) is the scalar $F_{qq} = 5$. Equation (5.139) then yields, for the scalar F_{qy} term,

$$\begin{aligned} F_{qy} &= f_1 = b_0(a_1 - b_0) - a_0 \\ &= 5(2 - 5) - 5 = -20. \end{aligned}$$

The transformation matrix is obtained from the matrices of Equation (5.130):

$$\boldsymbol{N} = \begin{bmatrix} 0 & 1 \\ 1 & 0 \end{bmatrix}; \quad \hat{\boldsymbol{N}} = \begin{bmatrix} 1 & -2 \\ 0 & 1 \end{bmatrix}$$

from which

$$\boldsymbol{T} = \begin{bmatrix} 1 & 0 \\ -2 & 1 \end{bmatrix}; \quad \boldsymbol{T}^{-1} = \begin{bmatrix} 1 & 0 \\ 2 & 1 \end{bmatrix}.$$

The G_q matrix–here a scalar–is obtained from Equation (5.141), with the matrix $\boldsymbol{P}$ evaluated from Equation (5.135):

$$G_q = [-5 \quad 1] \begin{bmatrix} 1 & 0 \\ 2 & 1 \end{bmatrix} \begin{bmatrix} 0 \\ \frac{1}{2} \end{bmatrix} = \tfrac{1}{2}.$$

The expression of Equation (5.140) is then

$$\dot{\hat{q}}(t) = -5\hat{q}(t) - 20y(t) + \tfrac{1}{2}u(t)$$

and gives the equation for the first-order observer. Finally, the transformation of Equation (5.142) yields the state estimate in the original system,

$$\begin{bmatrix} \hat{x}_1(t) \\ \hat{x}_2(t) \end{bmatrix} = \begin{bmatrix} 1 & 0 \\ -2 & 1 \end{bmatrix} \begin{bmatrix} 0 & 1 \\ 1 & 5 \end{bmatrix} \begin{bmatrix} \hat{q}(t) \\ y(t) \end{bmatrix} = \begin{bmatrix} y(t) \\ \hat{q}(t) - 3y(t) \end{bmatrix}.$$

Since $y(t) = x_1(t)$, the only state $x_2(t)$ is estimated, and the estimate is given by

$$\begin{aligned} \hat{x}_2(t) &= \hat{q}(t) - 3x_1(t) \\ \dot{\hat{q}}(t) &= -5\hat{q}(t) - 20x_1(t) + \tfrac{1}{2}u(t). \end{aligned}$$

The regulator system using this observer is shown in Figure 6-14. It is clear that

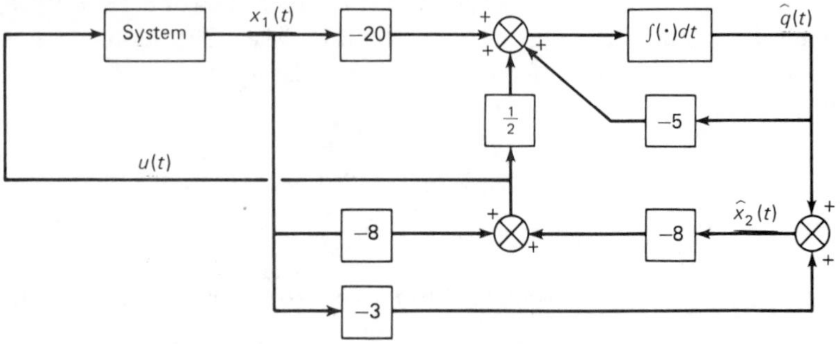

Figure 6-14. Optimal regulator using a reduced-order observer.

this system is a third-order system, while that of Figure 6-13 is a fourth-order system. The feedforward feature characteristic of the reduced-order observer can be clearly seen, since the estimate $\hat{x}_2(t)$ contains a term proportional to the measurement $y(t) = x_1(t)$.

This same example is treated in Chapter 7, but in the situation in which the measurement $y(t)$ is contaminated by additive noise, and there is driving noise within the system itself.

The reader should recognize that the use of a time invariant observer need not be limited to the regulator problem with its time invariant optimal filter gains. So long as the system itself is time invariant, the observer can be designed and used. In application to the general optimal feedback controller problem, however, the controller gains–and, in turn, the closed-loop response characteristics–vary with time, so that an appropriate relationship between closed-loop eigenvalues and observer eigenvalues may not exist for all values of time.

6.5.2. Optimal Control of Linear, Time Invariant Systems with Noise Contaminated Measurements

Physical measurements are never perfect–they are always contaminated with measurement noise to some extent. If the measurement noise is relatively small compared to the magnitude of the measured quantities, then it may be ignored. If a reduced-order observer is used, the feedforward characteristic may tend to enhance the noise effect. A full-order observer, on the other hand, would tend to filter out high frequency noise, because all state variable estimates are smoothed by the observer dynamics. If the noise is a slowly varying bias term, then it can be estimated by the observer with the system state variables, as discussed in Section 5.2.10. If the noise is appreciable and must be considered, then some state estimation technique based upon statistical estimation procedures must be used. Such an estimator is the topic of Chapter 7.

6.6. SOME FURTHER CONSIDERATIONS

The introduction to the calculus of variations contained in Appendix B, and the application of the calculus of variations to the optimal control problem contained in this chapter, serve as an introduction to the general optimal control theory. The resulting computational requirement is the solution of a two-point boundary value problem with its attendant difficulties. In the linear case, it was found possible to solve the two-point boundary value problem by solving the matrix Riccati differential equation; for the regulator problem, only the steady state solution of this equation was required. For the general nonlinear problem, very few solution methods were discussed. There are techniques for solving the nonlinear, two-point boundary value problem, but they are numerical and recursive in nature. The method of steepest descent, the use of second variations, and an iterative technique called quasilinearization are examples. These are specialized numerical techniques, and the reader is referred to the references at the end of this chapter for further information.

The concepts of sensitivity and robustness of optimal control systems is also an important question–one which is especially important when observer-controller combinations are used. Most of the development in this area relates to the linear system–and primarily to the time invariant linear system. The sensitivity of the system to variation in system parameters, as well as the effect of changes in the quadratic weighting matrices, is the principle concern in this area. Several of the references address this problem.

For the time-invariant linear system, the solution to the two-point boundary value problem is obtained by solving a nonlinear matrix differential equation–the Riccati equation. Since the Riccati equation is nonlinear, solution is no easy task. Efficient numerical techniques have been developed, some dependent upon particular traits of the problem. In certain cases, solution techniques using approximation methods have been developed. In others, equivalent problems with less computational complexity have been developed. For the regulator problem, only the steady state solution of the Riccati equation is required; this solution can be obtained from the algebraic Riccati equation. But the algebraic Riccati equation is actually a set of nonlinear simultaneous equations for which there is no general solution. In fact, in some instances solutions to such equations are most easily obtained as steady state values of associated differential equations. However, because of its special form, the algebraic Riccati equation has been the subject of several rather elegant methods of solution. Some of these methods are treated in the references listed at the end of this chapter.

Most of the optimal control problems for which solutions can be obtained in any kind of generalized form are linear and specify the optimality criterion in terms of a quadratic performance index. It is the designer's task to choose the quadratic weights to satisfy various design specifications. Suggestions vary from simple diagonal inverse square weighting to versions of a model following concept. Some of the examples of this chapter have shown how the selection of quadratic weights can affect closed-loop eigenvalues, and how such effects can have possibly serious consequences in observer-based control systems. The sensitivity of controller characteristics to variation in these quadratic weights is also an important consideration.

Discrete time systems are extremely important, simply because of the widespread use of digital controllers. In this text, developments are carried out in either the continuous time or the discrete time cases, depending upon which case has the more lucid development. This procedure often yields continuous time development followed by generalization to the discrete time version of the result. Although this technique is valid in most instances, especially if the time interval is suitably small, there does exist a class of interesting discrete time problems having no true continuous-time counterpart. These considerations, along with those relating to possible problem areas resulting from sampling time and roundoff errors, make the design of discrete time systems more than just a generalization of the continuous-time design process. Some of the references address this problem.

Finally, the design of state feedback controllers for systems with inaccessible state variables has been approached by using an observer to provide an estimate of the state vector for feedback purposes. This same technique was used in Chapter 5 for various state variable feedback schemes, but there the observer eigenvalues could be selected so that there was no interaction between the modal responses of the closed-loop system and those of the observer. In the case of optimal control, however, the values of the closed-loop eigenvalues are sensitive to the quadratic weights used in the objective function, so that observer design must also reflect the values of the quadratic weights. Perhaps a more important consideration is the fact that an optimal system using an observer for state variable estimation is no longer optimal; the best that can be said is that it is a suboptimal approximation to the truly optimal system. Just what degradation is produced by use of an observer is certainly of interest. Also of interest is the effect on the sensitivity and robustness of the control system due to the use of an observer. These concepts are addressed in some of the references.

6.6.1. Application of the Calculus of Variations

The general, nonlinear, two-point boundary value problem, produced by application of the calculus of variations to the optimal control problem with no control variable constraints, requires iterative numerical computation for its solution. Some of the iterative methods–the method of steepest descent, quasilinearization, methods based on the second variation, and invariant imbedding–can be found in Reference 13, 27, 42, and 80. A general treatment of these techniques as they are applied to the problem of guidance and control of spacecraft can be found in Reference 7. All of these techniques result in open-loop control functions.

6.6.2. Optimization with Control Variable Constraints

When control variable constraints are added to the problem, then the Maximum Principle must be used. The resulting two-point boundary value problem is even more restrictive in this case, because of the control variable constraints. The dynamic programming technique, discussed in Chapter 8, can be useful in relatively simple problems, but it rapidly fails as a practical solution method when the number of states grows large. Techniques for numerical solution of this problem are discussed in References 13, 27, and 80.

Application of the Maximum Principle to the minimum time problem shows that, in the event that the system is linear in the control function $\boldsymbol{u}(t)$, a bang-bang system results. These full deflection systems and their switching characteristics are an important application of optimal control theory. These problems are treated extensively in References 13 and 42. In Reference 79, the sensitivity of bang-bang systems to variations in system parameters is investigated.

6.6.3. Unconstrained Linear Problems

The primary emphasis in development has been in the area of linear controllers and regulators using a quadratic performance index. The reason, of course, is that general development in the area of nonlinear systems is fairly limited. The two concepts of primary interest in the design of linear systems with quadratic performance indices are the selection of the quadratic weighting term and the solution of the matrix Riccati equation. Again, for fairly obvious reasons, primary emphasis has been on the time invariant system. Techniques for the selection of the quadratic weighting terms vary from the inverse square weighting of Reference 13 to those based on asymptotic modal properties of linear regulators as the control input weights tend to zero (Reference 88). Other considerations are explored in References 5, 11, 21, and 30.

In the case of the linear regulator, an important consideration is the situation in which the quadratic weights associated with the control input tend to zero. In this event, it can be shown that the weighting matrix for the state vector is not unique—that is, more than one weighting matrix can produce the same optimal feedback control gains. This feature is discussed in References 34, 60, 61, and 82, where it is shown that, for single input–single output systems, selection of the weighting matrix is tantamount to a model following approach to system control. Also of interest in this special case is the location of the closed-loop eigenvalues, which tend to form in groups characterized by what is termed a Butterworth configuration, as shown in References 21, 69, 73, 78, and 84.

The solution of the matrix Riccati differential equation has received wide attention. Numerical techniques for solution of the Riccati equation are treated in References 22, 95, and 96. In References 47 and 86, a computationally effective algorithm for the solution of the Riccati equation by means of a technique known as Generalized Partitioned Solutions is developed. This technique constitutes a generalization of a set of algorithms known as Chandrasekhar or X-Y algorithms. In Reference 12, the particular problem involving a constrained terminal state is considered and the result obtained as the limit of certain solutions representing the optimal cost for the unconstrained problem with penalized terminal state.

The general concept of Chandrasekhar algorithms has received attention in References 17, 18, and 48. The basis of these algorithms is that the dimension of the Riccati equation is the same as that of the system state vector, regardless of the dimension of the input and output functions. This fact leads to certain redundancies in the components of the Riccati equation when the dimensions of the input and output functions are less than that of the state vector. These redundancies can be used to reformulate the Riccati equation as a pair of coupled, nonlinear matrix equations termed Generalized X-Y Functions. These equations constitute a set of $N(M+P)$ equations, in which N, M, and P are the dimensions of the state vector, the measurement function, and the input function, respectively. The Riccati equation, of course, constitutes a set of $N(N+1)/2$ equations. For the extreme example of the single input-single output system, the comparison is $N(N+1)/2$ for the Riccati equation and $2N$ for the Generalized X-Y Functions. For $N = 8$, for example, these values are 36 and 16, respectively, and the divergence increases with increasing N. Since numerical integration is often used in the practical solution to Riccati equations, the decrease in computational load can be significant. For the time invariant case, the matrix Riccati equation, from Equation (6.118), is

$$\dot{\boldsymbol{S}}(t) = -\boldsymbol{S}(t)\boldsymbol{F} - \boldsymbol{F}^T\boldsymbol{S}(t) + \boldsymbol{S}(t)\boldsymbol{G}\boldsymbol{B}^{-1}\boldsymbol{G}^T\boldsymbol{S}(t) - \boldsymbol{A}. \qquad (6.214)$$

The X-Y functions are defined as the matrices $\boldsymbol{L}(t)$ and $\boldsymbol{C}(t)$, and satisfy the equations

$$\dot{\boldsymbol{L}}(t) = -[\boldsymbol{F}^T - \boldsymbol{C}^T(t)\boldsymbol{B}^{-1/2}\boldsymbol{G}^T]\boldsymbol{L}(t) \qquad (6.215)$$

and

$$\dot{\boldsymbol{C}}(t) = \boldsymbol{B}^{-1}\boldsymbol{G}^T\boldsymbol{L}(t)\boldsymbol{L}^T(t)$$
$$\boldsymbol{C}(t_f) = \boldsymbol{0}. \qquad (6.216)$$

The matrix $\boldsymbol{L}(t)$ is of dimension $N \times M$, while the dimension of $\boldsymbol{C}(t)$ is $P \times N$.

These equations are integrated backward in time from the terminal time t_f. The initial condition for the $\boldsymbol{L}(t)$ equation is obtained from the special form assumed for the $\boldsymbol{A}$ matrix:

$$\boldsymbol{L}(t_f) = \boldsymbol{A}_M, \quad \boldsymbol{A} = \boldsymbol{A}_M \boldsymbol{A}_M^T \tag{6.217}$$

where $\boldsymbol{A}_M$ is an $N \times M$ full rank matrix. It is shown in References 60, 61, and 82 that there is no loss of generality in specifying the matrix $\boldsymbol{A}$ to be of this special form. With the X-Y functions thus defined, the optimal gain matrix is given by the matrix $\boldsymbol{C}(t)$. If the Riccati equation variable $\boldsymbol{S}(t)$ is desired, it may be computed from the differential equation

$$\dot{\boldsymbol{S}}(t) = -\boldsymbol{L}(t)\boldsymbol{L}^T(t). \tag{6.218}$$

Evaluation of the numerical efficiency of the above algorithm can be found in References 17 and 18. These procedures can be generalized to the time varying case.

An alternative approach to the solution of the matrix Riccati differential equation is to solve a linear matrix differential equation of twice the dimension of the Riccati equation. This concept follows from the developments of Section 6.3.1, and is formalized in References 1 and 17. The linear matrix differential equation is

$$\begin{bmatrix} \dot{\boldsymbol{S}}_1(t) \\ \dot{\boldsymbol{S}}_2(t) \end{bmatrix} = \left[\begin{array}{c|c} \boldsymbol{F} & \boldsymbol{G}\boldsymbol{B}^{-1}\boldsymbol{G}^T \\ \hline -\boldsymbol{A} & -\boldsymbol{F}^T \end{array}\right] \begin{bmatrix} \boldsymbol{S}_1(t) \\ \boldsymbol{S}_2(t) \end{bmatrix} \tag{6.219}$$

with terminal conditions

$$\boldsymbol{S}_1(t_f) = \boldsymbol{I}; \quad \boldsymbol{S}_2(t_f) = \boldsymbol{S}_f. \tag{6.220}$$

Then, the matrix $\boldsymbol{S}(t)$ of Equation (6.214) is obtained as

$$\boldsymbol{S}(t) = \boldsymbol{S}_2(t)\boldsymbol{S}_1^{-1}(t). \tag{6.221}$$

This result, while shown for the time invariant case, is also valid for time varying systems, but the utility in that instance is questionable.

The solution of the algebraic Riccati equation, as required in optimal regulator design, has also received wide attention. A general investigation into the properties of the covergence of the matrix Riccati differential equation to a steady state value is presented in Reference 36. Special cases of the algebraic Riccati equation are considered in References 15, 32, and 34. The most general method is that of eigenvector decomposition, related to the steady state solution of Equation (6.219). The matrix $\boldsymbol{M}$ is defined as

$$\boldsymbol{M} = \left[\begin{array}{c|c} \boldsymbol{F} & -\boldsymbol{G}\boldsymbol{B}^{-1}\boldsymbol{G}^T \\ \hline -\boldsymbol{A} & -\boldsymbol{F}^T \end{array}\right]. \tag{6.222}$$

Because of the special form of this matrix, it has no purely imaginary eigenvalues, and it also has the characteristic that if λ is an eigenvalue, then so also is $-\lambda$. If $\boldsymbol{T}$ is a matrix which transforms $\boldsymbol{M}$ to its Jordan canonical form, as discussed briefly in Appendix A, then

$$\boldsymbol{T} = \left[\begin{array}{c|c} \boldsymbol{T}_{11} & \boldsymbol{T}_{12} \\ \hline \boldsymbol{T}_{21} & \boldsymbol{T}_{22} \end{array}\right]; \qquad \boldsymbol{T}^{-1}\boldsymbol{M}\boldsymbol{T} = \boldsymbol{J} \tag{6.223}$$

where $\boldsymbol{J}$ is the Jordan canonical form. The desired matrix $\boldsymbol{S}$–the steady state value of $\boldsymbol{S}(t)$–is given by

$$\boldsymbol{S} = \boldsymbol{T}_{21}\boldsymbol{T}_{11}^{-1}. \tag{6.224}$$

This concept is developed in References 1, 17, and 33. An improvement on the computational method is presented in References 49 and 59. Other techniques are based upon the Chandrasekhar algorithms–for example, those in Reference 47. A direct approach to obtaining the steady state solution is to numerically integrate the matrix Riccati equation to its steady state condition. In Reference 46, the relationship of the algebraic Riccati equation to the Lyapunov stability condition of Equation (4.46) is exploited to yield bounds for the steady state solution. A direct iterative approach is presented in Reference 1.

The topic of sensitivity of the optimal control problem is treated in References 1 and 80. In Reference 23, the sensitivity of the general linear, time invariant system is considered, while Reference 43 deals specifically with the sensitivity of optimal linear regulators to variation in system parameters. The sensitivity of the performance index itself to variations in the system parameters is investigated in References 9, 10, and 20. The design of robust optimal control systems is the topic of References 43, 71, 74, and 93; a frequency domain approach to robust design is presented in Reference 47. In Reference 25, a practical design perspective of multivariable feedback control systems is developed, complete with the attendant concepts of sensitivity and robust design.

Examples of the application of optimal control techniques to practical problems can be found in References 7, 13, 16, 19, 56, 58, 63, 76, and 87. A critique of practical application of the optimal regulator in industrial control can be found in Reference 78.

6.6.4. Discrete Time Systems

While the discrete time optimal control problem is treated here as a generalization of the continuous time problem, it is by far the more practical. Modern digital control systems are by nature discrete time systems, and therefore they require the use of discrete time control algorithms. While it is, in general, permissible to generalize from continuous time to discrete time results, there are nevertheless some special considerations that must be observed. An approach to the discrete time control problem using dynamic programming is contained in Chapter 8. A complete development of the discrete time, linear control system can be found in Reference 90. A careful presentation of the discrete time regulator problem and a review of the basic results are given in Reference 24. In Reference 35, the stability of discrete time optimal systems is investigated, recognizing those areas where generalization from continuous time results may not be justifiable. The solution of the discrete time matrix Riccati differential equation is considered in Reference 75, while approaches to the solution of the algebraic equation are considered in References 24, 34, 49, and 72. The discrete time optimal control problem with a singular arc is treated in Reference 89.

The time-optimal discrete time problem is of particular interest because of its relationship to the concept of deadbeat control introduced in Section 5.3.4. A deadbeat controller produces a closed-loop system with all eigenvalues equal to zero, but the methods of Section 5.3.4 do not produce unique controllers. The optimal control approach permits the designer to shape the response in some sense; for example, the deadbeat controller which yields minimal control amplitudes is of fundamental importance. In Reference 50, the deadbeat controller resulting in minimum control gain is developed. This concept is developed further in Reference 83, which presents a minimum time controller that achieves additional optimal properties along the path by shaping the transient response of the closed-loop system. This feature obviously is related to the eigenstructure concepts of Section 5.1.4. A geometric approach to the deadbeat control problem is investigated in Reference 53. In Reference 35, the concept of a minimum energy deadbeat controller is introduced, and in Reference 26, numerical techniques for solving the feedback gains are developed.

The time-optimal, linear, discrete time problem has a special characteristic in that it is amenable to solution by standard linear programming techniques, discussed briefly in Appendix B. In Reference 6, the simplex method of linear programming is used to provide on-line computation for minimum time control processes.

Finally, the use of digital logic requires consideration of the implementation effects of quantization, computational roundoff effects, computational speed, word length, and precision. These concepts are considered in Reference 33, which also contains several examples of digital implementation of discrete time control logic. The sensitivity of discrete time linear regulators, and the possible loss of controllability due to sampling, is investigated in Reference 24. These, and other effects of digital implementation of control logic, are treated in Reference 41, which also contains several examples of microprocessor implementation of control logic. Design for optimal quantization is developed in Reference 90, in which the effect of quantization levels on system performance is considered. Coefficient word-length considerations are presented in Reference 62.

6.6.5. Inaccessible State Variables

The use of observers to obtain suboptimal state variable feedback control systems is considered in Reference 1. In Reference 77, it is shown that the linear, time invariant, dynamic compensator for systems with inaccessible state variables is a minimum order linear observer. The problem of finding the optimal compensator is shown to be mathematically identical to a steady state stochastic control problem whose optimal solution is known. The robustness of observer based control systems was discussed in Section 5.5.2.

When the problem of inaccessible state variables is coupled with those of a stochastic system, then the statistical concepts of Chapter 7 must be applied.

REFERENCES

1. Anderson, Brian D. O., and Moore, John B. *Linear Optimal Control.* Englewood Cliffs, N.J.: Prentice-Hall, 1971.
2. Athans, M., and Falb, P. L. *Optimal Control.* New York: McGraw-Hill, 1968.
3. Athans, Michael. The Role and Use of the Stochastic Linear–Quadratic–Gaussian Problem in Control System Design. *IEEE Transactions on Automatic Control*, **AC-16:** 529–551 (December 1971).
4. Bagchi, A., and Strijbos, R. C. W. Decoupled Decomposition of the Riccati Equation. *IEEE Transactions on Automatic Control*, **AC-27:** 696–698 (June 1982).
5. Barry, Patrick E. Optimal Control with Minimax Cost. *IEEE Transactions on Automatic Control*, **AC-16:** 194–196 (April 1971).
6. Bashein, G. A Simplex Algorithm for On-Line Computation of Time Optimal Controls. *IEEE Transactions on Automatic Control*, **AC-16:** 479–482 (October 1971).
7. Battin, R. H. *Astronomical Guidance.* New York: McGraw-Hill, 1964.
8. Becker, N. A Note on Performance Index Sensitivity of Time-Optimal Control Systems. *IEEE Transactions on Automatic Control*, **AC-25:** 819–821 (August 1980).
9. Bertele, U., and Gaurdabassi, G. Can the Terminal Condition of Time-Invariant Linear Control Systems be Made Immune Against Small Parameter Variations? *IEEE Transactions on Automatic Control*, **AC-16:** 460–462 (October 1971).
10. Bobrovsky, B. Z., and Graupe, D. Analysis of Optimal-Cost Sensitivity to Parameter Changes. *IEEE Transactions on Automatic Control*, **AC-16:** 487–489 (October 1971).
11. Broussard, John R. A Quadratic Weight Selection Algorithm. *IEEE Transactions on Automatic Control*, **AC-27:** 945–947 (August 1982).
12. Brunovsky, Pavol, and Komornik, Josef. The Riccati Equation Solution of the Linear–

Quadratic Problem with Constrained Terminal State. *IEEE Transactions on Automatic Control*, **AC-26:** 398–402 (April 1981).
13. Bryson, Arthur E., and Ho Yu-Chi. *Applied Optimal Control*. Waltham, Massachusetts: Blaisdell, 1969.
14. Callier, Frank M., and Willems, Jacques L. Criterion for the Convergence of the Solution of the Riccati Differential Equation. *IEEE Transactions on Automatic Control*, **AC-26:** 1232–1242 (December 1981).
15. Canno, Incertis F., and Torres, J. M. Martinez. An Extension on a Reformulation on the Algebraic Riccati Equation Problem. *IEEE Transactions on Automatic Control*, **AC-22:** 128–129 (February 1977).
16. Cassidy, John F., Athans, Michael and Wing-Hong, Lee. On the Design of Electronic Automotive Engine Controls Using Linear Quadratic Control Theory. *IEEE Transactions on Automatic Control*, **AC-25:** 901–912 (October 1980).
17. Casti, John L. *Dynamical Systems and Their Applications: Linear Theory*. New York: Academic Press, 1977.
18. —— and Kirschner, O. Numerical Experiments in Linear Control Theory Using Generalized *X–Y* Equations. *IEEE Transactions on Automatic Control*, **AC-21:** 792–795 (October 1976).
19. Cottrell, Ronald G. Optimal Intercept Guidance for Short-Range Tactical Missiles. *AIAA Journal*, **9:** 1414–1415 (July 1971).
20. Courtin, P., and Rootenberg, J. Performance Index Sensitivity of Optimal Control Systems. *IEEE Transactions on Automatic Control*, **AC-16:** 275–277 (June 1971).
21. Davison, E. J., and Maki, M. C. The Numerical Solution of the Matrix Riccati Equation. *IEEE Transactions on Automatic Control*, **AC-18:** 71–73 (February 1973).
22. D'Azzo, John J., and Houpis, Constantine H. *Linear Control System Analysis and Design*. New York: McGraw-Hill, 1981.
23. Denevy, Dallas G. Simplification in the Computation of the Sensitivity Functions for Constant Coefficient Linear Systems. *IEEE Transactions on Automatic Control*, **AC-16:** 348–350 (August 1971).
24. Dorato, Peter, and Levis, Alexander. Optimal Linear Regulators: The Discrete-Time Case. *IEEE Transactions on Automatic Control*, **AC-16:** 613–620 (December 1971).
25. Doyle, J. C., and Stein, G. Multivariable Feedback Design: Concepts for a Classical/Modern Synthesis. *IEEE Transactions on Automatic Control*, **AC-26:** 4–16 (February 1981).
26. Emami-Naeini, A., and Franklin, G. F. Deadbeat Control and Tracking of Discrete-Time Systems. *IEEE Transactions on Automatic Control*, **AC-27:** 176–181 (February 1982).
27. Filho, Antonio Salles Campos. Numerical Computation of Optimal Control Sequences and Trajectories. *IEEE Transactions on Automatic Control*, **AC-16:** 47–49 (February 1971).
28. Fischer, Thomas R. Optimal Quantized Control. *IEEE Transactions on Automatic Control*, **AC-27:** 996–998 (August 1982).
29. Geering, Hans P. Continuous-Time Optimal Control Theory for Cost Functionals Including Discrete State Penalty Terms. *IEEE Transactions on Automatic Control*, **AC-21:** 866–869 (December 1976).
30. Harvey, Charles A., and Stein, Gunter. Quadratic Weights for Asymptotic Regulator Properties. *IEEE Transactions on Automatic Control*, **AC-23:** 378–394 (June 1978).
31. Howerton, Robert D., and Hammond, Joseph L. A New Computational Solution of the Linear Optimal Regulator Problem. *IEEE Transactions on Automatic Control*, **AC-16:** 645–651 (December 1971).

32. Incertis, F. C. A New Formulation of the Algebraic Riccati Equation Problem. *IEEE Transaction on Automatic Control*, **AC-26:** 768–770 (June 1981).
33. Jacquot, Raymond G. *Modern Digital Control Systems.* New York: Marcel Dekker, 1981.
34. Jones, E. L. A Reformulation of the Algebraic Riccati Equation. *IEEE Transactions on Automatic Control*, **AC-21:** 113 (February 1976).
35. Jordan, David, and Korn, Jonathan. Deadbeat Algorithms for Multivariable Process Control. *IEEE Transactions on Automatic Control*, **AC-25:** 486–491 (June 1980).
36. Kailath, T., and Ljung, L. The Asymptotic Behavior of Constant-Coefficient Riccati Differential Equations. *IEEE Transactions on Automatic Control*, **AC-21:** 385–388 (June 1976).
37. ——. Some Chandrasekhar-Type Algorithms for Quadratic Regulators. *Proceedings of the IEEE Decision and Control Conference, December 1972.*
38. ——. Some New Algorithms for Recursive Estimation in Constant Linear Systems. *IEEE Transactions on Information Theory*, **IT-19:** 750–760 (November 1973).
39. Karanam, V. R. Lower Bounds on the Solution of Lyapunov Matrix and Algebraic Riccati Equations. *IEEE Transactions on Automatic Control*, **AC-26:** 1288–1290 (December 1981).
40. ——. Eigenvalue Bounds for Algebraic Riccati and Lyapunov Equations. *IEEE Transactions on Automatic Control*, **AC-27:** 461–463 (April 1982).
41. Katz, Paul. *Digital Control Using Microprocessors.* Englewood Cliffs, New Jersey: Prentice-Hall, 1981.
42. Kirk, Donald E. *Optimal Control Theory.* Englewood Cliffs, New Jersey: Prentice-Hall, 1970.
43. Krishnan, K. R., and Brzezowski, S. Design of Robust Linear Regulator with Prescribed Trajectory Insensitivity to Parameter Variations. *IEEE Transactions on Automatic Control*, **AC-23:** 474–478 (June 1978).
44. Kucera, V. The Structure Properties of Time-Optimal Discrete Linear Control. *IEEE Transactions on Automatic Control*, **AC-16:** 375–377 (August 1971).
45. Kuo, Benjiman C. *Automatic Control Systems.* Englewood Cliffs, New Jersey: Prentice-Hall, 1975.
46. Kwon, W. H., and Pearson, E. A Note on the Algebraic Matrix Riccati Equation. *IEEE Transactions on Automatic Control*, **AC-22:** 143–144 (February 1977).
47. Lainiotis, D. G. Generalized Chandrasekhar Algorithms: Time-Varying Models. *IEEE Transactions on Automatic Control*, **AC-21:** 728–732 (October 1976).
48. ——. Partitioned Riccati Solutions and Integration-Free Doubling Algorithms. *IEEE Transactions on Automatic Control*, **AC-21:** 677–698 (October 1976).
49. Laub, Alan J. A Schur Method for Solving Algebraic Riccati Equations. *IEEE Transactions on Automatic Control*, **AC-24:** 913–921 (December 1979).
50. Leden, B. Dead-Beat Control and the Riccati Equation. *IEEE Transactions on Automatic Control*, **AC-21:** 791–792 (October 1976).
51. Lee, E. B., and Markus, L. *Foundations of Optimal Control Theory.* New York: John Wiley & Sons, 1968.
52. Lehtomaki, Norman A., Sandell, Nils R., and Athans, Michael. Robustness Results in Linear Quadratic Gaussian Based Multivariable Control Designs. *IEEE Transactions on Automatic Control*, **AC-26:** 75–93 (February 1981).
53. Lewis, F. Riccati Equation Solution to the Minimum Time Output Control Problem. *IEEE Transactions on Automatic Control*, **AC-26:** 763–766 (June 1981).
54. Lewis, Frank L. A General Riccati Equation Solution to the Deadbeat Control Problem. *IEEE Transactions on Automatic Control*, **AC-27:** 186–188 (February 1982).

55. Luenberger, D. *Optimization by Vector Space Methods.* New York: John Wiley & Sons, 1969.
56. McDonald, John P., and Kwatny, Harry G. Design and Analysis of Boiler–Turbine-Generator Controls Using Optimal Linear Regulator Theory. *IEEE Transactions on Automatic Control*, **AC-18:** 202–209 (June 1973).
57. Meditch, J. S. *Stochastic Optimal Linear Estimation and Control.* New York: McGraw-Hill, 1969.
58. Merriam, C. W. *Optimization Theory and the Design of Feedback Control Systems.* New York: McGraw Hill, 1964.
59. Michelsen, Michael L. On the Eigenvalue–Eigenvector Method for Solution of the Stationary Discrete Matrix Riccati Equation. *IEEE Transactions on Automatic Control*, **AC-24:** 480–481 (June 1979).
60. Molinari, B. P. Redundancy in Linear Optimum Regulator Problems. *IEEE Transactions on Automatic Control*, **AC-16:** 83–85 (February 1971).
61. ——. The Stable Regulator Problem and Its Inverse. *IEEE Transactions on Automatic Control*, **AC-18:** 454–459 (October 1973).
62. Moroney, Paul, Willsky, Alan S., and Houpt, Paul K. The Digital Implementation of Control Compensators: The Coefficient Wordlength Issue. *IEEE Transactions on Automatic Control*, **AC-25:** 621–630 (August 1980).
63. Nazaroff, Gregory J. An Optimal Terminal Guidance Law. *IEEE Transactions on Automatic Control*, **AC-21:** 407–408 (June 1976).
64. Nedeljkovic, Nikola B. New Algorithms for Unconstrained Nonlinear Optimal Control Problems. *IEEE Transactions on Automatic Control*, **AC-26:** 868–884 (August 1981).
65. Nicholson, David W. Eigenvalue Bounds in the Lyapunov and Riccati Matrix Equations. *IEEE Transactions on Automatic Control*, **AC-26:** 1290–1291 (December 1981).
66. Nishimura, T., and Kano, H. Periodic Oscillations of Matrix Riccati Equations in Time-Invariant Systems. *IEEE Transactions on Automatic Control*, **AC-25:** 749–755 (August 1980).
67. Ogata, Katsuhiko. *Modern Control Engineering.* Englewood Cliffs, New Jersey: Prentice-Hall, 1967.
68. ——. *State Space Analysis of Control Systems.* Englewood Cliffs, New Jersey: Prentice-Hall, 1967.
69. Owens, D. H. On the Computation of Optimal System Asymptotic Root-Loci. *IEEE Transactions on Automatic Control*, **AC-25:** 100–102 (February 1980).
70. Pappas, Thrasyvoulos, Laub, Alan J., and Sandell, Nils R. On the Numerical Solution of the Discrete-Time Algebraic Riccati Equation. *IEEE Transactions on Automatic Control*, **AC-25:** 631–641 (August 1980).
71. Patel, R. V., Toda, M., and Sridhar, B. Robustness of Linear Quadratic State Feedback Designs in the Presence of System Uncertainty. *IEEE Transactions on Automatic Control*, **AC-22:** 945–949 (December 1977).
72. Payne, Harold J., and Silverman, Leonard M. On the Discrete Time Algebraic Riccati Equation. *IEEE Transactions on Automatic Control*, **AC-18:** 226–234 (June 1973).
73. Postlehwaite, I. A Note on the Characteristic Frequency Loci of Multivariable Linear Optimal Regulators. *IEEE Transactions on Automatic Control*, **AC-23:** 757–760 (August 1978).
74. Rao, Sira G., and Soudack, A. C. Synthesis of Optimal Control Systems with Near Sensitivity Feedback. *IEEE Transactions on Automatic Control*, **AC-16:** 194–196 (April 1971).
75. Rappaport, David, and Silverman, Leonard M. Structure and Stability of Discrete-

Time Optimal Systems. *IEEE Transactions on Automatic Control*, **AC-16**: 227–233 (June 1971).

76. Reid, R. E., and Mears, B. C. Design of the Steering Controller of a Supertanker Using Linear Quadratic Control Theory: A Feasibility Study. *IEEE Transactions on Automatic Control*, **AC-27**: 940–942 (August 1982).
77. Rom, Douglas B., and Sarachik, Philip E. Design of Optimal Compensators for Linear Constant Systems with Inaccessible States. *IEEE Transactions on Automatic Control*, **AC-18**: 509–512 (October 1973).
78. Rosenbrock, Howard H., and McMorran, Peter D. Good, Bad, or Optimal? *IEEE Transactions on Automatic Control*, **AC-16**: 552–553 (December 1971).
79. Ryan, E. P. On the Sensitivity of a Time-Optimal Switching Function. *IEEE Transactions on Automatic Control*, **AC-25**: 275–277 (April 1980).
80. Sage, Andrew P., and White, Chelsea C. *Optimum Systems Control*. Englewood Cliffs, New Jersey: Prentice-Hall, 1977.
81. Sakawa, Yoshiyuki, and Shindo, Yuji. On Global Convergence of an Algorithm for Optimal Control. *IEEE Transactions on Automatic Control*, **AC-25**: 1149–1153 (December 1980).
82. Schultz, Donald H., and Melsa, James L. *State Functions and Linear Control Systems*. New York: McGraw-Hill, 1967.
83. Sebakhy, O. A., and Abdel-Moneim, T. M. Design of Optimal Dead-Beat Controllers. *IEEE Transactions on Automatic Control*, **AC-25**: 604–606 (June 1980).
84. Shaked, U. The Asymptotic Behavior of the Root-Loci of Multivariable Optimal Regulators. *IEEE Transactions on Automatic Control*, **AC-23**: 425–430 (June 1978).
85. Shinners, Stanly B. *Techniques of System Engineering*. New York: McGraw-Hill, 1967.
86. Sidhu, Gursharan S., and Bierman, Gerald J. Integration-Free Interval Doubling for Riccati Equation Solutions. *IEEE Transactions on Automatic Control*, **AC-22**: 831–834 (October 1977).
87. Stallard, David V. Discrete Optimal Terminal Control, with Application to Terminal Guidance. *IEEE Transactions on Automatic Control*, **AC-18**: 373–377 (August 1973).
88. Stein, G. Generalized Quadratic Weights for Asymptotic Regulator Properties. *IEEE Transactions on Automatic Control*, **AC-24**: 559–565 (August 1979).
89. Tarn, Tzyh-Jong, Rao, Sudhakara Kumblekere, and Zaborsky, John. Singular Control in Linear-Discrete Systems. *IEEE Transactions on Automatic Control*, **AC-16**: 401–409 (October 1971).
90. Tou, Julius T. *Optimum Design of Digital Control Systems*. New York: Academic Press, 1963.
91. ——. *Modern Control Theory*. New York: McGraw-Hill, 1964.
92. Tsuchiya, Takeshi. Improved Direct Digital Control Algorithm for Microprocessor Application. *IEEE Transactions on Automatic Control*, **AC-27**: 295–306 (April 1982).
93. Vandelinde, V. David. Robust Properties of Solutions to Linear–Quadratic Estimation and Control Problems. *IEEE Transactions on Automatic Control*, **AC-22**: 138–139 (February 1977).
94. Williams, Jan C. Least Squares Optimal Control and the Algebraic Riccati Equation. *IEEE Transactions on Automatic Control*, **AC-16**: 621–634 (December 1971).
95. Womble, M. Edward, Potter, James E., and Speyer, Jason L. Approximations to Riccati Equations Having Slow and Fast Modes. *IEEE Transactions on Automatic Control*, **AC-21**: 846–855 (December 1976).
96. Yackel, Richard A. and Kokotovic, Peter V. A Boundary Layer Method for Matrix Riccati Equations. *IEEE Transactions on Automatic Control*, **AC-18**: 17–23 (February 1973).

97. Youla, Dante, Bongiorno, Joseph, and Jabr, Hamid. Modern Wiener-Hopf Design of Optimal Controllers. *IEEE Transactions on Automatic Control*, **AC-21:** 3–13, 319–338 (February and June 1976).

DEVELOPMENTAL EXERCISES

6.1. In the idealized orbit injection model of Section 6.1.2, assume that the thrust acceleration $a(t)$ is constant. Use $\beta(t)$ as the independent variable and solve the equations of motion. Generalize this development as far as you can.

6.2. Use the technique described in Exercise 6.1 to solve the equation of the example of Section 6.1.3.

6.3. Use the concept of the state transition matrix to show why the two-point boundary value problem can, in principle, be solved for linear systems. Discuss the difference between time varying and time invariant systems in this context.

6.4. Formulate the frequency domain representation of the linear regulator.

PROBLEMS

6.1. Consider a particle of mass m acted upon by thrust force at angle β, as shown in Figure P6-1.

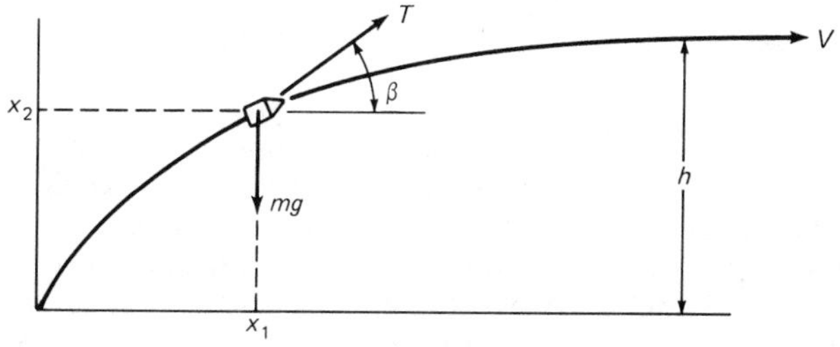

Figure P6-1.

(a) Show that the general minimum time problem results in a control law

$$\tan \beta = (-k_2 t + ku)/(-k_1 t + k_3).$$

This result is known as a *bilinear tangent law.*

(b) Show that the result of (a) reduces to a linear tangent law ($\tan \beta = k_1 + k_2 t$) when the terminal value of x_2 is specified to be h and the final velocity is specified to be V.

6.2. For the system of problem 6.1, find the thrust direction $\beta(t)$ which will take the particle from some initial point $x_1(t_0)$, $x_2(t_0)$ to the origin $x_1(t_f) = x_2(t_f) = 0$. The initial velocities are $x_3(t_0)$ and $x_4(t_0)$. Assume constant thrust acceleration a. Since the final velocities are not specified, this is an *intercept problem.*

6.3. Repeat Problem 6.2, but now with the final velocities specified to be zero. *This is a rendezvous problem.*

6.4. The roll channel of an aircraft autopilot can be considered to be a regulator which maintains the roll angle ϕ near zero, while staying within the physical limitations of aileron deflection and deflection rate. If $\delta(t)$ is the aileron deflection, the linearized equation of motion around the roll axis are

$$\dot{\phi}(t) = \omega(t)$$

$$\dot{\omega}(t) = \frac{1}{\tau}\,\omega(t) + k\delta(t)$$

$$\dot{\delta}(t) = u(t)$$

where τ is the roll time constant
k is the aileron torque constant
ω is the roll rate
u is the hydraulic flow rate to the aileron actuator.

Use the performance index

$$J = \tfrac{1}{2}\int_{t_0}^{t_f} (a_1\phi^2 + a_2\delta^2 + bu^2)\,dt$$

and derive the equations which define the optimal regulator. Carry your analysis as far as you can analytically and sketch the block diagram. Label blocks in terms of the elements of the $\boldsymbol{S}$ matrix if necessary.

6.5. Repeat Problem 6.4 for the case in which the aircraft is to be rotated so as to achieve a final angle near a desired value $\phi(t_f)$, while using minimal values of δ and u. Determine the equations which must be solved to specify the required terminal controller. Sketch the block diagram of the system and controller.

6.6. A spacecraft making a soft landing on a planet is modeled by the following equations:

$$\dot{h}(t) = v(t)$$

$$\dot{v}(t) = -g - [ku(t)/m]$$

$$\dot{m}(t) = u(t)$$

where $h(t)$ is the altitude, $v(t)$ the velocity, and $u(t)$ the thrust. Find the equations for determining the minimum time control function $u(t)$ to achieve a zero velocity landing ($h = 0$). Carry the solution as far as you can and describe the procedure to be followed from that point.

6.7. Repeat Problem 6.6 for the performance index corresponding to minimum fuel descent.

6.8. Consider the second-order system with equations

$$\dot{x}_1(t) = x_2(t)$$
$$\dot{x}_2(t) = u(t).$$

(a) Using the performance index

$$J = \frac{1}{2} \int_{t_0}^{t_f} [x_1^2(t) + 8x_1(t)x_2(t) + 4x_2^2(t) + bu^2(t)]\, dt$$

and determine the feedback gains for the optimum regulator for b values of 0.01, 1, and 100. Sketch a block diagram of the system and the controller.

(b) Find the control function $u(t)$ which will take the system from an initial condition $x_1(t_0) = 1$, $x_2(t_0) = 1$ to a terminal condition $x_1(t_f) = 0$, $x_2(t_f) = 0$, while minimizing the performance index

$$J = \frac{1}{2} \int_{t_0}^{t_f} u^2(t)\, dt$$

where $t_0 = 0$, $t_f = 1$.

6.9. Consider the system described by the scalar equation

$$\dot{x}(t) = -ax(t) + bu(t)$$

with performance index

$$J = \frac{1}{2}x^2(t_f) + \frac{1}{2} \int_{t_0}^{t_f} u^2(t)\, dt$$

and $x(t_0)$ fixed, $x(t_f)$ free. Find the optimal feedback control $u(t)$.

6.10. Consider the simple first-order system described by the equation

$$\dot{x}(t) = u(t)$$

which is to be transferred from some initial state $x(t_0)$ to the terminal state $x(t_f) = 0$ in a manner which minimizes the control effort. The performance index is

$$J = \int_{t_0}^{t_f} |x(t)|\, dt.$$

6.11. Consider the first-order linear system shown in Figure P6-2.

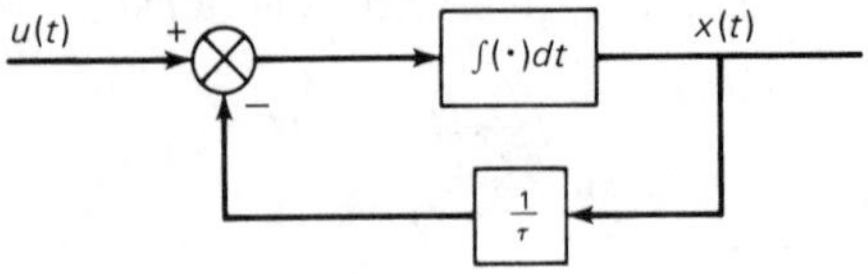

Figure P6-2.

(a) Determine the optimal regulator for this system, where optimality is defined by the performance index

$$J = \frac{1}{2} \int_{t_0}^{t_f} [ax^2(t) + bu^2(t)]\, dt.$$

Investigate the limiting cases $a \to \infty$ and $a \to 0$ and comment on the results.

(b) Consider this system from the terminal controller point of view. It is desired to move the system from some initial value $x(t_0)$ to a final value $x(t_f)$, where t_f is fixed, in such a manner as to minimize the performance index

$$J = \tfrac{1}{2} cx^2(t_f) + \tfrac{1}{2} \int_{t_0}^{t_f} u^2(t)\, dt.$$

Find the optimal feedback control $u(t)$ to accomplish this. Investigate the limiting case as $c \to \infty$, and comment on the physical interpretation of this limiting case.

(c) It is desired to find the control function $u(t)$ that will take the system from some initial state $x(t_0) = x_0$ to the final state $x(t_f) = 0$ while minimizing the performance index

$$J = \frac{1}{2} \int_{t_0}^{t_f} u^2(t)\, dt.$$

Show that this system yields the same result as (b) in the limiting case as $c \to \infty$.

6.12. Consider the first-order linear equation

$$\dot{x}(t) = x(t) + u(t).$$

(a) Using the performance index

$$J = \frac{1}{2} \int_{t_0}^{t_f} [2x^2(t) + u^2(t)]\, dt$$

find the optimum feedback regulator. Sketch the feedback system and identify the regulator portion of the diagram.

(b) Show the equivalent diagram of the regulated system (that is, system plus regulator) and find the response of the regulated system to some disturbance $x(t_0)$.

(c) Now design the optimum controller to bring the system from some initial state to a final state at time t_f, such that the performance index

$$J = \frac{1}{2}x^2(t_f) + \frac{1}{2} \int_{t_0}^{t_f} [2x^2(t) + u^2(t)]\, dt$$

is minimized. Show that, under the condition $t_f \to \infty$, this result reduces to that obtained in (a).

6.13. The pitch channel of an autopilot can be modeled by the following linearized equations.

$$\dot{\alpha}(t) = \frac{1}{T}\, \alpha(t) + \theta(t)$$

$$\dot{\theta}(t) = -\omega^2 [\alpha(t) - k\delta(t)]$$

where α is the angle of attack, θ the perturbation angle from the zero-lift angle of attack, and δ the elevator deflection (the control function here). The constants ω and k are the natural frequency of the pitch motion and the elevator effectiveness, respectively. Find the optimal regulator which minimizes the performance index

$$J = \frac{1}{2} \int_{t_0}^{t_f} [a_1 \alpha^2(t) + a_2 \dot{\theta}^2(t) + b\delta^2(t)]\, dt.$$

6.14. Consider the second-order system described by the state equation

$$\dot{\boldsymbol{x}}(t) = \boldsymbol{F}\boldsymbol{x}(t) + \boldsymbol{G}u(t)$$

where

$$\boldsymbol{x}(t) = \begin{bmatrix} x_1(t) \\ x_2(t) \end{bmatrix}; \quad \boldsymbol{F} = \begin{bmatrix} 0 & 1 \\ 2 & -1 \end{bmatrix}; \quad \boldsymbol{G} = \begin{bmatrix} 0 \\ 1 \end{bmatrix}.$$

A terminal controller is to be designed on the basis of the performance index

$$J = [x_1(t_f) - 1]^2 + \int_{t_0}^{t_f} \{[x_1(t) - 1]^2 + 0.04u^2(t)\}\, dt.$$

Determine the optimal control $u(t)$ in terms of the elements of the $\boldsymbol{S}$ matrix. Give the equations and boundary conditions for determining the elements of the $\boldsymbol{S}$ matrix.

6.15. Consider the undamped oscillator described by the state equation

$$\dot{\boldsymbol{x}}(t) = \boldsymbol{F}\boldsymbol{x}(t) + \boldsymbol{G}(t)\,\boldsymbol{u}(t)$$

where

$$\boldsymbol{x}(t) = \begin{bmatrix} x_1(t) \\ x_2(t) \end{bmatrix}; \quad \boldsymbol{F} = \begin{bmatrix} 0 & \omega \\ -\omega & 0 \end{bmatrix}; \quad \boldsymbol{G} = \begin{bmatrix} 0 \\ 1 \end{bmatrix}.$$

Determine the minimum time control which will take the system from some initial state $x_1(t_0), x_2(t_0)$ to the final state $x_1(t_f), x_2(t_f) = 0$.

6.16. Consider the discrete time system described by the equation

$$\boldsymbol{x}(l) = \quad \boldsymbol{x}(l - 1) + \quad u(l - 1)$$

where

$$\boldsymbol{x}(l) = \begin{bmatrix} x_1(l) \\ x_2(l) \end{bmatrix}; \quad \boldsymbol{\Phi} = \begin{bmatrix} 0 & 1 \\ -1 & 1 \end{bmatrix}; \quad \Gamma = \begin{bmatrix} 0 \\ 1 \end{bmatrix}; \quad \boldsymbol{x}(0) = \begin{bmatrix} 1 \\ 1 \end{bmatrix}$$

and $u(l)$ is a scalar control sequence. Find the opitmal control sequence $u(l)$ for $l = 0, 1$, and 2, which minimizes the performance index

$$J = \frac{1}{2} \sum_{l=0}^{2} [x_1^2(l + 1) + u^2(l)].$$

6.17. An attitude control system with viscous damping considered has the system equation

$$\ddot{x}(t) + \dot{x}(t) + x(t) = u(t).$$

(a) Determine the state space representation for this system.

(b) Find the minimum time control function $u(t)$ to take this system from some initial state $x(t_0), x(t_0)$ to a final state $x(t_f) = x(t_f) = 0$.

6.18. Determine the minimum time control function for the system described by the third-order equation

$$\dddot{x}(t) + 2\ddot{x}(t) + 6x(t) + 4x(t) = u(t).$$

The system is to transition from some arbitrary initial state to the final state $x(t_f) = \dot{x}(t_f) = \ddot{x}(t_f) = 0$.

6.19. Determine the optimal control for a simple regulator problem with the performance index

$$\int_{t_0}^{t_f} [x^2(t) + u^2(t)]\, dt$$

and system equation

$$\dot{x}(t) = -2x(t) + u(t).$$

There are no constraints on $u(t)$.

6.20. Determine the optimal control policy for the second-order system

$$\ddot{x}(t) + 4\dot{x}(t) + x(t) = u(t)$$

with the performance index

$$\int_{t_0}^{t_f} x^2(t)\, dt.$$

Assume that the magnitude of the input $u(t)$ is constrained to be less than one, and the control is to transfer the system from some initial value $x(t_0)$, $x(t_0)$ to a final state $x(t_f) = x(t_f) = 0$.

6.21. A second-order attitude control system is governed by the second-order equation

$$\ddot{x}(t) = u(t).$$

The control signal $u(t)$ must satisfy the constraint

$$|u(t)| \leqslant C.$$

Determine the minimum time control signal $u(t)$ which will take the system from some initial state $x(t_0)$, $x(t_0)$ to the final state $x(t_f) = x(t_f) = 0$.

6.22. For the second-order attitude control system described by the equation

$$\ddot{x}(t) = u(t)$$

find the control function $u(t)$ which will take the system from an initial state $x(t_0)$, $x(t_0)$ to the final state $x(t_f) = x(t_f) = 0$, while minimizing the energy expended

$$\int_{t_0}^{t_f} u^2(t)\, dt.$$

6.23. Consider the linear system with the system equations

$$\dot{x}_1(t) = x_2(t)$$
$$\dot{x}_2(t) = u(t).$$

Design the linear follower which minimizes the performance index

$$J = \frac{1}{2} \int [(x_1(t) - r(t))^2 + u^2]\, dt.$$

Sketch the block diagram of the resulting system.

6.24. Consider the system shown in Figure P6-3. It is desired to design a terminal

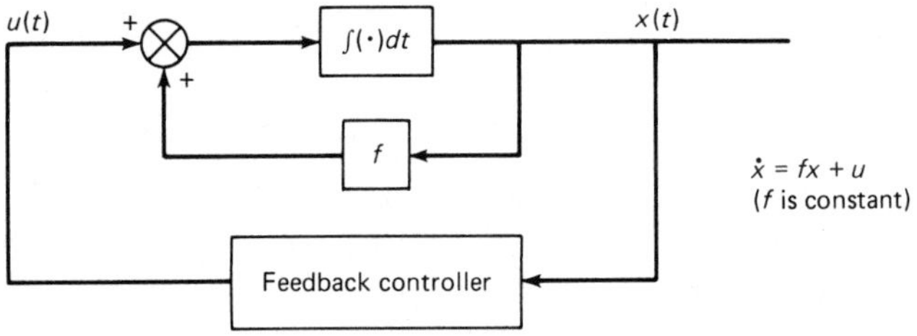

Figure P6-3.

controller which will take this system from an initial state $x(t_0)$ to a final state $x(t_f)$ near zero, such that the performance index

$$J = \frac{1}{2} a x^2(t_f) + \frac{1}{2} \int_{t_0}^{t_f} u^2(t)\, dt$$

is minimized.

(a) Comment on the performance index J.
(b) Determine the optimal feedback control law.
(c) Determine the closed-loop response with the optimal feedback controller in place and comment.
(d) Comment on the closed-loop system when $a = f$.

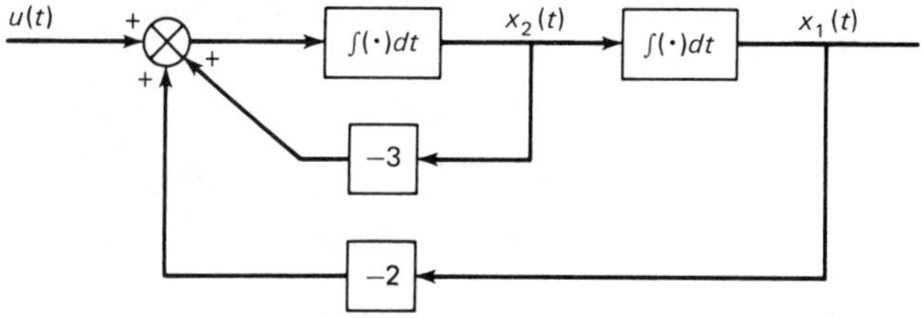

Figure P6-4.

6.25. For the system depicted in Figure P6-4, find the optimal regulator which minimizes the performance index

$$J = \frac{1}{2} \int_{t_0}^{t_f} [a x_1^2(t) + u^2(t)]\, dt.$$

Consider the values 1, 10, and 100 for a. Find the frequency domain transfer function of the regulated system.

6.26. A single input–single output system has the transfer function

$$V(s) = \frac{100}{s(s+1)(s+1)} = \frac{y(s)}{r(s)}.$$

Use the state vector

$$\mathbf{x}(t) = \begin{bmatrix} y(t) \\ \dot{y}(t) \\ \dot{y}(t) \end{bmatrix}$$

and performance index

$$J = \int_{t_0}^{t_f} [y^2(t) + 0.008y^2(t) + 0.06y^2(t) + u^2(t)]\, dt$$

and design the optimal regulator. Convert your answer to yield the closed loop transfer function $y(s)/r(s)$ for the controlled system. Refer to Appendix D.

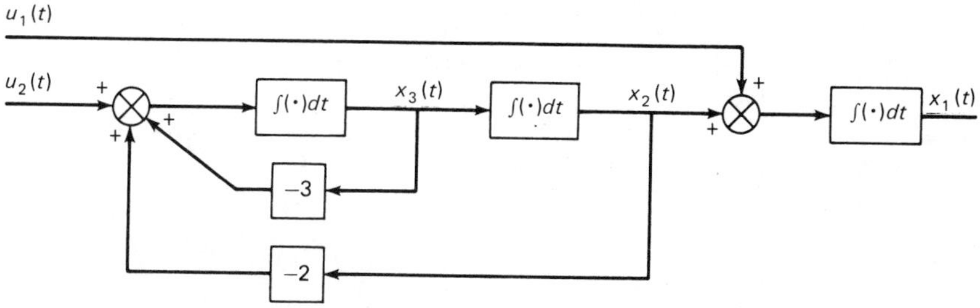

Figure P6-5.

6.27. For the system shown in Figure P6-5 find the optimal regulator which minimizes the performance index

$$J = \frac{1}{2} \int_{t_0}^{t_f} [x_1^2(t) + x_2^2(t) + u_1^2(t) + u_2^2(t)]\, dt.$$

6.28. Consider the general scalar system

$$\dot{x}(t) = fx(t) + gu(t).$$

Find the control function $u(t)$ which minimizes the performance index

$$J = \frac{1}{2} \int_{t_0}^{t_f} x^2(t)\, dt$$

when the control function has the constraint $u(t) \leqslant C$.

6.29. Consider the system described by the equations

$$\dot{x}_1(t) = x_2(t) + u_1(t)$$

$$\dot{x}_2(t) = u_2(t).$$

Use the performance index

$$J = \frac{1}{2} \int_{t_0}^{t_f} [x_1^2(t) + x_2^2(t) + u_1^2(t) + u_2^2(t)]\, dt$$

and design the optimal regulator. Sketch the resulting block diagram.

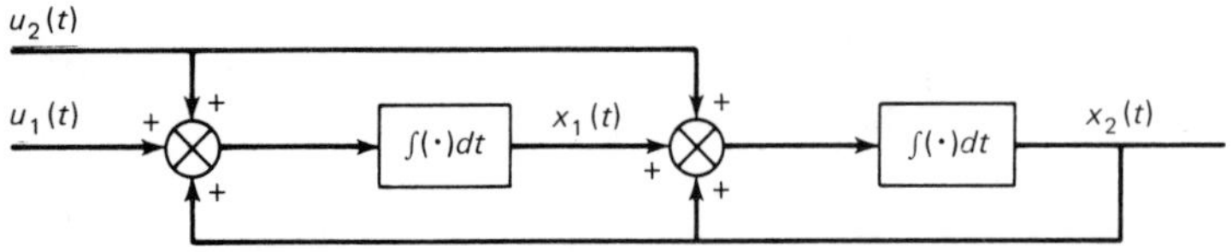

Figure P6-6.

6.30. For the system depicted in Figure P6-6, find the optimal regulator which minimizes the performance index

$$J = \frac{1}{2} \int_{t_0}^{t_f} [x_1^2(t) + (u_1(t) + u_2(t))^2 - 2u_1(t)u_2(t)]\, dt.$$

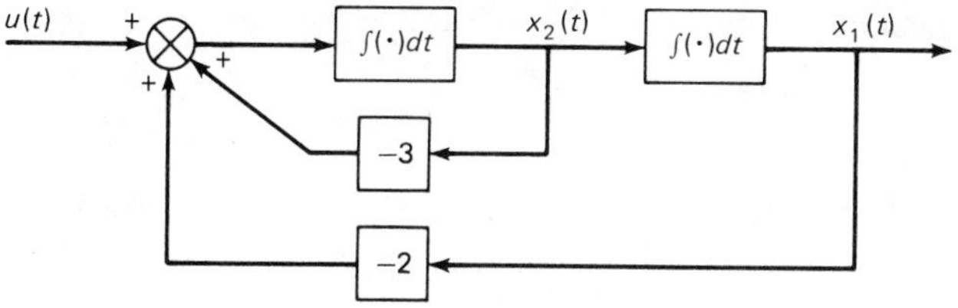

Figure P6-7.

6.31. A second-order system is described by the block diagram of Figure P6-7. Assume $t_0 = 0$, $t_f = 1$. For the performance index

$$J = \frac{1}{2} \int_{t_0}^{t_f} [x_1^2(t) + u^2(t)]\, dt$$

find the optimal control for $x_1(0) = x_2(0) = 1$.

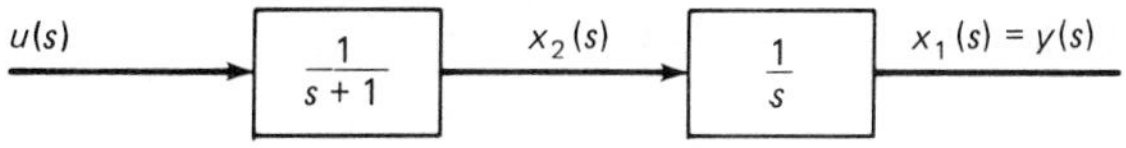

Figure P6-8.

6.32. Consider the second-order system described by the open-loop block diagram shown in Figure P6-8. Use the performance index

$$J = \frac{1}{2} \int_{t_0}^{t_f} [x_1^2(t) + 0.1x_2^2(t) + u^2(t)]\, dt$$

and find the optimal regulator for this system.

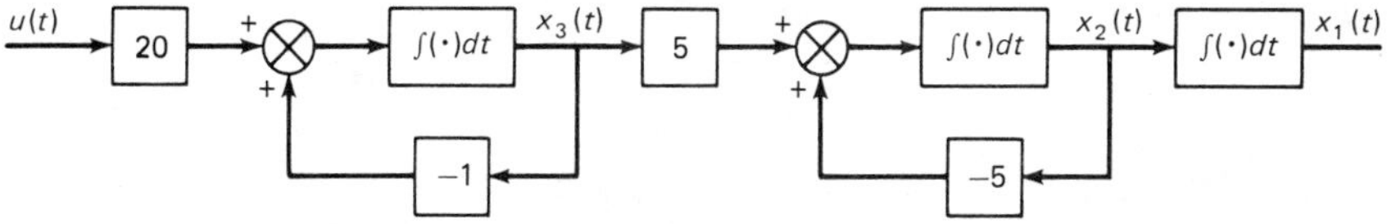

Figure P6-9.

6.33. A third-order system has the block diagram shown in Figure P6-9. Find the regulator which will minimize the performance index

$$J = \frac{1}{2} \int_{t_0}^{t_f} [x_1^2(t) + 0.008x_2^2(t) + x_3^3(t) + u^2(t)]\, dt.$$

Show the frequency domain representation of this system.

6.34. An undamped linear oscillator is described by the second order equation

$$\ddot{x} + \omega^2 x = u(t).$$

Find the driving function $u(t)$ that will bring this system from initial condition $x(t_0) = a$, $x(t_0) = b$, to the final state $x(t_f) = x(t_f) = 0$, with a minimum value of the performance index

$$J = \frac{1}{2} \int_{t_0}^{t_f} u^2 \, dt.$$

6.35. For the second-order attitude control system described by the equation

$$\ddot{x}(t) = u(t)$$

find the control function $u(t)$ which will take the system from an initial state $x(t_0)$, $\dot{x}(t_0)$ to the final state $x(t_f) = \dot{x}(t_f) = 0$. The system is to minimize total fuel consumption, given by

$$\int_{t_0}^{t_f} |u(t)| \, dt.$$

6.36. Repeat Problem 6.35 if the system equation is

$$\ddot{x}(t) + \dot{x}(t) + x(t) = u(t).$$

6.37. A control system is described by the equations

$$\dot{x}_1(t) = x_2(t)$$
$$\dot{x}_2(t) = -x_1(t) - x_2(t) + u(t).$$

If $u(t)$ is subject to the constraint

$$|u(t)| \leqslant C,$$

find the minimum time control function which will take the system from some initial state $x_1(t_0)$, $x_2(t_0)$ to the final state $x_1(t_f) = x_2(t_f) = 0$.

6.38. Repeat Problem 6.37 if the system equations are

$$\dot{x}_1(t) = x_2(t)$$
$$\dot{x}_2(t) = -4x_1(t) - 2x_2(t) + u(t).$$

6.39. Repeat Problem 6.37 if the system equations are

$$\dot{x}_1(t) = x_2(t)$$
$$\dot{x}_2(t) = -4x_1(t) - 4x_2(t) + u(t).$$

6.40. Consider the scalar system described by the equation

$$\dot{x}(t) = -2x(t) + u(t).$$

(a) Design a terminal controller with the performance index

$$J = \tfrac{1}{2}x^2(t_f) + \tfrac{1}{2}\int_{t_0}^{t_f} [x^2(t) + u^2(t)]\, dt.$$

(b) Plot the optimal gain as a function of time from the initial value t_0 to the final value $t_f = \infty$. Interpret these results in terms of regulator design.

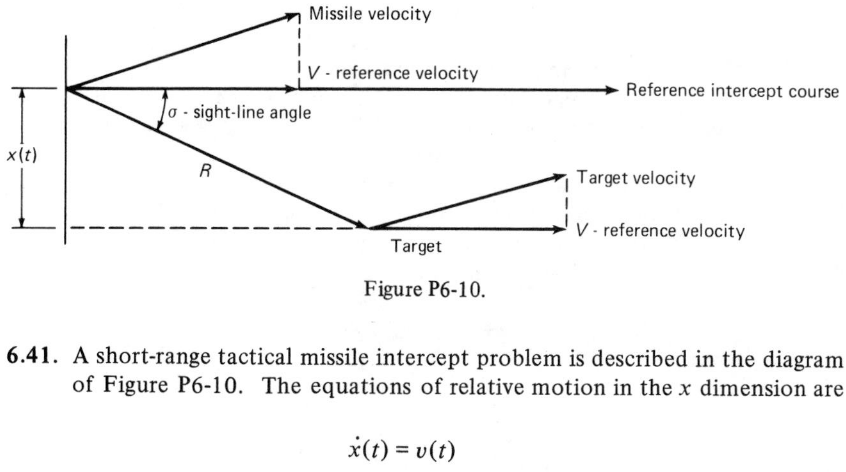

Figure P6-10.

6.41. A short-range tactical missile intercept problem is described in the diagram of Figure P6-10. The equations of relative motion in the x dimension are

$$\dot{x}(t) = v(t)$$
$$\dot{v}(t) = a_M(t) - a_T(t)$$
$$a_M(t) = \frac{1}{\tau}[u(t) - a_M(t)]$$

where v is the normal velocity, a_M and a_T are the missile and target normal accelerations, respectively, and $u(t)$ is the normal control acceleration. The missile dynamics are represented by the missile time constant τ. The guidance law for the missile is to be determined by designing a terminal controller with performance index

$$J = \frac{1}{2} \int_{t_0}^{t_f} u^2(t)\, dt.$$

(a) Assume target normal acceleration $a_T(t)$ is zero and design the optimum regulator with zero terminal miss distance, i.e., $x(t_f) = 0$. Terminal velocity is free. Express your result in terms of time-to-go defined as $t_g = t_f - t$.

(b) Show that in the limit as $\tau \to 0$ the guidance law reduces to

$$u(t) = -3\{[v(t)/(t_f - t)] + [x(t)/(t_f - t)^2]\}.$$

Also show that if the sight line angle σ is approximated by

$$\sigma(t) \approx x(t)/R(t) = x(t)/V(t - t_f)$$

Then, for $\tau \to 0$, the above expression can be written as

$$u(t) = -3V\dot{\sigma}(t).$$

This is the well known *proportional guidance law.*

(c) Find the expressions for $M(t)$, the predicted terminal miss distance if control is discontinued at time t. Use the limiting case $\tau \to 0$ to give intuitive appeal to your answer.

6.42. One of the requirements imposed on many control systems is that of fail-safe operation. Loss of one sensor in a system, for example, can result in degraded performance, but should not cause unstable operation. Consider a system described by the equation

$$\dot{\boldsymbol{x}}(t) = \boldsymbol{F}\boldsymbol{x}(t) + \boldsymbol{G}\boldsymbol{u}(t)$$

where

$$\boldsymbol{F} = \begin{bmatrix} 5 & 6 \\ -6 & -5 \end{bmatrix}; \qquad \boldsymbol{G} = \begin{bmatrix} 1 & 0 \\ 0 & 1 \end{bmatrix}$$

and both states x_1 and x_2 are available for measurement. The system is controlled by a regulator optimized for the performance index

$$J = \int_{t_0}^{t_f} [\tfrac{2}{3}x_1^2(t) + \tfrac{4}{3}x_2^2(t) + u_1^2(t) + u_2^2(t)]\, dt.$$

Determine the stability of this system when the measurement of x_1 is lost.

Chapter 7
Optimal Estimation and Filtering

7.1. POINT ESTIMATION

In previous chapters dealing with system control, it was shown that state variable feedback is an effective approach to the feedback control problem. Optimal control laws were also shown to result in requirements satisfied quite nicely by state variable feedback. But in most practical systems, observation of all state variables is not possible, and those that are available for observation may be contaminated with measurement noise. In those instances in which the noise contamination is negligible, a noise-free observer of the kind developed in Chapter 5 may be used to reconstruct the entire state vector from those observations that are available. On the other hand, in the presence of appreciable measurement noise other methods must be used. These considerations introduce the concept of estimation of the state vector of a system from noisy measurement data. The general concept applies to any situation in which an estimation process is required, regardless of whether or not a feedback control process is involved. From a broad viewpoint, however, most estimation processes provide information upon which some kind of decision is to be based, so that the feedback control analogy applies in most cases.

Estimation processes fall into three general catagories: point estimation, filtering, and smoothing. Point estimation involves a set of variables to be estimated, and the values of the variables do not change regardless of how many observations are made or what time interval occurs between observations. The filtering problem involves a time function which is to be estimated—so that the values of the variables to be estimated are not constant—and different relationships between the variables and the observations may exist at different times. The estimate at any time should include information contained in all measurements made up to that time. Smoothing is similar to filtering, except that the smoothed estimate for any particular time may include information contained in measurements made at subsequent times, implying a time delay in computing the smoothed estimate.

An example is provided by a radar tracking system which measures the position of an aircraft in two angular and one linear dimension—azimuth and elevation angles and slant range. In each case, the observations are contaminated with measurement noise, which is assumed to be additive. It is desired to estimate the state of the system as expressed by three position and three velocity coordinates—that is, the state vector is

$$\boldsymbol{x} = \begin{bmatrix} s_1 \\ s_2 \\ s_3 \\ v_1 \\ v_2 \\ v_3 \end{bmatrix} \tag{7.1}$$

The noise-free and noise-contaminated measurements are expressed in the general terms of Chapter 4:

$$\boldsymbol{y}(\boldsymbol{x}) = \boldsymbol{h}(\boldsymbol{x}) = \begin{bmatrix} \theta(\boldsymbol{x}) \\ \phi(\boldsymbol{x}) \\ R(\boldsymbol{x}) \end{bmatrix}; \qquad \boldsymbol{Z}(\boldsymbol{x}, \boldsymbol{\epsilon}) = \boldsymbol{y}(\boldsymbol{x}) + \boldsymbol{\epsilon} \tag{7.2}$$

where θ, ϕ, and R are the measured values of the azimuth angle, elevation angle, and slant range, respectively, and $\boldsymbol{\epsilon}$ is a vector of random error terms. The point estimation problem would involve the generation of estimates of s_1, s_2, and s_3, from the observed values of θ, ϕ, and R at a single point in time. Obviously, no velocity information can be obtained from measurements at a single time point, even if different measurements from different radars are used to form the estimate. On the other hand, if measurements are made at different points in time, and it is recognized that the system itself is dynamic in the sense that the states change with time in a known way, then both position and velocity coordinates can be estimated from the time sequence of available measurements. Such a situation represents a filtering problem, and the solution results in a set of time functions representing the estimates of the system states.

The point estimation problem itself falls into one of two different categories, each requiring its own particular development. In the first category, the variables to be estimated are not random but are completely unknown. That is, there is no information available concerning any probability distribution which might apply to the variable or variables to be estimated. The second category involves the problem of estimating a set of random variables which are again unknown but about which some information is available in the form of known probability distributions. In either event, the estimates are to be selected so that some kind of optimal criterion is satisfied, and, as has been the case in previous developments, the particular criterion is sometimes selected as much for mathematical convenience as for any other reason.

7.1.1. Parameter Estimation

When the variables to be estimated are unknown parameters, as contrasted here to random variables, then the problem is one of point estimation in which the only information available is the way in which the unknown parameters affect the noise-free measurements. That is, the measurement equation is

$$\mathbf{Z} = \boldsymbol{h}(\boldsymbol{x}) + \boldsymbol{\epsilon} \tag{7.3}$$

where the vector function $\boldsymbol{h}(\boldsymbol{x})$ is known. The problem is to estimate the parameter $\boldsymbol{x}$, based upon the noisy observations $\mathbf{Z}$, and the estimation process is to be performed in such a way that some prescribed optimization criterion is satisfied. Information about the probability distribution of the error terms may or may not be available, and this factor influences the criterion to be applied. The problem as stated above involves the general nonlinear measurement function $\boldsymbol{h}(\boldsymbol{x})$ and, as in most areas of analytical development, the nonlinear problem has no general solution. For this reason, and because linearization and perturbation techniques also apply to this situation, emphasis will be placed on the linear estimation problem, which has the general form

$$\mathbf{Z} = \boldsymbol{H}\boldsymbol{x} + \boldsymbol{\epsilon}. \tag{7.4}$$

Here the noise-free measurement is a linear function of the parameter to be estimated.

7.1.1.1. Least Squares Linear Estimation. Least squares estimation is probably familiar to the reader as a method of forming an estimate based upon redundant measurements of some parameter. It was originally used by Gauss in the late eighteenth century to estimate the orbital parameters of planets and comets. Although it has developed over the years as a method in its own right, it was originally conceived in the form of maximum likelihood estimation when error terms are assumed to be independent and normally distributed. In the general case, the measurement is of the form of Equation (7.3), and the distribution of the error terms $\boldsymbol{\epsilon}$ is not known. The principle of least squares states that nature normally tends to minimize the errors which are present in a measurement process, and therefore a logical estimate for the parameter $\boldsymbol{x}$ is that value which will yield a minimum sum of squares of the error terms themselves. That is, the error criterion specifies that the loss function defined as

$$q = \epsilon_1^2 + \epsilon_2^2 + \cdots + \epsilon_M^2 = [\mathbf{Z} - \boldsymbol{h}(\boldsymbol{x})]^T [\mathbf{Z} - \boldsymbol{h}(\boldsymbol{x})] \tag{7.5}$$

should have a minimum value. If q is to be minimized by choice of $\boldsymbol{x}$, the above expression results in a set of nonlinear, simultaneous algebraic equations which has no general solution. But if the measurement is linear and of the form given in Equation (7.4), then the loss function can be written as

$$q = [\mathbf{Z} - \boldsymbol{Hx}]^T [\mathbf{Z} - \boldsymbol{Hx}] = \mathbf{Z}^T\mathbf{Z} - 2\mathbf{Z}^T\boldsymbol{Hx} + \boldsymbol{x}^T\boldsymbol{H}^T\boldsymbol{Hx}. \tag{7.6}$$

The minimization of q is then accomplished by differentiating q with respect to the vector $\boldsymbol{x}$ and equating the result to zero. Following the methods outlined in Appendix A yields

$$\frac{\partial q}{\partial \boldsymbol{x}} = \begin{bmatrix} \frac{\partial q}{\partial x_1} & \frac{\partial q}{\partial x_2} & \cdots & \frac{\partial q}{\partial x_N} \end{bmatrix} = -2\mathbf{Z}^T\boldsymbol{H} + 2\boldsymbol{x}^T\boldsymbol{H}^T\boldsymbol{H} = \mathbf{0}. \tag{7.7}$$

Solving this equation for $\boldsymbol{x}$ yields the least squares estimate, denoted by $\hat{\boldsymbol{x}}$,

$$\hat{\boldsymbol{x}} = (\boldsymbol{H}^T\boldsymbol{H})^{-1}\boldsymbol{H}^T\mathbf{Z}. \tag{7.8}$$

This concise matrix equation yields a closed form solution for the value of the parameter $\hat{\boldsymbol{x}}$ which produces the minimum value of the scalar q. The reason for selecting the sum of squares as the optimization criterion is partly for mathematical convenience and partly to obtain the result of emphasizing the larger error terms. When the least squares method is used, it is important to recognize the basis for the optimization process.

7.1.1.2. Weighted Least Squares Linear Estimation. If the individual measurement error terms in Equation (7.4) are known to have variances which are not identical, then the argument for the error criterion function q in the particular form of Equation (7.5) cannot be made. Instead, a logical approach is to weight each squared error by a factor which somehow indicates the quality of the associated measurement, so that the loss function is now specified to be

$$q = c_1\epsilon_1^2 + c_2\epsilon_2^2 + \cdots + c_M\epsilon_M^2 \tag{7.9}$$

where c_i has a large value if the ith measurement is considered to have low noise content, and a small value if the ith measurement is considered to be very noisy. Such an effect can be incorporated into the previous result by letting

$$q = [\mathbf{Z} - \boldsymbol{Hx}]^T\boldsymbol{C}[\mathbf{Z} - \boldsymbol{Hx}] \tag{7.10}$$

where $\boldsymbol{C}$ is a diagonal matrix of the form

$$\boldsymbol{C} = \begin{bmatrix} c_1 & & & \\ & c_2 & & \\ & & \ddots & \\ & & & c_M \end{bmatrix}.$$

Then

$$q = \mathbf{Z}^T \boldsymbol{C} \mathbf{Z} - 2\mathbf{Z}^T \boldsymbol{C}\boldsymbol{H}\boldsymbol{x} - \boldsymbol{x}^T \boldsymbol{H}^T \boldsymbol{C}\boldsymbol{H}\boldsymbol{x} \tag{7.11}$$

and, from the matrix formalism of Appendix A,

$$\frac{\partial q}{\partial \boldsymbol{x}} = -2\mathbf{Z}^T \boldsymbol{C}\boldsymbol{H} + 2\boldsymbol{x}^T \boldsymbol{H}^T \boldsymbol{C}\boldsymbol{H}. \tag{7.12}$$

Solving this equation for the estimate $\hat{\boldsymbol{x}}$ yields

$$\hat{\boldsymbol{x}} = [\boldsymbol{H}^T \boldsymbol{C}\boldsymbol{H}]^{-1} \boldsymbol{H}^T \boldsymbol{C} \mathbf{Z} \tag{7.13}$$

which is the weighted least squares estimate of the parameter $\boldsymbol{x}$. The selection of the weighting functions c_i can be intuitive, although there is a systematic approach to their selection if the error variances are known, as will be shown below.

7.1.1.3. Minimum Variance Unbiased Linear Estimation. A basic characteristic of the least squares method is that the distribution of the error terms $\boldsymbol{\epsilon}$ is not known. When some information is available in the form of the mean value function $E[\boldsymbol{\epsilon}]$ and the covariance matrix

$$\boldsymbol{\Sigma} = E[\{\boldsymbol{\epsilon} - E[\boldsymbol{\epsilon}]\}\{\boldsymbol{\epsilon} - E[\boldsymbol{\epsilon}]\}^T]$$

then a much more systematic and intuitively satisfying approach to the linear estimation problem can be obtained. There will be no loss of generality if the error terms are assumed to have zero mean

$$E[\boldsymbol{\epsilon}] = \mathbf{0} \tag{7.14}$$

since this merely represents a correction term of the form

$$\mathbf{Z} - E[\boldsymbol{\epsilon}] = \boldsymbol{H}\boldsymbol{x} + \boldsymbol{\epsilon} - E[\boldsymbol{\epsilon}] \tag{7.15}$$

to generate a new set of measurements contaminated by zero mean noise terms. Thus, the measurement noise will be assumed to have the characteristics

$$E[\boldsymbol{\epsilon}] = \mathbf{0}$$
$$E[\boldsymbol{\epsilon}\boldsymbol{\epsilon}^T] = \boldsymbol{\Sigma}. \tag{7.16}$$

Consideration will now be limited to linear estimates of the form

$$\hat{\boldsymbol{x}} = \boldsymbol{BZ} \tag{7.17}$$

where $\boldsymbol{B}$ is, in general, a singular matrix. Furthermore, only unbiased estimates, for which

$$E[\hat{\boldsymbol{x}}] = \boldsymbol{x} \tag{7.18}$$

will be considered. Now, if one is considering all the unbiased linear estimates described by the two equations above, the most logical choice is that which yields the smallest estimation error variance. That is, if the loss function q is now defined as

$$q = \sum_{i=1}^{N} E[(x_i - \hat{x}_i)^2] \tag{7.19}$$

which is the sum of the variances of the estimation errors, then that unbiased linear estimate which yields the minimum value of q is a logical choice. This estimate is termed the *minimum variance, unbiased, linear estimate.*

In order to develop an expression for the minimum variance, unbiased, linear estimate, use is made of the estimation error covariance matrix, defined as

$$\boldsymbol{P} = E[(\boldsymbol{x} - \hat{\boldsymbol{x}})(\boldsymbol{x} - \hat{\boldsymbol{x}})^T] \tag{7.20}$$

where the particular form results from the zero mean assumption. In component form, this equation is

$$\boldsymbol{P} = \begin{bmatrix} E[(x_1 - \hat{x}_1)^2] & E[(x_1 - \hat{x}_1)(x_2 - \hat{x}_2)] & \cdots & E[(x_1 - \hat{x}_1)(x_N - \hat{x}_N)] \\ E[(\hat{x}_1 - \hat{x}_1)(x_2 - \hat{x}_2)] & E[(x_2 - \hat{x}_2)^2] & \cdots & E[(x_2 - \hat{x}_2)(x_N - \hat{x}_N)] \\ \cdots & \cdots & \cdots & \cdots \\ E[(x_1 - \hat{x}_1)(x_N - \hat{x}_N)] & E[(x_2 - \hat{x}_2)(x_N - \hat{x}_N)] & \cdots & E[(x_N - \hat{x}_N)^2] \end{bmatrix} \tag{7.21}$$

from which it can be seen that

$$q = \mathrm{Tr}\,\boldsymbol{P} \tag{7.22}$$

where Tr $\boldsymbol{P}$ is the trace of matrix $\boldsymbol{P}$. With the necessary terms and relationships thus defined, the development of a general closed form solution for the minimum variance, unbiased, linear estimate can be considered.

If the assumed linear form of Equation (7.17) is substituted into Equation (7.4), with the result

$$\hat{\boldsymbol{x}} = \boldsymbol{BHx} + \boldsymbol{B\epsilon} \tag{7.23}$$

then the estimation error is given by

$$\boldsymbol{x} - \hat{\boldsymbol{x}} = (\boldsymbol{I} - \boldsymbol{BH})\boldsymbol{x} - \boldsymbol{B\epsilon}. \tag{7.24}$$

The requirement for an unbiased estimate can then be expressed as

$$E[\boldsymbol{BHx} + \boldsymbol{B\epsilon}] = \boldsymbol{x}$$

or,

$$\boldsymbol{BHx} + \boldsymbol{B}E[\boldsymbol{\epsilon}] = \boldsymbol{x}. \tag{7.25}$$

The second term on the left-hand side is zero because of the zero mean assumption. By factoring terms, one obtains

$$(\boldsymbol{I} - \boldsymbol{BH})\boldsymbol{x} = \boldsymbol{0} \tag{7.26}$$

as the requirement for an unbiased estimate. In essence, Equation (7.26) represents a constraint on the choice of the matrix $\boldsymbol{B}$, in that it requires

$$\boldsymbol{I} - \boldsymbol{BH} = \boldsymbol{0}. \tag{7.27}$$

With this restriction on $\boldsymbol{B}$, the estimation error from Equation (7.24) is

$$\boldsymbol{x} - \hat{\boldsymbol{x}} = -\boldsymbol{B\epsilon}. \tag{7.28}$$

Substitution of this result into Equation (7.20) produces

$$\boldsymbol{P} = E[\boldsymbol{B\epsilon\epsilon}^T\boldsymbol{B}^T] = \boldsymbol{B\Sigma B}^T. \tag{7.29}$$

This expression gives the estimation error covariance matrix $\boldsymbol{P}$ in terms of the measurement error covariance matrix $\boldsymbol{\Sigma}$.

The problem can now be explicitly stated. It is desired to select the matrix $\boldsymbol{B}$ to yield the minimum value of the loss function

$$q = \text{Tr}\,(\boldsymbol{B\Sigma B}^T) \tag{7.30}$$

subject to the constraint

$$\boldsymbol{I} - \boldsymbol{BH} = \boldsymbol{0}.$$

The result is a constrained point optimization problem in which q is to be minimized with respect to each element of the matrix $\boldsymbol{B}$, subject to N constraints represented by the vector equation above. This minimization can be accomplished by defining a matrix of Lagrange multipliers $\boldsymbol{\Lambda}$, as discussed in Appendix B, such that the Lagrange function

$$\boldsymbol{F} = \text{Tr}\,[\boldsymbol{B\Sigma B}^T + (\boldsymbol{I} - \boldsymbol{BH})\boldsymbol{\Lambda}] \tag{7.31}$$

now becomes the objective function to be minimized. Two helpful relationships which can be obtained from Appendix A are

$$\frac{\partial}{\partial \boldsymbol{B}}\,\text{Tr}\,(\boldsymbol{B\Sigma B}^T) = 2\boldsymbol{\Sigma B}^T \tag{7.32}$$

and

$$\frac{\partial}{\partial \boldsymbol{B}}\,\text{Tr}\,(\boldsymbol{BH\Lambda}) = \boldsymbol{H\Lambda}.$$

Then, the necessary conditions are represented by the equation

$$\frac{\partial q}{\partial \boldsymbol{B}} = 2\boldsymbol{\Sigma B}^T - \boldsymbol{H\Lambda} = 0 \tag{7.33}$$

which, when solved for $\boldsymbol{B}$, yields

$$\boldsymbol{B} = \tfrac{1}{2}\boldsymbol{\Lambda}^T\boldsymbol{H\Sigma}^{-1}. \tag{7.34}$$

The value of $\boldsymbol{\Lambda}$, the matrix of Lagrange multipliers, can be obtained by substituting this expression into the constraint equation. Thus,

$$\boldsymbol{I} - \boldsymbol{BH} = \boldsymbol{I} + \tfrac{1}{2}\,\boldsymbol{\Lambda}^T\boldsymbol{H}^T\boldsymbol{\Sigma}^{-1}\boldsymbol{H} = \boldsymbol{0} \tag{7.35}$$

or, by multiplying on the right by $2(\boldsymbol{H}^T \boldsymbol{\Sigma}^{-1}\boldsymbol{H})^{-1}$,

$$\boldsymbol{\Lambda}^T = 2(\boldsymbol{H}^T \boldsymbol{\Sigma}^{-1}\boldsymbol{H})^{-1}.$$

Finally, substituting this expression into Equation (7.34) yields the desired result

$$\boldsymbol{B} = (\boldsymbol{H}^T \boldsymbol{\Sigma}^{-1}\boldsymbol{H})^{-1}\boldsymbol{H}^T \boldsymbol{\Sigma}^{-1}. \tag{7.36}$$

The optimal estimate $\hat{\boldsymbol{x}}$ is then obtained from Equation (7.17) as

$$\hat{\boldsymbol{x}} = (\boldsymbol{H}^T \boldsymbol{\Sigma}^{-1}\boldsymbol{H})^{-1}\boldsymbol{H}^T \boldsymbol{\Sigma}^{-1}\mathbf{Z}. \tag{7.37}$$

This expression gives the estimate $\hat{\boldsymbol{x}}$ which, of all possible unbiased linear estimates, yields the smallest value for the sum of the estimation error variances. These variances can be obtained from the estimation error covariance matrix $\boldsymbol{P}$, which, by virtue of Equation (7.29), can now be expressed as

$$\begin{aligned} \boldsymbol{P} = \boldsymbol{B}\boldsymbol{\Sigma}\boldsymbol{B}^T &= (\boldsymbol{H}^T \boldsymbol{\Sigma}^{-1}\boldsymbol{H})^{-1}\boldsymbol{H}^T \boldsymbol{\Sigma}^{-1}\boldsymbol{\Sigma}\boldsymbol{\Sigma}^{-1}\boldsymbol{H}(\boldsymbol{H}^T \boldsymbol{\Sigma}^{-1}\boldsymbol{H}) \\ &= (\boldsymbol{H}^T \boldsymbol{\Sigma}^{-1}\boldsymbol{H})^{-1} \end{aligned} \tag{7.38}$$

where the symmetry of the matrices $\boldsymbol{\Sigma}$ and $\boldsymbol{H}^T \boldsymbol{\Sigma}^{-1}\boldsymbol{H}$ has been used. It can be seen that this estimation process yields not only the estimate itself, but also the estimation error covariance matrix, which is quite useful in evaluating the quality of the estimate. In fact, it can be seen from Equations (7.17), (7.34), and (7.38) that

$$\hat{\boldsymbol{x}} = \boldsymbol{P}\boldsymbol{H}^T \boldsymbol{\Sigma}^{-1}\mathbf{Z} \tag{7.39}$$

which clearly indicates the necessity for determining $\boldsymbol{P}$ in the process of forming the estimate.

In the special event that the measurement error covariance matrix has the form

$$\boldsymbol{\Sigma} = \begin{bmatrix} \sigma_1^2 & & & \\ & \sigma_2^2 & & \\ & & \ddots & \\ & & & \sigma_M^2 \end{bmatrix} \tag{7.40}$$

then comparison of Equation (7.37) with Equation (7.13), the analogous expression for weighted least squares estimation, yields the equivalency

$$c_i = 1/\sigma_i^2. \tag{7.41}$$

The implication is that the least squares process should apply to each measurement a weighting factor which is inversely proportional to the variance of the error on that measurement, if such information is available. If the problem is further specialized so that

$$\boldsymbol{\Sigma} = \begin{bmatrix} \sigma^2 & & & \\ & \sigma^2 & & \\ & & \ddots & \\ & & & \sigma^2 \end{bmatrix} = \boldsymbol{I}\sigma^2 \tag{7.42}$$

then Equation (7.37) yields

$$\hat{\boldsymbol{x}} = \left(\frac{1}{\sigma^2}\boldsymbol{H}^T\boldsymbol{H}\right)^{-1}\boldsymbol{H}^T\left(\frac{1}{\sigma^2}\right)\mathbf{Z} = (\boldsymbol{H}^T\boldsymbol{H})^{-1}\boldsymbol{H}^T\mathbf{Z} \tag{7.43}$$

which is the solution to the least squares problem.

These relationships permit a clear comparison of the three estimation techniques. The least squares criterion, while intuitively satisfying in certain cases, is clearly not applicable if the quality of the measurements differs widely. The weighted least squares technique permits measurements of differing quality, but may fail if strong correlations exist among the different measurement errors.

7.1.1.4. Maximum Likelihood Estimation. The method of maximum likelihood has been mentioned briefly in Chapter 2. It is applicable to those situations in which the probability density function of the random variable constituting the measurement is known in terms of the parameter to be estimated. That is, if the measurement is the function

$$\mathbf{Z} = \boldsymbol{h}(\boldsymbol{x}) + \boldsymbol{\epsilon} \tag{7.44}$$

then $\mathbf{Z}$ is a random variable because of the randomness of $\boldsymbol{\epsilon}$, and its PDF will contain $\boldsymbol{x}$ as a parameter. That is,

$$f_{\mathbf{Z}}(\mathbf{z}) = f_{\mathbf{Z}}(\mathbf{z}, \boldsymbol{x}) \tag{7.45}$$

where the $\boldsymbol{x}$ dependency is not a functional dependency in the usual sense, but instead represents a parametric dependency in the sense that there is a family of PDFs $f_Z(\mathbf{z})$, characterized by the parameter $\boldsymbol{x}$. In the general sense, the method of maximum likelihood does not even require a measurement relationship as given in Equation (7.44). It is applicable to any situation in which there is a parametric dependence of $f_Z(\mathbf{z})$ on $\boldsymbol{x}$, as given in Equation (7.45), regardless of how this dependency arises. The only requirement is that the observation $\mathbf{Z}$ be a random variable with a probability distribution which depends to some extent on the value of the parameter $\boldsymbol{x}$. In light of this parametric dependency of $f_Z(\mathbf{z})$ on $\boldsymbol{x}$, one can envision a family of PDFs, $f_Z(\mathbf{z}, \boldsymbol{x})$, each member of the family being the result of a different value of the parameter $\boldsymbol{x}$. This parametric relationship is shown in Figure 7-1 for a scalar Z, where each individual curve represents the PDF, $f_Z(z)$, produced by the particular value of $\boldsymbol{x}$. The principle of maximum likelihod states that the estimate of the parameter $\boldsymbol{x}$ should be that which yields the maximum probability that the value of Z would be that which is actually observed. For example, in the problem illustrated in Figure 7-1, if the value of Z actually observed is z, then the probability of obtaining values near z, if the actual value of $\boldsymbol{x}$ is $\boldsymbol{x}_1$, is given by point (1) on the $f_Z(z)$ curve produced by parameter $\boldsymbol{x}_1$. Similarly, if x is actually $\boldsymbol{x}_2$, then the probability is given by point (2), and so on for the remaining curves in the family. The principle of maximum likelihood requires that this probability be maximized by selection of $\hat{\boldsymbol{x}}$, the estimate of $\boldsymbol{x}$. The previous discussion indicates that this selection can be accomplished by choosing $\hat{\boldsymbol{x}} = \boldsymbol{x}_2$ since this value yields the maximum probability that z, the actual value of the observation, would occur.

In this simple example, only discrete values of $\boldsymbol{x}$ were permitted, while in the more general case $\boldsymbol{x}$ may come from a continuum of values. This situation implies an infinite number of the curves in the parametric family and leads to the definition of the likelihood function

$$\ell(\mathbf{Z}, \boldsymbol{x}) = f_Z(\mathbf{Z}, \boldsymbol{x}) \tag{7.46}$$

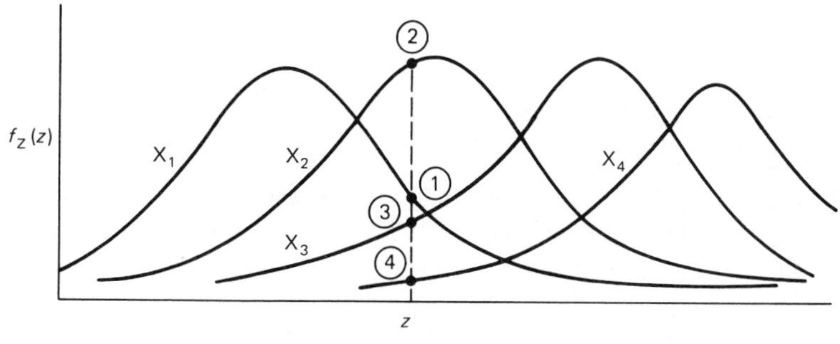

Figure 7-1. Family of PDFs $f_Z(z)$.

which is now considered to contain $\boldsymbol{x}$ and the actual observation $\mathbf{Z}$ as independent variables. The change in notation is intended to distinguish between the likelihood function, considered to be a function of $\boldsymbol{x}$ and $\mathbf{Z}$, and the PDF, which is a function of $\boldsymbol{z}$ and contains $\boldsymbol{x}$ only as a parameter. The principle of maximum likelihood requires that the estimate $\hat{\boldsymbol{x}}$ be chosen as that value of $\boldsymbol{x}$ which maximizes the function $\ell(\mathbf{Z}, \boldsymbol{x})$, and which can be obtained as the solution of the equation

$$\frac{\partial \ell(\mathbf{Z}, \boldsymbol{x})}{\partial \boldsymbol{x}} = \mathbf{0} \tag{7.47}$$

or, as may be the more widely used form,

$$\frac{\partial}{\partial \boldsymbol{x}} [\ln \ell(\mathbf{Z}, \boldsymbol{x})] = \mathbf{0}. \tag{7.48}$$

This latter expression arises because of the frequent occurrence of exponential terms, and because the logarithm is a nondecreasing function. The two expressions lead to identical results.

As an example of these principles, consider the problem of estimating the parameter $\boldsymbol{x}$ if the measurement is given by

$$\mathbf{Z} = \boldsymbol{x} + \boldsymbol{\epsilon} \tag{7.49}$$

where $\boldsymbol{\epsilon}$ has a multivariate normal distribution, given by

$$f_{\boldsymbol{\epsilon}}(\boldsymbol{\epsilon}) = \frac{1}{(2\pi)^{N/2}\sqrt{\det \boldsymbol{\Sigma}}} \exp\left[-\frac{1}{2}(\boldsymbol{\epsilon} - \boldsymbol{\mu})^T \boldsymbol{\Sigma}^{-1}(\boldsymbol{\epsilon} - \boldsymbol{\mu})\right]. \tag{7.50}$$

It is desired to find the maximum likelihood estimate of $\boldsymbol{x}$. Here, the family of PDFs is given by

$$f_{\mathbf{Z}}(\boldsymbol{z}) = \frac{1}{(2\pi)^{N/2}\sqrt{\det \boldsymbol{\Sigma}}} \exp\left[-\frac{1}{2}(\boldsymbol{z} - \boldsymbol{\mu} - \boldsymbol{x})\, \boldsymbol{\Sigma}^{-1}(\boldsymbol{z} - \boldsymbol{\mu} - \boldsymbol{x})\right] \tag{7.51}$$

which results directly from the fact that $\mathbf{Z}$ is a linear function of a normal random vector $\boldsymbol{\epsilon}$. The right-hand side of Equation (7.51), considered as a function of $\boldsymbol{x}$ and $\mathbf{Z}$, is the likelihood function $\ell(\mathbf{Z}, \boldsymbol{x})$, and

$$\ln \ell(\mathbf{Z}, \boldsymbol{x}) = \ln\left[\frac{1}{(2\pi)^{N/2}\sqrt{\det \boldsymbol{\Sigma}}}\right] - \frac{1}{2}(\mathbf{Z} - \boldsymbol{\mu} - \boldsymbol{x})^T \boldsymbol{\Sigma}^{-1}(\mathbf{Z} - \boldsymbol{\mu} - \boldsymbol{x}). \tag{7.52}$$

Then, Equation (7.48) yields

$$\frac{\partial}{\partial \boldsymbol{x}} [\ln \ell(\mathbf{Z}, \boldsymbol{x})] = (\mathbf{Z} - \boldsymbol{\mu} - \boldsymbol{x})^T \boldsymbol{\Sigma}^{-1} = (\mathbf{Z} - \boldsymbol{\mu})^T \boldsymbol{\Sigma}^{-1} - \boldsymbol{x}^T \boldsymbol{\Sigma}^{-1} = 0$$

or

$$\hat{\boldsymbol{x}} = \mathbf{Z} - \boldsymbol{\mu} \tag{7.53}$$

as the maximum likelihood estimate of $\boldsymbol{x}$. That is, if this value of $\boldsymbol{x}$ is substituted into Equation (7.51), the maximum value of $f_{\mathbf{Z}}(\mathbf{Z})$ will result. With respect to this particular example, note that the form of Equation (7.49) is linear; therefore, the results of the previous section can be used to determine the minimum variance unbiased linear estimate of $\boldsymbol{x}$. From Equation (7.37), with $\boldsymbol{H} = \boldsymbol{I}$,

$$\begin{aligned} \hat{\boldsymbol{x}} &= (\boldsymbol{I}^T \boldsymbol{\Sigma}^{-1} \boldsymbol{I})^{-1} \boldsymbol{I}^T \boldsymbol{\Sigma}^{-1} (\mathbf{Z} - \boldsymbol{\mu}) \\ &= \mathbf{Z} - \boldsymbol{\mu} \end{aligned} \tag{7.54}$$

which is identical to the previous result. The vector $\mathbf{Z} - \boldsymbol{\mu}$ on the right-hand side results from compensating for the fact that the measurement error in this case does not have a zero mean. The fact that these two methods–based on entirely different optimization criteria–yield the same result is a general characteristic when the noise processes are normal. This characteristic will be discussed further in the following section.

7.1.1.5. Equivalencies for Normal Distributions. If the measurements are linear in the sense of Equation (7.4), and the measurement noise has a multivariate normal distribution, then the observation vector $\mathbf{Z}$ also has a multivariate normal distribution with mean value vector

$$\boldsymbol{\mu}_{\mathbf{Z}} = \boldsymbol{H}\boldsymbol{x} \tag{7.55}$$

where $\boldsymbol{\mu}_{\boldsymbol{\epsilon}}$, the mean value vector of the distribution of $\boldsymbol{\epsilon}$, is assumed to be zero. The covariance matrix of the distribution of $\mathbf{Z}$ is $\boldsymbol{\Sigma}$, the covariance matrix of the distribution of $\boldsymbol{\epsilon}$. The likelihood function is then

$$\ell(\mathbf{Z}, \boldsymbol{x}) = f_{\mathbf{Z}}(\mathbf{Z}, \boldsymbol{x}) = \frac{1}{(2\pi)^{N/2}\sqrt{\det \boldsymbol{\Sigma}}} \exp\left[-\frac{1}{2}(\mathbf{Z} - \boldsymbol{H}\boldsymbol{x})^T \boldsymbol{\Sigma}^{-1} (\mathbf{Z} - \boldsymbol{H}\boldsymbol{x})\right]. \tag{7.56}$$

The logarithm of this expression is

$$\ln \ell(\mathbf{Z}, \boldsymbol{x}) = \ln\left[\frac{1}{(2\pi)^{N/2}\sqrt{\det \boldsymbol{\Sigma}}}\right] - \frac{1}{2}(\mathbf{Z} - \boldsymbol{H}\boldsymbol{x})^T \boldsymbol{\Sigma}^{-1} (\mathbf{Z} - \boldsymbol{H}\boldsymbol{x}). \tag{7.57}$$

This result, when substituted into Equation (7.48), yields

$$\frac{\partial}{\partial \boldsymbol{x}}[\ln \ell(\mathbf{Z}, \boldsymbol{x})] = -\frac{1}{2}\frac{\partial}{\partial \boldsymbol{x}}[\mathbf{Z}^T\boldsymbol{\Sigma}^{-1}\mathbf{Z} - 2\mathbf{Z}^T\boldsymbol{\Sigma}^{-1}\boldsymbol{H}\boldsymbol{x} + \boldsymbol{x}^T(\boldsymbol{H}^T\boldsymbol{\Sigma}^{-1}\boldsymbol{H})\boldsymbol{x}] \quad (7.58)$$

$$= \mathbf{0}$$

where the term within the brackets has been expanded by the indicated matrix multiplications. Taking the derivative with respect to $\boldsymbol{x}$ and equating to zero yield

$$-\mathbf{Z}^T\boldsymbol{\Sigma}^{-1}\boldsymbol{H} + \boldsymbol{x}^T(\boldsymbol{H}^T\boldsymbol{\Sigma}^{-1}\boldsymbol{H}) = \mathbf{0} \quad (7.59)$$

and, upon solution for $\boldsymbol{x}$, produce the maximum likelihood estimate

$$\hat{\boldsymbol{x}} = (\boldsymbol{H}^T\boldsymbol{\Sigma}^{-1}\boldsymbol{H})^{-1}\boldsymbol{H}^T\boldsymbol{\Sigma}^{-1}\mathbf{Z}. \quad (7.60)$$

This expression is identical to the minimum variance, unbiased, linear estimate given by Equation (7.37). This equivalency is a general result when the measurement is linear, and the contaminating noise is normally distributed.

7.1.1.6. The Perturbation Method Approach to Nonlinear Parameter Estimation. The reason for studying linear estimation techniques is similar to the reason for examining linear techniques in other areas. The nonlinear problem has no generally applicable method of solution, while the linear problem does; therefore, one frequently used approach to the nonlinear estimation problem involves a process of linearization, followed by application of the systematic linear theory. Specifically, if the measurement process is described by the expression

$$\mathbf{Z} = \boldsymbol{h}(\boldsymbol{x}) + \boldsymbol{\epsilon} \quad (7.61)$$

then the linear methods do not apply directly; however, if some prior estimate of $\boldsymbol{x}$, denoted by $\boldsymbol{x}^*$, is known, then the estimation problem can be linearized about this prior estimate in the following way. Define the perturbation $d\boldsymbol{x}$ as the difference between the actual value of $\boldsymbol{x}$ and the prior estimate $\boldsymbol{x}^*$,

$$d\boldsymbol{x} = \boldsymbol{x} - \boldsymbol{x}^* \quad (7.62)$$

and the perturbation $d\mathbf{Z}$ as

$$d\mathbf{Z} = \mathbf{Z} - \boldsymbol{h}(\boldsymbol{x}^*). \quad (7.63)$$

This expression gives the difference between the measurement actually observed and the noise free measurement which would be obtained if the prior estimate were actually the value of $\boldsymbol{x}$. Then,

$$d\mathbf{Z} = \boldsymbol{h}(x) - \boldsymbol{h}(x^*) + \boldsymbol{\epsilon} \tag{7.64}$$

and, if the first-order Taylor series is used to represent $\boldsymbol{h}(\boldsymbol{x})$, the result is

$$\boldsymbol{h}(x) = \boldsymbol{h}(x^*) + \left.\frac{\partial \boldsymbol{h}}{\partial x}\right|_{\boldsymbol{x}^*} d\boldsymbol{x}. \tag{7.65}$$

Then, substitution of this expression into Equation (7.64) yields, for the perturbation equation,

$$d\mathbf{Z} = \left.\frac{\partial \boldsymbol{h}}{\partial x}\right|_{\boldsymbol{x}^*} dx + \boldsymbol{\epsilon}. \tag{7.66}$$

The above expression constitutes a linear measurement process with additive noise $\boldsymbol{\epsilon}$, and the techniques of the previous sections can be used to determine an estimate of the perturbation $d\boldsymbol{x}$. The minimum variance, unbiased, linear estimate is

$$d\hat{\boldsymbol{x}} = \left[\left(\frac{\partial \boldsymbol{h}}{\partial x}\right)^T \boldsymbol{\Sigma}^{-1} \left(\frac{\partial \boldsymbol{h}}{\partial x}\right)\right]^{-1} \left(\frac{\partial \boldsymbol{h}}{\partial x}\right)^T \boldsymbol{\Sigma}^{-1}\, d\mathbf{Z}. \tag{7.67}$$

The nomenclature used here is based upon the matrix formalism defined in Appendix A; specifically,

$$\frac{\partial \boldsymbol{h}}{\partial x} = \begin{bmatrix} \dfrac{\partial h_1}{\partial x_1} & \dfrac{\partial h_1}{\partial x_2} & \cdots & \dfrac{\partial h_1}{\partial x_N} \\ \dfrac{\partial h_2}{\partial x_1} & \dfrac{\partial h_2}{\partial x_2} & \cdots & \dfrac{\partial h_2}{\partial x_N} \\ \cdots & \cdots & \cdots & \cdots \\ \dfrac{\partial h_M}{\partial x_1} & \dfrac{\partial h_M}{\partial x_2} & \cdots & \dfrac{\partial h_M}{\partial x_N} \end{bmatrix} \tag{7.68}$$

Such nomenclature is used extensively in the developments to follow. Once this estimate of the perturbation $d\boldsymbol{x}$ is obtained, the estimate for $\boldsymbol{x}$ is formed as

$$\hat{\boldsymbol{x}} = \boldsymbol{x}^* + d\hat{\boldsymbol{x}}. \tag{7.69}$$

This process is an approximation technique, and the primary source of error is the assumption that the higher-order terms in the Taylor series approximation of Equation (7.65) are small. This kind of error usually depends upon the quality of the prior estimate $\boldsymbol{x}^*$. Since it is reasonable to assume that the estimate $\hat{\boldsymbol{x}}$ will be closer to the actual value $\boldsymbol{x}$ than is the prior estimate $\boldsymbol{x}^*$, this value can be used as a revised prior estimate; that is,

$$\hat{\boldsymbol{x}} \longrightarrow \boldsymbol{x}^* \tag{7.70}$$

and the entire process repeated. This procedure results in a different value of the matrix

$$\left.\frac{\partial \boldsymbol{h}}{\partial \boldsymbol{x}}\right|_{\boldsymbol{x}^*}$$

since $\boldsymbol{x}^*$ now has a different value and will result in a revised perturbation estimate $d\hat{\boldsymbol{x}}$, which in turn will yield a revised estimate $\hat{\boldsymbol{x}}$. This process can be repeated as often as desired, and usually results in convergence of the sequence of $\hat{\boldsymbol{x}}$. Such a process is termed *linearization and successive approximation*.

As an example, consider the situation in which an aircraft flies a straight-line track from point A to point B as shown in Figure 7-2, during which time a number of angular measurements are made of the position of a radar station. These measurements are made at specific points along the line AB, and each point is identified by its distance from point A, labeled l_i in the figure. The problem is to estimate the position (x_1, x_2) of the radar station, using the complete set of measurements $\mathbf{Z}$, where

$$\mathbf{Z} = \begin{bmatrix} Z_1 \\ Z_2 \\ \vdots \\ Z_N \end{bmatrix} = \boldsymbol{h}(\boldsymbol{x}) + \boldsymbol{\epsilon}$$

and

$$\boldsymbol{x} = \begin{bmatrix} x_1 \\ x_2 \end{bmatrix}.$$

The nonlinear measurement equation can easily be determined from the geometry of the problem to be

$$Z_i = \arctan\left(\frac{x_2}{x_1 - l_i}\right) + \epsilon_i$$

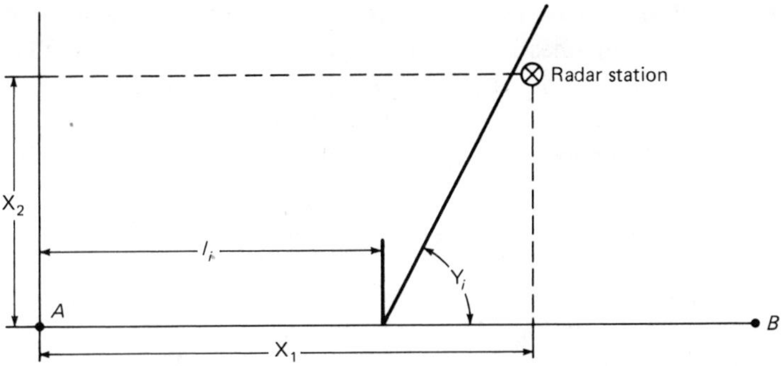

Figure 7-2. Geometry of estimation problem.

The measurement errors ϵ_i need not be uncorrelated, so that the minimum variance, unbiased, linear estimation process will be used; it is assumed that the measurement error covariance matrix is known. Since the measurement is non-linear, the perturbation technique must be used. In this case, the measurement matrix is

$$\frac{\partial \boldsymbol{h}}{\partial \boldsymbol{x}} = \begin{bmatrix} \frac{\partial h_1}{\partial x_1} & \frac{\partial h_1}{\partial x_2} \\ \frac{\partial h_2}{\partial x_1} & \frac{\partial h_2}{\partial x_2} \\ \vdots & \vdots \\ \frac{\partial h_M}{\partial x_1} & \frac{\partial h_M}{\partial x_2} \end{bmatrix} = \begin{bmatrix} \frac{-x_2^*}{(x_1^* - l_1) + x_2^{*2}} & \frac{x_1^* - l_1}{(x_1^* - l_1) + x_2^{*2}} \\ \frac{-x_2^*}{(x_1^* - l_2) + x_2^{*2}} & \frac{x_2^* - l_2}{(x_2^* - l_2) + x_2^{*2}} \\ \vdots & \vdots \\ \frac{-x_2^*}{(x_1^* - l_M)^2 + x_2^{*2}} & \frac{x_2^* - l_M}{(x_1^* - l_M)^2 + x_2^{*2}} \end{bmatrix}$$

where x_1^* and x_2^* are the coordinates describing the initial prior position estimate obtained from some appropriate procedure. The estimation process is then merely a straightforward application of Equation (7.67) to generate the estimated perturbation

$$d\hat{\boldsymbol{x}} = \begin{bmatrix} d\hat{x}_1 \\ d\hat{x}_2 \end{bmatrix}$$

followed by the computation of the estimated position

$$\hat{\boldsymbol{x}} = \begin{bmatrix} x_1^* + d\hat{x}_1 \\ x_2^* + d\hat{x}_2 \end{bmatrix}.$$

If the estimated perturbations $d\hat{x}$ are large enough so that some doubt as to the appropriateness of the linearization process exists, the successive approximation method can be applied, whereby a new prior estimate is obtained from $\hat{x} \rightarrow x^*$ and the linearization process is repeated. This successive approximation process is repeated until the desired convergence is obtained.

7.1.2. Estimation of Random Variables

When the quantities to be estimated are random variables, then the methods of the previous section do not apply, since the measurement equation can no longer be divided into random and nonrandom terms as in Equation (7.3). In the event that the form of the measurement is known to be linear in the sense of Equation (7.4), a minimum variance, unbiased, linear estimate can still be formulated, but the interpretation is slightly different in that the definition of an unbiased estimate of a random variable **X** requires that

$$E[\hat{\mathbf{X}}] = E[\mathbf{X}]. \tag{7.71}$$

Such a definition is required by the random character of **X** itself. From this point, the derivation of a minimum variance, unbiased, linear estimate follows that of Section 7.1.1.3, with a result identical to that represented by Equations (7.37) and (7.38).

If the precise relationship between the measurement and the random variable to be estimated is not known, the most general way in which the probabilistic relationship between these two random quantities can be described is by the joint probability distribution, most commonly defined by the joint probability density function

$$f_{\mathbf{XZ}}(\mathbf{x}, \mathbf{z})$$

where **X** and **Z** are the random vector to be estimated and the measurement vector, respectively. The problem, then, is to form an estimate of **X** which satisfies some specified error criterion. Since the quantity to be estimated is itself random, there can be no explicit deterministic error measure such as that used in parameter estimation. Instead, the error criterion must be expressed in probabilistic terms, the most common of which is the expected value of some error measure. This expectation is termed the risk and is given by

$$R = E[q(\mathbf{X} - \hat{\mathbf{X}})] \tag{7.72}$$

where q is a loss function in the same sense as in the developments of the previous sections. The error criterion then is the minimization of the risk function.

The expression for the risk is obtained by elementary application of the expectation operator:

$$R = \int_{-\infty}^{\infty} \int_{-\infty}^{\infty} q(\boldsymbol{x} - \hat{\boldsymbol{x}}) f_{\mathbf{XZ}}(\boldsymbol{x}, \boldsymbol{z})\, d\boldsymbol{x}\, d\boldsymbol{z}. \tag{7.73}$$

This expression results from the fact that $\hat{\mathbf{X}}$ is a function of $\mathbf{Z}$, since its value will be determined from $\mathbf{Z}$ in some way to be prescribed. When the error measure q is defined, this expression can be minimized to yield the estimate $\hat{\mathbf{X}}$.

7.1.2.1. Minimum Mean-Squared Error Estimation. If the error measure q is the sum of the squares of the estimation errors, then the risk is

$$R = E[(\mathbf{X} - \hat{\mathbf{X}})^T (\mathbf{X} - \hat{\mathbf{X}})] \tag{7.74}$$

and the resulting estimate yields the minimum expected sum of the squared errors. In this case, Equation (7.73) is

$$R = \int_{-\infty}^{\infty} \int_{-\infty}^{\infty} (\boldsymbol{x} - \hat{\boldsymbol{x}})^T (\boldsymbol{x} - \hat{\boldsymbol{x}}) f_{\mathbf{XZ}}(\boldsymbol{x}, \boldsymbol{z})\, d\boldsymbol{x}\, d\boldsymbol{z}. \tag{7.75}$$

The joint PDF $f_{\mathbf{XZ}}(\boldsymbol{x}, \boldsymbol{z})$ can be expressed as

$$f_{\mathbf{XZ}}(\boldsymbol{x}, \boldsymbol{z}) = f_{\mathbf{X}|\mathbf{Z}}(\boldsymbol{x}) f_{\mathbf{Z}}(\boldsymbol{z}) \tag{7.76}$$

which yields

$$R = \int_{-\infty}^{\infty} f_{\mathbf{Z}}(\boldsymbol{z}) \int_{-\infty}^{\infty} (\boldsymbol{x} - \hat{\boldsymbol{x}})^T (\boldsymbol{x} - \hat{\boldsymbol{x}}) f_{\mathbf{X}|\mathbf{Z}}(\boldsymbol{x})\, d\boldsymbol{x}\, d\boldsymbol{z}. \tag{7.77}$$

Since the quadratic form

$$(\boldsymbol{x} - \hat{\boldsymbol{x}})^T (\boldsymbol{x} - \hat{\boldsymbol{x}})$$

and the PDF $f_{\mathbf{Z}}(\mathbf{z})$ are nonnegative quantities, Equation (7.77) can be minimized by minimizing the inner integral. To accomplish this

$$\frac{\partial}{\partial \hat{\boldsymbol{x}}} \int_{-\infty}^{\infty} (\boldsymbol{x} - \hat{\boldsymbol{x}})^T (\boldsymbol{x} - \hat{\boldsymbol{x}}) f_{\mathbf{X}|\mathbf{Z}}(\boldsymbol{x})\, d\boldsymbol{x} = \mathbf{0}$$

or, after differentiating,

$$-\int_{-\infty}^{\infty} 2(\boldsymbol{x} - \hat{\boldsymbol{x}})^T f_{\mathbf{X}|\mathbf{Z}}(\boldsymbol{x})\, d\boldsymbol{x} = \mathbf{0}. \tag{7.78}$$

This integral can be expressed as the sum of two integrals

$$-2\int_{-\infty}^{\infty} \boldsymbol{x}^T f_{\mathbf{X}|\mathbf{Z}}(\boldsymbol{x})\, d\boldsymbol{x} + 2\hat{\boldsymbol{x}}^T \int_{-\infty}^{\infty} f_{\mathbf{X}|\mathbf{Z}}(\boldsymbol{x})\, d\boldsymbol{x} = \mathbf{0}. \tag{7.79}$$

The second integral has the value one, since it is the area under a PDF, in which case solution for the estimate yields

$$\hat{\mathbf{X}} = \int_{-\infty}^{\infty} \boldsymbol{x} f_{\mathbf{X}|\mathbf{Z}}(\boldsymbol{x})\, d\boldsymbol{x} = E_{\mathbf{X}|\mathbf{Z}}(\mathbf{X}) \tag{7.80}$$

which is the conditional mean of the random variable **X**, conditioned on the occurrence of the measurement **Z**. From an intuitive viewpoint this result makes sense, since it states that the estimate is the expected value of **X**, given the fact that **Z** has been observed to yield a particular value. Examples of this process will be deferred until the discussion of estimation relating to multivariate normal distributions in Section 7.1.2.3.

7.1.2.2. Maximum Posterior Probability Estimation. A second error measure which is sometimes used is the maximization of the conditional PDF

$$f_{\mathbf{X}|\mathbf{Z}}(\boldsymbol{x}) \tag{7.81}$$

which is the PDF of the quantity to be estimated conditioned on the measurement **Z**. This procedure results in an estimate which is the *mode* of the conditional distribution or, as it is commonly called, the *posterior distribution*. The minimum mean-squared error criterion, as is shown above, yields an estimate which is the *mean* of the conditional, or posterior, distribution. For those distributions in which the mode and the mean coincide–such as the normal, or Gaussian, distribution–these two criteria yield identical results. The estimate is obtained from maximizing Equation (7.81) with respect to **X**, accomplished by solution of the equation

$$\frac{\partial f_{\mathbf{X}|\mathbf{Z}}(\boldsymbol{x})}{\partial \boldsymbol{x}} = \mathbf{0}. \tag{7.82}$$

Note the similarity between this method and the maximum likelihood method of nonrandom parameter estimation developed in Section 7.1.1.

7.1.2.3. Estimation with Multivariate Normal Distributions. When the random quantity to be estimated and the measurement have a joint, multivariate normal distribution described by the mean vector

$$E\begin{bmatrix} \mathbf{X} \\ \hline \mathbf{Z} \end{bmatrix} = \begin{bmatrix} \mu_\mathrm{X} \\ \hline \mu_\mathrm{Z} \end{bmatrix} \tag{7.83}$$

and covariance matrix

$$K = \left[\begin{array}{c|c} K_{\mathrm{XX}} & K_{\mathrm{XZ}} \\ \hline K_{\mathrm{ZX}} & K_{\mathrm{ZZ}} \end{array}\right]$$

then the minimum mean squared-error estimate of $\mathbf{X}$ is readily available as the conditional mean

$$\hat{\mathbf{X}} = E_{\mathrm{X|Z}}[\mathbf{X}] = \mu_\mathrm{X} + K_{\mathrm{XZ}}K_{\mathrm{ZZ}}^{-1}[\mathbf{Z} - \mu_\mathrm{Z}]. \tag{7.84}$$

This expression also gives the maximum posterior probability estimate, since the distribution is normal. The expression of Equation (7.84) is a linear function of the measurement vector $\mathbf{Z}$, which implies the general conclusion that the minimum mean-squared error estimate, when the joint distribution of the vectors $\mathbf{X}$ and $\mathbf{Z}$ is normal, is a linear estimate. So also, of course, is the maximum posterior probability estimate. The conditional covariance matrix, which gives some indication of the quality of the estimate, is given by the expression

$$K_{\mathrm{X|Z}} = K_{\mathrm{XX}} - K_{\mathrm{XZ}}K_{\mathrm{ZZ}}^{-1}K_{\mathrm{XZ}}^{T}. \tag{7.85}$$

These expressions follow directly from general results of Chapter 2.

A particularly interesting case of minimum mean-squared error estimation occurs when the measurement vector $\mathbf{Z}$ is a linear function of the vector to be estimated, $\mathbf{X}$, contaminated by additive noise. This situation is represented by the measurement equation

$$\mathbf{Z} = H\mathbf{X} + \epsilon \tag{7.86}$$

where ϵ is a vector of measurement noise terms. Note that $\mathbf{X}$ here is a random quantity, as compared to x in Equation (7.4) which is an unknown parameter.

The matrices of Equation (7.84) in this case are

$$K_{XZ} = E[\mathbf{X}\mathbf{Z}^T] = E[\mathbf{X}(H\mathbf{X} + \boldsymbol{\epsilon})^T]$$
$$= E[\mathbf{X}\mathbf{X}^T]H + E[\mathbf{X}\boldsymbol{\epsilon}^T] \tag{7.87}$$

where zero mean values have been assumed for both $\mathbf{X}$ and $\boldsymbol{\epsilon}$. This restriction causes no loss of generality, since nonzero mean values merely produce additive terms such as $\boldsymbol{\mu}_X$ and $\boldsymbol{\mu}_Z$ in Equation (7.84). If a further assumption is made that the measurement noise $\boldsymbol{\epsilon}$ is uncorrelated with $\mathbf{X}$, then the second term of this expression is zero, with the result

$$K_{XZ} = E[\mathbf{X}\mathbf{X}^T]H^T = K_{XX}H^T. \tag{7.88}$$

The remaining matrix of Equation (8.84) is

$$K_{ZZ} = E[\mathbf{Z}\mathbf{Z}^T] = E[(H\mathbf{X} + \boldsymbol{\epsilon})(H\mathbf{X} + \boldsymbol{\epsilon})^T]$$
$$= HE[\mathbf{X}\mathbf{X}^T] + E[\boldsymbol{\epsilon}\mathbf{X}^T]H + HE[\mathbf{X}\boldsymbol{\epsilon}^T] + E[\boldsymbol{\epsilon}\boldsymbol{\epsilon}^T]. \tag{7.89}$$

Under the previous assumptions the two cross product terms are zero, yielding

$$K_{ZZ} = HK_{XX}H^T + \boldsymbol{\Sigma} \tag{7.90}$$

where $\boldsymbol{\Sigma}$ is the covariance matrix of the contaminating noise, in the same context as that given in Equation (7.16) for parameter estimation. Application of Equation (7.84) with the zero mean assumption to this situation yields

$$\hat{\mathbf{X}} = E_{X|Z}[\mathbf{X}] = K_{XZ}K_{ZZ}^{-1}\mathbf{Z} \tag{7.91}$$

which, upon substitution from Equations (7.88) and (7.90), finally gives

$$\hat{\mathbf{X}} = K_{XX}H^T(HK_{XX}H^T + \boldsymbol{\Sigma})^{-1}\mathbf{Z} \tag{7.92}$$

as the minimum mean-squared error estimate. The covariance matrix of this estimate is, from substitution into Equation (7.85),

$$K_{X|Z} = K_{XX} - K_{XZ}K_{ZZ}^{-1}K_{XZ}^T$$
$$= K_{XX} - K_{XX}H^T(HK_{XX}H^T + \boldsymbol{\Sigma}]^{-1}HK_{XX}. \tag{7.93}$$

Note the form of this equation; the covariance matrix of the estimate $\hat{\mathbf{X}}$ is equal to the covariance matrix of $\mathbf{X}$ minus a matrix term whose elements, in general, decrease as the quality of the measurement—as implied by small elements in $\boldsymbol{\Sigma}$—increases.

Further insight into the relationship between parameter estimation and estimation of random quantities can be obtained by using two matrix identities derived from expressions given in Appendix A, Section A.6:

$$K_{XX} - K_{XX}^T H^T (H K_{XX} H^T + \Sigma)^{-1} H K_{XX} = (K_{XX}^{-1} + H^T \Sigma H)^{-1} \quad (7.94)$$

and

$$K_{XX} H^T [H K_{XX} H^T + \Sigma]^{-1} = (K_{XX}^{-1} + H^T \Sigma^{-1} H)^{-1} H^T \Sigma^{-1}. \quad (7.95)$$

Applying the latter identity to Equation (7.92) and the former to Equation (7.93) yield an alternative set of expressions for the minimum mean-squared estimate and its covariance matrix:

$$\hat{\mathbf{X}} = (K_{XX}^{-1} + H^T \Sigma^{-1} H)^{-1} H^T \Sigma^{-1} \mathbf{Z} \quad (7.96)$$

$$K_{X|Z} = (K_{XX}^{-1} + H^T \Sigma^{-1} H)^{-1}. \quad (7.97)$$

The obvious substitution of Equation (7.97) into Equation (7.96) yields another alternative form of the estimation equation:

$$\hat{\mathbf{X}} = K_{X|Z} H^T \Sigma^{-1} \mathbf{Z}. \quad (7.98)$$

These expressions should be compared with Equations (7.37), (7.38), and (7.39), derived for minimum variance, unbiased, linear estimation of an unknown parameter. The difference between an unknown parameter and a random variable is that nothing at all is known about an unknown parameter, while some information in the form of a probability distribution is known about the value of a random variable. Just as a nonrandom constant can be considered to be a random variable with zero variance, an unknown parameter can be considered a random variable with infinite variance. Applied to the situation at hand, the random variable **X** approaches the unknown parameter $\boldsymbol{x}$ as the covariance matrix K_{XX} becomes very large. In the limit

$$K_{XX} \longrightarrow \infty; \qquad K_{XX}^{-1} \longrightarrow \mathbf{0}$$

and Equations (7.96) through (7.98) reduce to the expressions of Equations (7.37) through (7.39). Thus, the intuitive conclusion that minimum mean-squared estimation approaches minimum variance estimation as less and less is known about possible values of **X** is verified. One can further conclude that using minimum variance, unbiased, linear estimation, in the sense of Equation (7.71), essentially treats the random variable to be estimated as an unknown parameter. Such a procedure completely ignores the potentially useful information implicit in the prior probability distribution.

Before leaving this comparison of minimum mean-squared error and minimum variance estimation, one more point needs to be made. The covariance matrix $\boldsymbol{P}$ of Equation (7.38) is an estimation error covariance matrix, in the sense of Equation (7.20), while the matrix $\boldsymbol{K}_{X|Z}$ of Equations (7.93) and (7.97) is the conditional covariance matrix of $\mathbf{X}$, given the value of the measurement vector $\mathbf{Z}$. If the estimation error covariance matrix for the minimum mean-squared error estimate is defined in a manner analogous to that of Equation (7.20), then

$$\boldsymbol{P} = E[(\mathbf{X} - \hat{\mathbf{X}})(\mathbf{X} - \hat{\mathbf{X}})^T]. \tag{7.99}$$

The easiest approach to evaluation of this expression is to recognize that the estimate of Equation (7.91) is a linear estimate,

$$\hat{\mathbf{X}} = \boldsymbol{B}\mathbf{Z} = \boldsymbol{B}[\boldsymbol{H}\mathbf{X} + \boldsymbol{\epsilon}] \tag{7.100}$$

where

$$\boldsymbol{B} = \boldsymbol{K}_{XZ}\boldsymbol{K}_{ZZ}^{-1}.$$

Then, Equation (7.99) can be expressed as

$$\begin{aligned} \boldsymbol{P} &= E[(\mathbf{X} - \boldsymbol{BH}\mathbf{X} + \boldsymbol{B}\boldsymbol{\epsilon})(\mathbf{X} - \boldsymbol{BH}\mathbf{X} + \boldsymbol{B}\boldsymbol{\epsilon})^T] \\ &= (\boldsymbol{I} - \boldsymbol{BH})E[\mathbf{X}\mathbf{X}^T](\boldsymbol{I} - \boldsymbol{BH})^T + \boldsymbol{B}E[\boldsymbol{\epsilon}\boldsymbol{\epsilon}^T]\boldsymbol{B}^T \\ &= (\boldsymbol{I} - \boldsymbol{BH})\boldsymbol{K}_{XX}(\boldsymbol{I} - \boldsymbol{H}^T\boldsymbol{B}^T) + \boldsymbol{B}\boldsymbol{\Sigma}\boldsymbol{B}^T. \end{aligned} \tag{7.101}$$

Expanding the matrix multiplication terms yields

$$\boldsymbol{P} = \boldsymbol{K}_{XX} - \boldsymbol{BHK}_{XX} - \boldsymbol{K}_{XX}\boldsymbol{H}^T\boldsymbol{B}^T + \boldsymbol{BHK}_{XX}\boldsymbol{H}^T\boldsymbol{B}^T + \boldsymbol{B}\boldsymbol{\Sigma}\boldsymbol{B}^T. \tag{7.102}$$

The last two terms can be combined to yield

$$\boldsymbol{B}[\boldsymbol{HK}_{XX}\boldsymbol{H}^T + \boldsymbol{\Sigma}]\boldsymbol{B}^T = \boldsymbol{BK}_{ZZ}\boldsymbol{B}^T \tag{7.103}$$

where Equation (7.90) has been used. Then, further substitution for $\boldsymbol{B}$ yields the expression

$$\boldsymbol{K}_{XZ}\boldsymbol{K}_{ZZ}^{-1}\boldsymbol{K}_{ZZ}\boldsymbol{B}^T = \boldsymbol{K}_{XZ}\boldsymbol{B}^T. \tag{7.104}$$

In a similar manner, the second and third terms on the right-hand side of Equation (7.102) can be written as

$$\begin{aligned} \boldsymbol{BHK}_{XX} &= \boldsymbol{K}_{XZ}\boldsymbol{K}_{ZZ}^{-1}\boldsymbol{HK}_{XX} = \boldsymbol{K}_{XZ}\boldsymbol{K}_{ZZ}^{-1}\boldsymbol{K}_{XZ}^T = \boldsymbol{K}_{XZ}\boldsymbol{B}^T \\ \boldsymbol{K}_{XX}\boldsymbol{H}^T\boldsymbol{B}^T &= \boldsymbol{K}_{XZ}^T\boldsymbol{B}^T. \end{aligned} \tag{7.105}$$

Finally, substitution of Equations (7.103) through (7.105) into Equation (7.102) yields the result

$$P = K_{XX} - K_{XZ}B^T$$
$$= K_{XX} - K_{XX}H^T K_{ZZ}^{-1} H K_{XX}.$$

Further substitution from Equation (7.90) yields

$$P = K_{XX} - K_{XX}H^T(HK_{XX}H^T + \Sigma)^{-1}HK_{XX}. \qquad (7.106)$$

The fact that this expression is identical to that for $K_{X|Z}$ from Equation (7.93) shows that the covariance matrix of the conditional distribution of **X**, given the observed value of **Z**, is identical to the estimation error covariance matrix associated with the conditional mean, considered as the minimum mean-squared error estimate of **X**. This result is important to later developments.

As an example, consider the situation in which a random quantity X is measured M different times, such that the ith measurement is

$$Z_i = X + \epsilon_i, \qquad i = 1, 2, \ldots, M$$

where the ϵ_i are independent, normally distributed, zero mean random variables with variance σ_ϵ^2. That is, the random vector $\boldsymbol{\epsilon}$ has covariance matrix

$$\Sigma = \begin{bmatrix} \sigma_\epsilon^2 & & & \\ & \sigma_\epsilon^2 & & \\ & & \ddots & \\ & & & \sigma_\epsilon^2 \end{bmatrix} = I\sigma_\epsilon^2$$

Further, let X be a normal random variable, independent of $\boldsymbol{\epsilon}$, with zero mean and variance σ^2. Then, X and **Z** have a joint multivariate normal distribution with zero mean and covariance matrix

$$K = \left[\begin{array}{c|c} K_{XX} & K_{XZ} \\ \hline K_{ZX} & K_{ZZ} \end{array}\right]$$

where

$$K_{XX} = \sigma^2$$
$$K_{XZ} = E[X\mathbf{Z}^T] = E[(X^2 + X\epsilon_1) \quad (X^2 + X\epsilon_2) \quad \cdots \quad (X^2 + X\epsilon_M)]$$
$$= [\sigma^2 \quad \sigma^2 \quad \cdots \quad \sigma^2]$$
$$K_{ZX} = K_{XZ}^T$$

and

$$K_{ZZ} = E[\mathbf{ZZ}^T] = E\begin{bmatrix} (X+\epsilon_1)^2 & (X+\epsilon_1)(X+\epsilon_2) & \cdots \\ (X+\epsilon_1)(X+\epsilon_2) & (X+\epsilon_2)^2 & \cdots \\ \cdots & \cdots & \cdots \\ (X+\epsilon_1)(X+\epsilon_M) & (X+\epsilon_2)(X+\epsilon_M) & \cdots \end{bmatrix}$$

$$= \begin{bmatrix} \sigma_\epsilon^2 & & & \\ & \sigma_\epsilon^2 & & \\ & & \ddots & \\ & & & \sigma_\epsilon^2 \end{bmatrix} + \begin{bmatrix} \sigma \\ \sigma \\ \vdots \\ \sigma \end{bmatrix} [\sigma \quad \sigma \quad \cdots \quad \sigma].$$

The inverse of K_{ZZ} is required by Equation (7.84), and this can be obtained from the matrix identity

$$(A + B^T C)^{-1} = A^{-1} - A^{-1} B^T (I + CA^{-1} B^T)^{-1} CA^{-1}$$

with the association

$$A = I\sigma_\epsilon^2; \qquad B^T = C^T = \begin{bmatrix} \sigma \\ \sigma \\ \vdots \\ \sigma \end{bmatrix}.$$

Application of this equation yields

$$K_{ZZ}^{-1} = \frac{1}{\sigma_\epsilon^2} I - \frac{1}{\sigma_\epsilon^2(\sigma_\epsilon^2 + M\sigma^2)} \begin{bmatrix} \sigma \\ \sigma \\ \vdots \\ \sigma \end{bmatrix} [\sigma \quad \sigma \quad \cdots \quad \sigma].$$

Then the product

$$K_{XZ} K_{ZZ}^{-1} = \frac{1}{\sigma_\epsilon^2} [\sigma^2 \quad \sigma^2 \quad \cdots \quad \sigma^2] - \frac{M\sigma^2}{\sigma_\epsilon^2(\sigma_\epsilon^2 + M\sigma^2)} [\sigma^2 \quad \sigma^2 \quad \cdots \quad \sigma^2]$$

$$= \frac{1}{\sigma_\epsilon^2 + M\sigma^2} [\sigma^2 \quad \sigma^2 \quad \cdots \quad \sigma^2]$$

when substituted into Equation (7.84), with $\mu_X = \mu_{Z_i} = 0$, yields

$$\hat{X} = \frac{1}{\sigma_\epsilon^2 + M\sigma^2} [\sigma^2 \;\; \sigma^2 \;\; \cdots \;\; \sigma^2] \begin{bmatrix} Z_1 \\ Z_2 \\ \vdots \\ Z_M \end{bmatrix}$$

$$= \left[\frac{\sigma^2}{\frac{1}{M}\sigma_\epsilon^2 + \sigma^2} \right] \frac{1}{M} \sum_{i=1}^{M} Z_i.$$

This is the minimum mean-squared error estimate of X, and it can be seen to approach the value of the arithmetic average for large M. The variance of this estimate can be obtained from Equation (7.85) as

$$K_{X|Z} = \frac{\sigma^2 \sigma_\epsilon^2}{M\sigma^2 + \sigma_\epsilon^2}$$

which approaches zero for large M, as is anticipated. The solution to this problem can be simplified considerably by recognizing the measurement equation to be linear, in which case Equation (7.96) can be used directly:

$$\hat{X} = [K_{XX}^{-1} + \boldsymbol{H}^T \boldsymbol{\Sigma}^{-1} \boldsymbol{H}]^{-1} \boldsymbol{H}^T \boldsymbol{\Sigma}^{-1} \mathbf{Z}$$

$$= \left(\frac{1}{\sigma^2} + [1 \;\; 1 \;\; 1 \;\; \cdots \;\; 1] \left[\frac{1}{\sigma_\epsilon^2} \right] \begin{bmatrix} 1 \\ 1 \\ \vdots \\ 1 \end{bmatrix} \right)^{-1} [1 \;\; 1 \;\; 1 \;\; \cdots \;\; 1] \left[\frac{1}{\sigma_\epsilon^2} \right] \begin{bmatrix} Z_1 \\ Z_2 \\ \vdots \\ Z_M \end{bmatrix}.$$

The matrix inversion here is a scalar operation, yielding

$$\hat{X} = \left[\frac{\sigma^2}{\frac{1}{M}\sigma_\epsilon^2 + \sigma^2} \right] \frac{1}{M} \sum_{i=1}^{M} Z_i$$

which agrees with the previous result. The covariance matrix is given directly by Equation (7.97),

$$K_{X|Z} = \frac{\sigma^2 \sigma_\epsilon^2}{M\sigma_\epsilon^2 + \sigma^2}.$$

This example has been purposely complicated in order to illustrate the principles related to the general problem involving multivariate normal distributions. A more straightforward approach may be taken because of the simple form of the problem, whereby the conditional mean is computed in a more direct fashion using concepts discussed in Chapter 2.

If this same problem is approached from the point of view that a minimum variance unbiased linear estimate is desired, rather than the minimum mean-squared estimate just developed, the applicable equation is

$$\hat{\mathbf{X}} = (H^T \Sigma^{-1} H)^{-1} H^T \Sigma^{-1} \mathbf{Z}.$$

This expression results from the discussion preceding Equation (7.71), and the use of Equation (7.37). It is possible to use this method because the precise form of the measurement is known, and it is linear. That is, the measurement process can be represented in vector form as

$$\mathbf{Z} = H\mathrm{X} + \boldsymbol{\epsilon}$$

where

$$H = \begin{bmatrix} 1 \\ 1 \\ \vdots \\ 1 \end{bmatrix}$$

is an $M \times 1$ matrix, and X is a scalar. The estimate is then

$$\hat{\mathrm{X}} = \left([1 \;\; 1 \;\; \cdots \;\; 1] \frac{1}{\sigma_\epsilon^2} \begin{bmatrix} 1 \\ 1 \\ \vdots \\ 1 \end{bmatrix} \right)^{-1} [1 \;\; 1 \;\; \cdots \;\; 1] \frac{1}{\sigma_\epsilon^2} \begin{bmatrix} Z_1 \\ Z_2 \\ \vdots \\ Z_M \end{bmatrix} = \frac{1}{M} \sum_{i=1}^{N} \mathrm{Z}_i.$$

The difference between this estimate and that derived previously as the minimum mean-squared estimate results from the fact that the former does not consider the information available in the form of the probability distribution of **X**, while the latter makes explicit use of the fact that **X** is normally distributed with known mean and variance. Note that the minimum mean-squared error estimate approaches the minimum variance estimate as the variance of **X** increases, a result which agrees with the above interpretation.

7.1.2.4. Relationship to Linear Regression. The developments presented here in the form of estimation techniques for random and nonrandom quantities form the basis for a general method known as *regression analysis*. Usually an assumption of linear dependence is made in order to produce a set of tractable equations, in which case the process is called *linear regression*. In the event that the quantity to be estimated is not random, the results of Section 7.1.1.3 are directly applicable. For estimation of random variables, the linear assumption is

$$\hat{\mathbf{X}} = \boldsymbol{B}\mathbf{Z} \tag{7.107}$$

and the loss function is the sum of squares of estimation error, expressible as

$$q = (\mathbf{X} - \boldsymbol{B}\mathbf{Z})^T(\mathbf{X} - \boldsymbol{B}\mathbf{Z}) = \text{Tr}\ [(\mathbf{X} - \boldsymbol{B}\mathbf{Z})(\mathbf{X} - \boldsymbol{B}\mathbf{Z})^T]. \tag{7.108}$$

Then, the error criterion is to select $\boldsymbol{B}$ such that the risk $R = E[q]$ is minimized. This minimization can be obtained by writing

$$\begin{aligned} R &= E[\text{Tr}\ \{(\mathbf{X} - \boldsymbol{B}\mathbf{Z})(\mathbf{X} - \boldsymbol{B}\mathbf{Z})^T\}] \\ &= \text{Tr}\ \{E[\mathbf{X}\mathbf{X}^T - 2\boldsymbol{B}\mathbf{Z}\mathbf{X}^T + \boldsymbol{B}\mathbf{Z}\mathbf{Z}^T\boldsymbol{B}^T]\}. \end{aligned} \tag{7.109}$$

Again, for computational convenience the assumption of zero mean values will be made; for reasons previously discussed, this assumption causes no loss of generality. Performing the expectation operation yields

$$R = \text{Tr}\ (\boldsymbol{K}_{\mathbf{XX}} - 2\boldsymbol{B}\boldsymbol{K}_{\mathbf{ZX}} + \boldsymbol{B}\boldsymbol{K}_{\mathbf{ZZ}}\boldsymbol{B}^T). \tag{7.110}$$

This expression can now be differentiated with respect to $\boldsymbol{B}$, using the formalism developed in Appendix A. Equating the resulting derivative to zero produces the expression

$$-2\boldsymbol{K}_{\mathbf{ZX}} + 2\boldsymbol{K}_{\mathbf{ZZ}}\boldsymbol{B}^T = 0. \tag{7.111}$$

Then, solving for $\boldsymbol{B}$ yields

$$\boldsymbol{B} = \boldsymbol{K}_{\mathbf{XZ}}\boldsymbol{K}_{\mathbf{ZZ}}^{-1} \tag{7.112}$$

giving the matrix of regression coefficients which yield the minimum value for the risk of Equation (7.109) when used to form the linear estimate expressed by Equation (7.107). If the zero mean assumption is removed, the resulting equation is

$$\hat{\mathbf{X}} = \boldsymbol{\mu}_{\mathbf{X}} + \boldsymbol{K}_{\mathbf{XZ}}\boldsymbol{K}_{\mathbf{ZZ}}^{-1}(\mathbf{Z} - \boldsymbol{\mu}_{\mathbf{Z}}) \tag{7.113}$$

which is identical to Equation (7.84), developed for the multivariate normal distribution. This result is to be expected since the optimum estimate in that case was a linear function of the observation, and Equation (7.113) is the optimum linear estimate.

Linear regression analysis primarily consists of estimating the matrices K_{ZZ} and K_{XZ}, or more commonly, the matrix

$$K = \left[\begin{array}{c|c} K_{XX} & K_{XZ} \\ \hline K_{ZX} & K_{ZZ} \end{array}\right]$$

by use of a set of measured concurrent values of the quantities **X** and **Z**. From this matrix, the regression coefficients of Equation (7.112) are determined, and these coefficients in turn permit subsequent estimation of **X** based only upon measurements of **Z**.

7.1.3. Recursive Forms

A problem of considerable practical importance involves the situation in which the measurement data are obtained as a time sequence. The real-time processing problem then involves the formation of a revised estimate each time a new measurement is made. One approach to the solution of this problem is to consider each measurement as it is obtained as additional data–increasing the dimension of the measurement vector–and then to apply the techniques of the previous sections in a direct manner. This approach involves storage of all the data, which may be a limiting consideration in many practical situations, and computation involving matrices of dimensions which increase each time a new measurement is obtained. The dimension of the matrix inversion, however, remains fixed. As an alternative, a set of recursion relationships can be developed which permit computation of a revised estimate using only data from the previous estimate, together with the new measurement data.

7.1.3.1. Recursive, Minimum Variance, Unbiased, Linear Estimation. To develop the recursive equations for minimum variance, unbiased, linear estimation, assume that a minimum variance, unbiased, linear estimate has been obtained based upon measurement data up to and including the measurement $\mathbf{Z}_k$, given by

$$\mathbf{Z}_k = \boldsymbol{H}_k \boldsymbol{x} + \boldsymbol{\epsilon}_k \tag{7.114}$$

and further assume that the estimation error covariance matrix associated with this estimate has been determined:

$$\boldsymbol{P}_k = E[(\boldsymbol{x} - \hat{\boldsymbol{x}}_k)(\boldsymbol{x} - \hat{\boldsymbol{x}}_k)^T]. \tag{7.115}$$

Such an estimate could have been obtained by the methods of Section 7.1.1; specifically,

$$\hat{x}_k = (H_k^T \Sigma_k^{-1} H_k)^{-1} H_k^T \Sigma_k^{-1} Z_k$$
$$P_k = (H_k^T \Sigma_k^{-1} H_k)^{-1}. \qquad (7.116)$$

Now, assume that an additional measurement has been obtained of the form

$$Z_{k+1} = H_{k+1} x + \epsilon_{k+1} \qquad (7.117)$$

where the error terms ϵ_k and ϵ_{k+1} are uncorrelated, zero mean random vectors. That is

$$E[\epsilon_k \epsilon_{k+1}^T] = 0; \quad E[\epsilon_k] = E[\epsilon_{k+1}] = 0. \qquad (7.118)$$

It is now desired to form a revised estimate of x which will reflect the additional information available in the form of the new measurement data, and which can be accomplished by considering the two measurements as a single measurement vector of increased dimension, partitioned as follows:

$$\begin{bmatrix} Z_k \\ \hline Z_{k+1} \end{bmatrix} = \begin{bmatrix} H_k \\ \hline H_{k+1} \end{bmatrix} x + \begin{bmatrix} \epsilon_k \\ \hline \epsilon_{k+1} \end{bmatrix}. \qquad (7.119)$$

When the left-hand side of this equation is considered as a single measurement, the measurement error covariance matrix is

$$E\left[\begin{bmatrix} \epsilon_k \\ \hline \epsilon_{k+1} \end{bmatrix} [\epsilon_k^T \mid \epsilon_{k+1}^T]\right] = \left[\begin{array}{c|c} \Sigma_k & 0 \\ \hline 0 & \Sigma_{k+1} \end{array}\right] \qquad (7.120)$$

where zero submatrices result because of Equation (7.118). By direct application of Equation (7.37), the revised estimate is

$$\hat{x}_{k+1} = \left([H_k^T \mid H_{k+1}^T] \left[\begin{array}{c|c} \Sigma_k^{-1} & 0 \\ \hline 0 & \Sigma_{k+1}^{-1} \end{array}\right] \begin{bmatrix} H_k \\ \hline H_{k+1} \end{bmatrix} \right)^{-1} [H_k^T \mid H_{k+1}^T] \left[\begin{array}{c|c} \Sigma_k^{-1} & 0 \\ \hline 0 & \Sigma_{k+1}^{-1} \end{array}\right] \begin{bmatrix} Z_k \\ \hline Z_{k+1} \end{bmatrix}.$$

$$(7.121)$$

Performing the indicated multiplication yields

$$\hat{x}_{k+1} = (H_k^T \Sigma_k^{-1} H_k + H_{k+1}^T \Sigma_{k+1}^{-1} H_{k+1})^{-1} (H_k^T \Sigma_k^{-1} Z + H_{k+1}^T \Sigma_{k+1}^{-1} Z_{k+1})$$
$$= \hat{x}_k + (\hat{x}_{k+1} - \hat{x}_k) \qquad (7.122)$$

where the last expression is obtained by adding and subtracting the term $\hat{\mathbf{x}}_k$ from the left-hand side. Multiplication of this equation on the left by the matrix

$$\mathbf{H}_k^T \mathbf{\Sigma}^{-1} \mathbf{H}_k + \mathbf{H}_{k+1}^T \mathbf{\Sigma}_{k+1}^{-1} \mathbf{H}_{k+1}$$

yields

$$\begin{aligned} &\mathbf{H}_k^T \mathbf{\Sigma}_k^{-1} \mathbf{H}_k \hat{\mathbf{x}}_k + \mathbf{H}_{k+1}^T \mathbf{\Sigma}_{k+1}^{-1} \mathbf{H}_{k+1} \hat{\mathbf{x}}_k \\ &\quad + (\mathbf{H}_k^T \mathbf{\Sigma}_k^{-1} \mathbf{H}_k + \mathbf{H}_{k+1}^T \mathbf{\Sigma}_{k+1}^{-1} \mathbf{H}_{k+1})(\hat{\mathbf{x}}_{k+1} - \hat{\mathbf{x}}_k) \\ &\quad = \mathbf{H}_k^T \mathbf{\Sigma}^{-1} \mathbf{Z}_k + \mathbf{H}_{k+1}^T \mathbf{\Sigma}_{k+1}^{-1} \mathbf{Z}_{k+1}. \end{aligned} \tag{7.123}$$

The first term on the left-hand side and the first term on the right-hand side cancel, since multiplying Equation (7.116) on the left by the matrix $\mathbf{H}_k^T \mathbf{\Sigma}_k^{-1} \mathbf{H}_k$ yields

$$\mathbf{H}_k^T \mathbf{\Sigma}_k^{-1} \mathbf{H}_k \hat{\mathbf{x}}_k = \mathbf{H}_k^T \mathbf{\Sigma}_k^{-1} \mathbf{Z}_k. \tag{7.124}$$

Then, combining terms and multiplying on the left by the matrix

$$[\mathbf{H}_k^T \mathbf{\Sigma}_k^{-1} \mathbf{H}_k + \mathbf{H}_{k+1}^T \mathbf{\Sigma}_{k+1}^{-1} \mathbf{H}_{k+1}]^{-1}$$

yield

$$\hat{\mathbf{x}}_{k+1} = \hat{\mathbf{x}}_k + (\mathbf{H}_k^T \mathbf{\Sigma}_k^{-1} \mathbf{H}_k + \mathbf{H}_{k+1}^T \mathbf{\Sigma}_{k+1}^{-1} \mathbf{H}_{k+1})^{-1} \mathbf{H}_{k+1}^T \mathbf{\Sigma}_{k+1}^{-1} (\mathbf{Z}_{k+1} - \mathbf{H}_{k+1} \hat{\mathbf{x}}_k). \tag{7.125}$$

Substituting from Equation (7.116) yields finally

$$\hat{\mathbf{x}}_{k+1} = \hat{\mathbf{x}}_k + (\mathbf{P}_k^{-1} + \mathbf{H}_{k+1}^T \mathbf{\Sigma}_{k+1}^{-1} \mathbf{H}_{k+1})^{-1} \mathbf{H}_{k+1}^T \mathbf{\Sigma}_{k+1}^{-1} (\mathbf{Z}_{k+1} - \mathbf{H}_{k+1} \hat{\mathbf{x}}_k) \tag{7.126}$$

which is a recursive equation for $\hat{\mathbf{x}}_{k+1}$, yielding this revised estimate as a function of the previous estimate $\hat{\mathbf{x}}_k$ and the new measurement $\mathbf{Z}_{k+1}$. The estimation error covariance matrix is, from Equations (7.38) and (7.120),

$$\begin{aligned} \mathbf{P}_{k+1} &= \left([\mathbf{H}_k^T \,\vdots\, \mathbf{H}_{k+1}^T] \begin{bmatrix} \mathbf{\Sigma}_k^{-1} & \mathbf{0} \\ \mathbf{0} & \mathbf{\Sigma}_{k+1}^{-1} \end{bmatrix} \begin{bmatrix} \mathbf{H}_k \\ \mathbf{H}_{k+1} \end{bmatrix} \right)^{-1} \\ &= (\mathbf{P}_k^{-1} + \mathbf{H}_{k+1}^T \mathbf{\Sigma}_{k+1}^{-1} \mathbf{H}_{k+1})^{-1} \end{aligned} \tag{7.127}$$

which is a recursive equation for $\mathbf{P}_{k+1}$ involving the previous value $\mathbf{P}_k$ and the characteristics of the new measurement in terms of the matrices $\mathbf{H}_{k+1}$ and $\mathbf{\Sigma}_{k+1}$.

Combining Equations (7.126) and (7.127) yields a more compact form for the revised estimate

$$\hat{x}_{k+1} = \hat{x}_k + P_{k+1} H_{k+1}^T \Sigma_{k+1}^{-1} (Z_{k+1} - H_{k+1}\hat{x}_k). \tag{7.128}$$

The dimension of the matrix to be inverted in Equation (7.127) is the same as the dimension of the parameter to be estimated. This matrix inversion can be modified to be of the dimension of the new measurement Z_{k+1} by use of the matrix identity

$$(A + B^T C)^{-1} = A^{-1} - A^{-1} B^T (I + CA^{-1} B^T)^{-1} CA^{-1} \tag{7.129}$$

from Appendix A, and the associations

$$A \longrightarrow P_k^{-1}; \quad B \longrightarrow H_{k+1}; \quad C \longrightarrow \Sigma_{k+1}^{-1} H_{k+1}.$$

Application of this relationship yields, in lieu of Equation (7.127), the more common form of the recursive equation for P_k,

$$P_{k+1} = P_k - P_k H_{k+1}^T (\Sigma_{k+1} + H_{k+1} P_k H_{k+1}^T)^{-1} H_{k+1} P_k. \tag{7.130}$$

This equation and Equation (7.128) constitute the equations for recursive, minimum variance, unbiased, linear estimation. They apply equally to recursive least squares estimation, in the sense that the two methods are equivalent when

$$\Sigma = I\sigma^2. \tag{7.131}$$

In this event, the recursive equation for the covariance matrix P is, from Equation (7.130),

$$P_{k+1} = P_k - P_k H_{k+1}^T (I\sigma^2 + H_{k+1} P_k H_{k+1}^T)^{-1} H_{k+1} P_k \tag{7.132}$$

or, upon division by σ^2,

$$\frac{1}{\sigma^2} P_{k+1} = \frac{1}{\sigma^2} P_k + \frac{1}{\sigma_2} P_k H_{k+1}^T \left[I - H_{k+1} \left(\frac{1}{\sigma^2} P_k \right) H_{k+1}^T \right]^{-1} H_{k+1} \left(\frac{1}{\sigma^2} \right) P_k. \tag{7.133}$$

This result is a recursion relationship for the quantity

$$R_k = \frac{1}{\sigma^2} P_k = (H_k^T H_k)^{-1} \tag{7.134}$$

which represents a matrix of normalized coefficients. Thus, the recursive equations for least squares estimation are

$$R_{k+1} = R_k + R_k H_{k+1}^T (I - H_{k+1} R_k H_{k+1}^T)^{-1} H_{k+1} R_k \tag{7.135}$$

and

$$\hat{x}_{k+1} = \hat{x}_k + R_{k+1} H_{k+1}^T (Z_{k+1} - H_{k+1} \hat{x}_k).$$

If the actual error covariance matrix is desired, it is given by

$$P = \sigma^2 R. \tag{7.136}$$

In view of the equivalence of minimum variance and maximum likelihood estimates when the measurement noise is normally distributed, it should also be noted that the recursive minimum variance equations also represent a recursive form for maximum likelihood estimation, when the measurement noise is normally distributed.

7.1.3.2. Recursive, Minimum Mean-Squared Error Estimation. When the estimation criterion is that of minimum mean-squared error and the probability distribution involved is normal, or consideration is limited to a linear estimation and the measurement is linear, recursive forms of Equations (7.96) and (7.97) are desired. Following the basic procedure of the previous section yields, from Equation (7.96),

$$\hat{\mathbf{X}}_{k+1} = \left(K_{\mathbf{XX}}^{-1} + [H_k^T \mid H_{k+1}^T] \left[\begin{array}{c|c} \Sigma_k^{-1} & 0 \\ \hline 0 & \Sigma_{k+1}^{-1} \end{array} \right] \left[\begin{array}{c} H_k \\ \hline H_{k+1} \end{array} \right] \right)^{-1} \cdot [H_k^T \mid H_{k+1}^T] \left[\begin{array}{c|c} \Sigma_k^{-1} & 0 \\ \hline 0 & \Sigma_{k+1}^{-1} \end{array} \right] \left[\begin{array}{c} \mathbf{Z}_k \\ \hline \mathbf{Z}_{k+1} \end{array} \right]. \tag{7.137}$$

Expanding this expression yields

$$\begin{aligned} \hat{\mathbf{X}}_{k+1} &= (K_{\mathbf{XX}}^{-1} + H_k^T \Sigma_k^{-1} H_k + H_{k+1}^T \Sigma_{k+1}^{-1} H_{k+1})^{-1} (H_k^T \Sigma_k \mathbf{Z}_k + H_{k+1}^T \Sigma_{k+1}^{-1} \mathbf{Z}_{k+1}) \\ &= \hat{\mathbf{X}}_k + (\hat{\mathbf{X}}_{k+1} - \hat{\mathbf{X}}_k). \end{aligned} \tag{7.138}$$

In a manner analogous to that used to develop Equations (7.123) through (7.125), the following expression for $\hat{\mathbf{X}}_{k+1}$ can be obtained:

$$\begin{aligned} \hat{\mathbf{X}}_{k+1} = \hat{\mathbf{X}}_k + (K_{\mathbf{XX}}^{-1} &+ H_k^T \Sigma_k^{-1} H_k \\ &+ H_{k+1} \Sigma_{k+1}^{-1} H_{k+1})^{-1} H_{k+1}^T \Sigma_{k+1}^{-1} (\mathbf{Z}_{k+1} - H_{k+1} \hat{\mathbf{X}}_k). \end{aligned} \tag{7.139}$$

Since, from Equations (7.93), (7.97), and (7.106),

$$P_k = (K_{XX}^{-1} + H_k^T \Sigma_k^{-1} H_k)^{-1} \tag{7.140}$$

and

$$P_{k+1} = (K_{XX}^{-1} + H_k^T \Sigma_k H_k + H_{k+1}^T \Sigma_{k+1} H_{k+1})^{-1} \tag{7.141}$$

the final form of the recursive equations for minimum mean-squared error estimation are

$$\hat{X}_{k+1} = X_k + (P_k^{-1} + H_{k+1}^T \Sigma_{k+1}^{-1} H_{k+1})^{-1} H_{k+1}^T \Sigma_{k+1}^{-1} (Z_{k+1} - H_{k+1} \hat{X}_k) \tag{7.142}$$

and

$$P_{k+1} = (P_k^{-1} + H_{k+1}^T \Sigma_{k+1} H_{k+1})^{-1}.$$

These expressions are identical to those of Equations (7.126) and (7.127) for recursive, minimum variance, unbiased, linear estimation, leading to the following conclusion concerning recursive estimation. An algorithm based upon the expressions of Equation (7.142) can be used under the following circumstances involving linear measurements:

I. An unknown parameter is to be estimated by a minimum variance, unbiased, linear estimator.
II. An unknown parameter is to be estimated, the measurement is a linear function of the parameter contaminated by noise with a normal distribution, and maximum likelihood estimation is to be used.
III. A random variable is to be estimated, the probability distribution is normal, and the estimation criterion is minimum mean-squared error.
IV. A random variable is to be estimated using a linear estimate, and the estimation criterion is minimum mean-squared error.

The wide range of applicability of the recursive equations makes them generally useful and relieves the analyst of worrying too much about strict adherence to models described by normal probability distributions.

7.2. THE FILTERING PROBLEM

The discussion to this point has centered around point estimation, in which the quantity to be estimated does not change in any way as the measurements are taken. In the event that the quantity to be estimated changes with time in a known way, then the sequence of measurements still contains information from which an estimate of the time-varying quantity can be obtained. This situation describes the filtering problem, in which a time varying sequence of measurements is filtered to produce a time varying sequence of estimates. To illustrate such a situation, consider a sequence of radar measurements from which the track of an aircraft is to be obtained. In this case, the filter consists of the necessary equations which accept as input data the radar measurements and from these data generate the estimated system state in the form of position and velocity estimates. Again, the usual problem is nonlinear in nature with no general solution, while the linearized version of the problem can be solved using previously developed techniques.

As in any estimation technique, the filtering problem must be fully specified in terms of the criterion used. In view of the developments of the previous section, and particularly of their implications relating to linear estimation, much of this concern can be set aside. Because the expressions of Equation (7.142) are applicable to recursive estimation of either random variables or unknown parameters, under the various conditions outlined in their development, then any filtering techniques based upon these expressions will be equally applicable to a wide range of filtering problems. The approach to be taken here is to treat the general case of linear estimation of a random process. Estimation of a nonrandom time function follows as a special case. The particular estimation criterion used–minimum mean-squared error, minimum variance, etc.–is of no consequence because of the wide range of applicability of the recursive equations.

7.2.1. Linear Systems and the Kalman Filter

In the event that the time varying quantity to be estimated is determined by the time response of a linear system, it must satisfy the system state equations. In their most general form, the equations can be written as

$$\dot{\mathbf{X}}(t) = \boldsymbol{F}(t)\mathbf{X}(t) + \boldsymbol{D}(t)\mathbf{W}(t) + \boldsymbol{G}(t)\boldsymbol{u}(t) \tag{7.143}$$

or, in the discrete form, as

$$\mathbf{X}(t_l) = \boldsymbol{\Phi}(t_l, t_{l-1})\mathbf{X}(t_{l-1}) + \boldsymbol{\Delta}(t_l, t_{l-1})\mathbf{W}(t_{l-1}) + \boldsymbol{\Gamma}(t_l, t_{l-1})\boldsymbol{u}(t_{l-1}). \tag{7.144}$$

The distribution matrices $\boldsymbol{D}(t)$ and $\boldsymbol{\Delta}(t_l, t_{l-1})$ have been introduced here to distinguish between the effects of the random input functions $\mathbf{W}(t)$ and $\mathbf{W}(t_l)$, and those of the deterministic input functions $\boldsymbol{u}(t)$ and $\boldsymbol{u}(t_l)$. In the derivations to follow, the discrete form will be used because it more realistically represents the practical measurement situation. For ease of notation, the functional form of the previous equation will be reduced to

$$\mathbf{X}(l) = \boldsymbol{\Phi}(l, l-1)\mathbf{X}(l-1) + \boldsymbol{\Delta}(l, l-1)\mathbf{W}(l-1) \tag{7.145}$$

in conformance with the standards set in Chapter 4. Also, the deterministic input function $\boldsymbol{u}(l)$ has been assumed to be zero. The measurement equation is also considered to be linear and of the general form

$$\mathbf{Z}(l) = \boldsymbol{H}(l)\mathbf{X}(l) + \mathbf{V}(l) \tag{7.146}$$

where $\mathbf{V}(l)$ is a contaminating noise sequence. The problem, then, is to form a sequence of optimal estimates $\hat{\mathbf{X}}(l)$, given the sequence of measurements $\mathbf{Z}(l)$ and the state equation of Equation (7.145). The optimality criterion to be used is that of minimum variance, unbiased, linear estimation of a random time function. The results obtained are equally applicable to minimum mean-squared error estimation of a random function under the condition of normal probability distributions, to maximum likelihood estimation of a nonrandom time function when the measurement noise $\mathbf{V}(l)$ is normal, and, as a limiting case when the variance of the input noise $\mathbf{W}(l)$ approaches zero, to the minimum, variance, unbiased, linear estimation of an unknown time function.

Suppose that a minimum variance, unbiased, linear estimate of $\mathbf{X}(l-1)$ has been obtained, using measurements up to and including $\mathbf{Z}(l-1)$. This estimate and its associated estimation error covariance matrix will be represented by $\hat{\mathbf{X}}(l-1|l-1)$ and $\boldsymbol{P}(l-1|l-1)$ respectively. Then, without further measurement, an estimate of $\mathbf{X}(l)$ can be obtained from the expression

$$\hat{\mathbf{X}}(l|l-1) = \boldsymbol{\Phi}(l,l-1)\hat{\mathbf{X}}(l-1|l-1) + \boldsymbol{\Delta}(l,l-1)E[\mathbf{W}(l-1)] \tag{7.147}$$

where the notation on the left-hand side implies that $\mathbf{X}(l)$ is estimated, but measurements only through $\mathbf{Z}(l-1)$ have been used in forming this estimate. The estimation error covariance matrix for this estimate is denoted by $\boldsymbol{P}(l|l-1)$. That $\hat{\mathbf{X}}(l|l-1)$ is also an unbiased estimate can be seen by taking the expected value

$$E[\hat{\mathbf{X}}(l|l-1)] = \boldsymbol{\Phi}(l,l-1)E[\hat{\mathbf{X}}(l-1|l-1)] + \boldsymbol{\Delta}(l,l-1)E[\mathbf{W}(l-1)] \tag{7.148}$$

and comparing this result with $E[\mathbf{X}(l)]$ obtained from Equation (7.145),

$$E[\mathbf{X}(l)] = \boldsymbol{\Phi}(l,l-1)E[\mathbf{X}(l-1) + \boldsymbol{\Delta}(l,l-1)E[\mathbf{W}(l-1)]. \tag{7.149}$$

Because $\hat{\mathbf{X}}(l-1|l-1)$ is an unbiased estimate of $\mathbf{X}(l-1)$,

$$E[\hat{\mathbf{X}}(l-1|l-1)] = E[\mathbf{X}(l-1)] \tag{7.150}$$

and the equivalency of Equations (7.148) and (7.149) is established. The estimate $\hat{\mathbf{X}}(l|l-1)$ is therefore an unbiased estimate of $\mathbf{X}(l)$. Furthermore, it is intuitively apparent that $\hat{\mathbf{X}}(l|l-1)$ is the minimum variance, unbiased, linear estimate based on measurements up to and including $\mathbf{Z}(l-1)$, since no new information is obtained in extrapolating from the previous estimate. This conclusion is strengthened by examining the estimation error covariance matrix

$$\boldsymbol{P}(l|l-1) = E[\{\mathbf{X}(l) - \hat{\mathbf{X}}(l|l-1)\}\{\mathbf{X}(l) - \hat{\mathbf{X}}(l|l-1)\}^T] \\ - E[(\mathbf{X}(l) - \hat{\mathbf{X}}(l|l-1)]E[\mathbf{X}(l) - \hat{\mathbf{X}}(l|l-1)]^T. \quad (7.151)$$

Without loss of generality, this expression can be simplified by limiting consideration to that case in which

$$E[\mathbf{W}(l)] = 0 \quad \text{for all } l$$
$$E[\mathbf{X}(0)] = 0 \quad (7.152)$$

since, as discussed previously, the effect of nonzero means is treated as a deterministic factor and can thus be eliminated by compensation. These requirements further specify that

$$E[\mathbf{X}(l)] = 0 \quad (7.153)$$

since $\mathbf{X}(l)$ is a linear combination of $\mathbf{X}(0)$ and the $\mathbf{W}(l)$ terms. Then, Equation (7.147) reduces to

$$\hat{\mathbf{X}}(l|l-1) = \boldsymbol{\Phi}(l, l-1)\hat{\mathbf{X}}(l-1|l-1). \quad (7.154)$$

For convenience, define the estimation errors $\boldsymbol{\epsilon}(l|l-1)$ and $\boldsymbol{\epsilon}(l|l)$,

$$\boldsymbol{\epsilon}(l|l-1) = \mathbf{X}(l) - \hat{\mathbf{X}}(l|l-1)$$
$$\boldsymbol{\epsilon}(l|l) = \mathbf{X}(l) - \hat{\mathbf{X}}(l|l). \quad (7.155)$$

Then Equation (7.151) can be written as

$$\boldsymbol{P}(l|l-1) = E[\boldsymbol{\epsilon}(l|l-1)\boldsymbol{\epsilon}^T(l|l-1)] \quad (7.156)$$

since the second term of Equation (7.151) vanishes because the estimate $\hat{\mathbf{X}}(l|l-1)$ is unbiased. The estimation error $\boldsymbol{\epsilon}(l|l-1)$ is, from Equations (7.145) and (7.154),

$$\boldsymbol{\epsilon}(l|l-1) = \boldsymbol{\Phi}(l, l-1)\,\boldsymbol{\epsilon}(l-1|l-1) + \boldsymbol{\Delta}(l, l-1)\mathbf{W}(l-1). \quad (7.157)$$

Note that, whereas $\mathbf{W}(l)$ has a zero expected value, the values actually encountered are not zero. Then, from Equation (7.156)

$$\boldsymbol{P}(l|l-1)=E[\{\boldsymbol{\Phi}(l,l-1)\boldsymbol{\epsilon}(l-1|l-1) + \boldsymbol{\Delta}(l,l-1)\mathbf{W}(l-1)\}\{\boldsymbol{\Phi}(l,l-1)\boldsymbol{\epsilon}(l-1|l-1)+\boldsymbol{\Delta}(l,l-1)\mathbf{W}(l-1)\}^T] \tag{7.158}$$

This expression yields, upon expansion,

$$\begin{aligned}\boldsymbol{P}(l|l-1)=&\boldsymbol{\Phi}(l,l-1)E[\boldsymbol{\epsilon}(l-1|l-1)\boldsymbol{\epsilon}^T(l-1|l-1)]\boldsymbol{\Phi}^T(l,l-1)\\&+\boldsymbol{\Delta}(l,l-1)E[\mathbf{W}(l-1)\mathbf{W}^T(l-1)]\boldsymbol{\Delta}^T(l,l-1)\\&+\boldsymbol{\Phi}(l,l-1)E[\boldsymbol{\epsilon}(l-1|l-1)\mathbf{W}^T(l-1)]\boldsymbol{\Delta}^T(l,l-1)\\&+\boldsymbol{\Delta}(l,l-1)E[\mathbf{W}(l-1)\boldsymbol{\epsilon}^T(l-1|l-1)]\boldsymbol{\Phi}^T(l,l-1)\end{aligned} \tag{7.159}$$

If the input sequence $\mathbf{W}(l)$ is assumed to be a white noise sequence and is uncorrelated with $\hat{\mathbf{X}}(0|0)$ and $\mathbf{V}(l)$, then the expected values in the last two terms are zero, since $\mathbf{W}(l-1)$ is uncorrelated with all previous values, and therefore is uncorrelated with $\mathbf{X}(l-1)$. Furthermore, the expected value in the first term is

$$E[\boldsymbol{\epsilon}(l-1|l-1)\boldsymbol{\epsilon}^T(l-1|l-1)]=\boldsymbol{P}(l-1|l-1) \tag{7.160}$$

and that in the second term is

$$E[\mathbf{W}(l-1)\mathbf{W}^T(l-1)]=\boldsymbol{Q}(l-1) \tag{7.161}$$

where $\boldsymbol{Q}(l)$ is the covariance matrix function of the white noise process $\mathbf{W}(l)$. Finally, these results, when substituted into Equation (7.159), lead to an expression for $\boldsymbol{P}(l|l-1)$,

$$\boldsymbol{P}(l|l-1)=\boldsymbol{\Phi}(l,l-1)\boldsymbol{P}(l-1|l-1)\boldsymbol{\Phi}^T(l,l-1)+\boldsymbol{\Delta}(l,l-1)\boldsymbol{Q}(l-1)\boldsymbol{\Delta}^T(l,l-1) \tag{7.162}$$

which is the estimation error covariance matrix of the estimate $\hat{\mathbf{X}}(l|l-1)$ of $\mathbf{X}(l)$, based on measurements up to and including $\mathbf{Z}(l-1)$. That this expression must represent a minimum variance condition is evident from the fact that both covariance terms relate to time $l-1$, and the estimate at that time was of minimum variance.

To summarize, a minimum variance, unbiased, linear estimate at time $l-1$ was assumed available, together with its estimation error covariance matrix $\boldsymbol{P}(l-1|l-1)$, based on all measurements up to $\mathbf{Z}(l-1)$. Without further measurement, this estimate was extrapolated to time l by Equation (7.154), and, under the assumptions listed, the estimation error covariance matrix was determined as given in Equation (7.162). This estimate is also based on measurements through $\mathbf{Z}(l-1)$.

Now, a measurement is obtained at time l,

$$\mathbf{Z}(l) = \boldsymbol{H}(l)\mathbf{X}(l) + \mathbf{V}(l) \tag{7.163}$$

where $\mathbf{V}(l)$ is a zero mean white noise process with covariance matrix $\boldsymbol{R}(l)$, and it is desired to include this measurement in the estimate of $\mathbf{X}(l)$. The result will then be the estimate $\hat{\mathbf{X}}(l|l)$, since it will be based on measurements up to and including $\mathbf{Z}(l)$. The method for including this measurement has already been developed in Section 7.1.3, where recursive estimation was discussed. Specifically, Equations (7.128) and (7.130) apply, with the associations

$$\begin{aligned} &\hat{\mathbf{X}}(l|l-1) \longrightarrow \hat{\boldsymbol{x}}_k; \quad \boldsymbol{P}(l|l-1) \longrightarrow \boldsymbol{P}_k; \quad \boldsymbol{H}(l) \longrightarrow \boldsymbol{H}_{k+1};\\ &\hat{\mathbf{X}}(l|l) \longrightarrow \hat{\boldsymbol{x}}_{k+1}; \quad \boldsymbol{P}(l|l) \longrightarrow \boldsymbol{P}_{k+1}; \quad \boldsymbol{R}(l) \longrightarrow \boldsymbol{\Sigma}_k. \end{aligned} \tag{7.164}$$

When these substitutions are made, the resulting equations are

$$\begin{aligned} &\hat{\mathbf{X}}(l|l) = \hat{\mathbf{X}}(l|l-1) + \boldsymbol{P}(l|l)\boldsymbol{H}^T(l)\boldsymbol{R}^{-1}(l)[\mathbf{Z}(l) - \boldsymbol{H}(l)\hat{\mathbf{X}}(l|l-1)]\\ &\boldsymbol{P}(l|l) = \boldsymbol{P}(l|l-1) - \boldsymbol{P}(l|l-1)\boldsymbol{H}^T(l)[\boldsymbol{R}(l) + \boldsymbol{H}(l)\boldsymbol{P}(l|l-1)\boldsymbol{H}^T(l)]^{-1}\boldsymbol{H}(l)\boldsymbol{P}(l|l-1) \end{aligned} \tag{7.165}$$

and together with Equations (7.154) and (7.162), repeated here,

$$\begin{aligned} &\hat{\mathbf{X}}(l|l-1) = \boldsymbol{\Phi}(l, l-1)\hat{\mathbf{X}}(l-1|l-1)\\ &\boldsymbol{P}(l|l-1) = \boldsymbol{\Phi}(l, l-1)\boldsymbol{P}(l-1|l-1)\boldsymbol{\Phi}^T(l, l-1) + \boldsymbol{\Delta}(l, l-1)\boldsymbol{Q}(l-1)\boldsymbol{\Delta}^T(l, l-1) \end{aligned} \tag{7.166}$$

constitute a set of recursion relationships for $\hat{\mathbf{X}}(l|l)$ and $\boldsymbol{P}(l|l)$. Given an initial estimate $\hat{\mathbf{X}}(0|0)$ and the error covariance matrix $\boldsymbol{P}(0|0)$ associated with that initial estimate, a recursive computation of $\hat{\mathbf{X}}(l|l)$ and $\boldsymbol{P}(l|l)$ is provided each time a measurement $\mathbf{Z}(l)$ becomes available. Thus, these equations constitute a filter as described at the beginning of this chapter.

The recursive expressions of Equations (7.165) and (7.166) constitute an algorithm known as the Kalman Filter and represent a systematic approach to real-time, linear filtering with a well-established optimality criterion. The terms $\hat{\mathbf{X}}(l|l-1)$ and $\boldsymbol{P}(l|l-1)$ are known as the *a priori* or *prior* estimate and estimation error covariance matrix, while $\hat{\mathbf{X}}(l|l)$ and $\boldsymbol{P}(l|l)$ are known as the *posterior* quantities. For rather obvious reasons the term

$$\boldsymbol{K}(l) = \boldsymbol{P}(l|l)\boldsymbol{H}^T(l)\boldsymbol{R}^{-1}(l) \tag{7.167}$$

is known as the *Kalman gain matrix*.

7.2.1.1. Alternate Forms of the Kalman Filter Equations. Alternative forms of the basic Kalman filter equations may be derived by use of certain matrix identities derived from expressions in Appendix A, Section A.6. By use of the matrix identity

$$(\boldsymbol{P}^{-1} + \boldsymbol{H}^T\boldsymbol{R}^{-1}\boldsymbol{H})^{-1} = \boldsymbol{P} - \boldsymbol{P}\boldsymbol{H}^T(\boldsymbol{H}\boldsymbol{P}\boldsymbol{H}^T + \boldsymbol{R})^{-1}\boldsymbol{H}\boldsymbol{P} \tag{7.168}$$

the second expression of Equation (7.165) can be written as

$$\boldsymbol{P}(l|l) = [\boldsymbol{P}^{-1}(l|l-1) + \boldsymbol{H}^T(l)\boldsymbol{R}^{-1}(l)\boldsymbol{H}(l)]^{-1}. \tag{7.169}$$

A second identity,

$$\boldsymbol{P}\boldsymbol{H}^T(\boldsymbol{H}\boldsymbol{P}\boldsymbol{H}^T + \boldsymbol{R})^{-1} = (\boldsymbol{P}^{-1} + \boldsymbol{H}^T\boldsymbol{R}^{-1}\boldsymbol{H})^{-1}\boldsymbol{H}^T\boldsymbol{R}^{-1} \tag{7.170}$$

can be used to simplify the expression for the Kalman gain matrix, when written with Equation (7.169) substituted for $\boldsymbol{P}(l|l)$,

$$\boldsymbol{K}(l) = [\boldsymbol{P}^{-1}(l|l-1) + \boldsymbol{H}^T(l)\boldsymbol{R}^{-1}(l)\boldsymbol{H}(l)]^{-1}\boldsymbol{H}^T(l)\boldsymbol{R}^{-1}(l). \tag{7.171}$$

By use of the matrix identity of Equation (7.170), this expression can be written as

$$\boldsymbol{K}(l) = \boldsymbol{P}(l|l-1)\boldsymbol{H}^T(l)[\boldsymbol{H}(l)\boldsymbol{P}(l|l-1)\boldsymbol{H}^T(l) + \boldsymbol{R}(l)]^{-1}. \tag{7.172}$$

Then also, from Equation (7.165),

$$\begin{aligned}\boldsymbol{P}(l|l) &= \boldsymbol{P}(l|l-1) - \boldsymbol{K}(l)\boldsymbol{H}(l)\boldsymbol{P}(l|l-1)\\ &= [\boldsymbol{I} - \boldsymbol{K}(l)\boldsymbol{H}(l)]\boldsymbol{P}(l|l-1).\end{aligned} \tag{7.173}$$

There are several ways of combining these results to form alternative expressions to those of Equations (7.165) and (7.166), but the most logical one is

$$\begin{aligned}\hat{\mathbf{X}}(l|l) &= \hat{\mathbf{X}}(l|l-1) + \boldsymbol{K}(l)[\mathbf{Z}(l) - \boldsymbol{H}(l)\hat{\mathbf{X}}(l|l-1)]\\ \boldsymbol{P}(l|l) &= [\boldsymbol{I} - \boldsymbol{K}(l)\boldsymbol{H}(l)]\boldsymbol{P}(l|l-1)\\ \boldsymbol{K}(l) &= \boldsymbol{P}(l|l-1)\boldsymbol{H}^T(l)[\boldsymbol{H}(l)\boldsymbol{P}(l|l-1)\boldsymbol{H}^T(l) + \boldsymbol{R}(l)]^{-1}\end{aligned} \tag{7.174}$$

which, together with Equation (7.166), is a form found commonly in the literature.

An alternative view of the Kalman filter equations can be obtained by combining the prior and the posterior equations. Thus, by substituting the first expression of Equation (7.174) into that of Equation (7.166) and incrementing l, one obtains

$$\begin{aligned}\hat{\mathbf{X}}(l+1|l) &= \boldsymbol{\Phi}(l+1,l)\{\hat{\mathbf{X}}(l|l-1) + \boldsymbol{K}(l)[\mathbf{Z}(l) - \boldsymbol{H}(l)\hat{\mathbf{X}}(l|l-1)]\} \\ &= \boldsymbol{\Phi}(l+1,l)[\boldsymbol{I} - \boldsymbol{K}(l)\boldsymbol{H}(l)]\hat{\mathbf{X}}(l|l-1) + \boldsymbol{\Phi}(l+1,l)\boldsymbol{K}(l)\mathbf{Z}(l).\end{aligned} \tag{7.175}$$

This form of the estimation equation shows that the prior estimate is effectively produced by a linear discrete time system with transition matrix

$$\boldsymbol{\Phi}(l+1,l)[\boldsymbol{I} - \boldsymbol{K}(l)\boldsymbol{H}(l)]. \tag{7.176}$$

The response characteristics of the filter are determined by the eigenvalues of this matrix, and this information can be quite useful in design considerations. The recursion equations for the estimation error covariance matrix can also be combined by incrementing l by one in Equation (7.166) and substituting for $\boldsymbol{P}(l|l)$ from Equation (7.173) to yield

$$\begin{aligned}\boldsymbol{P}(l+1|l) = \boldsymbol{\Phi}(l+1,l)[\boldsymbol{I} - \boldsymbol{K}(l)\boldsymbol{H}(l)]\boldsymbol{P}(l|l-1)\boldsymbol{\Phi}^T(l+1,l) \\ + \boldsymbol{\Delta}(l+1,l)\boldsymbol{Q}(l)\boldsymbol{\Delta}^T(l+1,l).\end{aligned} \tag{7.177}$$

The Kalman gain matrix $\boldsymbol{K}(l)$ is obtained from Equation (7.172). This particular form of the Kalman filter equations has been termed the *time update form*, while that given in terms of $\hat{\mathbf{X}}(l|l)$ is called the *measurement update form*. The time update form consists of Equations (7.172), (7.175), and (7.177),

$$\begin{aligned}\hat{\mathbf{X}}(l+1|l) &= \boldsymbol{\Phi}(l+1,l)[\boldsymbol{I} - \boldsymbol{K}(l)\boldsymbol{H}(l)]\hat{\mathbf{X}}(l|l-1) + \boldsymbol{\Phi}(l+1,l)\boldsymbol{K}(l)\mathbf{Z}(l) \\ \boldsymbol{P}(l+1|l) &= \boldsymbol{\Phi}(l+1,l)[\boldsymbol{I} - \boldsymbol{K}(l)\boldsymbol{H}(l)]\boldsymbol{P}(l|l-1)\boldsymbol{\Phi}^T(l+1,l) \\ &\quad + \boldsymbol{\Delta}(l+1,l)\boldsymbol{Q}(l)\boldsymbol{\Delta}^T(l+1,l) \\ \boldsymbol{K}(l) &= \boldsymbol{P}(l|l-1)\boldsymbol{H}^T(l)[\boldsymbol{H}(l)\boldsymbol{P}(l|l-1)\boldsymbol{H}^T(l) + \boldsymbol{R}(l)]^{-1}.\end{aligned} \tag{7.178}$$

While this form presents the filter as a readily identifiable linear system, it has the disadvantage of an inherent time delay of one sample period. In some applications such a delay may not be detrimental.

7.2.1.2. The Structure of the Kalman Filter. The basic concept behind the Kalman filter results from the fact that the time function being estimated is the state vector of a linear system, and that the value of this state vector at any time can be related to the value at some other time by the state transition matrix. This viewpoint implies that the filter itself must contain some mechanism for determining this relationship. The estimate equations from Equations (7.165) and (7.166) are shown in block diagram form in Figure 7-3, together with the system for which the state is to be estimated. The inclusion of a system model in the filter equations is apparent. Another feature of the Kalman filter equations is also apparent from this figure. The recursive computation of the estimation error covariance matrix, as described by the second expression of Equations (7.165) and (7.166), is completely independent of the calculation of the estimate itself. This feature is important from two points of view. First, the $\boldsymbol{P}(l|l)$ calculation, and the resulting Kalman gain matrix $\boldsymbol{K}(l)$, can be precomputed and stored for later use. This possibility may be important in computation-sensitive application. Second and perhaps more important, an error analysis of the filtering process can be conducted by simply implementing the $\boldsymbol{P}(l|l)$ equations on a computer and solving them. In very simple cases, closed-form solutions may be obtainable.

7.2.1.3. Restrictions Imposed on the Kalman Filter Equations. In the derivation of the Kalman filter equations as a recursive, minimum variance, unbiased, linear estimator, three limiting assumptions were made with regard to eliminating the cross product terms in Equation (7.159). The first of these assumptions requires that the input function $\mathbf{W}(l)$ be a zero mean white noise sequence; that is,

$$E[\mathbf{W}(l)] = 0$$

$$E[\mathbf{W}(l)\mathbf{W}^T(m)] = \begin{cases} \boldsymbol{Q}(l), & \text{if } l = m \\ 0, & \text{if } l \neq m. \end{cases} \tag{7.179}$$

The second assumption requires that $\mathbf{W}(l)$ be uncorrelated with the random vector $\hat{\mathbf{X}}(0|0)$:

$$E[\hat{\mathbf{X}}(0|0)\mathbf{W}^T(l)] = 0 \quad \text{for all } l. \tag{7.180}$$

Finally, the third assumption requires that $\mathbf{W}(l)$ also be uncorrelated with the measurement noise function $\mathbf{V}(l)$:

$$E[\mathbf{W}(l)\mathbf{V}^T(m)] = 0 \quad \text{for all } l \text{ and } m. \tag{7.181}$$

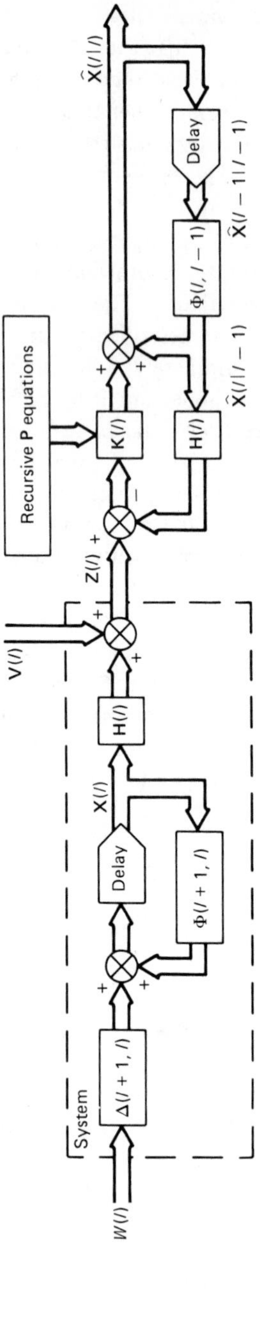

Figure 7-3. Discrete time Kalman filter.

A fourth assumption is required in the application of Equation (7.163). This assumption requires $\mathbf{V}(l)$ to also be a zero-mean, white noise sequence,

$$E[\mathbf{V}(l)] = 0$$
$$E[\mathbf{V}(l)\mathbf{V}^T(m)] = \begin{cases} \boldsymbol{R}(l), & \text{if } l = m \\ 0, & \text{if } l \neq m. \end{cases} \tag{7.182}$$

These assumptions constitute restrictions on the class of estimation problems to which the Kalman filter operations may be applied. They are not, however, as severe as they appear, and in the next section some of these restrictions will be relaxed to a point where the filter equations are widely applicable.

7.2.2. Relaxing Restrictions on the Kalman Filter Equations

While some filtering applications may fit into the general form of the filter equations derived in the previous sections, many do not because they violate assumptions basic to the derivation. Here, the filter equations are modified to permit a more realistic class of applications.

7.2.2.1. Nonzero Mean Value Function. The zero mean assumptions for the various functions are not really limiting, in the sense that a mean value function can be treated as a deterministic function, for which the linearity of the system and filter equations ensure that compensation can be made.

First, if the measurement noise process $\mathbf{V}(l)$ has known nonzero mean value sequence $\boldsymbol{\mu}_{\mathbf{V}}(l)$, then compensation can be made directly in the measurement process:

$$\mathbf{Z}(l) = \boldsymbol{H}(l)\mathbf{X}(l) + \mathbf{V}(l) - \boldsymbol{\mu}_{\mathbf{V}}(l). \tag{7.183}$$

This kind of compensation is effectively what occurs when sensors are adjusted to eliminate bias errors. If $\boldsymbol{\mu}_{\mathbf{V}}(l)$ represents an unknown bias, then it can be estimated along with the system state vector in a manner similar to that described for use in linear observers in Chapter 5. This situation can be treated as a special case of the concepts discussed in Section 7.2.2.3.

If the input function $\mathbf{W}(l)$ has a nonzero mean value function $\boldsymbol{\mu}_{\mathbf{W}}(l)$, it can be treated as a deterministic input function. In fact, the provision for a deterministic input function is desirable whether or not $\mathbf{W}(l)$ is a zero mean process. The system model which accounts for a deterministic input function is obtained by resorting to the general situation represented by Equation (7.144):

$$\mathbf{X}(l) = \boldsymbol{\Phi}(l, l-1)\mathbf{X}(l-1) + \boldsymbol{\Gamma}(l, l-1)\boldsymbol{u}(l-1) + \boldsymbol{\Delta}(l, l-1)\mathbf{W}(l-1) \tag{7.184}$$

where $\boldsymbol{u}(l)$ is the deterministic input function. Since all operations used in deriving the Kalman filter estimate equations are linear, the effect of including the $\boldsymbol{\Gamma}(l, l-1)\boldsymbol{u}(l-1)$ term is obtained immediately from Equation (7.166) as

$$\hat{\mathbf{X}}(l|l-1) = \boldsymbol{\Phi}(l, l-1)\hat{\mathbf{X}}(l-1|l-1) + \boldsymbol{\Gamma}(l, l-1)\boldsymbol{u}(l-1). \qquad (7.185)$$

Furthermore, since the term $\boldsymbol{\Gamma}(l, l-1)\boldsymbol{u}(l-1)$ is deterministic and additive, it does not affect the covariance equations. Thus, Equation (7.185) represents the only modification necessary to include consideration of the deterministic input function $\boldsymbol{u}(l)$. On the basis of the arguments above, if the noise function $\mathbf{W}(l)$ has mean value function $\boldsymbol{\mu}_{\mathbf{W}}(l)$, consideration of this term it can be included by the same process. Thus, in the most general case the estimate is given by

$$\hat{\mathbf{X}}(l|l-1) = \boldsymbol{\Phi}(l, l-1)\hat{\mathbf{X}}(l-1|l-1) + \boldsymbol{\Gamma}(l, l-1)\boldsymbol{u}(l-1) + \boldsymbol{\Delta}(l, l-1)\boldsymbol{\mu}_{\mathbf{W}}(l-1) \qquad (7.186)$$

where both a deterministic input function $\boldsymbol{u}(l)$ and mean value function $\boldsymbol{\mu}_{\mathbf{W}}(l)$ are considered.

7.2.2.2. Correlated Input and Measurement Noise. The requirement of Equations (7.180) and (7.181) were imposed to facilitate simplification of Equation (7.159). The restriction of Equation (7.180) essentially requires that the input noise function $\mathbf{W}(l)$ be uncorrelated with the initial state of the system, which is reasonable under most circumstances. The restriction imposed by Equation (7.181), however, may be violated in a number of practical applications. Fortunately, this restriction can be relaxed at the expense of additional complexity in the filter equations. In particular, consider the third term of Equation (7.159):

$$\boldsymbol{\Phi}(l, l-1)E[\boldsymbol{\epsilon}(l-1|l-1)\mathbf{W}^T(l-1)]\,\boldsymbol{\Delta}^T(l, l-1). \qquad (7.187)$$

By using Equations (7.146), (7.155), and (7.174), the expectation operation in this expression can be written as

$$\begin{aligned} &E[\{\mathbf{X}(l-1) - \hat{\mathbf{X}}(l-1|l-2) - \boldsymbol{K}(l-1)[\boldsymbol{H}(l-1)\mathbf{X}(l-1) \\ &\quad + \mathbf{V}(l-1) - \boldsymbol{H}(l-1)\hat{\mathbf{X}}(l-1|l-2)]\}\mathbf{W}^T(l-1)] \\ &\quad = E[\{[\boldsymbol{I} - \boldsymbol{K}(l-1)\boldsymbol{H}(l-1)]\,[\mathbf{X}(l-1) - \hat{\mathbf{X}}(l-1|l-2)] \\ &\qquad - \boldsymbol{K}(l-1)\mathbf{V}(l-1)\}\mathbf{W}^T(l-1)]. \end{aligned} \qquad (7.188)$$

When the expectation is taken, the only nonzero term is

$$-\boldsymbol{K}(l-1)E[\mathbf{V}(l-1)\mathbf{W}^T(l-1)] \qquad (7.189)$$

since the condition established by Equation (7.180) assures that $\mathbf{W}(l-1)$ is uncorrelated with $\mathbf{X}(l-1)$ and $\hat{\mathbf{X}}(l-1|l-1)$.

An equivalent result is obtained when the fourth term of Equation (7.159) is considered. The matrix $\boldsymbol{M}(l)$ is defined as

$$\boldsymbol{M}(l) = E[\mathbf{V}(l)\mathbf{W}^T(l)]$$
$$\boldsymbol{M}^T(l) = E[\mathbf{W}(l)\mathbf{V}^T(l)] \tag{7.190}$$

resulting in the revised form of Equation (7.162):

$$\begin{aligned} \boldsymbol{P}(l|l-1) = \boldsymbol{\Phi}(l, l-1)\boldsymbol{P}(l-1|l-1)\boldsymbol{\Phi}^T(l, l-1) \\ + \boldsymbol{\Delta}(l, l-1)\boldsymbol{Q}(l-1)\boldsymbol{\Delta}^T(l, l-1) \\ - \boldsymbol{\Phi}(l, l-1)\boldsymbol{K}(l-1)\boldsymbol{M}(l-1)\boldsymbol{\Delta}^T(l, l-1) \\ - \boldsymbol{\Delta}(l, l-1)\boldsymbol{M}^T(l-1)\boldsymbol{K}^T(l-1)\boldsymbol{\Phi}^T(l, l-1). \end{aligned} \tag{7.191}$$

This equation then replaces Equation (7.162) to provide Kalman filter equations which consider correlation between the input function $\mathbf{W}(l)$ and the measurement noise function $\mathbf{V}(l)$. Note that the requirement that both of these functions be white noise sequences has not been altered, in that Equation (7.190) considers correlation between $\mathbf{W}(l)$ and $\mathbf{V}(l)$ only at the same time instant.

7.2.2.3. Sequentially Correlated Input Noise. When the input function $\mathbf{W}(l)$ is not a white noise sequence, then the condition of Equation (7.179) is violated and the Kalman filter equations are no longer strictly valid. Typically, the input noise is a natural phenomenom of some kind and quite often will approximate a Markov process, as discussed in Chapter 3. A Markov process can be considered to be the output of a linear system with a white noise input, and can therefore be modeled in the same manner as the system for which the state vector is to be estimated. In those instances in which the input noise function $\mathbf{W}(l)$ is not a white noise process, but can be modeled as a Markov random sequence, the noise model is

$$\mathbf{W}(l) = \boldsymbol{\Phi}_{\mathbf{W}}\mathbf{W}(l-1) + \boldsymbol{\Delta}_{\mathbf{W}}\boldsymbol{\nu}(l-1) \tag{7.192}$$

where $\boldsymbol{\nu}(l)$ is a zero-mean white noise sequence with covariance matrix

$$E[\boldsymbol{\nu}(l)\boldsymbol{\nu}^T(m)] = \begin{cases} \boldsymbol{I}, & \text{if } l = m \\ 0, & \text{if } l \neq m \end{cases} \tag{7.193}$$

and $\boldsymbol{\Delta}_{\mathbf{W}}$ is a diagonal matrix whose elements determine the variances of the white noise terms. The elements of the constant matrix $\boldsymbol{\Phi}_{\mathbf{W}}$ are selected to provide the required correlation. Once this model has been determined to be a mechanism by which $\mathbf{W}(l)$ is produced, the noise model may be incorporated into the system model by using the vector sequence $\mathbf{W}(l)$ to augment the system state vector, giving the augmented system model

$$\begin{bmatrix} \mathbf{X}(l) \\ \hline \mathbf{W}(l) \end{bmatrix} = \left[\begin{array}{c|c} \mathbf{\Phi}(l, l-1) & \Delta(l, l-1) \\ \hline \mathbf{0} & \mathbf{\Phi}_\mathbf{W} \end{array}\right] \begin{bmatrix} \mathbf{X}(l-1) \\ \hline \mathbf{W}(l-1) \end{bmatrix} + \begin{bmatrix} \mathbf{0} \\ \hline \Delta_\mathbf{W} \end{bmatrix} \boldsymbol{\nu}(l-1)$$

$$\mathbf{Z}(l) = [\boldsymbol{H}(l) \mid \mathbf{0}] \begin{bmatrix} \mathbf{X}(l) \\ \hline \mathbf{W}(l) \end{bmatrix} + \mathbf{V}(l) \tag{7.194}$$

where $\mathbf{V}(l)$ is a white noise sequence. Several aspects of this system should be noted: the $\mathbf{X}(l)$ equations are identical to the model of Equation (7.145); the $\mathbf{W}(l)$ equations are identical to Equation (7.192); and, finally, the order of the system has been increased by P, the dimension of the input function.

The system model of Equation (7.194) is a discrete time, linear system with a white noise input, and its state can therefore be estimated by the Kalman filter equations as developed. Note that the Markov input noise $\mathbf{W}(l)$ is estimated along with the state vector $\mathbf{X}(l)$, and that correlation between components of $\mathbf{W}(l)$ are modeled by the off-diagonal terms of $\mathbf{\Phi}_\mathbf{W}$. It is also possible for only a portion of the input noise components to be modeled as Markov sequences, while the remainder are white noise sequences. The modifications to the model of Equation (7.194) to accommodate these changes are not complex.

A further generalization can be obtained by selectively populating the zero submatrix of the first expression of Equation (7.194) with ones so that certain components of the input noise are modeled as Markov noise with an additive white noise component. Individual components can also be modeled as general Nth order Markov processes by further augmenting the state vector. In general, N additional states are required to produce an Nth-order scalar Markov process. In most instances, first-order Markov properties suitably model the input noise process.

7.2.2.4. Sequentially Correlated Measurement Noise. In principle, sequentially correlated measurement noise can be considered by the same approach used for sequentially correlated input noise in the previous section. The state vector can be augmented to form the system

$$\begin{bmatrix} \mathbf{X}(l) \\ \hline \mathbf{V}(l) \end{bmatrix} = \left[\begin{array}{c|c} \mathbf{\Phi}(l, l-1) & \mathbf{0} \\ \hline \mathbf{0} & \mathbf{\Phi}_\mathbf{V} \end{array}\right] \begin{bmatrix} \mathbf{X}(l-1) \\ \hline \mathbf{V}(l-1) \end{bmatrix} + \left[\begin{array}{c|c} \Delta(l, l-1) & \mathbf{0} \\ \hline \mathbf{0} & \Delta_\mathbf{V} \end{array}\right] \begin{bmatrix} \mathbf{W}(l-1) \\ \hline \boldsymbol{\nu}(l-1) \end{bmatrix}$$

$$\mathbf{Z}(l) = [\boldsymbol{H}(l) \mid \boldsymbol{I}] \begin{bmatrix} \mathbf{X}(l) \\ \hline \mathbf{V}(l) \end{bmatrix} \tag{7.195}$$

where $\boldsymbol{\nu}(l)$ is a zero mean white noise sequence with covariance matrix as given by Equation (7.193), and the constant matrix $\boldsymbol{\Phi}_\mathbf{V}$ is selected to produce the required correlation. Again, the $\mathbf{X}(l)$ equations are unaltered, and the $\mathbf{V}(l)$ equations satisfy those of a Markov random sequence. Now, however, the measurement is noise free, implying that $\boldsymbol{R}(l)$ is zero in Equation (7.165). Such a zero value presents a computational difficulty, even if the particular equations do not require use of $\boldsymbol{R}^{-1}$, since the filter tends to a condition where the $\boldsymbol{P}(l|l)$ matrix is singular. This result is to be expected since, if the dimension of the measurement process is P, then P linear combinations of the $(N+P)$ states can be determined precisely from the noise-free measurements. This fact in turn implies that these states need not be estimated, and the problem should revert to one of N dimensions. In practice, not all measurement noise terms are modeled as Markov processes, so that the $\boldsymbol{R}(l)$ matrix is not zero. In fact, it is not uncommon to model a noise process as the sum of a Markov process and a white noise process. A simple extension of this model can provide for this characteristic.

The conceptual description of the anticipated computational problem arising from modeling measurement noise as a Markov process can be used to advantage by the following technique. Consider the measurement noise to be modeled as in Equation (7.195) as

$$\mathbf{V}(l+1) = \boldsymbol{\Phi}_\mathbf{V}\mathbf{V}(l) + \boldsymbol{\Delta}_\mathbf{V}\boldsymbol{\nu}(l). \tag{7.196}$$

Then define the following linear combination of successive measurement vectors:

$$\boldsymbol{\eta}(l) = \mathbf{Z}(l+1) - \boldsymbol{\Phi}_\mathbf{V}\mathbf{Z}(l). \tag{7.197}$$

Substituting from Equations (7.145) and (7.146) yields

$$\begin{aligned}\boldsymbol{\eta}(l) &= \boldsymbol{H}(l+1)\mathbf{X}(l+1) + \boldsymbol{\Phi}_\mathbf{V}\mathbf{V}(l) + \boldsymbol{\Delta}_\mathbf{V}\boldsymbol{\nu}(l) - \boldsymbol{\Phi}_\mathbf{V}[\boldsymbol{H}(l)\mathbf{X}(l) + \mathbf{V}(l)]\\ &= \boldsymbol{H}(l+1)[\boldsymbol{\Phi}(l+1,l)\mathbf{X}(l) + \boldsymbol{\Delta}(l+1,l)\mathbf{W}(l)] + \boldsymbol{\Phi}_\mathbf{V}\mathbf{V}(l)\\ &\quad + \boldsymbol{\Delta}_\mathbf{V}\boldsymbol{\nu}(l) - \boldsymbol{\Phi}_\mathbf{V}[\boldsymbol{H}(l)\mathbf{X}(l) + \mathbf{V}(l)].\end{aligned} \tag{7.198}$$

The $\mathbf{V}(l)$ terms cancel, leaving

$$\boldsymbol{\eta}(l) = [\boldsymbol{H}(l+1)\boldsymbol{\Phi}(l+1,l) - \boldsymbol{\Phi}_\mathbf{V}\boldsymbol{H}(l)]\mathbf{X}(l) + \boldsymbol{H}(l+1)\boldsymbol{\Delta}(l+1,l)\mathbf{W}(l) + \boldsymbol{\Delta}_\mathbf{V}\boldsymbol{\nu}(l). \tag{7.199}$$

This equation is of the form

$$\boldsymbol{\eta}(l) = \tilde{\boldsymbol{H}}(l)\mathbf{X}(l) + \tilde{\boldsymbol{\epsilon}}(l)$$

where

$$\tilde{H}(l) = H(l+1)\Phi(l+1, l) - \Phi_V H(l)$$
$$\tilde{\epsilon}(l) = H(l+1)\Delta(l+1, l)\mathbf{W}(l) + \Delta_V \nu(l). \tag{7.200}$$

The vector sequence $\tilde{\epsilon}(l)$ is a white noise sequence, and now Equation (7.200) can be considered the system measurement process with additive white noise. Then the Kalman filter equations as previously derived can be used.

It should be noted that correlation exists between the newly defined measurement noise process $\tilde{\epsilon}(l)$ and the input noise process $\mathbf{W}(l)$ even if none existed between $\mathbf{V}(l)$ and $\mathbf{W}(l)$. Specifically,

$$E[\tilde{\epsilon}(l)\mathbf{W}^T(l)] = E[\{H(l+1)\Delta(l+1, l)\mathbf{W}(l) + \Delta_V \nu(l)\}\mathbf{W}^T(l)]$$
$$= H(l+1)\Delta(l+1, l)Q(l) \tag{7.201}$$

where the fact that $\mathbf{W}(l)$ and $\nu(l)$ are uncorrelated has been used to eliminate the cross product term. This situation can be accommodated by the methods discussed in Section 7.2.2.2. In addition, the pseudo-measurement $\eta(l)$ lags one time interval behind the actual measurement $\mathbf{Z}(l)$. This lag may or may not present a problem, depending upon the particular application.

As an example of the application of the Kalman filter to the problem of estimating the state of a linear stochastic system, consider the water brake problem of Sections 4.2 and 4.4.1. The physical system under consideration is a rocket sled which encounters a water brake at time $t = 0$. The rocket thrust $\mathrm{U}(t)$ is a white noise process with

$$E[\mathrm{U}(t)] = 62{,}100 \text{ pounds}$$
$$K_\mathrm{U}(t, \sigma) = 3.86 \times 10^7 \,\delta(t - \sigma).$$

Other system parameters are

Mass of sled = 2000 pounds

Coefficient of drag force = 124.2 pounds sec/ft.

From the results of Section 4.2, the continuous time system equations have the solution

$$\begin{bmatrix} \mathrm{X}(t) \\ \mathrm{V}(t) \end{bmatrix} = \begin{bmatrix} 1 & \frac{1}{2} - \frac{1}{2}e^{-2(t-t_0)} \\ 0 & e^{-2(t-t_0)} \end{bmatrix} \begin{bmatrix} \mathrm{X}(t_0) \\ \mathrm{V}(t_0) \end{bmatrix} + \int_{t_0}^{t} \begin{bmatrix} 1 & \frac{1}{2} - \frac{1}{2}e^{-2(t-\sigma)} \\ 0 & e^{-2(t-\sigma)} \end{bmatrix} \begin{bmatrix} 0 \\ 1 \end{bmatrix} \mathrm{U}(\sigma)\, d\sigma.$$

which is a direct application of Equation (4.20). Now suppose that position measurements are obtained at one-tenth second intervals, and that these measurements are contaminated by a white noise sequence $N(t_l)$ with

$$E[N(t_l)] = 0$$

$$K_N(t_l, t_m) = \begin{cases} 100 \text{ ft}, & \text{if } t_l = t_m \\ 0, & \text{otherwise.} \end{cases}$$

It is desired to design a Kalman filter for this system so that position and velocity estimates can be obtained from the noisy position measurements.

The first consideration here is that a discrete time measurement process is to be applied to a continuous time physical process. The system differential equations must be converted to difference equations so that the discrete time Kalman filter equations can be applied. The techniques of Section 4.2.2 apply here, and Equation (4.24) yields

$$\mathbf{\Gamma}(l, l-1) \int_{t_{l-1}}^{t_l} \begin{bmatrix} 1 & \frac{1}{2} - \frac{1}{2}e^{-2(t_l-\sigma)} \\ 0 & e^{-2(t_l-\sigma)} \end{bmatrix} \begin{bmatrix} 0 \\ 1 \end{bmatrix} d\sigma.$$

In this example, the time interval $(t_l - t_{l-1})$ is one-tenth second. The designated integration is

$$\int_0^{0.1} \begin{bmatrix} 1 & \frac{1}{2} - \frac{1}{2}e^{-2(0.1-\sigma)} \\ 0 & e^{-2(0.1-\sigma)} \end{bmatrix} \begin{bmatrix} 0 \\ 1 \end{bmatrix} d\sigma = \begin{bmatrix} 0.00468 \\ 0.09063 \end{bmatrix}.$$

The transition matrix evaluated for $(t_l - t_{l-1})$ equal one-tenth second is

$$\mathbf{\Phi}(l, l-1) = \begin{bmatrix} 1 & 0.09063 \\ 0 & 0.81873 \end{bmatrix}.$$

Thus, the difference equations based on a time interval of one-tenth second are expressed as

$$\begin{bmatrix} X(l) \\ V(l) \end{bmatrix} = \begin{bmatrix} 1 & 0.09063 \\ 0 & 0.81873 \end{bmatrix} \begin{bmatrix} X(l-1) \\ V(l-1) \end{bmatrix} + \begin{bmatrix} 0.00468 \\ 0.09063 \end{bmatrix} U(l-1).$$

This expression is analogous to Equation (4.23).

With the system now in discrete time form, the Kalman filter equations of Equation (7.174) can be applied. There is one difficulty, however, and that is the fact that the input function U(t) is not a zero-mean process. This problem can be handled quite easily, as shown in Section 7.2.2.1, by considering the input process to be composed of the sum of two components–a deterministic input sequence equal to the mean value sequence of the white noise input process, and a zero-mean white noise sequence with covariance kernel equal to that of the actual input process. The result is the system equation

$$\begin{bmatrix} X(l) \\ V(l) \end{bmatrix} = \begin{bmatrix} 1 & 0.09063 \\ 0 & 0.81873 \end{bmatrix} \begin{bmatrix} X(l-1) \\ V(l-1) \end{bmatrix} + \begin{bmatrix} 0.00468 \\ 0.09063 \end{bmatrix} 10^3 + \begin{bmatrix} 0.00468 \\ 0.09063 \end{bmatrix} W(l-1)$$

where W(l - 1) is the zero-mean white noise sequence, and the mean value sequence is constant at the value

$$E[U(l)] = (62{,}100)(32.2/2000) = 10^3 \text{ ft/sec.}$$

The measurement process is described by the expression

$$Z(l) = \boldsymbol{H}\mathbf{X}(l) + V(l)$$

where the measurement vector $\boldsymbol{H}$ is

$$\boldsymbol{H} = [0 \quad 1]$$

and where V(l) is the zero-mean white noise sequence used to model the measurement noise.

The Kalman filter described by Equation (7.174) can now be applied, with the modification of Equation (7.186), by the following associations:

$$\boldsymbol{\Phi} = \begin{bmatrix} 1 & 0.09063 \\ 0 & 0.81873 \end{bmatrix}; \quad \boldsymbol{\Gamma} = \boldsymbol{\Delta} = \begin{bmatrix} 0.00468 \\ 0.09063 \end{bmatrix}; \quad Q = 10^3; \quad R = 10^2$$

and with the prior estimate of Equation (7.186) given by

$$\begin{bmatrix} \hat{X}(l|l-1) \\ \hat{V}(l|l-1) \end{bmatrix} = \begin{bmatrix} 1 & 0.09063 \\ 0 & 0.81873 \end{bmatrix} \begin{bmatrix} X(l-1|l-1) \\ V(l-1|l-1) \end{bmatrix} + \begin{bmatrix} 0.00468 \\ 0.09063 \end{bmatrix} 10^3$$

The value of Q in the proper units is obtained from the expression

$$Q = (3.86 \times 10^7)(32.2/2000)^2 \Delta t = 10^4 \Delta t = 10^3$$

as required by the discussion of the relationship between white noise processes and white noise sequences in Chapter 3, Section 3.6.4. If the initial estimate and covariance matrix are

$$\begin{bmatrix} X(0|0) \\ V(0|0) \end{bmatrix} = \begin{bmatrix} 1000 \\ 2000 \end{bmatrix}; \qquad \boldsymbol{P}(0|0) = \begin{bmatrix} 10{,}000 & 0 \\ 0 & 20{,}000 \end{bmatrix}$$

then the application of the algorithm begins with determination of

$$\begin{bmatrix} \hat{X}(1|0) \\ \hat{V}(1|0) \end{bmatrix} \quad \text{and} \quad \boldsymbol{P}(1|0).$$

Then, when $Z(1)$ becomes available,

$$\boldsymbol{P}(1|1) \quad \text{and} \quad \begin{bmatrix} X(1|1) \\ V(1|1) \end{bmatrix}$$

are determined, which completes the first iteration. With these values, the new prior estimate and error covariance matrix are determined:

$$\begin{bmatrix} \hat{X}(2|1) \\ \hat{V}(2|1) \end{bmatrix} \quad \text{and} \quad \boldsymbol{P}(2|1)$$

followed by the posterior values

$$\boldsymbol{P}(2|2) \quad \text{and} \quad \begin{bmatrix} \hat{X}(2|2) \\ \hat{V}(2|2) \end{bmatrix}$$

when the second measurement $Z(2)$ becomes available. This iterative process continues, providing a filtered estimate $\mathbf{X}(l|l)$ for each value of l.

The reader should note that the equations for the estimation error covariance matrix $\boldsymbol{P}(l|l)$ do not require the measurement $Z(l)$ for its computation. Thus, the sequence for $\boldsymbol{P}(l|l)$ can be determined without actually making any measurements at all. As discussed previously, this process can be used to provide an analysis of filter performance.

To further exploit this example, assume now that the thrust is not white noise sequence, but instead is a Markov random sequence which can be modeled by the scalar equation

$$W(l) = -0.1W(l-1) + \Delta_W \nu(l-1)$$

corresponding to Equation (7.192). In accordance with the development of

Section 7.2.2.3, the system equations are augmented, in the form of Equation (7.194), to produce the following characteristics for the augmented system

$$\Phi = \begin{bmatrix} 1 & 0.09063 & 0.00468 \\ 0 & 0.81873 & 0.09063 \\ 0 & 0 & 0.10000 \end{bmatrix}; \qquad \Gamma = \begin{bmatrix} 0 \\ 0 \\ \Delta_W \end{bmatrix}; \qquad Q = I; \qquad R = 10^2.$$

The value of the scalar Δ_W is chosen so that the input Markov sequence has the required variance when driven by a zero-mean white noise sequence with unit variance.

7.2.3. Optimality of the Kalman Filter Equations

The derivation of the Kalman filter equations presented here is based upon the concept of the minimum variance, unbiased, linear estimate as discussed in Section 7.1.1.3, and upon the considerations leading to Equation (7.71). The estimate was restricted to be a linear function of the measured variables; therefore it is clear why the Kalman filter equations constitute a system of linear difference equations. It seems immediately apparent that, since a linear estimate is a rather restricted class of all possible estimates which might be used, such limitation might not be wise. The answer to this question usually depends upon what knowledge is available about the random processes involved. The development of the Kalman filter equations presented here required only that the estimate be linear and that the noise processes be described through second-order statistics, i.e., by the mean value and covariance kernel functions. If more information is available–perhaps in the form of the probability density functions–then this additional information might possibly be used to produce a different set of estimation equations. Some of the conclusions of Section 7.1 bear directly upon this question.

In Section 7.1, the question of unknown parameter estimation as distinct from estimation of a random variable was broached, with the conclusion that random variable estimation should use all prior information available in terms of the known distribution characteristics of the variable to be estimated. For this reason, the minimum mean-squared error (MMSE) and maximum posterior probability (MAP) estimates were considered. Even in parameter estimation, the distribution of the measurement noise, if known, was considered in the development of the maximum likelihood (ML) estimate. Only in the worst case, in which the distribution of the measurement noise was known only through second-order statistics, was the minimum variance, unbiased, linear estimate even considered. But for a very important class of problems–those involving normal distributions–the maximum likelihood estimate; the minimum variance, unbiased, linear estimate; the minimum mean-squared error estimate; and the maximum posterior probability estimate were all given by the same linear function of the measured variables. This same feature carries over to the Kalman filter equations, with the following results:

I. If $\mathbf{X}(0)$ is a normally distributed random vector and $\mathbf{W}(l)$ and $\mathbf{V}(l)$ are normal white noise sequences, then $\mathbf{X}(l)$ and $\mathbf{Z}(l)$ are normal random processes and the Kalman filter equations are optimal in the sense of minimum mean-squared error estimation.

II. If $\mathbf{V}(l)$ is a normal white noise process, but nothing is known about the distribution of $\mathbf{W}(l)$, then $\mathbf{X}(l)$ can be considered an unknown parameter, and the Kalman filter equations are optimal in the sense of maximum likelihood estimation.

III. If the distribution of $\mathbf{W}(l)$ and $\mathbf{V}(l)$ are known only through second-order statistics, then the Kalman filter equations are optimal in the sense of minimum variance, unbiased, linear estimation.

Statement I can be made even more encompassing by introducing the concept of an admissible loss function. Recall from Section 7.1.2 that the error criterion in estimation of random variables is the risk R, defined by Equation (7.72). In the filtering problem,

$$R = E[q(\mathbf{X}(t) - \hat{\mathbf{X}}(t))] \tag{7.202}$$

and the loss function q in minimum mean-squared error estimation is the sum of the estimation error variances. These results can be extended to include all loss functions $q(\mathbf{X}(t) - \hat{\mathbf{X}}(t))$ which satisfy the following criteria:

$$\begin{gathered} q(\mathbf{0}) = 0 \\ q(\boldsymbol{a}) = q(-\boldsymbol{a}) \\ q(\boldsymbol{a}) > q(\boldsymbol{b}) \quad \text{when} \quad \rho(\boldsymbol{b}) > \rho(\boldsymbol{a}) > 0 \end{gathered} \tag{7.203}$$

where $\rho(\mathbf{X})$ is a nonnegative, convex function. The minimum mean-squared error criterion fits this description, as do many others. For example, the absolute value function

$$q(\mathrm{X}(t) - \hat{\mathrm{X}}(t)) = |\mathrm{X}(t) - \hat{\mathrm{X}}(t)| \tag{7.204}$$

is an admissible loss function in the above context. It happens that statement I is valid in the sense of any admissible loss function, not just the minimum mean-squared error criterion.

Statement III can also be extended somewhat since, if the estimate is restricted to be linear and only the second-order statistics of the random process are known, the Kalman filter equations are optimal in the sense of any admissible loss function. These results are important in that they give some indication of the wide applicability of the Kalman filter equations. In typical applications, the system designer does not worry too much about the Gaussian character of the input or measurement noise processes. He is dealing with a linear filter which is fairly straightforward in its implementation and well suited to real-time, digital computation because of its recursive nature. If the noise processes are not Gaussian, he is usually willing to accept a linear filter which is optimal, even though it may be suboptimal in the class of all possible filters which might apply.

It is interesting to note here that the original work of Kalman was based upon the strictly Gaussian case and considered admissible loss functions in the sense of Equation (7.203).

7.2.4. The Continuous Time Kalman Filter Equations

The development of the Kalman filter equations has used the discrete time description of linear processes. There are two reasons for this. The first is the fact that discrete processes permit a much more lucid description of the features of the filter than do continuous time processes. This feature is especially true when considering white noise. The second reason is a pragmatic one, in that the modern world of digital computation normally requires a discrete time approach to filtering problems. Nevertheless, the continuous time problem is important and is therefore briefly discussed here.

The system model to be considered is the standard linear system described by the equation

$$\begin{aligned} \dot{\mathbf{X}}(t) &= \boldsymbol{F}(t)\mathbf{X}(t) + \boldsymbol{D}(t)\mathbf{W}(t) + \boldsymbol{G}(t)\boldsymbol{u}(t) \\ \mathbf{Z}(t) &= \boldsymbol{H}(t)\mathbf{X}(t) + \mathbf{V}(t) \end{aligned} \tag{7.205}$$

where $\mathbf{W}(t)$ and $\mathbf{V}(t)$ are white noise processes. For reasons discussed previously, the deterministic input $\boldsymbol{u}(t)$ can be ignored. The white noise processes $\mathbf{W}(t)$ and $\mathbf{V}(t)$ are described by their second-order statistics:

$$\begin{aligned} E[\mathbf{W}(t)] &= \mathbf{0} \\ E[\mathbf{W}(t)\mathbf{W}^T(\sigma)] &= \boldsymbol{Q}(t)\delta(t-\sigma) \end{aligned} \tag{7.206}$$

and

$$\begin{aligned} E[\mathbf{V}(t)] &= \mathbf{0} \\ E[\mathbf{V}(t)\mathbf{V}^T(\sigma)] &= \boldsymbol{R}(t)\delta(t-\sigma) \end{aligned} \tag{7.207}$$

where $\delta(t - \sigma)$ is the Dirac delta function, discussed in Chapter 3. The zero mean requirement here is not limiting, since mean value functions can be accommodated in a manner entirely analogous to that considered for discrete time processes in Section 7.2.2.1.

Perhaps the easiest approach to obtaining the continuous time equations is to determine the limiting behavior of the discrete time equations as the time interval approaches zero. Consider the time interval Δt to be fixed, so that the discrete time equations of Equation (7.144) can be expressed as

$$\mathbf{X}(t_l) = \mathbf{\Phi}(t_l, t_l - \Delta t)\mathbf{X}(t_l - \Delta t) + \mathbf{\Delta}(t_l, t_l - \Delta t)\mathbf{W}(t_l - \Delta t). \quad (7.208)$$

By subtracting $\mathbf{X}(t_l - \Delta t)$ from both sides and dividing by Δt, one obtains

$$\frac{\mathbf{X}(t_l) - \mathbf{X}(t_l - \Delta t)}{\Delta t} = \frac{[\mathbf{\Phi}(t_l, t_l - \Delta t) - I]\mathbf{X}(t_l - \Delta t)}{\Delta t} + \frac{\mathbf{\Delta}(t_l, t_l - \Delta t)\mathbf{W}(t_l - \Delta t)}{\Delta t}. \quad (7.209)$$

Then, taking the limit as t approaches zero, yields

$$\frac{[\mathbf{\Phi}(t_l, t_l - \Delta t) - \boldsymbol{I}]}{\Delta t} \longrightarrow \boldsymbol{F}(t_l) \longrightarrow \boldsymbol{F}(t)$$

$$\frac{\mathbf{\Delta}(t_l, t_l - \Delta t)}{\Delta t} \longrightarrow \boldsymbol{D}(t_l) \longrightarrow \boldsymbol{D}(t). \quad (7.210)$$

The first result can be readily seen from the matrix expansion of the transition matrix $\mathbf{\Phi}(t_l, t_l - \Delta t)$. The second can be considered a definition of $\boldsymbol{D}(t)$. These results then lead to Equation (7.205) with $\boldsymbol{u}(t) = \mathbf{0}$, as was expected. Applying this same technique to the Kalman filter equations as given by Equation (7.165) results in

$$\hat{\mathbf{X}}(t_l|t_l) = \hat{\mathbf{X}}(t_l|t_l - \Delta t) + \boldsymbol{P}(t_l)\boldsymbol{H}^T(t_l)[\Delta t \boldsymbol{R}(t_l)]^{-1}[\mathbf{Z}(t_l) - \boldsymbol{H}(t)\hat{\mathbf{X}}(t_l|t_l - \Delta t)]\Delta t \quad (7.211)$$

where

$$\hat{\mathbf{X}}(t_l|t_l - \Delta t) = \hat{\mathbf{X}}(l|l - 1).$$

The measurement noise covariance term has been multiplied by Δt, as required by the limiting process described in Chapter 3. As described there, the continuous time white noise process is a mathematical fiction exhibiting an infinite variance; nevertheless, it is useful in analysis of systems which only approach this condition. Substituting from Equation (7.166) for $\hat{\mathbf{X}}(t_l|t_l - \Delta t)$ and then subtracting this result from both sides of Equation (7.211) and dividing by Δt yield

$$\frac{\hat{\mathbf{X}}(t_l|t_l) - \hat{\mathbf{X}}(t_l - \Delta t|t_l - \Delta t)}{\Delta t} = \frac{[\boldsymbol{\Phi}(t_l, t_l - \Delta t) - \boldsymbol{I}]\hat{\mathbf{X}}(t_l - \Delta t|t_l - \Delta t)}{\Delta t} + \boldsymbol{P}(t_l)\boldsymbol{H}^T(t_l)[\boldsymbol{R}(t_l)\Delta t]^{-1}[\mathbf{Z}(t_l) - \boldsymbol{H}(t_l)\hat{\mathbf{X}}(t_l|t_l - \Delta t)] \tag{7.212}$$

Taking the limit of this equation as Δt approaches zero yields, after dropping the subscript l,

$$\dot{\hat{\mathbf{X}}}(t) = \boldsymbol{F}(t)\hat{\mathbf{X}}(t) + \boldsymbol{P}(t)\boldsymbol{H}^T(t)\boldsymbol{R}^{-1}(t)[\mathbf{Z}(t) - \boldsymbol{H}(t)\hat{\mathbf{X}}(t)]. \tag{7.213}$$

Note that $\boldsymbol{R}(t)$ is no longer the covariance matrix of the measurement process, but is the matrix used in Equation (7.207) to describe the strengths of the Dirac delta functions constituting the covariance matrix. This expression is the continuous time Kalman filter estimate equation.

This same process can be applied to the covariance equation given by Equations (7.165) and (7.166). By substituting from Equation (7.166) into Equation (7.165) and subtracting $\boldsymbol{P}(t_l - \Delta t|t_l - \Delta t)$ from both sides, one obtains

$$\begin{aligned} &\boldsymbol{P}(t_l|t_l) - \boldsymbol{P}(t_l - \Delta t|t_l - \Delta t) \\ &\quad = [\boldsymbol{\Phi}(t_l, t_l - \Delta t)\boldsymbol{P}(t_l - \Delta t|t_l - \Delta t)\boldsymbol{\Phi}^T(t_l, t_l - \Delta t) \\ &\quad - \boldsymbol{P}(t_l - \Delta t|t_l - \Delta t)] + \frac{\boldsymbol{\Delta}(t_l, t_l - \Delta t)\boldsymbol{Q}(t_l - \Delta t)\Delta t\,\boldsymbol{\Delta}^T(t_l, t_l - \Delta t)}{\Delta t} \\ &\quad - \boldsymbol{P}(t_l|t_l - \Delta t)\boldsymbol{H}^T(t_l)[\boldsymbol{H}(t_l)\boldsymbol{P}(t_l|t_l - \Delta t)\boldsymbol{H}^T(t_l) \\ &\quad + \boldsymbol{R}^{-1}(t_l)]^{-1}\boldsymbol{H}(t_l)\boldsymbol{P}(t_l|t_l - \Delta t) \end{aligned} \tag{7.214}$$

where $\boldsymbol{Q}(t_l - \Delta t)$ has been multiplied by Δt in accordance with procedure for handling white noise processes. After dividing this equation by Δt and manipulating, one can obtain the following expression

$$\frac{\boldsymbol{P}(t_l|t_l - \Delta t) - \boldsymbol{P}(t_l - \Delta t|t_l - \Delta t)}{\Delta t} = \frac{[\boldsymbol{\Phi}(t_l, t_l - \Delta t) - \boldsymbol{I}]}{\Delta t}\boldsymbol{P}(t_l - \Delta t|t_l - \Delta t)\frac{[\boldsymbol{\Phi}^T(t_l, t_l - \Delta t) - \boldsymbol{I}]}{\Delta t}$$

$$
\begin{aligned}
&+\frac{\boldsymbol{P}(t_l - \Delta t|t_l - \Delta t)[\boldsymbol{\Phi}^T(t_l, t_l - \Delta t) - \boldsymbol{I}]}{\Delta t} \\
&+\frac{[\boldsymbol{\Phi}(t_l, t_l - \Delta t) - \boldsymbol{I}]\boldsymbol{P}(t_l - \Delta t|t_l - \Delta t)}{\Delta t} \\
&+\frac{\boldsymbol{\Delta}(t_l, t_l - \Delta t)\boldsymbol{Q}(t_l - \Delta t)\Delta t\boldsymbol{\Delta}^T(t_l, t_l - \Delta t)}{\Delta t^2} \\
&-\boldsymbol{P}(t_l|t_l - \Delta t)\boldsymbol{H}^T(t_l)[\boldsymbol{H}(t_l)\boldsymbol{P}(t_l|t_l - \Delta t)\boldsymbol{H}^T(t_l)\Delta t \\
&+\boldsymbol{R}(t_l)\Delta t]^{-1}\boldsymbol{H}(t_l)\boldsymbol{P}(t_l|t_l - \Delta t).
\end{aligned}
\tag{7.215}
$$

As Δt approaches zero in this expression, the following limits apply:

$$
\begin{aligned}
&\boldsymbol{R}(t_l)\Delta t \longrightarrow \boldsymbol{R}(t_l) \longrightarrow \boldsymbol{R}(t) \\
&\boldsymbol{P}(t_l|t_l) \longrightarrow \boldsymbol{P}(t_l|t_l - \Delta t) \longrightarrow \boldsymbol{P}(t) \\
&\boldsymbol{Q}(t_l - \Delta t)\Delta t \longrightarrow \boldsymbol{Q}(t_l) \longrightarrow \boldsymbol{Q}(t) \\
&\frac{\boldsymbol{\Delta}(t_l, t_l - \Delta t)}{\Delta t} \longrightarrow \boldsymbol{D}(t_l) \longrightarrow \boldsymbol{D}(t) \\
&\frac{\boldsymbol{\Phi}(t_l, t_l - \Delta t) - \boldsymbol{I}}{\Delta t} \longrightarrow \boldsymbol{F}(t_l) \longrightarrow \boldsymbol{F}(t) \\
&\frac{\boldsymbol{P}(t_l|t_l) - \boldsymbol{P}(t_l - \Delta t|t_l - \Delta t)}{\Delta t} \longrightarrow \dot{\boldsymbol{P}}(t_l) \longrightarrow \dot{\boldsymbol{P}}(t).
\end{aligned}
\tag{7.216}
$$

This process gives the final result

$$
\dot{\boldsymbol{P}}(t) = \boldsymbol{P}(t)\boldsymbol{F}^T(t) + \boldsymbol{F}(t)\boldsymbol{P}(t) + \boldsymbol{D}(t)\boldsymbol{Q}(t)\boldsymbol{D}^T(t) - \boldsymbol{P}(t)\boldsymbol{H}^T(t)\boldsymbol{R}^{-1}(t)\boldsymbol{H}(t)\boldsymbol{P}(t).
\tag{7.217}
$$

This expression is a matrix differential equation for the estimation error covariance matrix $\boldsymbol{P}(t)$. It can be solved forward in time from an initial condition $\boldsymbol{P}(t_0)$, and has the specific form of a matrix Riccati equation as discussed in Chapter 6. In fact, this expression should be compared to that of Equation (6.118) relating to optimal control of a linear, continuous time system, which is solved backward in time from some terminal condition. This fact leads to the conclusion that the optimal control problem and the optimal filtering problem are duals of one another.

In summary then, the continuous time Kalman filter equations are given by

$$\dot{\hat{\mathbf{X}}}(t) = \boldsymbol{F}(t)\hat{\mathbf{X}}(t) + \boldsymbol{P}(t)\boldsymbol{H}^T(t)\boldsymbol{R}^{-1}(t)[\mathbf{Z}(t) - \boldsymbol{H}(t)\hat{\mathbf{X}}(t)]$$

$$\dot{\boldsymbol{P}}(t) = \boldsymbol{P}(t)\boldsymbol{F}^T(t) + \boldsymbol{F}(t)\boldsymbol{P}(t) + \boldsymbol{D}(t)\boldsymbol{Q}(t)\boldsymbol{D}^T(t) - \boldsymbol{P}(t)\boldsymbol{H}^T(t)\boldsymbol{R}^{-1}(t)\boldsymbol{H}(t)\boldsymbol{P}(t). \tag{7.218}$$

The block diagram of the continuous time filter is shown in Figure 7-4, where the Kalman gain matrix $\boldsymbol{K}(t)$ has been defined as

$$\boldsymbol{K}(t) = \boldsymbol{P}(t)\boldsymbol{H}^T(t)\boldsymbol{R}^{-1}(t). \tag{7.219}$$

Again, it can be seen that the $\boldsymbol{P}(t)$ equation is solved independently of the estimation equations, thus providing the analyst with a means of error analysis. This feature also permits off-line computation of the Kalman gain matrix $\boldsymbol{K}(t)$.

For obvious reasons, the restrictions on the continuous time Kalman filter equations are the same as those imposed on the discrete time version. Some of these restrictions can be relaxed, as was the case for the discrete time equations. Results analogous to those of Section 7.2.2 can be developed for the continuous time filter in terms of augmenting the state vector. The modification for sequentially correlated measurement noise considered in Section 7.2.2.3 would result in a differential equation obtained by a limiting process similar to that used in this section.

An obvious analogy exists between the Kalman estimate of Equation (7.218) and the observer estimate of Equation (5.85). In fact, if the Kalman estimate is modified to account for a deterministic input function, then considerations analogous to those of Section 7.2.2.1 relating to discrete time systems lead to the estimate equation

$$\dot{\hat{\mathbf{X}}}(t) = \boldsymbol{F}(t)\hat{\mathbf{X}}(t) + \boldsymbol{G}(t)\boldsymbol{u}(t) + \boldsymbol{K}(t)[\mathbf{Z}(t) - \boldsymbol{H}(x)\hat{\mathbf{X}}(t)] \tag{7.220}$$

which reduces to Equation (5.85) in the time invariant case. This analogy is discussed further in Section 7.4, where stochastic control is considered.

7.2.5. Time Invariant Kalman Filters

In many instances of practical importance, the Kalman filter equations are applied to time invariant systems in both the discrete and continuous time forms. Some important results obtained from this special situation are discussed below.

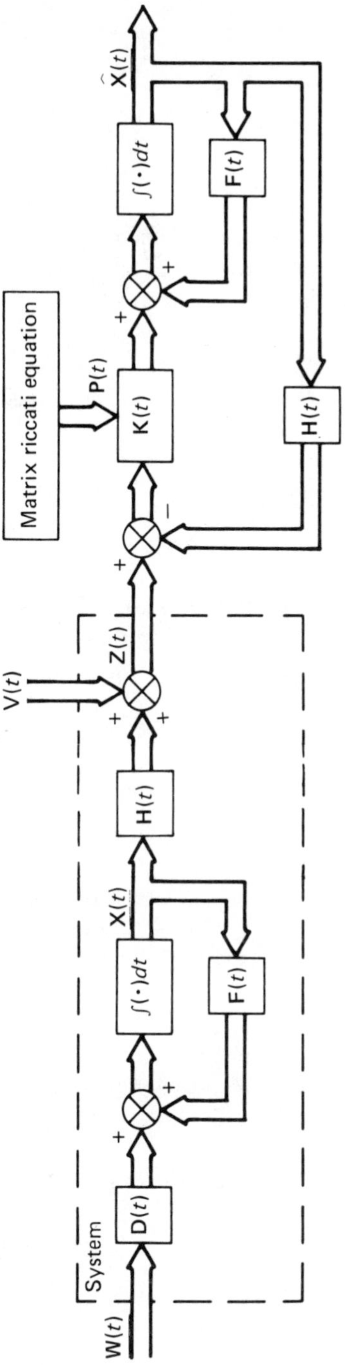

Figure 7-4. Continuous time Kalman filter.

7.2.5.1. Discrete Time Filters. The time invariant, discrete time linear system is described by the equations

$$\mathbf{X}(l) = \mathbf{\Phi X}(l-1) + \mathbf{\Delta W}(l-1) + \mathbf{\Gamma} \boldsymbol{u}(l-1)$$
$$\mathbf{Z}(l) = \boldsymbol{H}\mathbf{X}(l) + \mathbf{V}(l) \tag{7.221}$$

where a fixed interval $(t_l - t_{l-1})$ has been used. The deterministic input $\boldsymbol{u}(l)$ can be ignored without loss of generality, as discussed previously. The discrete time Kalman filter equations for this system are obtained from Equations (7.165) and (7.166) as

$$\hat{\mathbf{X}}(l|l) = \hat{\mathbf{X}}(l|l-1) + \boldsymbol{P}(l|l)\boldsymbol{H}^T\boldsymbol{R}^{-1}[\mathbf{Z}(l) - \boldsymbol{H}\hat{\mathbf{X}}(l|l-1)]$$
$$\hat{\mathbf{X}}(l|l-1) = \mathbf{\Phi}\hat{\mathbf{X}}(l-1|l-1)$$
$$\boldsymbol{P}(l|l) = \boldsymbol{P}(l|l-1) - \boldsymbol{P}(l|l-1)\boldsymbol{H}^T[\boldsymbol{R} + \boldsymbol{HP}(l|l-1)\boldsymbol{H}^T]^{-1}\boldsymbol{HP}(l|l-1)$$
$$\boldsymbol{P}(l|l-1) = \mathbf{\Phi}\boldsymbol{P}(l-1|l-1)\mathbf{\Phi}^T + \mathbf{\Delta}\boldsymbol{Q}\mathbf{\Delta}^T. \tag{7.222}$$

Note that the covariance matrices $\boldsymbol{Q}$ and $\boldsymbol{R}$ are constant, as are the system matrices $\mathbf{\Phi}$, $\mathbf{\Delta}$, and $\boldsymbol{H}$. Substituting the third expression into the fourth and rearranging yield

$$\boldsymbol{P}(l|l-1) = \mathbf{\Phi}\{\boldsymbol{P}(l-1|l-2)$$
$$- \boldsymbol{P}(l-1|l-2)\boldsymbol{H}^T[\boldsymbol{R} + \boldsymbol{HP}(l-1|l-2)\boldsymbol{H}^T]^{-1}\boldsymbol{HP}(l-1|l-2)\}\mathbf{\Phi}^T + \mathbf{\Delta}\boldsymbol{Q}\mathbf{\Delta}^T. \tag{7.223}$$

In many applications involving time invariant systems, this matrix difference equation will approach a steady state solution independent of the initial conditions. If this steady state behavior does occur, then

$$\boldsymbol{P}(l|l-1) = \boldsymbol{P}(l-1|l-2) = \bar{\boldsymbol{P}} \tag{7.224}$$

where $\bar{\boldsymbol{P}}$ is the steady state value of the prior error covariance matrix, obtainable from the expression of Equation (7.223) as the solution of the algebraic equation

$$\bar{\boldsymbol{P}} = \mathbf{\Phi}\{\bar{\boldsymbol{P}} - \bar{\boldsymbol{P}}\boldsymbol{H}^T[\boldsymbol{R} + \boldsymbol{H}\bar{\boldsymbol{P}}\boldsymbol{H}^T]^{-1}\boldsymbol{H}\bar{\boldsymbol{P}}\}\mathbf{\Phi}^T + \mathbf{\Delta}\boldsymbol{Q}\mathbf{\Delta}^T. \tag{7.225}$$

The Kalman gain matrix, as given by Equation (7.172), also approaches a steady state value, given by

$$\boldsymbol{K} = \bar{\boldsymbol{P}}\boldsymbol{H}^T(\boldsymbol{H}\bar{\boldsymbol{P}}\boldsymbol{H}^T + \boldsymbol{R})^{-1} \tag{7.226}$$

and the posterior covariance is obtained from Equation (7.174):

$$\boldsymbol{P} = (\boldsymbol{I} - \boldsymbol{KH})\bar{\boldsymbol{P}}. \tag{7.227}$$

The estimate $\hat{\boldsymbol{X}}(l|l)$ is then

$$\hat{\mathbf{X}}(l|l) = \hat{\mathbf{X}}(l-1|l-1) + \boldsymbol{K}[\mathbf{Z}(l) - \boldsymbol{H}\hat{\mathbf{X}}(l|l-1)]. \tag{7.228}$$

Alternatives to these particular expressions exist, but the basic idea is clear. The Kalman filter expression may approach steady state values for large l, and these values can be obtained by solution of algebraic equations, a great simplification over the general time varying case. It is important to recognize the limitation of this approach, however, since the steady state values are not valid during the interval when the transient behavior of the filter is dominant.

7.2.5.2. Continuous Time Filters. The time invariant, continuous time system is described by the equations

$$\begin{aligned} \dot{\mathbf{X}}(t) &= \boldsymbol{F}\mathbf{X}(t) + \boldsymbol{D}\mathbf{W}(t) + \boldsymbol{G}\boldsymbol{u}(t) \\ \mathbf{Z}(t) &= \boldsymbol{H}\mathbf{X}(t) + \mathbf{V}(t). \end{aligned} \tag{7.229}$$

The deterministic input $\boldsymbol{u}(t)$ can be ignored without loss of generality. The continuous time Kalman filter equations are then obtained from Equation (7.218) as

$$\begin{aligned} \dot{\hat{\mathbf{X}}}(t) &= \boldsymbol{F}\hat{\mathbf{X}}(t) + \boldsymbol{P}(t)\boldsymbol{H}^T\boldsymbol{R}^{-1}[\mathbf{Z}(t) - \boldsymbol{H}\hat{\mathbf{X}}(t)] \\ \dot{\boldsymbol{P}}(t) &= \boldsymbol{P}(t)\boldsymbol{F}^T + \boldsymbol{F}\boldsymbol{P}(t) + \boldsymbol{D}\boldsymbol{Q}\boldsymbol{D}^T - \boldsymbol{P}(t)\boldsymbol{H}^T\boldsymbol{R}^{-1}\boldsymbol{H}\boldsymbol{P}(t). \end{aligned} \tag{7.230}$$

In many instances of practical import, the matrix Riccati equation for $\boldsymbol{P}(t)$ will reach a steady state condition described by

$$\dot{\boldsymbol{P}}(t) = 0$$

in which case the matrix $\boldsymbol{P}$ can be found from the algebraic equation

$$\boldsymbol{P}\boldsymbol{F}^T + \boldsymbol{F}\boldsymbol{P} + \boldsymbol{D}\boldsymbol{Q}\boldsymbol{D} - \boldsymbol{P}\boldsymbol{H}^T\boldsymbol{R}^{-1}\boldsymbol{H}\boldsymbol{P} = 0. \tag{7.231}$$

The estimate $\hat{\mathbf{X}}(t)$ is then obtained from Equations (7.218) and (7.219),

$$\dot{\hat{\mathbf{X}}}(t) = \boldsymbol{F}\hat{\mathbf{X}}(t) + \boldsymbol{K}[\mathbf{Z}(t) - \boldsymbol{H}\hat{\mathbf{X}}(t)] \tag{7.232}$$

where the Kalman gain matrix is constant and is given by Equation (7.219),

$$\boldsymbol{K} = \boldsymbol{P}\boldsymbol{H}^T\boldsymbol{R}^{-1}. \tag{7.233}$$

Again, alternative forms may be obtained.

Some interesting aspects of this problem can be exploited. For example, Equation (7.232) can be rewritten as

$$\dot{\hat{\mathbf{X}}}(t) = (\boldsymbol{F} - \boldsymbol{K}\boldsymbol{H})\hat{\mathbf{X}}(t) + \boldsymbol{K}\mathbf{Z}(t) \tag{7.234}$$

which very clearly identifies the filter as a time invariant, linear system driven by a linear function of the observed variables. The relationship to the state variable feedback control problem should be evident by comparing this expression with results obtained in Chapter 5. This system of equations produces the transition matrix

$$\boldsymbol{\Phi}(t - t_0) = \exp - [(\boldsymbol{F} - \boldsymbol{K}\boldsymbol{H})(t - t_0)] \tag{7.235}$$

so that the system response is

$$\hat{\mathbf{X}}(t) = \int_{t_0}^{t} [\exp - (\boldsymbol{F} - \boldsymbol{K}\boldsymbol{H})(\sigma - t_0)]\boldsymbol{K}\mathbf{Z}(\sigma)\, d\sigma. \tag{7.236}$$

It can be shown that this expression is equivalent to that describing a multivariable Wiener filter, which is a classical result in statistical estimation theory.

The concept of steady state filter operation can be illustrated by a simple, one-dimensional problem in which a particle moves along a single dimension in a region where viscous damping is present. It is desired to estimate the velocity of the particle, based upon noisy velocity measurements. The random acceleration acting on the particle is considered to be a zero mean white noise process. The equation of motion is then

$$\dot{\mathrm{X}}(t) = -c\mathrm{X}(t) + \mathrm{W}(t)$$

where $\mathrm{X}(t)$ is the velocity, and c is the coefficient of viscous damping. The acceleration $\mathrm{W}(t)$ is a white noise process with zero mean and covariance kernel function

$$k_{\mathrm{W}}(t, \sigma) = E[\mathrm{W}(t)\mathrm{W}(\sigma)] = q\delta(t - \sigma).$$

The observation equation is

$$Z(t) = X(t) + V(t)$$

where $V(t)$ is a zero mean, white noise process with covariance kernel function

$$k_V(t, \sigma) = E[V(t)V(\sigma)] = r\delta(t - \sigma).$$

The solution to the filtering problem follows directly from application of Equation (7.230). The Ricatti equation is

$$\dot{p}(t) = -2cp(t) + q - (1/r)p^2(t)$$

which can be shown to have the solution

$$p(t) = r[\sqrt{c^2 + (q/r)} - c] + \frac{2r\sqrt{c^2 + (q/r)}}{\dfrac{p(t_0) + r[\sqrt{c^2 + (q/r)} + c]}{p(t_0) - r[\sqrt{c^2 + (q/r)} - c]} \exp\,[2\sqrt{c^2 + (q/r)}\,t - 1]}$$

Once $p(t)$ is determined, the optimal estimate can be found from solution of the differential equation

$$\dot{\hat{X}}(t) = -c\hat{X}(t) + (1/r)p(t)[Z(t) - \hat{X}(t)].$$

Even in this simple case, solution of the nonlinear differential equation is difficult. Certain limiting cases are informative, however, and serve to illustrate the characteristics of the statistical filter. For t very large, the solution for $p(t)$ reaches a steady state condition, given by

$$p = r[\sqrt{c^2 + (q/r)} - c]$$

which yields the stationary filter which would result from application of the Wiener theory to this problem. The optimal estimate of X under these conditions is the solution of the equation

$$\dot{\hat{X}}(t) = -c\hat{X}(t) + [\sqrt{c^2 + (q/r)} - c][Z(t) - \hat{X}(t)].$$

This nonhomogeneous, first-order, linear equation can be readily solved for $\hat{X}(t)$, if the driving function $Z(t)$ is known. The block diagram of such a system is shown in Figure 7-5. In the limiting case in which r gets very large, indi-

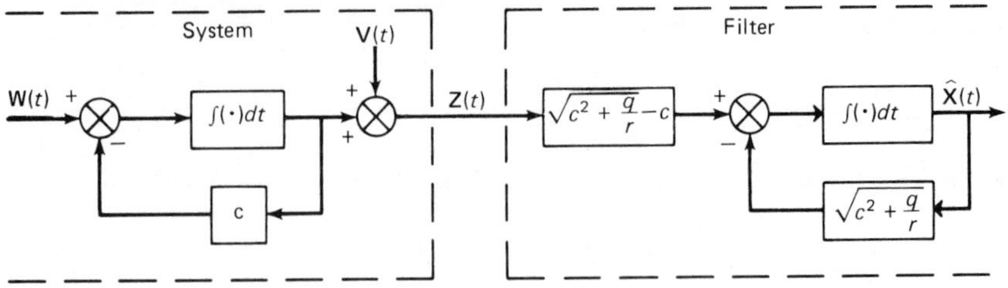

Figure 7-5. Steady state Kalman filter.

cating that the measurements are extremely noisy, the estimation equation is

$$\dot{\hat{X}}(t) = -c\hat{X}(t)$$

which implies that the filter is merely predicting $X(t)$ based upon the previous estimate, thus ignoring the poor quality measurement data.

7.3. THE SMOOTHING PROBLEM

The Kalman filter equations present a systematic approach to the real time filtering problem of determining an estimate for the state of a linear system at time t_l, given measurements up to and including $\mathbf{Z}(t_l)$, i.e., the estimate $\hat{\mathbf{X}}(l|l)$ is determined very nicely by the Kalman filter equations. For real time estimation, where the estimate $\hat{\mathbf{X}}(l|l)$ is needed at time t_l or shortly thereafter, the filtering algorithms provide the best available information in the form of the filtered estimate. There are other applications, however, in which the estimate at time t_l is not required in a real time manner, so now one must examine the filtering algorithms to determine if they actually produce the best possible estimate obtainable from the available measurement data. If a series of measurements are made,

$$\mathbf{Z}(1), \mathbf{Z}(2), \mathbf{Z}(3), \ldots, \mathbf{Z}(l), \ldots, \mathbf{Z}(L) \tag{7.237}$$

then the filtering algorithms yield, for time t_l, the estimate $\hat{\mathbf{X}}(l|l)$, the estimate based upon the l measurements

$$\mathbf{Z}(1), \mathbf{Z}(2), \mathbf{Z}(3), \ldots, \mathbf{Z}(l). \tag{7.238}$$

For real time estimation, this estimate is the best that can be accomplished. Suppose that instead of the estimates being required in a real time fashion, the estimate at time t_l is not required until some time after the last measurement $\mathbf{Z}(L)$ has been obtained. In this case, the estimate of $\mathbf{X}(l)$ can be based on all measurements up to and including $\mathbf{Z}(L)$. That is, it is possible to obtain $\hat{\mathbf{X}}(l|L)$ based upon all the measurements of Equation (7.237), rather than only those given by Equation (7.238). In this manner, estimates based upon all measurements up to and including $\mathbf{Z}(L)$ are obtained for the entire sequence $\mathbf{X}(l)$. This general concept of including measurements made both prior to and subsequent to the time of the estimate is called *smoothing*. Because more data is used for each estimate–except for the last–in smoothing than in filtering, one would expect the estimates to be better.

There are three general smoothing problems which will be considered, each with a different objective. These are:

I. *Fixed point smoothing.* Here it is desired to use the entire sequence of measurements given by Equation (7.237) to estimate the state vector at a single time point t_0. That is, the estimate $\hat{\mathbf{X}}(0|L)$ is desired.

II. *Fixed lag smoothing.* Here it is desired to estimate the sequence of state vectors $\mathbf{X}(l)$ over the interval $(0, L)$; each estimate is to be based upon all measurements prior to time t_l and also upon M measurements made subsequently to time t_l, i.e., the estimate $\hat{\mathbf{X}}(l|l+M)$ is desired.

III. *Fixed interval smoothing.* This case is the general situation described above, in which it is desired to estimate the sequence of state vectors $\mathbf{X}(l)$ over the interval $(0, L)$, and in which each estimate is to be based upon all measurements made over the interval. Here, it is $\hat{\mathbf{X}}(l|L)$ that is desired.

As might be expected, these three smoothing problems are related to each other and to the filtering problem. The approach to be used here is to modify the Kalman filtering problem so that it actually accomplishes smoothing. Since this approach uses the principles of Kalman filtering already considered, no new conceptual development is required. Instead, the smoothing problem is made to look like a filtering problem, and then the filtering techniques are applied directly. This approach is not the only one that may be used, but it is easily developed and intuitively satisfying.

7.3.1. Fixed Point Smoothing

In fixed point smoothing, the objective is to estimate the state vector of the linear system described by the equations

$$\mathbf{X}(l) = \mathbf{\Phi}(l, l-1)\mathbf{X}(l-1) + \mathbf{\Delta}(l, l-1)\mathbf{W}(l-1) + \mathbf{\Gamma}(l, l-1)\boldsymbol{u}(l-1)$$

$$\mathbf{Z}(l) = \boldsymbol{H}(l)\mathbf{X}(l) + \mathbf{V}(l) \tag{7.239}$$

where $\mathbf{W}(l)$ and $\mathbf{V}(l)$ are zero mean, white noise sequences with covariance kernel functions

$$E[\mathbf{W}(l)\mathbf{W}^T(m)] = \begin{cases} Q(l), & \text{if } l = m \\ 0, & \text{if } l \neq m \end{cases}$$

$$E[\mathbf{V}(l)\mathbf{V}^T(m)] = \begin{cases} R(l), & \text{if } l = m \\ 0, & \text{if } l \neq m. \end{cases} \tag{7.240}$$

The deterministic input $\boldsymbol{u}(l)$ can be ignored without loss of generality. Nonzero mean value functions for $\mathbf{W}(l)$ and $\mathbf{V}(l)$ and correlation between $\mathbf{W}(l)$ and $\mathbf{V}(l)$ can be considered by the methods discussed in Section 7.2.2. The approach here is simple. The system of Equation (7.239) is augmented by defining a second vector as

$$\mathbf{X}_1(l) = \mathbf{X}(l-1), \qquad \text{for } 0 < l < L \tag{7.241}$$

and initializing it to the value $\mathbf{X}(0)$. That is, the augmented state equation is

$$\begin{bmatrix} \mathbf{X}_0(l) \\ \hline \mathbf{X}_1(l) \end{bmatrix} = \left[\begin{array}{c|c} \boldsymbol{\Phi}(l, l-1) & \mathbf{0} \\ \hline \mathbf{0} & I \end{array}\right] \begin{bmatrix} \mathbf{X}_0(l-1) \\ \hline \mathbf{X}_1(l-1) \end{bmatrix} + \begin{bmatrix} \boldsymbol{\Delta}(l, l-1) \\ \hline \mathbf{0} \end{bmatrix} \mathbf{W}(l-1) \tag{7.242}$$

where the zero subscript refers to the original variables $\mathbf{X}(l)$, and $\mathbf{X}_1(0) = \mathbf{X}(0)$. The behavior of the augmenting vector can be seen by examining the sequence of values

$$\mathbf{X}_1(0) = \mathbf{X}(0)$$
$$\mathbf{X}_1(1) = \mathbf{X}_1(0) = \mathbf{X}(0)$$
$$\mathbf{X}_1(2) = \mathbf{X}_1(1) = \mathbf{X}_1(0) = \mathbf{X}(0).$$

Thus, $\mathbf{X}_1(l)$ is constant at the value $\mathbf{X}(0)$. The measurement equation is

$$\mathbf{Z}(l) = [\mathbf{H}(l) \,\vert\, \mathbf{0}] \begin{bmatrix} \mathbf{X}_0(l) \\ \hline \mathbf{X}_1(l) \end{bmatrix} + \mathbf{V}(l). \tag{7.243}$$

The expressions of Equations (7.242) and (7.243) represent a linear system of the form of Equations (7.145) and (7.146); therefore, the Kalman filter equations apply directly. From Equations (7.166) and (7.174),

$$\left[\begin{array}{c}\hat{\mathbf{X}}_0(l|l-1)\\ \hline \hat{\mathbf{X}}_1(l|l-1)\end{array}\right]=\left[\begin{array}{c|c}\boldsymbol{\Phi}(l,l-1) & \mathbf{0}\\ \hline \mathbf{0} & \boldsymbol{I}\end{array}\right]\left[\begin{array}{c}\hat{\mathbf{X}}_0(l-1|l-1)\\ \hline \hat{\mathbf{X}}_1(l-1|l-1)\end{array}\right]=\left[\begin{array}{c}\boldsymbol{\Phi}(l,l-1)\hat{\mathbf{X}}_0(l-1|l-1)\\ \hline \hat{\mathbf{X}}_1(l-1|l-1)\end{array}\right]$$

$$\left[\begin{array}{c}\hat{\mathbf{X}}_0(l|l)\\ \hline \hat{\mathbf{X}}_1(l|l)\end{array}\right]=\left[\begin{array}{c}\hat{\mathbf{X}}_0(l|l-1)\\ \hline \hat{\mathbf{X}}_1(l|l-1)\end{array}\right]+\left[\begin{array}{c}\boldsymbol{K}_0(l)\\ \hline \boldsymbol{K}_1(l)\end{array}\right][\mathbf{Z}(l)-\boldsymbol{H}(l)\hat{\mathbf{X}}_0(l|l-1)] \tag{7.244}$$

The smoothing result is produced by the smoothing gain matrix $\boldsymbol{K}_1(l)$, which modifies $\hat{\mathbf{X}}_1(l|l)$ after each measurement, in effect updating the estimate $\hat{\mathbf{X}}(0)$. The Kalman gain equation, from Equation (7.174), is

$$\left[\begin{array}{c}\boldsymbol{K}_0(l)\\ \hline \boldsymbol{K}_1(l)\end{array}\right]=\left[\begin{array}{c|c}\boldsymbol{P}_{00}(l|l-1) & \boldsymbol{P}_{10}^T(l|l-1)\\ \hline \boldsymbol{P}_{10}(l|l-1) & \boldsymbol{P}_{11}(l|l-1)\end{array}\right]\left[\begin{array}{c}\boldsymbol{H}^T(l)\\ \hline \mathbf{0}\end{array}\right]\left[[\boldsymbol{H}(l)\,|\,\mathbf{0}]\right.$$

$$\left.\cdot\left[\begin{array}{c|c}\boldsymbol{P}_{00}(l|l-1) & \boldsymbol{P}_{10}^T(l|l-1)\\ \hline \boldsymbol{P}_{10}(l|l-1) & \boldsymbol{P}_{11}(l|l-1)\end{array}\right]\left[\begin{array}{c}\boldsymbol{H}^T(l)\\ \hline \mathbf{0}\end{array}\right]+\boldsymbol{R}(l)\right]^{-1} \tag{7.245}$$

Performing the indicated multiplication in this equation yields

$$\left[\begin{array}{c}\boldsymbol{K}_0(l)\\ \hline \boldsymbol{K}_1(l)\end{array}\right]=\left[\begin{array}{c}\boldsymbol{P}_{00}(l|l-1)\boldsymbol{H}^T(l)\\ \hline \boldsymbol{P}_{10}(l|l-1)\boldsymbol{H}^T(l)\end{array}\right][\boldsymbol{H}(l)\boldsymbol{P}_{00}(l|l-1)\boldsymbol{H}^T(l)+\boldsymbol{R}(l)]^{-1}. \tag{7.246}$$

Two important points are seen here. First, the $\boldsymbol{K}_0(l)$ equation is identical to the Kalman filter equation, as is the case with Equation (7.244). Secondly, no additional matrix inversion is required to determine the smoothing gain matrix $\boldsymbol{K}_1(l)$.

Finally, from Equations (7.162) and (7.173), the covariance equations are

$$\left[\begin{array}{c|c}\boldsymbol{P}_{00}(l|l-1) & \boldsymbol{P}_{10}^T(l|l-1)\\ \hline \boldsymbol{P}_{10}(l|l-1) & \boldsymbol{P}_{11}(l|l-1)\end{array}\right]$$

$$=\left[\begin{array}{c|c}\boldsymbol{\Phi}(l,l-1) & \mathbf{0}\\ \hline \mathbf{0} & \boldsymbol{I}\end{array}\right]\left[\begin{array}{c|c}\boldsymbol{P}_{00}(l-1|l-1) & \boldsymbol{P}_{10}^T(l-1|l-1)\\ \hline \boldsymbol{P}_{10}(l-1|l-1) & \boldsymbol{P}_{11}(l-1|l-1)\end{array}\right]\left[\begin{array}{c|c}\boldsymbol{\Phi}^T(l,l-1) & \mathbf{0}\\ \hline \mathbf{0} & \boldsymbol{I}\end{array}\right]$$

$$+\left[\begin{array}{c}\boldsymbol{\Delta}(l,l-1)\\ \hline \mathbf{0}\end{array}\right]\boldsymbol{Q}(l-1)[\boldsymbol{\Delta}^T(l,l-1)\,|\,\mathbf{0}] \tag{7.247}$$

and

$$\left[\begin{array}{c:c} \boldsymbol{P}_{00}(l|l) & \boldsymbol{P}_{10}^T(l|l) \\ \hdashline \boldsymbol{P}_{10}(l|l) & \boldsymbol{P}_{11}(l|l) \end{array}\right] = \left[\boldsymbol{I} - \left[\begin{array}{c} \boldsymbol{K}_0(l) \\ \hdashline \boldsymbol{K}_1(l) \end{array}\right] [\boldsymbol{H}(l) \mid \boldsymbol{0}]\right] \left[\begin{array}{c:c} \boldsymbol{P}_{00}(l|l-1) & \boldsymbol{P}_{10}^T(l|l-1) \\ \hdashline \boldsymbol{P}_{10}(l|l-1) & \boldsymbol{P}_{11}(l|l-1) \end{array}\right]. \tag{7.248}$$

Expanding these equations yields

$$\begin{aligned} &\left[\begin{array}{c:c} \boldsymbol{P}_{00}(l|l-1) & \boldsymbol{P}_{10}^T(l|l-1) \\ \hdashline \boldsymbol{P}_{10}(l|l-1) & \boldsymbol{P}_{11}(l|l-1) \end{array}\right] \\ &\quad = \left[\begin{array}{c:c} \boldsymbol{\Phi}(l, l-1)\boldsymbol{P}_{00}(l-1|l-1)\boldsymbol{\Phi}^T(l, l-1) & \boldsymbol{\Phi}(l, l-1)\boldsymbol{P}_{10}^T(l-1|l-1) \\ \hdashline \boldsymbol{P}_{10}(l-1|l-1)\boldsymbol{\Phi}^T(l, l-1) & \boldsymbol{P}_{11}(l-1|l-1) \end{array}\right] \\ &\quad + \left[\begin{array}{c:c} \boldsymbol{\Delta}(l, l-1)\boldsymbol{Q}(l-1)\boldsymbol{\Delta}^T(l, l-1) & \boldsymbol{0} \\ \hdashline \boldsymbol{0} & \boldsymbol{0} \end{array}\right] \end{aligned} \tag{7.249}$$

and

$$\begin{aligned} &\left[\begin{array}{c:c} \boldsymbol{P}_{00}(l|l) & \boldsymbol{P}_{10}^T(l|l) \\ \hdashline \boldsymbol{P}_{10}(l|l) & \boldsymbol{P}_{11}(l|l) \end{array}\right] \\ &= \left[\begin{array}{c:c} [\boldsymbol{I} - \boldsymbol{K}_0(l)\boldsymbol{H}(l)]\boldsymbol{P}_{00}(l|l-1) & [\boldsymbol{I} - \boldsymbol{K}_0(l)\boldsymbol{H}(l)]\boldsymbol{P}_{10}^T(l|l-1) \\ \hdashline \boldsymbol{K}_1(l)\boldsymbol{H}(l)\boldsymbol{P}_{00}(l|l-1) + \boldsymbol{P}_{10}(l|l-1) & \boldsymbol{K}_1(l)\boldsymbol{H}(l)\boldsymbol{P}_{10}^T(l|l-1) + \boldsymbol{P}_{11}(l|l-1) \end{array}\right]. \end{aligned} \tag{7.250}$$

The apparent asymmetry in this expression can be resolved by substituting for $\boldsymbol{K}_0(l)$ and $\boldsymbol{K}_1(l)$ from Equation (7.246) into the off-diagonal submatrices. Identical expressions will result, so that Equation (7.250) can be rewritten as

$$\begin{aligned} &\left[\begin{array}{c:c} \boldsymbol{P}_{00}(l|l) & \boldsymbol{P}_{10}^T(l|l) \\ \hdashline \boldsymbol{P}_{10}(l|l) & \boldsymbol{P}_{11}(l|l) \end{array}\right] \\ &\quad = \left[\begin{array}{c:c} [\boldsymbol{I} - \boldsymbol{K}_0(l)\boldsymbol{H}(l)]\boldsymbol{P}_{00}(l|l-1) & [\boldsymbol{I} - \boldsymbol{K}_0(l)\boldsymbol{H}(l)]\boldsymbol{P}_{10}^T(l|l-1) \\ \hdashline \boldsymbol{P}_{10}(l|l-1)[\boldsymbol{I} - \boldsymbol{H}^T(l)\boldsymbol{K}_0^T(l)] & \boldsymbol{K}_1(l)\boldsymbol{H}(l)\boldsymbol{P}_{10}^T(l|l-1) + \boldsymbol{P}_{11}(l|l-1) \end{array}\right]. \end{aligned} \tag{7.251}$$

Examination of Equations (7.246), (7.249), and (7.251) reveals that the matrix $\boldsymbol{P}_{11}$ does not affect the computation of any other quantities required by the algorithm and therefore need not itself be computed. The algorithm can then be summarized by the expressions

$$\boldsymbol{K}_0(l) = \boldsymbol{P}_{00}(l|l-1)\boldsymbol{H}^T(l)[\boldsymbol{H}(l)\boldsymbol{P}_{00}(l|l-1)\boldsymbol{H}^T(l) + \boldsymbol{R}(l)]^{-1}$$
$$\boldsymbol{K}_1(l) = \boldsymbol{P}_{10}(l|l-1)\boldsymbol{H}^T(l)[\boldsymbol{H}(l)\boldsymbol{P}_{00}(l|l-1)\boldsymbol{H}^T(l) + \boldsymbol{R}(l)]^{-1} \tag{7.252}$$

$$\boldsymbol{P}_{00}(l|l-1) = \boldsymbol{\Phi}(l, l-1)\boldsymbol{P}_{00}(l-1|l-1)\boldsymbol{\Phi}^T(l, l-1) + \boldsymbol{\Delta}(l, l-1)\boldsymbol{Q}(l-1)\boldsymbol{\Delta}^T(l, l-1)$$
$$\boldsymbol{P}_{10}(l|l-1) = \boldsymbol{P}_{10}(l-1|l-1)\boldsymbol{\Phi}^T(l, l-1) \tag{7.253}$$

$$\boldsymbol{P}_{00}(l|l) = [\boldsymbol{I} - \boldsymbol{K}_0(l)\boldsymbol{H}(l)]\boldsymbol{P}_{00}(l|l-1)$$
$$\boldsymbol{P}_{10}(l|l) = \boldsymbol{P}_{10}(l|l-1)[\boldsymbol{I} - \boldsymbol{H}^T(l)\boldsymbol{K}_0^T(l)] \tag{7.254}$$

$$\hat{\mathbf{X}}_0(l|l-1) = \boldsymbol{\Phi}(l, l-1)\hat{\mathbf{X}}_0(l-1|l-1)$$
$$\hat{\mathbf{X}}_1(l|l-1) = \hat{\mathbf{X}}_1(l-1|l-1) \tag{7.255}$$

$$\hat{\mathbf{X}}_0(l|l) = \hat{\mathbf{X}}_0(l|l-1) + \boldsymbol{K}_0(l)[\mathbf{Z}(l) - \boldsymbol{H}(l)\hat{\mathbf{X}}_0(l|l-1)]$$
$$\hat{\mathbf{X}}_1(l|l) = \hat{\mathbf{X}}_1(l-1|l-1) + \boldsymbol{K}_1(l)[\mathbf{Z}(l) - \boldsymbol{H}(l)\mathbf{X}_0(l|l-1)]. \tag{7.256}$$

The fact is immediately obvious that the zero-subscripted variables constitute the Kalman filter results, so it is concluded that the smoothing algorithm generates the filtered estimates as part of the smoothing process. The initial covariance matrices $\boldsymbol{P}_{00}(0|0)$ and $\boldsymbol{P}_{10}(0|0)$ have the value $\boldsymbol{P}(0|0)$, the initial value of the Kalman filter covariance matrix.

The computation proceeds as in the Kalman filter algorithm, with recursive application of Equations (7.252) through (7.254) used to produce the gain matrices and covariance matrices, and of Equations (7.255) and (7.256) to produce the filtered estimate $\hat{\mathbf{X}}(l|l)$ and the smoothed estimate $\hat{\mathbf{X}}(0|l)$. The relationships are

$$\hat{\mathbf{X}}_0(l|l) = \hat{\mathbf{X}}(l|l)$$
$$\hat{\mathbf{X}}_1(l|l) = \hat{\mathbf{X}}(0|l). \tag{7.257}$$

That is, $\hat{\mathbf{X}}_0(l|l)$ is the Kalman filtered estimate of $\mathbf{X}(l)$, and $\hat{\mathbf{X}}_1(l|l)$ the smoothed estimate of $\mathbf{X}(0)$ based upon measurements up to and including $\mathbf{Z}(l)$.

If the estimation error covariance matrix associated with the smoothed estimate $\hat{\mathbf{X}}(0|l)$ is desired, it can be recursively computed using the appropriate submatrices of Equations (7.249) and (7.251), resulting in the recursion relation

$$P_{11}(l|l) = K_1(l)H(l)P_{10}^T(l|l-1) + P_{11}(l-1|l-1). \qquad (7.258)$$

Computation of this matrix is not required by the smoothing algorithm. The only reason for including this computation in the smoothing processing would be to obtain the statistical information which it contains.

If a smoothed point estimation is desired at some time instant t_m other than t_0, where $0 < m < L$, then it is only necessary to modify the definition of $\mathbf{X}_1(l)$ to

$$\mathbf{X}_1(l) = \mathbf{X}(l-1) \quad \text{for} \quad m < l < L \qquad (7.259)$$

with initialization at time t_m

$$\mathbf{X}_1(m) = \mathbf{X}(m).$$

Prior to time t_m, only the filtering algorithm is used; however, subsequent to time t_m, the filtering and smoothing algorithm described above is used. The fact that this procedure produces the smoothed estimate $\hat{\mathbf{X}}(m|L)$ can be seen by recognizing that the filtered estimate $\hat{\mathbf{X}}(m|m)$ contains information from all measurements up to and including $\mathbf{Z}(m)$, while the smoothing algorithm refines this estimate by including information contained in subsequent measurements $\mathbf{Z}(m+1)$ through $\mathbf{Z}(L)$.

A block diagram depicting fixed point smoothing is shown in Figure 7-6. The

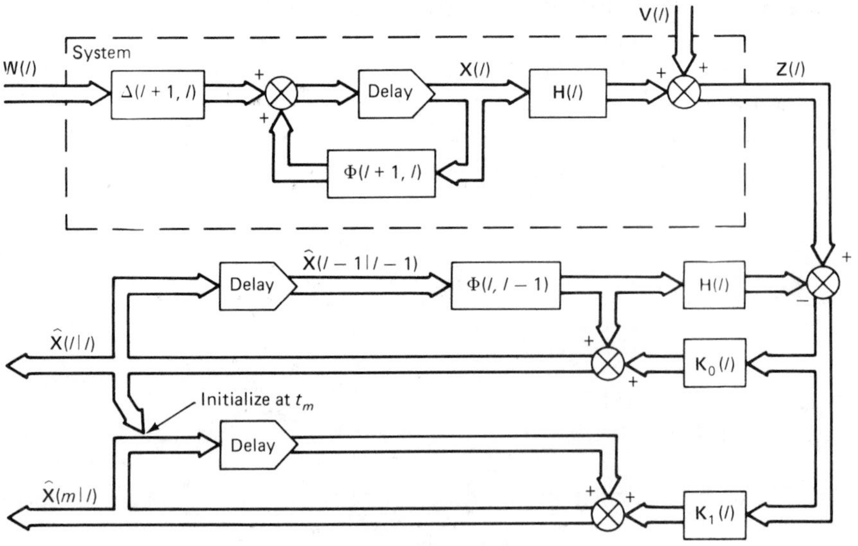

Figure 7-6. Fixed point smoothing.

diagram merely shows implementation of Equation (7.244). Computation of the Kalman gain matrices $\boldsymbol{K}_0(l)$ and $\boldsymbol{K}_1(l)$, and of the covariance matrices is not shown. These computations are accomplished according to Equations (7.247) and (7.251). As was the case with the filtering algorithms, computation of these matrices can be done off-line in order to relieve the real-time computational burden or to yield an error analysis of the smoothing process. From Figure 7-6 and Equations (7.247) and (7.251), it is also apparent that the zero subscripted quantities directly implement the Kalman filter equations. Finally, it should be noted that the smoothed estimate $\hat{\mathbf{X}}(m|l)$ is not defined until time t_{m+1}, since the smoothing delay loop is not initialized until time t_m. The estimate $\hat{\mathbf{X}}(m|m)$ is available at the Kalman filter output at time t_m.

7.3.2. Fixed Lag Smoothing

The preceding development involved the use of smoothing to produce an estimate of the state at a single time point. Fixed lag smoothing consists of estimating the entire sequence of state vectors $\mathbf{X}(l)$ over the interval $(0, L)$. Each estimate is to be based on all previous measurements and the M succeeding measurements, resulting in a smoothed estimate of the state vector $\mathbf{X}(l)$. This smoothed estimate is based upon measurements $\mathbf{Z}(1)$ through $\mathbf{Z}(l+M)$, and the computation of this estimate must obviously lag the measurement process by a factor of M. The system model is again given by Equations (7.239) and (7.240), and again the deterministic input $\boldsymbol{u}(l)$ can be ignored. The technique of state vector augmentation will be used again to obtain the fixed lag smoothing equations, with the number of augmenting vectors now equal to M. Specifically, the augmented state equation is

$$\begin{bmatrix} \mathbf{X}_0(l) \\ \mathbf{X}_1(l) \\ \mathbf{X}_2(l) \\ \vdots \\ \mathbf{X}_M(l) \end{bmatrix} = \begin{bmatrix} \boldsymbol{\Phi}(l, l-1) & \mathbf{0} & \cdots & \mathbf{0} & \mathbf{0} \\ I & \mathbf{0} & \cdots & \mathbf{0} & \mathbf{0} \\ \mathbf{0} & I & \cdots & \mathbf{0} & \mathbf{0} \\ \vdots & \vdots & & \vdots & \vdots \\ \mathbf{0} & \mathbf{0} & \cdots & I & \mathbf{0} \end{bmatrix} \begin{bmatrix} \mathbf{X}_0(l-1) \\ \mathbf{X}_1(l-1) \\ \mathbf{X}_2(l-1) \\ \vdots \\ \mathbf{X}_M(l-1) \end{bmatrix} + \begin{bmatrix} \boldsymbol{\Delta}(l, l-1) \\ \mathbf{0} \\ \mathbf{0} \\ \vdots \\ \mathbf{0} \end{bmatrix} W(l-1)$$

$$\mathbf{Z}(l) = [H(l) \;\; \mathbf{0} \;\; \mathbf{0} \;\; \cdots \;\; \mathbf{0}] \begin{bmatrix} \mathbf{X}_0(l) \\ \mathbf{X}_1(l) \\ \mathbf{X}_2(l) \\ \vdots \\ \mathbf{X}_M(l) \end{bmatrix} + \mathbf{V}(l). \qquad (7.260)$$

The behavior of the augmenting state vectors can be seen from this equation to be

$$\begin{aligned} \mathbf{X}_1(l) &= \mathbf{X}_0(l-1) \\ \mathbf{X}_2(l) &= \mathbf{X}_1(l-1) \\ &\vdots \\ \mathbf{X}_M(l) &= \mathbf{X}_{M-1}(l-1) \end{aligned} \tag{7.261}$$

so that these vectors always contain the M previous values of the state vector $\mathbf{X}(l)$. The zero subscript, as before, refers to the original values; that is, by definition

$$\mathbf{X}_0(l) = \mathbf{X}(l). \tag{7.262}$$

Once this system model has been defined, the fixed lag smoothing process is accomplished by applying the Kalman filter equations to the expanded model of Equation (7.260). The estimate is obtained from Equation (7.174):

$$\begin{bmatrix} \hat{\mathbf{X}}_0(l|l) \\ \hat{\mathbf{X}}_1(l|l) \\ \hat{\mathbf{X}}_2(l|l) \\ \vdots \\ \hat{\mathbf{X}}_M(l|l) \end{bmatrix} = \begin{bmatrix} \hat{\mathbf{X}}_0(l|l-1) \\ \hat{\mathbf{X}}_1(l|l-1) \\ \hat{\mathbf{X}}_2(l|l-1) \\ \vdots \\ \hat{\mathbf{X}}_M(l|l-1) \end{bmatrix} + \begin{bmatrix} K_0(l) \\ K_1(l) \\ K_2(l) \\ \vdots \\ K_M(l) \end{bmatrix} [\mathbf{Z}(l) - H(l)\hat{\mathbf{X}}_0(l|l-1)]. \tag{7.263}$$

The prior estimates are given by Equation (7.166); when these expressions are substituted using the transition matrix of Equation (7.260), the result is

$$\begin{bmatrix} \hat{\mathbf{X}}_0(l|l) \\ \hat{\mathbf{X}}_1(l|l) \\ \hat{\mathbf{X}}_2(l|l) \\ \vdots \\ \hat{\mathbf{X}}_M(l|l) \end{bmatrix} = \begin{bmatrix} \mathbf{\Phi}(l, l-1)\hat{\mathbf{X}}_0(l-1|l-1) \\ \hat{\mathbf{X}}_0(l-1|l-1) \\ \hat{\mathbf{X}}_1(l-1|l-1) \\ \vdots \\ \hat{\mathbf{X}}_{M-1}(l-1|l-1) \end{bmatrix}$$

$$+\begin{bmatrix} \boldsymbol{K}_0(l) \\ \hdashline \boldsymbol{K}_1(l) \\ \hdashline \boldsymbol{K}_2(l) \\ \hdashline \vdots \\ \hdashline \boldsymbol{K}_M(l) \end{bmatrix} [\mathbf{Z}(l) - \boldsymbol{H}(l)\boldsymbol{\Phi}(l, l-1)\hat{\mathbf{X}}_0(l-1|l-1)]. \quad (7.264)$$

The character of the smoothing algorithm is obvious. The zero-subscripted terms are the Kalman filter equations. The filtered estimate $\hat{\mathbf{X}}_0(l|l)$ is based on measurements up to and including $\mathbf{Z}(l)$. The previous value of this estimate, $\hat{\mathbf{X}}_0(l-1|l-1)$ is updated by gain matrix $\boldsymbol{K}_1(l)$ to become

$$\hat{\mathbf{X}}_1(l|l) = \hat{\mathbf{X}}(l-1|l). \quad (7.265)$$

The previous value of this estimate is updated by gain matrix $\boldsymbol{K}_2(l)$ to become

$$\hat{\mathbf{X}}_2(l|l) = \hat{\mathbf{X}}(l-2|l) \quad (7.266)$$

and so on, until the updated term

$$\hat{\mathbf{X}}_M(l|l) = \hat{\mathbf{X}}(l-M|l) \quad (7.267)$$

is obtained. This expression, the fixed lag estimate of $\mathbf{X}(l-M)$ based upon information contained in measurements $\mathbf{Z}(1)$ through $\mathbf{Z}(M)$, is the desired smoother output and, as predicted, is delayed by a factor of M behind the observation process.

The smoothing filter gain matrices are obtained from Equation (7.174) as applied to the augmented system of Equation (7.260). The result is

$$\begin{bmatrix} \boldsymbol{K}_0(l) \\ \hdashline \boldsymbol{K}_1(l) \\ \hdashline \boldsymbol{K}_2(l) \\ \hdashline \vdots \\ \hdashline \boldsymbol{K}_M(l) \end{bmatrix} = \begin{bmatrix} \boldsymbol{P}_{00}(l|l-1) & \boldsymbol{P}_{10}^T(l|l-1) & \boldsymbol{P}_{20}^T(l|l-1) & \cdots & \boldsymbol{P}_{M0}^T(l|l-1) \\ \boldsymbol{P}_{10}(l|l-1) & \boldsymbol{P}_{11}(l|l-1) & \boldsymbol{P}_{21}^T(l|l-1) & \cdots & \boldsymbol{P}_{M1}^T(l|l-1) \\ \boldsymbol{P}_{20}(l|l-1) & \boldsymbol{P}_{21}(l|l-1) & \boldsymbol{P}_{22}(l|l-1) & \cdots & \boldsymbol{P}_{M2}^T(l|l-1) \\ \cdots & \cdots & \cdots & \cdots & \cdots \\ \boldsymbol{P}_{M0}(l|l-1) & \boldsymbol{P}_{M1}(l|l-1) & \boldsymbol{P}_{M2}(l|l-1) & \cdots & \boldsymbol{P}_{MM}(l|l-1) \end{bmatrix}$$

$$
\cdot \begin{bmatrix} H^T(l) \\ \hdashline 0 \\ \hdashline 0 \\ \hdashline \vdots \\ \hdashline 0 \end{bmatrix} \Bigg[[H(l) \,\vdots\, 0 \,\vdots\, 0 \,\vdots\, \cdots \,\vdots\, 0]
$$

$$
\cdot \begin{bmatrix} P_{00}(l|l-1) & P_{10}^T(l|l-1) & P_{20}^T(l|l-1) & \cdots & P_{M0}^T(l|l-1) \\ P_{10}(l|l-1) & P_{11}(l|l-1) & P_{21}^T(l|l-1) & \cdots & P_{M1}^T(l|l-1) \\ P_{20}(l|l-1) & P_{21}(l|l-1) & P_{22}(l|l-1) & \cdots & P_{M2}^T(l|l-1) \\ \cdots & \cdots & \cdots & \cdots & \cdots \\ P_{M0}(l|l-1) & P_{M1}(l|l-1) & P_{M2}(l|l-1) & \cdots & P_{MM}(l|l-1) \end{bmatrix}
$$

$$
\cdot \begin{bmatrix} H^T(l) \\ \hdashline 0 \\ \hdashline 0 \\ \hdashline \vdots \\ \hdashline 0 \end{bmatrix} + R(l) \Bigg]^{-1} . \tag{7.268}
$$

Performing the indicated multiplication yields

$$
\begin{bmatrix} K_0(l) \\ \hdashline K_1(l) \\ \hdashline K_2(l) \\ \hdashline \vdots \\ \hdashline K_M(l) \end{bmatrix} = \begin{bmatrix} P_{00}(l|l-1)H^T(l) \\ \hdashline P_{10}(l|l-1)H^T(l) \\ \hdashline P_{20}(l|l-1)H^T(l) \\ \hdashline \vdots \\ \hdashline P_{M0}(l|l-1)H^T(l) \end{bmatrix} [H(l)P_{00}(l|l-1)H^T(l) + R(l)]^{-1}. \tag{7.269}
$$

Here again, the zero-subscripted terms represent the Kalman filter equations, and no further matrix inversion is required to determine the smoothing gain matrices.

The covariance equations are obtained from Equations (7.166) and (7.174), as applied to the system of Equation (7.260). The result is

$$\left[\begin{array}{c|c|c|c} \boldsymbol{P}_{00}(l|l-1) & \boldsymbol{P}_{10}^T(l|l-1) & \cdots & \boldsymbol{P}_{M0}^T(l|l-1) \\ \hline \boldsymbol{P}_{10}(l|l-1) & \boldsymbol{P}_{11}(l|l-1) & \cdots & \boldsymbol{P}_{M1}^T(l|l-1) \\ \hline \boldsymbol{P}_{20}(l|l-1) & \boldsymbol{P}_{21}(l|l-1) & \cdots & \boldsymbol{P}_{M2}^T(l|l-1) \\ \hline \cdots & \cdots & \cdots & \cdots \\ \hline \boldsymbol{P}_{M0}(l|l-1) & \boldsymbol{P}_{M1}(l|l-1) & \cdots & \boldsymbol{P}_{MM}(l|l-1) \end{array}\right]$$

$$= \left[\begin{array}{c|c|c|c|c} \boldsymbol{\Phi}(l, l-1) & \mathbf{0} & \cdots & \mathbf{0} & \mathbf{0} \\ \hline \boldsymbol{I} & \mathbf{0} & \cdots & \mathbf{0} & \mathbf{0} \\ \hline \mathbf{0} & \boldsymbol{I} & \cdots & \mathbf{0} & \mathbf{0} \\ \hline \cdots & \cdots & \cdots & \cdots & \cdots \\ \hline \mathbf{0} & \mathbf{0} & \cdots & \boldsymbol{I} & \mathbf{0} \end{array}\right]$$

$$\cdot \left[\begin{array}{c|c|c|c} \boldsymbol{P}_{00}(l-1|l-1) & \boldsymbol{P}_{10}^T(l-1|l-1) & \cdots & \boldsymbol{P}_{M0}^T(l-1|l-1) \\ \hline \boldsymbol{P}_{10}(l-1|l-1) & \boldsymbol{P}_{11}(l-1|l-1) & \cdots & \boldsymbol{P}_{M1}^T(l-1|l-1) \\ \hline \boldsymbol{P}_{20}(l-1|l-1) & \boldsymbol{P}_{21}(l-1|l-1) & \cdots & \boldsymbol{P}_{M2}^T(l-1|l-1) \\ \hline \cdots & \cdots & \cdots & \cdots \\ \hline \boldsymbol{P}_{M0}(l-1|l-1) & \boldsymbol{P}_{M1}(l-1|l-1) & \cdots & \boldsymbol{P}_{MM}(l-1|l-1) \end{array}\right]$$

$$\cdot \left[\begin{array}{c|c|c|c|c} \boldsymbol{\Phi}^T(l, l-1) & \boldsymbol{I} & \mathbf{0} & \cdots & \mathbf{0} \\ \hline \mathbf{0} & \mathbf{0} & \boldsymbol{I} & \cdots & \mathbf{0} \\ \hline \mathbf{0} & \mathbf{0} & \mathbf{0} & \cdots & \mathbf{0} \\ \hline \cdots & \cdots & \cdots & \cdots & \cdots \\ \hline \mathbf{0} & \mathbf{0} & \mathbf{0} & \cdots & \boldsymbol{I} \\ \hline \mathbf{0} & \mathbf{0} & \mathbf{0} & \cdots & \mathbf{0} \end{array}\right]$$

$$+ \left[\begin{array}{c} \boldsymbol{\Delta}(l, l-1) \\ \hline \mathbf{0} \\ \hline \mathbf{0} \\ \hline \vdots \\ \hline \mathbf{0} \end{array}\right] Q(l-1)[\boldsymbol{\Delta}^T(l, l-1) \mid \mathbf{0} \mid \mathbf{0} \mid \cdots \mid \mathbf{0}]. \tag{7.270}$$

Performing the indicated multiplications in this equation results in

$$
\left[\begin{array}{c|c|c|c}
P_{00}(l|l-1) & P_{10}^T(l|l-1) & \cdots & P_{M0}^T(l|l-1) \\ \hline
P_{10}(l|l-1) & P_{11}(l|l-1) & \cdots & P_{M1}^T(l|l-1) \\ \hline
P_{20}(l|l-1) & P_{21}(l|l-1) & \cdots & P_{M2}^T(l|l-1) \\ \hline
\cdots & \cdots & \cdots & \cdots \\ \hline
P_{M0}(l|l-1) & P_{M1}(l|l-1) & \cdots & P_{MM}(l|l-1)
\end{array}\right]
$$

$$
= \left[\begin{array}{c|c|c|c}
\Phi(l,l-1)P_{00}(l-1|l-1)\Phi^T(l,l-1) & \Phi(l,l-1)P_{00}(l-1|l-1) & \cdots & \Phi(l,l-1)P_{M-1,0}^T(l-1|l-1) \\ \hline
P_{00}(l-1|l-1)\Phi^T(l,l-1) & P_{00}(l-1|l-1) & \cdots & P_{M-1,0}^T(l-1|l-1) \\ \hline
P_{10}(l-1|l-1)\Phi^T(l,l-1) & P_{10}(l-1|l-1) & \cdots & P_{M-1,1}^T(l-1|l-1) \\ \hline
\cdots & \cdots & \cdots & \cdots \\ \hline
P_{M-1,0}(l-1|l-1)\Phi^T(l,l-1) & P_{M-1,0}(l-1|l-1) & \cdots & P_{M-1,M-1}(l-1|l-1)
\end{array}\right]
$$

$$
+ \left[\begin{array}{c|c|c|c}
\Delta(l,l-1)Q(l-1)\Delta^T(l,l-1) & \mathbf{0} & \cdots & \mathbf{0} \\ \hline
\mathbf{0} & \mathbf{0} & \cdots & \mathbf{0} \\ \hline
\mathbf{0} & \mathbf{0} & \cdots & \mathbf{0} \\ \hline
\cdots & \cdots & \cdots & \cdots \\ \hline
\mathbf{0} & \mathbf{0} & \cdots & \mathbf{0}
\end{array}\right]. \tag{7.271}
$$

The second expression, from Equation (7.174) is,

$$
\left[\begin{array}{c|c|c|c}
P_{00}(l|l) & P_{10}^T(l|l) & \cdots & P_{M0}^T(l|l) \\ \hline
P_{10}(l|l) & P_{11}(l|l) & \cdots & P_{M1}^T(l|l) \\ \hline
P_{20}(l|l) & P_{21}(l|l) & \cdots & P_{M2}^T(l|l) \\ \hline
\cdots & \cdots & \cdots & \cdots \\ \hline
P_{M0}(l|l) & P_{M1}(l|l) & \cdots & P_{MM}^T(l|l)
\end{array}\right]
$$

$$
= \left[I - \left[\begin{array}{c}
K_0(l) \\ \hline
K_1(l) \\ \hline
K_2(l) \\ \hline
\vdots \\ \hline
K_M(l)
\end{array}\right]
[H(l) \,|\, \mathbf{0} \,|\, \mathbf{0} \,|\, \cdots \,|\, \mathbf{0}]
\right.
$$

$$\cdot \left[\begin{array}{c|c|c|c} \boldsymbol{P}_{00}(l|l-1) & \boldsymbol{P}_{10}^T(l|l-1) & \cdots & \boldsymbol{P}_{M0}^T(l|l-1) \\ \hline \boldsymbol{P}_{10}(l|l-1) & \boldsymbol{P}_{11}(l|l-1) & \cdots & \boldsymbol{P}_{M1}^T(l|l-1) \\ \hline \boldsymbol{P}_{20}(l|l-1) & \boldsymbol{P}_{21}(l|l-1) & \cdots & \boldsymbol{P}_{M2}^T(l|l-1) \\ \hline \cdots & \cdots & \cdots & \cdots \\ \hline \boldsymbol{P}_{M0}(l|l-1) & \boldsymbol{P}_{M1}(l|l-1) & \cdots & \boldsymbol{P}_{MM}(l|l-1) \end{array}\right]\Bigg]. \tag{7.272}$$

Performing the indicated multiplication in this expression produces

$$\left[\begin{array}{c|c|c|c} \boldsymbol{P}_{00}(l|l) & \boldsymbol{P}_{10}^T(l|l) & \cdots & \boldsymbol{P}_{M0}^T(l|l) \\ \hline \boldsymbol{P}_{10}(l|l) & \boldsymbol{P}_{11}(l|l) & \cdots & \boldsymbol{P}_{M1}^T(l|l) \\ \hline \boldsymbol{P}_{20}(l|l) & \boldsymbol{P}_{21}(l|l) & \cdots & \boldsymbol{P}_{M2}^T(l|l) \\ \hline \cdots & \cdots & \cdots & \cdots \\ \hline \boldsymbol{P}_{M0}(l|l) & \boldsymbol{P}_{M1}(l|l) & \cdots & \boldsymbol{P}_{MM}(l|l) \end{array}\right]$$

$$= \left[\begin{array}{c|c|c} [\boldsymbol{I} - \boldsymbol{K}_0(l)\boldsymbol{H}(l)]\boldsymbol{P}_{00}(l|l-1) & [\boldsymbol{I} - \boldsymbol{K}_0(l)]\boldsymbol{P}_{00}(l|l-1) & \cdots \\ \hline \boldsymbol{P}_{10}(l|l-1) - \boldsymbol{K}_1(l)\boldsymbol{H}(l)\boldsymbol{P}_{00}(l|l-1) & \boldsymbol{P}_{11}(l|l-1) - \boldsymbol{K}_1(l)\boldsymbol{H}(l)\boldsymbol{P}_{10}^T(l|l-1) & \cdots \\ \hline \boldsymbol{P}_{20}(l|l-1) - \boldsymbol{K}_2(l)\boldsymbol{H}(l)\boldsymbol{P}_{00}(l|l-1) & \boldsymbol{P}_{22}(l|l-1) - \boldsymbol{K}_2(l)\boldsymbol{H}(l)\boldsymbol{P}_{10}^T(l|l-1) & \cdots \\ \hline \cdots & \cdots & \cdots \\ \hline \boldsymbol{P}_{M-1,0}(l|l-1) - \boldsymbol{K}_M(l)\boldsymbol{H}(l)\boldsymbol{P}_{00}(l|l-1) & \boldsymbol{P}_{M-1,1}(l|l-1) - \boldsymbol{K}_M(l)\boldsymbol{H}(l)\boldsymbol{P}_{10}^T(l|l-1) & \cdots \end{array}\right]. \tag{7.273}$$

While these equations seem complex, it can be seen from Equations (7.263) and (7.269) that only the left-hand column of submatrices in this expression need actually be computed. Furthermore, the apparent asymmetry in Equation (7.273) can be resolved by substitution for the smoothing gain matrices from Equation (7.269). Then, the equations resulting from the left-hand column of submatrices are

$$\begin{aligned} \boldsymbol{P}_{00}(l|l) &= [\boldsymbol{I} - \boldsymbol{K}_0(l)\boldsymbol{H}(l)]\boldsymbol{P}_{00}(l|l-1) \\ \boldsymbol{P}_{10}(l|l) &= \boldsymbol{P}_{10}(l|l-1)[\boldsymbol{I} - \boldsymbol{H}^T(l)\boldsymbol{K}_0^T(l)] \\ \boldsymbol{P}_{20}(l|l) &= \boldsymbol{P}_{20}(l|l-1)[\boldsymbol{I} - \boldsymbol{H}^T(l)\boldsymbol{K}_0^T(l)] \\ &\vdots \\ \boldsymbol{P}_{M0}(l|l) &= \boldsymbol{P}_{M0}(l|l-1)[\boldsymbol{I} - \boldsymbol{H}^T(l)\boldsymbol{K}_0^T(l)]. \end{aligned} \tag{7.274}$$

From Equation (7.271), the left-hand column of submatrices produces the equations

$$\begin{aligned}
P_{00}(l|l-1) &= \Phi(l, l-1)P_{00}(l-1|l-1)\Phi^T(l, l-1)\\
P_{10}(l|l-1) &= P_{00}(l-1|l-1)\Phi^T(l, l-1)\\
P_{20}(l|l-1) &= P_{10}(l-1|l-1)\Phi^T(l, l-1)\\
&\vdots\\
P_{M0}(l|l-1) &= P_{M-1,0}(l-1|l-1)\,\Phi^T(l, l-1).
\end{aligned} \tag{7.275}$$

These two groups of equations are used recursively to form the smoothing covariance matrices. The zero-subscripted equations are those constituting the Kalman filter algorithm.

The smoothing gain matrices are obtained from Equation (7.269), repeated here in component form:

$$\begin{aligned}
K_0(l) &= P_{00}(l|l-1)H^T(l)[H(l)P_{00}(l|l-1)H^T(l) + R(l)]^{-1}\\
K_1(l) &= P_{10}(l|l-1)H^T(l)[H(l)P_{00}(l|l-1)H^T(l) + R(l)]^{-1}\\
K_2(l) &= P_{20}(l|l-1)H^T(l)[H(l)P_{00}(l|l-1)H^T(l) + R(l)]^{-1}\\
&\vdots\\
K_M(l) &= P_{M0}(l|l-1)H^T(l)[H(l)P_{00}(l|l-1)H^T(l) + R(l)]^{-1}.
\end{aligned} \tag{7.276}$$

Finally, the smoothed estimates themselves are given by Equation (7.263), repeated here in component form and in the notation of Equations (7.265) through (7.267).

$$\begin{aligned}
\hat{\mathbf{X}}(l|l) &= \hat{\mathbf{X}}(l|l-1) + K_0(l)[\mathbf{Z}(l) - H(l)\hat{\mathbf{X}}(l|l-1)]\\
\hat{\mathbf{X}}(l-1|l) &= \hat{\mathbf{X}}(l-1|l-1) + K_1(l)[\mathbf{Z}(l) - H(l)\hat{\mathbf{X}}(l|l-1)]\\
\hat{\mathbf{X}}(l-2|l) &= \hat{\mathbf{X}}(l-2|l-1) + K_2(l)[\mathbf{Z}(l) - H(l)\hat{\mathbf{X}}(l|l-1)]\\
&\vdots\\
\hat{\mathbf{X}}(l-M|l) &= \hat{\mathbf{X}}(l-M|l-1) + K_M(l)[\mathbf{Z}(l) - H(l)\hat{\mathbf{X}}(l|l-1)].
\end{aligned} \tag{7.277}$$

Notice that the prior estimate on the right-hand side of each equation is the old value of the posterior estimate from the left-hand side of the equation immediately above.

The Kalman fixed lag smoothing algorithm is then summarized by Equations

(7.274) through (7.277). The complexity is increased considerably over that of the Kalman filter algorithm, and storage requirements have increased by a multiplicative factor of nearly M. A block diagram implementation of the smoothing algorithm is shown in Figure 7-7, where it is indicated that the algorithm consists

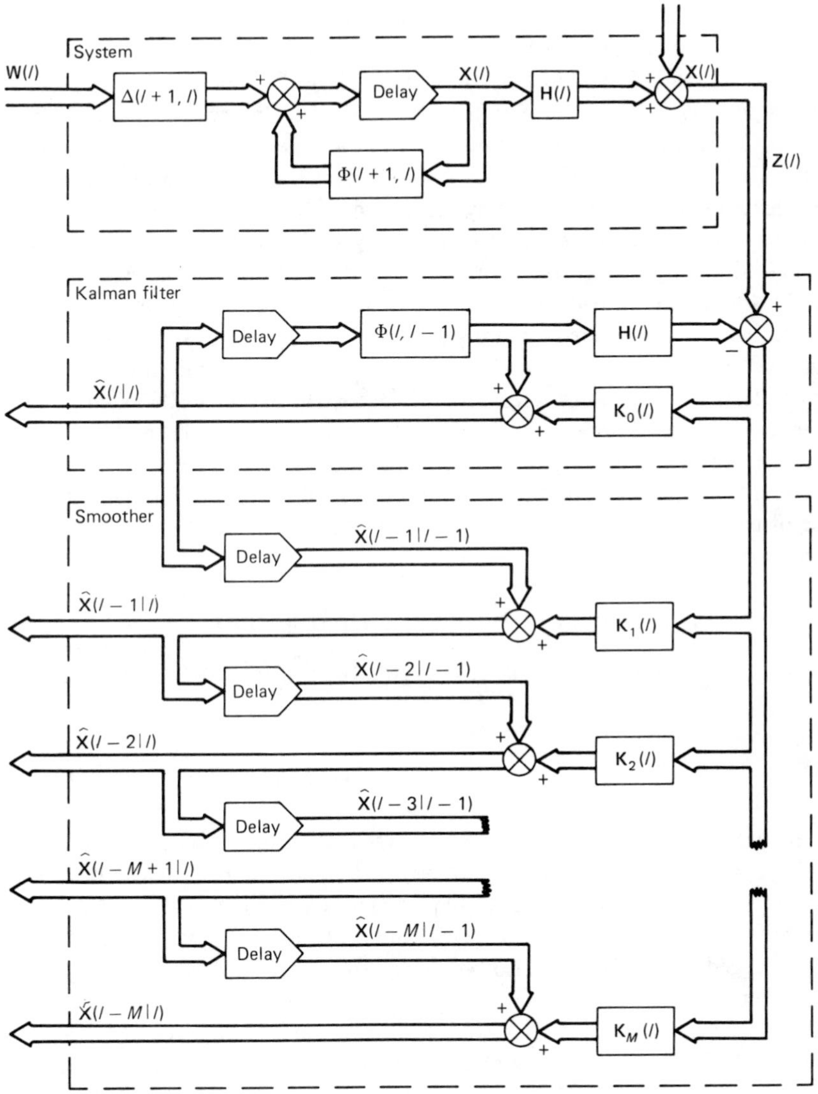

Figure 7-7. Fixed lag smoothing.

of a Kalman filter and a set of smoothing computations which are driven by the Kalman filter output and the measurement residuals, $\mathbf{Z}(l) - \boldsymbol{H}(l)\hat{\mathbf{X}}(l|l-1)$. The computation of the smoother gain matrices is not shown; these matrices are obtained from implementation of the recursive algorithm given by Equations (7.274) and (7.275). Normally, the only output of interest is the lagged estimate $\hat{\mathbf{X}}(l-M|l)$, but it can be seen that the estimates at all other lag values are also available without additional computation.

There are other approaches to the fixed lag smoothing problem which yield only the Kalman filter estimate $\hat{\mathbf{X}}(l|l)$ and the lagged estimate $\hat{\mathbf{X}}(l-M|l)$. These techniques involve two correction terms applied to the prior estimate. One is related to the Kalman filter gain matrix, and the other to the combination of the various gain matrices and time delays of Figure 7-7. The approach presented here is fairly lucid and presents a systematic view of the smoothing process.

There is one other aspect of the smoothing algorithm presented by Equations (7.274) through (7.277) which should be considered. One of the characteristics of the Kalman filter algorithm is that the estimation error covariance matrix is determined in the estimation process, a very useful feature which permits real time error analysis and evaluation. Examination of the smoothing algorithm indicates that an analogous feature does not exist, except in the Kalman filter component. In fact, except for $\boldsymbol{P}_{00}$, the matrices in Equations (7.274) and (7.275) are not even symmetric. If estimation error covariance matrices are desired, they can be computed independently of the smoothing algorithm by use of expressions obtained from Equations (7.271) and (7.273). The diagonal terms other than $\boldsymbol{P}_{00}$ yield

$$\begin{aligned}
\boldsymbol{P}_{11}(l|l) &= \boldsymbol{P}_{00}(l-1|l-1) - \boldsymbol{K}_1(l)\boldsymbol{H}(l)\boldsymbol{P}_{10}^T(l|l-1)\\
\boldsymbol{P}_{22}(l|l) &= \boldsymbol{P}_{11}(l-1|l-1) - \boldsymbol{K}_2(l)\boldsymbol{H}(l)\boldsymbol{P}_{20}^T(l|l-1)\\
&\vdots\\
\boldsymbol{P}_{MM}(l|l) &= \boldsymbol{P}_{M-1,M-1}(l-1|l-1) - \boldsymbol{K}_M(l)\boldsymbol{H}(l)\boldsymbol{P}_{M0}^T(l|l-1). \qquad (7.278)
\end{aligned}$$

That these matrices are symmetric can be verified by substituting from Equation (7.276) for the smoothing gain matrices. This recursion relation can be used to determine the estimation error covariance matrix associated with the lagged estimate $\hat{\mathbf{X}}(l-M|l)$, which is $\boldsymbol{P}_{MM}(l|l)$. Note that the covariance matrices for all other lagged estimates must also be computed in order to provide the required recursion.

Before leaving the fixed lag smoothing algorithm, it is appropriate to discuss the effect produced on the estimation error variances as the lag interval M is varied. While it is difficult to assess this effect in general, some insight can be obtained by substituting the prior smoothing covariance matrices, as given by

Equation (7.275), into the expression of Equation (7.274). The result is

$$\begin{aligned}
\boldsymbol{P}_{10}(l|l) &= \boldsymbol{P}_{00}(l-1|l-1)\boldsymbol{\Phi}^T(l,l-1)[\boldsymbol{I}-\boldsymbol{H}^T(l)\boldsymbol{K}_0^T(l)]\\
\boldsymbol{P}_{20}(l|l) &= \boldsymbol{P}_{10}(l-1|l-1)\boldsymbol{\Phi}^T(l,l-1)[\boldsymbol{I}-\boldsymbol{H}^T(l)\boldsymbol{K}_0^T(l)]\\
&\vdots\\
\boldsymbol{P}_{M0}(l|l) &= \boldsymbol{P}_{M-1,0}(l-1|l-1)\boldsymbol{\Phi}^T(l,l-1)[\boldsymbol{I}-\boldsymbol{H}^T(l)\boldsymbol{K}_0^T(l)].
\end{aligned} \tag{7.279}$$

From the recursion relationship implied by these equations, it can be seen that the matrices on the left-hand side tend to diminish at a rate determined by the dominant eigenvalues of the matrix

$$\boldsymbol{\Phi}^T(l,l-1)[\boldsymbol{I}-\boldsymbol{H}^T(l)\boldsymbol{K}_0^T(l)] \tag{7.280}$$

or, equivalently, of its transpose

$$[\boldsymbol{I}-\boldsymbol{K}_0(l)\boldsymbol{H}(l)]\boldsymbol{\Phi}(l,l-1). \tag{7.281}$$

Since difference equations are involved here, eigenvalues with real parts near 1 tend to produce slow time responses, while those with real parts near 0 tend to produce rapid responses. This effect can be roughly described by defining the time constants of these equations by

$$\tau_i = 1/\ln \lambda_i \tag{7.282}$$

where the λ_i are the eigenvalues of the matrix of Equation (7.281). As a general rule, smoothing over intervals longer than three times the dominant time constant yields little additional information. The same general conclusion can be reached with respect to fixed point smoothing, based upon use of Equations (7.252) through (7.254). More specific consideration is given to this concept in Section 7.3.4, where time invariant smoothing is discussed.

7.3.3. Fixed Interval Smoothing

When estimates of the state vector $\mathbf{X}(l)$ over the interval $(0, L)$ are desired, and every estimate is to be based on all measurements $\mathbf{Z}(1)$ through $\mathbf{Z}(L)$, then the estimate desired is $\hat{\mathbf{X}}(l|L)$. The most direct approach is to consider the problem as a fixed lag smoothing problem with $M = L$. After processing the entire set of

L measurements, the smoothed estimates given by Equation (7.277) would be

$$\begin{gathered}\hat{\mathbf{X}}(L|L)\\ \hat{\mathbf{X}}(L-1|L)\\ \hat{\mathbf{X}}(L-2|L)\\ \vdots\\ \hat{\mathbf{X}}(0|L)\end{gathered} \tag{7.283}$$

which is precisely the desired result. The difficulty with this approach is that L may well be very large, so that storage of the intermediate estimates and covariance matrices may become a problem.

An alternative approach to the fixed interval smoothing problem is to use a Kalman filter to produce the filtered estimates $\hat{\mathbf{X}}(l|l)$ over the interval $(0, L)$, and then a backward pass through the data to adjust each filtered estimate to account for subsequent measurements. The algorithm is given here without development. After a forward pass through the data using the Kalman filter algorithm and storage of all filtered estimates $\hat{\mathbf{X}}(l|l)$, the smoothing equation is

$$\begin{aligned}\hat{\mathbf{X}}(l|L) &= \hat{\mathbf{X}}(l|l) + \boldsymbol{A}(l)[\hat{\mathbf{X}}(l+1|L) - \hat{\mathbf{X}}(l+1|l)\\ \boldsymbol{A}(l) &= \boldsymbol{P}(l|l)\boldsymbol{\Phi}^T(l+1, l)\boldsymbol{P}^{-1}(l+1|l).\end{aligned} \tag{7.284}$$

This equation is initialized at t_L with the final result of the Kalman filter equations, $\hat{\mathbf{X}}(L|L)$. The smoothing estimation error covariance matrix is not required in the above equation, but it can be computed recursively from the expression

$$\boldsymbol{P}(l|L) = \boldsymbol{P}(l|l) + \boldsymbol{A}(l)[\boldsymbol{P}(l+1|L) - \boldsymbol{P}(l+1|l)]\boldsymbol{A}^T(l) \tag{7.285}$$

which is an expression for the estimation error covariance matrix associated with the smoothed estimate $\hat{\mathbf{X}}(l|L)$.

The use of these two equations requires the matrix $\boldsymbol{P}(l|l)$, which is computed recursively in the forward pass through the data using the Kalman filter algorithm. This requirement implies that this sequence of matrices be either stored along with the corresponding estimates $\hat{\mathbf{X}}(l|l)$ or recomputed during the backward pass through the data. Storage is often impractical when L is large, for the same reasons that the fixed lag smoothing algorithm is impractical, so that recomputation is usually the preferred approach. The applicable equations for such recomputation are

$$P(l|l-1) = P(l|l) + P(l|l)H^T(l)[H(l)P(l|l)H^T(l) - P(l|l)]^{-1}H(l)P(l|l)$$

$$P(l-1|l-1) = \Phi(l-1,l)[P(l|l-1) - \Delta(l,l-1)Q(l-1)\Delta^T(l,l-1)]\Phi^T(l-1,l) \tag{7.286}$$

and are merely the Kalman filter equations solved for $P(l|l-1)$ and $P(l-1|l-1)$, given $P(l|l)$. These equations must be used with caution, however, since they tend toward instability. A third and perhaps more practical approach to the fixed interval smoothing problem is the use of a suboptimal smoothing algorithm based on a fixed lag smoother using a lag interval M much smaller than L. The reasoning is heuristic and is based on the concept of the time constant of the Kalman filter discussed in the previous section. If M is chosen so that it corresponds to a lag interval which is several times larger than the dominant time constant of the filter, then little additional benefit is obtained by using the full observation interval $(0, L)$ as the lag interval. Quite often, the filter time constant is much smaller than the available observation interval, and this approach can be effectively used.

7.3.4. Time Invariant Smoothing

When the system equations are time invariant, and the input and measurement noise statistics are stationary, the system model is

$$\begin{aligned} \mathbf{X}(l) &= \Phi\mathbf{X}(l-1) + \Delta\mathbf{W}(l) + \Gamma u(l) \\ \mathbf{Z}(l) &= H\mathbf{X}(l) + \mathbf{V}(l) \\ Q(l) &= Q; \qquad R(l) = R. \end{aligned} \tag{7.287}$$

Thus, the matrices Φ, Δ, H, Q, and R in the smoothing equations are constant. As before, the deterministic input $u(l)$ can be ignored without loss of generality.

The fixed point smoothing algorithm contains a recursive computation of the covariance matrices, as described by Equations (7.252) through (7.256). When the system is time invariant, the Kalman filter covariance equations may reach steady state conditions with increasing l, so that in these equations

$$\begin{aligned} P_{00}(l|l) &= P_{00}(l-1|l-1) = P_{00} = [I - K_0H]\bar{P}_{00} \\ K_0(l) &= K_0; \qquad P_{00}(l|l-1) = P_{00}(l-1,l-2) = \bar{P}_{00} \end{aligned} \tag{7.288}$$

where P_{00}, $\bar{P}_{00}$ and K_0 are constant matrices. This situation has been discussed previously in Section 7.2.5. In the event that this steady state condition in the Kalman filter equations occurs, then the $\boldsymbol{P}_{10}$ expressions from Equations (7.253) and (7.254) are

$$\boldsymbol{P}_{10}(l|l-1) = \boldsymbol{P}_{10}(l-1|l-1)\boldsymbol{\Phi}^T$$
$$\boldsymbol{P}_{10}(l|l) = \boldsymbol{P}_{10}(l|l-1)(\boldsymbol{I} - \boldsymbol{H}^T\boldsymbol{K}_0^T). \tag{7.289}$$

Combining these equations yields

$$\boldsymbol{P}_{10}(l|l) = \boldsymbol{P}_{10}(l-1|l-1)\boldsymbol{\Phi}^T(\boldsymbol{I} - \boldsymbol{H}^T\boldsymbol{K}_0^T) \tag{7.290}$$

from which it can be seen that $\boldsymbol{P}_{10}(l|l)$ is not necessarily constant with varying l, even though $\boldsymbol{P}_{00}(l|l)$ and $\boldsymbol{K}_0(l)$ are. This conclusion also applies to the smoothing gain matrix $\boldsymbol{K}_1(l)$ since, from Equation (7.252),

$$\boldsymbol{K}_1(l) = \boldsymbol{P}_{10}(l|l-1)\boldsymbol{H}^T(\boldsymbol{H}\bar{\boldsymbol{P}}_{00}\boldsymbol{H}^T + \boldsymbol{R})^{-1}. \tag{7.291}$$

Since Equation (7.290) is a time invariant difference equation, its response characteristics are determined by the dominant eigenvalues of the matrix

$$\boldsymbol{\Phi}^T(\boldsymbol{I} - \boldsymbol{H}^T\boldsymbol{K}_0^T). \tag{7.292}$$

This result in turn implies that the smoothing efficiency decreases with l in such a manner that essentially no smoothing is obtained after two or three dominant time constants.

For fixed lag smoothing, which also includes the Kalman filter algorithm involving zero-subscripted variables, steady state conditions may also be reached by the Kalman filter covariance and gain matrices, resulting in a condition analogous to that implied by Equation (7.289). The effect which this situation has on the smoothing covariance matrices can be seen by examining Equations (7.274) and (7.275). For example, the $\boldsymbol{P}_{10}$ matrices are governed by the recursion equations

$$\boldsymbol{P}_{10}(l|l-1) = \boldsymbol{P}_{00}\boldsymbol{\Phi}^T = \bar{\boldsymbol{P}}_{10} \tag{7.293}$$

$$\boldsymbol{P}_{10}(l|l) = \boldsymbol{P}_{10}(l|l-1)(\boldsymbol{I} - \boldsymbol{H}^T\boldsymbol{K}_0^T)$$
$$= \bar{\boldsymbol{P}}_{10}(\boldsymbol{I} - \boldsymbol{H}^T\boldsymbol{K}_0^T) = \boldsymbol{P}_{10} \tag{7.294}$$

where $\bar{\boldsymbol{P}}_{10}$ and $\boldsymbol{P}_{10}$ are constant matrices. This result implies that $\boldsymbol{P}_{10}(l|l-1)$ and $\boldsymbol{P}_{10}(l|l)$ reach steady state conditions along with $\boldsymbol{P}_{00}(l|l)$. Then applying this same concept to the $\boldsymbol{P}_{20}$ matrices yields

$$\boldsymbol{P}_{20}(l|l-1) = \boldsymbol{P}_{10}\boldsymbol{\Phi}^T = \bar{\boldsymbol{P}}_{10}(\boldsymbol{I} - \boldsymbol{H}^T\boldsymbol{K}_0^T)\boldsymbol{\Phi}^T = \bar{\boldsymbol{P}}_{20} \tag{7.295}$$

$$\boldsymbol{P}_{20}(l|l) = \bar{\boldsymbol{P}}_{20}(\boldsymbol{I} - \boldsymbol{H}^T\boldsymbol{K}_0^T) = \boldsymbol{P}_{20} \tag{7.296}$$

where $\bar{\boldsymbol{P}}_{20}$ and $\boldsymbol{P}_{20}$ are constant matrices.

The same reasoning applies to all smoothing covariance matrices of Equation (7.275), further implying that the smoothing gain matrices of Equation (7.276) also reach a constant value. Furthermore, this constant value is entirely determined by the steady state value $\bar{\boldsymbol{P}}_{00}$ reached by the Kalman filter covariance matrix, which can be seen by substituting from Equations (7.292) through (7.296) into the expressions for the filter gains contained in Equation (7.276). These results, extended to the general case, are

$$\begin{aligned}
\boldsymbol{K}_0 &= \bar{\boldsymbol{P}}_{00}\boldsymbol{H}^T(\boldsymbol{H}\bar{\boldsymbol{P}}_{00}\boldsymbol{H}^T + \boldsymbol{R})^{-1}\\
\boldsymbol{K}_1 &= \bar{\boldsymbol{P}}_{10}\boldsymbol{H}^T(\boldsymbol{H}\bar{\boldsymbol{P}}_{00}\boldsymbol{H}^T + \boldsymbol{R})^{-1}\\
\boldsymbol{K}_2 &= \bar{\boldsymbol{P}}_{20}\boldsymbol{H}^T(\boldsymbol{H}\bar{\boldsymbol{P}}_{00}\boldsymbol{H}^T + \boldsymbol{R})^{-1}\\
\boldsymbol{K}_3 &= \bar{\boldsymbol{P}}_{30}\boldsymbol{H}^T(\boldsymbol{H}\bar{\boldsymbol{P}}_{00}\boldsymbol{H}^T + \boldsymbol{R})^{-1}\\
&\vdots\\
\boldsymbol{K}_M &= \bar{\boldsymbol{P}}_{M0}\boldsymbol{H}^T(\boldsymbol{H}\bar{\boldsymbol{P}}_{00}\boldsymbol{H}^T + \boldsymbol{R})^{-1}
\end{aligned} \tag{7.297}$$

where the smoothing covariance matrices are given by the recursion relation

$$\begin{aligned}
\bar{\boldsymbol{P}}_{10} &= \boldsymbol{P}_{00}\boldsymbol{\Phi}^T = (\boldsymbol{I} - \boldsymbol{K}_0\boldsymbol{H})\bar{\boldsymbol{P}}_{00}\boldsymbol{\Phi}^T = \bar{\boldsymbol{P}}_{00}(\boldsymbol{I} - \boldsymbol{H}^T\boldsymbol{K}_0^T)\boldsymbol{\Phi}^T\\
\bar{\boldsymbol{P}}_{20} &= \bar{\boldsymbol{P}}_{10}(\boldsymbol{I} - \boldsymbol{H}^T\boldsymbol{K}_0^T)\boldsymbol{\Phi}^T\\
\bar{\boldsymbol{P}}_{30} &= \bar{\boldsymbol{P}}_{20}(\boldsymbol{I} - \boldsymbol{H}^T\boldsymbol{K}_0^T)\boldsymbol{\Phi}^T\\
&\vdots\\
\bar{\boldsymbol{P}}_{M0} &= \bar{\boldsymbol{P}}_{M-1,0}(\boldsymbol{I} - \boldsymbol{H}^T\boldsymbol{K}_0^T)\boldsymbol{\Phi}^T.
\end{aligned} \tag{7.298}$$

Thus, starting with only the value of $\bar{\boldsymbol{P}}_{00}$, all filter gain matrices can be computed. The covariance equations are time invariant, linear, matrix difference equations whose response characteristics are determined by the eigenvalues of the matrix

$$(\boldsymbol{I} - \boldsymbol{H}^T\boldsymbol{K}_0^T)\boldsymbol{\Phi}^T. \tag{7.299}$$

The dominant eigenvalue of this matrix determines the reduction of error variance attained with various lag intervals. Virtually all reduction will be obtained when the lag interval corresponds to three time constants as determined from the eigenvalues of this matrix.

The smoothing estimation error covariance matrices can be obtained from Equation (7.278), which, by the same arguments presented above, in this case yield the recursion relations

$$\begin{aligned}
\boldsymbol{P}_{11} &= \boldsymbol{P}_{00} - \boldsymbol{K}_1 \boldsymbol{H} \bar{\boldsymbol{P}}_{10}^T \\
\boldsymbol{P}_{22} &= \boldsymbol{P}_{11} - \boldsymbol{K}_2 \boldsymbol{H} \bar{\boldsymbol{P}}_{20}^T \\
\boldsymbol{P}_{33} &= \boldsymbol{P}_{22} - \boldsymbol{K}_3 \boldsymbol{H} \bar{\boldsymbol{P}}_{30}^T \\
&\vdots \\
\boldsymbol{P}_{MM} &= \boldsymbol{P}_{M-1,\,M-1} - \boldsymbol{K}_M \boldsymbol{H} \bar{\boldsymbol{P}}_{M0}^T .
\end{aligned} \tag{7.300}$$

From consideration of Equations (7.297) and (7.298), it is evident that these matrices are determined completely by the steady state Kalman filter covariance matrix $\bar{\boldsymbol{P}}_{00}$. The error reducing efficiency of the smoothing algorithm can be seen quite readily by examining the smoothing estimation error covariance matrix $\boldsymbol{P}_{MM}$.

The fact that the fixed lag smoothing equations are time invariant is of practical importance, since these equations then provide a systematic approach to smoother design for systems which operate over extended time periods.

7.4. THE LINEAR STOCHASTIC CONTROL PROBLEM

In the introductory remarks of Section 7.1, it was noted that estimation processes are important to feedback control. The state variable feedback concepts of Chapter 5 and the optimal control concepts of Chapter 6 result in feedback control functions obtained from the state vector by linear transformations. When the state vector itself is not accessible in a deterministic system, then an observer of the kind considered in Chapter 5 can be used. In a stochastic system, the analogous result is the use of a state estimator, or filter, to provide an estimate of the system state vector for feedback purposes. Following the concept suggested for deterministic systems, the stochastic control system would consist of a stochastic state estimator, followed by a controller designed as though the system were deterministic and the entire state vector were available for state variable feedback. The controller itself may be an optimal controller, as discussed in Chapter 6, or it may be designed for some specific purpose–such as decoupling as described in Chapter 5. If the Kalman filter is used as the state

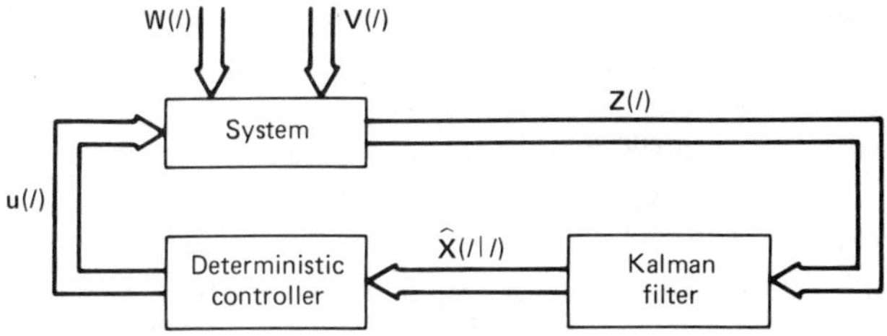

Figure 7-8. A stochastic control system using the Kalman filter as the state estimator.

estimator, then the control system appears something like that depicted in Figure 7-8.

The question of whether a system should be treated as deterministic or stochastic is one of degree. Since all systems are stochastic to some extent, the question is whether or not the benefit derived from stochastic modeling is justified in the light of the increased complexity and computational requirements inherent in state estimators such as the Kalman filter. The same kind of question arises in considering optimal control–is the increased complexity justified by the gain in the performance? In many cases, the answer is yes. In fact, in some instances effective control may be impossible to achieve without a stochastic state estimator. In others, the use of stochastic techniques serves to improve the performance of a system already functional using only a deterministic controller design. It is also possible–in those systems in which the random influence is relatively small–that stochastic state estimation may produce only minor improvements in performance, or may even be counterproductive because of higher susceptibility to numerical problems.

Some conceptual appreciation of the designer's choice in whether a system is treated as deterministic or stochastic can be gained by considering a time invariant linear regulator problem. In both deterministic and stochastic treatment of the problem, the state feedback gain matrix is implied by Equation (6.128)

$$\boldsymbol{C} = -\boldsymbol{B}^{-1}\boldsymbol{G}^T\boldsymbol{S} \tag{7.301}$$

where $\boldsymbol{S}$ is the steady state solution of the Riccati equation. In the deterministic treatment of the problem, the optimal control input vector is obtained as

$$\boldsymbol{u}(t) = \boldsymbol{C}\hat{\boldsymbol{x}}(t) \tag{7.302}$$

and $\hat{\boldsymbol{x}}(t)$ is the observer estimate obtained from Equation (5.85),

$$\dot{\hat{\boldsymbol{x}}}(t) = \boldsymbol{F}\hat{\boldsymbol{x}}(t) + \boldsymbol{G}\boldsymbol{u}(t) + \boldsymbol{K}[\boldsymbol{y}(t) - \boldsymbol{H}\hat{\boldsymbol{x}}(t)]. \tag{7.303}$$

In this expression, $\boldsymbol{K}$ is the matrix of observer gains, and is completely determined by the system parameters and the assigned eigenstructure of the observer. As shown in Section 5.2.1, the observer response characteristics can be assigned independently of the closed-loop system response characteristics produced by the optimum feedback gain matrix $\boldsymbol{C}$.

On the other hand, if the system is treated as being stochastic, then the optimal control input vector is still obtained from Equation (7.302), but now in the form

$$\boldsymbol{u}(t) = \boldsymbol{C}\hat{\mathbf{X}}(t) \tag{7.304}$$

where the upper case letter denotes an estimate of the random process $\mathbf{X}(t)$. The estimate $\hat{\mathbf{X}}(t)$, in this case, is obtained from a Kalman filter. The appropriate expression is obtained from Equation (7.220) applied to the time invariant case under steady state conditions, or from Equation (7.232) modified to account for a deterministic input function $\boldsymbol{u}(t)$,

$$\dot{\hat{\mathbf{X}}} = \boldsymbol{F}\hat{\mathbf{X}}(t) + \boldsymbol{G}\boldsymbol{u}(t) + \boldsymbol{K}[\mathbf{Z}(t) - \boldsymbol{H}\hat{\mathbf{X}}(t)]. \tag{7.305}$$

The differences between this expression and that of Equation (7.303) are that the gain matrix $\boldsymbol{K}$ in this expression is the steady state Kalman gain matrix, and that the measurement function is the noise contaminated measurement $\mathbf{Z}(t)$. While the observer gain matrix of Equation (7.303) is determined by the designers assignment of observer eigenstructure, the Kalman gain matrix of Equation (7.305) is determined solely by the system parameters and the covariance matrices of the input and measurement noise. This fact is clearly established by examination of equations (7.231) and (7.233).

In this particular case, if the system is considered deterministic, then the observer gain matrix $\boldsymbol{K}$ of Equation (7.303) can be prescribed to ensure an observer eigenstructure compatible with that produced by the optimal regulator gain matrix $\boldsymbol{C}$. If appreciable driving noise and/or measurement noise are present, however, observer performance may be far from that implied by the theory. On the other hand, if the system is considered stochastic, then the Kalman filter estimate is the best obtainable under the existing circumstances, but filter eigenstructure may not be entirely compatible with that of the optimal closed loop system. For this reason, systems with optimal designs may sometimes exhibit undesirable transient characteristics.

To illustrate these concepts, the spring-mass-damper system treated in Sections 4.2 and 6.3.2 will be considered. The system is described by the matrices

$$\boldsymbol{F} = \begin{bmatrix} 0 & 1 \\ -5 & -2 \end{bmatrix}; \quad \boldsymbol{G} = \begin{bmatrix} 0 \\ \frac{1}{2} \end{bmatrix}.$$

In addition, suppose that the measured quantity is the position of the mass, yielding for the measurement equation

$$y(t) = x_2(t)$$

or

$$\boldsymbol{H} = [0 \quad 1].$$

This is a practical problem, since certain inertial measurement devices can be adequately represented by the spring-mass-damper model. If the measurement noise and driving noise can be considered negligible, then a state variable feedback control system can be implemented by use of an observer, designed in accordance with the developments of Section 5.2. While a reduced-order observer can obviously be developed, a full-order observer will be used here for illustrative purposes.

Suppose that it is desired to use an optimal linear regulator with performance index

$$J = \int_{t_0}^{t_f} [56x_1^2 + 24x_2^2 + (u^2/4)]\ dt.$$

This performance index is chosen by the designer to produce some desired characteristic. From the performance index specifications, the $\boldsymbol{A}$ and $\boldsymbol{B}$ matrices of Equation (6.125) are

$$\boldsymbol{A} = \begin{bmatrix} 56 & 0 \\ 0 & 24 \end{bmatrix}; \qquad B = [\tfrac{1}{4}].$$

This optimal regulator problem was solved in Section 6.3.2, where the resulting optimal gain matrix was determined to be

$$\boldsymbol{C} = [-8 \quad -8]$$

The resulting closed-loop system was critically damped with eigenvalues of -3.

The optimal regulator design presumes the availability of all state variables for feedback purposes, but in this case only the position variable $x_1(t)$ can be measured. Suppose that the observer of Equation (5.85) is to be used to produce an estimate of the state vector, and that the observer response is to be critically damped with eigenvalues of -5. The procedures of Section 5.2 can be used tc determine the observer gain matrix $\boldsymbol{K}$ of Equation (5.85), with the result

$$\boldsymbol{K} = \begin{bmatrix} -4 \\ 8 \end{bmatrix}.$$

Since the observer response is much faster than that of the closed-loop system, the argument is that near optimal response is produced. The overall system block diagram is shown in Figure 7-9, where the alternative form of Equation (5.85),

$$\dot{\hat{\boldsymbol{x}}}(t) = [\boldsymbol{F} - \boldsymbol{KH}]\hat{\boldsymbol{x}}(t) + \boldsymbol{G}\boldsymbol{u}(t) + \boldsymbol{K}\boldsymbol{y}(t)$$

has been used. The input noise $\mathbf{W}(t)$ and measurement noise $\mathbf{V}(t)$ are presumed negligible.

On the other hand, if these noise processes are significant, then the estimation of the state vector by means of an observer is no longer a reasonable approach, and statistical estimation must be used. Since this is a regulator problem, the steady state Kalman filter equations–together with the algebraic Riccati equation of Equation (7.231)–can be used. Suppose that the variance of the input noise is 5.64, and that of the measurement noise is 1. Then the terms in Equation (7.231) are

$$\boldsymbol{F} = \begin{bmatrix} 0 & 1 \\ -5 & -2 \end{bmatrix}; \quad Q = [5.64]; \quad R = [1]; \quad \boldsymbol{D} = \begin{bmatrix} 0 \\ 1 \end{bmatrix}$$

and the Ricatti equation is

$$\begin{bmatrix} p_1 & p_2 \\ p_2 & p_3 \end{bmatrix}\begin{bmatrix} 0 & -5 \\ 1 & -2 \end{bmatrix} + \begin{bmatrix} 0 & 1 \\ -5 & -2 \end{bmatrix}\begin{bmatrix} p_1 & p_2 \\ p_2 & p_3 \end{bmatrix} + \begin{bmatrix} 0 \\ 1 \end{bmatrix}[5.64][0 \quad 1]$$
$$- \begin{bmatrix} p_1 & p_2 \\ p_2 & p_3 \end{bmatrix}\begin{bmatrix} 1 \\ 0 \end{bmatrix}[1][1 \quad 0]\begin{bmatrix} p_1 & p_2 \\ p_2 & p_3 \end{bmatrix} = \mathbf{0}.$$

The resulting equations are

$$2p_2 - p_1^2 = 0$$
$$-5p_1 - 2p_2 + p_3 - p_1 p_2 = 0$$
$$-10p_2 - 4p_3 - p_2 p_3 + 5.64 = 0$$

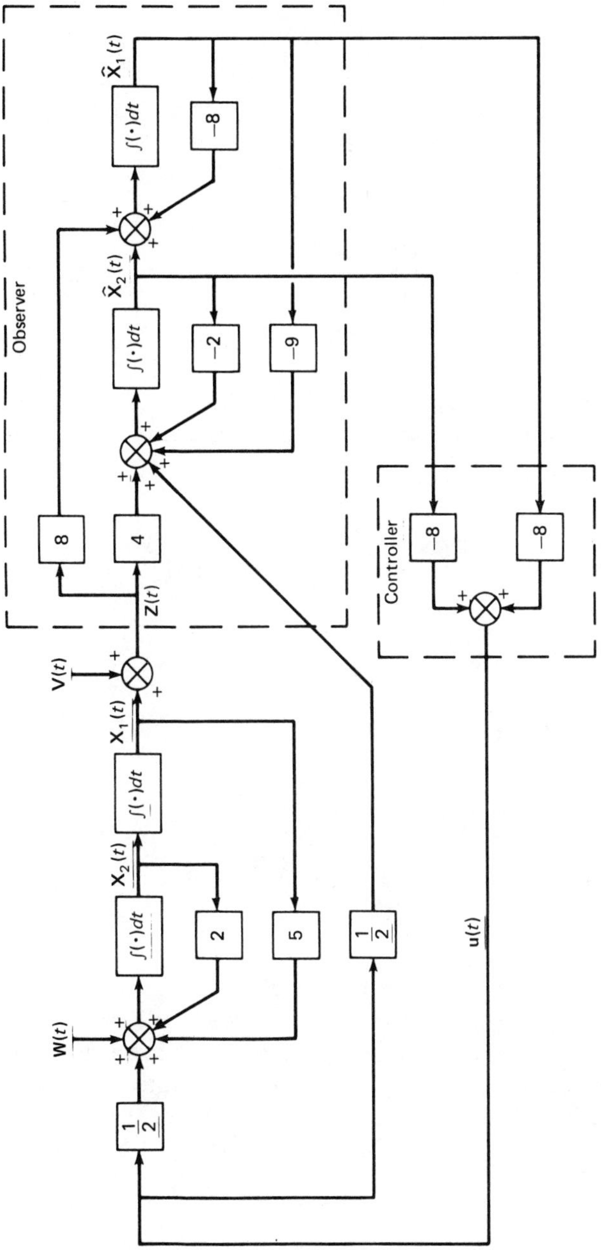

Figure 7-9. Optimal regulator synthesis using a full-order observer.

with solution

$$P = \begin{bmatrix} 0.250 & 0.031 \\ 0.031 & 1.320 \end{bmatrix}.$$

Then, from Equation (7.233), the Kalman gain matrix is

$$K = \begin{bmatrix} 0.250 & 0.031 \\ 0.031 & 1.320 \end{bmatrix} \begin{bmatrix} 1 \\ 0 \end{bmatrix} [1]$$

$$= \begin{bmatrix} 0.250 \\ 0.031 \end{bmatrix}.$$

The form of the resulting filter is identical to that obtained using the observer, but the gain terms are different. The eigenvalues of the filter are those of the matrix

$$\boldsymbol{F} - \boldsymbol{KH} = \begin{bmatrix} 0 & 1 \\ -5 & -2 \end{bmatrix} - \begin{bmatrix} 0.250 \\ 0.031 \end{bmatrix} [1 \quad 0]$$

$$= \begin{bmatrix} -0.250 & 1 \\ -5.031 & -2 \end{bmatrix}.$$

These eigenvalues are

$$\lambda_1 = -1.12 + i2.07; \qquad \lambda_2 = -1.12 - i2.07$$

implying oscillating response characteristics very near those of the system itself. The system open-loop eigenvalues are

$$\lambda_1 = -1 + i2; \qquad \lambda_2 = -1 - i2.$$

In general, as the driving noise variance increases, the Kalman filter gains will increase, causing the filter to weight the observed values more heavily. Large values of the Kalman gain terms, in turn, make the filter response faster. The point to be made is that the response characteristics of the filter are not under control of the designer, but rather are uniquely determined by the noise variances. Whatever the transient response, it is optimal with respect to the selected performance index.

This example has considered a very special case—but nevertheless an important one—of linear stochastic control. Steady state conditions were used for both the controller and filter, permitting use of the algebraic Ricatti equation in both instances. This steady state operating condition for the Kalman filter permitted comparison of the Kalman filter and observer estimate equations. In the general filter problem, in which the Kalman gain terms are time varying, such comparison is not possible. Also, for the general linear controller problem, the optimal controller gain terms are time varying, making it difficult to assign observer eigenvalues which are appropriate for all time. It may be possible, however, to devise appropriate time varying observer gains.

7.4.1. The Separation Principle: Continuous Time Systems

The discussion leading to the example above has implied that a stochastic control system should consist of a stochastic state estimator, such as the Kalman filter, followed by a deterministic state variable feedback controller designed for some specific purpose. This conclusion resulted from the same reasoning process which lead to the use of linear observers in system synthesis. That is, on the one hand powerful systematic techniques are available for the design of deterministic state variable feedback controllers with specific characteristics, while on the other hand the Kalman filter equations provide an optimal estimate of the system state variables. The cascading of the optimal filter and deterministic controller permits the use of two well known systematic procedures in a manner which is certainly intuitively satisfying. In the case of optimal stochastic control, however, it is possible to specifically define an optimization criterion which considers the random nature of the control problem. While some general theory has been developed in this area, consideration here will be limited to linear systems and additive normal or Gaussian noise processes. Specifically, consideration is limited to the system described by the equations

$$\dot{\mathbf{X}}(t) = \boldsymbol{F}(t)\mathbf{X}(t) + \boldsymbol{D}(t)\mathbf{W}(t) + \boldsymbol{G}(t)\mathbf{U}(t)$$
$$\mathbf{Z}(t) = \boldsymbol{H}(t)\mathbf{X}(t) + \mathbf{V}(t). \tag{7.306}$$

These expressions are identical to those of Equation (7.205) in Section 7.2.4, where the continuous time Kalman filter is developed. Here, $\mathbf{W}(t)$ and $\mathbf{V}(t)$ are zero-mean, normal, white noise processes with covariance kernels described by the matrices $\boldsymbol{Q}(t)$ and $\boldsymbol{R}(t)$, respectively, as specified by Equations (7.206) and (7.207). The purpose here is to consider an optimum controller with a quadratic performance such as that specified by Equation (6.95) for use in the development of deterministic controllers. In this case, however, the state variable function $\mathbf{X}(t)$ is a random variable. Also, since the control function $\mathbf{U}(t)$ is to be provided by feedback of the measurement vector $\mathbf{Z}(t)$, it is also a random function. Since the functions are random processes, the performance index must be modified to represent a quantity that can actually be minimized. As was done when estimation criteria were considered in Section 7.1, the expectation operator is used, and the performance index becomes

$$J = E\left[\tfrac{1}{2}[\mathbf{X}^T(t_f)\boldsymbol{S}_f\mathbf{X}(t_f)] + \tfrac{1}{2}\int_{t_0}^{t_f}[\mathbf{X}^T(t)\boldsymbol{A}(t)\mathbf{X}(t) + \mathbf{U}^T(t)\boldsymbol{B}(t)\mathbf{U}(t)]\,dt\right]. \tag{7.307}$$

The objective is to find the general feedback control function

$$\mathbf{U}(t) = \boldsymbol{f}(t, \mathbf{Z}(t)) \tag{7.308}$$

which minimizes J. No stipulation as to the specific form of the control function is made.

The development of the solution to this problem is rather complex and is not presented here, but the solution itself is surprisingly simple. The optimum controller which minimizes the performance index of Equation (7.307) is a cascade of the Kalman filter–producing the estimated state variables–and the optimum deterministic linear controller designed in accordance with the developments of Section 6.3 to minimize the performance index of Equation (6.95). That is, the optimal control function is

$$\mathbf{U}(t) = \boldsymbol{C}(t)\hat{\mathbf{X}}(t) \tag{7.309}$$

where $\boldsymbol{C}(t)$ is the deterministic controller gain matrix and $\hat{\mathbf{X}}(t)$ is the Kalman filter state estimate. Thus, the intuitive procedures discussed at the beginning of this section result in a truly optimal feedback control system in the case of normal or Gaussian noise processes. The system, in general form, is as depicted in Figure 7-8, and in more specific form is illustrated in Figure 7-10. The matrix Riccati equation in the Kalman filter is solved forward in time starting from the initial value of the estimation error covariance matrix $\boldsymbol{P}(0|0)$, while that in the deterministic controller is solved backward in time from terminal condition $\boldsymbol{S}_f$, as discussed in Section 6.3.1. Off-line computation is possible in both instances.

The solution to the optimal stochastic regulator problem follows directly from the discussion above for those instances in which the matrix Ricatti equations reach steady state conditions. These conditions imply steady state conditions in both the filter and the controller, and the resulting optimal regulator is exemplified by the example problem considered at the beginning of this section.

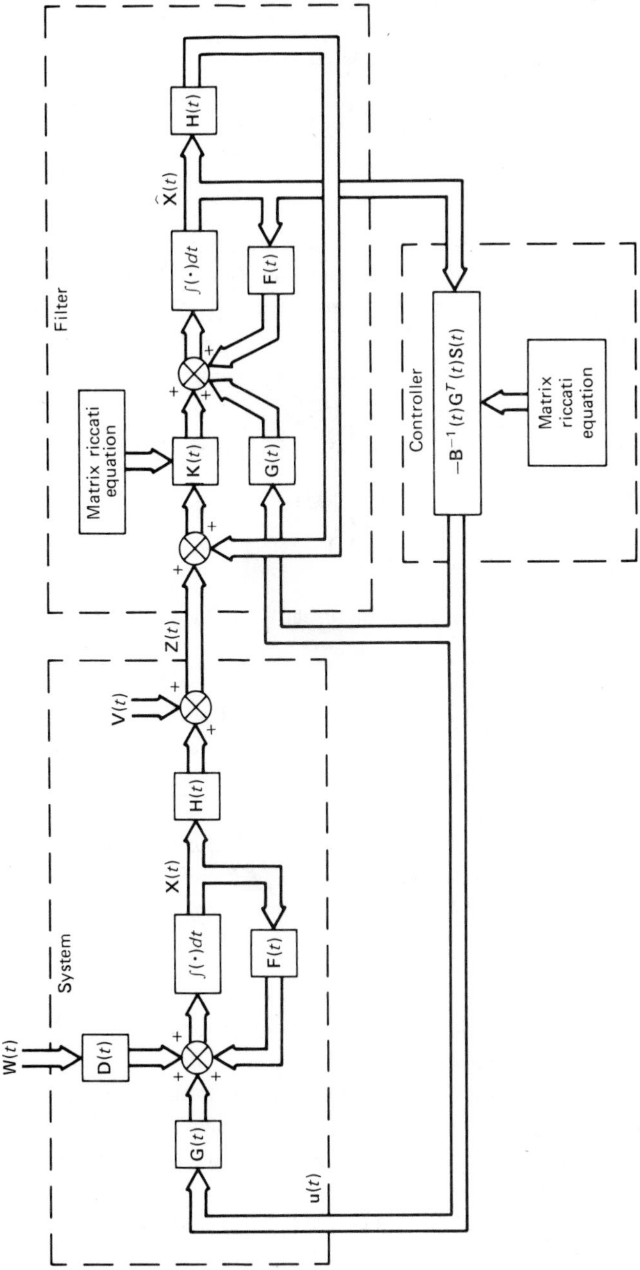

Figure 7-10. Optimal stochastic control systems.

7.4.2. The Separation Principle: Discrete Time Systems

For a discrete time system described by the equations

$$\begin{aligned}\mathbf{X}(l) &= \mathbf{\Phi}(l, l-1)\mathbf{X}(l-1) + \mathbf{\Delta}(l, l-1)\mathbf{W}(l-1) + \mathbf{\Gamma}(l, l-1)\mathbf{U}(l-1)\\ \mathbf{Z}(l) &= \boldsymbol{H}(l)\mathbf{X}(l) + \mathbf{V}(l)\end{aligned} \tag{7.310}$$

a result similar to the separation principle for continuous time systems can be obtained. The driving noise sequence $\mathbf{W}(l)$ and the measurement noise sequence $\mathbf{V}(l)$ are zero-mean, normal, white noise processes described by covariance matrices $\boldsymbol{Q}(l)$ and $\boldsymbol{R}(l)$. That is,

$$\begin{aligned}E[\mathbf{W}(l)\mathbf{W}^T(m)] &= \begin{cases}\boldsymbol{Q}(l), & l = m\\ \mathbf{0}, & l \neq m\end{cases}\\ E[\mathbf{V}(l)\mathbf{V}^T(m)] &= \begin{cases}\boldsymbol{R}(l), & l = m\\ \mathbf{0}, & l \neq m\end{cases}\end{aligned} \tag{7.311}$$

and the two processes themselves are uncorrelated. The performance index to be minimized is

$$J = E\left[\tfrac{1}{2}\mathbf{X}^T(L)\boldsymbol{S}_L\mathbf{X}(L) + \tfrac{1}{2}\sum_{l=0}^{L-1}[\mathbf{X}^T(l)\boldsymbol{A}(l)\mathbf{X}(l) + \mathbf{U}^T(l)\boldsymbol{B}(l)\mathbf{U}(l)]\right]. \tag{7.312}$$

Analogously to the result expressed by Equation (7.309), the optimal control sequence $\mathbf{U}(l)$ is given by the expression

$$\mathbf{U}(l) = \boldsymbol{C}(l)\hat{\mathbf{X}}(l|l) \tag{7.313}$$

where $\boldsymbol{C}(l)$ is the deterministic controller gain matrix and $\hat{\mathbf{X}}(l|l)$ is the state estimate provided by the Kalman filter.

One of the more striking features of the separation principle is that the optimum feedback control gain matrix $\boldsymbol{C}(l)$ or $\boldsymbol{C}(t)$ is independent of all the statistical parameters of the problem, while the Kalman gain matrix $\boldsymbol{K}(l)$ or $\boldsymbol{K}(t)$ is independent of the parameters in the performance index. One must recall, however, that there is also a performance criterion associated with the Kalman filter, and that is one of minimum estimation error variance. The validity of the separation principle is dependent on this characteristic.

7.5. AN APPROACH TO NONLINEAR FILTERING: THE EXTENDED KALMAN FILTER

When the process to be estimated is the state vector of a nonlinear system, then the state and measurement equations have the general form

$$\begin{aligned} \mathbf{X}(l) &= \boldsymbol{f}(\mathbf{X}(l-1), \boldsymbol{u}(l-1), \mathbf{W}(l-1), l) \\ \mathbf{Z}(l) &= \boldsymbol{h}(\mathbf{X}(l), l) + \mathbf{V}(l). \end{aligned} \tag{7.314}$$

Obviously, linear estimation techniques do not apply in this case, but the perturbation method of Section 4.5 can be applied to produce a linearized version of these equations. These linearized equations then permit linear estimation techniques to be applied to the perturbation variables. In this case, Equation (7.314) is to be linearized about some reference solution $\mathbf{X}^*(l)$, obtained with reference deterministic input $\boldsymbol{u}^*(l)$. The function $\boldsymbol{u}^*(l)$ is assumed to contain any bias terms present in $\mathbf{W}(l)$, and $\mathbf{W}^*(l)$ is assumed equal to zero. The following perturbation variables are then defined,

$$\begin{aligned} \delta\mathbf{X}(l) &= \mathbf{X}(l) - \mathbf{X}^*(l) \\ \delta\boldsymbol{u}(l) &= \boldsymbol{u}(l) - \boldsymbol{u}^*(l) \\ \delta\mathbf{W}(l) &= \mathbf{W}(l) \end{aligned} \tag{7.315}$$

where the third expression results from the fact that $\mathbf{W}^*(l)$ is zero in the reference solution. Applying the techniques of Section 4.5.2 yields the perturbation equations

$$\begin{aligned} \delta\mathbf{X}(l) &= \left[\frac{\partial \boldsymbol{f}}{\partial \mathbf{X}}\right]_{\mathbf{X}^*(l),\boldsymbol{u}^*(l)} \delta\mathbf{X}(l-1) + \left[\frac{\partial \boldsymbol{f}}{\partial \boldsymbol{u}}\right]_{\mathbf{X}^*(l),\boldsymbol{u}^*(l)} \delta\boldsymbol{u}(l-1) \\ &\quad + \left[\frac{\partial \boldsymbol{f}}{\partial \mathbf{W}}\right]_{\mathbf{X}^*(l),\boldsymbol{u}^*(l)} \delta\mathbf{W}(l-1) \\ \delta\mathbf{Z}(l) &= \left[\frac{\partial \boldsymbol{h}}{\partial \mathbf{X}}\right]_{\mathbf{X}^*(l)} \delta\mathbf{X}(l) + V(l) \end{aligned} \tag{7.316}$$

where the nomenclature implies that the partial derivative matrices are evaluated along the reference solution. Since these equations are linear, the $\delta\boldsymbol{u}$ term can be ignored without loss of generality, and the Kalman filter equations of Section 7.2.1 applied directly in estimation of the state perturbation vector $\delta\mathbf{X}(l)$. The resulting equations are simply those given in Equations (7.165) and

(7.166), or equivalently, Equation (7.174), with the associations

$$\Phi(l, l-1) \longrightarrow \left[\frac{\partial f}{\partial \mathbf{X}}\right]_{\mathbf{X}^*(l), u^*(l)}; \quad \Delta(l, l-1) \longrightarrow \left[\frac{\partial f}{\partial \mathbf{W}}\right]_{\mathbf{X}^*(l), u^*(l)}$$

$$H(l) \longrightarrow \left[\frac{\partial h}{\partial \mathbf{X}}\right]_{\mathbf{X}^*(l)}. \tag{7.317}$$

The estimate produced by the Kalman filter is $\delta\hat{\mathbf{X}}(l|l)$, which in turn is used to estimate the system state vector by inverting the relationship of Equation (7.315),

$$\hat{\mathbf{X}}(l|l) = \mathbf{X}^*(l) + \delta\hat{\mathbf{X}}(l|l). \tag{7.318}$$

Since the reference solution $\mathbf{X}^*(l)$ is known, the filtered state estimate can be obtained.

The continuous time version of these equations follows by direct analogy. Based upon the continuous time system equations,

$$\dot{\mathbf{X}}(t) = f(\mathbf{X}(t), u(t), \mathbf{W}(t), t)$$

$$\mathbf{Z}(t) = h(\mathbf{X}(t), t) + \mathbf{V}(t) \tag{7.319}$$

the perturbation equations are

$$\delta\mathbf{X}(t) = \mathbf{X}(t) - \mathbf{X}^*(t)$$

$$\delta\dot{\mathbf{X}}(t) = \left[\frac{\partial f}{\partial \mathbf{X}}\right]_{\mathbf{X}^*(t), u^*(t)} \delta\mathbf{X}(t) + \left[\frac{\partial f}{\partial u}\right]_{\mathbf{X}^*(t), u^*(t)} \delta u(t) + \left[\frac{\partial f}{\partial \mathbf{W}}\right]_{\mathbf{X}^*(t), u^*(t)} \delta\mathbf{W}(t)$$

$$\delta\mathbf{Z}(t) = \left[\frac{\partial h}{\partial \mathbf{X}}\right]_{\mathbf{X}^*(t)} \delta\mathbf{X}(t) + \mathbf{V}(t). \tag{7.320}$$

The continuous time Kalman filter algorithm of Equation (7.218) is then applied to this linear system, with the associations

$$F(t) \longrightarrow \left[\frac{\partial f}{\partial \mathbf{X}}\right]_{\mathbf{X}^*(t), u^*(t)}; \quad D(t) \longrightarrow \left[\frac{\partial f}{\partial \mathbf{W}}\right]_{\mathbf{X}^*(t), u^*(t)}$$

$$H(t) \longrightarrow \left[\frac{\partial h}{\partial \mathbf{X}}\right]_{\mathbf{X}^*(t)}. \tag{7.321}$$

The filtered state estimate is obtained from $\delta\mathbf{X}(t)$ by the equation

$$\hat{\mathbf{X}}(t) = \mathbf{X}^*(t) + \delta\mathbf{X}(t). \tag{7.322}$$

An obvious characteristic of the perturbation equations is the dependence of the coefficient matrices on t, since they are evaluated along the reference trajectory. Thus, even though $\boldsymbol{f}$ in either Equation (7.314) or (7.319) may not be explicitly time dependent, the partial derivative matrices are. This means that time invariant systems are seldom encountered in nonlinear estimation by means of the perturbation techniques.

When such linearization techniques are used to obtain approximate solutions to nonlinear equations, the primary concern over the validity of the method is the requirement for maintaining the perturbations within a region where the approximations are valid. This requirement normally implies that the perturbation variables must be small, since a first-order multivariable Taylor series expansion serves as the basis for the perturbation technique. Such consideration is of special importance in filtering, since the system is not necessarily controlled in a manner which maintains the value of the state vector near the reference value. In the case of filtering, a method analogous to that of linearization and successive approximation, used for nonlinear point estimation in Section 7.1.1.6, can be used. This technique, which has been termed the *extended Kalman filter*, uses the currently derived estimate of the perturbation state vector to obtain a revised reference solution–by means of Equation (7.318) or (7.322)–and is most easily demonstrated for the discrete time filter. If the estimate $\delta\hat{\mathbf{X}}(l|l)$ has been obtained using reference solution $\mathbf{X}^*(l)$, then the estimate of the state vector at time l is given by Equation (7.318). If this value is used as the new reference solution, then it is obvious that the prior estimate $\delta\hat{\mathbf{X}}(l+1|l)$ is

$$\delta\hat{\mathbf{X}}(l+1|l) = \boldsymbol{\Phi}(l+1, l)\delta\hat{\mathbf{X}}(l|l) = 0 \tag{7.323}$$

since $\delta\hat{\mathbf{X}}(l|l)$ based on the revised reference solution is zero. The rationale behind this procedure is that the estimate $\hat{\mathbf{X}}(l|l)$ is probably closer to the true value of $\mathbf{X}(l)$ than is the reference solution value $\mathbf{X}^*(l)$. If each stage in the discrete time filter is treated in this manner, then the filter estimate equations are

$$\begin{aligned}
\delta\hat{\mathbf{X}}(l|l) &= \boldsymbol{P}(l|l)\boldsymbol{H}(l)\boldsymbol{R}^{-1}(l)\delta\mathbf{Z}(l)\\
\hat{\mathbf{X}}(l) &= \mathbf{X}^*(l) + \delta\hat{\mathbf{X}}(l|l)\\
&= \boldsymbol{f}(\hat{\mathbf{X}}(l-1|l-1), \boldsymbol{u}^*(l-1), l) + \delta\hat{\mathbf{X}}(l|l)
\end{aligned} \tag{7.324}$$

where the function $\boldsymbol{f}$ is that occurring in Equation (7.314).

If this same concept is applied to the continuous time problem, the analogous results are

$$\begin{aligned}
\delta\hat{\mathbf{X}}(t) &\approx \mathbf{0}\\
\delta\dot{\hat{\mathbf{X}}}(t) &= \dot{\hat{\mathbf{X}}}(t) - \dot{\mathbf{X}}^*(t)
\end{aligned} \tag{7.325}$$

which, when substituted into Equation (7.218) as it applies here, yields

$$\dot{\hat{\mathbf{X}}}(t) = \boldsymbol{f}(\hat{\mathbf{X}}(t), \boldsymbol{u}^*(t), t) + \boldsymbol{P}(t)\boldsymbol{H}(t)\boldsymbol{R}^{-1}(t)[\mathbf{Z}(t) - \boldsymbol{h}(\hat{\mathbf{X}}(t), t)] \quad (7.326)$$

where the $\boldsymbol{f}$ function is obtained from Equation (7.319). Both the discrete and continuous versions of this form of the extended Kalman filter entail solution of the nonlinear equation; such solutions may or may not be plausible. In addition there are still approximations involved in the use of the equations, as is evident from the above derivations, so that there are some grounds for concern over the convergence properties of the technique. This same problem is encountered in nonlinear point estimation by the method of linearization and successive approximation discussed in Section 7.1.1.6, but to a somewhat lesser degree because of the time invariant nature of the quantity to be estimated. It can only be said here that there are intuitive reasons for the use of the method.

Other variations of the extended Kalman filter exist, all intended to maintain the perturbation variables within an acceptable region or to extend the acceptable region. In certain instances, only some of the state variables are updated, in which case suitable adjustments in the equations are required. Any of these variations may produce better or worse results than any other, depending upon the specific application. Usually, some sort of simulation analysis is required to evaluate the effectiveness of any particular technique. The following examples will serve to illustrate the use of extended Kalman filtering techniques.

A system may involve linear system equations and non-linear measurement equations, in which case a variation of the extended Kalman filter approach can be used. As an example consider a situation similar to the example of Section 7.1.1.6, except that now the radar station makes position measurements of the aircraft, from which it is desired to obtain the filtered values of position and velocity. The geometry of the problem is shown in Figure 7-11, in which the position measurements are contaminated with noise so that they do not fall on the actual trajectory. This is a four-dimensional problem (two position and two

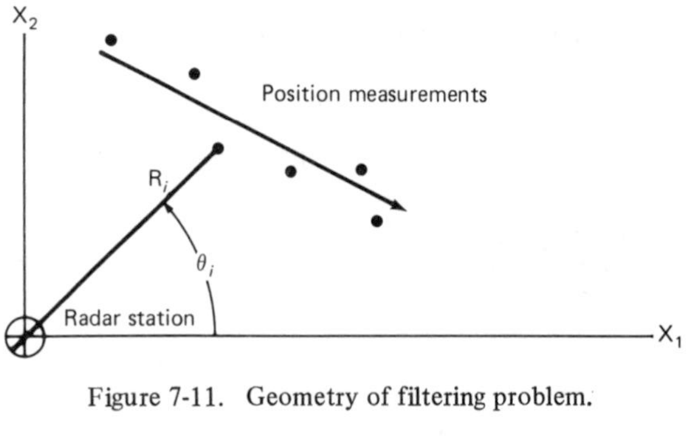

Figure 7-11. Geometry of filtering problem.

velocity coordinates), with state equations

$$\dot{X}_1(t) = X_3(t)$$
$$\dot{X}_2(t) = X_4(t)$$
$$\dot{X}_3(t) = W_1(t)$$
$$\dot{X}_4(t) = W_2(t)$$

where the assumption of straight-line motion has been made. The accelerations $W_1(l)$ and $W_2(l)$ are considered to be zero mean, white noise, random processes. The measurement equation is

$$\boldsymbol{h}(\mathbf{X}(t)) = \begin{bmatrix} R(t) \\ \theta(t) \end{bmatrix} = \begin{bmatrix} \sqrt{X_1^2(t) + X_2^2(t)} \\ \arctan\,[X_2(t)/X_1(t)] \end{bmatrix}.$$

Here, the state equation is linear but the measurement equation is nonlinear, thus requiring the use of nonlinear filtering techniques. The system matrices are

$$\boldsymbol{F} = \begin{bmatrix} 0 & 0 & 1 & 0 \\ 0 & 0 & 0 & 1 \\ 0 & 0 & 0 & 0 \\ 0 & 0 & 0 & 0 \end{bmatrix}; \quad \boldsymbol{D} = \begin{bmatrix} 0 & 0 \\ 0 & 0 \\ 1 & 0 \\ 0 & 1 \end{bmatrix}$$

$$\boldsymbol{H} = \begin{bmatrix} \dfrac{\partial R}{\partial X_1} & \dfrac{\partial R}{\partial X_2} & 0 & 0 \\ \dfrac{\partial \theta}{\partial X_1} & \dfrac{\partial \theta}{\partial X_2} & 0 & 0 \end{bmatrix} = \begin{bmatrix} \dfrac{X_1}{\sqrt{X_1^2 + X_2^2}} & \dfrac{X_2}{\sqrt{X_1^2 + X_2^2}} & 0 & 0 \\ -\dfrac{1}{2X_1}\left(\dfrac{X_2}{1 + \left(\dfrac{X_1}{X_2}\right)^2}\right) & -\dfrac{1}{X_1}\left(\dfrac{1}{1 + \left(\dfrac{X_1}{X_2}\right)^2}\right) & 0 & 0 \end{bmatrix}$$

and the filter equations follow directly from the discussion of this section. The linearized perturbation equations are

$$\delta\dot{\mathbf{X}} = \boldsymbol{F}\delta\mathbf{X} + \boldsymbol{D}\mathbf{W}$$
$$\delta\mathbf{Z} = \boldsymbol{H}\delta\mathbf{X} + \mathbf{V}$$

and the filtered state is

$$\hat{\mathbf{X}}(t) = \mathbf{X}^*(t) + \delta\hat{\mathbf{X}}(t)$$

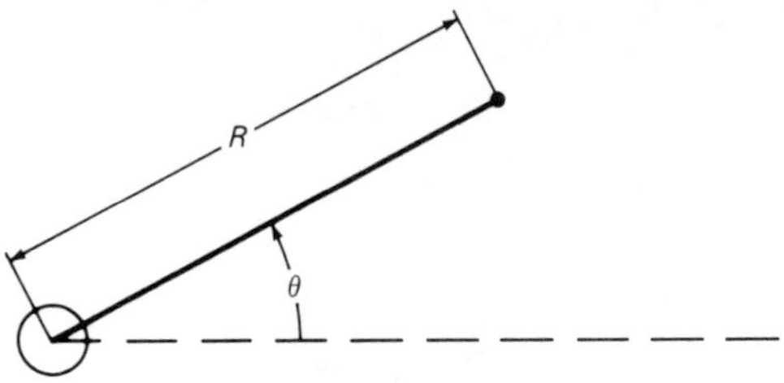

Figure 7-12. Satellite motion diagram.

where the estimate of $\delta\mathbf{X}(t)$ is obtained from Equation (7.213). If the extended Kalman filter is used, then an equation for $\hat{\mathbf{X}}$ itself is obtained, using Equation (7.326), which in this case is

$$\dot{\hat{\mathbf{X}}}(t) = \boldsymbol{F}\hat{\mathbf{X}}(t) + \boldsymbol{P}(t)\boldsymbol{H}(t)\boldsymbol{R}^{-1}[\mathbf{Z}(t) - \boldsymbol{h}(\hat{\mathbf{X}}(t))].$$

It should be mentioned that both methods are approximations, with the extended Kalman filter giving the user an extra degree of confidence in the validity of the linear approximations.

As a second, and perhaps more illustrative example, consider a satellite of mass m in an orbit around a planet of mass M, as depicted in Figure 7-12. Here, the reference solution is some general elliptical orbit, and the motion is described in spherical coordinates by the equations

$$\ddot{\mathrm{R}}(t) = \mathrm{R}\dot{\theta}^2(t) - [GM/\mathrm{R}^2(t)]$$
$$\ddot{\theta}(t) = -2\dot{\mathrm{R}}(t)\dot{\theta}(t)/\mathrm{R}(t)$$

where G is the universal gravitational constant. Suppose that noisy R and θ measurements are obtained from radar beacons located on the planet's surface, and from this information the satellite state vector must be specified. For present purposes the state variables

$$\mathbf{X}(t) = \begin{bmatrix} \mathrm{R}(t) \\ \dot{\mathrm{R}}(t) \\ \theta(t) \\ \dot{\theta}(t) \end{bmatrix} = \begin{bmatrix} \mathrm{X}_1(t) \\ \mathrm{X}_2(t) \\ \mathrm{X}_3(t) \\ \mathrm{X}_4(t) \end{bmatrix}$$

are chosen. Then the measurement equation is linear:

$$\mathbf{Z}(t) = \boldsymbol{H}\mathbf{X}(t) + \mathbf{V}(t)$$

where

$$H = [1 \quad 0 \quad 1 \quad 0]$$

and $\mathbf{V}(t)$ is the measurement noise process. Then the state equation corresponding to the first expression of Equation (7.319) is

$$\begin{bmatrix} \dot{X}_1(t) \\ \dot{X}_2(t) \\ \dot{X}_3(t) \\ \dot{X}_4(t) \end{bmatrix} = \begin{bmatrix} X_2(t) \\ X_1(t)X_4^2(t) - [GM/X_1^2(t)] + W_1(t) \\ X_4(t) \\ -[2X_2(t)X_4(t)/X_1(t)] + W_2(t) \end{bmatrix}$$

where the driving noise

$$\mathbf{W}(t) = \begin{bmatrix} W_1(t) \\ W_2(t) \end{bmatrix}$$

has been included. These noise terms have the units of tangential and radial acceleration. Further definition in terms of actual forces could be made if such information were available. The state equation is of the general nonlinear form of Equation (7.319):

$$\dot{\mathbf{X}}(t) = \boldsymbol{f}(\mathbf{X}(t), \mathbf{W}(t))$$

which must be linearized about some reference solution $\mathbf{X}^*(t)$. The partial derivative matrix of Equation (7.320) is

$$\boldsymbol{F}(t) = \frac{\partial f}{\partial \mathbf{X}} = \begin{bmatrix} 0 & 1 & 0 & 0 \\ X_4^{*2}(t) + 2GM/X_1^{*3}(t) & 0 & 0 & 2X_1^*(t)X_4^*(t) \\ 0 & 0 & 0 & 0 \\ 2X_2^*(t)X_4^*(t)/X_1^{*2}(t) & -2X_4^*(t)/X_1^*(t) & 0 & -2X_2^*(t)/X_1^*(t) \end{bmatrix}$$

The measurement equation is linear, with

$$H = \begin{bmatrix} 1 & 0 & 0 & 0 \\ 0 & 0 & 1 & 0 \end{bmatrix}.$$

These $\boldsymbol{F}(t)$ and $\boldsymbol{H}$ matrices are used in the linearized Kalman filter. The $\boldsymbol{F}(t)$ matrix is a function of time because some of its elements are functions of the four state variables evaluated along the reference solution. The equations

$$\delta\dot{\mathbf{X}}(t) = \boldsymbol{F}(t)\delta\mathbf{X}(t) + \mathbf{W}(t)$$
$$\delta\mathbf{Z}(t) = \boldsymbol{H}\delta\mathbf{X}(t) + \mathbf{V}(t)$$

represent the linearized system to which the Kalman filter equations are to be applied. So long as the input disturbances $\mathbf{W}(t)$ are small enough so that the perturbation variables $\delta\mathbf{X}(t)$ remain small, the linearized Kalman filter will function properly. In the event that it cannot be assured that these perturbations remain small, then updating of the reference solution by means of some form of an extended Kalman filter is required. In the event that sufficient computational capacity is available for solution of the nonlinear equations at the required rate, then the extended Kalman filter of Equation (7.326) can be used.

A particularly simple linearized system results when the reference solution is a circular orbit, for then

$$\dot{R}^*(t) = X_2^*(t) = 0; \qquad R(t) = X_1^*(t) = r, \text{ a constant}$$
$$\dot{\theta}^*(t) = X_4^*(t) = \alpha, \text{ a constant}; \qquad \theta^*(t) = X_3^*(t) = \alpha t.$$

The $\boldsymbol{F}(t)$ matrix then has the simple form

$$\boldsymbol{F}(t) = \boldsymbol{F} = \begin{bmatrix} 0 & 1 & 0 & 0 \\ \alpha^2 + (2GM/r) & 0 & 0 & 2r\alpha \\ 0 & 0 & 0 & 0 \\ 0 & -2\alpha/r & 0 & 0 \end{bmatrix}$$

and the resulting linearized system is time invariant.

Practical implementation of a linearized or extended Kalman filter for a system like the one described in this example would require discrete time techniques—which would then require use of Equations (7.314) through (7.318)—but this modification presents no difficulty.

It is important to note that linearization and the extended Kalman filter are not the only approaches to the nonlinear filtering problem. More direct approaches, however, are fraught with theoretical difficulties which have limited the development of practical optimal filtering techniques in all but severely restricted special cases. Even in those instances in which the problem can be solved, the computational load tends to be excessive, and almost all require normal or Gaussian noise statistics. As a result, most applications have used some form of the extended Kalman filter.

7.6. FURTHER CONSIDERATIONS

As with state variable feedback, the use of observers, and optimal control, filtering and smoothing concepts must be considered in light of the practical aspects of their implementation. Because of the dual nature of the control and filtering problems, the same kinds of concerns that were raised in the previous chapter must also be considered here. The questions of sensitivity and robustness are pertinent, as is the concern over the computational burden of implementation. The sensitivity of the Kalman filter is particularly important for two reasons. First, filter performance is sensitive not only to model parameters, as in the case of feedback control, but also to the statistical parameters of the noise processes. Second, the estimation errors may eventually grow entirely out of proportion to the elements of the estimation error covariance matrix, after an extended period of filter operation—a phenomenon known as *divergence*. The mechanism behind divergence is clear—the computed estimation error covariance matrix becomes unrealistically small, so that undue confidence is placed in the prior estimate and subsequent measurements are effectively ignored. The causes of divergence, on the other hand, are not easy to identify and may vary from one application to another. Some identified causes are incorrect modeling of the system, incorrect values for the noise variances, unidentified biases in the system or in the measurement process, and numerical effects which cause the estimation error covariance matrix to lose its positive semidefinite character. The identification of divergence in filter operation can often be obtained by examining the statistics of the term $(\mathbf{Z} - \boldsymbol{H}\mathbf{X})$—the so-called innovations process—which should be a zero mean white noise process. Or, the estimation error covariance matrix can be monitored to determine if the diagonal elements tend toward zero with increasing time.

Methods of coping with divergence constitute an art as much as a science, and they result in a suboptimal filter. Such measures include artificially increasing the variances of the input noise, use of a finite length memory, and arbitrarily placing a lower bound on the diagonal elements of the estimation error covariance matrix. The technique of exponentially weighting older data—what is termed a *fading memory*—can be shown to be equivalent to increasing both the system and measurement noise variances by some multiplicative factor. Divergence problems due to computational factors, which cause the estimation error covariance matrix to lose its positive semidefinite character, can be countered by use of what is termed *square root filtering*. This technique carries on the sequential computation in terms of the square roots of the covariance matrices—usually in triangular form—and has the advantage that the product of the square root matrix with its transpose is always positive semidefinite. An additional advantage acrues in the fact that the square root matrix is less sensitive to computational errors than the covariance matrix itself.

The reader should note that divergence is not a manifestation of filter instability, for such instability would result not only in unbounded errors, but also in an indication of those unbounded errors in terms of the estimation error covariance matrix. It is entirely possible, of course, to estimate the state of an unstable system with a Kalman filter which is itself stable, and indeed it has been noted that such a situation is typically associated with divergence effects in the filter. Filter stability, on the other hand, is more closely related to the observability and controllability of the system. For the time invariant case, the stability of the filter is equivalent to that of the matrix Riccati differential equation, for which the stability requirements can be formulated in terms of the more general characteristics of detectability and stabilizability. For a detectable and stabilizable system, the filter is stable and the Riccati equation reaches a steady state value. Under the stronger conditions of observability and controllability, the system exhibits a more robust stability with respect to the positive semidefinite character of the estimation error covariance matrix. Further details can be found in the references.

The sensitivity of the Kalman filter to parameter variations in both the model and noise variances is also of interest even if some form of divergence is not present. Various error analysis algorithms have been developed and are presented in some of the references. Since the Kalman filter has several successively less stringent optimal characteristics–minimum mean-squared error, maximum likelihood, and minimum variance unbiased linear estimate–the sensitivity analyses are usually oriented toward one specific case. If Gaussian or normal noise statistics and minimum mean-squared error estimation are considered, sensitivity of filter performance to deviation from a Gaussian or normal distribution is also significant. Several of the references at the end of this chapter deal with the error analysis of the Kalman filter.

The design of robust forms of the Kalman filter have also been developed, some of which estimate values of uncertain parameters by state augmentation techniques. This concept requires some form of nonlinear filtering, for which the extended Kalman filter is a likely candidate.

Computational requirements for the discrete time Kalman filter are important when implementation is considered. The computation time per iteration and the required storage determine the sampling rate and memory size. The sampling rate must be compatible with the quantization requirements of the filter, and with the characteristics of the sensors used to provide the measurement data. The computational load can be minimized by recognizing the specific structure of the matrices involved, since they often are rather sparse–i.e., they often contain many zeros. A design feature that exploits the sparse nature of the matrices is obtained at the cost of larger storage requirements. It is also possible to realize rather large reductions in computation time by a technique known as *sequential processing*, in which the measurement vector obtained at any particular time is partitioned into groups of statistically independent measurements, and then each group is processed as though it were a separate measurement. The computational efficiency of this technique results from the fact that the matrix inversion required in the Kalman filter equations is of the order of the dimension of the measurement vector. The ultimate effect is obtained when independent measurements are processed as one-dimensional vectors, in which case the matrix inversion is of order one—a simple reciprocation. The reader should be cautioned, however, that there is a tradeoff between the reduction in the complexity per iteration offered by sequential processing and the additional number of iterations required by its implementation.

Just about everything said about practical implementation of the Kalman filter applies equally well to smoothing algorithms based on the Kalman filter. There are additional stability problems involved in the smoothing algorithms which must be considered. In fact, the algorithm based on forward and backward passes through the measured data has been shown to be unstable under the very conditions under which the Kalman filter is stable. The state augmentation approach to smoothing, as developed in Section 7.3, is stable so long as the filter itself is stable.

The computational requirements of smoothing algorithms grow very rapidly with the size of the lag used. The dimension of the fixed lag smoothing problem is in general NM, where N is the dimension of the system and M is the size of the lag. In certain cases, it is possible to devise reduced-order smoothing algorithms. The reader is referred to the references at the end of this chapter for further detail.

The subject of nonlinear estimation and filtering has received little attention in this chapter, except for its relationship to the extended Kalman filter. This consideration has been limited not so much by the fact that nonlinear estimation, filtering, and smoothing techniques would not be of practical use, but rather by the fact that useful results of a general nature are difficult to obtain. The conditional mean, for example, is the minimum mean-squared error estimate in any case, but useful results based upon this fact in other than the linear case are few. Those results that can be obtained are often first- or second-order approximations to the desired estimate. Other optimization criteria, such as Maximum Posterior Probability (MAP), have also been used. In light of the fact that linearized filters of some sort are usually required, and the fact that the Kalman filter is an elegant, systematic, and much studied technique of linear filtering, it is not surprising that many of the practical applications of nonlinear filtering have been made through the vehicle of the extended Kalman filter.

Iterative techniques have been applied to the various nonlinear filtering approximate methods, including the extended Kalman filter. The so-called *locally iterated extended Kalman filter* is an improved version of the extended Kalman filter in which, at each step of the iterative procedure, the measurement is linearized about the most recent estimate obtained from the extended Kalman filter. Yet another method–the Gaussian sum approach–involves a bank of extended Kalman filters to provide a better estimate of the posterior density function, at the expense of increased computational and storage requirements.

One topic that has not been discussed in this chapter, but that relates to Kalman filtering in general–and nonlinear filtering in particular–is that of *system identification*. In many instances, the precise form of a time invariant system is not known, so that this information must be inferred from the observed data at the same time, perhaps, that the state is being estimated. The result, of course, is a nonlinear problem to which the extended Kalman filter can be applied with some useful results. Techniques which identify practically all aspects of a linear, time invariant system–even the order of the system–from the measured data have been developed. One particular concept that relates to the discrete time problem is the *autoregressive moving average* model (ARMA) in which the process of interest is the solution of an nth-order linear difference equation driven by a Markov noise process. The application to single input-single output discrete systems is immediately obvious, since the ARMA process relates to the control canonical form of the system. This concept also supplies a link with the statitician's concept of time-series analysis, since the ARMA model is a standard model of time-series analysis. The vector generalization of the ARMA model has proven very useful in development of discrete time filtering and identification results.

As a final note, it is necessary to call attention to a general approach to linear estimation theory known as the *innovations approach*. The basic concept is to first convert the measurement to a white noise process called the *innovations process*, and then to solve the easier estimation problem with white noise measurements. The solution to the simplified problem can then be converted to yield the desired estimate. The innovations approach has been applied to almost every aspect of linear estimation theory, and to nonlinear filtering as well.

7.6.1. Estimation and Filtering

A very complete treatise on the discrete time Kalman filter is contained in Reference 3. The sensitivity of the Kalman filter is treated in References 3, 23, 27, 64, and 69. The discussion in Reference 55 deals specifically with the sensitivity of the Kalman filter as a minimum variance, unbiased, linear estimator, as opposed to the minimum mean-squared error estimate. The particular problem of divergence is considered in References 3, 27, and 61. In Reference 22, divergence is studied in detail and the distinction made between *true divergence*, in which the errors actually become unbounded, and *apparent divergence*, in which there are finite degradations due to modeling inaccuracies. Conditions contributing to divergence of both kinds are identified and investigated. The development of Reference 33 considers a fixed-length memory version of the Kalman filter as a means of preventing divergence.

When divergence is due to computational effects, the square root formulation of the Kalman filter is sometimes used in an effort to reduce the effects of the problem. Square root filters are treated in References 3, 11, 27, 42, and 68 for the discrete time filter, and in Reference 49 for the continuous time filter. The stability of the Kalman filter is treated in Reference 3, and that of the matrix Riccati differential equation in Reference 40. The solution to the algebraic Riccati equation is considered in References 4, 16, 27, 32, 35, and 60, and in other references given at the end of Chapter 6. The solution to the differential matrix Riccati equation is treated in References 4, 27, and 58.

The design of robust Kalman filters is addressed and treated in Reference 49 for the minimum mean-squared error form, in the sense that the filter is made insensitive to deviations of the probability distributions from the normal distribution in certain characteristic ways. Other techniques combine some form of parameter estimation with the filtering process, such as those described in References 17, 46, 51, and 67. In References 45 and 67, techniques are developed which estimate the steady-state Kalman gain matrix for systems with unknown noise covariances. A method developed in Reference 19 deals with uncertainties in the system parameters and employs the extended Kalman filter in the joint estimation process.

The use of Kalman filter techniques in estimating the state of large scale systems–in which the number of states makes the computational load prohibitive–is treated in References 6, 30, and 56. The basic approach is to use the Kalman filter on subsystems and thus reduce the computational load by implementing a number of lower-dimensional filters. In References 3 and 25, alternatives to the Kalman filter as a discrete time linear estimation technique are considered. In Reference 43, it is shown that the Kalman filter has a minimax feature in that it minimizes the maximum of a linear functional of the estimation error.

In many high-reliability designs, redundant measurements are made for use in fault detection and backup mode operation. Although the measured quantity is the same in these redundancies, the actual measurements are independent and thus can be used to advantage in the estimation process. This concept is developed in Reference 77, in which redundancy is exploited in order to reduce the computational load of the filter. In Reference 24, the special situation in which both discrete time and continuous time measurement data are available in a steady state Kalman filter is considered.

The computational requirements of a Kalman filter implementation are considered in References 3 and 27, and a detailed breakdown of the individual operations and their computational requirements is given in Reference 52. The information in this reference is somewhat dated with respect to computer cycle times, but the basic information and conclusions are still valid. The use of sequential processing as a means of reducing computational load is considered, and some general rules are developed for its use. In Reference 70, quantization effects are considered, and optimal quantization criteria are established. The effect of quantization errors, multiplication errors, finite word length, and other digital implementation characteristics are treated in References 21 and 32. Finally, the three major aspects of sensor placement–location, type, and scheduling–are investigated in Reference 8. It is shown there that proper choices of sensor placement can greatly improve the quality of the estimate.

7.6.2. The Smoothing Problem

The use of smoothing has been fairly limited in the design of feedback systems because of the inherent lag characteristic of smoothing algorithms. But smoothing techniques continue to be of interest in those instances in which a reconstruction of a state trajectory is desired. A general approach to the smoothing problem is detailed in Reference 10, while the approach through the Kalman filter is considered in References 3, 18, and 50 for discrete time systems. In Reference 13, the stability of the fixed lag filter derived by state augmentation is proven. The concept of nonlinear smoothing is treated in Reference 64, using maximum posterior probability (MAP) techniques.

In Reference 31, the performance of Kalman smoothers at lag values other than the design lag is investigated, with some interesting results relating to suboptimal smoothers. One result indicates that the filter itself can sometimes be used as a suboptimal smoother.

7.6.3. The Linear Stochastic Control Problem

The linear stochastic control design problem is treated with some detail in Reference 9, in which the concepts of deterministic perturbation control, stochastic state estimation, linearized stochastic control, and the separation principle are considered. The off-line nature of the design process is stressed. The discrete time control problem is also treated in Reference 50, and the separation principle developed. In Reference 70, dynamic programming techniques are used to develop the solution to the discrete time control problem in terms of the separation principle. For continuous time systems, the linear stochastic control problem is developed in Reference 71. All of these treatments involve an assumption of Gaussian or normal noise statistics. The reader must be aware of the fact that the separation principle is often applied to systems in which the Gaussian character of the noise cannot be established. In such cases, suboptimal controllers result, but such is the price paid for use of the systematic procedures of the Kalman filter in linear optimal control. In Reference 53, the effect of word length in microprocessor implementations of the filter/controller stochastic control algorithm is investigated.

7.6.4. Nonlinear Filtering

The general concept of nonlinear filtering is treated in Reference 64, where the conditional mean and maximum posterior probability (MAP) estimates are considered. The extended Kalman filter is treated in References 3, 15, 27, 34, and 64. The robustness of the extended Kalman filter is treated in Reference 63. This reference also introduces the concept of a *constant gain extended Kalman filter*, designed to be optimal for a stochastic, linear, time invariant model crudely approximating the actual nonlinear system. The robustness of regulators based upon constant gain filters and the separation principle is also investigated. The *locally iterated extended Kalman filter* is treated in Reference 34, and a critical evaluation is given in Reference 5.

The nonlinear filtering problem is closely related to the robust design of Kalman filters, because robust design usually incorporates some form of parameter estimation along with state estimation. Such a problem is inherently nonlinear. This concept is treated in References 2, 46, 51, 59, and 72. Several of these developments use the extended Kalman filter as the nonlinear estimator.

Applications of the *autoregressive moving average* (ARMA) model are presented in References 1, 12, 44, and 76.

Finally, the application of the *innovations approach* to linear estimation theory is developed in References 1, 3, 28, 37, 38, 39, and 40. Applications to nonlinear filtering is treated in Reference 26.

REFERENCES

1. Aasnaes, Hans Bert, and Kailth, Thomas. An Innovations Approach to Least-Squares Estimation–Part VII: Some Applications of Vector Autoregressive Moving Average Models. *IEEE Transactions on Automatic Control*, **AC-18:** 601–607 (December 1973); **AC-24:** 511–512 (June 1978).
2. Aidala, V. J. Parameter Estimation via the Kalman Filter. *IEEE Transactions on Automatic Control*, **AC-22:** 471–472 (June 1977).
3. Anderson, Brian D. O., and Moore, John B. *Optimal Filtering*. Englewood Cliffs, New Jersey: Prentice Hall, 1979.
4. —— and ——. *Linear Optimal Control*. Englewood Cliffs, New Jersey: Prentice-Hall, 1971.
5. Andrade Netto, M. L., Gimeno, L., and Mendes, M. J. On the Optimal and Suboptimal Nonlinear Filtering Problem for Discrete-Time Systems. *IEEE Transactions on Automatic Control*, **AC-19:** 1062–1067 (December 1978).
6. Angel, Edward, and Jain, Anil K. A Dimensionality Reducing Model for Distributed Filtering. *IEEE Transactions on Automatic Control*, **AC-18:** 59–62 (February 1973).
7. Aoki, M. *Optimization of Stochastic Systems*. New York: Academic Press, 1967.

8. Arbel, Ami. Sensor Placement in Optimal Filtering and Smoothing Problems. *IEEE Transactions on Automatic Control*, **AC-27:** 94–98 (February 1982).
9. Athans, Michael. The Role and Use of the Stochastic–Linear–Quadratic–Gaussian Problem in Control System Design. *IEEE Transactions on Automatic Control*, **AC-16:** 529–551 (December 1971).
10. Badawi, Faris A., Lindquist, Anders, and Pavon, Michello. A Stochastic Realization Approach to the Smoothing Problem. *IEEE Transactions on Automatic Control*, **AC-24:** 878–888 (December 1979).
11. Battin, R. H. *Astronomical Guidance*. New York: McGraw-Hill, 1964.
12. Benveniste, Albert, and Choure, Christian. AR and ARMA Identification Algorithms of Levinson Type: An Innovations Approach. *IEEE Transactions on Automatic Control*, **AC-26:** 1243–1261 (December 1981).
13. Biswas, K. K., and Mahalanabis, A. K. On the Stability of a Fixed Lag Smoother. *IEEE Transactions on Automatic Control*, **AC-18:** 63–64 (February 1973).
14. Brotherton, T. W., and Caines, P. E. ARMA System Identification via the Cholesky Least Squares Method. *IEEE Transactions on Automatic Control*, **AC-24:** 698–702 (August 1978).
15. Bryson, Arthur E., and Ho Yu-Chi. *Applied Optimal Control*. Waltham, Massachusetts: Blaisdell, 1969.
16. Canno, F. Incertis, and Torres, J. M. Martinez. An Extension on a Reformulation of the Algebraic Riccati Equation Problem. *IEEE Transactions on Automatic Control*, **AC-22:** 128–129 (February 1977).
17. Carew, Burian, and Belanger, Piere. Identification of Optimum Filter Steady-State Gain for Systems with Unknown Noise Covariances. *IEEE Transactions on Automatic Control*, **AC-18:** 582–587 (December 1973).
18. Catlin, D. E. The Independence of Forward and Backward Estimation Errors in the Two-Filter Form of the Fixed Interval Kalman Smoother. *IEEE Transactions on Automatic Control*, **AC-25:** 1111–1115 (December 1980).
19. Chung, Richard C., and Belanger, Piere R. Minimum Sensitivity Filter for Linear Time-Invariant Stochastic Systems with Uncertain Parameters. *IEEE Transactions on Automatic Control*, **AC-21:** 98–100 (February 1976).
20. Fagin, S. L. Recursive Linear Regression Theory, Optimal Filter Theory, and Error Analysis of Optimal Systems. *IEEE Convention Record* **12:** 216–240 (1964).
21. Fam, Adly T. Word Length and Memory Requirements of the Integer Parts of Some Digital Control Parameters. *IEEE Transactions on Automatic Control*, **AC-27:** 496–498 (April 1982).
22. Fitzgerald, R. J. Divergence in the Kalman Filters. *IEEE Transactions on Automatic Control*, **AC-16:** 736–747 (December 1971).
23. Friedland, Bernard. On the Effect of Incorrect Gain in Kalman Filters. *IEEE Transaction on Automatic Control*, **AC-12:** 610–611 (October 1967).
24. ——. Steady-State Behavior of the Kalman Filter with Discrete- and Continuous-Time Observations. *IEEE Transactions on Automatic Control*, **AC-25:** 988–992 (October 1980).
25. Friedlander, B., Kailath, T., Morf, M., and Ljung, L. Extended Levinson and Chandrasekhar Equations for General Discrete-Time Linear Estimation Problems. *IEEE Transactions on Automatic Control*, **AC-23:** 653–659 (August 1978).
26. Frost, P., and Kailath, T. An Innovations Approach to Least-Squares Estimation–Part III: Non-Linear Estimation in White Gaussian Noise. *IEEE Transactions on Automatic Control*, **AC-16:** 217–226 (June 1971).
27. Gelb, Arthur. *Applied Optimal Estimation*. Cambridge, Massachusetts: M.I.T. Press, 1974.

28. Gevers, Michel R., and Kailath, T. An Innovations Approach to Least-Squares Estimation–Part VI: Discrete-Time Innovations Representations and Recursive Estimation. *IEEE Transactions on Automatic Control*, **AC-18:** 588–600 (December 1973).
29. Harvey, Charles, A., and Stein, Gunter. Quadratic Weights for Asymptotic Regulator Properties. *IEEE Transactions on Automatic Control*, **AC-23:** 378–387 (June 1978).
30. Hassan, M. F., Salut, G., Singh, M. G., and Titli, A. A Decentralized Computational Algorithm for the Global Kalman Filter. *IEEE Transactions on Automatic Control*, **AC-23:** 262–267 (April 1978).
31. Hedelin, Per, and Jonsson, Ingvar. Applying a Smoothing Criterion to the Kalman Filter. *IEEE Transactions on Automatic Control*, **AC-23:** 916–921 (October 1978).
32. Jacquot, Raymond G. *Modern Digital Control Systems.* New York: Marcel Dekker, 1981.
33. Jazwinski, A. H. Limited Memory Optimal Filtering. *IEEE Transactions on Automatic Control*, **AC-13:** 558–563 (October 1968).
34. ——. *Stochastic Processes and Filtering Theory.* New York: Academic Press, 1970.
35. Jones, E. L. A Reformulation of the Algebraic Riccati Equation. *IEEE Transactions on Automatic Control*, **AC-21:** 113 (February 1976).
36. Kalaith, Thomas. A View of Three Decades of Linear Filtering Theory. *IEEE Transactions on Information Theory*, **IT-20:** 146–180 (March 1974).
37. ——. An Innovations Approach to Least-Squares Estimation–Part I: Linear Filtering in Additive White Noise. *IEEE Transactions on Automatic Control*, **AC-13:** 646–655 (December 1968).
38. —— and Frost, P. An Innovations Approach to Least-Squares Estimation–Part II: Linear Smoothing in Additive White Noise. *IEEE Transactions on Automatic Control*, **AC-13:** 655–660 (December 1968).
39. —— and Geesey, R. An Innovations Approach to Least-Squares Estimation–Part IV: Recursive Estimation Given Lumped Covariance Functions. *IEEE Transactions on Automatic Control*, **AC-16:** 720–727 (December 1971).
40. —— and ——. An Innovations Approach to Least-Squares Estimation–Part V: Innovations Representations and Recursive Estimation in Colored Noise. *IEEE Transactions on Automatic Control*, **AC-18:** 435–453 (October 1973).
41. —— and Ljung, L. The Asymptotic Behavior of Constant-Coefficient Riccati Differential Equations. *IEEE Transactions on Automatic Control*, **AC-21:** 385–388 (June 1976).
42. Kaminiski, P. G., Bryson, A. E., and Schmidt, S. F. Discrete Square Root Filtering: A Survey of Current Techniques. *IEEE Transactions on Automatic Control*, **AC-16:** 727–735 (December 1971).
43. Krener, Authur J. Kalman-Bucy and Minimax Filtering. *IEEE Transactions on Automatic Control*, **AC-25:** 291–292 (April 1980).
44. Lee, Daniel T., Friedlander, Benjamin, and Morf, Marlin. Recursive Ladder Algorithms for ARMA Modeling. *IEEE Transactions on Automatic Control*, **AC-27:** 753–764 (August 1982).
45. Lee, Tony T. A Direct Approach to Identify the Noise Covariances of Kalman Filtering. *IEEE Transactions on Automatic Control*, **AC-25:** 841–842 (August 1980).
46. Ljung, L. Asymptotic Behavior of the Extended Kalman Filter as a Parameter Estimator for Linear Systems. *IEEE Transactions on Automatic Control*, **AC-24:** 36–50 (February 1979).
47. ——. Convergence Analysis of Parameter Identification Methods. *IEEE Transactions on Automatic Control*, **AC-23:** 770–783 (October 1978).
48. Madrid, G. A. and Bierman, G. J. Application of Kalman Filtering to Spacecraft Range

Residual Prediction. *IEEE Transactions on Automatic Control*, **AC-23:** 430–433 (June 1978).

49. Masreliez, C. J., and Martin, R. D. Robust Bayesian Estimation for the Linear Model and Robustifying the Kalman Filter. *IEEE Transactions on Automatic Control*, **AC-22:** 361–371 (June 1977).
50. Meditch, J. S. *Stochastic Optimal Linear Estimation and Control.* New York: McGraw-Hill, 1969.
51. Mehra, Raman K. On-Line Identification of Linear Dynamic Systems with Applications to Kalman Filtering. *IEEE Transactions on Automatic Control*, **AC-16:** 12–21 (February 1971).
52. Mendel, J. M. Computational Requirements for a Discrete Kalman Filter. *IEEE Transactions on Automatic Control*, **AC-16:** 748–758 (December 1971).
53. Morf, M., Levy, B. and Kailath, T. Square-Root Algorithms for the Continuous-Time Linear Least-Square Estimation Process. *IEEE Transactions on Automatic Control*, **AC-23:** 907–911 (October 1978).
54. Moroney, Paul, Willsky, Alan S., and Houpt, Paul K. The Digital Implementation of Control Compensators: The Coefficient Wordlength Issue. *IEEE Transactions on Automatic Control*, **AC-25:** 621–630 (August 1980).
55. Neal, S. R. Linear Estimation in the Presence of Errors in Assumed Plant Dynamics. *IEEE Transactions on Automatic Control*, **AC-12:** 592–594 (October 1967).
56. Noton, A. R. M. A Two-Level Form of the Kalman Filter. *IEEE Transactions on Automatic Control*, **AC-18:** 128–133 (April 1971).
57. Nuyan, Seyhan, and Carrol, Robert L. Minimal Order Arbitrarily Fast Adaptive Observers and Identifiers. *IEEE Transactions on Automatic Control*, **AC-24:** 289–297 (April 1979).
58. Orfandis, Sophocles J. An Exact Solution of the Time-Invariant Discrete Kalman Filter. *IEEE Transactions on Automatic Control*, **AC-27:** 240–242 (February 1982).
59. Panuska, V. A New Form of the Extended Kalman Filter for Parameter Estimation in Linear Systems with Correlated Noise. *IEEE Transactions on Automatic Control*, **AC-25:** 229–235 (April 1980).
60. Payne, H. J., and Silverman, L. M. On the Discrete Time Algebraic Riccati Equation. *IEEE Transactions on Automatic Control*, **AC-18:** 226–234 (June 1973).
61. Price, C. F. An Analysis of the Divergence Problem in the Kalman Filter. *IEEE Transactions on Automatic Control*, **AC-13:** 699–701 (December 1968).
62. Rhodes, Ian. A Tutorial Introduction to Estimation and Filtering. *IEEE Transactions on Automatic Control*, **AC-16:** 688–706 (December 1971).
63. Safanov, Michael G., and Athens, Michael. Robustness and Computational Aspects of Nonlinear Stochastic Estimators and Regulators. *IEEE Transactions on Automatic Control*, **AC-23:** 717–725 (August 1978).
64. Sage, Andrew P., and Melsa, James L. *Estimation Theory with Applications to Communications and Control.* New York: McGraw-Hill, 1971.
65. Sage, Andrew P., and White, Chelsea C. *Optimum Systems Control*, Second Edition. Englewood Cliffs, N.J.: Prentice-Hall, 1977.
66. Sorenson, H. W. Least Squares Estimation: From Gauss to Kalman. *IEEE Spectrum* **7:** 63–68 (July 1970).
67. Tajima, K. Estimation of Steady-State Kalman Filter Gain. *IEEE Transactions on Automatic Control*, **AC-23:** 944–945 (October 1978).
68. Tapley, B. D., and Choe, C. Y. An Algorithm for Propagating the Square-Root Covariance Matrix in Triangular Form. *IEEE Transactions on Automatic Control*, **AC-21:** 122–123 (February 1976).

69. Toda, M., and Patel, R. V. Bounds on Estimation Errors of Discrete-Time Filters Under Modeling Uncertainty. *IEEE Transactions on Automatic Control*, **AC-25**: 1115–1121 (December 1980).
70. Tou, Julius T. *Optimum Design of Digital Control Systems.* New York: Academic Press, 1963.
71. Tse, Edison. On the Optimum Control of Stochastic Linear Systems. *IEEE Transactions on Automatic Control*, **AC-16**: 776–785 (December 1971).
72. Ursin, Bjorn. Asymptotic Convergence Properties of the Extended Kalman Filter Using Filtered State Estimates. *IEEE Transactions on Automatic Control*, **AC-25**: 1207–1211 (December 1980).
73. Van Trees, Harry L. *Detection, Estimation, and Modulation Theory.* New York: John Wiley & Sons, 1968.
74. Weiss, H., and Moore, J. B. Improved Extended Kalman Filter Design for Passive Tracking. *IEEE Transactions on Automatic Control*, **AC-25**: 807–811 (August 1980).
75. Westphall, L. C. An Adaptive Observer Incorporating A Priori System Data. *IEEE Transactions on Automatic Control*, **AC-24**: 124–125 (February 1979).
76. Westerlund, Tapio. A Digital Quality Control System for an Industrial Dry Process Rotary Cement Kiln. *IEEE Transactions on Automatic Control*, **AC-26**: 885–890 (August 1981).
77. Yonezawa, Katsuo. Suboptimal Design of Discrete Kalman Filter and Smoother with Redundant Measurements. *IEEE Transactions on Automatic Control*, **AC-26**: 561–562 (April 1981).

DEVELOPMENTAL EXERCISES

7.1. A fundamental precept of estimation theory is the *orthogonality principle*, which states that, for minimum mean-squared error estimation, the estimation error must be orthogonal, in the statistical sense, to the data. That is,

$$E[(X - \hat{X})\mathbf{Z}] = \mathbf{0}$$

where X is a random variable and **Z** is a vector of observations. Use the orthogonality principle to develop the minimum mean-squared error linear estimate.

7.2. Generalize Exercise 7.1 to the multivariate case, for which the orthogonality principle yields

$$E[(\mathbf{X} - \hat{\mathbf{X}})\mathbf{Z}^T] = \mathbf{0}.$$

Derive the minimum mean-squared error linear estimate.

7.3. Find the frequency domain representation of the steady state Kalman filter.

7.4. If the measurement vector is processed one component at a time, considerable decrease in complexity is obtained because the matrix inversion is of the same dimension as the measurement vector. Other practical advantages also may accrue from use of this concept. Investigate sequential processing, showing where computational complexity is decreased and where it is increased.

7.5. One of the characteristics of the Kalman filter is that all measurements, no matter how far in the past, affect the current estimate. With perfect modeling, this feature is appropriate, but in many practical instances it produces near-singular covariance matrices for large t. One suggested remedy for this problem is exponential weighting of the covariance matrix, so that the effect of older data is gradually diminished. Such weighting can be achieved by arbitrarily multiplying the prior covariance matrix $\boldsymbol{P}(l|l-1)$ by a factor $\alpha > 1$ at each filter iteration. Investigate this concept and show that the suggested weighting is equivalent to arbitrarily increasing the noise.

PROBLEMS

7.1. The parameters x_1 and x_2 are to be estimated by the least squares method. The measurements

$$Z_1 = x_1 + x_2 + \epsilon_1 = 2$$

$$Z_2 = x_1 + \epsilon_2 \qquad = 1$$

are obtained.

(a) Find the estimates for x_1 and x_2.

(b) An additional measurement

$$Z_3 = x_1 + 2x_2 + \epsilon_3 = 4$$

is obtained. Find the revised estimates of x_1 and x_2.

(c) Repeat (b) using the recursive estimation algorithm.

7.2. Two observations of the parameter

$$\boldsymbol{x} = \begin{bmatrix} x_1 \\ x_2 \end{bmatrix}$$

are obtained as

$$Z_1 = 2x_1 + x_2 + \epsilon_1 = 2$$

$$Z_1 = x_2 \qquad + \epsilon_2 = 1.$$

(a) Using Z_1 and Z_2 as a single vector observation

$$\mathbf{Z} = \begin{bmatrix} Z_1 \\ Z_2 \end{bmatrix}$$

find the least squares estimate of $\boldsymbol{x}$.

(b) Use the result of (a) and a third measurement

$$Z_3 = x_1 + 2x_2 + \epsilon_3 = 4$$

to generate a least squares estimate based on all three measurements.

(c) Is the quality of the x_1 estimate improved by the new measurement? Is the quality of the x_2 estimate improved by the new measurement?

7.3. Given the observations

$$3x_1 + 4x_2 = 6$$

$$2x_1 + 3x_2 = 5$$

$$4x_1 + 3x_2 = 8$$

$$2x_1 + x_2 = 4.$$

If the measurement errors have a multivariate normal distribution with mean zero and covariance matrix.

$$E[\boldsymbol{\epsilon}\boldsymbol{\epsilon}^T] = \begin{bmatrix} 1 & 0 & 0 & 0 \\ 0 & \frac{1}{2} & 0 & 0 \\ 0 & 0 & 1 & 0 \\ 0 & 0 & 0 & \frac{1}{2} \end{bmatrix}$$

Find the minimum variance unbiased linear estimate of $\begin{bmatrix} x_1 \\ x_2 \end{bmatrix}$.

7.4. The position of a ship is to be determined with respect to a specified (x_1, x_2) coordinate grid. Theodolite measurements are made to determine the angular position of the ship with respect to three tracking stations. An assumed position of $x_1^* = 1000$ yards, $x_2^* = 1000$ yards, is used as a reference solution about which the equations are linearized. If the measurement perturbations are dy_1, dy_2, and dy_3, and the linearized equations evaluated at the reference solution are

$$dy_1 = dx_1 + 2dx_2$$

$$dy_2 = 4dx_1 - 6dx_2$$

$$dy_3 = 4dx_1 - 4dx_2$$

and the differences (in degrees) between the observed angles and those predicted by the assumed position are

$$dy_1 = 29$$

$$dy_2 = 21$$

$$dy_3 = 2$$

find a suitable estimate of x_1 and x_2. The three measurement errors are independent with variances of 1, 2, and 4, respectively. Assume correct units have been used in all equations. Once a solution is obtained, comment on how a refinement in the estimate might be determined.

7.5. Consider the following example of least-squares estimation. Given

$$6 = 3x_1 + 2x_2$$
$$1 = x_1 + x_2$$
$$-3 = 4x_1 - 6x_2$$
$$4 = 2x_1 + x_2$$

show that the solution is

$$\hat{x} = \begin{bmatrix} 1.11 \\ 1.23 \end{bmatrix}.$$

Suppose an additional measurement

$$3 = 2x_1 + 2x_2$$

is obtained. Find the revised estimate by two methods:
(a) Directly, by the equation $\hat{x} = (H^T H)^{-1} H^T Z$.
(b) Indirectly, by the recursive equations.

7.6. A weapon calibration range is arranged in the form of an equilateral triangle, with a theodolite station at each apex. The objective is to precisely determine the position of a ship within the range by use of the angular measurements obtained from the theodolites. The geometry is shown in Figure P7-1.

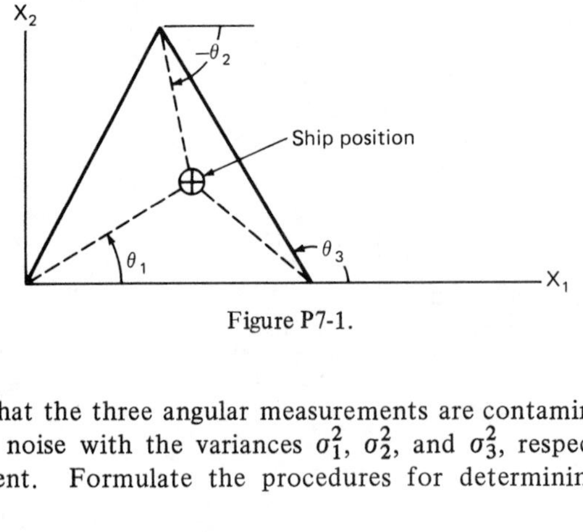

Figure P7-1.

Assume that the three angular measurements are contaminated with measurement noise with the variances σ_1^2, σ_2^2, and σ_3^2, respectively, and are independent. Formulate the procedures for determining an optimum estimate of the ship's position. List the elements of the various matrices involved.

7.7. Three tracking stations are used to determine the range and altitude of an aircraft target. The tracking devices are such that the individual stations measure only the following data:

Station 1	$Z_1 = 2r + h + \epsilon_1$
Station 2	$Z_2 = r + h + \epsilon_2$
Station 3	$Z_3 = r + \epsilon_3$

where r is range, h is altitude, and the ϵ's are measurement noise terms.

(a) Find $\boldsymbol{B}$ such that the linear estimate $\hat{x} = \boldsymbol{B}Z$ is the minimum variance, unbiased, linear estimate.

(b) If

$$Z_1 = 30{,}000$$

$$Z_2 = 21{,}000$$

$$Z_3 = 9{,}000$$

what are the estimates of r and h if the individual measurement noises are uncorrelated with variances of $\frac{1}{4}$, 1, and 2, respectively?

(c) What is the covariance between the errors on the r and h estimates? How is this to be interpreted?

(d) If a fourth measurement

$$Z_4 = r + 2h + \epsilon_4$$

is obtained, comment on the two possible approaches which may be used to include this measurement into the estimate.

7.8. In an experiment designed to measure the acceleration of gravity, the position of a test mass is made at four different times, $t_1 = 1$, $t_2 = 2$, $t_3 = 3$, and $t_4 = 4$. The position measurements are $x_1 = 25.3$ ft, $x_2 = 85.7$ ft, $x_3 = 174.3$ ft, and $x_4 = 298.4$ ft. The mass possesses some unknown initial velocity v_0 as it passes the zero reference point of the position measurement, so that the distance x is given by

$$x = v_0 t + \tfrac{1}{2} g t^2.$$

Find the least square estimates of v_0 and g. Comment on the relative probable errors in the v_0 and g estimates.

7.9. A discrete time system is described by the scalar equation

$$X(l) = 0.5X(l-1) + W(l)$$

with the measurement process

$$Z(l) = X(l) + V(l).$$

W(l) and V(l) are white noise sequences with covariance kernels

$$k_W(l, m) = \begin{cases} 1, & l = m \\ 0, & l \neq m \end{cases} \qquad k_V(l, m) = \begin{cases} 2, & l = m \\ 0, & l \neq m. \end{cases}$$

Formulate the Kalman filter equations and solve for $\hat{X}(l|l)$, $K(l)$, and $P(l|l)$ for $l = 1, 2, 3, 4$, and 5. Assume $E[X(0)]$ and $P(0|0) = 1$.

7.10. A scalar discrete time process is defined as

$$\begin{aligned} X(l+1) &= X(l) + W(l) \\ Z(l) &= X(l) + V(l) \end{aligned} \qquad l = 1, 2, 3$$

where W(l) and V(l) are zero mean white noise sequences with covariance kernels

$$k_W(l, m) = \begin{cases} 1, & l = m \\ 0, & l \neq m \end{cases} \qquad k_V(l, m) = \begin{cases} 1, & l = m \\ 0, & l \neq m. \end{cases}$$

Use point estimation techniques to find the minimum variance, unbiased, linear estimate of the vector

$$\mathbf{X} = \begin{bmatrix} X(1) \\ X(2) \\ X(3) \end{bmatrix}$$

Assume $E[X(0)] = 0$, VAR$[X(0)] = 0$.

7.11. A message is transmitted over a noisy communications channel. The message model is

$$\dot{X}(t) = -AX(t) + W(t)$$

where W(t) is a scalar white noise process with zero mean and covariance kernel $q\delta(t - \sigma)$. The received message is

$$Z(t) = X(t) + V(t)$$

where the channel noise V(t) is white noise sequence with covariance kernel $r\delta(t - \sigma)$.

(a) Find the minimum variance, unbiased, linear filter to extract an estimate of X(t) from the measurement Z(t).

(b) Suppose that measurement data $Z(t)$ is obtained until time t_1, after which it is not available. Find the prediction of $X(t)$ for $t > t_1$ and the estimation error covariance matrix associated with the prediction.

(c) Show and discuss the limiting behavior in (b) for $t \gg t_1$.

7.12. As a simple example of a continuous time Kalman filter, consider the system described by the scalar differential equation

$$\dot{X}(t) = 0$$

and measurement process

$$Z(t) = X(t) + V(t).$$

This is a model of estimation of the constant $X(0)$ by means of the time function $Z(t)$. If $X(0)$ has zero mean and variance 10, and if $V(t)$ is a zero mean white noise process with covariance kernel $\delta(t - \sigma)$, find the Kalman filter estimate equation. Does the result make intuitive sense? Relate this problem to bias estimation considered in Chapters 5 and 7.

7.13. Consider the continuous time system described by the equation

$$\dot{\mathbf{X}}(t) = \boldsymbol{F}\mathbf{X}(t) + \boldsymbol{D}W(t)$$

where

$$\boldsymbol{F} = \begin{bmatrix} 0 & 1 \\ -1 & 0 \end{bmatrix}; \qquad \mathbf{X}(t) = \begin{bmatrix} X_1(t) \\ X_2(t) \end{bmatrix}; \qquad \boldsymbol{D} = \begin{bmatrix} 0 \\ 1 \end{bmatrix}$$

The measurement equation is

$$Z(t) = X_1(t) + V(t).$$

$W(t)$ and $V(t)$ are zero mean white noise processes with covariance kernels $\delta(t - \sigma)$ and $3\delta(t - \sigma)$, respectively. Find and sketch the steady state Kalman filter for this system.

7.14. Consider the continuous time system described by the scalar system and measurement equations

$$\dot{X}(t) = -X(t) + W(t)$$

$$Z(t) = X(t) + V(t)$$

where $W(t)$ and $V(t)$ are zero mean white noise sequences with covariance kernels $2q\delta(t - \sigma)$ and $q\delta(t - \sigma)$, respectively. Find and sketch the steady state Kalman filter.

7.15. A continuous time system is described by the scalar equation

$$\dot{X}(t) = -2X(t) + W(t)$$

where $W(t)$ is a white noise process with covariance kernel $\delta(t - \sigma)$. The measurement process is

$$Z(t) = X(t) + V(t).$$

Use the Kalman filter predictor

$$\hat{X}(t + h) = \Phi(t, t + h)\hat{X}(t)$$

to design a steady state prediction filter to yield the prediction $X(t + 2|t)$, i.e., the estimate of $X(t + 2)$ based upon measurements up to $Z(t)$.

7.16. A reentry vehicle is tracked with a measurement and computation every ten seconds. Consider the one-dimensional problem, in which the equations of motion are

$$\dot{S} = V$$
$$\dot{V} = A$$

and A is a zero mean white noise sequence with covariance kernel $k_A = \delta(t - \sigma)$.

(a) At time t_l the prior estimates are $\hat{S}(l|l)$ and $\hat{V}(l|l)$ with estimation error variances of 100 ft^2 and 4 ft^2/sec^4, respectively, and the estimation errors are independent. Consider the acceleration A constant during the interval (t_l, t_{l+1}), and determine the prior estimates $\hat{S}(l + 1|l)$ and $\hat{V}(l + 1|l)$.

(b) If the position $S(t)$ is observed by radar, yielding the measurement equation

$$Z(t_l) + S(t_l) + V(t_l)$$

where $V(t)$ is a white noise sequence with covariance kernel $4\delta(t - \sigma)$, formulate the Kalman filter equations for estimating $S(l)$ and $V(l)$. (*Note:* Convert the continuous time problem to a discrete time problem.)

7.17. Consider the system shown in the diagram of Figure P7-2, where $W(t)$ and

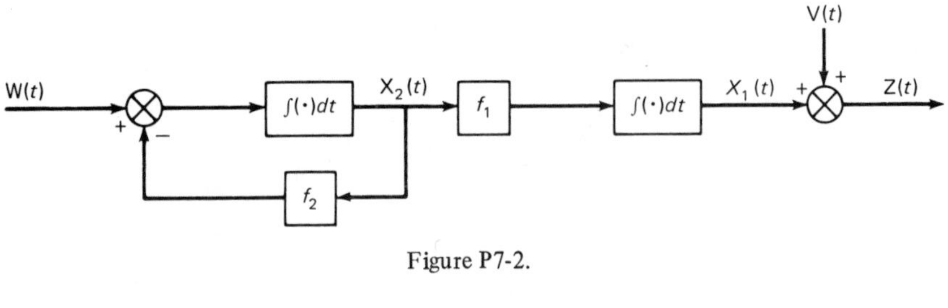

Figure P7-2.

V(t) are independent white noise processes with covariance kernels $q\delta(t - \sigma)$ and $r\delta(t - \sigma)$. Develop the Kalman filter equations for this system. Determine the Kalman gain matrix for the steady state filter, and the steady state value of the estimation error covariance matrix. Sketch the block diagram of the filter.

7.18. In the system of Problem 7.17, the measurement noise consists of the sum of a white noise term and Markov random process,

$$Z(t) = X(t) + V(t)$$

where V(t) is a Markov random process with zero mean and covariance kernel $q_c e^{-|t|/\tau}$. Develop the Kalman filter equations for this system.

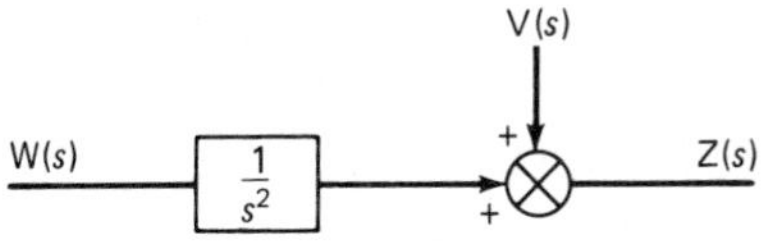

Figure P7-3.

7.19. Consider the simple second-order system shown in Figure P7-3. Formulate the Kalman filter for this system. W(t) and V(t) are independent white noise processes with covariance kernels $q\delta(t - \sigma)$ and $r\delta(t - \sigma)$, respectively.

7.20. Consider the system given by the difference equation

$$\mathbf{X}(l + 1) = \mathbf{\Phi X}(l) + \mathbf{\Delta W}(l)$$

where

$$\mathbf{\Phi} = \begin{bmatrix} 1 & 1 \\ 0 & 1 \end{bmatrix}; \quad \mathbf{\Delta} = \begin{bmatrix} 0 \\ 1 \end{bmatrix}; \quad \mathbf{X}(l) = \begin{bmatrix} X_1(l) \\ X_2(l) \end{bmatrix}$$

and W(l) is a scalar, zero mean, white noise sequence with covariance kernel

$$k_W(l, m) = \begin{cases} 1, & \text{if } l = m \\ 0, & \text{if } l \neq m. \end{cases}$$

The measurement process is

$$Z(l) = X_1(l) + V(l)$$

where V(l) is a scalar, zero mean, white noise sequence, independent of

W(l), with covariance kernel

$$k_V(l, m) = \begin{cases} 2 + (-1)^l, & l = m \\ 0, & l \neq m \end{cases}$$

(a) Develop the Kalman filter equations for this system.
(b) Determine and plot the diagonal elements of the estimation error covariance matrix $\boldsymbol{P}(l|l)$ for $l = 1, 2, 3, 4,$ and 5, assuming that the initial value is

$$\boldsymbol{P}(0|0) = \begin{bmatrix} 10 & 0 \\ 0 & 10 \end{bmatrix}.$$

(c) Determine the Kalman gain matrix $\boldsymbol{K}(l)$ for $l = 1, 2, 3, 4,$ and 5, and explain the behavior noted for odd and even values of l.

7.21. Consider the scalar system described by the equations

$$\dot{X}(t) = -f\,X(t) + W(t)$$

$$Z(t) = X(t) + V(t)$$

where W(t) and V(t) are white noise processes with covariance kernels $q\delta(t-\sigma)$ and $r\delta(t-\sigma)$. Find the steady state Kalman filter for this system. Sketch the block diagram of the filter.

7.22. The Wiener random process can be represented in differential equation form as

$$\dot{X}(t) = aW(t); \qquad X(0) = 0$$

where W(t) is a normal white noise process with zero mean and covariance kernel $q\delta(t-\sigma)$. If a noisy observation of the random process X(t) is obtained as

$$Z(t) = X(t) + V(t)$$

where V(t) is a white noise process with zero mean and variance $r\delta(t-\sigma)$, find the time varying Kalman filter for estimating X(t) from the measurement Z(t).

7.23. The system equation is

$$\dot{\mathbf{X}}(t) = \boldsymbol{F}\mathbf{X}(t) + \boldsymbol{D}W(t)$$

where

$$\boldsymbol{F} = \begin{bmatrix} 0 & 1 & 0 \\ 0 & 0 & 1 \\ -1 & 0 & -2 \end{bmatrix}; \quad \boldsymbol{D} = \begin{bmatrix} 0 \\ 0 \\ 1 \end{bmatrix}; \quad \mathbf{X}(t) = \begin{bmatrix} X_1(t) \\ X_2(t) \\ X_3(t) \end{bmatrix}$$

It is desired to formulate the Kalman filter equations for this system, when the measurement equation is

$$\mathbf{Z}(t) = \boldsymbol{H}\mathbf{X}(t) + \mathbf{V}(t)$$

where

$$\boldsymbol{H} = \begin{bmatrix} 1 & 0 & 0 \\ 0 & 1 & 0 \end{bmatrix}; \qquad \mathbf{V}(t) = \begin{bmatrix} V(t) \\ 0 \end{bmatrix}.$$

W(t) and V(t) are zero mean, white noise processes. It is immediately noted that $X_2(t)$ is measured noise free, and that this characteristic presents a problem for the continuous time filter because the matrix $\boldsymbol{R}(t)$ is singular. Show, by differentiating the measurement $Z_2(t)$, that $X_3(t)$ can also be estimated without error, and that by differentiating a second time, a pseudo-measurement $Z_3(t) = \ddot{Z}_2(t)$ is defined, containing the noise term W(t). Find the Kalman filter for the remaining system involving the system state $X_1(t)$ and noise contaminated measurements $Z_1(t)$ and $Z_4(t) = \dot{Z}_2(t)$. (This approach is representative of a general technique for dealing with noise-free measurements in continuous time systems.)

7.24. Repeat Problem 7.23 for the case in which W(t) and V(t) are white noise processes but are correlated.

7.25. Repeat Problem 7.23 for the case in which W(t) and V(t) are uncorrelated, but V(t) is sequentially correlated and modeled as a Markov process.

7.26. A satellite in earth orbit makes observation of a fixed target on the earth's surface. These observations consist of angular measurements as shown in Figure P7-4. The measurement is contaminated with white noise.

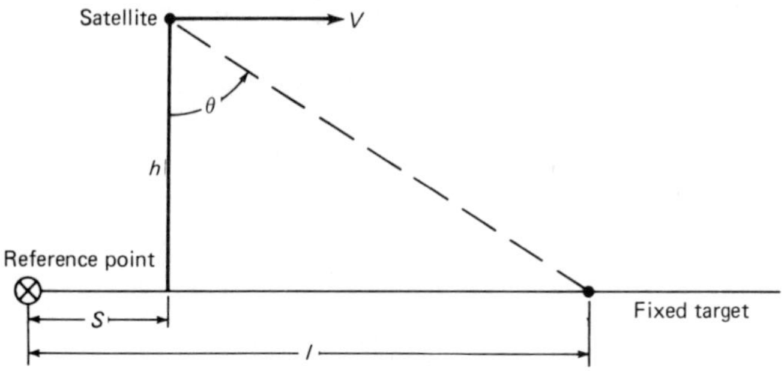

Figure P7-4.

Formulate the Kalman filter equations to provide estimates for the position s, altitude h, and velocity v.

7.27. Consider the scalar linear system with nonlinear measurement

$$\dot{X}(t) = -X(t) + u(t) + W(t)$$

$$Z(t) = X^3(t) + V(t)$$

where $u(t)$ is a deterministic input function. Derive an extended Kalman filter for this system. W(t) and V(t) are uncorrelated, and both are white noise sequences.

7.28. Repeat Problem 7.27, but now use the concept of a second-order Taylor series expansion around the reference solution.

7.29. Consider the scalar nonlinear system with linear measurement

$$\dot{X}(t) = -X^3(t) + W(t)$$

$$Z(t) = X(t) + V(t)$$

where W(t) and V(t) are white noise processes. Derive an extended Kalman filter for this system.

7.30. Repeat Problem 7.29, but now use the concept of a second-order Taylor series expansion around the reference solution.

7.31. Consider the nonlinear system with nonlinear measurement equation

$$\dot{X}(t) = [1 + aX^2(t)]\,W(t)$$

$$Z(t) = \arctan\,[bX(t)] + V(t)]$$

where W(t) and V(t) are white noise processes. Derive an extended Kalman filter for this system.

7.32. Consider the second-order linear system with nonlinear measurement

$$\dot{\mathbf{X}}(t) = \boldsymbol{F}\,\mathbf{X}(t) + \boldsymbol{D}\,\mathbf{W}(t)$$

$$Z(t) = \arctan\,[X_1(t)] + V(t)].$$

where **W**(t) and V(t) are white noise processes. Derive an extended Kalman filter for this system.

7.33. The equation describing a second-order nonlinear system is

$$\dot{\mathbf{X}} = \boldsymbol{f}(\mathbf{X}(t), W(t)$$

where

$$\boldsymbol{f} = \begin{bmatrix} X_2(t) \\ [W(t) - X_1(t)]^2 \end{bmatrix}.$$

The measurement equation is linear and noise free:

$$Z(t) = X_1(t).$$

Show, by differentiating Z(t), that the system state can be estimated without error.

7.34. Consider a rocket of mass M in vertical flight as shown in Figure P7-5.

Assume that the rocket is subject to a drag force $d = M\rho(\dot{h})^2 e^{-h}$, where $\rho > 0$. (The model used here is that of an exponential atmosphere wherein the air density, and therefore the drag, decreases exponentially with altitude.) Assume further that the rocket is also acted upon by a gravitational force

$$g(t) = kM/[h_0 + h(t)]^2$$

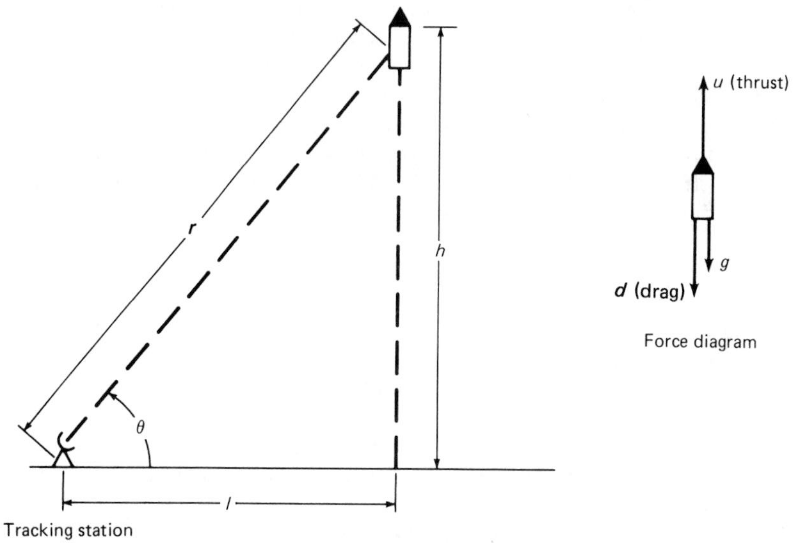

Figure P7-5.

where k and h_0 are positive constants, and by a thrust force $u(t)$ (upward). For simplicity, assume also that the mass of the rocket may be considered constant over the region of flight which is of interest.

(a) Determine the linearized equations of motion for a nominal $h(t)$, $\dot{h}(t)$, and $u(t)$.

(b) Assuming that the rocket is tracked during its vertical flight by measuring its range r and its elevation angle θ at a tracking station located as shown ($l > 0$ is a known distance), determine the linearized measurement equations. Assume that measurement errors are additive, i.e.,

$$\mathbf{Z}(t) = \begin{bmatrix} r(t) + V_r(t) \\ \theta(t) + V_\theta(t) \end{bmatrix}$$

where $V_r(t)$ = error in range measurement and $V_\theta(t)$ = error in elevation angle measurement.

7.35. Consider the scalar system described by the equation

$$\dot{X}(t) = X(t) + W(t)$$

with the measurement equation

$$Z(t) = X(t) + V(t)$$

where $W(t)$ and $V(t)$ are zero mean white noise processes with covariance kernels $q\delta(t - \sigma)$ and $r\delta(t - \sigma)$, respectively.

(a) Determine the Kalman filter equations for this system.

(b) The $X(t)$ equation resulting from (a) is a first-order, nonhomogeneous, time varying differential equation. Solve this equation for a $\hat{X}(t)$ as a function of time. The limit of this value as $t \to \infty$ is the time invariant filter resulting from application of the Weiner theory to this system.

7.36. Consider the bias estimation problem for which

$$\dot{X}(t) = 0; \qquad E[X] = 1; \qquad \text{VAR}[X] = 10$$

and the measurement equation is

$$Z(t) = 2X(t) + V(t)$$

where $V(t)$ is a zero mean white noise process with covariance kernel $\delta(t - \sigma)$. Find the equation for the estimate of $X(t)$ and sketch the resulting system.

7.37. Consider the first-order system described by the following equation:

$$\dot{X}(t) = -\tfrac{1}{2}X(t) + W(t)$$

where $W(t)$ is a white noise process with covariance kernel $2\delta(t - \sigma)$. The measurement equation is

$$Z(t) = X(t) + V(t)$$

where $V(t)$ is a white noise process with covariance kernel $0.25\delta(t - \sigma)$.

(a) Convert these continuous time equations to their discrete time equivalent, using time interval Δt.

(b) Using the results of (a), form the discrete time Kalman filter for the system.

7.38. The one-dimensional equation of motion of an inertial mass is simply

$$\dot{X}_1(t) = X_2(t)$$

$$\dot{X}_2(t) = A(t)$$

where X_1 is the position, X_2 the velocity, and A the acceleration of the mass.

(a) Model the acceleration as a Markov process and determine the augmented state transition matrix.
(b) Determine the transition matrix $\mathbf{\Phi}(\Delta t)$, and write the discrete version of the equation,

$$X(l) = \mathbf{\Phi}X(l - 1).$$

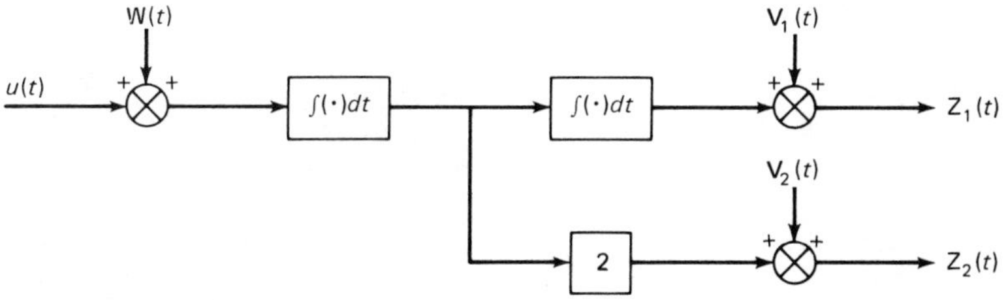

Figure P7-6.

7.39. Consider the system described by the block diagram of Figure P7-6. Find the optimal steady state Kalman filter if $W(t)$, $V_1(t)$ and $V_2(t)$ are uncorrelated, zero mean, white noise processes with covariance kernels $\delta(t - \sigma)$, $16\delta(t - \sigma)$, and $\delta(t - \sigma)$, respectively.

7.40. Repeat Problem 7.39, but now consider the case in which only the $Z_1(t)$ output is available.

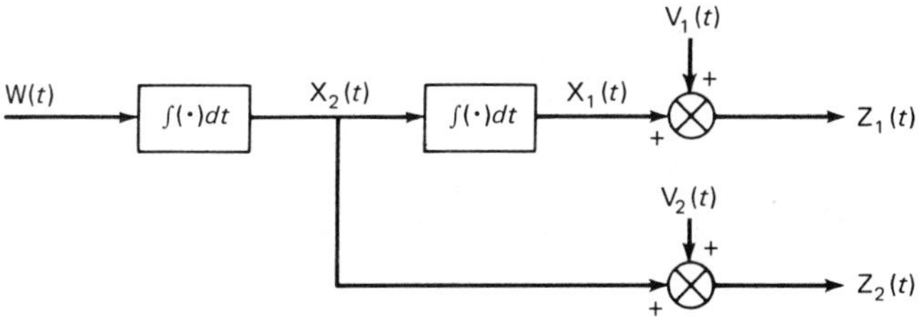

Figure P7-7.

7.41. Consider the Kalman filter for the system shown in Figure P7-7. The input function $W(t)$ is a zero mean white noise process with covariance kernel $\delta(t - \sigma)$, and the measurement noises are independent, zero mean white noise processes with covariance kernel $2\delta(t - \sigma)$. Assume the initial states X_1 and X_2 are uncorrelated, zero mean random variables with variance 4. Sketch the block diagram of the filter and relate this problem to a physical system. Solve the estimation error covariance matrix equation.

7.42. Repeat Problem 7.41 for the case in which only Z_1 is available for measurement.

7.43. Repeat Problem 7.41 for the case in which fixed biases b_1 and b_2 are present in the measurements Z_1 and Z_2, respectively.

7.44. Repeat Problem 7.41 for the case in which the input noise is a zero mean, first-order Markov process with covariance kernel $k_W(t, \sigma) = e^{-|t-\sigma|}$.

7.45. Consider the discrete-time system described by the equations

$$X_1(l) = X_1(l - 1) + X_2(l - 2) + W_1(l - 1)$$

$$X_2(l) = X_2(l - 1) + W_2(l - 1).$$

The measurement equation is

$$Z(l) = X_1(l) + V(l).$$

The input sequences $W_1(l)$ and $W_2(l)$ are uncorrelated, zero mean white noise sequences with covariance kernels q_1 and q_2, respectively, and $V(l)$ is a zero mean white noise sequence with covariance kernel r. Determine the Kalman filter equations for this system.

7.46. Kalman filtering is to be used to track an aircraft flying at constant altitude. The system model is to have four states, two velocity coordinates and two position coordinates. Acceleration is to be modeled as white noise. Range and azimuth measurements are converted to x and y position coordinates prior to filtering. The time between measurements t is not fixed, but varies with each measurement.

Formulate the Kalman filter equations for this system in terms of the time between measurements Δt, the range and azimuth error covariances σ_r^2 and σ_a^2, and the acceleration noise error covariances σ_x^2 and σ_y^2. This will require determination of the state transition matrix. Discuss how these equations would be used to produce the filtered estimates.

Chapter 8
Dynamic Programming as a System Optimization Technique

8.1. DYNAMIC PROGRAMMING

The system optimization problems encountered in Chapter 6 were based upon the classical theory of the calculus of variations, as generalized by the work of Pontryagin. In the general case, the result was a two-point boundary value problem with its inherent solution difficulty under all but linear, time invariant conditions. Dynamic programming is a mathematical programming technique, developed as a result of the computational power provided by the modern digital computer, which provides potential for numeric solutions to system optimization problems. The basic premise underlying the concept of dynamic programming also permits development of alternative forms of the necessary conditions for system optimization.

Dynamic programming is applicable to those problems which can be represented as a multistage decision process. Since the stages usually consist of time increments, the systems to which the technique is applied are usually dynamic systems. Lest this be misleading, however, it should be noted that many decision processes which do not actually involve time can be considered to be sequences of decisions and thus can be treated as a dynamic problem. This concept should become clear as some examples are considered.

Because of its wide range of capabilities, dynamic programming is applicable in many areas pertinent to the system analysis problem. In optimal control theory it permits approximate solutions to the nonlinear problem; in inventory theory it is applicable to those problems with time varying demand; it can be used in reliability analysis to determine optimum redundancy configurations. There are many other applications, some of which will undoubtably come to the reader's mind as the discussion progresses.

Before proceeding, it is necessary to temper the present discussion by pointing out two primary difficulties encountered in the application of dynamic programming. The first of these is an operational problem in that dynamic programming is not a neat, easily applied collection of algorithms, such as is the case with linear programming. It is instead a general approach to a class of problems which involves a progression through a number of stages, each of which requires a decision of some sort. Secondly, the process itself generates a large number of variables, which will become apparent as the discussion progresses, and therefore suffers from the issue of dimensionality as the problem itself grows large. In what follows, the general concept is developed, followed by specific examples of the application of that concept to various types of problems.

8.2. THE GENERAL PROBLEM

The method of dynamic programming is applicable to those problems which can be represented as multistage decision processes. As an introduction to this type of problem, consider the single-stage decision process represented by Figure 8-1. Here, the output y is determined by the input x, the decision d, and the transition process T,

$$y = T(x, d). \tag{8.1}$$

In effecting this transition, a return function is generated which may depend on x, d, and y. Thus,

$$j = j(x, y, d) = j(x, d) \tag{8.2}$$

where the second form results from application of Equation (8.1). That this representation can be used to describe many physical processes encountered in analysis is evident. Usually the return function j is to be optimized by proper selection of the decision d.

The multistage process is obtained when the output of a single-stage process serves as the input to a succeeding stage, and the output of that stage is the input to yet another stage, and so on. Such an arrangement is shown in Fig-

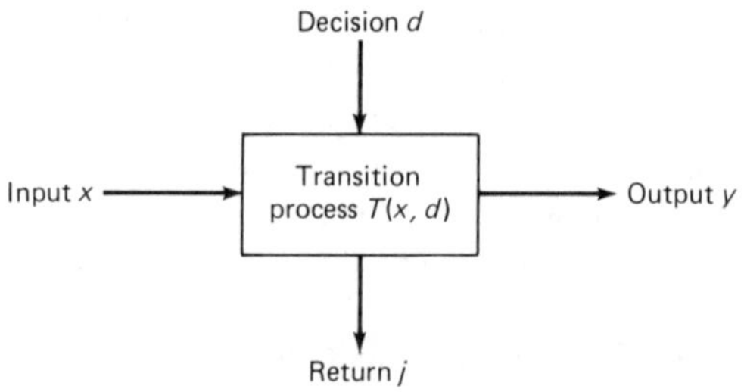

Figure 8-1. A single-stage decision process.

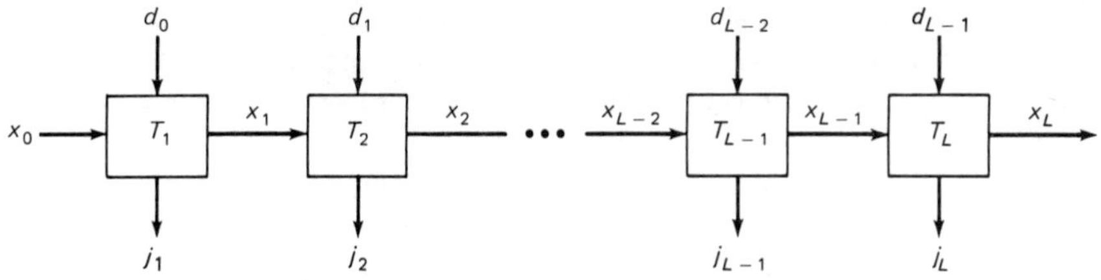

Figure 8-2. A multistage decision process.

ure 8-2, where the process consists of L stages. In this situation the process return

$$J = J(j_1, j_2, \ldots, j_L) \tag{8.3}$$

is to be optimized by proper selection of the decisions $d_0, d_1, \ldots, d_{L-1}$, subject to the constraints imposed by the individual stage transitions

$$x_l = T_l(d_{l-1}, x_{l-1}). \tag{8.4}$$

(The similarity to the calculus of variations problem considered in Appendix B should be obvious here.)

It is apparent that the decision d_{l-1} cannot be chosen on the basis of only the input to the lth stage, for the decision d_{l-1} not only determines the stage return j_l, but also the input to the following stage, which in turn affects all the subsequent stage returns. The sole exception is the last stage, since the output of the last stage does not affect any other stage. If the output of a particular stage is termed the state at that stage, then the underlying concept of dynamic programming can be expressed as follows:

> An optimal policy has the property that no matter what the previous decisions and states have been, the remaining decisions must constitute an optimal policy with regard to the states resulting from those previous decisions.

This is the so-called *principle of optimality*, and its repeated application, starting with the last stage and working backward, is the essence of the dynamic programming technique. It will be illustrated by a general example, to be followed by more specific application.

Consider the process described by Figure 8-2, for which the process return is defined as

$$J(j_1, j_2, \ldots, j_L) = \sum_{l=1}^{L} j_l \tag{8.5}$$

This form makes the problem even more representative of the calculus of variations, where J represents the functional to be optimized. Consider also that the optimization process is one of maximization. The problem can then be stated as

$$\max[j_1(x_0, d_0) + j_2(x_1, d_1) + \cdots + j_L(x_{L-1}, d_{L-1})] \tag{8.6}$$

subject to the constraints

$$x_l = T_l(x_{l-1}, d_{l-1}), \quad \text{for} \quad l = 1, 2, \ldots, L.$$

The maximization is with respect to the decision variables $d_0, d_1, \ldots, d_{L-1}$. It should be apparent that the optimum value of J is a function only of the input to the first stage x_0, assuming that all the decisions are made in an optimal manner. This fact can be represented by defining the function

$$J_1(x_0) = \max[j_1(x_0, d_0) + j_2(x_1, d_1) + \cdots] \tag{8.7}$$

which in turn can be expressed as

$$J_1(x_0) = \max(\text{w.r.t. } d_0)\{\max(\text{w.r.t. } d_1, d_2, \ldots, d_{L-1})[j_1(x_0, d_0) + j_2(x_1, d_1) + \cdots]\}. \tag{8.8}$$

But the first term in the summation, $j_1(x_0, d_0)$, is not dependent upon $d_1, \ldots, d_{L-1}$, so that Equation (8.8) can also be represented as

$$J_1(x_0) = \max(\text{w.r.t. } d_0)\{j_1(x_0, d_0) + \max(\text{w.r.t. } d_1, d_2, \ldots, d_{L-1})[j_2(x_1, d_1) + j_3(x_2, d_2) + \cdots]\}. \tag{8.9}$$

If the term

$$J_2(x_1) = \max(\text{w.r.t. } d_1, d_2, \ldots, d_{L-1})[j_2(x_1, d_1) + j_3(x_2, d_2) + \cdots] \tag{8.10}$$

is defined, then Equation (8.9) reduces to

$$J_1(x_0) = \max(\text{w.r.t. } d_0)[j_1(x_0, d_0) + J_2(x_1)]. \tag{8.11}$$

Note the implication of this result—the optimum value of J is given by the maximum, with respect to d_0, of the sum of the return from the current stage and the maximum of the return from the following stages, if optimal policies are followed there. The second term within the brackets, $J_2(x_1)$, is dependent upon d_0 because x_1 is a function of d_0, as implied by the transition equation

$$x_1 = T_1(x_0, d_0). \tag{8.12}$$

Then Equation (8.10) can be represented as

$$\begin{aligned} J_2(x_1) &= \max(\text{w.r.t. } d_1)\{\max(\text{w.r.t. } d_2, d_3, \ldots, d_{L-1})[\,j_2(x_1, d_1) \\ &\quad + j_3(x_2, d_2) + \cdots]\} \\ &= \max(\text{w.r.t. } d_1)\{\,j_2(x_1, d_1) \\ &\quad + \max(\text{w.r.t. } d_2, d_3, \ldots, d_{L-1})[\,j_3(x_2, d_2) + j_4(x_3, d_3) + \cdots]\} \end{aligned} \tag{8.13}$$

As before, a new function is defined,

$$J_3(x_2) = \max(\text{w.r.t. } d_2, d_3, \ldots, d_{L-1})[\,j_3(x_2, d_2) + j_4(x_3, d_3) + \cdots] \tag{8.14}$$

which permits Equation (8.13) to be expressed as

$$J_2(x_1) = \max(\text{w.r.t. } d_1)[\,j_2(x_1, d_1) + J_3(x_2)]. \tag{8.15}$$

This equation implies that $J_2(x_1)$ is the maximum, with respect to d_1, of the sum of the return from stage 2 and the maximum of the return from the following stages, if optimal policies are followed there. Proceeding in this manner results in the sequence of functions

$$\begin{aligned} J_3(x_2) &= \max(\text{w.r.t. } d_2)[\,j_3(x_2, d_2) + J_4(x_3)] \\ J_4(x_3) &= \max(\text{w.r.t. } d_3)[\,j_4(x_3, d_3) + J_5(x_4)] \\ &\;\;\vdots \\ J_{L-1}(x_{L-2}) &= \max(\text{w.r.t. } d_{L-2})[\,j_{L-1}(x_{L-2}, d_{L-2}) + J_L(x_{L-1})] \\ J_L(x_{L-1}) &= \max(\text{w.r.t. } d_{L-1})\, j_L(x_{L-1}, d_{L-1}). \end{aligned} \tag{8.16}$$

It is here that the principle of optimality is used, since the last function can be readily determined, being the result of a single-stage decision process. For every possible value of x_{L-1}, the optimum return $J_L(x_{L-1})$ can be determined. Then, the solution process proceeds in the reverse order, as follows:

I. Determine $J_L(x_{L-1})$, a function of x_{L-1}, as

$$\max(\text{w.r.t. } d_{L-1})[\,j_L(x_{L-1}, d_{L-1})].$$

II. Determine $J_{L-1}(x_{L-2})$ as

$$\max(\text{w.r.t. } d_{L-2})[j_{L-1}(x_{L-2}, d_{L-2}) + J_L(x_{L-1})].$$

III. Continue this process until $J_1(x_0)$ is determined as

$$\max(\text{w.r.t. } d_0)[j_1(x_0, d_0) + J_2(x_1)].$$

Normally x_0 will be known, but if it is not, or if a functional relationship is desired, $J_1(x_0)$ may be obtained for any desired range of x_0.

Note that the dynamic programming solution begins with the terminal condition and proceeds backward through the problem, a characteristic of dynamic optimization procedures.

While the preceding development has tacitly assumed that there were no restrictions on either the system state $x(l)$ or the decision d_l, it is apparent from the development that such restrictions can be easily incorporated by considering only admissible values in each stage of optimization. In fact, in application to control problems the restrictions on the control variables is almost a necessity in order to limit the dimensionality of the overall problem.

The general discussion above concerns the particular case in which the objective function to be optimized, as given by Equation (8.5), is the sum of the individual stage returns. While this arrangement is representative of a large class of problems to which the method can be applied, it is certainly not a practical restriction. In fact, the only requirement imposed on the objective function is that it be separable in some way into the individual stage returns. Certainly, the objective function of Equation (8.5) satisfies this criterion, as do a wide range of other combinations of stage returns. One form which will be encountered in the subsequent discussion is the product

$$L = \prod_{l=1}^{L} j_l \tag{8.17}$$

which also satisfies the separability criterion. This form arises quite often when the objective function is a probability of some sort. An example of an objective function which is not separable is

$$J(j_1, j_2, \ldots, j_L) = \left[\sum_{l=1}^{M} j_l\right]\left[\sum_{l=M+1}^{L} j_l\right]. \tag{8.18}$$

That this function fails the test of separability can be seen from the presence of product terms such as $j_1 j_l$, which prevent the stage-by-stage assessment of optimum values characteristic of the general approach. The determination of the separability of the objective function must be evaluated for each particular problem, based upon whether or not the principle of optimality applies.

As a final note before proceeding, it should be pointed out that uniformity in the separation is not a requirement. That is, each stage return does not necessarily have to relate to the objective function in the same manner. For example, the objective function

$$J(j_1, j_2, \ldots, j_L) = \sum_{l=1}^{M} j_l + \prod_{l=M+1}^{L} j_l \tag{8.19}$$

is separable in terms of the concepts just discussed.

Although all problems to which dynamic programming technqiues can be applied do not necessarily fit the mold of the general problem presented above, the approach is indicative of that to be applied to any particular problem. Particular problems will vary according to such characteristics as discrete or continuous states, deterministic or probabilistic transitions, dynamic or static transitions, and the like, but the concept of the application of the principle of optimality to a progression of stages, then working backward through the problem, is common to all.

8.3. DISCRETE STATE PROBLEMS

As is usual in situations such as this, a discrete state problem is more easily understood than one involving continuous states. For this reason, the discrete state problem is considered first, in the form of two examples. The first is a deterministic problem which is not dynamic in nature but can be considered so for application of the dynamic programming technqiues. The second problem involves a probabilistic transition which is dynamic in nature. These two examples serve to illustrate the general approach.

8.3.1. Static Discrete State Examples

In Section 8.1, it was noted that certain multiple decision processes can be considered to occur sequentially in time, in which case the optimization problem can be considered one of dynamic programming even though the dynamic characteristic of the system is completely suppressed. The example to be considered here is of that type.

Consider a situation in which six technicians are available for duty on four different ships, and the return from such assignments can be represented by a return matrix given in Table 8-1. The elements of the matrix might be dollars

Table 8-1. Return Matrix for Example Problem, Money Saved ($1000).

Men Assigned \ Ship	1	2	3	4
0	0	0	0	0
1	2	3	4	1
2	5	6	4	3
3	7	8	4	8
4	8	9	3	8
5	9	10	2	7
6	10	11	1	6

saved by the repair of some critical equipment. Notice that on ships 1 and 2, the return is a nondecreasing function, while for ships 3 and 4 the return increases and then decreases as the number of men assigned is increased. Such a situation could easily exist in the event that the labor cost of idle time becomes a major factor. In any event, the problem is to determine the optimum assignment of men to ships so that the return is maximized.

To consider this a dynamic programming problem, one can think of the assignments being made sequentially, in which case the situation is described as in Figure 8-3, where the initial state has the value of six and subsequent states are the number of men left to be assigned. The decisions are the number of men assigned to each ship, and the transition process is merely a subtraction of the number of men assigned from the number available. Here, the process return is the sum of the individual stage returns.

In accordance with the general approach, the solution begins at the last stage. In terms of the general developments of Section 8.2, the optimum return is $J_4(x_3)$, which, as a function of x_3, can be presented in tabular form as shown in Table 8-2, where d_3^* is the value of d_3 producing J_4. Clearly, $J_4(x_3)$ as given here represents the return from stage 4 if there are x_3 men available for assignment to ship 4, and the optimum assignment is made. The next step in the solu-

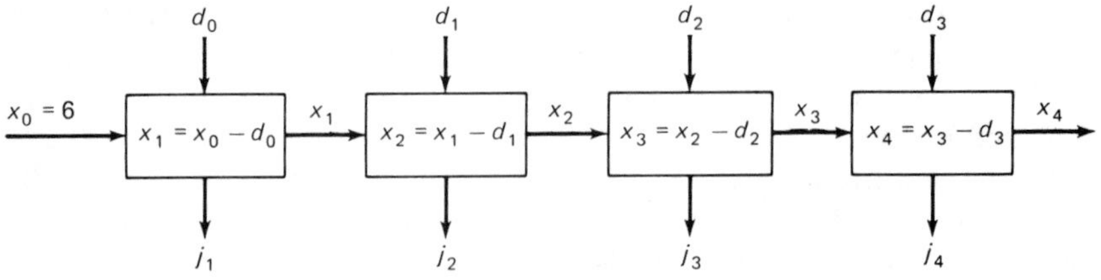

Figure 8-3. Example problem as a multistage decision process.

Table 8-2. Last-Stage Return Function $J_4(x_3)$.

x_3	$J_4(x_3)$	$d_3^*(x_3)$	x_3	$J_4(x_3)$	$d_3^*(x_3)$
0	0	0	4	8	3, 4
1	1	1	5	8	3, 4
2	3	2	6	8	3, 4
3	8	3			

tion is to determine

$$J_3(x_2) = \max(\text{w.r.t. } d_2)[j_3(x_2, d_2) + J_4(x_3)]$$

which represents the maximum return from stages 3 and 4 if there are x_2 men available for assignment to ships 3 and 4, and optimum assignments are made. Since a large number of possibilities must be considered and the solution is to be presented here in its clearest form, this information is given in tabular format in Table 8-3, showing all possible combinations. For example, if $x_2 = 3$, then

Table 8-3. Third-Stage Optimization.

x_2	d_3	x_3	$j_3(x_2, d_2)$	$+ J_4(x_3)$	$= S_3(x_2, x_3, d_2)$	$J_3(x_2)$	$d_2^*(x_2)$
0	0	0	0	0	0	0	0
1	0	1	0	1	1		
	1	0	4	0	4	4	1
2	0	2	0	3	3		
	1	1	4	1	5	5	1
	2	0	4	0	4		
3	0	3	0	8	8	8	0
	1	2	4	3	7		
	2	1	4	1	5		
	3	0	4	0	4		
4	0	4	0	8	8		
	1	3	4	8	12	12	1
	2	2	4	3	7		
	3	1	4	1	5		
	4	0	3	0	3		
5	0	5	0	8	8		
	1	4	4	8	12	12	1
	2	3	4	8	12	12	2
	3	2	4	3	7		
	4	1	3	1	4		
	5	0	2	0	2		
6	0	6	0	8	8		
	1	5	4	8	12	12	1
	2	4	4	8	12	12	2
	3	3	4	8	12	12	3
	4	2	3	3	6		
	5	1	2	1	3		
	6	0	1	0	1		

Table 8-4. Third-Stage Return Function $J_3(x_3)$.

x_2	$J_3(x_2)$	$d_2^*(x_2)$	x_2	$J_3(x_2)$	$d_2^*(x_2)$
0	0	0	4	12	1
1	4	1	5	12	1, 2
2	5	1	6	12	1, 2, 3
3	8	0			

there are four possible values of the decision d_2; these values are 0, 1, 2, and 3. These decision values in turn produce x_3 values of 3, 2, 1, and 0, respectively, as obtained from the transition equation. For each value of x_3, there is a corresponding value of J_3, as obtained from the calculations pertaining to stage 4, and these are 8, 3, 1, and 0, respectively. By summing j_3 and J_4 for each value of d_2, one obtains 8, 5, 7, and 4, respectively, for d_2 values of 3, 2, 1, and 0. The maximum of these is 8, obtained with d_2 equal to zero, and this result becomes $J_3(x_2)$ for the value $x_2 = 3$. Similar results are obtained for the remaining values of x_2, so that the function $J_3(x_2)$ can be represented in tabular form as shown in Table 8-4. This table represents the return from stages 3 and 4, assuming that optimal decisions are made, as a function of the input to stage 3, which is x_2. The d_2^* is the value of d_2 which produces this optimum return.

The next step consists of determining the function

$$J_2(x_1) = \max(\text{w.r.t. } d_1)[j_2(x_1, d_2) + J_3(x_2)]$$

which is the maximum return from stages 2, 3, and 4, assuming optimum assignments are made in these stages. Following the example of the preceding steps produces Table 8-5, showing the optimization of stage 2. Again, the pertinent data from this table can be summarized as shown in Table 8-6. This table represents the return from stages 2, 3, and 4, assuming optimum decisions are made in these stages.

The final computation involves the determination of the first-stage return

$$J_1(x_0) = J_{\max} = \max(\text{w.r.t. } d_0)[j_1(x_0, d_0) + J_2(x_1)]$$

where x_0 can take only the single value of 6. The result is given in Table 8-7. The pertinent information from this table is

$$J_1(x_0) = 18 = J_{\max}$$

$$d_1^* = 0$$

which is the solution to the problem. The optimum decisions can be obtained by progressing forward through the system in the following manner. Since

Table 8-5. Second-Stage Optimization.

x_1	d_1	x_2	$j_2(x_1, d_2) + J_3(x_2) = S_2(x_1, x_2, d_1)$			$J_2(x_1)$	$d_1^*(x_1)$
0	0	0	0	0	0	0	0
1	0	1	0	4	4	4	1
	1	0	3	0	3		
2	0	2	0	5	5		
	1	1	3	4	7	7	1
	2	0	6	0	6		
3	0	3	0	8	8		
	1	2	3	5	8		
	2	1	6	4	10	10	2
	3	0	8	0	8		
4	0	4	0	12	12	12	0
	1	3	3	8	11		
	2	2	6	5	11		
	3	1	8	4	12	12	3
	4	0	9	0	9		
5	0	5	0	12	12		
	1	4	3	12	15	15	1
	2	3	6	8	14		
	3	2	8	5	13		
	4	1	9	4	13		
	5	0	10	0	10		
6	0	6	0	12	12		
	1	5	3	12	15		
	2	4	6	12	18	18	2
	3	3	8	8	16		
	4	2	9	5	14		
	5	1	10	4	14		
	6	0	11	0	11		

$$d_1^* = 0$$

then

$$x_1 = 6 - 0 = 6.$$

From the table of $J_2(x_1)$, with the above value of $x_1 = 6$,

Table 8-6. Second-Stage Return Function $J_2(x_1)$.

x_1	$J_2(x_1)$	$d_1^*(x_1)$	x_1	$J_2(x_1)$	$d_1^*(x_1)$
0	0	0	4	12	0, 3
1	4	1	5	15	1
2	7	1	6	18	2
3	10	2			

Table 8-7. First-Stage Optimization.

x_0	d_1	x_1	$j_1(x_0, d_0)$	$+\ J_2(x_1)$	$= S_1(x_0, x_1, d_0)$	$J_1(x_0)$	$d_1^*(x_1)$
6	0	6	0	18	18	18	0
	1	5	2	15	17		
	2	4	5	12	17		
	3	3	7	10	17		
	4	2	8	7	15		
	5	1	9	4	13		
	6	0	0	10	10		

$$d_2^* = 2$$

$$x_2 = 6 - 2 = 4.$$

From the table of $J_3(x_2)$, with the above value of $x_2 = 4$,

$$d_3^* = 3$$

$$x_3 = 4 - 1 = 3.$$

From the table of $J_4(x_3)$, with the above value of $x_3 = 3$,

$$d_4^* = 3$$

$$x_4 = 0.$$

Thus, the optimum return is 18, obtained with the following assignments:

Ship	Number Assigned
1	0
2	2
3	1
4	3

Any other combination of assignments will result in a smaller value of the process return.

The detailed representation of the problem solution in terms of the stage optimization tables are given to aid the reader in understanding the application of the technique. In certain cases there are short-cut methods which permit direct determination of the summarized return function tables. Although these methods are not too important, because practical application of the dynamic

Table 8-8. Alternative Derivation of Table 8-4.

x_2 \ d_2	$j_3(x_2, d_2) + J_4(x_2 - d_2)$ 0	1	2	3	4	5	6	$J_3(x_2)$	$d_2^*(x_2)$
0	0							0	0
1	1	4						4	1
2	3	5	4					5	1
3	8	7	5	4				8	0
4	8	12	7	5	3			12	1
5	8	12	12	7	4	2		12	1, 2
6	8	12	12	12	6	3	1	12	1, 2, 3

programming technique almost always entails the use of a digital computer, they do aid in the manual generation of the solution. For example, in those problems in which the stage transition is

$$x_l = x_{l-1} - d_{l-1}$$

and a zero decision at any stage results in a zero return, a simple algorithm permits rapid determination of the return function tables. The procedure can be demonstrated in the solution to the current example problem by constructing Table 8-8 as shown. The entries along the main diagonal are the entries in column 3 of the return matrix given by Table 8-1. Entries in the first column of the table are the results of the $J_4(x_3)$ table determined previously and given in Table 8-2. A term below the main diagonal is determined as the sum of the corresponding first column entry in the diagonal containing that term with the main diagonal entry in the column containing the term. Once the off-diagonal terms are obtained, the $J_3(x_2)$ value is determined as the maximum occurring in each row, and $d_2^*(x_2)$ is the d_2 value producing that maximum. The analogous table for $J_2(x_1)$ is shown in Table 8-9. Finally, the first stage with its

Table 8-9. Alternative Derivation of Table 8-5.

x_1 \ d_1	$j_2(x_1, d_1) + J_3(x_1 - d_1)$ 0	1	2	3	4	5	6	$J_2(x_1)$	$d_1^*(x_1)$
0	0							0	0
1	4	3						4	0
2	5	7	6					7	1
3	8	8	10	8				10	2
4	12	11	11	12	9			12	0, 3
5	12	15	14	13	13	10		15	1
6	12	15	18	16	14	14	11	18	2

Table 8-10. Alternative Derivation of Table 8-6.

x_0 \ d_0	$j_1(x_0) + J_2(x_0 - d_0)$								
	0	1	2	3	4	5	6	$J_1(x_0)$	$d_0^*(x_0)$
6	18	17	17	17	15	13	10	18	0

single possible input state yields Table 8-10. These entries are easily obtainable as the sum of the $J_2(x_1)$ entries of the previous table with the complementary entries from the first column of the return table.

Although it may not be evident at this point, the calculations carried out in the solution of this problem also provide for the determination of optimal policies in the event that one of the decisions is improperly made or a stage transition does not occur in the prescribed manner. For example, suppose that an arbitrary decision is made to assign three men to ship 2, rather than two men as prescribed by the optimal policy. Even though it is not possible to change the assignment to ship 1, one can still maximize the return by proper assignments to ships 3 and 4. This solution can be obtained from Table 8-4 ($J_3(x_2)$), entering with the argument

$$x_2 = 6 - 3 = 3$$

for which

$$J_3 = 8; \quad d_2^* = 0.$$

Then, from Table 8-6 ($J_4(x_3)$), with

$$x_3 = 3 - 0 = 3$$

the optimum stage 3 decision is

$$d_3^* = 3$$

which, in turn, yields

$$x_4 = 0.$$

This result implies that the optimum value of the process return, with three men arbitrarily assigned to ship 2 and none to ship 1, is

$$\begin{aligned} J_{max} &= 8 + J_3(3) \\ &= 8 + 8 = 16. \end{aligned}$$

This feature, in effect, provides a feedback capability to dynamic programming, and it is this characteristic which makes the method applicable to optimal control problems.

Generally speaking, when dynamic programming is applied to static problems of this sort, the analyst replaces a single multivariate optimization problem in a space of dimension N by a sequence of N one-dimensional, interdependent optimization problems. The analogy to the representation of a single Nth-order equation by a set of N first-order equations is obvious. As N gets large, the complexity of the N-dimensional optimization grows much faster than that of the sequence of one-dimensional optimizations, and hence the usefulness of dynamic programming for application to this type of problem is clear. Even so, the issue of dimensionality eventually becomes a limiting factor as the complexity of the problem grows. It should also be noted that where specialized methods such as linear programming can be applied the specialized methods are usually more computationally efficient.

Finally, it should be pointed out that the ordering here is artificial, since the individual assignments are actually made simultaneously. Obviously, then, the ships can be reordered in any desired way without changing the results. There are many problems in which the ordering is prescribed, one notable example being the ordering dictated by the passage of time in a discrete time optimal control problem. It is in the solution of this type of problem that the full advantage of the dynamic programming method is realized. The example problem of the following section is of this type.

8.3.2. A Dynamic Discrete State Example with Probabilistic Effects

One of the advantages of the dynamic programming method is its ability to consider multistage processes in which the stage transitions exhibit a random characteristic rather than being deterministic in nature. This capability is extremely important in stochastic control problems, in which random noise affects system dynamics. The following example, in addition to demonstrating the true dynamic nature of the process, also provides an insight into its usefulness in this critical area.

Suppose that a certain missile has two acceleration commands, labeled +1 and -1, which can be selected at discrete, equally spaced time points. The acceleration produces an instantaneous velocity change ΔV, which is a random variable having the distribution shown in Table 8-11. The numbers in the table represent the probabilities of obtaining the listed velocity changes with the acceleration settings shown. This table then represents the probabilistic transition process for each stage, with the acceleration setting of +1 or -1 serving as the decision process. As a standard multistage decision process, the situation would appear

Table 8-11. Acceleration Probabilities, $P(\Delta V|d)$ (velocity change ΔV in a single stage).

ΔV \ d	-1	$+1$
-1	0.2	0
0	0.8	0.1
1	0	0.7
2	0	0.2

as in Figure 8-4, in which the stage transitions are random in nature, and the pertinent distribution depends upon whether the decision variable d_l is chosen as +1 or -1.

So far, the stage return j_l has not been defined. Suppose that the desired objective is a velocity change of four or five units during a time period consisting of five intervals. Since the system is probabilistic in nature, a probabilistic measure of effectiveness must be adopted, which in this case is simply the probability of achieving the desired goal. To accomplish this, the individual stage return can be specified as

$$j_l(V_l, V_{l-1}, d_{l-1}) = P[V_l|V_{l-1}, d_{l-1}] = p(V_l - V_{l-1}|d_{l-1}) \qquad (8.20)$$

where $p(V_l - V_{l-1}|d_{l-1})$ is taken from Table 8-11. Equation (8.20) gives the conditional probability of obtaining velocity V_l, given the value of V_{l-1}. The decision variable d_{l-1} is also included, since the conditional distribution of V_{l-1} will depend on d_{l-1} in a manner such that d_{l-1} can be considered to be a parameter of the distribution. Also, for the sake of clarity in what is to follow, the following definitions are provided:

l—the number of the stage (time interval in this case), $l = 0, 1, \ldots, 5$.
d_{l-1}—the acceleration setting chosen at time l, $d_l = +1$ or -1

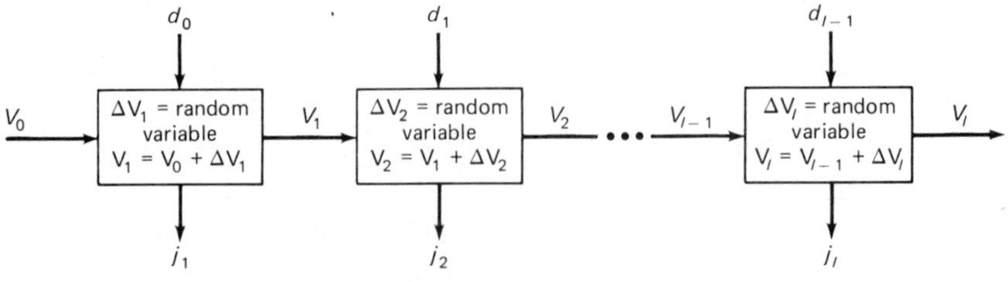

Figure 8-4. Probabilistic example as a multistage decision process.

ΔV_l–the velocity change produced by the acceleration setting d_l. It is a random variable.

V_l–the velocity at time l. It is a random variable.

$j_l(V_l|V_{l-1}, d_{l-1})$–the conditional probability of V_l given V_{l-1}, with d_{l-1} as a parameter (from Equation 8.20)

$J_{l+1}(V_l)$–the probability of obtaining the desired goal (4 or 5 velocity units increase in 5 time intervals) if the velocity is V_l at time l, and the optimum procedure is followed from stage $l+1$ to the final stage.

$P_l(V_l, V_{l-1}, d_{l-1})$–the probability of obtaining the desired goal (4 or 5 velocity units increase in 5 time intervals) given the velocity V_{l-1}, the application of decision d_{l-1} at time l and the resulting velocity V_l, and assuming that optimal decisions are made in subsequent stages.

d_l^*–the optimum decision (+1 or −1) at stage l.

With these definitions, it can be shown that

$$P_l(V_l, V_{l-1}, d_{l-1}) = p(V_l - V_{l-1}|d_{l-1})J_{l+1}(V_l). \tag{8.21}$$

The right-hand side of this expression is the product of the maximum probability J_{l+1} of obtaining the desired goal, given velocity V_l, with the probability of producing velocity V_l, given V_{l-1} and d_{l-1}. Then the optimum probability of obtaining the desired goal, given V_{l-1}, is obtained as

$$J_{l+1}(V_l) = \max(\text{w.r.t. } d_l) \sum P_{l+1}(V_{l+1}, V_l, d_l). \tag{8.22}$$

The summation is over the probabilities of the various possible outcomes of applying decision d_l. If the quantity $J_{l+1}(V_l)$ is determined for the final stage, producing $J_5(V_4)$, then the values for the preceeding stages can be determined by recursive application of the above equations in accordance with the diagram of Figure 8-4. As in the previous example, this process is most easily demonstrated by use of tables. Before proceeding, however, it is advisable to consider limitations imposed on the values of the states V_l. Without loss of generality, V_0 can be taken as zero, in which case Table 8-12 lists the states attainable at

Table 8-12. Attainable States.

Time l	Possible States V_l
0	0
1	−1 through +2
2	−2 through +4
3	−3 through +6
4	−4 through +8
5	−5 through +10

Table 8-13. State Permitting Final Goal of 4 or 5 Velocity Units.

Time l	Permissible States V_l
5	4 or 5
4	2 through 6
3	0 through 7
2	−2 through +8
1	−4 through +9
0	−6 through +10

each stage. In addition, only certain states at each stage will permit the desired goal of a change of 4 or 5 velocity increments to be realized. These restrictions are shown in Table 8-13 and clearly relate to the general concept of controllability discussed in Chapter 4. More generally, at stage l the smallest value of V_l which need be considered is given by the expression

$$\max[-l, 4 - 2(5 - l)] = \max(-l, 2l - 6)$$

as shown in Table 8-12. The maximum value requiring consideration is shown in Table 8-13 and is given by the expression

$$\min(2l, 5 + l).$$

These results yield Table 8-14, showing the states which must be considered. With these preliminaries now established, the optimization table for stage 5 can now be determined. In its complete form, this table appears as Table 8-15. From this point, the table for stage 4 can be determined, using the information from Table 8-15, in accordance with Equation (8.21). In the stage 4 table, the

Table 8-14. Attainable States Permitting Final Goal of 4 or 5 Velocity Units.

0	0
1	−1 through +2
2	−2 through +4
3	0 through 6
4	2 through 6
5	4 or 5

Table 8-15. Fifth-Stage Optimization.

V_4	d_4	ΔV_5	V_5	$p(\Delta V_5 \mid d_4)$	$P_5(V_5, V_4, d_4)$	$J_5(V_4)$	$d_4^*(V_4)$
2	+1	2	4	0.20*	0.20	0.2	+1
		1	3	0.70			
		0	2	0.10			
	−1	−1	1	0.20			
		0	2	0.80	0		
3	+1	2	5	0.20*			
		1	4	0.70*	0.90	0.90	+1
		0	0	0.10			
	−1	−1	2	0.20			
		0	3	0.80	0		
4	+1	2	6	0.20			
		1	5	0.70*			
		0	4	0.10*	0.80		+1
	−1	−1	3	0.20		0.80	
		0	4	0.80*	0.80		−1
5	+1	2	7	0.20			
		1	6	0.70			
		0	5	0.10*	0.10		
	−1	−1	4	0.20*			
		0	5	0.80*	1.00	1.00	−1
6	+1	2	8	0.20			
		1	7	0.70			
		0	6	0.10	0		
	−1	−1	5	0.20*	0.20	0.20	−1
		0	6	0.80			

*Only those situations resulting in V_5 values of 4 or 5 are of interest.

quantity $P_4(V_4, V_3, d_3)$ is the probability of obtaining the desired goal of an increase of 4 or 5 velocity units at the end of the fifth stage if the velocity into stage four is V_3, the decision is d_3 and the resulting velocity is V_4. Also, $J_4(V_3)$ is the maximum sum of probabilities, as indicated by Equation (8.22). Proceeding in this fashion for every possible value of V_3, and for the ΔV_4's capable of being produced by decision d_3 of +1 or −1, yields the results presented as Table 8-16. The reader should note that this table contains the information from Table 8-15 necessary for optimum decision making. For example, if the velocity at time $l = 3$ has the observed value of 1 then the optimum decision is $d = +1$, with a resulting value of 0.32 as the probability of obtaining the desired result after the fifth stage, assuming, of course, that optimum decisions will be applied in subsequent stages. It is important to recognize that the probabilistic nature of the stage transitions has been considered in this development.

The table describing the third-stage transitions can be determined in a similar fashion. In the interest of brevity, only a summarized third-stage table is shown

Table 8-16. Fourth-Stage Optimization.

V_3	d_3	ΔV_4	V_4	$p(\Delta V_4 \mid d_3)$	$J_5(V_4)$	$P_4(V_4, V_3, d_3)$	$J_4(V_3)$	$d_3^*(V_3)$
0	1	2	2	0.2	0.2	0.04	0.04	+1
		1	1	0.7	0			
		0	0	0.1	0			
	−1	−1	−1	0.2	0			
		0	0	0.8	0			
1	1	2	3	0.2	0.9	0.18	0.32	+1
		1	2	0.7	0.2	0.14		
		0	1	0.1	0			
	−1	−1	0	0.2	0			
		0	1	0.8	0			
2	1	2	4	0.2	0.8	0.16		
		1	3	0.7	0.9	0.63	0.81	+1
		0	2	0.1	0.2	0.02		
	−1	0	2	0.8	0.2	0.16		
3	1	2	5	0.2	1.0	0.20		
		1	4	0.7	0.8	0.56	0.85	+1
		0	3	0.1	0.9	0.09		
	−1	−1	2	0.2	0.2	0.04		
		0	3	0.8	0.9	0.72		
4	1	2	6	0.2	0.2	0.04		
		1	5	0.7	1.0	0.70	0.82	+1
		0	4	0.1	0.8	0.08		
	−1	−1	3	0.2	0.9	0.18	0.82	−1
		0	4	0.8	0.8	0.64		
5	1	2	7	0.2	0			
		1	6	0.7	0.2	0.14		
		0	5	0.1	1.0	0.10		
	−1	−1	4	0.2	0.8	0.16		
		0	5	0.8	1.0	0.80	0.96	−1
6	1	2	8	0.2	0			
		1	7	0.7	0			
		0	6	0.1	0.2	0.02		
	−1	−1	5	0.2	1.0	0.20	0.36	−1
		0	6	0.8	0.2	0.16		

as Table 8-17, in which Σ denotes the sum of the entries in the numbered columns to its left. The reader may verify these results by use of the larger, more complete table. Repeating the process for the $J_2(V_1)$ and $J_1(V_0)$ tables produces the summarized data shown as Tables 8-18 and 8-19. Table 8-19 has only one V_0 argument, since the initial value of V_0 is specified to be zero. The implication of the $J_1(V_0)$ term is that the overall probability of obtaining the desired result is 0.81, assuming the optimal decisions are made at each stage. The optimum first stage decision is $d_1^*(V_0) = +1$.

Table 8.17. Third-Stage Optimization.

		$p(V_3 - V_2 \mid d_2) \times J_4(V_3)$									
V_2	d_2	$V_3 = 0$	1	2	3	4	5	6	$\sum$	$J_3(V_2)$	$d_2^*(V_2)$
−2	1	0.01							0.01	0.01	1
−1	1	0.03	0.06						0.09	0.09	1
0	1		0.22	0.08					0.30	0.30	1
1	1		0.03	0.57	0.17				0.77	0.77	1
2	1			0.08	0.58	0.16			0.82	0.82	1
	−1		0.06	0.65					0.71		
3	1				0.08	0.57	0.19		0.84	0.84	1
	−1			0.16	0.67				0.83		
4	1					0.08	0.67	0.07	0.82		
	−1				0.17	0.66			0.83	0.83	−1

It should be emphasized that this result, 0.81, is a prior probability, indicating the probability of success before any action is taken. For example, if the optimum decision is made at the first stage, then the setting of +1 is used. Suppose that the result is a gain of two velocity increments, in which case V_1 has the value 2. Examination of the $J_2(V_1)$ value in Table 8-18 shows a probability of success of 0.84, which is a conditional probability based on the known outcome of the previous stage. The reader should also note that another possible outcome of the first stage is a velocity loss of one unit, in which case V_1 is −1. The probability of success in this event is only 0.37, but the $d_2^*(V_1)$ column still gives the optimum decision to achieve this probability. This feature is an illustration of the principle of optimality, in that current decisions must be made on the basis of the current situation, without regard for previous decisions which have produced the current situation. Since the decisions to be made are determined by observation of the current situation at each stage of the process, this

Table 8-18. Second-Stage Optimization.

		$p(V_2 - V_1 \mid d_1) \times J_3(V_2)$									
V_1	d_1	$V_2 = -2$	−1	0	1	2	3	4	$\sum$	$J_2(V_1)$	$d_1^*(V_1)$
−1	1		0.01	0.21	0.15				0.37	0.37	1
0	1			0.03	0.54	0.16			0.73	0.73	1
1	1				0.08	0.57	0.17		0.82	0.82	1
2	1					0.08	0.59	0.17	0.84	0.84	1
	−1				0.15	0.66			0.81		

Table 8-19. First-Stage Optimization.

		$p(V_1 - V_0 \mid d_0) \times J_2(V_1)$							
V_0	d_0	$V_1 =$	−1	0	1	2	Σ	$J_1(V_0)$	$d_0^*(V_0)$
0	1			0.07	0.57	0.17	0.81	0.81	1
	−1		0.07	0.58			0.65		

case also serves to demonstrate the feedback capability of dynamic programming. Furthermore, the feedback control law specified by these results is optimal in the sense that it maximizes the probability of obtaining the desired goal.

A graphic representation of the results obtained above is shown in Figure 8-5, in which the possible transitions from one state to the next are represented by directed lines. For the interest of clarity, only the optimal decisions are represented. For example, at time 0 the optimal decision is +1, which has the three possible outcomes shown, occurring with the indicated probabilities. Similarly, at stage 2 only a +1 decision is indicated, with the overall possibilities of the stage at time 2 being 0, 1, 2, 3, and 4. If it is 4, a −1 decision is indicated, as shown, while +1 decisions are required for the remaining four states. In essence, this figure shows the possible results which can be produced by following the optimal policy. The reader should note that it is possible, even using the optimal policy, to produce states 2, 3, or 6 at time 5. This conclusion is in agreement with the tabular results, which show a value 0.81 for the probability of achieving states 4 or 5, given that the optimal policy is followed at each stage. This result in turn implies a 0.19 probability of obtaining states other than 4 or 5, even though the optimal policy is followed at each stage.

It is obvious that the dynamic programming method represents a systematic approach to optimization which is well adapted to computer implementation.

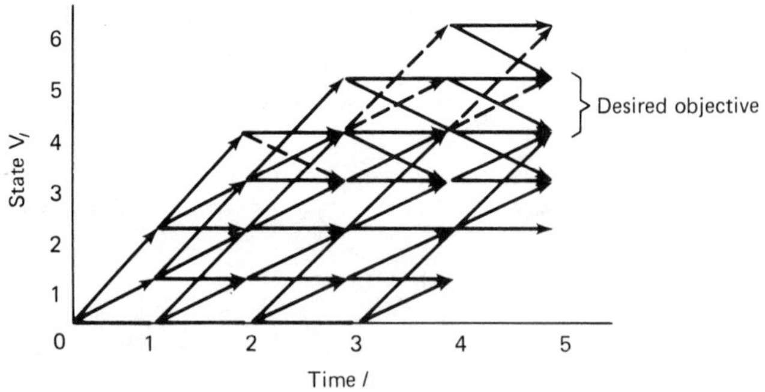

Figure 8-5. Optimal strategy.

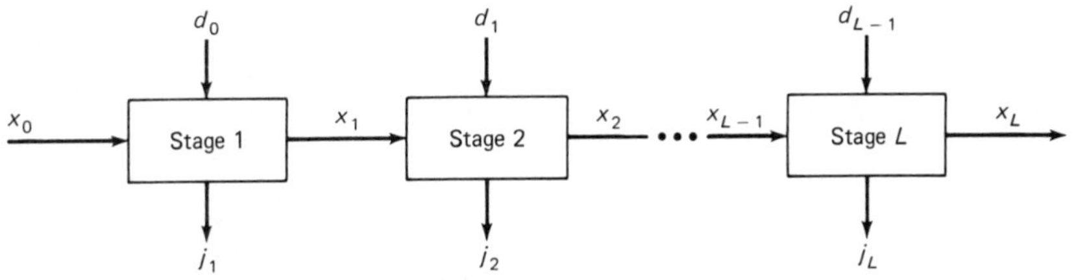

Figure 8-6. Dynamic programming as a multistage decision process.

8.4. DYNAMIC PROGRAMMING APPROACH TO OPTIMAL CONTROL

The relationship between dynamic programming and the concept of system optimization considered in Chapter 6 may not be readily apparent to the reader, but some reflection will reveal that these are actually two different approaches to the same general type of problem. Dynamic programming involves a multistage decision process in which the state at the output of a given stage is dependent upon the decision made in the current stage, the state at the output of the previous stage, and the particular process determining the transition between stages. This situation can be represented as shown in Figure 8-6, where the decision at the lth stage is d_{l-1}, and the scalar return function from each state is j_l. Although this model can fit any number of situations because of the very general nature of the decision process and of the mechanics of the interstage transitions, it can be specifically tailored to fit the feedback control problem by defining the decision process as the result obtained by applying a given control vector $\boldsymbol{u}$ and by letting the stage transition be defined by a general difference equation

$$\boldsymbol{x}(l) = \boldsymbol{f}(\boldsymbol{x}(l-1), \boldsymbol{u}(l-1), l). \tag{8.23}$$

The optimization involves the sum of the individual stage returns

$$J = \sum_{l=1}^{L} j_l. \tag{8.24}$$

Such a situation is shown in Figure 8-7 and clearly represents a discrete feedback control system. Further similarities can be seen in the methods of solution, since both involve examination of the process in reverse time order. For example, the starting point for computation in dynamic programming is the last stage, whereas the optimum control problem involves integration backward in time from some terminal condition, as illustrated by the matrix Riccati equation method for linear systems developed in Chapter 6. It is not surprising, then, that the control problem can be approached by treating it as a special case of dynamic pro-

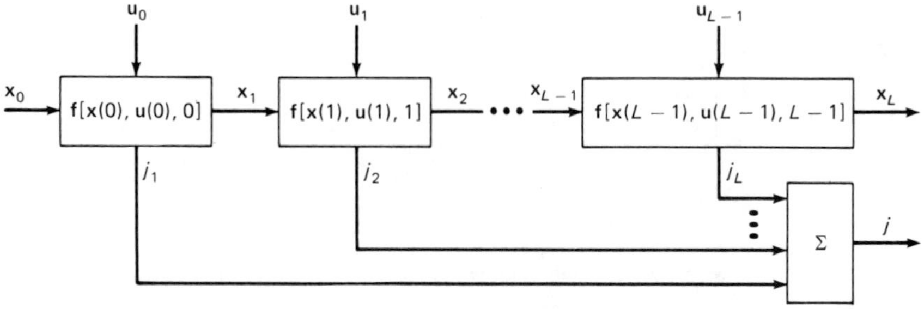

Figure 8-7. Feedback control as a multistage decision process.

gramming. Being a feedback method, the dynamic programming approach can actually be superior in the presence of modeling errors and/or input noise. It can also provide an approach to the solution of nonlinear control problems in certain situations in which the two-point boundary value problem resulting from application of the techniques of Chapter 6 proves to be intractable. To further demonstrate the general equivalency of optimum control and dynamic programming, the general optimum control problem will now be developed using dynamic programming concepts and the principle of optimality.

The expressions of Equations (8.23) and (8.24) should be compared with those of Equations (6.164) and (6.165) to illustrate the fact that the problem depicted in Figure 8-7 is indeed the discrete version of the optimal control problem considered in Chapter 6. To formalize this concept, let the j_l of Equation (8.24) be written in the functional form, and the stage return modified slightly to include a function of the final state $\phi(\boldsymbol{x}(L))$. The result is

$$J = \phi(\boldsymbol{x}(L)) + \sum_{l=1}^{L} j_l(\boldsymbol{x}(l-1), \boldsymbol{u}(l-1)). \tag{8.25}$$

The problem can now be stated as one of minimizing the performance index of Equation (8.25), subject to the constraints of Equation (8.23). This problem is clearly analogous to that considered in Chapter 6. The application of dynamic programming to the solution of this problem entails the use of the recurrence relationships of Equation (8.16), for which the general expression is

$$J_l(\boldsymbol{x}(l-1)) = \min(\text{w.r.t. } \boldsymbol{u}(l-1))[j_l(\boldsymbol{x}(l-1), \boldsymbol{u}(l-1)) + J_{l+1}(\boldsymbol{x}(l))]. \tag{8.26}$$

beginning with the terminal stage

$$J_L(\boldsymbol{x}(L-1)) = \min(\text{w.r.t. } \boldsymbol{u}(L-1))[j_L(\boldsymbol{x}(L-1), \boldsymbol{u}(L-1)) + \phi(\boldsymbol{x}(L))]. \tag{8.27}$$

The stage transitions are governed by Equation (8.23).

Solution of this problem proceeds in a manner similar to that of previous examples. Notice that control variable constraints such as

$$C_p^l \leqslant u_p(l) \leqslant C_p^u \tag{8.28}$$

and state variable constraints represented by

$$V_i^l \leqslant x_i(l) \leqslant V_i^u \tag{8.29}$$

are easily considered by this process, since only permitted values are considered in forming the tables of cumulative return functions J_l. The solution of Equation (8.26) yields the optimal control law $\boldsymbol{u}(l)$, obtained by essentially trying all admissible control values $\boldsymbol{u}(l)$ at each possible state value $\boldsymbol{x}(l)$. The formulation and storage of tables of the cumulative return functions make it necessary that the control function $\boldsymbol{u}(l)$ and the state variables $\boldsymbol{x}(l)$ be quantized into a reasonable number of possible values, which requires that the system be modeled as a discrete value, discrete parameter random process. Here, the concept of dimensionality and its effect upon dynamic programming solutions becomes apparent, for there are L time intervals, N state variables quantized into K_N discrete values, and P control variables quantized into K_P discrete values. These three types of variables are used to formulate L tables of cumulative return functions. Typical problems may sometimes tax the storage capabilities of even the larger computers.

8.4.1. The Discrete Time Linear System

When the general system described by Equation (8.23) is linear, the resulting difference equation is

$$\boldsymbol{x}(l) = \boldsymbol{\Phi}(l, l-1)\boldsymbol{x}(l-1) + \boldsymbol{\Gamma}(l, l-1)\boldsymbol{u}(l-1). \tag{8.30}$$

If the performance measure is specified as the discrete time version of the quadratic performance index of Section 6.4,

$$J = \tfrac{1}{2}[\boldsymbol{x}^T(L)\boldsymbol{S}_L\boldsymbol{x}(L)] + \tfrac{1}{2}\sum_{l=0}^{L-1}[\boldsymbol{x}^T(l)\boldsymbol{A}(l)\boldsymbol{x}(l) + \boldsymbol{u}^T(l)\boldsymbol{B}(l)\boldsymbol{u}(l)] \tag{8.31}$$

then the table of cumulative return functions is constructed as implied by Equations (8.26) and (8.27) as follows. The term

$$\begin{aligned} J_L(\boldsymbol{x}(L-1)) = \min(\text{w.r.t. } \boldsymbol{u}(L-1))[\tfrac{1}{2}\boldsymbol{x}^T(L-1)\boldsymbol{A}(L-1)\boldsymbol{x}(L-1) \\ + \tfrac{1}{2}\boldsymbol{u}^T(L-1)\boldsymbol{B}(L-1)\boldsymbol{u}(L-1) + \tfrac{1}{2}\boldsymbol{x}^T(L)\boldsymbol{S}_L\boldsymbol{x}(L)] \end{aligned} \tag{8.32}$$

is the return from the final stage, including the terminal cost function. From stage $L - 1$, the accumulated return is

$$J_{L-1}(\boldsymbol{x}(L-2)) = \min(\text{w.r.t. } \boldsymbol{u}(L-2))[\tfrac{1}{2}\boldsymbol{x}^T(L-2)\boldsymbol{A}(L-2)\boldsymbol{x}(L-2) + \tfrac{1}{2}\boldsymbol{u}^T(L-2)\boldsymbol{B}(L-2)\boldsymbol{u}(L-2) + J_L(\boldsymbol{x}(L-1))] \quad (8.33)$$

and so on, yielding the general term

$$J_l(\boldsymbol{x}(l-1)) = \min(\text{w.r.t. } \boldsymbol{u}(l-1))[\tfrac{1}{2}\boldsymbol{x}^T(l-1)\boldsymbol{A}(l-1)\boldsymbol{x}(l-1) + \tfrac{1}{2}\boldsymbol{u}^T(l-1)\boldsymbol{B}(l-1)\boldsymbol{u}(l-1) + J_{l+1}(\boldsymbol{x}(l))] \quad (8.34)$$

as implied by Equation (8.26). This technique will then lead to tabular functions of the accumulated return functions for each stage, similar to those produced for the examples of Section 8.3. Then, the $\boldsymbol{u}(l)$ sequence producing minimum J can be readily obtained.

If there are no bounds on the control sequence $\boldsymbol{u}(l)$ or state vector sequence $\boldsymbol{x}(l)$, the general approach to the solution of this problem can be carried even further. Since the return function from the last stage, as given by Equation (8.32), is the minimum with respect to $\boldsymbol{u}(l)$, then this minimum can be obtained by formal differentiation. That is, if the function

$$J_L(\boldsymbol{x}(L-1), \boldsymbol{u}(L-1)) = \tfrac{1}{2}\boldsymbol{x}^T(L-1)\boldsymbol{A}(L-1)\boldsymbol{x}(L-1) + \tfrac{1}{2}\boldsymbol{u}^T(L-1)\boldsymbol{B}(L-1)\boldsymbol{u}(L-1) + \tfrac{1}{2}\boldsymbol{x}^T(L)\boldsymbol{S}_L\boldsymbol{x}(L) \quad (8.35)$$

is defined, then

$$J_L(\boldsymbol{x}(L-1)) = \min(\text{w.r.t. } \boldsymbol{u}(L-1))\, J_L(\boldsymbol{x}(L-1), \boldsymbol{u}(L-1)) \quad (8.36)$$

can be obtained from the equation

$$\partial J_L/\partial \boldsymbol{u} = \boldsymbol{0} \quad (8.37)$$

so long as the resulting value of $\boldsymbol{u}(L-1)$ does not violate some constraint relating to the control sequence $\boldsymbol{u}(l)$. Before performing this differentiation, however, the dependence of $\boldsymbol{x}(L)$ on $\boldsymbol{u}(L-1)$ must be incorporated by substituting from Equation (8.30), which produces

$$
\begin{aligned}
J_L(\boldsymbol{x}(L-1), \boldsymbol{u}(L-1)) = {} & \tfrac{1}{2}\boldsymbol{x}^T(L-1)\boldsymbol{A}(L-1)\boldsymbol{x}(L-1) \\
& + \tfrac{1}{2}\boldsymbol{u}^T(L-1)\boldsymbol{B}(L-1)\boldsymbol{u}(L-1) \\
& + \tfrac{1}{2}\boldsymbol{x}^T(L-1)\boldsymbol{\Phi}^T(L, L-1)\boldsymbol{S}_L\boldsymbol{\Phi}(L, L-1)\boldsymbol{x}(L-1) \\
& + \boldsymbol{x}^T(L-1)\boldsymbol{\Phi}^T(L, L-1)\boldsymbol{S}_L\boldsymbol{\Gamma}(L, L-1)\boldsymbol{u}(L-1) \\
& + \tfrac{1}{2}\boldsymbol{u}^T(L-1)\boldsymbol{\Gamma}^T(L, L-1)\boldsymbol{S}_L\boldsymbol{\Gamma}(L, L-1)\boldsymbol{u}(L-1).
\end{aligned} \tag{8.38}
$$

Here, the symmetry of the $\boldsymbol{S}_L$ matrix, and the fact that all terms are scalars, have been used in combining the cross product terms. Then, performing the differentiation implied by Equation (8.37) yields

$$
\begin{aligned}
\boldsymbol{u}^T(L-1)\boldsymbol{B}(L-1) + \boldsymbol{x}^T(L-1)\boldsymbol{\Phi}^T(L, L-1)\boldsymbol{S}_L\boldsymbol{\Gamma}(L, L-1) \\
+ \boldsymbol{u}^T(L-1)\boldsymbol{\Gamma}^T(L, L-1)\boldsymbol{S}_L\boldsymbol{\Gamma}(L, L-1) = \boldsymbol{0}.
\end{aligned} \tag{8.39}
$$

By transposing and solving for $\boldsymbol{u}(L-1)$, one obtains

$$
\begin{aligned}
\boldsymbol{u}(L-1) = -[\boldsymbol{B}(L-1) \\
+ \boldsymbol{\Gamma}^T(L, L-1)\boldsymbol{S}_L(L, L-1)]^{-1}\boldsymbol{\Gamma}^T(L, L-1)\boldsymbol{S}_L\boldsymbol{\Phi}(L, L-1)\boldsymbol{x}(L-1).
\end{aligned} \tag{8.40}
$$

Again, the optimum control $\boldsymbol{u}(L-1)$ is a linear function of the state $\boldsymbol{x}(L-1)$.

This result from the last stage can then be substituted into the preceeding stage equation, as given by Equation (8.33). That is, the return function $J_L(\boldsymbol{x}(L-1))$, as given by Equation (8.32), has now been determined by performing the indicated minimization, and the specific expression can be obtained by substituting for $\boldsymbol{u}(L-1)$ from Equation (8.40) into Equation (8.38) and collecting terms. Since this process involves very complex expressions, let the feedback gain matrix be represented by $\boldsymbol{C}(L-1)$, defined as

$$
\begin{aligned}
\boldsymbol{C}(L-1) = -[\boldsymbol{B}(L-1) \\
+ \boldsymbol{\Gamma}^T(L, L-1)\boldsymbol{S}_f\boldsymbol{\Gamma}(L, L-1)]^{-1}\boldsymbol{\Gamma}^T(L, L-1)\boldsymbol{S}_L\boldsymbol{\Phi}(L, L-1)
\end{aligned} \tag{8.41}
$$

in which case

$$
\boldsymbol{u}(L-1) = \boldsymbol{C}(L-1)\boldsymbol{x}(L-1). \tag{8.42}
$$

Then, by substituting Equations (8.42) and (8.30) into Equation (8.35) and collecting terms, one obtains

$$\begin{aligned} J_L(\boldsymbol{x}(L-1)) = \tfrac{1}{2}\boldsymbol{x}^T(L-1)[&\boldsymbol{A}(L-1) + \boldsymbol{C}^T(L-1)\boldsymbol{B}(L-1)\boldsymbol{C}(L-1) \\ &+ [\boldsymbol{\Phi}(L, L-1) + \boldsymbol{\Gamma}(L, L-1)\,\boldsymbol{C}(L-1)]^T\boldsymbol{S}_L\,[\boldsymbol{\Phi}(L, L-1) \\ &+ \boldsymbol{\Gamma}(L, L-1)\boldsymbol{C}(L-1)]\,]\boldsymbol{x}(L-1) \end{aligned} \tag{8.43}$$

as the minimum return from the last stage. Let the matrix of this quadratic form be denoted as $\boldsymbol{S}(L-1)$. That is

$$\begin{aligned} \boldsymbol{S}(L-1) = \boldsymbol{A}(L-1) + \boldsymbol{C}^T(L-1)\boldsymbol{B}(L-1)\boldsymbol{C}(L-1) + [\boldsymbol{\Phi}(L, L-1) \\ + \boldsymbol{\Gamma}(L, L-1)\boldsymbol{C}(L-1)]^T\boldsymbol{S}_L\,[\boldsymbol{\Phi}(L, L-1) + \boldsymbol{\Gamma}(L, L-1)\boldsymbol{C}(L-1)] \end{aligned} \tag{8.44}$$

in which case

$$J_L(\boldsymbol{x}(L-1)) = \tfrac{1}{2}\boldsymbol{x}^T(L-1)\boldsymbol{S}(L-1)\boldsymbol{x}(L-1). \tag{8.45}$$

Now this form of the return function is substituted into the equation for the preceeding stage, which from Equation (8.33) is

$$\begin{aligned} J_{L-1}(\boldsymbol{x}(L-2)) = \min(\text{w.r.t. } \boldsymbol{u}(L-2))[&\tfrac{1}{2}\boldsymbol{x}^T(L-2)\boldsymbol{A}(L-2)\boldsymbol{x}(L-2) \\ &+ \tfrac{1}{2}\boldsymbol{u}(L-2)\boldsymbol{B}(L-2)\boldsymbol{u}(L-2) \\ &+ \tfrac{1}{2}\boldsymbol{x}^T(L-1)\boldsymbol{S}(L-1)\boldsymbol{x}(L-1)]. \end{aligned} \tag{8.46}$$

The significant characteristic of this equation is that it is identical to Equation (8.32), with L indexed down by one and $\boldsymbol{S}_L = \boldsymbol{S}(L-1)$. The minimization is then identical to that developed for the last stage, with the analogous result

$$\begin{aligned} \boldsymbol{u}(L-2) &= \boldsymbol{C}(L-2)\boldsymbol{x}(L-2) \\ J_{L-1}(\boldsymbol{x}(L-2)) &= \tfrac{1}{2}\boldsymbol{x}^T(L-2)\boldsymbol{S}(L-2)\boldsymbol{x}(L-2) \end{aligned} \tag{8.47}$$

where

$$\begin{aligned} \boldsymbol{C}(L-2) &= [\boldsymbol{B}(L-2) + \boldsymbol{\Gamma}^T(L-1, L-2)\boldsymbol{S}(L-1)\boldsymbol{\Gamma}(L-1, L-2)]^{-1} \\ &\quad \cdot \boldsymbol{\Gamma}^T(L-1, L-2)\boldsymbol{S}(L-1)\boldsymbol{\Phi}(L-1, L-2) \\ \boldsymbol{S}(L-2) &= \boldsymbol{A}(L-2) + \boldsymbol{C}^T(L-2)\boldsymbol{B}(L-2)\boldsymbol{C}(L-2) + [\boldsymbol{\Phi}(L-1, L-2) \\ &\quad + \boldsymbol{\Gamma}(L-1, L-2)\boldsymbol{C}(L-2)]^T\boldsymbol{S}(L-1)[\boldsymbol{\Phi}(L-1, L-2) \\ &\quad + \boldsymbol{\Gamma}(L-1, L-2)\boldsymbol{C}(L-2)] \end{aligned} \tag{8.48}$$

Continuing in this manner yields the general recursion equations

$$\begin{aligned}
\boldsymbol{u}(l) &= \boldsymbol{C}(l)\boldsymbol{x}(l)\\
J_l(\boldsymbol{x}(l-1)) &= \tfrac{1}{2}\boldsymbol{x}^T(l-1)\,\boldsymbol{S}(l-1)\boldsymbol{x}(l-1)\\
\boldsymbol{C}(l) &= -[\boldsymbol{B}(l) + \boldsymbol{\Gamma}^T(l+1,l)\boldsymbol{S}(l+1)\boldsymbol{\Gamma}(l+1,l)]^{-1}\\
&\quad \cdot \boldsymbol{\Gamma}^T(l+1,l)\boldsymbol{S}(l+1)\boldsymbol{\Phi}(l+1,l)\\
\boldsymbol{S}(l) &= \boldsymbol{A}(l) + \boldsymbol{C}^T(l)\boldsymbol{B}(l)\boldsymbol{C}(l) + [\boldsymbol{\Phi}(l+1,l)\\
&\quad + \boldsymbol{\Gamma}(l+1,l)\boldsymbol{C}(l)]^T\boldsymbol{S}(l+1)[\boldsymbol{\Phi}(l+1,l)\\
&\quad + \boldsymbol{\Gamma}(l+1,l)\boldsymbol{C}(l)]
\end{aligned} \tag{8.49}$$

where the equations must be solved backward in time from the initial condition

$$\boldsymbol{S}(L) = \boldsymbol{S}_L. \tag{8.50}$$

The difference equation for $\boldsymbol{S}(l)$ in Equation (8.49) is a discrete time version of the matrix Riccati equation. The equivalent result of Section 6.4.1 can be obtained by expanding the equation to yield

$$\begin{aligned}
\boldsymbol{S}(l) &= \boldsymbol{A}(l) + \boldsymbol{C}^T(l)[\boldsymbol{B}(l) + \boldsymbol{\Gamma}^T(l+1,l)\boldsymbol{S}(l+1)\,\boldsymbol{\Gamma}(l+1,l)]\,\boldsymbol{C}(l)\\
&\quad + \boldsymbol{\Phi}^T(l+1,l)\boldsymbol{S}(l+1)\,\boldsymbol{\Phi}(l+1,l)\\
&\quad + 2\boldsymbol{\Phi}^T(l+1,l)\boldsymbol{S}(l+1)\,\boldsymbol{\Gamma}(l+1,l)\boldsymbol{C}(l).
\end{aligned} \tag{8.51}$$

The factor 2 in the last term results from the symmetry of $\boldsymbol{S}(l)$, and can be readily verified by substituting for $\boldsymbol{C}(l)$. By substituting for $\boldsymbol{C}^T(l)$ in the second term of this expression and applying the matrix identity of Equation (A.61) of Appendix A, the second term can be shown to reduce to

$$-\boldsymbol{\Phi}^T(l+1,l)\boldsymbol{S}(l+1)\boldsymbol{\Gamma}(l+1,l)\boldsymbol{C}(l) \tag{8.52}$$

which is identical in form to the last term of Equation (8.51). Thus, Equation (8.51) reduces to

$$\begin{aligned}
\boldsymbol{S}(l) &= \boldsymbol{A}(l) + \boldsymbol{\Phi}^T(l+1,l)\boldsymbol{S}(l+1)\,\boldsymbol{\Phi}(l+1,l)\\
&\quad + \boldsymbol{\Phi}^T(l+1,l)\boldsymbol{S}(l+1)\,\boldsymbol{\Gamma}(l+1,l)\boldsymbol{C}(l)\\
&= \boldsymbol{A}(l) + \boldsymbol{\Phi}^T(l+1,l)\boldsymbol{S}(l+1)[\boldsymbol{\Phi}(l+1,l) + \boldsymbol{\Gamma}(l+1,l)\boldsymbol{C}(l)]
\end{aligned} \tag{8.53}$$

which is the result of Section 6.4.1.

These results imply that, when the system difference equations are linear and the optimization criterion is a quadratic performance index as described by Equation (8.31), the discrete time, optimal feedback control sequence $\boldsymbol{u}(l)$ is obtained as a linear function of the state vector $\boldsymbol{x}(l)$ from Equation (8.49). The recursive solution for $\boldsymbol{C}(l)$ and $\boldsymbol{S}(l)$ can be carried out off line, and the values of $\boldsymbol{C}(l)$ stored for use in real-time processing. Furthermore, the value of the performance index over the entire interval $(0, L)$ is obtained as

$$J_1(\boldsymbol{x}(0)) = \tfrac{1}{2}\boldsymbol{x}^T(0)\boldsymbol{S}(0)\boldsymbol{x}(0) \tag{8.54}$$

where $\boldsymbol{S}(0)$ is the value obtained from backward solution of the $\boldsymbol{S}(l)$ equation. Some reflection on the development above will also reveal that this result is actually a feedback system, in that the control always considers only the optimum path from the current state $\boldsymbol{x}(l)$ to the final state $\boldsymbol{x}(L)$, and is independent of any costs incurred in reaching the present state. This is merely a restatement of the principle of optimality, upon which the whole concept of dynamic programming is based. Along with this realization goes the fact that the minimum return incurred in going from the present state $\boldsymbol{x}(l)$ to the final state is always given by the expression

$$J_{l+1}(\boldsymbol{x}(l)) = \tfrac{1}{2}\boldsymbol{x}^T(l)\boldsymbol{S}(l)\boldsymbol{x}(l) \tag{8.55}$$

which is a convenient way of monitoring the efficacy of the control process.

Finally, sufficient conditions for a minimum at each stage can be obtained from the second partial derivative matrix of the general J_l term. This fact follows from the generalized version of Equations (8.37) and (8.39) as

$$\frac{\partial^2 J}{\partial \boldsymbol{u}^2(l-1)} = \boldsymbol{B}(l-1) + \boldsymbol{\Gamma}^T(l, l-1)\boldsymbol{S}(l)\boldsymbol{\Gamma}(l, l-1) \tag{8.56}$$

where the matrix formalism of Appendix A has been used. From considerations of Appendix B, it can be determined that a minimum is reached if this matrix is positive definite. The definition of $\boldsymbol{B}(l)$ and $\boldsymbol{S}_f$ as positive definite, symmetric matrices suffices to ensure that the matrix of Equation (8.56) is positive definite. Note that this matrix is also computed and then inverted in computation leading to evaluation of $\boldsymbol{C}(l-1)$.

8.4.2. The Discrete Time Linear Regulator

When the system under investigation is time invariant, and the time interval $(t_l - t_{l-1})$ is constant for all l, then the system equations are

$$\boldsymbol{x}(l) = \boldsymbol{\Phi}\boldsymbol{x}(l-1) + \boldsymbol{\Gamma}\boldsymbol{u}(l-1). \tag{8.57}$$

Furthermore, if the matrices $\boldsymbol{A}$ and $\boldsymbol{B}$ are constant, then the recursive expression for $\boldsymbol{C}(l)$ and $\boldsymbol{S}(l)$ given by Equation (8.49) may reach a steady state condition independent of the value of $\boldsymbol{S}_L$, much in the manner described in Section 6.3.2 for the continuous time, linear regulator. These steady state values can be obtained by solving Equation (8.49) recursively until a steady state condition is reached. Alternatively, an analytical approach can be used by recognizing that, under steady state conditions,

$$\begin{aligned} \boldsymbol{S}(l+1) &= \boldsymbol{S}(l) = \boldsymbol{S} \\ \boldsymbol{C}(l+1) &= \boldsymbol{C}(l) = \boldsymbol{C} \end{aligned} \tag{8.58}$$

in which case the algebraic equations

$$\begin{aligned} \boldsymbol{C} &= -[\boldsymbol{B} + \boldsymbol{\Gamma}^T \boldsymbol{S} \boldsymbol{\Gamma}]^{-1} \boldsymbol{\Gamma}^T \boldsymbol{S} \boldsymbol{\Phi} \\ \boldsymbol{S} &= \boldsymbol{A} + \boldsymbol{C}^T \boldsymbol{B} \boldsymbol{C} + [\boldsymbol{\Phi} + \boldsymbol{\Gamma} \boldsymbol{C}]^T \boldsymbol{S} [\boldsymbol{\Phi} + \boldsymbol{\Gamma} \boldsymbol{C}] \\ &= \boldsymbol{A} + \boldsymbol{\Phi}^T \boldsymbol{S} [\boldsymbol{\Phi} + \boldsymbol{\Gamma} \boldsymbol{C}] \end{aligned} \tag{8.59}$$

can be solved for the constant values of $\boldsymbol{C}$ and $\boldsymbol{S}$. Alternative expressions for the algebraic Riccati equation can be obtained from the developments of Section 6.4.1. The regulator is then implemented by the feedback equation

$$\boldsymbol{u}(l) = \boldsymbol{C}\boldsymbol{x}(l) \tag{8.60}$$

in a manner completely analogous to the continuous time problem developed in Section 6.3.2.

8.4.3. The Continuous Time Equations

By considering the limiting case of the discrete time feedback control developed by dynamic programming, the Maximum Principle of Pontryagin can be shown, and certain useful interpretations placed upon variables of the control problem. The general optimal control problem treated in Chapter 6 requires selection of the control function $\boldsymbol{u}(t)$ to minimize or maximize the functional

$$J = \phi[\boldsymbol{x}(t_f)] + \int_{t_0}^{t_f} L(\boldsymbol{x}(t), \boldsymbol{u}(t), t)\, dt \tag{8.61}$$

subject to the constraint

$$\dot{\boldsymbol{x}}(t) = \boldsymbol{f}(\boldsymbol{x}(t), \boldsymbol{u}(t), t). \tag{8.62}$$

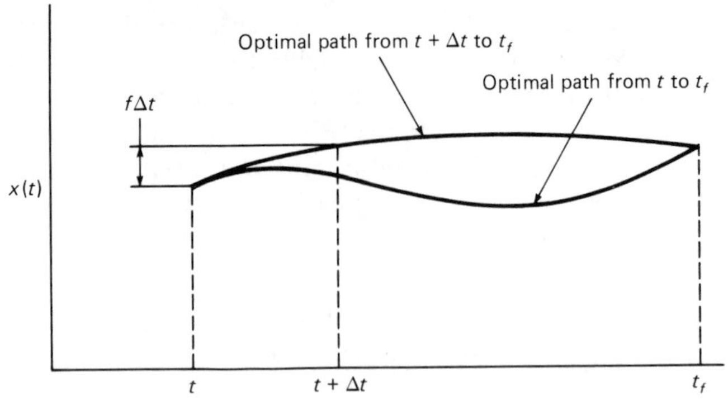

Figure 8-8. Symbolic representation of optimum and nonoptimum paths.

In theory, an optimum function $\boldsymbol{u}^*(t)$ can be found which, when applied to the system, produces an optimum path $\boldsymbol{x}^*(t)$ and an optimum value $\boldsymbol{J}^*$. The actual path $\boldsymbol{x}(t)$ may differ from the optimum, however, because of modeling errors, input disturbances, or other factors. If some means is provided for observing the state $\boldsymbol{x}(t)$ and its divergence from the optimum, appropriate action can be taken to determine a revised optimum control function. This method is the one used in defining the optimum linear controller in Section 6.3.1. The object here will be to investigate such a process as the time interval between observations, or stages, becomes small and as the limiting case of a continuous time system is reached.

Suppose that the system state at time t is $\boldsymbol{x}(t)$, and that it then proceeds to time $t + \Delta t$, where Δt is small. From the constraint equation, if $\boldsymbol{f}$ is a continuous function, it can be determined that the first-order approximation of the state at time $t + \Delta t$ is

$$\boldsymbol{x}(t + \Delta t) = \boldsymbol{x}(t) + \boldsymbol{f}(\boldsymbol{x}, \boldsymbol{u}, t)\,\Delta t \tag{8.63}$$

where the control $\boldsymbol{u}(t)$ is not necessarily optimal. There are two optimal paths defined–that existing from t to t_f and that from $t + \Delta t$ to t_f. These paths are shown symbolically in Figure 8-8 for a scalar $x(t)$. If $\boldsymbol{u}(t)$ corresponds to the optimal value, then the two paths are identical. The performance index evaluated for $t_0 = t$ along the non-optimal path of Figure 8-8 is

$$J(\boldsymbol{x}, t) = J^*(\boldsymbol{x} + \boldsymbol{f}\Delta t, t + \Delta t) + L(\boldsymbol{x}, \boldsymbol{u}, t)\,\Delta t. \tag{8.64}$$

The first term is the value of J obtained along the optimum path from $t + \Delta t$ to t_f; the second is that contribution obtained from the nonoptimal path from t to $t + \Delta t$. These conditions clearly imply that the optimum value of J is a function only of the initial state $\boldsymbol{x}(t)$ and perhaps t. The objective is to select $\boldsymbol{u}(t)$ such that $J(\boldsymbol{x}, t)$ is optimized, or, in the case of minimization,

$$\begin{aligned} J^*(\boldsymbol{x}, t) &= \min(\text{w.r.t. } \boldsymbol{u}(t))\; J(\boldsymbol{x}, t) \\ &= \min(\text{w.r.t. } \boldsymbol{u}(t))\; [J^*(\boldsymbol{x} + \boldsymbol{f}\Delta t, t + \Delta t) + L(\boldsymbol{x}, \boldsymbol{u}, t)\,\Delta t]. \end{aligned} \tag{8.65}$$

If the term within the brackets is expanded in a Taylor series, retaining only first-order terms, then

$$J^*(\boldsymbol{x}, t) = \min(\text{w.r.t. } \boldsymbol{u}(t)) \left[J^*(\boldsymbol{x}, t) + \frac{\partial J^*}{\partial t}\,\Delta t + \frac{\partial J^*}{\partial \boldsymbol{x}}\,\boldsymbol{f}\Delta t + L(\boldsymbol{x}, \boldsymbol{u}, t)\,\Delta t \right]. \tag{8.66}$$

Since the minimization is with respect to $\boldsymbol{u}(t)$, those terms not dependent upon $\boldsymbol{u}(t)$ can be removed from the brackets, with the result

$$J^*(\boldsymbol{x}, t) = J^*(\boldsymbol{x}, t) + \frac{\partial J^*}{\partial t}\,\Delta t + \min(\text{w.r.t. } \boldsymbol{u}(t)) \left[\frac{\partial J^*}{\partial \boldsymbol{x}}\,\boldsymbol{f}\Delta t + L(\boldsymbol{x}, \boldsymbol{u}, t)\,\Delta t \right]$$

or, upon factoring of the Δt term,

$$\frac{\partial J^*}{\partial t} = -\min(\text{w.r.t. } \boldsymbol{u}(t)) \left[\frac{\partial J^*}{\partial \boldsymbol{x}}\,\boldsymbol{f} + L(\boldsymbol{x}, \boldsymbol{u}, t) \right]. \tag{8.67}$$

The interpretation of the quantities $\partial J^*/\partial \boldsymbol{x}$ and $\partial J^*/\partial t$ can be obtained by considering the calculus of variations problem with variable initial time. This procedure follows the development leading up to Equation (6.28), except that now, in addition to the small change in terminal time, dt_f, a small change dt_0 is permitted in the initial time. This result is due to an argument analogous to that presented in support of Equation (6.25). The modified performance index is given by Equation (6.26), and when perturbations $\delta \boldsymbol{x}(t)$, $\delta \boldsymbol{u}(t)$, and small changes dt_0 and dt_f are considered, the total change in $\bar{J}$ is

$$\begin{aligned} \Delta \bar{J} = {} & \left[\frac{\partial \phi}{\partial \boldsymbol{x}} + \boldsymbol{\nu}^T \frac{\partial \boldsymbol{\psi}}{\partial \boldsymbol{x}} \right] d\boldsymbol{x}(t_f) - \boldsymbol{\lambda}^T \delta \boldsymbol{x} \Big|_{t=t_f} + \int_{t_0}^{t_f} \left\{ \left[\frac{\partial H}{\partial \boldsymbol{x}} + \dot{\boldsymbol{\lambda}}^T \right] \delta \boldsymbol{x} \right. \\ & \left. + \frac{\partial H}{\partial \boldsymbol{u}}\,\delta \boldsymbol{u} \right\} dt + \boldsymbol{\lambda}^T \delta \boldsymbol{x} \Big|_{t=t_0} + \frac{d}{dt_f} \left[\int_{t_0}^{t_f} [H - \boldsymbol{\lambda}^T \dot{\boldsymbol{x}}]\, dt \right] dt_f \\ & + \frac{d}{dt_0} \left[\int_{t_0}^{t_f} [H - \boldsymbol{\lambda}^T \dot{\boldsymbol{x}}]\, dt \right] dt_0. \end{aligned} \tag{8.68}$$

The last term is due to permitted changes in t_0, which was not a factor in the development leading to Equation (6.28). The last two terms can be differentiated by Leibniz' rule, yielding the result

$$\Delta \bar{J} = \left[\frac{\partial \phi}{\partial \boldsymbol{x}} + \boldsymbol{\nu}^T \frac{\partial \boldsymbol{\psi}}{\partial \boldsymbol{x}} - \boldsymbol{\lambda}^T\right] d\boldsymbol{x}(t_f) + \int_{t_0}^{t_f} \left\{\left[\frac{\partial H}{\partial \boldsymbol{x}} + \dot{\boldsymbol{\lambda}}^T\right] \delta \boldsymbol{x} + \frac{\partial H}{\partial \boldsymbol{u}} \delta \boldsymbol{u}\right\} dt$$

$$+ \left[H + \frac{\partial \phi}{\partial t}\right]_{t=t_f} dt_f + \boldsymbol{\lambda}^T \delta \boldsymbol{x}\,|_{t=t_0} - [H - \boldsymbol{\lambda}^T \dot{\boldsymbol{x}}]_{t=t_0}\, dt_0. \quad (8.69)$$

If a solution to this problem is obtained with the conditions of Equation (6.29) satisfied, then the only remaining term in this expression is

$$\Delta \bar{J} = [\boldsymbol{\lambda}^T \dot{\boldsymbol{x}} - H]\big|_{t=t_0}\, dt_0 = \boldsymbol{\lambda}^T\, d\boldsymbol{x}(t_0) - H\big|_{t=t_0}\, dt_0. \quad (8.70)$$

Since t_0 corresponds to t in the current discussion, this expression identifies $\boldsymbol{\lambda}^{*T}(t)$ as the row vector of partial derivatives

$$\boldsymbol{\lambda}^{*T}(t) = \frac{\partial J^*}{\partial \boldsymbol{x}} \quad (8.71)$$

and $-H(\boldsymbol{x}^*(t), \boldsymbol{u}^*(t), \boldsymbol{\lambda}^*(t), t)$ as the partial derivative

$$-H(\boldsymbol{x}^*(t), \boldsymbol{u}^*(t), \boldsymbol{\lambda}^*(t), t) = \partial J^*/\partial t \quad (8.72)$$

where $\boldsymbol{x}^*(t)$, $\boldsymbol{u}^*(t)$, and $\boldsymbol{\lambda}^*(t)$ are the values obtained along the optimal path. That is, the Lagrange multipliers and the Hamiltonian imply the influence on the optimal performance index of small changes in the state variables and initial time, respectively.

With these interpretations and the fact that the Hamiltonian function is given by

$$H(\boldsymbol{x}(t), \boldsymbol{u}(t), \boldsymbol{\lambda}(t), t) = L(\boldsymbol{x}(t), \boldsymbol{u}(t), t) + \boldsymbol{\lambda}^T(t) \boldsymbol{f}(\boldsymbol{x}(t), \boldsymbol{u}(t), t) \quad (8.73)$$

Equation (8.67) can be written as

$$\begin{aligned} H^* &= H(\boldsymbol{x}^*(t), \boldsymbol{u}^*(t), \boldsymbol{\lambda}^*(t), t) \\ &= \min(\text{w.r.t. } \boldsymbol{u}(t))\ [L(\boldsymbol{x}(t), \boldsymbol{u}(t), t) + \boldsymbol{\lambda}^{*T}(t) \boldsymbol{f}(\boldsymbol{x}(t) \boldsymbol{u}(t), t)] \\ &= \min(\text{w.r.t. } \boldsymbol{u}(t))\, H(\boldsymbol{x}(t), \boldsymbol{u}(t), \boldsymbol{\lambda}(t), t) \end{aligned} \quad (8.74)$$

where $\boldsymbol{\lambda}^*(t)$ is defined as that value produced by the necessary condition

$$\dot{\boldsymbol{\lambda}}^T(t) = -\partial H/\partial \boldsymbol{x}. \tag{8.75}$$

This result is merely a statement of the Pontrygagin Maximum Principle discussed in Chapter 6. For $\boldsymbol{u}(t)$ unconstrained, the above equation is equivalent to selecting $u(t)$ so that

$$\partial H/\partial \boldsymbol{u} = 0 \tag{8.76}$$

in which case the necessary conditions of Equation (6.29) are obtained.

8.4.4. Application to the Continuous Time Linear Problem

These techniques can be applied to the continuous time linear problem with quadratic performance index by specializing to the system equations

$$\dot{\boldsymbol{x}}(t) = \boldsymbol{F}(t)\boldsymbol{x}(t) + \boldsymbol{G}(t)\boldsymbol{u}(t) \tag{8.77}$$

and the performance index

$$J = \tfrac{1}{2}[\boldsymbol{x}^T(t_f)\boldsymbol{S}_f\boldsymbol{x}(t)] + \int_{t_0}^{t_f} \tfrac{1}{2}[\boldsymbol{x}^T(t)\boldsymbol{A}(t)\boldsymbol{x}^T(t) + \boldsymbol{u}^T(t)\boldsymbol{B}(t)\boldsymbol{u}(t)]\,dt. \tag{8.78}$$

The result is a solution by way of the matrix Riccati equations as developed in Section 6.3.1. Implicit in this particular approach to the problem is the relationship

$$J(\boldsymbol{x}(t), t) = \tfrac{1}{2}[\boldsymbol{x}^T\boldsymbol{S}(t)\boldsymbol{x}(t)] \tag{8.79}$$

relating the value of the return generated by progressing from the current time t to the final time t_f to the current value of $\boldsymbol{S}(t)$ as generated by the Riccati equation. This result is analogous to the expression of Equation (8.55) for discrete time linear systems.

8.4.5. Direct Application of Dynamic Programming to Optimal Control

The principles of dynamic programming are useful to the control designer from two points of view. The first of these considers dynamic programming an analytical tool, useful in developing important relationships of optimal control. This view is exemplified by the developments of the previous two sections, in which several of the more important concepts of both discrete and continuous time optimal control are derived. The second view considers dynamic programming, in the form of Equation (8.26), as a technique that can be applied directly—in the manner of the developments of Section 8.3—to the discrete time optimal control problem. Such an approach has a number of attractive features, as listed below:

- Very general kinds of system equations, optimization criteria, and constraints can be easily considered.
- The control obtained is a true feedback control, as opposed to the open-loop control functions produced by other general approaches.
- Questions of existence and uniqueness are, in a sense, avoided.
- Since the procedure is numeric in nature, tabular functions with no analytic representation are easily considered.

These are very real advantages, but they come at an often overwhelming computational burden in terms of computer operations and required storage. The problems encountered in the practical application of dynamic programming principles will be considered in more detail after a brief description of the general process is given.

The direct dynamic programming approach to the discrete time optimal control problem is applied to the general system of Equation (8.23), and the optimization criterion is the maximization or minimization of some objective function satisfying the separability criterion discussed in Section 8.3—perhaps that of Equation (8.24). The solution procedure is to establish a table, similar to those established in the solution of the example problem of Section 8.3, for each discrete time point in the interval of interest. Furthermore, unless they are already in discrete form because of the nature of the problem, the state variable and the control functions must be quantized. This quantization is a sensitive procedure, since one is driven to use small quantization intervals in order to better approximate the continuous functions, but at the same time would like a large interval to reduce the computational load.

Once the quantization of the state variables and control functions has been established, the procedure consists of alternate application of Equations (8.24) and (8.26). The results are stored in tables of the symbolic form shown in Figure 8-9. These tables are represented in an abstract fashion in Figure 8-9, since in the general case the entries $\boldsymbol{x}(l)$ and $\boldsymbol{u}(l)$ are vectors. The procedure used in determining the table entries parallels that of the ship assignment example of Section 8.3, as represented by Tables 8-2 through 8-7. The general procedure can be illustrated by considering the general case of Table l. For each admissible value of $\boldsymbol{x}(l-1)$, the admissible values of $\boldsymbol{u}(l-1)$ are listed, together with the resulting values of $\boldsymbol{x}(l)$ and the sum $(j_l + J_{l+1})$. The general entries appear as in

Table L

$\mathbf{x}(L-1)$	$\mathbf{u}(L-1)$	$\mathbf{x}(L)$	$j_L(\mathbf{x}(L-1)) + (\mathbf{x}(L))$
— — —	— — —	— — —	— — —
— — —	— — —	— — —	— — —
— — —	— — —	— — —	— — —

Table L^*

$\mathbf{x}(L-1)$	$j_L(\mathbf{x}(L-1))$	$\mathbf{u}^*(L-1)$
— — —	— — —	— — —
— — —	— — —	— — —
— — —	— — —	— — —

Table $(L-1)$

$\mathbf{x}(L-2)$	$\mathbf{u}(L-2)$	$\mathbf{x}(L-1)$	$j_{L-1}[\mathbf{x}(L-2), \mathbf{u}(L-2)) + J_L(\mathbf{x}(L-1))]$
— — —	— — —	— — —	— — —
— — —	— — —	— — —	— — —
— — —	— — —	— — —	— — —

Table $(L-1)^*$

$\mathbf{x}(L-2)$	$J_{L-1}(\mathbf{x}(L-2))$	$\mathbf{u}^*(L-2)$
— — —	— — —	— — —
— — —	— — —	— — —
— — —	— — —	— — —

⋮

Table l

$\mathbf{x}(l-1)$	$\mathbf{u}(l-1)$	$\mathbf{x}(1)$	$j_l(\mathbf{x}(l-1), \mathbf{u}(l-1)) + J_{l+1}(\mathbf{x}(l))$
— — —	— — —	— — —	— — —
— — —	— — —	— — —	— — —
— — —	— — —	— — —	— — —

Table l^*

$\mathbf{x}(l-1)$	$J_l(\mathbf{x}(l-1)$	$\mathbf{u}^*(l-1)$
— — —	— — —	— — —
— — —	— — —	— — —
— — —	— — —	— — —

⋮

Table 1

$\mathbf{x}(0)$	$\mathbf{u}(0)$	$\mathbf{x}(1)$	$j_1(\mathbf{x}(0), \mathbf{u}(0)) + J_2(\mathbf{x}(1))$
— — —	— — —	— — —	— — —
— — —	— — —	— — —	— — —
— — —	— — —	— — —	— — —

Table 1*

$\mathbf{x}(0)$	$J_1(\mathbf{x}(0))$	$\mathbf{u}^*(0)$
— — —	— — —	— — —
— — —	— — —	— — —
— — —	— — —	— — —

$$J = \phi(\mathbf{x}(L)) + \sum_{l=1}^{L} j_l(\mathbf{x}(l-1), \mathbf{u}(l-1))$$

J_1 = optimum value of J

Figure 8-9. Tabular functions of dynamic programming.

Table *l*

x(*l* − 1)	u(*l* − 1)	$j_l(x(l-1), u(l-1)) + J_{l+1}(x(\ ))$	
− − −	− − −	− − − ⎫ optimum = J_l	
	− − −	− − − ⎬	
	− − −	− − − ⎭ u = u*	
− − −	− − −	− − − ⎫ optimum = J_l	
	− − −	− − − ⎬	
	− − −	− − − ⎭ u = u*	
− − −	− − −	− − − ⎫ optimum = J_l	
	− − −	− − − ⎬	
	− − −	− − − ⎭ u = u*	

Table *l**

x(*l* − 1)	$J_l(x(l-1))$	u*(*l* − 1)
− − −	− − −	− − −
− − −	− − −	− − −
− − −	− − −	− − −

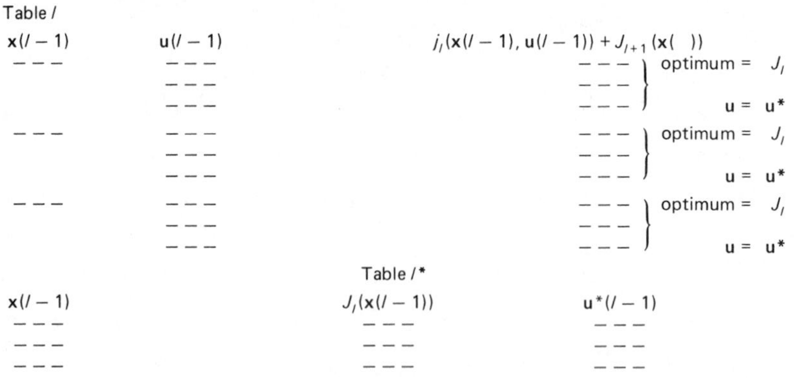

Figure 8-10. Details of Table *l*.

Figure 8-10, illustrating the fact that, for each admissible value of $x(l-1)$, there may be many admissible values of $u(l-1)$.

For each value of $x(l-1)$, $u*(l-1)$—that value of $u(l-1)$ producing the optimum sum—is listed in Table *l**, together with the resulting optimal value, now designated as $J_l(x(l-1))$ in accordance with Equation (8.26). This procedure starts at the terminal stage $x(L)$, represented by Table L, and proceeds to the initial stage, represented by Table 1. Terminal constraints are met at either or both ends by proper definition of admissible states, as are general state variable constraints. Control variable constraints are easily considered by limiting the control functions to admissible values.

It is easily seen that the advantages outlined above are inherent in the process. But there are disadvantages as well—so many, in fact, that the use of dynamic programming as a practical method of numerically solving any reasonably complex control problem is severely limited. Some of these disadvantages are:

- If the state vector or control vector contains many components, generation and storage of the tables becomes computationally difficult.
- As the tables grow in complexity, the search for the minimizing $u(l)$ becomes a computationally difficult problem.
- It is not known in advance how many, or which, of the values of $x(l-1)$ need be included in Table *l*. It may well happen that in the development of Table $(l-1)$, the admissible values of $u(l-2)$ may produce an $x(l-1)$ value not included in Table *l*. On the other hand, values of $x(l-1)$ may be included in Table *l* which are not produced by any admissible values of $u(l-2)$ Table $(l-1)$.
- Interpolation between quantized values of the states is usually required. When Table $(l-1)$ is developed, the $x(l-2)$ and $u(l-2)$ values—which are quantized—are unlikely to produce precisely one of the quantized values of

$\boldsymbol{x}(l-1)$ included in Table l. Thus, some kind of interpolation is required which, when the number of states is large or when j_l is a discontinuous function, can be a severe computational problem.

- Most of the above characteristics are directly affected by the quantization of the continuous time state and control vectors in such a manner that smaller intervals only make the problem more severe.

These difficulties are very real—even in this era of very fast computers and inexpensive memory—to the extent that there have been few practical developments in the use of dynamic programing techniques. Even with the introduction of new procedures with greatly reduced computational requirements, practical applications in the control area have been limited.

8.4.6. Application to the Stochastic Problem

If the system equation of Equation (8.23) is modified to include a driving noise function $\mathbf{W}(l)$, then the state vector is a random process,

$$\mathbf{X}(l) = \boldsymbol{f}(\mathbf{X}(l-1), \boldsymbol{u}(l-1), \mathbf{W}(l-1), l). \tag{8.80}$$

The objective function must then be modified to recognize the random character of $\mathbf{W}(l)$ and $\mathbf{X}(l)$, and the most logical choice is

$$J = E\left[\sum_{l=1}^{L} j_l(\mathbf{X}(l-1), \boldsymbol{u}(l-1)\mathbf{W}(l-1))\right]. \tag{8.81}$$

Then, if $\mathbf{W}(l)$ is a white noise process, recurrence relations analogous to Equation (8.16) can be developed, resulting in the general expression

$$J_l(\mathbf{X}(l-1) = \min(\text{w.r.t. } \boldsymbol{u}(l-1)) \cdot \{E[j_l(\mathbf{X}(l-1), \boldsymbol{u}(l-1), \mathbf{W}(l-1)) + J_{l+1}(\mathbf{X}(l))]\}. \tag{8.82}$$

The limitation to white noise driving functions can be relaxed by augmenting the state vector, in a manner similar to that discussed in Section 7.2.2.2. Since the recurrence relation of Equation (8.82) is of the same form as that of Equation (8.26), the procedures of the previous section are directly applicable. The random function $\mathbf{W}(l)$ must be quantized and the corresponding probability density functions converted to probability functions. The expectation operator is then implemented by a summation.

If the more general problem entailing noisy measurements as well as a random driving function is considered, the result is a statement of the Separation

Principle—that is, the resulting optimum stochastic control system consists of a Kalman filter cascaded with the deterministic controller.

8.5. FURTHER COMMENTS

Dynamic programming is an elegant and powerful technique for the solution of optimization problems in engineering, economics, operations research, and other areas. It is a product of the computer age, because of the enormity of the computational burden imposed by even relatively simple problems. The standard method of direct application outlined in Section 8.4.4, while rather easily developed and understood, is prohibitively demanding of computational capacity. A number of other methods which retain the desirable features of dynamic programming, yet substantially reduce computational requirements, have been developed. Some of these include:

- Polynomial approximation of the minimum cost function $J_l(\boldsymbol{x}(l - 1))$. Then only the polynomial coefficients need be stored.
- Search procedures other than direct comparison. It may be possible to use some efficient search procedure, such as the simplex method of linear programming—briefly discussed in Appendix B—to find the minimum of Equation (8.26). Ordinarily, direct comparison would be used.
- Use of known characteristics of the solution as obtained from analytical considerations. Even though a solution may not be easily obtainable from an analytical approach, certain characteristics helpful to the designer may be inferred. For example, for a system linear in the control function, it is known that the solution is a bang-bang—or full deflection—control. Such information is useful in defining the set of admissible control functions.
- State increment dynamic programming. This technique is useful in alleviating the quantization problem in discrete time problems that have been transformed from continuous time problems. The time interval is adjusted so that the number of quantized states required to determine the minimum cost function is reduced.
- Forward dynamic programming. The problem is reformulated as one in which calculations proceed from the first stage to the last, rather than backward from the terminal stage. In some cases the forward solution is more desirable.

These techniques offer alternatives to the standard approach of Section 8.4.5, and therefore yield a complete feedback control solution. These procedures reduce the required storage space, but the processing time does not change drastically.

8.5.1. Discrete State Problems

Problems with discrete-valued state variables are rare in control theory, but they abound in fields such as operations research. Examples can be found in References 9, 10, and 11. Normally, the application deals with resource allocation of some kind. The discrete time, discrete state random process is known as a *chain.* A particular type of chain, the Markov chain, has been analyzed extensively and is essentially the output of a discrete-valued linear system driven by white noise. The reader should consider this fact with regard to the quantization aspects discussed in Section 8.4.5. Application to the pattern recognition problem is discussed in Reference 8, while Reference 22 attempts to show the value of dynamic programming in image processing.

8.5.2. Application to Control Systems

The application of dynamic programming in the development of basic optimal control results is presented in References 2, 4, 13, 14, 16, 21, 25, and 26. In References 16 and 26, the stochastic problem is considered and the Separation Principle derived. The direct application of dynamic programming and the attendant computational difficulties are considered in References 7, 12, 13, 14, and 21. In Reference 15, a dated but useful compendium of computational procedures is presented. Specific techniques for overcoming the dimensionality problem are presented in Reference 1–in which extrapolation and interpolation in state space is considered–and in Reference 20, where specific application to bang-bang systems is considered. Differential dynamic programming, which applies the principle of optimality in the neighborhood of a nominal, possibly nonoptimal, solution is treated in Reference 19. Application of differential dynamic programming to bang-bang systems is the subject of Reference 12. In Reference 6, specific application to the linear problem with a quadratic performance index is considered, and a simplified algorithm is developed. The application of distributed processing–using multiple central processing units–to the dynamic programming problem is treated in Reference 3. This concept is an interesting one, and one which may hold the key to overcoming the computational burden of dynamic programming.

In References 17 and 27, some practical applications of dyanmic programming are given, while Reference 23 presents some results comparing dynamic programming to other optimization techniques, for one specific application.

Application of dynamic programming to a class of nonlinear stochastic systems is considered in Reference 18.

REFERENCES

1. Bard, Yonathan. Interpolation and Extrapolation Schemes in Dynamic Programming. *IEEE Transactions on Automatic Control,* **AC-12:** 97–99 (February 1967).
2. Bellman, Richard. *Dynamic Programming.* Princeton, New Jersey: Princeton University Press, 1957.
3. Bertsekas, Dimitri. Distributed Dynamic Programming. *IEEE Transactions on Automatic Control,* AC-27: 611–616 (June 1978).

4. —— *Dynamic Programing and Stochastic Control.* New York, New York: Academic Press, Inc. 1976.
5. —— and Shreve, Steven N. *Stochastic Optimal Control: The Discrete-Time Case.* New York, New York: Academic Press, Inc. 1978.
6. Dreyfus, S. E., and Kan, Y. C. A General Dynamic Programming Solution of Discrete-Time Linear Optimal Control Problems. *IEEE Transactions on Automatic Control,* **AC-18:** 286–289 (June 1973).
7. Dyer, Peter, and McReynolds, Stephen R. *The Computation and Theory of Optimal Control.* New York, New York: Academic Press, Inc. 1970.
8. Fu, K. S. and Cardillo, G. P. An Optimum Finite Sequential Procedure for Feature Selection and Pattern Classification. *IEEE Transactions on Automatic Control,* **AC-12:** 588–591 (October 1967).
9. Gauer, Donald P., and Thompson, Gerald P. *Programming and Probability Models in Operation Research.* Monterey, California: Brookes/Cole, 1973.
10. Gue, Ronald L., and Thomas, Michael E. *Mathematical Methods in Operations Research.* New York: Macmillan, 1968.
11. Hillier, Frederick S., and Lieberman, Gerald J. *Introduction to Operations Research.* San Francisco: Holden-Day, 1967.
12. Jacobsen, David H. Differential Dynamic Programming Methods for Solving Bang-Bang Control Problems. *IEEE Transactions on Automatic Control,* **AC-13:** 661–675 (December 1968).
13. Kirk, Donald E. *Optimal Control Theory.* Englewood Cliffs, New Jersey: Prentice-Hall, 1970.
14. Koppel, Lowell B. *Introduction to Control Theory.* Englewood Cliffs, New Jersey: Prentice-Hall, 1968.
15. Larson, Robert E. A Survey of Dynamic Programming Computational Procedures. *IEEE Transactions on Automatic Control,* **AC-12:** 767–774 (December 1967).
16. Meditch, J. S. *Stochastic Optimal Linear Estimation and Control.* New York, New York: McGraw-Hill, 1969.
17. Meier, Lewis, Larson, Robert E., and Tether, Anthony J. Dynamic Programming for Stochastic Control of Discrete Systems. *IEEE Transactions on Automatic Control,* **AC-16:** 767–775 (December 1971).
18. Mohler, Ronald R., and Kolodziej, Wojciech J. Optimal Control of a Class of Non-Linear Stochastic Systems. *IEEE Transactions on Automatic Control,* **AC-26:** 1048–1054 (October 1981).
19. Ohno, Katsuhisa. A New Approach to Differential Dynamic Programming for Discrete-Time Systems. *IEEE Transactions on Automatic Control,* **AC-23:** 37–47 (February 1978).
20. Rillings, James H., and Roy, Rob J. Bang-Bang Control Using a Tabular Adaptive Model. *IEEE Transactions on Automatic Control,* **AC-12:** 310–312 (June 1967).
21. Sage, Andrew, P., and White, Chelsea C. *Optimum Systems Control.* Englewood Cliffs, New Jersey: Prentice-Hall, 1977.
22. Scharf, L. L., and Elliot, H. Aspects of Dynamic Programming in Signal and Image Processing. *IEEE Transactions on Automatic Control,* **AC-26:** 1018–1021 (October 1981).
23. Shoemaker, Christine A. Applications of Dynamic Programming and Other Optimization Methods in Pest Management. *IEEE Transactions on Automatic Control,* **AC-26:** 1125–1132 (October 1981).
24. Tou, Julius T. *Modern Control Theory.* New York, New York: McGraw-Hill, 1964.
25. —— *Optimum Design of Digital Control Systems.* New York, New York: Academic Press, Inc., 1963.

26. Wang, Peter J., and Larson, Robert E. Optimization of Natural-Gas Pipeline Systems Via Dynamic Programming. *IEEE Transactions on Automatic Control*, AC-13: 475–481 (October 1968).

PROBLEMS

8.1. In the four by four grid shown in Figure P8-1, find the minimum length path from point A to point B, moving only to the right from one grid point to the next. There are 70 different paths. How many are actually calculated?

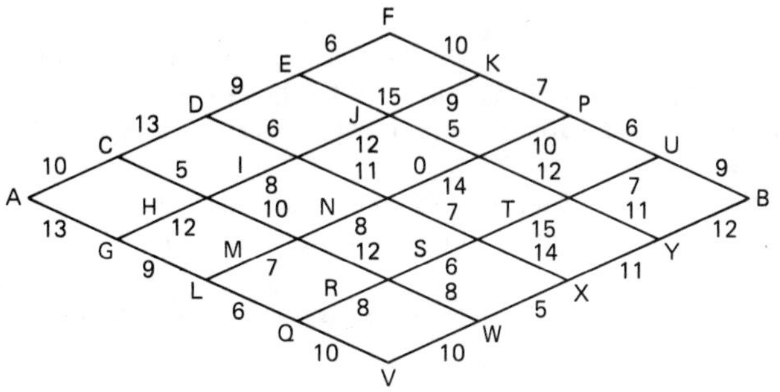

Figure P8-1.

8.2. Find the maximum length path in the grid of Problem 8.1.

8.3. Dynamic programming can be used to solve resource allocation problems. Suppose three units of a resource are available for allocation to three different activities. The return from each activity as a function of the resources allocated to that activity is given below. Allocation is made in single units only. Use dynamic programming to determine the optimal allocation policy.

allocation	activity 1	2	3
0	0	0	0
1	2	1	3
2	4	5	5
3	6	6	6

8.4. A first-order discrete time system is described by the difference equation

$$x(l) = 0.5x(l-1) + u(l-1).$$

The performance index

$$J = \sum_{l=0}^{4} x(l)$$

is to be minimized. The control variable and state variable are constrained by the inequalities

$$-2 \leqslant x(l) \leqslant 2$$

$$-1 \leqslant u(l) \leqslant 1$$

Use dynamic programming to determine the optimal control sequence $u(l)$, $l = 0, 1, 2, 3$. Use quantization of 1.0 for both x and u.

8.5. Repeat Problem 8.4 for the case in which a terminal state $x(4) = 1$ is to be obtained.

8.6. A truck can carry a 10-ton load. Three types of product are available for shipment. Their weights and values are tabulated below. Consider the loading of the truck a sequential operation of loading type A items, then type B items, and finally, type C items. Determine the loading for maximum value.

Type	Value	Weight (tons)
A	20	1
B	50	2
C	60	2

8.7. A production line contains a machine which is t years old and has no salvage value. The annual income I from this machine is

$$I = \begin{cases} 25 - t^2, & \text{if } 0 \leqslant t \leqslant 4 \\ 0 & \text{otherwise.} \end{cases}$$

The cost of replacement is 21, and the machine has an expected life of 5 years. If the present machine is 2 years old and annual decisions are made on the basis of maximizing profit over the next five years, find the optimal replacement policy. Model the process as a multistage decision process with a replace or keep decision at each stage.

8.8. Use dynamic programming to maximize the function

$$J = \sum_{i=1}^{3} (4x_i - ix_i^2)$$

subject to the constraint

$$\sum_{i=1}^{3} x_i = 4.$$

Assume that x_i can take on only the values 0 through 4.

8.9. A manufacturing process contains four units, through which the product is sequentially passed. Raw material enters the first unit, and the product from one unit serves as the raw material for the next. The product from any particular unit may be withdrawn and sold, or passed to the next unit. At each stage, the profit of withdrawn products is given by

$$6x_i - ix_i^2$$

where x_i is the withdrawn product flow. If the input flow is 10 units, find the flow to be withdrawn at each stage in order to maximize the return from the process.

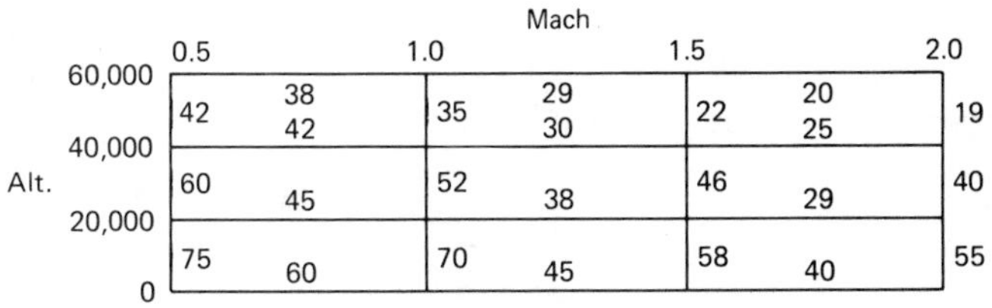

Figure P8-2.

8.10. An aircraft is limited to constant velocity climbs or constant altitude acceleration. The desired goal is 60,000 ft altitude at Mach 2 in minimum time, starting from sealevel at Mach 0.5. The climb must correspond to altitude–Mach number points shown in Figure P8-1. The time to climb or accelerate from one point to the next is shown. Find the minimum-time climb profile.

8.11. Determine the sequence of control variables $u(l)$ which maximize the performance index

$$J = \sum_{l=0}^{2} u^2(l)$$

subject to the constraint

$$x(l) = x(l-1) + u(l-1), \qquad l = 1, 2, 3.$$

The sequence $x(l)$ can take on only the values 0 through 5, and $x(3)$ is constrained to have the value 5.

8.12. A scalar, discrete time system is described by the difference equation

$$x(l) = \phi x(l-1) + \gamma u(l-1).$$

The control sequence $u(l)$ is constrained by the inequality

$$|u(l)| \leqslant 1.$$

Use dynamic programming to determine the feedback control which minimizes the performance index

$$J = \sum_{l=0}^{2} x^2(l)$$

The initial state is $x(0)$, and the final state is free.

8.13. A discrete time system is described by the equation

$$x(l) = x(l-1) + u(l-1).$$

The performance index is specified as

$$J = x^2(L) + \sum_{l=0}^{L-1} u^2(l)$$

and there are state and control constraints specified as

$$0 \leqslant x(l) \leqslant 1.5$$

$$-1.0 \leqslant u(l) \leqslant 1.0.$$

Consider $x(l)$ and $u(l)$ to be quantitized in steps of 0.5, and determine the optimal control sequence for $L = 2$.

Appendix A
Matrix Operations

A.1. INTRODUCTION

The use of vector–matrix notation and certain formal definitions for representing rather complex vector–matrix manipulation is important to a succinct representation of multivariable operations.

This appendix provides the basic definitions used throughout the text. Vectors, in general, are considered to be column vectors, with the transposed form used for row vectors. The dimension of a matrix is $N \times M$, where N is the number of rows and M the number of columns. The identity matrix is indicated by $\boldsymbol{I}$, and the inverse of the matrix $\boldsymbol{A}$ is represented by $\boldsymbol{A}^{-1}$. The transpose of $\boldsymbol{A}$ is represented by $\boldsymbol{A}^T$.

A.2. PARTITIONED MATRICES

In many instances, the physical transformations producing the definition of certain matrices provide for the natural partitioning of the matrices into smaller matrices. For example, if the matrix $\boldsymbol{A}$ is subdivided as

$$\boldsymbol{A} = \left[\begin{array}{cc|ccc} a_{11} & a_{12} & a_{13} & a_{14} & a_{15} \\ a_{21} & a_{22} & a_{23} & a_{24} & a_{25} \\ \hline a_{31} & a_{32} & a_{33} & a_{34} & a_{35} \\ a_{41} & a_{42} & a_{43} & a_{44} & a_{45} \end{array}\right] = \left[\begin{array}{c|c} \boldsymbol{A}_{11} & \boldsymbol{A}_{12} \\ \hline \boldsymbol{A}_{21} & \boldsymbol{A}_{22} \end{array}\right] \tag{A.1}$$

then the definition of the submatrices $\boldsymbol{A}_{11}$, $\boldsymbol{A}_{12}$, $\boldsymbol{A}_{21}$, and $\boldsymbol{A}_{22}$ follows directly. Such partitioning also results in some computational convenience because of the following easily verified relationship. If $\boldsymbol{B}$ is a matrix partitioned as

$$\boldsymbol{B} = \left[\begin{array}{c|cc} b_{11} & b_{12} & b_{13} \\ b_{21} & b_{22} & b_{23} \\ \hline b_{31} & b_{32} & b_{33} \\ b_{41} & b_{42} & b_{43} \\ b_{51} & b_{52} & b_{53} \end{array}\right] = \left[\begin{array}{c|c} \boldsymbol{B}_{11} & \boldsymbol{B}_{12} \\ \hline \boldsymbol{B}_{21} & \boldsymbol{B}_{22} \end{array}\right] \tag{A.2}$$

then the product $\boldsymbol{AB}$ can be formed from

$$\left[\begin{array}{c:c} A_{11} & A_{12} \\ \hdashline A_{21} & A_{22} \end{array}\right]\left[\begin{array}{c:c} B_{11} & B_{12} \\ \hdashline B_{21} & B_{22} \end{array}\right] = \left[\begin{array}{c:c} A_{11}B_{11} + A_{12}B_{21} & A_{11}B_{12} + A_{12}B_{22} \\ \hdashline A_{21}B_{11} + A_{22}B_{21} & A_{21}B_{12} + A_{22}B_{22} \end{array}\right]. \tag{A.3}$$

That is, the product matrix is obtained by considering each submatrix as though it were a matrix element. Generalization of this concept to any matrix product is easily accomplished, but it must be recognized that the partitioning must be done in such a way that the individual matrix products are defined.

An example of the usefulness of partitioned matrices can be obtained from consideration of the conditional distributions derived from multivariate normal distributions. If the random vector $\mathbf{X}$ is partitioned as

$$\mathbf{X} = \left[\begin{array}{c} \mathbf{X}_1 \\ \hdashline \mathbf{X}_2 \end{array}\right]$$

and the corresponding subdivisions made in the covariance matrix

$$\boldsymbol{\Sigma} = \left[\begin{array}{c:c} \boldsymbol{\Sigma}_{11} & \boldsymbol{\Sigma}_{12} \\ \hdashline \boldsymbol{\Sigma}_{21} & \boldsymbol{\Sigma}_{12} \end{array}\right]$$

then $\boldsymbol{\Sigma}_{11}$ constitutes the covariance matrix of $\mathbf{X}_1$ and $\boldsymbol{\Sigma}_{22}$ is the covariance matrix of $\mathbf{X}_2$. Multivariate normal distributions are special in that conditional probability density functions derived from such distributions are also multivariate normal. In particular, the conditional probability density function of $\mathbf{X}_1$ given $\mathbf{X}_2$ has covariance matrix

$$\boldsymbol{\Sigma}_{12}\boldsymbol{\Sigma}_{22}^{-1}\boldsymbol{\Sigma}_{21}$$

and mean vector

$$\boldsymbol{\mu}_1 + \boldsymbol{\Sigma}_{12}\boldsymbol{\Sigma}_{22}^{-1}(\mathbf{X}_2 - \boldsymbol{\mu}_2)$$

where the mean vector $\boldsymbol{\mu}$ is also partitioned as

$$\boldsymbol{\mu} = \left[\begin{array}{c} \boldsymbol{\mu}_1 \\ \hdashline \boldsymbol{\mu}_2 \end{array}\right].$$

Thus, the individual submatrices of the covariance matrix have meanings germane to the physical problem at hand, that of determining the distribution of one set of random variables conditioned on observation of the remaining variables.

A.3. DEFINED FORMALISM

There are certain fundamental operations involving multivariate functions which can most succinctly be represented in terms of matrices and matrix manipulation. For this reason, a certain set of basic definitions is made here which represents only a mathematical formalism, in the sense that they exist only in terms of the definitions themselves. For example, a matrix exponential of the form $e^{\boldsymbol{A}}$ has no meaning in strict mathematical terms, but here it will be defined as the infinite series

$$e^{\boldsymbol{A}} = I + \boldsymbol{A} + \frac{1}{2!}\boldsymbol{AA} + \frac{1}{3!}\boldsymbol{AAA} + \cdots . \tag{A.4}$$

In terms of the formalism of Equation (A.4), the matrix exponential does have meaning and can be used whenever the series of Equation (A.4) is implied. A group of such definitions will now be made.

A.3.1. Differentiation and Integration of a Matrix

It is often desired to indicate formally the differentiation of each element of a matrix with respect to some independent variable, such as time. The notation used to designate this operation is

$$\frac{dA}{dt} = \begin{bmatrix} \dfrac{da_{11}}{dt} & \dfrac{da_{12}}{dt} & \cdots & \dfrac{da_{1N}}{dt} \\ \dfrac{da_{21}}{dt} & \dfrac{da_{22}}{dt} & \cdots & \dfrac{da_{2N}}{dt} \\ \cdots & \cdots & \cdots & \cdots \\ \dfrac{da_{N1}}{dt} & \dfrac{da_{N2}}{dt} & \cdots & \dfrac{da_{NN}}{dt} \end{bmatrix}. \tag{A.5}$$

Similarly, the integral of a matrix is used to denote the matrix with elements that are the integrals of the elements of the original matrix.

$$\int A\,dt = \begin{bmatrix} \int a_{11}\,dt & \int a_{12}\,dt & \cdots & \int a_{1N}\,dt \\ \int a_{21}\,dt & \int a_{22}\,dt & \cdots & \int a_{2N}\,dt \\ \cdots & \cdots & \cdots & \cdots \\ \int a_{N1}\,dt & \int a_{N2}\,dt & \cdots & \int a_{NN}\,dt \end{bmatrix}. \tag{A.6}$$

If $f(\boldsymbol{x})$ is a scalar function of the vector $\boldsymbol{x}$, then the row vector consisting of elements which are the partial derivatives of $f(\boldsymbol{x})$ with respect to the elements of $\boldsymbol{x}$ is called the partial derivative of the scalar $f(\boldsymbol{x})$ with respect to the vector $\boldsymbol{x}$. Thus,

$$\frac{\partial f(\boldsymbol{x})}{\partial \boldsymbol{x}} = \begin{bmatrix} \frac{\partial f}{\partial x_1} & \frac{\partial f}{\partial x_2} & \cdots & \frac{\partial f}{\partial x_N} \end{bmatrix}. \tag{A.7}$$

The reason for this particular definition is related to the expression for the total differential of $f(\boldsymbol{x})$,

$$df(\boldsymbol{x}) = \frac{\partial f}{\partial x_1} dx_1 + \frac{\partial f}{\partial x_2} dx_2 + \cdots + \frac{\partial f}{\partial x_N} dx_N \tag{A.8}$$

which now can be written as

$$df(\boldsymbol{x}) = \frac{\partial f}{\partial \boldsymbol{x}} d\boldsymbol{x} \tag{A.9}$$

where

$$d\boldsymbol{x} = \begin{bmatrix} dx_1 \\ dx_2 \\ \vdots \\ dx_N \end{bmatrix}.$$

The shorthand notation provided by definitions such as Equation (A.7) proves very valuable when dealing with multivariate expressions.

If $\boldsymbol{f}(\boldsymbol{x})$ is a vector function of the vector $\boldsymbol{x}$, then the following formal definition of the partial derivative of $\boldsymbol{f}(\boldsymbol{x})$ with respect to $\boldsymbol{x}$ is made:

$$\frac{\partial \boldsymbol{f}(\boldsymbol{x})}{\partial \boldsymbol{x}} = \begin{bmatrix} \frac{\partial f_1}{\partial x_1} & \frac{\partial f_1}{\partial x_2} & \cdots & \frac{\partial f_1}{\partial x_N} \\ \frac{\partial f_2}{\partial x_1} & \frac{\partial f_2}{\partial x_2} & \cdots & \frac{\partial f_2}{\partial x_N} \\ \cdots & \cdots & \cdots & \cdots \\ \frac{\partial f_M}{\partial x_1} & \frac{\partial f_M}{\partial x_2} & \cdots & \frac{\partial f_M}{\partial x_N} \end{bmatrix}. \tag{A.10}$$

Again, this particular form is used because of the convenience in writing an expression for the vector of total differentials as

$$df(x) = \begin{bmatrix} df_1(x) \\ df_z(x) \\ \vdots \\ df_M(x) \end{bmatrix} = \begin{bmatrix} \frac{\partial f_1}{\partial x_1} & \frac{\partial f_1}{\partial x_2} & \cdots & \frac{\partial f_1}{\partial x_N} \\ \frac{\partial f_2}{\partial x_1} & \frac{\partial f_2}{\partial x_2} & \cdots & \frac{\partial f_2}{\partial x_N} \\ \cdots & \cdots & \cdots & \cdots \\ \frac{\partial f_M}{\partial x_1} & \frac{\partial f_M}{\partial x_2} & \cdots & \frac{\partial f_M}{\partial x_N} \end{bmatrix} \begin{bmatrix} dx_1 \\ dx_1 \\ \vdots \\ dx_N \end{bmatrix} \tag{A.11}$$

or, more compactly, as

$$df(x) = \frac{\partial f}{\partial x}\, dx. \tag{A.12}$$

Note the similarity between Equations (A.9) and (A.12), and the corresponding expression for the total differential.

With these formal definitions, the second partial derivative of the scalar $f(\boldsymbol{x})$ with respect to the vector $\boldsymbol{x}$ is defined as

$$\frac{\partial}{\partial x}\left[\frac{\partial f(x)}{\partial x}\right]^T = \frac{\partial}{\partial x}\begin{bmatrix} \frac{\partial f}{\partial x_1} \\ \frac{\partial f}{\partial x_2} \\ \vdots \\ \frac{\partial f}{\partial x_N} \end{bmatrix} \tag{A.13}$$

or

$$\frac{\partial^2 f}{\partial x^2} = \begin{bmatrix} \frac{\partial^2 f}{\partial x_1^2} & \frac{\partial^2 f}{\partial x_1 \partial x_2} & \cdots & \frac{\partial^2 f}{\partial x_1 \partial x_N} \\ \frac{\partial^2 f}{\partial x_1 \partial x_2} & \frac{\partial^2 f}{\partial x_2^2} & \cdots & \frac{\partial^2 f}{\partial x_2 \partial x_N} \\ \cdots & \cdots & \cdots & \cdots \\ \frac{\partial^2 f}{\partial x_N \partial x_1} & \frac{\partial^2 f}{\partial x_N \partial x_2} & \cdots & \frac{\partial^2 f}{\partial x_N^2} \end{bmatrix}. \tag{A.14}$$

In most cases of interest

$$\frac{\partial^2 f}{\partial x_i \partial x_j} = \frac{\partial^2 f}{\partial x_j \partial x_i}.$$

so that the matrix of Equation (A.14) is symmetric.

An example of the usefulness of these definitions is provided by the concise form in which the Taylor series expansion through second order terms can be expressed. In terms of these matrices,

$$f(\boldsymbol{x}+\boldsymbol{h}) = f(\boldsymbol{x}) + \frac{\partial f}{\partial \boldsymbol{x}}\boldsymbol{h} + \frac{1}{2!}\boldsymbol{h}^T \frac{\partial^2 f}{\partial \boldsymbol{x}^2}\boldsymbol{h} + \text{h.o.t.}$$

From this expression, the necessary condition for a maximum or minimum is

$$\partial f/\partial \boldsymbol{x} = 0$$

and the sufficient condition for a minimum is

$$\boldsymbol{h}^T \frac{\partial^2 f}{\partial \boldsymbol{x}^2}\boldsymbol{h} > 0$$

for all $\boldsymbol{h}$. This topic is treated in Appendix B.

Finally, the differentiation of a scalar with respect to the elements of a matrix is defined as

$$\frac{\partial f}{\partial \boldsymbol{A}} = \begin{bmatrix} \frac{\partial f}{\partial a_{11}} & \frac{\partial f}{\partial a_{12}} & \cdots & \frac{\partial f}{\partial a_{1N}} \\ \frac{\partial f}{\partial a_{21}} & \frac{\partial f}{\partial a_{22}} & \cdots & \frac{\partial f}{\partial a_{2N}} \\ \cdots & \cdots & \cdots & \cdots \\ \frac{\partial f}{\partial a_{N1}} & \frac{\partial f}{\partial a_{N2}} & \cdots & \frac{\partial f}{\partial a_{NN}} \end{bmatrix}. \tag{A.15}$$

Special cases of this operation will be discussed in the next section.

A.3.2. Differentiation of Matrix Products

In many instances it is desired to differentiate a linear form represented by matrix products with respect to variables that are elements of some of the matrices. For example, if the function $\boldsymbol{f}(\boldsymbol{x})$ is the linear form

$$\boldsymbol{f}(\boldsymbol{x}) = \boldsymbol{A}\boldsymbol{x} \tag{A.16}$$

then the definition of Equation (A.10) leads directly to the expression,

$$\frac{\partial}{\partial \boldsymbol{x}}[\boldsymbol{A}\boldsymbol{x}] = \boldsymbol{A}. \tag{A.17}$$

By similar application of the definitions of the previous section, the partial derivative of the scalar $\boldsymbol{x}^T\boldsymbol{A}\boldsymbol{x}$ is, when $\boldsymbol{A}$ is symmetric,

$$\frac{\partial}{\partial \boldsymbol{x}}[\boldsymbol{x}^T\boldsymbol{A}\boldsymbol{x}] = 2\boldsymbol{x}^T\boldsymbol{A} \tag{A.18}$$

and the second derivative is

$$\begin{aligned}\frac{\partial^2}{\partial \boldsymbol{x}^2}[\boldsymbol{x}^T\boldsymbol{A}\boldsymbol{x}] &= \frac{\partial}{\partial \boldsymbol{x}}[2\boldsymbol{x}^T\boldsymbol{A}]^T \\ &= 2\frac{\partial}{\partial \boldsymbol{x}}[\boldsymbol{A}^T\boldsymbol{x}] = 2\boldsymbol{A}.\end{aligned} \tag{A.19}$$

The particular form

$$\boldsymbol{x}^T\boldsymbol{A}\boldsymbol{x}$$

is called a *quadratic form of the matrix A* and will be considered in detail in the subsequent discussion. The particular case for $\boldsymbol{A}$ symmetric is of fundamental importance, and for this reason such a condition has been specified in the development of Equations (A.18) and (A.19).

A *bilinear form* of a symmetric matrix $\boldsymbol{A}$ is defined as the scalar

$$\boldsymbol{y}^T\boldsymbol{A}\boldsymbol{x}$$

and its derivative is

$$\frac{\partial}{\partial \boldsymbol{x}}[\boldsymbol{y}^T\boldsymbol{A}\boldsymbol{x}] = \boldsymbol{y}^T\boldsymbol{A}. \tag{A.20}$$

The second derivative is, of course,

$$\frac{\partial^2}{\partial \boldsymbol{x}^2}[\boldsymbol{y}^T\boldsymbol{A}\boldsymbol{x}] = \frac{\partial}{\partial \boldsymbol{x}}[\boldsymbol{y}^T\boldsymbol{A}]^T = 0. \tag{A.21}$$

If the derivative with respect to $\boldsymbol{y}$ is desired, the form is transposed and the relationship

$$\boldsymbol{y}^T\boldsymbol{A}\boldsymbol{x} = [\boldsymbol{y}^T\boldsymbol{A}\boldsymbol{x}]^T = \boldsymbol{x}^T\boldsymbol{A}^T\boldsymbol{y} \tag{A.22}$$

is easily established, since the bilateral form itself is a scalar. Then, from previous developments

$$\frac{\partial}{\partial \boldsymbol{y}}[\boldsymbol{y}^T\boldsymbol{A}\boldsymbol{x}] = \frac{\partial}{\partial \boldsymbol{y}}[\boldsymbol{x}^T\boldsymbol{A}^T\boldsymbol{y}] = \boldsymbol{x}^T\boldsymbol{A}^T. \tag{A.23}$$

These equations for derivatives of scalar forms are useful in certain optimization problems in which the quantity to be optimized is expressed in terms of quadratic or bilinear forms.

A.4. CHARACTERISTICS OF MATRICES

There are certain properties of matrices which are important to the application of matrix analysis to practical problems. Included are the concepts of singular and nonsingular matrices and their relationship to the solution of linear algebraic equations, the concept of the rank of a matrix, and the theory of eigenvalues and eigenvectors associated with a square matrix. Two properties—the trace of a matrix and the definiteness of a matrix—are so important to developments in the text that they are outlined here.

A.4.1. Definiteness of a Matrix

The quadratic form has been discussed previously in relation to its properties under differentiation. A quadratic form of the square matrix $\boldsymbol{A}$ is any scalar of the form

$$q = \boldsymbol{x}^T \boldsymbol{A} \boldsymbol{x} \tag{A.24}$$

where $\boldsymbol{x}$ is any nonzero vector of the proper dimension. If the matrix $\boldsymbol{A}$ is of such a character that the quadratic form q is positive for every possible vector $\boldsymbol{x}$ of the proper dimension, then the matrix is said to be *positive definite*. It can be shown, using the techniques to be developed in the subsequent discussion, that matrices which are positive definite have only positive eigenvalues.

If the quadratic form q is positive or zero, regardless of the vector $\boldsymbol{x}$, then the matrix $\boldsymbol{A}$ is said to be *positive semidefinite*, and its eigenvalues are either zero or positive. Similarly, those matrices producing quadratic forms which are all negative are termed *negative definite* matrices, while *negative semidefinite* matrices are those which produce only nonpositive quadratic forms. The eigenvalues of these matrices are all negative, or nonpositive, respectively.

Since it is not feasible to test all possible vectors $\boldsymbol{x}$ to determine whether or not a matrix is positive definite, some practical way of making this determination is required. One method is to solve for the eigenvalues, but a more direct way is provided by *Sylvester's criterion*, which states that a necessary and sufficient condition for a symmetric matrix to be positive definite is that the determinant and all leading principal minors of the determinant be positive. That is,

$$a_{11} > 0; \quad \begin{vmatrix} a_{11} & a_{12} \\ a_{21} & a_{22} \end{vmatrix} > 0; \quad \begin{vmatrix} a_{11} & a_{12} & a_{13} \\ a_{21} & a_{22} & a_{23} \\ a_{31} & a_{32} & a_{33} \end{vmatrix} > 0; \quad \ldots; \quad \det \boldsymbol{A} > 0. \tag{A.25}$$

The fact that Sylvester's criterion applies only to symmetric matrices is not a limiting factor, for the following reason. Every real square matrix $\boldsymbol{A}$ can be written as the sum of a symmetric matrix $\boldsymbol{A}_s$ and an antisymmetric matrix $\boldsymbol{A}_a$

$$\boldsymbol{A} = \boldsymbol{A}_s + \boldsymbol{A}_a \tag{A.26}$$

where

$$\begin{aligned} \boldsymbol{A}_s &= (\boldsymbol{A} + \boldsymbol{A}^T) \\ \boldsymbol{A}_a &= (\boldsymbol{A} - \boldsymbol{A}^T). \end{aligned} \tag{A.27}$$

The quadratic form q is then

$$\begin{aligned} q = \boldsymbol{x}^T\boldsymbol{A}\boldsymbol{x} &= \boldsymbol{x}^T[\boldsymbol{A}_s + \boldsymbol{A}_a]\boldsymbol{x} \\ &= \boldsymbol{x}^T\boldsymbol{A}_s\boldsymbol{x} + \boldsymbol{x}^T\boldsymbol{A}_a\boldsymbol{x}. \end{aligned} \tag{A.28}$$

The second term on the right-hand side is

$$[x_1 \quad x_2 \quad \cdots \quad x_N] \begin{bmatrix} 0 & (a_{12} - a_{21}) & \cdots & (a_{1N} - a_{N1}) \\ (a_{21} - a_{12}) & 0 & \cdots & (a_{2N} - a_{N2}) \\ \cdots & \cdots & \cdots & \cdots \\ (a_{N1} - a_{1N}) & (a_{N2} - a_{2N}) & \cdots & 0 \end{bmatrix} \begin{bmatrix} x_1 \\ x_2 \\ \vdots \\ x_N \end{bmatrix} \tag{A.29}$$

indicating that any quadratic form of an antisymmetric matrix is zero. Thus, a quadratic form of the generally nonsymmetric matrix $\boldsymbol{A}$ can be expressed in terms of the symmetric part of $\boldsymbol{A}$, as follows

$$q = \boldsymbol{x}^T\boldsymbol{A}\boldsymbol{x} = \boldsymbol{x}^T\boldsymbol{A}_s\boldsymbol{x}. \tag{A.30}$$

From these considerations, Sylvester's criterion can be applied to any square matrix $\boldsymbol{A}$ by examining the determinant and leading principle minors of its symmetric part. Also for this reason, consideration of quadratic forms is usually limited to symmetric matrices.

There is also a form of Sylvester's criterion for negative definiteness. It requires that the leading term be negative, and that leading principal minors alternate in sign. The determinant is positive or negative, depending upon whether the dimension of the matrix is odd or even. That is,

$$a_{11} < 0; \quad \begin{vmatrix} a_{11} & a_{12} \\ a_{21} & a_{22} \end{vmatrix} > 0; \quad \begin{vmatrix} a_{11} & a_{12} & a_{13} \\ a_{21} & a_{22} & a_{23} \\ a_{31} & a_{32} & a_{33} \end{vmatrix} < 0; \quad \ldots; \quad \det \boldsymbol{A}. \tag{A.31}$$

Table A-1. Definiteness of Matrices.

Characteristic	Quadratic Forms	Eigenvalues	Determinant	Leading Principal Minors	Principal Minors
Positive definite	$q > 0$	>0	>0	>0	–
Negative definite	$q < 0$	<0	<0 if N odd >0 if N even	<0 if odd >0 if even	–
Positive semidefinite	$q \geqslant 0$	$\geqslant 0$	$\geqslant 0$	–	$\geqslant 0$
Negative semidefinite	$q \leqslant 0$	$\leqslant 0$	$\leqslant 0$ if N odd $\geqslant 0$ if N even	–	$\leqslant 0$ if N odd $\geqslant 0$ if N even
Indefinite	$\pm$	$\pm$	$\pm$	$\pm$	$\pm$

This criterion can be applied only to symmetric matrices, or to the symmetric part of non-symmetric square matrices.

Tests for positive semidefinite and negative semidefinite matrices are slightly more complex, but they follow the general principles outlined above. The requirement for positive semidefiniteness is that all principal minors be nonnegative. For negative semidefiniteness, the principal minors must alternate in sign. These results are summarized in Table A-1, where it is indicated that those matrices which do not fall into one of the four categories are termed *indefinite matrices*.

An example of the usefulness of these concepts can be obtained by again considering the Taylor series approach to multivariate optimization presented in Appendix B. There it is shown that a sufficient condition for a minimum of a function is the vanishing of the first derivative term, and the requirement that the quadratic form

$$\boldsymbol{h}^T \frac{\partial^2 f}{\partial \boldsymbol{x}^2} \boldsymbol{h} > 0$$

for all $\boldsymbol{h}$. This requirement is independent of $\boldsymbol{h}$, and it can be seen to depend on the positive definiteness of the matrix $\partial^2 f/\partial \boldsymbol{x}^2$, which can easily be tested by Sylvester's criterion.

A.4.2. Trace of a Matrix

In some forms of analysis, the sum of the diagonal terms of a square matrix is an important quantity, and for this reason it is called the trace of the matrix. Thus,

$$\operatorname{Tr} \boldsymbol{A} = \sum_{i=1}^{N} a_{ii}. \tag{A.32}$$

Relationships involving the trace of a matrix which are easily seen are

$$\mathrm{Tr}\,(\boldsymbol{A} + \boldsymbol{B}) = \mathrm{Tr}\,\boldsymbol{A} + \mathrm{Tr}\,\boldsymbol{B}$$
$$\mathrm{Tr}\,(\boldsymbol{AB}) = \mathrm{Tr}\,(\boldsymbol{BA}). \qquad \text{(A.33)}$$

A second set of relationships which are not so easily proven, but which are useful in certain kinds of problems, is

$$\frac{\partial}{\partial \boldsymbol{A}} \mathrm{Tr}\,(\boldsymbol{ABA}^T) = 2\boldsymbol{B}^T\boldsymbol{A}^T, \quad \text{for symmetric } \boldsymbol{B}$$
$$\frac{\partial}{\partial \boldsymbol{A}} \mathrm{Tr}\,(\boldsymbol{ABC}) = \boldsymbol{BC} \qquad \text{(A.34)}$$

where $\boldsymbol{A}$ and $\boldsymbol{C}$ are matrices which are not necessarily square, but where the product $\boldsymbol{ABC}$ is a square matrix. Note that the trace of a matrix is a scalar, and consequently the derivative of the trace with respect to the matrix $\boldsymbol{B}$ is a matrix of the same dimension as $\boldsymbol{B}$, in accordance with the definition of Equation (A.15).

A.5. COORDINATE TRANSFORMATIONS

One of the most useful applications of matrix theory is in the establishment of the mathematical relationships involved in coordinate transformation. For example, suppose that a linear operator is represented by the matrix $\boldsymbol{A}$ in a specified coordinate system, such that two vectors $\boldsymbol{x}$ and $\boldsymbol{y}$ are related by the expression

$$\boldsymbol{y} = \boldsymbol{Ax} \qquad \text{(A.35)}$$

in the specified coordinate system. Now consider a transformation to a second coordinate system, represented by the transformation matrix $\boldsymbol{T}$. That is, any vector in the original coordinate system is obtained from a vector in the second coordinate system by multiplying on the left by $\boldsymbol{T}$. Thus

$$\boldsymbol{x} = \boldsymbol{Tx}'$$
$$\boldsymbol{y} = \boldsymbol{Ty}' \qquad \text{(A.36)}$$

and the relationship between the transformed vectors $\boldsymbol{x}'$ and $\boldsymbol{y}'$ is

$$\boldsymbol{y}' = \boldsymbol{A}'\boldsymbol{x}' \qquad \text{(A.37)}$$

in the new coordinate system. The relationship between $\boldsymbol{A}'$ and $\boldsymbol{A}$ is obtained simply by substituting Equation (A.36) into Equation (A.35). The result is

$$\boldsymbol{Ty}' = \boldsymbol{ATx}' \qquad \text{(A.38)}$$

or, from multiplying on the left by T^{-1},

$$y' = T^{-1}ATx'. \tag{A.39}$$

From this expression, it can be seen that the linear operator represented by A in the initial coordinate system is represented by the matrix

$$A' = T^{-1}AT \tag{A.40}$$

in the second coordinate system. A transformation of the kind represented by Equation (A.40) is called a *similarity transformation*. Matrices which are related to one another through similarity transformations are called *similar matrices*. It can be shown, using techniques to be developed in the subsequent discussion, that certain matrix characteristics are invariant under similarity transformations. These characteristics are the determinant, the characteristic or determinantal equation, the eigenvalues, and the trace. Specifically,

$$\det\,[T^{-1}AT] = \det A$$
$$\mathrm{Tr}\,[T^{-1}AT] = \mathrm{Tr}\,A. \tag{A.41}$$

The equality of the characteristic equations and eigenvalues follow directly from these considerations.

A.5.1. Diagonalization of Matrices

If the $N \times N$ matrix A has N linearly independent eigenvectors, $x^1, x^2, \ldots, x^N$, then a particularly interesting transformation results when the transformation matrix is formed by letting the columns of the matrix be the eigenvectors. That is, consider the transformation matrix P, where

$$P = [x^1 \mid x^2 \mid \cdots \mid x^N]. \tag{A.42}$$

Since the eigenvectors x^i must obey the determinantal equation

$$Ax = \lambda x \tag{A.43}$$

the product AP can be written as

$$AP = A[x^1 \mid x^2 \mid \cdots \mid x^N] = [\lambda_1 x^1 \mid \lambda_2 x^2 \mid \cdots \mid \lambda_N x^N]. \tag{A.44}$$

The right-hand side of the above equation can be expressed as

$$[\lambda_1 \boldsymbol{x}^1 \,\vdots\, \lambda_2 \boldsymbol{x}^2 \,\vdots\, \cdots \,\vdots\, \lambda_n \boldsymbol{x}^N] = [\boldsymbol{x}^1 \,\vdots\, \boldsymbol{x}^2 \,\vdots\, \cdots \,\vdots\, \boldsymbol{x}^N] \begin{bmatrix} \lambda_1 & 0 & 0 & \cdots & 0 \\ 0 & \lambda_2 & 0 & \cdots & 0 \\ \cdots & \cdots & \cdots & \cdots & \cdots \\ 0 & 0 & 0 & \cdots & \lambda_N \end{bmatrix} \tag{A.45}$$

$$= \boldsymbol{PD}$$

where the diagonal matrix $\boldsymbol{D}$ is defined as

$$\boldsymbol{D} = \begin{bmatrix} \lambda_1 & 0 & 0 & \cdots & 0 \\ 0 & \lambda_2 & 0 & \cdots & 0 \\ \cdots & \cdots & \cdots & \cdots & \cdots \\ 0 & 0 & 0 & \cdots & \lambda_N \end{bmatrix}.$$

Substituting this expression into Equation (A.44), and multiplying on the left by $\boldsymbol{P}^{-1}$, yield

$$\boldsymbol{P}^{-1}\boldsymbol{AP} = \boldsymbol{D} \tag{A.46}$$

a similarity transformation which transforms $\boldsymbol{A}$ into a diagonal matrix in which the eigenvalues of $\boldsymbol{A}$ are the diagonal elements. The matrix $\boldsymbol{P}$ is not unique, of course, because the eigenvectors are not unique.

The diagonalization of matrices is particularly important in identifying the most basic characteristics of systems of equations. A very useful example is provided by the linear dynamic system which can be described by a set of homogeneous, time invariant, first-order, linear differential equations of the form

$$\dot{\boldsymbol{x}}(t) = \boldsymbol{Fx}(t) \tag{A.47}$$

where $\boldsymbol{F}$ is a constant matrix. If the transformation

$$\boldsymbol{x}(t) = \boldsymbol{Pz}(t); \qquad \boldsymbol{z}(t) = \boldsymbol{P}^{-1}\boldsymbol{x}(t) \tag{A.48}$$

is applied, then Equation (A.47) is

$$\boldsymbol{P}\dot{\boldsymbol{z}}(t) = \boldsymbol{FPz}(t) \tag{A.49}$$

or, from multiplication on the left by $\boldsymbol{P}^{-1}$,

$$\dot{\boldsymbol{z}}(t) = \boldsymbol{P}^{-1}\boldsymbol{FPz}(t) \tag{A.50}$$

where $\boldsymbol{P}^{-1}\boldsymbol{FP} = \boldsymbol{D}$ is a diagonal matrix with the eigenvalues of $\boldsymbol{F}$ as the diagonal elements. Thus,

$$\dot{\boldsymbol{z}}(t) = \boldsymbol{D}\boldsymbol{z}(t)$$

or, in component form,

$$\begin{aligned} \dot{z}_1(t) &= \lambda_1 z_1(t) \\ \dot{z}_2(t) &= \lambda_2 z_2(t) \\ &\vdots \\ \dot{z}_N(t) &= \lambda_N z_N(t). \end{aligned} \qquad \text{(A.51)}$$

This set of uncoupled, linear differential equations can be readily solved to yield

$$\begin{aligned} z_1(t) &= z_1(0)\,e^{\lambda_1 t} \\ z_2(t) &= z_2(0)\,e^{\lambda_2 t} \\ &\vdots \\ z_N(t) &= z_N(0)\,e^{\lambda_N t}. \end{aligned} \qquad \text{(A.52)}$$

Once the solution for the z_i have been obtained, transformation back to the $\boldsymbol{x}$ equation can be accomplished by the transformation of Equation (A.48). Notice that the solution to Equation (A.47), a set of coupled, linear equations, has been obtained by transforming the equations into their most basic form, solving the resulting uncoupled equations, and then transforming this solution back into the original system. The solutions for the x_i will consist of linear combinations of the basic solutions given by Equation (A.52), and for this reason the basic solutions are called the modes of response of the system. The importance that each mode plays in the solutions for $\boldsymbol{x}(t)$ depends upon the initial condition $z_i(0)$ and the elements of the transformation matrix $\boldsymbol{P}$. The reader should note the possibility of complex eigenvalues, which must occur in conjugate pairs. Each pair is associated with a single oscillatory mode of response, obtained by combining the individual modal responses of the members of the pair by use of Euler's formula. This so-called *modal analysis* can be quite useful in understanding the response of linear systems under various initial excitations, and is exploited fully in Chapter 5.

Modal analysis is closely related to the classical method of solution for linear systems, as might be expected. In fact, if Equation (A.47) were changed to a single Nth order equation of the form

$$a_N x^{(N)}(t) + a_{N-1} x^{(N-1)}(t) + \cdots + a_0 x(t) = 0 \qquad \text{(A.53)}$$

the characteristic equation obtained by applying a differential operator to this equation would be the determinantal equation of the matrix $\boldsymbol{F}$; therefore, the roots of the characteristic equation of Equation (A.53) and the eigenvalues of $\boldsymbol{F}$ are identical.

A.5.2. The Jordan Canonical Form

An explicit assumption of the previous section was the existence of N linearly independent eigenvectors. If this is not the case, then the matrix $\boldsymbol{P}$, composed of the eigenvectors themselves, is a singular matrix. If $\boldsymbol{P}$ is singular, then the transformation of Equation (A.46) does not exist. Thus, in analysis involving the reduction of matrices to their most elementary form, consideration must be given to two questions. First, does the matrix have linearly independent eigenvectors, and is it therefore amenable to diagonalization? Second, if it does not have linearly independent eigenvectors, what is the most fundamental form to which the matrix may be transformed?

In attempting to answer the first question, it might be noted that matrices with distinct eigenvalues will have linearly independent eigenvectors, and that Hermitian and symmetric matrices will have distinct eigenvalues. If a matrix does not have distinct eigenvalues, then the multiplicity of eigenvectors associated with each multiple eigenvalue must be investigated to determine the total number of linearly independent eigenvectors.

If the number of linearly independent eigenvalues is less than N, then the matrix cannot be diagonalized, and the most fundamental form to which it can be transformed is the *Jordan canonical form*, in which the elements along the main diagonal are the eigenvalues, the elements immediately above the main diagonal are either one or zero, and all other elements are zero. The particular form of the matrix depends on which eigenvalues are associated with the linearly dependent eigenvectors. For example, consider the following matrices:

$$\overset{\text{I}}{\begin{bmatrix} \lambda_1 & 1 & 0 & 0 \\ 0 & \lambda_1 & 0 & 0 \\ 0 & 0 & \lambda_2 & 0 \\ 0 & 0 & 0 & \lambda_3 \end{bmatrix}}; \quad \overset{\text{II}}{\begin{bmatrix} \lambda_1 & 0 & 0 & 0 \\ 0 & \lambda_1 & 0 & 0 \\ 0 & 0 & \lambda_3 & 0 \\ 0 & 0 & 0 & \lambda_4 \end{bmatrix}}; \quad \overset{\text{III}}{\begin{bmatrix} \lambda_1 & 0 & 0 & 0 \\ 0 & \lambda_2 & 1 & 0 \\ 0 & 0 & \lambda_2 & 0 \\ 0 & 0 & 0 & \lambda_4 \end{bmatrix}}$$

$$\overset{\text{IV}}{\begin{bmatrix} \lambda_1 & 1 & 0 & 0 \\ 0 & \lambda_1 & 1 & 0 \\ 0 & 0 & \lambda_1 & 1 \\ 0 & 0 & 0 & \lambda_1 \end{bmatrix}}; \quad \overset{\text{V}}{\begin{bmatrix} \lambda_1 & 1 & 0 & 0 \\ 0 & \lambda_1 & 0 & 0 \\ 0 & 0 & \lambda_1 & 1 \\ 0 & 0 & 0 & \lambda_1 \end{bmatrix}}; \quad \overset{\text{VI}}{\begin{bmatrix} \lambda_1 & 1 & 0 & 0 \\ 0 & \lambda_1 & 0 & 0 \\ 0 & 0 & \lambda_2 & 1 \\ 0 & 0 & 0 & \lambda_2 \end{bmatrix}}.$$

The eigenvalue–eigenvector relationship of these matrices is as follows:

I. This matrix and all matrices similar to this matrix have multiple eigenvalues λ_1, and the corresponding eigenvectors are not linearly independent.
II. This matrix and all matrices similar to this matrix have multiple eigenvalues λ_1, but the two eigenvectors corresponding to these eigenvalues are linearly independent.
III. This matrix and all matrices similar to this matrix have multiple eigenvalues λ_2, and the associated eigenvectors are not linearly independent.
IV. This matrix and all matrices similar to this matrix have multiple eigenvalues λ_1, and only one linearly independent eigenvector.
V. This matrix and all matrices similar to this matrix have multiple eigenvalues λ_1, but there are two linearly independent eigenvectors.
VI. This matrix and all matrices similar to this matrix have two sets of multiple eigenvalues λ_1 and λ_2. Each set has associated with it only one linearly independent eigenvector.

A thorough discussion of Jordan forms is beyond the scope of the current development, but a general concept of their existence is important.

A.6. A USEFUL MATRIX IDENTITY

In the development of several useful relationships in estimation theory, a particular matrix identity is used. This identity is of the general form

$$(\boldsymbol{A} + \boldsymbol{B}^T\boldsymbol{C})^{-1} = \boldsymbol{A}^{-1} - \boldsymbol{A}^{-1}\boldsymbol{B}^T(\boldsymbol{I} + \boldsymbol{C}^T\boldsymbol{A}^{-1}\boldsymbol{B})^{-1}\boldsymbol{C}\boldsymbol{A}^{-1}. \tag{A.54}$$

To prove this identity, the matrix $(\boldsymbol{A} + \boldsymbol{B}^T\boldsymbol{C})$ is multiplied on the left by its own inverse, to yield

$$\begin{aligned} \boldsymbol{I} &= (\boldsymbol{A} + \boldsymbol{B}^T\boldsymbol{C})^{-1}(\boldsymbol{A} + \boldsymbol{B}^T\boldsymbol{C}) \\ &= (\boldsymbol{A} + \boldsymbol{B}^T\boldsymbol{C})^{-1}\boldsymbol{A} + (\boldsymbol{A} + \boldsymbol{B}^T\boldsymbol{C})^{-1}\boldsymbol{B}^T\boldsymbol{C} \end{aligned} \tag{A.55}$$

Multiplying on the right by $\boldsymbol{A}^{-1}$ yields

$$\boldsymbol{A}^{-1} = (\boldsymbol{A} + \boldsymbol{B}^T\boldsymbol{C})^{-1} + (\boldsymbol{A} + \boldsymbol{B}^T\boldsymbol{C})^{-1}\boldsymbol{B}^T\boldsymbol{C}\boldsymbol{A}^{-1} \tag{A.56}$$

or,

$$(\boldsymbol{A} + \boldsymbol{B}^T\boldsymbol{C})^{-1}\boldsymbol{B}^T\boldsymbol{C}\boldsymbol{A}^{-1} = \boldsymbol{A}^{-1} - (\boldsymbol{A} + \boldsymbol{B}^T\boldsymbol{C})^{-1}. \tag{A.57}$$

Now, both sides of Equation (A.56) are multiplied on the right by $\boldsymbol{B}^T$, yielding

$$\begin{aligned} \boldsymbol{A}^{-1}\boldsymbol{B}^T &= (\boldsymbol{A} + \boldsymbol{B}^T\boldsymbol{C})^{-1}\boldsymbol{B}^T + (\boldsymbol{A} + \boldsymbol{B}^T\boldsymbol{C})^{-1}\boldsymbol{B}^T\boldsymbol{C}\boldsymbol{A}^{-1}\boldsymbol{B}^T \\ &= (\boldsymbol{A} + \boldsymbol{B}^T\boldsymbol{C})^{-1}\boldsymbol{B}^T(\boldsymbol{I} + \boldsymbol{C}\boldsymbol{A}^{-1}\boldsymbol{B}^T). \end{aligned} \tag{A.58}$$

Then, by multiplying on the right by the matrix $(\boldsymbol{I} + \boldsymbol{C}\boldsymbol{A}^{-1}\boldsymbol{B}^T)^{-1}$, one obtains

$$\boldsymbol{A}^{-1}\boldsymbol{B}^T(\boldsymbol{I} + \boldsymbol{C}\boldsymbol{A}^{-1}\boldsymbol{B}^T)^{-1} = (\boldsymbol{A} + \boldsymbol{B}^T\boldsymbol{C})^{-1}\boldsymbol{B}^T \tag{A.59}$$

or, after multiplying on the right by $\boldsymbol{C}\boldsymbol{A}^{-1}$,

$$\boldsymbol{A}^{-1}\boldsymbol{B}^T(\boldsymbol{I} + \boldsymbol{C}\boldsymbol{A}^{-1}\boldsymbol{B}^T)^{-1}\boldsymbol{C}\boldsymbol{A}^{-1} = (\boldsymbol{A} + \boldsymbol{B}^T\boldsymbol{C})\boldsymbol{B}^T\boldsymbol{C}\boldsymbol{A}^{-1}. \tag{A.60}$$

Now, Equation (A.56) is used to substitute for the right-hand side of this expression, yielding finally

$$(\boldsymbol{A} + \boldsymbol{B}^T\boldsymbol{C})^{-1} = \boldsymbol{A}^{-1} - \boldsymbol{A}^{-1}\boldsymbol{B}^T(\boldsymbol{I} + \boldsymbol{C}\boldsymbol{A}^{-1}\boldsymbol{B}^T)^{-1}\boldsymbol{C}\boldsymbol{A}^{-1} \tag{A.61}$$

which is the identity of Equation (A.54).

The usefulness of this identity derives from the fact that there is no general relationship between the inverse of the sum of two matrices and the inverses of the individual matrices. When one of the matrices is expressible as a product of two matrices, then the identity of Equation (A.61) can be used. Note that $\boldsymbol{B}$ and $\boldsymbol{C}$ need not be square matrices, and in most useful applications they are not.

An example of the usefulness of this identity is given in Chapter 7, where the inversion

$$(\boldsymbol{P}^{-1} + \boldsymbol{H}^T\boldsymbol{\Sigma}^{-1}\boldsymbol{H})^{-1} \tag{A.62}$$

is required. Here, $\boldsymbol{P}$ is an $N \times N$ matrix, $\boldsymbol{\Sigma}$ is an $M \times M$ matrix, and $\boldsymbol{H}$ is a nonsquare matrix of dimension $M \times N$. By making the associations

$$\boldsymbol{A} \longrightarrow \boldsymbol{P}^{-1}; \quad \boldsymbol{B} \longrightarrow \boldsymbol{H}; \quad \boldsymbol{C} \longrightarrow \boldsymbol{\Sigma}^{-1}\boldsymbol{H}$$

the result

$$(\boldsymbol{P}^{-1} + \boldsymbol{H}^T\boldsymbol{\Sigma}^{-1}\boldsymbol{H})^{-1} = \boldsymbol{P} - \boldsymbol{P}\boldsymbol{H}^T(\boldsymbol{I} + \boldsymbol{\Sigma}^{-1}\boldsymbol{H}\boldsymbol{P}\boldsymbol{H}^T)^{-1}\boldsymbol{\Sigma}^{-1}\boldsymbol{H}\boldsymbol{P} \tag{A.63}$$

is obtained. Using the general matrix property

$$(\boldsymbol{A}\boldsymbol{B})^{-1} = \boldsymbol{B}^{-1}\boldsymbol{A}^{-1}$$

Equation (A.63) can be expressed as

$$(\boldsymbol{P}^{-1} + \boldsymbol{H}^T\boldsymbol{\Sigma}^{-1}\boldsymbol{H})^{-1} = \boldsymbol{P} - \boldsymbol{P}\boldsymbol{H}^T(\boldsymbol{\Sigma} + \boldsymbol{H}\boldsymbol{P}\boldsymbol{H}^T)^{-1}\boldsymbol{H}\boldsymbol{P}. \tag{A.64}$$

The noteworthy thing about this expression is that the matrix inversion is of dimension M, whereas that of Equation (A.62) is of dimension N. In estimation problems, N is generally greater than M, so that this result produces a reduction in problem complexity.

DEVELOPMENTAL EXERCISES

A.1. Show that $\operatorname{Tr} \boldsymbol{AB} = \operatorname{Tr} \boldsymbol{BA}$.

A.2. Show that $\operatorname{Tr} \boldsymbol{AB}^T = \operatorname{Tr} \boldsymbol{BA}^T = \operatorname{Tr} \boldsymbol{A}^T\boldsymbol{B}$.

A.3. Show that

$$\left[\begin{array}{c:c} \boldsymbol{A}_{11} & \boldsymbol{A}_{12} \\ \hdashline \mathbf{0} & \boldsymbol{A}_{22} \end{array}\right]^{-1} = \left[\begin{array}{c:c} \boldsymbol{A}_{11}^{-1} & -\boldsymbol{A}_{11}^{-1}\boldsymbol{A}_{12}\boldsymbol{A}_{22}^{-1} \\ \hdashline \mathbf{0} & \boldsymbol{A}_{22}^{-1} \end{array}\right].$$

A.4. Show that

$$\left[\begin{array}{c:c} \boldsymbol{A}_{11} & \mathbf{0} \\ \hdashline \boldsymbol{A}_{21} & \boldsymbol{A}_{22} \end{array}\right]^{-1} = \left[\begin{array}{c:c} \boldsymbol{A}_{11}^{-1} & \mathbf{0} \\ \hdashline -\boldsymbol{A}_{22}^{-1}\boldsymbol{A}_{22}\boldsymbol{A}_{11}^{-1} & \boldsymbol{A}_{22}^{-1} \end{array}\right].$$

A.5. Show that

$$\left[\begin{array}{c:c} \boldsymbol{A}_{11} & \mathbf{0} \\ \hdashline \mathbf{0} & \boldsymbol{A}_{22} \end{array}\right]^{-1} = \left[\begin{array}{c:c} \boldsymbol{A}_{11}^{-1} & \mathbf{0} \\ \hdashline \mathbf{0} & \boldsymbol{A}_{22}^{-1} \end{array}\right].$$

A.6. Prove the assertion of Equation (A.3).

A.7. Prove the assertion of Equation (A.17).

A.8. Prove the assertion of Equation (A.18).

A.9. Prove the assertion of Equation (A.19).

A.10. Prove the assertions of Equation (A.33).

A.11. Prove the assertions of Equation (A.34).

PROBLEMS

A.1. Determine the matrix product

$$[\boldsymbol{A}_1 \,\vdots\, \boldsymbol{A}_2] \left[\begin{array}{c} \boldsymbol{B}_1 \\ \hdashline \boldsymbol{B}_2 \end{array}\right] = \boldsymbol{AB}$$

where

$$\boldsymbol{A} = \left[\begin{array}{cc:cc} 2 & 2 & 1 & 4 \\ -3 & 1 & 1 & -1 \end{array}\right]; \quad \boldsymbol{B} = \left[\begin{array}{cc} 3 & 1 \\ 1 & 0 \\ \hdashline 2 & 0 \\ 0 & 1 \end{array}\right].$$

A.2. Find the inverse of the matrix $\boldsymbol{A}$ using the characteristics of partitioned matrices:

$$\boldsymbol{A} = \begin{bmatrix} 1 & 2 & 0 & 0 \\ 2 & 1 & 0 & 0 \\ 0 & 0 & 2 & 1 \\ 0 & 0 & 1 & 2 \end{bmatrix}.$$

A.3. Find the inverse of the matrix $\boldsymbol{A}$ using the characteristics of partitioned matrices:

$$\boldsymbol{A} = \begin{bmatrix} 1 & 2 & 0 & 0 \\ 2 & 1 & 0 & 0 \\ 3 & 2 & 2 & 1 \\ 4 & 3 & 1 & 2 \end{bmatrix}.$$

A.4. Find a closed form expression for the matrix exponential $e^{\boldsymbol{A}}$, where

$$\boldsymbol{A} = \begin{bmatrix} 0 & 1 \\ -2 & -3 \end{bmatrix}.$$

A.5. Determine the matrix product $\boldsymbol{AB}$, where

$$\boldsymbol{A} = \begin{bmatrix} 1 & 2 & 0 & 0 \\ 2 & 1 & 0 & 0 \\ 1 & 2 & 0 & 1 \\ 1 & 2 & 1 & 0 \end{bmatrix}; \quad \boldsymbol{B} = \begin{bmatrix} 1 & 2 & 0 & 1 \\ 2 & 1 & 1 & 0 \\ 1 & 0 & 0 & 0 \\ 2 & 1 & 0 & 0 \end{bmatrix}.$$

A.6. Find $\int_0^T \boldsymbol{A}(t)\,dt$ and $d\boldsymbol{A}(t)/dt$ for

$$\boldsymbol{A} = \begin{bmatrix} \sin \omega t & t^2 + t \\ 2 & \arctan(2t) \end{bmatrix}.$$

A.7. Find the ranks of the following matrices:

$$\text{(a)} \begin{bmatrix} -2 & 2 & 3 \\ -4 & 10 & 5 \\ -6 & 5 & 6 \end{bmatrix} \qquad \text{(b)} \begin{bmatrix} 7 & 4 & -4 \\ 4 & 7 & -4 \\ -1 & -1 & 4 \end{bmatrix}$$

$$\text{(c)} \begin{bmatrix} 1 & 2 & 1 \\ 0 & 0 & 1 \\ 3 & 5 & 1 \end{bmatrix} \qquad \text{(d)} \begin{bmatrix} 0 & 0 & 1 \\ 0 & 0 & 6 \\ 1 & 1 & 0 \end{bmatrix}.$$

A.8. Determine the definiteness of the following matrices:

$$\text{(a)} \begin{bmatrix} 1 & -2 \\ 2 & 3 \end{bmatrix} \qquad \text{(b)} \begin{bmatrix} 2 & 1 & 1 \\ 1 & 3 & 0 \\ 1 & 0 & 1 \end{bmatrix}$$

$$\text{(c)} \begin{bmatrix} 2 & 1 & 1 \\ 1 & 3 & 2 \\ 1 & 2 & 1 \end{bmatrix} \qquad \text{(d)} \begin{bmatrix} 1 & 1 & 1 \\ 1 & 3 & 1 \\ 1 & 1 & 1 \end{bmatrix}.$$

A.9. Find the sequence $\boldsymbol{A}, \boldsymbol{A}^2, \boldsymbol{A}^3, \ldots$ for

$$\boldsymbol{A} = \begin{bmatrix} (1 - (2/e)] & 1/e & 1/e \\ (1 - (1/e)] & 1/e & 0 \\ (1 - (2/e)] & 1/e & 1/e \end{bmatrix}.$$

Appendix B
Optimization Theory

B.1. APPLICATIONS OF CALCULUS TO OPTIMIZATION PROBLEMS

Any student of fundamental calculus recognizes the necessary condition for a maximum or minimum of a function to be the vanishing of the first derivative. A concept of such generality can be quite misleading in certain cases, however, and care must be taken to stipulate the assumptions concerning such things as continuity and differentiability of the functions in question. As an example, the functions shown in Figure B-1 both have a minimum at the point $x = x^*$, but the function in (b) certainly does not have a zero first derivative at that point. In fact, the derivative is infinitely large. The difference, of course, is in the differentiability of the function at the point $x = x^*$. The most general requirement for a minimum is that the function must increase immediately to the right and left of the point $x = x^*$. Sufficient conditions sometimes are misinterpreted also, especially when one makes the broad statement that the second derivative can be zero at a point where the first derivative vanishes and that the function can still exhibit a minimum at that point. Some of these problems arise from the way in which minima and maxima are identified in basic calculus. A preferable method, which can easily be generalized to problems involving more than one variable, is that utilizing the Taylor series expansion of a function. Basically, Taylor's theorem states that a function can be expressed in terms of its value and those of its derivatives at some point x^* as follows:

$$f(x^* + h) = f(x^*) + \left.\frac{df}{dx}\right|_{x=x^*} h + \frac{1}{2!}\left.\frac{d^2f}{dx^2}\right|_{x=x^*} h^2 + \cdots. \tag{B.1}$$

It is important to note that Equation (B.1) as it stands is not an approximation, but rather is an exact expression when the infinite number of terms is considered. Only when the series is truncated is the expression an approximation. Also note that h need not be small. Of course, in most cases the error produced by truncating the series after a fixed number of terms increases with h. For small h, only first-order terms are important and a useful approximation is

$$f(x^* + dx) \approx f(x^*) + \left.\frac{df}{dx}\right|_{x=x^*} dx \tag{B.2}$$

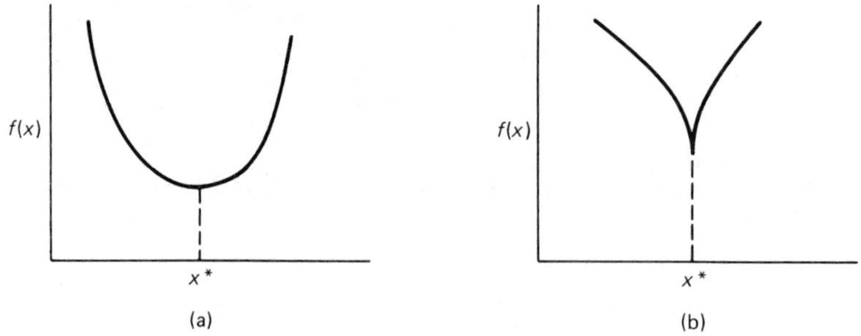

Figure B-1. Two functions exhibiting minima.

where use of the differential dx implies the smallness of h. Equation (B.2) is used in the linearization of nonlinear functions.

Examination of Equation (B.2) permits immediate identification of necessary and sufficient conditions for a minimum or a maximum to exist, at least in the event that the characteristics of the function $f(x)$ permit expansion in the form of a Taylor series. If $f(x^*)$ is a minimum, for example, then the term

$$\left.\frac{df}{dx}\right|_{x=x^*} dx$$

must be positive or zero if dx is positive, and it must be negative or zero if dx is negative. If $f(x^*)$ is actually a minimum, then for any dx there must be no decrease in $f(x^* + dx)$, regardless of the algebraic sign of dx; therefore a requirement for a minimum is

$$\left.\frac{df}{dx}\right|_{x=x^*} = 0. \tag{B.3}$$

A similar approach yields this same requirement for a maximum. Thus, Equation (B.3) represents a necessary but not sufficient condition for a maximum or minimum.

If Equation (B.3) is satisfied, then the Taylor series expansion takes the form

$$f(x^* + dx) \approx f(x^*) + \frac{1}{2!}\frac{d^2f}{dx^2}dx^2. \tag{B.4}$$

Now, if $f(x^*)$ is a minimum, then the second-derivative term must be positive, since the squared term represents a positive number regardless of the algebraic sign of dx. If not, then a small change dx would produce an even smaller value of $f(x^* + dx)$, a contradiction of the premise that $f(x^*)$ is a minimum. Similarly, the condition existing when $f(x^*)$ is a maximum requires that the second derivative be negative. In this way, the necessary and sufficient conditions can easily be determined. In the event that

$$\left.\frac{d^2f}{dx^2}\right|_{x=x^*} = 0 \quad \text{and} \quad \left.\frac{df}{dx}\right|_{x=x^*} = 0 \tag{B.5}$$

then the Taylor series expansion is

$$f(x^* + dx) \approx f(x^*) + \frac{1}{3!}\left.\frac{d^3f}{dx^3}\right|_{x=x^*} dx^3. \tag{B.6}$$

Here, positive values of dx produce positive values of dx^3, and negative values of dx produce negative values of dx^3. For this reason, the argument put forth for the requirement that the first derivative vanish at $x = x^*$ also requires that the third derivative vanish for a maximum or minimum of $f(x)$ to exist at x^*. If the third derivative is positive, then the first derivative is positive on either side of x^*; if it is negative the first derivative is negative on either side of x^*. The function $f(x)$ is then said to exhibit an inflection point at $x = x^*$.

In the event that the third derivative does vanish at $x = x^*$, then the Taylor series expansion is

$$f(x^* + dx) \approx f(x^*) + \frac{1}{4!}\left.\frac{d^4f}{dx^4}\right|_{x=x^*} dx^4. \tag{B.7}$$

Again, dx^4 is positive for both positive and negative values of dx, resulting in the requirement for a positive fourth derivative if $x = x^*$ produces a minimum, and a negative fourth derivative if $x = x^*$ produces a maximum. Continuing this process indefinitely leads to the following general conclusion: if $f(x)$ and its first n derivatives are continuous, then $f(x)$ has a maximum or a minimum at the point $x = x^*$ if and only if n is even, where n is the order of the first nonvanishing derivative at $x = x^*$. If the first nonvanishing derivative is of even order and is negative, then $f(x)$ exhibits a maximum at $x = x^*$. If it is positive, $f(x)$ exhibits a minimum at $x = x^*$. If an inflection point occurs, as indicated by the order of the first nonvanishing derivative being odd, then $f(x)$ is nondecreasing in the region near $x = x^*$ if this first nonvanishing derivative is positive; if it is negative, then $f(x)$ is nonincreasing in the region near $x = x^*$.

Admittedly, these principles can be developed from elementary ideas concerning the behavior of functions, but the use of the Taylor series sets the stage for the generalization to functions of more than one variable.

B.1.1. Optimization of Functions of More than One Variable

When a function of more than one variable is to be minimized with respect to those variables, the necessary conditions can be determined immediately from the multivariate form of the Taylor series expansion:

$$f(x_1^* + h_1, x_2^* + h_2, \ldots, x_n^* + h_n) = f(x_1^*, x_2^*, \ldots, x_n^*) + \sum_{i=1}^{N} \frac{\partial f}{\partial x_i} h_i + \frac{1}{2!} \sum_{i=1}^{N} \sum_{j=1}^{N} \frac{\partial^2 f}{\partial x_i \partial x_j} h_i h_j + \cdots. \quad \text{(B.8)}$$

Note that the second-order terms involve not only expressions such as

$$\frac{\partial^2 f}{\partial x_i^2} h_i^2$$

but also expressions involving the mixed partial derivatives

$$\frac{\partial^2 f}{\partial x_i \partial x_j} h_i h_j.$$

The third-order terms would involve a summation

$$\sum_{i=1}^{N} \sum_{j=1}^{N} \sum_{k=1}^{N} \frac{\partial^3 f}{\partial x_i \partial x_j \partial x_k} h_i h_j h_k$$

and so on. For terms through second order, Equation (B.8) can be expressed in matrix form as

$$f(\boldsymbol{x}^* + \boldsymbol{h}) \approx f(\boldsymbol{x}^*) + \frac{\partial f}{\partial \boldsymbol{x}} + \frac{1}{2!} \boldsymbol{h}^T \frac{\partial^2 f}{\partial \boldsymbol{x}^2} \boldsymbol{h} \quad \text{(B.9)}$$

where the following definitions have been assumed:

$$\boldsymbol{x} = \begin{bmatrix} x_1 \\ x_2 \\ \vdots \\ x_N \end{bmatrix}; \quad \boldsymbol{h} = \begin{bmatrix} h_1 \\ h_2 \\ \vdots \\ h_N \end{bmatrix}; \quad \frac{\partial f}{\partial \boldsymbol{x}} = \begin{bmatrix} \frac{\partial f}{\partial x_1} & \frac{\partial f}{\partial x_2} & \cdots & \frac{\partial f}{\partial x_N} \end{bmatrix};$$

$$\frac{\partial^2 f}{\partial \boldsymbol{x}^2} = \begin{bmatrix} \frac{\partial^2 f}{\partial x_1^2} & \frac{\partial^2 f}{\partial x_1 \partial x_2} & \cdots & \frac{\partial^2 f}{\partial x_1 \partial x_N} \\ \frac{\partial^2 f}{\partial x_1 \partial x_2} & \frac{\partial^2 f}{\partial x_2^2} & \cdots & \frac{\partial^2 f}{\partial x_2 \partial x_N} \\ \cdots & \cdots & \cdots & \cdots \\ \frac{\partial^2 f}{\partial x_1 \partial x_N} & \frac{\partial^2 f}{\partial x_2 \partial x_N} & \cdots & \frac{\partial^2 f}{\partial x_N^2} \end{bmatrix}.$$

It can be easily verified that the matrix products of Equation (B.9) yield the summations of Equation (B.8). By the same reasoning which leads to the necessary conditions in the single-variable case, it can be seen that if $f(\boldsymbol{x}^*)$ is a minimum, then infinitesimal movement away from the point $\boldsymbol{x}^*$ along any of the coordinates must yield zero change in $f(\boldsymbol{x})$. Thus, the necessary condition is that

$$\frac{\partial f}{\partial \boldsymbol{x}} = 0$$

or, equivalently,

$$\begin{aligned} \frac{\partial f}{\partial x_1} &= 0 \\ \frac{\partial f}{\partial x_2} &= 0 \\ &\vdots \\ \frac{\partial f}{\partial x_N} &= 0. \end{aligned} \tag{B.10}$$

These equations must be satisfied simultaneously, since a zero net change in $f(\boldsymbol{x})$ must be observed for arbitrary changes in the coordinates. As before, these relationships constitute necessary conditions for both minima and maxima. The sufficiency condition can be determined in a manner analogous to the univariate case by examining the Taylor series expansion when Equation (B.10) is satisfied. That is,

$$f(\boldsymbol{x}^* + d\boldsymbol{x}) \approx f(\boldsymbol{x}^*) + \frac{1}{2!}\, d\boldsymbol{x}^T \frac{\partial^2 f}{\partial \boldsymbol{x}^2}\, d\boldsymbol{x} \tag{B.11}$$

from which it can be seen that the term

$$d\boldsymbol{x}^T \frac{\partial^2 f}{\partial \boldsymbol{x}^2}\, d\boldsymbol{x} \tag{B.12}$$

must be positive for all the values of $d\boldsymbol{x}$ if $f(\boldsymbol{x}^*)$ is a minimum. By similar reasoning, this quantity must be negative for arbitrary $d\boldsymbol{x}$ if $f(\boldsymbol{x}^*)$ is a maximum. The form of Equation (B.12) can be recognized as a quadratic form of the matrix $\partial^2 f/\partial \boldsymbol{x}^2$. The property of a matrix which requires a positive value for the quadratic form, independent of the value of $d\boldsymbol{x}$, is that of positive definiteness. Similarly, a negative definite matrix has a negative value for the quadratic form. Tests such as Sylvester's criterion from Appendix A can be used to determine these characteristics.

If the quadratic form of Equation (B.12) can be either positive or zero, but not negative, then the matrix is said to be positive semidefinite, in which case further examination is required to determine whether or not $f(\boldsymbol{x}^*)$ is a minimum. A similar statement can be made regarding maxima and the property of negative semidefiniteness. If both positive and negative values of the quadratic form can exist, then the point $f(\boldsymbol{x}^*)$ represents a saddle point, which is the multivariate counterpart of an inflection point. In two-dimensional space, this condition implies a point where a maximum occurs with respect to one variable, and a minimum occurs with respect to the other, resulting in a surface similar in form to the middle of a saddle.

B.1.2. Constrained Problems in Two Dimensions

A problem of considerable importance is that in which it is desired to:

maximize or minimize the function

$$f(x_1, x_2)$$

subject to the constraint

$$g(x_1, x_2) = 0. \tag{B.13}$$

That is, the objective is to maximize or minimize $f(x_1, x_2)$, but arbitrary choices of x_1 and x_2 are not permitted. Instead, the values of x_1 and x_2 must come from pairs which satisfy the constraint equation, so that the problem is to find, from all values of x_1 and x_2 which satisfy the constraint equation, those values which maximize or minimize $f(x_1, x_2)$. The procedure to be followed here is known as the method of Lagrange multipliers, and is a fundamental tool in analysis of constrained optimization problems. The first step is to define a new function, the *Lagrange function*:

$$F(x_1, x_2, \lambda) = f(x_1, x_2) + \lambda g(x_1, x_2) \tag{B.14}$$

where λ is some constant, known as the *Lagrange multiplier*, the value of which may be determined later. Obviously, if $F(x_1, x_2, \lambda)$ is maximized or minimized with respect to x_1, x_2, and λ, and the constraint equation is also satisfied, then $f(x_1, x_2)$ is also maximized or minimized. Now, the necessary conditions for maximization or minimization of Equation (B.14) with respect to x_1, x_2, and λ are

$$\frac{\partial F}{\partial x_1} = \frac{\partial f}{\partial x_1} + \lambda \frac{\partial g}{\partial x_1} = 0$$

$$\frac{\partial F}{\partial x_2} = \frac{\partial f}{\partial x_2} + \lambda \frac{\partial g}{\partial x_2} = 0$$

$$\frac{\partial F}{\partial \lambda} = g(x_1, x_2) = 0. \tag{B.15}$$

The last expression is merely a restatement of the constraint equation. If these three equations are solved simultaneously, then $F(x_1, x_2, \lambda)$ is maximized or minimized with respect to x_1, x_2, and λ, but satisfaction of the third equation insures that the constraint is also satisfied. Satisfaction of the constraint in turn implies that $f(x_1, x_2)$ has been maximized or minimized, since the second term on the right-hand side of Equation (B.14) is zero. Thus, $f(x_1, x_2)$ has been maximized or minimized, with the simultaneous satisfaction of the constraint equation. Since Equation (B.15) is a set of three simultaneous equations in the three unknowns x_1, x_2, and λ, solution for x_1 and x_2 implies also a solution for λ, although its precise value is only of academic interest. The mathematical meaning of λ can be inferred from Equation (B.15) as

$$\lambda = -\frac{\partial f/\partial x_1}{\partial g/\partial x_1} = -\frac{\partial f/\partial x_2}{\partial g/\partial x_2} \tag{B.16}$$

evaluated at the point (x_1^*, x_2^*) which maximizes or minimizes $f(x_1, x_2)$.

As an example of the constrained and unconstrained optimization problems, consider the following exercise. Suppose it is desired to minimize the function

$$f(x_1, x_2) = x_1^2 + x_2^2$$

subject to the constraint

$$x_1 - x_2 = 5$$

or, equivalently,

$$g(x_1, x_2) = x_1 - x_2 - 5 = 0.$$

Here, the function $f(x_1, x_2)$ is a paraboloid, and if it were minimized subject to no constraint, the obvious answer would be

$$x_1 = x_2 = 0$$

$$\min f(x_1, x_2) = 0.$$

The constrained problem limits consideration of values of x_1 and x_2 to those which lie on the line represented by the constraint equation. This situation is shown in Figure B-2, which depicts the projection of the paraboloid on the (x_1, x_2) plane, together with the constraint line $x_1 - x_2 = 5$. This problem is relatively simple, and for this reason two different approaches may be used. First, the constraint equation can be solved for x_2 in terms of x_1:

$$x_2 = x_1 - 5$$

and this result substituted into the objective function (the function to be minimized):

$$f(x_1, x_2) = x_1^2 + (x_1 - 5)^2$$
$$= 2x_1^2 - 10x_1 + 25 = f(x_1).$$

This procedure has reduced the problem to an unconstrained minimization in one dimension. The solution is straightforward:

$$\frac{\partial f}{\partial x_1} = 4x_1 - 10 = 0$$
$$x_1 = \tfrac{5}{2}.$$

Then, substitution into the constraint equation yields

$$x_2 = -\tfrac{5}{2}.$$

Direct substitution such as this can be used when the constraint equation permits solution for one variable in terms of the other. Such substitution is not always possible, and for this reason the method of Lagrange multipliers is the more general solution. To solve the present problem by this method, the Lagrange function is defined as

$$F(x_1, x_2, \lambda) = x_1^2 + x_2^2 + \lambda(x_1 - x_2 - 5).$$

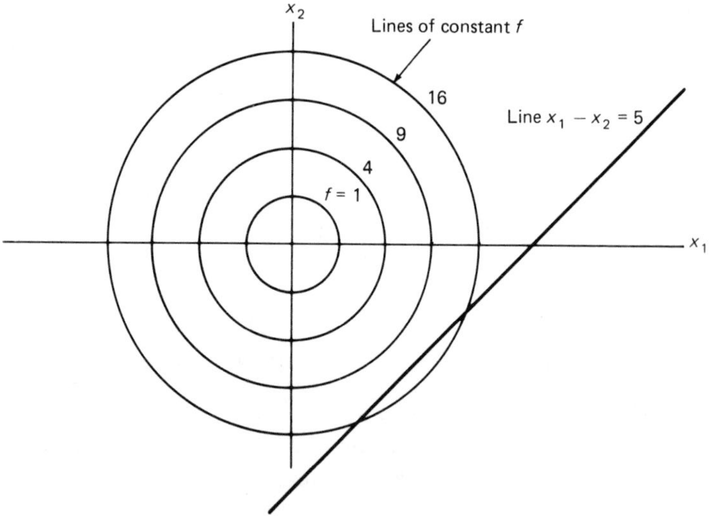

Figure B-2. Minimization of $x_1^2 + x_2^2$, subject to equality constraints.

The necessary conditions are

$$\frac{\partial f}{\partial x_1} = 2x_1 + \lambda = 0$$

$$\frac{\partial f}{\partial x_2} = 2x_2 - \lambda = 0$$

together with the constraint equation

$$x_1 - x_2 - 5 = 0.$$

Solved simultaneously, these three equations yield

$$x_1 = \tfrac{5}{2}; \qquad x_2 = -\tfrac{5}{2}; \qquad \lambda = -5$$

which agrees with the previous result. Note that solution for λ, while not required, results from the simultaneous solution of the three equations.

B.1.3. Constrained Problems in More than Two Dimensions

When the method of Lagrange multipliers is applied to a multivariate problem involving multiple constraints, the procedure to be followed is a direct generalization of those developed in the previous section. Since vector-matrix notation provides for compact expression of the problem, it will be used here. The problem is to maximize the scalar function

$$f(\boldsymbol{x}) = f(x_1, x_2, \ldots, x_N) \tag{B.17}$$

subject to the constraints

$$\boldsymbol{g}(\boldsymbol{x}) = \boldsymbol{0}.$$

The above vector equation is actually a set of M constraint equations

$$\begin{aligned} g_1(x_1, x_2, \ldots, x_N) &= 0 \\ g_2(x_1, x_2, \ldots, x_N) &= 0 \\ &\vdots \\ g_M(x_1, x_2, \ldots, x_N) &= 0. \end{aligned} \tag{B.18}$$

For each constraint equation, a Lagrange multiplier is introduced, yielding the vector of Lagrange multipliers

$$\boldsymbol{\lambda} = \begin{bmatrix} \lambda_1 \\ \lambda_2 \\ \vdots \\ \lambda_M \end{bmatrix}. \tag{B.19}$$

Now a Lagrange function is formed as

$$\begin{aligned} F(\boldsymbol{x}, \boldsymbol{\lambda}) &= F(x_1, x_2, \ldots, x_N, \lambda_1 \lambda_2, \ldots, \lambda_M) \\ &= f(x) + \lambda_1 g_1(x) + \lambda_2 g_2(x) + \cdots + \lambda_M g_M(x) \\ &= f(\boldsymbol{x}) + \boldsymbol{\lambda}^T \boldsymbol{g}(\boldsymbol{x}) \end{aligned} \tag{B.20}$$

where $\boldsymbol{\lambda}^T$ indicates the transpose of the vector $\boldsymbol{\lambda}$. The final expression above is the most compact way of representing the Lagrange function. Now if $F(\boldsymbol{x}, \boldsymbol{\lambda})$ is maximized or minimized with respect to the components of $\boldsymbol{x}$ and $\boldsymbol{\lambda}$, then the necessary conditions are compactly stated as

$$\frac{\partial F}{\partial \boldsymbol{x}} = 0; \quad \frac{\partial F}{\partial \boldsymbol{\lambda}} = 0 \tag{B.21}$$

where the implied component equations are

$$\begin{aligned} \frac{\partial F}{\partial x_1} &= 0; & \frac{\partial F}{\partial \lambda_1} &= 0 \\ \frac{\partial F}{\partial x_2} &= 0; & \frac{\partial F}{\partial \lambda_2} &= 0 \\ &\vdots & &\vdots \\ \frac{\partial F}{\partial x_N} &= 0; & \frac{\partial F}{\partial \lambda_M} &= 0. \end{aligned}$$

It can easily be seen that equations involving the derivatives with respect to the Lagrange multipliers merely reproduce the original M constraint equations. The remaining equations constitute N equations in the $(N + M)$ variables $x_1, x_2, \ldots, x_N, \lambda_1, \lambda_2, \ldots, \lambda_M$. The general equation is

$$\frac{\partial F}{\partial x_i} = \frac{\partial f}{\partial x_i} + \lambda_1 \frac{\partial g_1}{\partial x_i} + \lambda_2 \frac{\partial g_2}{\partial x_i} + \cdots + \lambda_M \frac{\partial g_M}{\partial x_i} = 0, \qquad i = 1, 2, \ldots, N.$$

In vector notation, these equations can be expressed as

$$\frac{\partial F}{\partial \boldsymbol{x}} = \frac{\partial f}{\partial \boldsymbol{x}} + \boldsymbol{\lambda}^T \frac{\partial \boldsymbol{g}}{\partial \boldsymbol{x}} = \boldsymbol{0}. \tag{B.22}$$

Thus, there are $(N+M)$ equations in $(N+M)$ unknowns. Simultaneous solution of these equations yields the desired values of $\boldsymbol{x}$, as well as $\boldsymbol{\lambda}$.

The sufficiency condition for this case is more complex than those considered previously. It is still necessary for the quadratic form

$$\boldsymbol{h}^T \frac{\partial^2 F}{\partial \boldsymbol{x}^2} \boldsymbol{h} = \sum_{i=1}^{N} \sum_{j=1}^{N} \frac{\partial^2 F}{\partial x_i \partial x_j} h_i h_j$$

to be positive for a minimum and negative for a maximum, but now this requirement must be met only for values of $\boldsymbol{h}$ and $\boldsymbol{x}$ which are permitted by the constraint equation. It can be shown that the sufficient conditions for a minimum are met if the determinantal equation

$$\det \left[\begin{array}{c|c} \dfrac{\partial^2 F}{\partial \boldsymbol{x}^2} - I\sigma & \dfrac{\partial \boldsymbol{g}^T}{\partial \boldsymbol{x}} \\ \hline \dfrac{\partial \boldsymbol{g}}{\partial \boldsymbol{x}} & \boldsymbol{0} \end{array}\right] = 0 \qquad \text{(B.23)}$$

has only positive roots σ_i, where the definition of the derivative of the vector $\boldsymbol{g}$ with respect to the vector $\boldsymbol{x}$ is implied as described in Appendix A,

$$\frac{\partial \boldsymbol{g}}{\partial \boldsymbol{x}} = \begin{bmatrix} \dfrac{\partial g_1}{\partial x_1} & \dfrac{\partial g_1}{\partial x_2} & \cdots & \dfrac{\partial g_1}{\partial x_N} \\ \dfrac{\partial g_2}{\partial x_1} & \dfrac{\partial g_2}{\partial x_2} & \cdots & \dfrac{\partial g_2}{\partial x_N} \\ \cdots & \cdots & \cdots & \cdots \\ \dfrac{\partial g_M}{\partial x_1} & \dfrac{\partial g_M}{\partial x_2} & \cdots & \dfrac{\partial g_M}{\partial x_N} \end{bmatrix}. \qquad \text{(B.24)}$$

Other definitions are in accordance to those made in association with Equation (B.9). The sufficient condition for a maximum is the negativeness of the roots of Equation (B.23).

To illustrate the satisfaction of the sufficient condition, consider again the example problem of the previous section, where the function

$$f(x_1, x_2) = x_1^2 + x_2^2$$

was minimized, subject to the constraint

$$g(x_1, x_2) = x_1 - x_2 - 5 = 0.$$

The solution obtained there was $x_1 = \frac{5}{2}$, $x_2 = -\frac{5}{2}$. Now, the matrix of Equation (B.23) will be determined for this case. First, from their definitions,

$$\frac{\partial^2 F}{\partial \boldsymbol{x}^2} = \begin{bmatrix} \dfrac{\partial^2 F}{\partial x_1^2} & \dfrac{\partial^2 F}{\partial x_1 \partial x_2} \\ \dfrac{\partial^2 F}{\partial x_1 \partial x_2} & \dfrac{\partial^2 F}{\partial x_2^2} \end{bmatrix} = \begin{bmatrix} 2 & 0 \\ 0 & 2 \end{bmatrix}$$

and

$$\frac{\partial g}{\partial \boldsymbol{x}} = \begin{bmatrix} \dfrac{\partial g}{\partial x_1} & \dfrac{\partial g}{\partial x_2} \end{bmatrix} = [1 \quad -1].$$

Then, the determinantal equation is

$$\det \left[\begin{array}{c:c} \dfrac{\partial^2 F}{\partial \boldsymbol{x}^2} - \boldsymbol{I}\sigma & \dfrac{\partial g^T}{\partial \boldsymbol{x}} \\ \hdashline \dfrac{\partial g}{\partial \boldsymbol{x}} & \boldsymbol{0} \end{array}\right] = \det \left[\begin{array}{cc:c} 2-\sigma & 0 & 1 \\ 0 & 2-\sigma & -1 \\ \hdashline 1 & -1 & 0 \end{array}\right] = 0$$

If the roots of this equation are positive, then the solution obtained represents a minimum. Expanding this determinant yields

$$(2-\sigma)\begin{vmatrix} (2-\sigma) & -1 \\ -1 & 0 \end{vmatrix} + \begin{vmatrix} 0 & (2-\sigma) \\ 1 & -1 \end{vmatrix} = -4 + 2\sigma = 0$$

from which $\sigma = +2$ is determined to be the only root. Thus, the solution $x_1 = 5$, $x_2 = -5$, is a minimum.

Note that, in the present example, x_1, x_2, and λ do not appear in the equation for σ, indicating that the solution to the necessary condition equations represents a global, rather than a local minimum, since all points satisfying the necessary conditions must be minima. In the more general case, Equation (B.23) will contain components of $\boldsymbol{x}$ and perhaps $\boldsymbol{\lambda}$, so that some points which satisfy the necessary conditions may be maxima, while others may be minima. There is also the possibility that both positive and negative roots to Equation (B.23) will be obtained, in which case other means must be used to identify the point as a maximum, minimum, or saddle point. To illustrate, consider the problem of minimizing the function

$$f(x_1, x_2) = x_1 - x_2^2$$

subject to the constraint

$$x_1^2 + x_2^2 = 1.$$

The Lagrange function is

$$F(x_1, x_2, \lambda) = x_1 - x_2^2 + \lambda(x_1^2 + x_2^2 - 1)$$

and the resulting necessary conditions are

$$\frac{\partial F}{\partial x_1} = 1 + 2\lambda x_1 = 0; \qquad \frac{\partial F}{\partial x_2} = 2x_2(\lambda - 1) = 0.$$

If these equations are solved simultaneously with the constraint equation, the following solutions result:

x_1	x_2	λ
1	0	$-\frac{1}{2}$
-1	0	$\frac{1}{2}$
$-\frac{1}{2}$	$\sqrt{3}/2$	1
$-\frac{1}{2}$	$-\sqrt{3}/2$	1

The question now arises as to which of these points represent minima. To apply the test of Equation (B.23), the following matrices are needed:

$$\frac{\partial^2 F}{\partial \boldsymbol{x}^2} = \begin{bmatrix} \dfrac{\partial^2 F}{\partial x_1^2} & \dfrac{\partial^2 F}{\partial x_1 \partial x_2} \\ \dfrac{\partial^2 F}{\partial x_1 \partial x_2} & \dfrac{\partial^2 F}{\partial x_2^2} \end{bmatrix} = \begin{bmatrix} 2 & 0 \\ 0 & 2\lambda - 2 \end{bmatrix}$$

and

$$\frac{\partial g}{\partial \boldsymbol{x}} = \begin{bmatrix} \dfrac{\partial g}{\partial x_1} & \dfrac{\partial g}{\partial x_2} \end{bmatrix} = [2x_1 \quad 2x_2].$$

Then, Equation (B.23) as applied to this problem is

$$\det \begin{bmatrix} 2\lambda - \sigma & 0 & 2x_1 \\ 0 & 2\lambda - 2 - \sigma & 2x_2 \\ 2x_1 & 2x_2 & 0 \end{bmatrix} = 0.$$

If this equation is solved for values of σ for each of the four points representing a solution, the following results are obtained:

x_1	x_2	λ	σ	Conclusion
1	0	$-\frac{1}{2}$	-3	maximum
-1	0	$\frac{1}{2}$	-1	maximum
$-\frac{1}{2}$	$\sqrt{3}/2$	1	$\frac{3}{2}$	minimum
$-\frac{1}{2}$	$-\sqrt{3}/2$	1	$\frac{3}{2}$	minimum

There are, therefore, two points yielding a minimum, and this minimum value is

$$x_1 - x_2^2 = -\tfrac{5}{4}.$$

B.1.4. Inequality Constraints

A problem of considerable interest involves inequality constraints. In its general form, it can be stated as the problem of maximizing or minimizing the function

$$f(\mathbf{x}) = f(x_1, x_2, \ldots, x_N)$$

subject to a set of constraint inequalities

$$\begin{aligned} g_1(x) &= g_1(x_1, x_2, \ldots, x_N) \leqslant b_1 \\ g_2(x) &= g_2(x_1, x_2, \ldots, x_N) \leqslant b_2 \\ &\vdots \\ g_M(x) &= g_M(x_1, x_2, \ldots, x_N) \leqslant b_M \end{aligned} \tag{B.25}$$

or, more compactly,

$$\mathbf{g}(\mathbf{x}) \leqslant \mathbf{b}.$$

The approach to this kind of problem is to introduce new variables, called *slack variables* because they take up the slack between the two sides of the constraint inequality, such that

$$g_i(x) + \mu_i^2 = b_i, \qquad i = 1, 2, \ldots, M. \tag{B.26}$$

The squared form μ_i^2 is used to ensure that the slack variable is never negative, as required by the less than or equal to constraint. Introduction of the slack variables transforms the problem into one involving equality constraints, which can then be treated according to the procedures of the previous section. That is, it is now desired to maximize or minimize $f(\mathbf{x})$ subject to the M constraints of Equation (B.26). The Lagrange function is now

$$\begin{aligned} F(\mathbf{x}, \boldsymbol{\lambda}, \boldsymbol{\mu}) = f(\mathbf{x}) + \lambda_1[g_1(\mathbf{x}) + \mu_1^2 - b_i] + \lambda_2[g_2(\mathbf{x}) - \mu_2^2 - b_2] \\ + \cdots + \lambda_M[g_M(\mathbf{x}) + \mu_M^2 - b_M] \end{aligned} \tag{B.27}$$

where $\boldsymbol{\mu}$ is the vector of the nonnegative slack variables μ_i. The Lagrange function is now differentiated with respect to $\mathbf{x}$, $\boldsymbol{\lambda}$, and the slack variables $\boldsymbol{\mu}$, and the results equated to zero:

$$\left.\begin{aligned}&\frac{\partial F}{\partial x_i}=0, \qquad i=1,2,\ldots,N\\ &\begin{aligned}&\frac{\partial F}{\partial \mu_i}=0,\\ &\frac{\partial F}{\partial \lambda_i}=0,\end{aligned}\right\} \quad i=1,2,\ldots,M.\end{aligned} \tag{B.28}$$

The first of these equations yields

$$\frac{\partial F}{\partial x_i}=\frac{\partial f}{\partial x_i}+\sum_{j=1}^{M}\lambda_j\frac{\partial g_j}{\partial x_i}=0 \tag{B.29}$$

in accordance with Equation (B.22). The second equation yields

$$\frac{\partial F}{\partial \mu_i}=2\lambda_i\mu_i=0, \qquad i=1,\ldots,M \tag{B.30}$$

and the third yields

$$\mu_i^2=b_i-g_i(x), \qquad i=1,\ldots,M. \tag{B.31}$$

If Equation (B.30) is multiplied by μ_i, and Equation (B.31) by λ_i, then the results combine to yield

$$\lambda_i\mu_i^2=\lambda_i[b_i-g_i(x)]=0. \tag{B.32}$$

This equation states that either $\lambda_i=0$ or the ith constraint is satisfied as an equality, in accordance with the following intuitive reasoning. If the optimum solution occurs at a point where the ith constraint is satisfied as an inequality, then the problem could have been solved without consideration of the ith constraint and λ_i need not be defined. On the other hand, if the optimum solution occurs with the ith constraint satisfied as an equality, then the problem could have been solved with the ith constraint considered as an equality. Now, suppose that λ_j is not necessarily zero, and therefore the jth constraint is satisfied as an equality:

$$g_j(x)-b_j=0$$

and further suppose that the problem is one of maximization. If $\boldsymbol{x}$ is the point which produces the maximum, then consider a small change Δx_i in the ith coordinate x_i which causes the jth constraint above to be satisfied as an inequality. Now, if Δx_i is positive, then

$$\Delta x_i > 0; \quad \frac{\partial f}{\partial x_i} < 0; \quad \frac{\partial g_j}{\partial x_i} < 0 \tag{B.33}$$

and if Δx_i is negative,

$$\Delta x_i < 0; \quad \frac{\partial f}{\partial x_i} > 0; \quad \frac{\partial g_j}{\partial x_i} > 0.$$

These relationships result from the particular form of the inequality constraint, and from the fact that $f(\boldsymbol{x})$ represents a maximum. In either case, Equation (B.33), when considered in conjunction with Equation (B.29), requires that

$$\lambda_j < 0. \tag{B.34}$$

Similar consideration of the case where $f(\boldsymbol{x})$ is to be minimized results in the condition

$$\lambda_j > 0. \tag{B.35}$$

These results can also be obtained from Equation (B.16). In summary, there are four conditions which must be met by the solution to the optimization problem with inequality constraints, as posed by Equation (B.25):

$$\begin{aligned}
&\text{I.} \quad \frac{\partial f(\boldsymbol{x})}{\partial x_i} + \sum_{j=1}^{M} \lambda_j \frac{\partial g_j(\boldsymbol{x})}{\partial x_i} = 0, && i = 1, 2, \ldots, N \\
&\text{II.} \quad \lambda_i [g_i(\boldsymbol{x}) - b_i] = 0, && i = 1, 2, \ldots, M \\
&\text{III.} \quad \left.\begin{array}{l} \lambda_i \leqslant 0 \ \text{ for a maximum,} \\ \lambda_i \geqslant 0 \ \text{ for a minimum,} \end{array}\right\} && i = 1, 2, \ldots, M \\
&\text{IV.} \quad g_i(\boldsymbol{x}) \leqslant b_i, && i = 1, 2, \ldots, M
\end{aligned} \tag{B.36}$$

Note that these results are listed as conditions, since they do not constitute a set of equations from which the solution can be obtained. These relationships are called the *Kuhn–Tucker conditions* for optimization with inequality constraints.

The Kuhn–Tucker conditions are necessary conditions, and, in order to discuss the conditions for sufficiency, it is imperative to consider the concept of *convexity*. Simply speaking, a function is said to be *convex* if a straight line drawn between any two points on the surface generated by the function lies completely above or on the surface. If the line lies above the surface, the function is said to be *strictly convex*. As examples of these concepts, Figures B-3(a), (c), and (e) present some convex and strictly convex functions. The property of *concavity* of a function is defined in a similar manner, except that a *concave* function has the property that a straight line drawn between any two points on the surface generated by the function lies on or below the surface. If the line lies below the

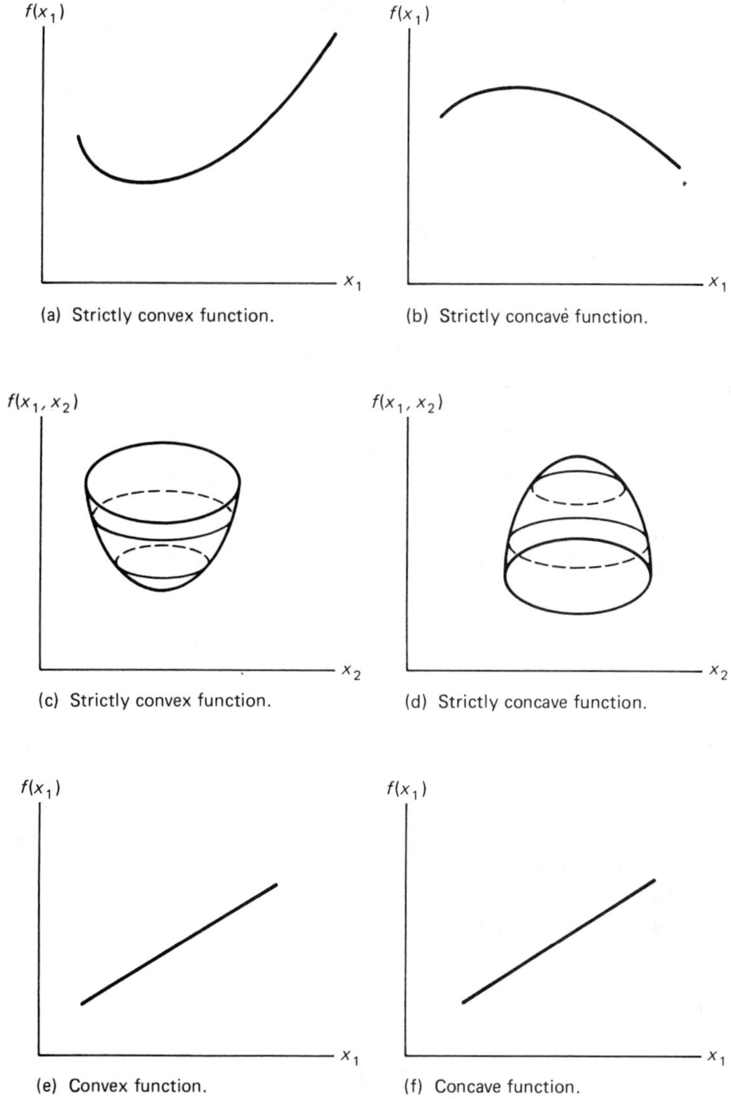

Figure B-3. Convex and concave functions.

surface the function is said to be *strictly concave*. Examples are shown in Figures B-3(b), (d), and (f). Notice that some functions may satisfy the conditions of convexity and concavity simultaneously, as does the straight line in Figures B-3(e) and (f). No function may be strictly concave and simultaneously be convex, or vice versa. The property of convexity can be determined by use of the

Table B-1. Convexity and Concavity of Functions.

	$\frac{\partial^2 f(x)}{\partial x^2}$	Eigenvalues
Strictly convex	positive definite	$\lambda_i > 0$
Convex	positive semidefinite	$\lambda_i \geqslant 0$
Strictly concave	negative definite	$\lambda_i < 0$
Concave	negative semidefinite	$\lambda_i \leqslant 0$
No classification	indefinite	some $\lambda_i > 0$ some $\lambda_i < 0$

matrix defined in conjunction with Equation (B.9):

$$\frac{\partial^2 f(x)}{\partial x^2} = \begin{bmatrix} \frac{\partial^2 f}{\partial x_1^2} & \frac{\partial^2 f}{\partial x_1 \partial x_2} & \cdots & \frac{\partial^2 f}{\partial x_1 \partial x_N} \\ \frac{\partial^2 f}{\partial x_1 \partial x_2} & \frac{\partial^2 f}{\partial x_2^2} & \cdots & \frac{\partial^2 f}{\partial x_2 \partial x_N} \\ \cdots & \cdots & \cdots & \cdots \\ \frac{\partial^2 f}{\partial x_1 \partial x_N} & \frac{\partial^2 f}{\partial x_2 \partial x_N} & \cdots & \frac{\partial^2 f}{\partial x_N^2} \end{bmatrix} \tag{B.37}$$

If this matrix is positive definite, then the function $f(x)$ is strictly convex at the point x. If it is positive semidefinite, then the function is convex. Similar relationships hold for concavity, with the results tabulated as Table B-1. The matrix properties have also been stated in terms of the eigenvalues, although definiteness determination by use of Sylvester's criterion is probably an easier method of investigating these properties.

Now the sufficiency of the Kuhn–Tucker conditions can be examined. Suppose that $f(x)$ of Equation (B.25) is strictly concave, and each of the $g_i(x)$ are convex. Consider a maximization process so that the Kuhn–Tucker conditions require that

$$\lambda_i < 0, \quad \text{for all } i$$

in accordance with Equation (B.36). Then, the terms

$$\lambda_i g_i(x), \quad \text{for all } i$$

are all concave, and the sum

$$f(x) + \sum_{i=1}^{M} \lambda_i g_i(x)$$

must be strictly concave, since it is the sum of a strictly concave function with other concave functions. The Lagrange function, from Equation (B.27) is

$$F(\boldsymbol{x}, \boldsymbol{\lambda}, \boldsymbol{\mu}) = f(\boldsymbol{x}) + \sum_{i=1}^{M} \lambda_i g_i(\boldsymbol{x}) + \sum_{i=1}^{M} \lambda_i \mu_i^2 - \sum_{i=1}^{M} \lambda_i b_i \tag{B.38}$$

and, since the optimization process yielded

$$\lambda_i \mu_i^2 = 0 \tag{B.39}$$

and the products $\lambda_i b_i$ are constants, the Lagrange function itself must be strictly concave. This result indicates that $F(\boldsymbol{x}, \boldsymbol{\lambda}, \boldsymbol{\mu})$ reaches a maximum at the point $\boldsymbol{x}$, in turn implying that $f(\boldsymbol{x})$ also reaches a maximum, with the concurrent satisfaction of the inequality constraints. A similar development shows that the sufficiency of the Kuhn–Tucker conditions for a minimum is established by the convexity of the $g_i(\boldsymbol{x})$ function together with a strictly convex property for $f(\boldsymbol{x})$. The concepts are summarized in Table B-2, which shows the stipulations under which the Kuhn–Tucker conditions are both necessary and sufficient. If these stipulations do not hold, then the Kuhn–Tucker conditions are necessary only, and extreme care must be taken in interpreting their satisfaction in any particular instance.

As an example of these principles, consider again the problem presented in Section B.1.2, and illustrated in Figure B-2. Only now consider the minimization of the function

$$f(\boldsymbol{x}) = x_1^2 + x_2^2$$

subject to the two inequality constraints

$$x_1 - x_2 \leqslant 5$$

$$x_1 - x_2 \geqslant 1.$$

The inequality constraints can be put into the standard form of Equation (B.25) as follows:

$$g_1 = x_1 - x_2 - 5 \leqslant 0$$

$$g_2 = -x_1 + x_2 + 1 \leqslant 0.$$

Table B-2. Stipulations Under Which Kuhn–Tucker Conditions Are Necessary and Sufficient.

	$f(\boldsymbol{x})$	$g_i(\boldsymbol{x})$
Maximum	strictly concave	convex
Minimum	strictly convex	convex

The Lagrange function in this instance is

$$F(\mathbf{x}, \boldsymbol{\lambda}) = x_1^2 + x_2^2 + \lambda_1(x_1 - x_2 - 5) + \lambda_2(-x_1 + x_2 + 1).$$

The Kuhn–Tucker conditions are

$$\text{I.}\quad \frac{\partial F}{\partial x_1} = 2x_1 + \lambda_1 - \lambda_2 = 0$$

$$\frac{\partial F}{\partial x_2} = 2x_2 - \lambda_1 + \lambda_2 = 0$$

$$\text{II.}\quad \lambda_1(x_1 - x_2 - 5) = 0$$
$$\lambda_2(-x_1 + x_2 + 1) = 0$$

$$\text{III.}\quad \lambda_1 \geqslant 0$$
$$\lambda_2 \geqslant 0$$

$$\text{IV.}\quad x_1 - x_2 \leqslant 5$$
$$x_1 - x_2 \geqslant 1.$$

Since these expressions are conditions which must be satisfied, rather than equations, all possible solutions must be investigated to determine whether or not they satisfy the above conditions. First, the condition I equations can be combined to yield

$$x_2 = -x_1$$

which must hold regardless of the values of λ_1 and λ_2. Now, for $\lambda_1 = 0$, the second equation of condition II yields

$$-x_1 + x_2 = -1 = -2x_1$$

or

$$x_1 = \tfrac{1}{2}; \qquad x_2 = -\tfrac{1}{2}$$

and, from the first equation of condition I,

$$2x_1 = \lambda_2$$

$$\lambda_2 = 1.$$

These results satisfy conditions III and IV and therefore satisfy the Kuhn–Tucker conditions. However, they may not be unique, and other solutions must be considered. Suppose $\lambda_1 = 0$ and $\lambda_2 = 0$. Then, the only solution to the equations of condition I is

$$x_1 = 0$$

$$x_2 = 0$$

which violates condition IV. For $\lambda_2 = 0$, the first equation of condition II yields

$$x_1 - x_2 = 5 = 2x_1$$

$$x_1 = \frac{5}{2}; \qquad x_2 = -\frac{5}{2}$$

and, from the first equation of condition I,

$$\lambda_1 = -2x_1 = -5$$

which violates condition III, so that this solution is not a minimum. For λ_1 and λ_2 nonzero, condition II yields the equations

$$x_1 - x_2 = 5$$

$$-x_1 + x_2 = -1$$

which have no solution, since they represent parallel lines. In review, then, all possibilities have been examined, with the following tabular results:

Possibility	Result
$\lambda_1 = 0, \lambda_2 \neq 0$	Satisfies all four conditions
$\lambda_1 = 0, \lambda_2 = 0$	Violates condition IV
$\lambda_1 \neq 0, \lambda_2 = 0$	Violates condition III
$\lambda_1 \neq 0, \lambda_2 \neq 0$	Violates condition II

Thus, the solution

$$\lambda_1 = 0; \qquad \lambda_2 = 1$$

$$x_1 = \frac{1}{2}; \qquad x_2 = -\frac{1}{2}$$

uniquely satisfies the Kuhn–Tucker conditions. Furthermore, since the two constraint inequalities are linear, the $g_1(\boldsymbol{x})$ and $g_2(\boldsymbol{x})$ functions are convex. This fact, together with the convexity of the function

$$f(\boldsymbol{x}) = x_1^2 + x_2^2$$

describes the situation in which the Kuhn–Tucker conditions are both necessary and sufficient for a minimum. Furthermore, since the matrix

$$\frac{\partial^2 f}{\partial \boldsymbol{x}^2} = \begin{bmatrix} 2 & 0 \\ 0 & 2 \end{bmatrix}$$

does not depend upon the values of x_1 and x_2, the point in question is a global minimum.

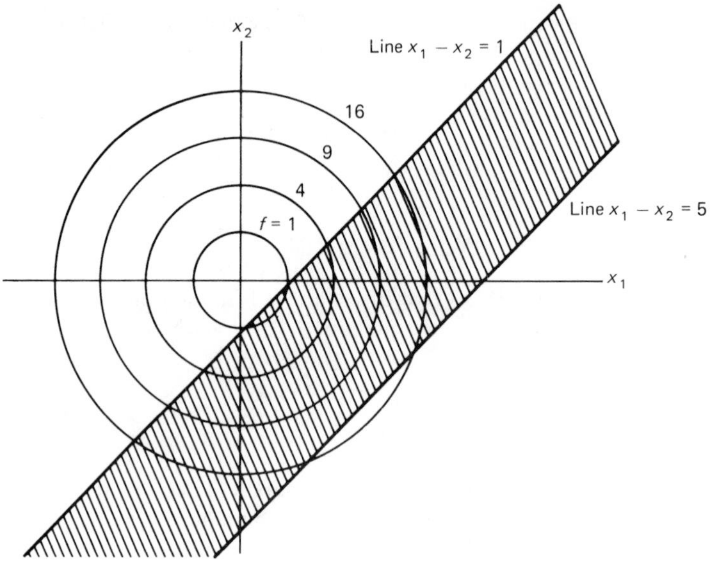

Figure B-4. Minimization of $x_1^2 + x_2^2$ subject to inequality constraints.

It should be noted that the solution

$$\lambda_1 = -5; \qquad \lambda_2 = 0$$
$$x_1 = \tfrac{5}{2}; \qquad x_2 = -\tfrac{5}{2}$$

satisfies conditions I, II, and IV, and also satisfies condition III for a maximum, as outlined in Equation (B.36). The Kuhn–Tucker conditions are not sufficient for a maximum, however, because the $f(\boldsymbol{x})$ term is not strictly concave. In fact, the above solution does not constitute a maximum, as can be seen from the form of $f(\boldsymbol{x})$, which has its maximum at infinity. From Figure B-4, the above solution can be seen to be a saddle point. Also from Figure B-4, where the cross-hatched region represents the inequality constraints, it can be seen that the point $(\frac{1}{2}, -\frac{1}{2})$ gives the minimum of all values lying within this region.

B.1.5. The Linear Problem

If the optimization problem is specified in terms of a linear function to be optimized, of linear inequality and/or equality constraints, and of a requirement that all variables be nonnegative, the particular form of the problem is to optimize

$$f(\boldsymbol{x}) = c_1 x_1 + c_2 x_2 + \cdots + c_N x_N \tag{B.40}$$

subject to the constraints

$$
\begin{aligned}
a_{11}x_1 + a_{12}x_2 + \cdots + a_{1N}x_N &\gtrless b_1 \\
a_{21}x_1 + a_{22}x_2 + \cdots + a_{2N}x_N &\gtrless b_2 \\
&\vdots \\
a_{M1}x_1 + a_{22}x_2 + \cdots + a_{MN}x_N &\gtrless b_M.
\end{aligned} \tag{B.41}
$$

For ease of development, consider the situation where $N = 3$ and $M = 2$, with less than or equal to constraints:

$$
\begin{aligned}
f(\boldsymbol{x}) &= c_1x_1 + c_2x_2 + c_3x_3 \\
a_{11}x_1 + a_{12}x_2 + a_{13}x_3 &\leqslant b_1 \\
a_{12}x_1 + a_{22}x_2 + a_{23}x_3 &\leqslant b_2 \\
x_1 \geqslant 0; \quad x_2 \geqslant 0;& \quad x_3 \geqslant 0.
\end{aligned} \tag{B.42}
$$

Applying the Kuhn–Tucker conditions to this problem yields the Lagrange function

$$
\begin{aligned}
F(x_1, x_2, x_3, \lambda_1, \lambda_2, \lambda_3, \lambda_4, \lambda_5) = {} & c_1x_1 + c_2x_2 + c_3x_3 \\
& + \lambda_1(a_{11}x_1 + a_{12}x_2 + a_{13}x_3) \\
& + \lambda_2(a_{21}x_1 + a_{22}x_2 + a_{23}x_3) \\
& + \lambda_3x_1 + \lambda_4x_2 + \lambda_5x_3.
\end{aligned} \tag{B.43}
$$

The Kuhn–Tucker conditions are then

$$
\text{I.} \quad \frac{\partial F}{\partial \boldsymbol{x}} = 0; \qquad
\begin{aligned}
c_1 + a_{11}\lambda_1 + a_{21}\lambda_2 + \lambda_3 &= 0 \\
c_2 + a_{12}\lambda_1 + a_{22}\lambda_2 + \lambda_4 &= 0 \\
c_3 + a_{13}\lambda_1 + a_{23}\lambda_2 + \lambda_5 &= 0
\end{aligned}
$$

$$
\text{II.} \quad \lambda_i[g_i(\boldsymbol{x}) - b_i] = 0; \qquad
\begin{aligned}
\lambda_1(a_{11}x_1 + a_{12}x_2 + a_{13}x_3) &= 0 \\
\lambda_2(a_{21}x_1 + a_{22}x_2 + a_{23}x_3) &= 0 \\
\lambda_3x_1 &= 0 \\
\lambda_4x_2 &= 0 \\
\lambda_5x_3 &= 0
\end{aligned}
$$

$$
\text{III.} \quad
\begin{aligned}
\lambda_i &< 0 \quad \text{for} \quad \max i = 1, \ldots, 5 \\
\lambda_i &> 0 \quad \text{for} \quad \min i = 1, \ldots, 5
\end{aligned}
$$

$$
\text{IV.} \quad g_i(\boldsymbol{x}) \leqslant b_i; \qquad
\begin{aligned}
&a_{11}x_1 + a_{12}x_2 + a_{13}x_3 \leqslant b_1 \\
&a_{21}x_1 + a_{22}x_2 + a_{23}x_3 \leqslant b_2 \\
&x_1 \geqslant 0; \quad x_2 \geqslant 0; \quad x_3 \geqslant 0.
\end{aligned}
$$

By examination of condition I, it can be seen that there are no values of the λ's which can satisfy this equation, without two or more of them being zero From condition II, this fact in turn implies that the solution must lie on at least two boundaries. Since the constraints are linear and in the form of planes in the three-dimensional (x_1, x_2, x_3) space, it is further implied that the solution must lie at the intersection of constraint planes. In general, solutions to problems of this type are constrained to lie at intersections of constraint planes in the multidimensional $\boldsymbol{x}$ space, and this fact serves as the basis for a specialized solution technique known as the *simplex method.* A problem of the general form given by Equations (B.40) and (B.41) is known as a *linear programming* problem and can be solved most efficiently by the simplex method, which systematically searches through all constraint plane intersections for a minimum or a maximum.

Up to this point in the discussion, various problems involving optimization—in the sense of maximization or minimization of a function of a variable or a set of variables—have been considered. In this sense, these developments represent point optimization, in that the point or points at which the function is optimized have been determined. There is another class of optimization problem in which an integral function of a quantity, in turn a function of some independent variable, is to be optimized. A particular branch of calculus—the calculus of variations—was developed to provide for solutions of this kind of problem and is the topic of the remainder of this appendix.

B.2. THE CALCULUS OF VARIATIONS

The fundamental problem of the calculus of variations involves the minimization or maximization of an integral

$$J = \int_a^b L[x(t), \dot{x}(t), t]\, dt \tag{B.44}$$

where the dependent variable x is a function of the independent variable t, $\dot{x} = dx/dt$, and L is some prescribed function. The integral J is dependent upon the value of the function L over the entire interval (a, b) and as such can be considered a function of the function $x(t)$. Such terms are called *functionals*, and the basic problem of the calculus of variations is the selection of the function $x(t)$ defined over the interval (a, b) which maximizes or minimizes the functional J. In previous work involving point optimization, it was assumed that some general point achieved the desired maximum or minimum, and then the Taylor series expansion around this point was examined to determine what characteristics an optimum point must exhibit. In the present situation, this technique can be used, but now, instead of a single point, an entire function must be examined to determine what characteristics it must exhibit at every point in the interval (a, b). In order to accomplish this, a new term is defined as the *variation in* $x(t)$ and denoted by $\delta x(t)$,

$$\delta x(t) = x(t) - x^*(t) \tag{B.45}$$

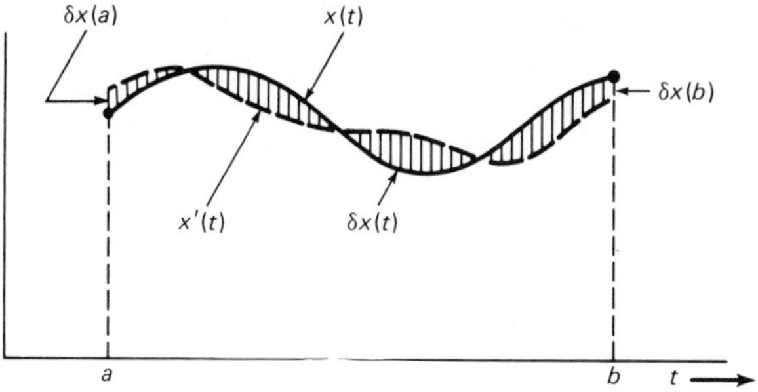

Figure B-5. Definition of the variation $\delta x(t)$.

where $x^*(t)$ is any function defined over the interval (a, b) and $x(t)$ is some other function lying close to $x^*(t)$ everywhere in the interval. This concept is illustrated in Figure B-5, where the difference between the two curves represents the variation $\delta x(t)$. Note that this concept is just a generalization of that of a differential and, as such, can be considered an operator defining certain mathematical procedures. For example, differentiation of $\delta x(t)$ with respect to the independent variable t yields

$$\begin{aligned} \frac{d}{dt}[\delta x(t)] &= \frac{d}{dt}[x(t) - x^*(t)] = \frac{dx(t)}{dt} - \frac{dx^*(t)}{dt} \\ &= \delta \dot{x}(t) \end{aligned} \tag{B.46}$$

which shows that the derivative of the variation is the variation of the derivative; thus, the operations of differentiating and taking variation commute. Similarly,

$$\begin{aligned} \int_{t_0}^{t} \delta x(\sigma)\, d\sigma &= \int_{t_0}^{t} [x(\sigma) - x^*(\sigma)]\, d\sigma \\ &= \int_{t_0}^{t} x(\sigma)\, d\sigma - \int_{t_0}^{t} x^*(\sigma)\, d\sigma \\ &= \delta\left[\int_{t_0}^{t} x(\sigma)\, d\sigma\right] \end{aligned} \tag{B.47}$$

so that integrating and taking variation commute.

Now assume that the function $x^*(t)$ over the entire interval (a, b) has been found such that the integral of Equation (B.44) has been maximized or minimized, and consider some adjacent function $x(t)$ which defines the variation

$\delta x(t)$. The resulting change in the functional J is then

$$\Delta J = J - J^* = \int_a^b L(x, \dot{x}, t)\, dt - \int_a^b L(x^*, \dot{x}^*, t)\, dt. \tag{B.48}$$

The right-hand side can be written as

$$\int_a^b [L(x, \dot{x}, t) - L(x^*, \dot{x}^*, t)]\, dt$$

where the integrand is δL, the variation in L due to variation in $x^*(t)$ and the resulting variation $\delta x(t)$. Then,

$$\Delta J = \int_a^b \delta L\, dt. \tag{B.49}$$

At every point t, the variation δL can be expressed as

$$\begin{aligned} \delta L &= L(x, \dot{x}, t) - L(x^*, \dot{x}^*, t) \\ &= L(x^* + \delta x, \dot{x}^* + \delta \dot{x}, t) - L(x^*, x^*, t). \end{aligned} \tag{B.50}$$

The first term on the right-hand side can be expanded in a Taylor series to yield

$$\delta L = L(x^*, \dot{x}^*, t) + \frac{\partial L}{\partial x}\delta x + \frac{\partial L}{\partial \dot{x}}\delta \dot{x} + \text{h.o.t.} - L(x^*, \dot{x}^*, t). \tag{B.51}$$

If the usual assumptions of differentiability are made, and if $x(t)$ is permitted to approach very closely to $x^*(t)$ at every point in the interval (a, b), then the first-order terms dominate, and the above equation is closely approximated by

$$\delta L = \frac{\partial L}{\partial x}\delta x + \frac{\partial L}{\partial \dot{x}}\delta \dot{x} \tag{B.52}$$

which gives an expression for the variation in L caused by the variation in $x^*(t)$. If Equation (B.52) is applied at every point t in the interval (a, b), then δL is defined over the interval, and Equation (B.49) is

$$\Delta J = \int_a^b \left[\frac{\partial L}{\partial x}\delta x + \frac{\partial L}{\partial \dot{x}}\delta \dot{x}\right] dt \tag{B.53}$$

yielding the change in J caused by the variation $\delta x(t)$. By the same reasoning leading to the necessary conditions for point optimization, it can be argued that

ΔJ must be zero for arbitrary variations δx in order for $x^*(t)$ to be the function which minimizes J. For ΔJ to vanish for arbitrary δx, the integrand of Equation (B.53) must be zero. This procedure would lead to an equation involving δx and $\delta \dot{x}$, but there is no general functional relationship between these two quantities. For this reason, the $\delta \dot{x}$ term is eliminated in the following manner. The integral of Equation (B.53) is written as

$$\Delta J = \int_a^b \frac{\partial L}{\partial x} \delta x \, dt + \int_a^b \frac{\partial L}{\partial \dot{x}} \delta \dot{x} \, dt. \qquad \text{(B.54)}$$

The second term on the right-hand side can be expressed as

$$\int_a^b \frac{\partial L}{\partial \dot{x}} \frac{d(\delta x)}{dt} dt = \int_{t=a}^{t=b} \frac{\partial L}{\partial \dot{x}} d(\delta x). \qquad \text{(B.55)}$$

The change of variable from t to δx has resulted in a change in the limits of integration, where the limit $t = a$ implies the value of δx when $t = a$, and so on. The right-hand side of Equation (B.55) can be integrated by parts to yield

$$\int_{t=a}^{t=b} \frac{\partial L}{\partial \dot{x}} d(\delta x) = \frac{\partial L}{\partial \dot{x}} \delta x \Bigg|_{t=a}^{t=b} - \int_{t=a}^{t=b} \delta x \, d\left[\frac{\partial L}{\partial \dot{x}}\right]. \qquad \text{(B.56)}$$

The variable of integration in the second term can now be changed back to time with the following result

$$\int_a^b \frac{\partial L}{\partial \dot{x}} \delta \dot{x} \, dt = \frac{\partial L}{\partial \dot{x}} \Bigg|_{t=a}^{t=b} - \int_a^b \frac{d}{dt}\left[\frac{\partial L}{\partial \dot{x}}\right] \delta x \, dt. \qquad \text{(B.57)}$$

Now, by substituting Equation (B.57) for the second term of Equation (B.54), and combining the integrals with respect to t, the following result is obtained:

$$\Delta J = \int_a^b \left\{\frac{\partial L}{\partial x} - \frac{d}{dt}\left[\frac{\partial L}{\partial \dot{x}}\right]\right\} \delta x \, dt - \frac{\partial L}{\partial \dot{x}} \delta x \Bigg|_{t=a}^{t=b}. \qquad \text{(B.58)}$$

The necessary condition for a maximum or minimum is the vanishing of ΔJ for any arbitrary $\delta x(t)$, which can occur only if the following conditions are simultaneously met:

$$\frac{\partial L}{\partial x} - \frac{d}{dt}\left[\frac{\partial L}{\partial \dot{x}}\right] = 0$$

$$\frac{\partial L}{\partial \dot{x}} \delta x \Bigg|_{t=a}^{t=b} = \frac{\partial L}{\partial \dot{x}} \delta x \Bigg|_{t=b} - \frac{\partial L}{\partial \dot{x}} \delta x \Bigg|_{t=a} = 0. \qquad \text{(B.59)}$$

The first equation is the *Euler–Lagrange equation* and must be met regardless of the end conditions—the values of $x(t)$ at the points $t = a$ and $t = b$. The second equation involves only end conditions, and may be automatically satisfied by the problem specification. For example, if $x(a)$ and $x(b)$ are specified, then $\delta x(a)$ and $\delta x(b)$ are both zero since no variation is permitted at these points, in which case the second equation is satisfied without yielding any information. If, as is commonly the case, $x(a)$ is specified while $x(b)$ is not, then $\delta x(a)$ is zero, but $\delta x(b)$ can take on arbitrary values. In this event, the second equation reduces to

$$\left.\frac{\partial L}{\partial \dot{x}}\,\delta x\right|_{t=b} = 0. \tag{B.60}$$

Similarly, if $x(a)$ is not specified but $x(b)$ is, the second equation reduces to

$$\left.\frac{\partial L}{\partial \dot{x}}\,\delta x\right|_{t=a} = 0. \tag{B.61}$$

In summary, the second equation of Equation (B.59) is always satisfied, but the amount of information it contains will vary with the boundary conditions on $x(t)$.

As an example of these principles, consider the problem of minimizing the functional

$$J = \int_0^1 (\dot{x}^2 + x)\,dt$$

with the boundary conditions

$$x(0) = 2$$
$$x(1) = 3.$$

Application of the Euler–Lagrange equation from Equation (B.59) yields

$$\begin{aligned}\frac{\partial L}{\partial x} - \frac{d}{dt}\left[\frac{\partial L}{\partial \dot{x}}\right] &= 1 - \frac{d}{dt}\,[2\dot{x}] \\ &= 1 - 2\ddot{x} = 0\end{aligned}$$

a second-order differential equation which, together with the two boundary conditions on $x(t)$, can be solved to yield

$$x(t) = \tfrac{1}{4}t^2 + \tfrac{3}{4}t + 2.$$

This result is the function $x(t)$ which minimizes J, while simultaneously meeting the required boundary conditions.

Now, suppose that the requirement that $x(1) = 3$ is removed. Then, the second expression of Equation (B.59) yields

$$\left.\frac{\partial L}{\partial \dot{x}}\,\delta x\right|_{t=1} = 0$$

which requires that

$$\left.\frac{\partial L}{\partial \dot{x}}\right|_{t=1} = 0 = 2\dot{x}\big|_{t=1}.$$

Note that this equation provides the additional boundary condition required for solution of the Euler–Lagrange equation. In this case, the solution is

$$x(t) = \tfrac{1}{4}t^2 - \tfrac{1}{2}t + 2.$$

This function $x(t)$ minimizes J, while simultaneously meeting the boundary condition $x(0) = 2$. Since the second case is the less restrictive, the minimum obtained should be less than or equal to that obtained in the first case. This fact can be shown by evaluation of J in the two cases, which yield

$$J = 3.48 \quad \text{and} \quad J = 1.63$$

respectively, for the first and second cases.

B.2.1. The Constrained Variational Problem

A problem of considerable interest is the optimization of a functional, subject to a differential equation of constraint. That is, it is desired to maximize or minimize the functional

$$J = \int_a^b L[x(t), u(t), t]\; dt \tag{B.62}$$

subject to the constraint

$$\dot{x} = f[x, u, t].$$

This problem is analogous to that of the constrained point optimization problem in two dimensions, which was approached using the method of Lagrange multipliers. In this case, however, the constraint equation must be satisfied at every point in the interval (a, b). The solution to this problem requires the definition of a Lagrange multiplier function $\lambda(t)$, which is used to form a new functional, analogous to the Lagrange function, as

$$\bar{J} = \int_a^b (L(x, u, t) + \lambda(t)[f(x, u, t) - \dot{x}])\; dt. \tag{B.63}$$

By the same reasoning as previously used, if $\bar{J}$ is optimized, with the simultaneous satisfaction of the constraint equation, then J is also optimized because the term within the square brackets is zero. If the change in $\bar{J}$ due to variations δx, δu, and $\delta\lambda$ is equated to zero, the equation resulting from the $\delta\lambda$ coefficient is

$$f(x, u, t) - \dot{x} = 0$$

which is the constraint equation. Thus, $\bar{J}$ can be optimized with respect to x and u alone, if the resulting equations are solved simultaneously with the constraint equation. Before examining the change $\Delta\bar{J}$, it is convenient to rearrange the equation as

$$\bar{J} = \int_a^b \{[L(x, u, t) + \lambda(t) f(x, u, t)]\}\, dt - \int_a^b \lambda(t)\dot{x}(t)\, dt. \tag{B.64}$$

The second term on the right-hand side can be integrated by parts as follows:

$$\int_a^b \lambda \frac{dx}{dt}\, dt = \int_{t=a}^{t=b} \lambda\, dx = \lambda x \Big|_{t=a}^{t=b} - \int_{t=a}^{t=b} x\, d\lambda. \tag{B.65}$$

The variable of integration in the integral term can then be changed back to t, with the result

$$\int_a^b \lambda\dot{x}\, dt = \lambda x \Big|_{t=a}^{t=b} - \int_a^b x\dot{\lambda}\, dt. \tag{B.66}$$

Then, substitution of this expression into Equation (B.64) yields

$$\bar{J} = \int_a^b [L(x, u, t) + \lambda(t) f(x, u, t) + x(t)\dot{\lambda}(t)]\, dt - \lambda(t) x(t)\Big|_{t=a}^{t=b}. \tag{B.67}$$

Now that the $\dot{x}$ term has been eliminated, the change in $\bar{J}$ due to variations δu and δx can be obtained as

$$\Delta\bar{J} = \int_a^b \left[\left(\frac{\partial L}{\partial u} + \lambda\frac{\partial f}{\partial u}\right)\delta u + \left(\frac{\partial L}{\partial x} + \lambda\frac{\partial f}{\partial x} + \dot{\lambda}\right)\delta x\right] dt - \lambda(b)\,\delta x(b) + \lambda(a)\,\delta x(a). \tag{B.68}$$

For this quantity to vanish for arbitrary values of δu and δx, the following conditions must be met:

$$\frac{\partial L}{\partial u} + \lambda \frac{\partial f}{\partial u} = 0$$

$$\frac{\partial L}{\partial x} + \lambda \frac{\partial f}{\partial x} = -\dot{\lambda}$$

$$\lambda(b)\,\delta x(b) = 0$$

$$\lambda(a)\,\delta x(a) = 0. \tag{B.69}$$

The first two equations constitute the Euler–Lagrange equations for this problem, while the remaining two determine boundary conditions for the problem, as required. Note that these latter equations may or may not yield useful information, depending upon the boundary conditions specified for x. The first equation is an algebraic equation which can be solved for u in terms of x. Substituting this result into the second equation yields a first-order differential equation involving λ and x. Boundary conditions are supplied by the problem statement and the latter two expressions of Equation (B.69), as required.

As an example, consider the following problem. It is desired to minimize the functional

$$J = \int_0^1 (u^2 + x)\, dt$$

subject to the constraint

$$\dot{x} = u$$

and the boundary conditions

$$x(0) = 2$$

$$x(1) = 3.$$

An immediate solution can be obtained by noting that substitution of the constraint equation into the functional yields the example previously solved. Here, solution as a constrained problem will be accomplished. The necessary conditions from the Euler–Lagrange equations of Equation (B.69) are

$$\frac{\partial L}{\partial u} + \lambda \frac{\partial f}{\partial u} = 2u + \lambda = 0$$

and

$$\frac{\partial L}{\partial x} + \lambda \frac{\partial f}{\partial x} = -\dot{\lambda} = 1.$$

After substitution from the first equation above, the constraint equation is

$$\dot{x} = u = -\frac{\lambda}{2}.$$

This equation and the second equation from above constitute a set of two simultaneous, first-order differential equations which, when combined with the prescribed boundary conditions, yield the solution

$$x(t) = \tfrac{1}{4}t^2 + \tfrac{3}{4}t + 2.$$

This result agrees with that obtained previously. If the boundary conditions are changed so that $x(1)$ is free, then $\delta x(1)$ is not necessarily zero and the third expression of Equation (B.69) yields $\lambda(1) = 0$ since $\delta x(1)$ is arbitrary. This new boundary condition modifies the previous solution to yield

$$x(t) = \tfrac{1}{4}t^2 - \tfrac{1}{2}t + 2$$

which also agrees with a previous result.

Note that problems such as this characteristically yield a set of simultaneous differential equations with split boundary conditions. That is, one boundary condition will be known at the initial time and one at the final time. When the differential equations are linear, as was the case in the previous example, this type of problem can usually be solved. The situation involving nonlinear equations presents a severe computational handicap since, not only does the nonlinear character of the equations usually require numerical solutions, but iterative techniques are usually required in order to satisfy the boundary conditions.

B.2.2. The Variational Problem in More than One Dimension

Generalization of the previously derived principles to more than one dimension is most easily accomplished using vector–matrix notation. The unconstrained problem requires optimization of a functional of the following form

$$J = \int_a^b L(\boldsymbol{x}(t), \dot{\boldsymbol{x}}(t), t)\, dt \tag{B.70}$$

where L is a scalar function of the vectors

$$\boldsymbol{x}(t) = \begin{bmatrix} x_1(t) \\ x_2(t) \\ \vdots \\ x_N(t) \end{bmatrix} \quad \text{and} \quad \dot{\boldsymbol{x}}(t) = \begin{bmatrix} \dot{x}_1(t) \\ \dot{x}_2(t) \\ \vdots \\ \dot{x}_N(t) \end{bmatrix}.$$

That is,

$$L(\boldsymbol{x}, \dot{\boldsymbol{x}}, t) = L(x_1, x_2, \ldots, x_N, \dot{x}_1, \dot{x}_2, \ldots, \dot{x}_N, t).$$

Now the vector variation $\delta\boldsymbol{x}(t)$ must be considered, together with the resulting vector variation $\delta\dot{\boldsymbol{x}}(t)$. The resulting change in the functional J is

$$\Delta J = \int_a^b \left[\frac{\partial L}{\partial \boldsymbol{x}}\,\delta\boldsymbol{x} - \frac{\partial L}{\partial \dot{\boldsymbol{x}}}\,\delta\dot{\boldsymbol{x}}\right] dt \tag{B.71}$$

where the vectors $\partial L/\partial \boldsymbol{x}$ and $\partial L/\partial \dot{\boldsymbol{x}}$ are defined as the row vectors

$$\frac{\partial L}{\partial \boldsymbol{x}} = \left[\frac{\partial L}{\partial x_1}\;\frac{\partial L}{\partial x_2}\;\cdots\;\frac{\partial L}{\partial x_N}\right]$$
$$\frac{\partial L}{\partial \dot{\boldsymbol{x}}} = \left[\frac{\partial L}{\partial \dot{x}_1}\;\frac{\partial L}{\partial \dot{x}_2}\;\cdots\;\frac{\partial L}{\partial \dot{x}_N}\right]. \tag{B.72}$$

The integration by parts of the second term of the integral of Equation (B.71) proceeds in a manner similar to that used previously:

$$\int_a^b \frac{\partial L}{\partial \dot{\boldsymbol{x}}}\,\delta\dot{\boldsymbol{x}}\,dt = \int_a^b \frac{\partial L}{\partial \dot{\boldsymbol{x}}}\,\frac{d(\delta\boldsymbol{x})}{dt}\,dt = \int_{t=a}^{t=b} \frac{\partial L}{\partial \dot{\boldsymbol{x}}}\,d(\delta\boldsymbol{x}). \tag{B.73}$$

Integration of this expression yields

$$\frac{\partial L}{\partial \dot{\boldsymbol{x}}}\,\delta\boldsymbol{x}\Bigg|_{t=a}^{t=b} - \int_a^b \frac{d}{dt}\left[\frac{\partial L}{\partial \dot{\boldsymbol{x}}}\right]\delta\boldsymbol{x}\,dt.$$

Then, substitution of this expression into Equation (B.71) yields

$$\Delta J = \int_a^b \left\{\frac{\partial L}{\partial \boldsymbol{x}} - \frac{d}{dt}\left[\frac{\partial L}{\partial \dot{\boldsymbol{x}}}\right]\right\}\delta\boldsymbol{x}\,dt + \frac{\partial L}{\partial \dot{\boldsymbol{x}}}\,\delta\boldsymbol{x}\Bigg|_{t=a}^{t=b}. \tag{B.74}$$

These equations are just the multidimensional generalization of Equation (B.58), resulting in the set of Euler–Lagrange equations given in vector–matrix form as

$$\frac{\partial L}{\partial \boldsymbol{x}} - \frac{d}{dt}\left[\frac{\partial L}{\partial \dot{\boldsymbol{x}}}\right] = 0$$

with boundary conditions

$$\frac{\partial L}{\partial \dot{\boldsymbol{x}}}\,\delta\boldsymbol{x}\Bigg|_{t=a}^{t=b} = 0. \tag{B.75}$$

The first of the above vector equations is actually the set of scalar equations

$$\frac{\partial L}{\partial x_1} - \frac{d}{dt}\left[\frac{\partial L}{\partial \dot{x}_1}\right] = 0$$

$$\frac{\partial L}{\partial x_2} - \frac{d}{dt}\left[\frac{\partial L}{\partial \dot{x}_2}\right] = 0$$

$$\vdots$$

$$\frac{\partial L}{\partial x_N} - \frac{d}{dt}\left[\frac{\partial L}{\partial \dot{x}_N}\right] = 0 \tag{B.76}$$

which must be solved simultaneously. Similarly, the vector equation describing the boundary conditions is actually the set of scalar equations

$$\left.\frac{\partial L}{\partial \dot{x}_1}\,\delta x_1\right|_{t=a}^{t=b} = 0$$

$$\left.\frac{\partial L}{\partial \dot{x}_2}\,\delta x_2\right|_{t=a}^{t=b} = 0$$

$$\vdots$$

$$\left.\frac{\partial L}{\partial \dot{x}_N}\,\delta x_N\right|_{t=a}^{t=b} = 0. \tag{B.77}$$

Note how the use of vector-matrix notation greatly reduces the effort required to express a set of equations.

The multidimensional constrained problem is approached in a similar manner. The problem involves optimization of the functional

$$J = \int_a^b L(\boldsymbol{x}(t), \boldsymbol{u}(t), t)\, dt \tag{B.78}$$

subject to the set of constraint equations

$$\dot{\boldsymbol{x}} = \boldsymbol{f}(\boldsymbol{x}, \boldsymbol{u}, t) \tag{B.79}$$

where $\boldsymbol{f}(\boldsymbol{x}, \boldsymbol{u}, t)$ is an $N \times 1$ vector function of the vectors $\boldsymbol{x}$ and $\boldsymbol{u}$, and of t. In this instance there are N constraint equations, so that N Lagrange multiplier functions $\lambda_1(t), \lambda_2(t), \ldots, \lambda_N(t)$ are required. These are represented by the vector function

$$\boldsymbol{\lambda}(t) = \begin{bmatrix} \lambda_1(t) \\ \lambda_2(t) \\ \vdots \\ \lambda_N(t) \end{bmatrix}. \tag{B.80}$$

Then, the modified functional is formed as

$$\bar{J} = \int_a^b \{L(\boldsymbol{x}, \boldsymbol{u}, t) + \boldsymbol{\lambda}^T(t)[\boldsymbol{f}(\boldsymbol{x}, \boldsymbol{u}, t) - \dot{\boldsymbol{x}}]\}\, dt. \tag{B.81}$$

By the same reasoning applied in the scalar case, if $\bar{J}$ is optimized and the constraint equations simultaneously satisfied, then J is also optimized. As before, the third term of the integral is considered separately and integrated by parts,

$$\int_a^b \boldsymbol{\lambda}^T \dot{\boldsymbol{x}}\, dt = \int_a^b \dot{\boldsymbol{\lambda}}^T \frac{d\boldsymbol{x}}{dt}\, dt = \int_{t=a}^{t=b} \dot{\boldsymbol{\lambda}}^T\, d\boldsymbol{x}. \tag{B.82}$$

Integrating this term by parts yields

$$\int_a^b \boldsymbol{\lambda}^T \dot{\boldsymbol{x}}\, dt = \boldsymbol{\lambda}^T \boldsymbol{x}\Big|_{t=a}^{t=b} - \int_a^b \dot{\boldsymbol{\lambda}}^T \boldsymbol{x}\, dt \tag{B.83}$$

and substituting this expression into Equation (B.81) yields, for $\bar{J}$,

$$\bar{J} = \int [L(\boldsymbol{x}, \boldsymbol{u}, t) + \boldsymbol{\lambda}^T(t)\boldsymbol{f}(\boldsymbol{x}, \boldsymbol{u}, t) + \dot{\boldsymbol{\lambda}}^T(t)\boldsymbol{x}(t)]\, dt - \boldsymbol{\lambda}^T(t)\boldsymbol{x}(t)\Big|_{t=a}^{t=b}. \tag{B.84}$$

Then, the change in $\bar{J}$ due to variations $\delta\boldsymbol{x}$ and $\delta\boldsymbol{u}$ is

$$\Delta\bar{J} = \int_a^b \left[\left(\frac{\partial L}{\partial \boldsymbol{u}} + \lambda^T \frac{\partial \boldsymbol{f}}{\partial \boldsymbol{u}}\right)\delta\boldsymbol{u} + \left(\frac{\partial L}{\partial \boldsymbol{x}} + \boldsymbol{\lambda}^T \frac{\partial L}{\partial \boldsymbol{x}} + \dot{\boldsymbol{\lambda}}^T\right)\delta\boldsymbol{x}\right] dt$$
$$- \boldsymbol{\lambda}^T(b)\,\delta\boldsymbol{x}(b) + \boldsymbol{\lambda}^T(a)\,\delta\boldsymbol{x}(a) \tag{B.85}$$

where the matrix $\partial\boldsymbol{f}/\partial\boldsymbol{u}$ is

$$\frac{\partial \boldsymbol{f}}{\partial \boldsymbol{u}} = \begin{bmatrix} \frac{\partial f_1}{\partial u_1} & \frac{\partial f_1}{\partial u_2} & \cdots & \frac{\partial f_1}{\partial u_P} \\ \frac{\partial f_2}{\partial u_1} & \frac{\partial f_2}{\partial u_2} & \cdots & \frac{\partial f_2}{\partial u_P} \\ \cdots & \cdots & \cdots & \cdots \\ \frac{\partial f_N}{\partial u_1} & \frac{\partial f_N}{\partial u_2} & \cdots & \frac{\partial f_N}{\partial u_P} \end{bmatrix} \tag{B.86}$$

and the row vectors $\partial L/\partial\boldsymbol{x}$ and $\partial L/\partial\boldsymbol{u}$ are defined as in Equation (B.72).

The Euler–Lagrange equations resulting from equating $\Delta\bar{J}$ in Equation (B.85) to zero are

$$\frac{\partial L}{\partial \boldsymbol{u}} + \boldsymbol{\lambda}^T \frac{\partial f}{\partial \boldsymbol{u}} = 0$$

$$\frac{\partial L}{\partial \boldsymbol{x}} + \boldsymbol{\lambda}^T \frac{\partial f}{\partial \boldsymbol{x}} = -\dot{\boldsymbol{\lambda}}^T \tag{B.87}$$

and the boundary conditions are

$$\boldsymbol{\lambda}^T(b)\,\delta\boldsymbol{x}(b) = 0$$

$$\boldsymbol{\lambda}^T(a)\,\delta\boldsymbol{x}(a) = 0. \tag{B.88}$$

Again, these equations are just the multidimensional generalizations of Equation (B.69). The first expression of Equation (B.87) is an algebraic equation which can be used to eliminate $\boldsymbol{u}$ from the set of $2N$ simultaneous, first-order differential equations formed by the constraint equation and the second expression of Equation (B.87). That is

$$\dot{\boldsymbol{x}} = \boldsymbol{f}(\boldsymbol{x}, \boldsymbol{u}, t)$$

$$\dot{\boldsymbol{\lambda}} = -\left[\frac{\partial L}{\partial \boldsymbol{x}}\right]^T - \left[\frac{\partial f}{\partial \boldsymbol{x}}\right]^T \boldsymbol{\lambda} \tag{B.89}$$

constitute a set of $2N$ simultaneous, differential equations in $\boldsymbol{\lambda}$ and $\boldsymbol{x}$, from which $\boldsymbol{u}$ can be eliminated by use of the algebraic equation

$$\frac{\partial L}{\partial \boldsymbol{u}} + \boldsymbol{\lambda}^T \frac{\partial f}{\partial \boldsymbol{u}} = 0.$$

Boundary conditions for these equations are either specified by the problem statement as initial and/or final conditions on $\boldsymbol{x}(t)$, or from the $\boldsymbol{\lambda}(t)$ boundary conditions specified by Equation (B.88). In either case, a split boundary condition problem results.

The general usefulness of the calculus of variations approach to optimization problems is limited by the severe computational difficulty imposed by the split boundary conditions. In those cases in which linear equations occur as a result of application of Equation (B.89), however, analytical solutions can be obtained and the approach can be quite useful. The general approach also serves to build a solid foundation upon which special techniques are based.

REFERENCES

1. Beveridge, Gordon S. G., and Schechter, Robert S. *Optimization: Theory and Practice.* New York: McGraw-Hill, 1970.
2. Gue, Ronald L., and Thomas, Michael E. *Mathematical Methods in Operations Research.* New York: Macmillan, 1968.

3. Schultz, Donald G., and Melsa, James L. *State Functions and Linear Control Systems.* New York: McGraw-Hill, 1967.

DEVELOPMENTAL EXERCISES

B.1. Show that the multivariable Taylor series expansion through the second-order terms is given by Equation (B.9).

B.2. Investigate the form of the third-order terms of the Taylor series expansion of Equation (B.9).

B.3. Develop a general approach to the constrained optimization problem described by Equation (B.13) using direct substitution.

B.4. Interpret the implication of Equation (B.16).

B.5. Prove the assertions of Table B-1.

B.6. Use a graphical approach to the following problem:

$$\begin{aligned} \text{Minimize} \quad & 4x_1 - x_2 \\ \text{Subject to} \quad & -x_1 + x_2 \leqslant 2 \\ & x_1 + x_2 \leqslant 4 \\ & x_1 \geqslant 1 \\ & x_1 - 2x_2 \leqslant 2. \end{aligned}$$

PROBLEMS

B.1. The relationship between battery energy (E), power (P), time (T), speed (S), auxilliary power (A) and distance (D), for an electrically powered submarine is given by the expressions

$E = PT$ — E is constant for a given battery

$P = KS^{\lambda} + A$ — K and λ are constants characterizing the submarine.

$D = ST.$

Find the expression for the maximum distance in terms of E, K, A, and λ.

B.2. Consider the function

$$f(x_1, x_2, x_3) = x_1^2 + x_2^2 + x_3^2 - 4x_1 - 8x_2 - 12x_3 + 56.$$

Investigate this function for extreme points and identify them as maxima or minima.

B.3. Find the maximum of the function

$$f(x_1, x_2) = 1/[(x_1 - 1)^2 + (x_2 - 1)^2 + 1].$$

Sketch the function to verify your answer.

B.4. Find the dimensions of the cylindrical container which has the maximum volume for a fixed surface area.

B.5. A rectangle is to be inscribed in an ellipse. Find the particular rectangle with the maximum perimeter. The equation of an ellipse is

$$(x_1^2/a^2) + (x_2^2/b^2) = 1.$$

B.6. Use the Lagrange multiplier method to find the maximum value of the function

$$f(x) = x^2 - x + 1$$

subject to the constraint

$$g(x) = x^2 - 4 \leqslant 0.$$

B.7. Use the Lagrange multiplier method to find the maximum of the function

$$f(\mathbf{x}) = 10x^3 - 4x^2 + 3x - 1$$

subject to the constraint

$$-3 \leqslant x \leqslant 11.$$

B.8. Investigate the stationary points of the function

$$f(x_1, x_2) = 3x_1^2 + x_2^3 + 3x_1x_2 + 3x_2^2.$$

B.9. Investigate the stationary points of the function

$$f(x_1, x_2) = x_1^3 - 3x_1x_2^2.$$

B.10. Investigate the stationary points of the function

$$f(x_1, x_2, x_3) = 3x_1^2 + 4x_1 + x_2^2 + x_1x_2 + 2x_3 + 2x_3^2 + x_2x_3.$$

B.11. What should be the proportions of a cone-roofed tank to obtain the greatest volume of storage for a fixed surface area? The diameter and height of the lower cylindrical section are $2R$ and I, respectively, and the height of the upper conical section is S. (The constraint here is fixed surface area A).

B.12. Find the global minimum and maximum of the function

$$f(x_1, x_2) = x_2 - x_1^2$$

if it is subject to the restriction that

$$g(x_1, x_2) = 1 - x_1^2 - x_2^2 = 0.$$

Use the Lagrange multiplier method, and then graph the constraint and various constant $f(x)$ contours. Correlate the analytical and graphical results.

B.13. Find the maximum and minimum distances from the origin to the curve

$$5x_1^2 + 6x_1x_2 + 5x_2^2 = 8.$$

Use the Lagrange multiplier method and verify your result by a graphical solution.

B.14. Find the point nearest the origin of a three-dimensional coordinate system which satisfies the constraint equation

$$x_1 + x_2 + x_3 = 5$$
$$x_1^2 + x_2^2 + x_3^2 = 9.$$

B.15. Find the values of x_1 and x_2 which yield a minimum for the function

$$f(x_1, x_2) = (x_1^2/a^2) + (x_2^2/b^2)$$

subject to the constraint

$$g(x_1, x_2) = x_1 + mx_2 - c = 0.$$

Interpret your result in a geometric sense.

B.16. Find the point nearest to the origin in the (x, y, z) rectangular coordinate system lying on the line which is the intersection of the planes described by the equations

$$2x + y + 3z = 10, \qquad 3x - y + 2z = 1.$$

This point is found by minimizing the function

$$f(x, y, z) = x^2 + y^2 + z^2$$

subject to the constraints above.

B.17. Find the minimum of the function

$$f(x_1, x_2) = x_1^4 + 3x_2^2 + 2x_2^4.$$

B.18. Find the minima and maxima of the function

$$f(x_1, x_2) = x_2 - x_1^2$$

subject to the constraint

$$g(x_1, x_2) = 1 - x_1^2 - x_2^2 = 0.$$

B.19. Find the minima and maxima of the function

$$f(x_1, x_2) = 4x_1^2 + 5x_2^2$$

subject to the constraint

$$g(x_1, x_2) = 2x_1 + 3x_2 - 6 = 0.$$

B.20. Find the dimensions of a closed cylindrical tank of fixed volume V which has minimum surface area A.

B.21. Find the minimum of the function

$$f(x_1, x_2) = 4x_1^2 + 5x_2^2$$

subject to the constraint $x_1 \geqslant 1$.

B.22. Find the maximum of the function

$$f(x_1, x_2) = x_2^2 - 2x_1 - x_1^2$$

subject to the constraint

$$x_1^2 + x_2^2 - 1 \leqslant 0.$$

B.23. Solve the following problem using the Kuhn–Tucker conditions. (Check by linear programming if you wish.) Check for sufficiency if possible.

$$\begin{array}{ll} \text{Maximize} & f(x_1, x_2) = 2x_1 - x_2 \\ \text{Subject to} & -3x_1 + 2x_2 \leqslant 2 \\ & -2x_1 - 4x_2 \leqslant 3 \\ & x_1 + x_2 \leqslant 6 \\ & x_1, x_2 \geqslant 0. \end{array}$$

B.24. Find the values of the x_1 and x_2 which minimize the function

$$f(x_1, x_2) = (x_1^2/a^2) + (x_2^2/b^2)$$

subject to the constraint

$$x_1 x_2 = c$$

where a, b, and c are positive constants. Interpret your results geometrically.

B.25. Find the minimum of the function:

$$f(x_1, x_2) = 4x_1^2 - 2x_1 + 3x_1 x_2 + 5x_2^2 - 4x_2$$

subject to the constraint

$$x_1 + x_2 \geqslant 1.$$

Check your result using the Kuhn–Tucker conditions.

B.26. Find the possible minima of the function:

$$f(x_1, x_2) = x_1 + 3x_2 - 6x_3^2$$

subject to the constraint

$$x_1^2 + x_2^2 + x_3^2 \leqslant 1.$$

B.27. Find the maximum of the function

$$f(x_1, x_2) = 10x_1^2 - 4x_1x_2 + 3x_2^2 + 5x_2x_3$$

subject to the constraints

$$\begin{aligned} x_1 + 2x_2 \qquad &\leqslant 3 \\ x_2 - x_3 &\geqslant 2 \\ x_1 \qquad\qquad &\geqslant 1 \end{aligned}$$

Check your solution using the Kuhn–Tucker conditions.

B.28. Consider the following function:

$$f(x_1, x_2) = 2x_1 + 4x_2 + x_1^2 + 3x_2^2 + 3x_1x_2$$

which is to minimized subject to the constraints

$$\begin{aligned} -x_1 + 4x_2 - 4 &\leqslant 0 \\ 6x_1 + 2x_2 - 1 &\leqslant 0 \\ x_1 - x_2 - 3 &\leqslant 0 \\ x_1 &\geqslant 0 \\ x_2 &\geqslant 0. \end{aligned}$$

Show that the solution is $x_1 = x_2 = 0$.

B.29. Investigate the minima and maxima of the function

$$f(x_1, x_2) = 4x_1^2 + 5x_2^2$$

subject to the constraint

$$g(x_1, x_2) = 2x_1 + 3x_2 - 6 = 0.$$

(a) Use the Lagrange multiplier method and investigate both necessary and sufficient conditions.
(b) Use direct substitution to solve this problem.

B.30. Consider the function

$$f(x_1, x_2, x_3) = x_1^2 + x_2^2 + x_3^2 - 4x_1 - 8x_1 - 12x_3 + 56.$$

Find the minimum of this function, subject to the constraints

$$x_1 \leqslant 3$$
$$x_2 \leqslant 5$$
$$x_3 \leqslant 10.$$

Show that your answer satisfies the Kuhn–Tucker conditions.

B.31. Find the function $x(t)$ which minimizes the functional

$$J = \int_0^1 [x^2(t) + 2tx + \dot{x}^2(t)]\, dt, \qquad x(0) = 2.$$

B.32. Find the function $u(t)$ which will minimize the functional

$$J = \int_1^2 u^2(t)\, dt$$

subject to the constraint

$$\dot{x}(t) = x(t) + u(t).$$

The boundary conditions are $x(1) = 1$, $x(2) = 0$.

B.33. Find the function $x(t)$ which minimizes the functional

$$J = \int_0^{\pi/2} [\dot{x}^2(t) - x^2(t)]\, dt.$$

Can you prove that your result is a minimum?

B.34. Find the function $u(t)$ which minimizes the functional

$$J = \int_0^1 [x^2(t) + u^2(t)]\, dt$$

subject to the constraint

$$\dot{x}(t) = u(t) - 4x(t), \qquad x(0) = 1.$$

Also find $x(t)$ and J.

B.35. Find the function $u(t)$ which minimizes the functional

$$J = \int_0^1 [x^2(t) + u^2(t)]\, dt$$

subject to the constraint

$$\dot{x}(t) = u(t) - x(t), \qquad x(0) = 1, \qquad x(1) = 0.$$

B.36. Find the function $x(t)$ which minimizes the functional

$$J = \int_0^2 [\tfrac{1}{2}\dot{x}^2(t) + x(t)\dot{x}(t) + 2x^2(t)]\, dt, \qquad x(0) = 1.$$

B.37. Find the function $x(t)$ and the final time t_f which minimizes the functional

$$J = \int_{t_0}^{t_f} [4x(t) + \dot{x}^2(t)]\, dt$$

for $t_0 = 1$, where $x(t_0) = 4$, $x(t_f) = 4$.

B.38. Find the functions $x_1(t)$ and $x_2(t)$ which minimize the functional

$$J = \int_0^{\pi/4} [\tfrac{1}{2}x_1^2(t) + 2x_2^2(t) + \tfrac{1}{2}\dot{x}_1(t)\dot{x}_2(t)]\, dt$$

with initial conditions $x_1(0) = 0$, $x_2(0) = 1$ and terminal conditions $x_1(\pi/4) = 1$, $x_2(\pi/4) = 0$.

B.39. Find the functions $x_1(t)$ and $x_2(t)$ which minimize the functional

$$J = \int_0^{\pi/4} [\tfrac{1}{2}x_1^2(t) + \tfrac{1}{2}\dot{x}_1(t)\dot{x}_2(t) + \tfrac{1}{2}\dot{x}_2^2(t)]\, dt$$

with initial conditions $x_1(0) = 1$, $x_2(0) = \frac{3}{2}$ and terminal condition $x_1(\pi/4) = 2$.

B.40. Consider the functional

$$J = \frac{1}{2} \int_0^{t_f} [\dot{x}^2(t) + t\dot{x}(t)] \, dt.$$

Find the function $x(t)$ which minimizes this functional under the following conditions:
(a) $t_f = 3$; $x(0) = 1$; $x(2) = 2$.
(b) $t_f = 5$; $x(0) = 1$; $x(2)$ free.
(c) t_f free; $x(0) = 1$; $x(t_f) = 2$.

B.41. Find the functions $x_1(t)$ and $x_2(t)$ which minimize the functional

$$J = \frac{1}{2} \int_{t_0}^{t_f} u^2(t) \, dt$$

subject to the constraint equations

$$\dot{x}_1(t) = x_2(t)$$
$$\dot{x}_2(t) = -x_2(t) + u(t).$$

To evaluate the arbitrary constants use the following conditions.

$$t_0 = 0; \quad t_f = 2; \quad x(0) = \begin{bmatrix} 0 \\ 0 \end{bmatrix}; \quad x(2) = \begin{bmatrix} 2 \\ 0 \end{bmatrix}.$$

Appendix C
The Solution of Linear Equations

C.1. THE CANONICAL FORMS FOR *N*TH-ORDER LINEAR DIFFERENTIAL EQUATIONS

There exists a general equivalency between an Nth order linear differential equation and a set of N coupled, first-order, linear differential equations. That is, any equation in the form

$$a_N(t)\,x^{(N)}(t) + a_{N-1}(t)\,x^{(N-1)}(t) + \cdots + a_1(t)\,\dot{x}(t) + a_0(t)\,x(t) = 0 \quad \text{(C.1)}$$

can also be expressed as a set of first order equations

$$\begin{aligned}
\dot{x}_1(t) &= f_{11}(t)\,x_1(t) + f_{12}(t)\,x_2(t) + \cdots + f_{1N}(t)\,x_N(t) \\
\dot{x}_2(t) &= f_{21}(t)\,x_1(t) + f_{22}(t)\,x_2(t) + \cdots + f_{2N}(t)\,x_N(t) \\
&\vdots \\
\dot{x}_N(t) &= f_{N1}(t)\,x_1(t) + f_{N2}(t)\,x_2(t) + \cdots + f_{NN}(t)\,x_N(t)
\end{aligned} \quad \text{(C.2)}$$

or, in the more convenient matrix form,

$$\dot{\boldsymbol{x}}(t) = \boldsymbol{F}(t)\,\boldsymbol{x}(t). \quad \text{(C.3)}$$

It can be shown that this matrix has a solution in terms of an associated matrix called the *transition matrix*, which is of paramount importance to the generalized solution of linear differential equations.

C.2. THE TRANSITION MATRIX

A completely rigorous development of the transition matrix is beyond the scope of the present discussion, but a heuristic argument for its existence can be provided as follows. The scalar equivalent of Equation (C.3) is

$$\dot{x} = f(t)\,x(t). \quad \text{(C.4)}$$

Solving this equation by the separation of variables yields the well known solution

$$x(t) = x(t_0)\, e^{\int_{t_0}^{t} f(\sigma)\, d\sigma} \tag{C.5}$$

or, when $f(\sigma) = f$, a constant

$$x(t) = x(t_0)\, e^{f(t - t_0)}. \tag{C.6}$$

Note the general form of Equation (C.5):

$$x(t) = \phi(t, t_0)\, x(t_0) \tag{C.7}$$

expressing $x(t)$ as a linear function of $x(t_0)$. It is reasonable to assume that the multivariate equation as presented in Equation (C.3) will have the same general form of solution, but modified to the extent that one component of the vector function $\boldsymbol{x}(t)$ will be a linear combination of all the components evaluated at t_0. That is,

$$\begin{aligned}
x_1(t) &= \phi_{11}(t, t_0)x_1(t_0) + \phi_{12}(t, t_0)x_2(t_0) + \cdots + \phi_{1N}(t, t_0)x_N(t_0) \\
x_2(t) &= \phi_{21}(t, t_0)x_1(t_0) + \phi_{22}(t, t_0)x_2(t_0) + \cdots + \phi_{2N}(t, t_0)x_N(t_0) \\
&\vdots \\
x_N(t) &= \phi_{N1}(t, t_0)x_1(t_0) + \phi_{N2}(t, t_0)x_2(t_0) + \cdots + \phi_{NN}(t, t_0)x_N(t_0)
\end{aligned} \tag{C.8}$$

or, in more concise matrix form

$$\boldsymbol{x}(t) = \boldsymbol{\Phi}(t, t_0)\boldsymbol{x}(t_0). \tag{C.9}$$

If this expression is to be a solution to Equation (C.3), then the requirements on $\boldsymbol{\Phi}(t, t_0)$ can be inferred by direct substitution. Thus, by differentiating Equation (C.9), one obtains

$$\dot{\boldsymbol{x}}(t) = \frac{d}{dt}\boldsymbol{\Phi}(t, t_0)\boldsymbol{x}(t_0)$$

which, upon substitution into Equation (C.3) yields

$$\frac{d}{dt}\boldsymbol{\Phi}(t, t_0)\boldsymbol{x}(t_0) = F(t)\boldsymbol{\Phi}(t, t_0)\boldsymbol{x}(t_0). \tag{C.10}$$

This equation is satisfied if

$$\frac{d}{dt}\boldsymbol{\Phi}(t, t_0) = F(t)\boldsymbol{\Phi}(t, t_0) \tag{C.11}$$

which establishes the requirement placed on $\mathbf{\Phi}(t, t_0)$ in order that Equation (C.9) represent the desired solution. Equation (C.11) is a matrix differential equation comprising a set of N^2 first order, linear differential equations.

As an example, consider the equation of the spring–mass–damper system introduced in Section 4.1 and described by the equation

$$2\ddot{x} + 4\dot{x} + 10x = 0.$$

In order to reduce this second-order equation to a set of two first-order equations, let

$$x_1 = x$$
$$x_2 = \dot{x}$$

which, upon substitution, yields

$$\dot{x}_1 = x_2$$
$$\dot{x}_2 = -5x_1 - 2x_2.$$

In matrix form, these equations are

$$\begin{bmatrix} \dot{x}_1(t) \\ \dot{x}_2(t) \end{bmatrix} = \begin{bmatrix} 0 & 1 \\ -5 & -2 \end{bmatrix} \begin{bmatrix} x_1(t) \\ x_2(t) \end{bmatrix}$$

which corresponds to Equation (C.3). This result implies that a solution of the form

$$\begin{bmatrix} x_1(t) \\ x_2(t) \end{bmatrix} = \begin{bmatrix} \phi_{11}(t, t_0) & \phi_{12}(t, t_0) \\ \phi_{21}(t, t_0) & \phi_{22}(t, t_0) \end{bmatrix} \begin{bmatrix} x_1(t_0) \\ x_2(t_0) \end{bmatrix}$$

exists. To prove this assertion, the original equation can be solved by the standard techniques to yield

$$\begin{aligned} x(t) &= e^{-(t-t_0)}[\cos 2(t - t_0) + \tfrac{1}{2}\sin 2(t - t_0)]\, x(t_0) \\ &\quad + e^{-(t-t_0)}[\tfrac{1}{2} \sin 2(t - t_0)]\, \dot{x}(t_0) \\ \dot{x}(t) &= e^{-(t-t_0)}[-\tfrac{5}{2} \sin 2(t - t_0)]\, x(t_0) \\ &\quad + e^{-(t-t_0)}[-\tfrac{1}{2} \sin 2(t - t_0) + \cos 2(t - t_0)]\, \dot{x}(t_0). \end{aligned}$$

The second equation is obtained by differentiating the first. Now, making the substitutions for x and $\dot{x}$ and writing the above expression in matrix form yield

$$\begin{bmatrix} x_1(t) \\ x_2(t) \end{bmatrix} = \begin{bmatrix} e^{-(t-t_0)}[\cos 2(t-t_0) + \frac{1}{2}\sin 2(t-t_0)] & e^{-(t-t_0)}[\frac{1}{2}\sin 2(t-t_0) \\ e^{-(t-t_0)}[-\frac{5}{2}\sin 2(t-t_0)] & e^{-(t-t_0)}[-\frac{1}{2}\sin 2(t-t_0) + \cos 2(t-t_0)] \end{bmatrix} \begin{bmatrix} x_1(t_0) \\ x_2(t_0) \end{bmatrix}$$

which is the form of Equation (C.9). The transition matrix $\boldsymbol{\Phi}(t, t_0)$ is immediately identified, and the reader can verify that this matrix satisfies Equation (C.11). In the particular case in which

$$t_0 = 0; \qquad x_1(0) = 2; \qquad x_2(0) = -4$$

for example, the above equation yields

$$x_1(t) = e^{-t}[2 \cos 2t - \sin 2t]$$
$$x_2(t) = e^{-t}[-4 \cos 2t - 3 \sin 2t]$$

which agrees with classical results.

C.3. METHODS OF DETERMINING THE TRANSITION MATRIX

The previous section has introduced the concept of the transition matrix and its relationship to the solution of a set of first-order, linear, differential equations. In many instances of theoretical development the existence of the transition matrix is the most important aspect, since it provides a closed-form expression for the solution to the set of equations. There may also be a requirement for the determination of the transition matrix, which is, of course, tantamount to solution of the equations. While there is no generally applicable method for determination of the transition matrix when $\boldsymbol{F}$ is a time varying matrix, there are quite general methods which can be applied when $\boldsymbol{F}$ is constant. This fact is most easily seen by considering a transformation of variable

$$\boldsymbol{x}(t) = \boldsymbol{P}\boldsymbol{z}(t) \tag{C.12}$$

where $\boldsymbol{P}$ is a transformation matrix composed of eigenvectors of $\boldsymbol{F}$. It is shown in Appendix A that such a transformation yields a diagonal matrix composed of the eigenvalues of $\boldsymbol{F}$, with the resulting solution for $\boldsymbol{z}(t)$ being

$$\boldsymbol{z}(t) = \begin{bmatrix} z_1(t_0)e^{\lambda_1(t-t_0)} \\ z_2(t_0)e^{\lambda_2(t-t_0)} \\ \vdots \\ z_N(t_0)e^{\lambda_N(t-t_0)} \end{bmatrix} = \begin{bmatrix} e^{\lambda_1(t-t_0)} & & & \\ & e^{\lambda_2(t-t_0)} & & \\ & & \ddots & \\ & & & e^{\lambda_N(t-t_0)} \end{bmatrix} \begin{bmatrix} z_1(t_0) \\ z_2(t_0) \\ \vdots \\ z_N(t_0) \end{bmatrix}.$$

Then, utilizing the transformation of Equation (C.12) in the form

$$z(t) = P^{-1}x(t)$$
$$z(t_0) = P^{-1}x(t_0)$$

this equation can be written as

$$P^{-1}x(t) = \begin{bmatrix} e^{\lambda_1(t-t_0)} & & & \\ & e^{\lambda_2(t-t_0)} & & \\ & & \ddots & \\ & & & e^{\lambda_N(t-t_0)} \end{bmatrix} P^{-1}x(t_0). \tag{C.13}$$

Or, if this expression is multiplied on the left by P, the result is

$$x(t) = P \begin{bmatrix} e^{\lambda_1(t-t_0)} & & & \\ & e^{\lambda_2(t-t_0)} & & \\ & & \ddots & \\ & & & e^{\lambda_N(t-t_0)} \end{bmatrix} P^{-1}x(t_0). \tag{C.14}$$

Comparing this expression with Equation (C.9) immediately identifies $\Phi(t, t_0)$ as

$$\Phi(t, t_0) = P \begin{bmatrix} e^{\lambda_1(t-t_0)} & & & \\ & e^{\lambda_2(t-t_0)} & & \\ & & \ddots & \\ & & & e^{\lambda_N(t-t_0)} \end{bmatrix} P^{-1}. \tag{C.15}$$

The combination of Equations (C.9) and (C.15) provides a systematic method of closed-form solution for a set of constant coefficient, first-order, homogeneous, linear differential equations. It involves no integration and no special techniques, and results from the fact that the general equation has already been solved in the form of Equation (C.9).

The form of Equation (C.14) implies that the time response of the set of homogeneous equations given by Equation (C.2), or equivalently by Equation (C.3), is a linear combination of exponential terms, each of which involves eigenvalues of the matrix F. If F cannot actually be reduced to a diagonal form, as often occurs, the resulting Jordan canonical form (see Appendix A) will contain some off-diagonal terms of value one. In any case, the eigenvalues with positive real parts imply an unstable mode of response, since infinite values of the components of the vector $x(t)$ result for sufficiently large t. Furthermore, the damping characteristics of each mode can be inferred from the magnitude of the real part of the eigenvalues producing that mode. Large-magnitude negative values produce rapid damping of the particular mode of response, while small-magnitude values imply lightly damped responses. In fact, it is entirely correct to view the reciprocal of the real part of the eigenvalue as the time constant of the associated mode of response.

If the exponential terms of Equation (C.15) are expanded in a Taylor series, then the equation can be expressed as

$$\boldsymbol{\Phi}(t, t_0) = \boldsymbol{PIP}^{-1} + \boldsymbol{PDP}^{-1} + \frac{1}{2!}\boldsymbol{PDP}^{-1}\boldsymbol{PDP}^{-1} + \cdots \tag{C.16}$$

where

$$\boldsymbol{D} = \begin{bmatrix} \lambda_1(t - t_0) & & & \\ & \lambda_2(t - t_0) & & \\ & & \ddots & \\ & & & \lambda_N(t - t_0) \end{bmatrix}.$$

From developments contained in Appendix A it is known that

$$\boldsymbol{F} = \boldsymbol{P}\begin{bmatrix} \lambda_1 & & & \\ & \lambda_2 & & \\ & & \ddots & \\ & & & \lambda_N \end{bmatrix}\boldsymbol{P}^{-1} \tag{C.17}$$

which, when substituted into Equation (C.16), yields

$$\boldsymbol{\Phi}(t, t_0) = \boldsymbol{I} + \boldsymbol{F}(t - t_0) + \frac{1}{2!}\boldsymbol{FF}(t - t_0)^2 + \frac{1}{3!}\boldsymbol{FFF}(t - t_0)^3 + \cdots \tag{C.18}$$

This matrix expansion is denoted as $e^{\boldsymbol{F}(t-t_0)}$ for obvious reasons, with the resulting formalism

$$\boldsymbol{\Phi}(t, t_0) = e^{\boldsymbol{F}(t-t_0)}. \tag{C.19}$$

This relationship is useful when the series can be truncated after a small number of terms to obtain an approximation for $\boldsymbol{\Phi}(t, t_0)$.

A less useful but more general result applicable to time varying equations can be obtained as

$$\boldsymbol{\Phi}(t, t_0) = e^{\int_{t_0}^{t_f} \boldsymbol{F}(\sigma)\, d\sigma} \tag{C.20}$$

where the integral of the matrix $\boldsymbol{F}(t)$ is to be interpreted as indicated in Appendix A. This result reduces to Equation (C.19) when $\boldsymbol{F}(t)$ is the constant matrix $\boldsymbol{F}$.

As a final note, attention is called to the result that, for constant $\boldsymbol{F}$, the transition matrix is a function of $(t - t_0)$ rather than t and t_0 independently. This conclusion is reasonable, especially in light of the implications of Equations (C.5) and (C.6), and is a general result for constant $\boldsymbol{F}$.

As an example of these concepts, consider the equations

$$\dot{x}_1 = -2x_2$$

$$\dot{x}_2 = x_1 - 3x_2$$

where

$$\boldsymbol{F} = \begin{bmatrix} 0 & -2 \\ 1 & -3 \end{bmatrix}.$$

The eigenvalues of $\boldsymbol{F}$ are

$$\lambda_1 = -1; \qquad \lambda_2 = -2$$

and the associated eigenvectors are

$$\boldsymbol{x}^1 = \begin{bmatrix} 1 \\ \frac{1}{2} \end{bmatrix}; \qquad \boldsymbol{x}^2 = \begin{bmatrix} 1 \\ 1 \end{bmatrix}$$

yielding

$$\boldsymbol{P} = \begin{bmatrix} 1 & 1 \\ \frac{1}{2} & 1 \end{bmatrix}; \qquad \boldsymbol{P}^{-1} = \begin{bmatrix} 2 & -2 \\ -1 & 2 \end{bmatrix}.$$

Then,

$$\boldsymbol{\Phi}(t, t_0) = \boldsymbol{P} \begin{bmatrix} e^{-(t-t_0)} & 0 \\ 0 & e^{-2(t-t_0)} \end{bmatrix} \boldsymbol{P}^{-1}$$

$$= \begin{bmatrix} [2e^{-(t-t_0)} - e^{-2(t-t_0)}] & [-2e^{-(t-t_0)} + 2e^{-2(t-t_0)}] \\ [-e^{-(t-t_0)} - e^{-2(t-t_0)}] & [-e^{-(t-t_0)} + 2e^{-2(t-t_0)}] \end{bmatrix}.$$

The solution for $\boldsymbol{x}(t)$ can be obtained, for any specified initial condition $\boldsymbol{x}(t_0)$, from the expression

$$\boldsymbol{x}(t) = \boldsymbol{\Phi}(t - t_0)\boldsymbol{x}(t_0).$$

This same result can be obtained by use of the expression of the matrix exponential $e^{F(t-t_0)}$ but one must be alert to recognize the resulting scalar series constituting the terms of the matrix as expansions of known forms.

In the event that $\boldsymbol{F}(t)$ is not time invariant, numerical techniques are usually required for evaluation of $\boldsymbol{\Phi}(t, t_0)$. These techniques are based upon the following special feature of linear differential equations. If Equation (C.3) is solved, using the particular initial condition

$$\boldsymbol{x}(t_0) = \begin{bmatrix} 1 \\ 0 \\ 0 \\ \vdots \\ 0 \end{bmatrix} \tag{C.21}$$

then the resulting solution can also be expressed in terms of Equation (C.9) as

$$\boldsymbol{x}(t) = \boldsymbol{\Phi}(t, t_0) \begin{bmatrix} 1 \\ 0 \\ 0 \\ \vdots \\ 0 \end{bmatrix} = \begin{bmatrix} \phi_{11}(t, t_0) \\ \phi_{21}(t, t_0) \\ \phi_{31}(t, t_0) \\ \vdots \\ \phi_{N1}(t, t_0) \end{bmatrix}. \tag{C.22}$$

The right-hand side of Equation (C.22) is the first column of the transition matrix $\boldsymbol{\Phi}(t, t_0)$. Similarly, the initial condition

$$\boldsymbol{x}(t_0) = \begin{bmatrix} 0 \\ 1 \\ 0 \\ \vdots \\ 0 \end{bmatrix} \tag{C.23}$$

produces, as the solution to Equation (C.3), the second column of the transition matrix, and so on. Thus, by solving Equation (C.3) N times with initial conditions of the special form of Equations (C.22) and (C.23), all N columns of the transition matrix can be determined. Although this technique is normally used when numerical solution of time varying equations is required, the procedure is general and can also be used in analytic solution for simple time varying and time invariant equations.

The numerical technique described above is also useful in the analysis of nonlinear systems by the perturbation method discussed in Chapter 4. If the general nonlinear equations are solved with initial conditions such as those specified in Equations (C.22) and (C.23), the resulting transition matrix describes the response of the perturbation variable $\delta \boldsymbol{x}(t)$ to changes in the initial condition $\delta \boldsymbol{x}(t_0)$. Selection of 1 as the initial perturbation normalizes the results very nicely, but care must be taken to ensure that the resulting perturbations do not violate the linearity assumptions.

DEVELOPMENTAL EXERCISES

C.1. Solve Equation (C.11) by use of the Laplace transform, for the case in which $\boldsymbol{\Phi}(t, t_0) = \boldsymbol{\Phi}(t - t_0)$.

C.2. Relate the expression of Equation (C.19) to the result of Exercise C.1.

C.3. The system of equations with state transition matrix $\boldsymbol{\Phi}^{-1}(t, t_0)$ is called *adjoint* of the system with state transition matrix $\boldsymbol{\Phi}(t, t_0)$.

(a) If the original system is described by the equation

$$\dot{\boldsymbol{x}}(t) = \boldsymbol{F}(t)\boldsymbol{x}(t)$$

show that the adjoint system is described by the equation

$$\dot{\boldsymbol{y}}(t) = -\boldsymbol{F}^T(t)\boldsymbol{y}(t).$$

(b) Show that

$$\dot{\boldsymbol{\Phi}}^{-1}(t, t_0) = -\boldsymbol{\Phi}^{-1}(t, t_0)\boldsymbol{F}(t).$$

(c) Show that the adjoint system can be used in numerical calculation, solving backward in time from appropriate terminal conditions, to yield the state transition matrix $\boldsymbol{\Phi}(t, t_0) = \boldsymbol{\Phi}^{-1}(t_0, t)$.

C.4. Apply the concept of the state transition matrix to difference equations. Take the z-transform of the resulting matrix.

PROBLEMS

C.1. Consider the time dependent system described by the equations

$$\dot{x}_1(t) = tx_1(t) + x_2$$

$$\dot{x}_2(t) = x_1 + tx_2(t).$$

Use the general form of the matrix exponential given by Equation (C.56) to determine the state transition matrix.

C.2. Find the state transition matrix for the following systems:

(a) $\dot{x}_1(t) = x_1(t) - 2x_2(t)$
$\dot{x}_2(t) = 4x_1(t) - 5x_2(t)$

(b) $\dot{x}_1(t) = 5x_2(t)$
$\dot{x}_2(t) = -4x_2(t)$

(c) $\dot{x}_1(t) = x_2(t)$
$\dot{x}_2(t) = -2x_1(t) - 3x_2(t)$

(d) $\dot{x}_1(t) = -4x_1(t) + 2x_2(t)$
$\dot{x}_2(t) = -2x_1(t) + x_2(t)$

C.3. For the second-order system matrix

$$F = \begin{bmatrix} 0 & -2 \\ 1 & -3 \end{bmatrix}$$

find the state transition matrix by use of the matrix expansion e^{Ft} and by identifying the resulting series expansions.

C.4. Consider the homogeneous equations

$$\dot{x}_1(t) = -x_1(t) + x_2(t)$$
$$\dot{x}_2(t) = -2x_2(t).$$

Use the matrix expansion e^{Ft} to determine the state transition matrix.

C.5. Consider the coupled, homogeneous, first-order, linear differential equations

$$\dot{x}_1(t) = -3x_1(t) - 2x_2(t) - x_3(t)$$
$$\dot{x}_2(t) = x_1(t) - x_3(t)$$
$$\dot{x}_3(t) = -4x_3(t).$$

Solve these equations by use of the state transition matrix.

C.6. Consider the homogeneous, third-order, linear differential equation

$$\dddot{x}(t) + 7\ddot{x}(t) + 14\dot{x}(t) + 8x(t) = 0.$$

Solve this equation by use of the state transition matrix.

C.7. Consider the homogeneous, third-order, linear differential equation

$$\dddot{x}(t) + 6\ddot{x}(t) + 11\dot{x}(t) + 6x(t) = 0.$$

Solve this equation by use of the state transition matrix.

C.8. Consider the coupled, homogeneous, first-order, linear differential equations

$$\dot{x}_1(t) = x_2(t)$$
$$\dot{x}_2(t) = x_3(t)$$
$$\dot{x}_3(t) = -6x_1(t) - 11x_2(t) - 6x_3(t).$$

Solve these equations by use of the state transition matrix.

C.9. Consider the coupled, homogeneous, first-order, linear differential equations

$$\dot{x}_1(t) = x_1(t) + x_2(t) + x_3(t)$$
$$\dot{x}_2(t) = -x_1(t) - 2x_2(t) - x_3(t)$$
$$\dot{x}_3(t) = x_1(t) + 4x_2(t) + 9x_3(t).$$

Solve these equations by use of the state transition matrix.

C.10. Consider the set of coupled, homogeneous, first-order, linear differential equations

$$\dot{x}_1(t) = 2x_1(t) + x_2(t) + x_3(t)$$
$$\dot{x}_2(t) = x_1(t) + 2x_2(t) + x_3(t)$$
$$\dot{x}_3(t) = x_1(t) + x_2(t) + 2x_3(t).$$

Use the matrix exponential $e^{\boldsymbol{F}t}$ to solve these equations.

Appendix D
A Comparison of Modern and Classical Control Theory

D.1. HISTORICAL PERSPECTIVE

Automatic control has developed from rather poorly organized principles in the mid-1930s to the well-defined systematic concepts presented in Chapters 4–7. While the analyst today has available the awesome computational power of the modern digital computer, early workers in the field did not enjoy this capability, and their development of such frequency domain techniques as the Bode, Nyquist, and root-locus methods to overcome this handicap constitutes inventive genius in its finest form. The emphasis which World War II placed on the design of efficient servomechanisms served as an impetus to development, and the state of the art progressed rapidly during that period.

The work on feedback controllers during this time concentrated on the single input–single output system, and the design criteria used were rather nebulous and arbitrary. Stability and response time were the primary measures of system performance. Since that time, the stress has been more on systems with specific performance criteria, in the context of adaptive or optimal systems. In addition, the requirement for controlling multiple input–multiple output systems became an important factor, as process control developed as an important application. The extension of frequency domain techniques to the optimal and multiple input–multiple output systems is possible and has been developed, but the advent of the modern digital computer has placed emphasis on the more computationally sensitive state space techniques known as modern control theory.

The classical methods are powerful, as evidenced by their success over the years, and are complemented rather than replaced by state space techniques in many applications. For this reason, the two methods are considered and compared in this appendix.

D.2. CONTINUOUS TIME SYSTEMS

Continuous time, or analog, systems have the longest history of development and, until recently, were representative of nearly all practical control systems. Because time functions are Laplace transformable, and because differential equations in the time domain transform to algebraic equations in the Laplace– or frequency–domain, the general differential operator techniques such as the

Laplace transform were used extensively in analysis of such systems. The basic tool developed for describing system input–output characteristics is the transfer function, defined as the ratio of the Laplace transform of the output time function to that of the input time function. The discussion to follow considers the relationship of the transfer function to the state space representation of the system.

D.2.1. Single Input–Single Output Systems

The single input–single output closed-loop system of classical control theory is shown in Figure D-1, where $V(s)$ and $W(s)$ are the forward and feedback transfer functions. The closed-loop transfer function relating the output $y(s)$ to the input $r(s)$ is given by the familiar equation

$$\frac{y(s)}{r(s)} = \frac{V(s)}{1 + V(s)\,W(s)}. \tag{D.1}$$

In general, the transfer functions $V(s)$ and $W(s)$ involve polynomials in s. The order of the denominator polynomial is equal to the order of the differential equation describing the process.

To see how this representation relates to the state space approach, consider the transfer function $V(s)$ from Figure D-1, which implies the transformed equation

$$y(s) = V(s)\,u(s). \tag{D.2}$$

If $V(s)$ is of order N, then an Nth order linear differential equation is implied:

$$a_N \frac{d^N y}{dt^N} + a_{N-1} \frac{d^{N-1} y}{dt^{N-1}} + \cdots + a_1 \frac{dy}{dt} + a_0 y = u(t) \tag{D.3}$$

where $y(t)$ and $u(t)$ are the inverse Laplace transforms of $y(s)$ and $u(s)$, respectively. The representation of this equation in terms of state space techniques would be by way of the N first-order, coupled, linear differential equations, which take the standard form

$$\dot{\boldsymbol{x}}(t) = \boldsymbol{F}\boldsymbol{x}(t) + \boldsymbol{g}u(t). \tag{D.4}$$

The $y(t)$ function is then a linear combination of the $\boldsymbol{x}(t)$ vector function,

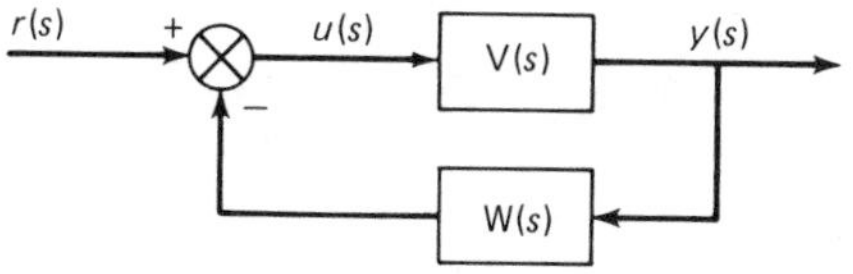

Figure D-1. Classical single input–single output feedback system.

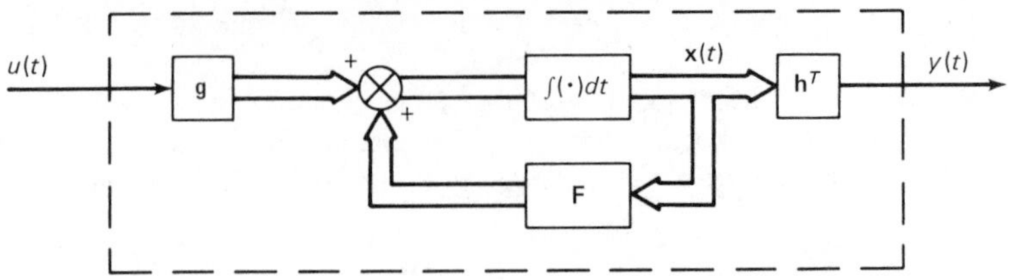

Figure D-2. State variable representation of single input–single output linear system.

$$y(t) = \boldsymbol{h}^T \boldsymbol{x}(t). \tag{D.5}$$

The vector $\boldsymbol{x}(t)$, which is not unique in its selection, is termed the *state vector* of the system. The transfer function $V(s)$ is thus seen to be the frequency domain representation of the input–output relationship of the system described by Equations (D.4) and (D.5). The block diagram of this system is shown in Figure D-2, where the dashed line indicates the input–output description of the system. By multiplying Equation (D.4) on the left by the vector $\boldsymbol{h}^T$, one obtains

$$\boldsymbol{h}^T \dot{\boldsymbol{x}}(t) = \dot{y}(t) = \boldsymbol{h}^T \boldsymbol{F} \boldsymbol{x}(t) + \boldsymbol{h}^T \boldsymbol{g} u(t) \tag{D.6}$$

which illustrates the dependence of $y(t)$ on the system states.

The Laplace transform of Equation (D.4) with $\boldsymbol{x}(0) = \boldsymbol{0}$ is

$$s\boldsymbol{x}(s) = \boldsymbol{F}\boldsymbol{x}(s) + \boldsymbol{g}u(s) \tag{D.7}$$

which, when solved for $\boldsymbol{x}(s)$, yields

$$\boldsymbol{x}(s) = (s\boldsymbol{I} - \boldsymbol{F})^{-1} \boldsymbol{g} u(s). \tag{D.8}$$

When this expression is substituted into the transform of Equation (D.5), the result is the input–output equation

$$y(s) = \boldsymbol{h}^T (s\boldsymbol{I} - \boldsymbol{F})^{-1} \boldsymbol{g} u(s). \tag{D.9}$$

Thus, the frequency domain transfer function $V(s)$ can be obtained from the state space parameters as the bilinear form

$$V(s) = \boldsymbol{h}^T (s\boldsymbol{I} - \boldsymbol{F})^{-1} \boldsymbol{g}. \tag{D.10}$$

Note that this is a scalar polynomial of degree N in s.

The inverse matrix in this equation can be expressed as

$$(s\boldsymbol{I} - \boldsymbol{F})^{-1} = \frac{1}{\det (s\boldsymbol{I} - \boldsymbol{F})} \operatorname{Adj} (s\boldsymbol{I} - \boldsymbol{F}) \tag{D.11}$$

where Adj $(s\boldsymbol{I} - \boldsymbol{F})$ is the adjoint matrix. Since the adjoint matrix has no poles, the poles of $V(s)$ must occur at the zeros of the determinant. These are found from the equation

$$\det (s\boldsymbol{I} - \boldsymbol{F}) = 0. \tag{D.12}$$

When this equation is compared to the eigenvalue equation for the matrix $\boldsymbol{F}$,

$$\det (\boldsymbol{F} - \lambda \boldsymbol{I}) = 0 \tag{D.13}$$

it is obvious that the poles of $V(s)$ are the eigenvalues of the matrix $\boldsymbol{F}$. Furthermore, since stability and response characteristics of $V(s)$ are a function of pole location, the stability and response characteristics of Equation (D.4) depend upon the eigenvalues of the matrix $\boldsymbol{F}$.

The resolvent matrix $\boldsymbol{\Phi}(s)$ is defined as

$$\boldsymbol{\Phi}(s) = (s\boldsymbol{I} - \boldsymbol{F})^{-1} \tag{D.14}$$

and is an $N \times N$ matrix function of the frequency domain variable s. Use of this notation in Equations (D.8), (D.9), and (D.10) yields

$$\begin{aligned} \boldsymbol{x}(s) &= \boldsymbol{\Phi}(s)\, \boldsymbol{g}u(s) \\ y(s) &= \boldsymbol{h}^T \boldsymbol{\Phi}(s)\, \boldsymbol{g}u(s) \\ V(s) &= \boldsymbol{h}^T \boldsymbol{\Phi}(s)\, \boldsymbol{g}. \end{aligned} \tag{D.15}$$

The resolvent matrix has a very special property which can be seen by returning to Equation (D.8) and including the effect of the initial condition $\boldsymbol{x}(0)$,

$$\begin{aligned} \boldsymbol{x}(s) &= (s\boldsymbol{I} - \boldsymbol{F})^{-1} [\boldsymbol{g}u(s) + \boldsymbol{x}(0)] \\ &= \boldsymbol{\Phi}(s)\, \boldsymbol{g}u(s) + \boldsymbol{\Phi}(s)\, \boldsymbol{x}(0). \end{aligned} \tag{D.16}$$

If $\boldsymbol{\Phi}(t)$ is the inverse Laplace transform of $\boldsymbol{\Phi}(s)$, then by use of the convolution integral form for the inverse transform of a product, Equation (D.16) transforms to

$$\boldsymbol{x}(t) = \boldsymbol{\Phi}(t)\, \boldsymbol{x}(0) + \int_0^t \boldsymbol{\Phi}(t - \sigma)\, \boldsymbol{g}u(\sigma)\, d\sigma. \tag{D.17}$$

Comparison of this result with the state transition matrix described in Chapter 4 and Appendix C clearly identifies $\boldsymbol{\Phi}(t)$ as the state transition matrix of the time invariant system described by the system matrix $\boldsymbol{F}$. That is,

$$\boldsymbol{\Phi}(t, \sigma) = \boldsymbol{\Phi}(t - \sigma) \tag{D.18}$$

since the matrix F is not a function of t. Thus, the resolvent matrix is the Laplace transform of the state transition matrix.

D.2.2. Multiple Input-Multiple Output Systems

The applications of frequency domain techniques to multiple input–multiple output systems results in a generalization of the transfer function known as the transfer matrix $V(s)$, defined by the equation

$$y(s) = V(s)\,u(s). \tag{D.19}$$

The matrix $V(s)$ need not be square, since in general the dimension of $y(s)$ and $u(s)$ need not be the same. In expanded form, Equation (D.19) is

$$\begin{bmatrix} y_1(s) \\ y_2(s) \\ \vdots \\ y_M(s) \end{bmatrix} = \begin{bmatrix} V_{11}(s) & V_{12}(s) & \cdots & V_{1P}(s) \\ V_{21}(s) & V_{22}(s) & \cdots & V_{2P}(s) \\ \cdots & \cdots & \cdots & \cdots \\ V_{M1}(s) & V_{M2}(s) & \cdots & V_{MP}(s) \end{bmatrix} \begin{bmatrix} u_1(s) \\ u_2(s) \\ \vdots \\ u_P(s) \end{bmatrix} \tag{D.20}$$

so that elements are seen to be transfer functions relating an output $y_i(s)$ to an input $u_j(s)$.

The time domain representation of this system is given by the general state space equations

$$\begin{aligned} \dot{x}(t) &= Fx(t) + Gu(t) \\ y(t) &= Hx(t). \end{aligned} \tag{D.21}$$

The Laplace transforms of these equations with $x(0) = \mathbf{0}$ are

$$\begin{aligned} sx(s) &= Fx(s) + Gu(s) \\ y(s) &= Hx(s). \end{aligned} \tag{D.22}$$

Following the procedure of the previous section yields

$$x(s) = \Phi(s)\,Gu(s) \tag{D.23}$$

and

$$y(s) = H\Phi(s)\,Gu(s) \tag{D.24}$$

where the resolvent matrix is again defined as

$$\Phi(s) = (sI - F)^{-1}. \tag{D.25}$$

The transfer matrix $V(s)$ is obtained as

$$V(s) = \boldsymbol{H}\boldsymbol{\Phi}(s)\,\boldsymbol{G}. \tag{D.26}$$

Thus, the multiple input–multiple output system is a simple generalization when viewed from the state variable approach.

D.2.3. Closed-Loop Response

The closed-loop response given by Equation (D.1) serves as the basis for most feedback system analysis of single input–single output systems. It gives the closed-loop transfer function as a ratio of polynomials in s, and the location of the poles and zeros of that ratio of polynomials determines the response characteristics of the closed-loop system. Most of the emphasis in classical control theory lies in specifying the location of these roots or inferring their approximate locations without actually determining their values. To relate state variable feedback to the closed-loop transfer function, consider a linear, time invariant system with a single input and state variable feedback. The system equations are

$$\begin{aligned} \dot{\boldsymbol{x}}(t) &= \boldsymbol{F}\boldsymbol{x}(t) + \boldsymbol{g}u(t) \\ y(t) &= \boldsymbol{h}^T\boldsymbol{x}(t) \end{aligned} \tag{D.27}$$

where a scalar output $y(t)$ has been specified. The fact that the output is $y(t)$— while the entire state vector has been assumed available for feedback purposes— should cause no concern, since the specific point under discussion is the equivalence of state variable feedback to transfer function feedback through the dynamic element described by $W(s)$ in Figure D-1. Using the input equation

$$u(t) = r(t) + \boldsymbol{c}^T\boldsymbol{x}(t) \tag{D.28}$$

and substituting into Equation (D.27) yield the closed-loop equation

$$\dot{\boldsymbol{x}}(t) = (\boldsymbol{F} + \boldsymbol{g}\boldsymbol{c}^T)\,\boldsymbol{x}(t) + \boldsymbol{g}r(t). \tag{D.29}$$

Taking the Laplace transform with $\boldsymbol{x}(0) = 0$ and solving for $\boldsymbol{x}(s)$ yield

$$\boldsymbol{x}(s) = (s\boldsymbol{I} - \boldsymbol{F} + \boldsymbol{g}\boldsymbol{c}^T)^{-1}\boldsymbol{g}r(s). \tag{D.30}$$

The closed-loop resolvent matrix $\tilde{\boldsymbol{\Phi}}(s)$ is now defined as

$$\tilde{\boldsymbol{\Phi}}(s) = (s\boldsymbol{I} - \boldsymbol{F} + \boldsymbol{g}\boldsymbol{c}^T)^{-1} \tag{D.31}$$

which allows Equation (D.30) to be written as

$$\boldsymbol{x}(s) = \tilde{\boldsymbol{\Phi}}(s)\,\boldsymbol{g}r(s). \tag{D.32}$$

By transforming $y(t)$ of Equation (D.27), multiplying Equation (D.31) on the left by $\boldsymbol{h}^T$, and substituting, one obtains the closed-loop input–output relationship

$$y(s) = \boldsymbol{h}^T \tilde{\boldsymbol{\Phi}}(s) \boldsymbol{g} r(s). \tag{D.33}$$

This expression immediately yields the closed-loop transfer function of Equation (D.1)

$$\frac{y(s)}{r(s)} = \boldsymbol{h}^T \tilde{\boldsymbol{\Phi}}(s) \boldsymbol{g} r(s). \tag{D.34}$$

By arguments made in the previous section, closed-loop stability and response characteristics are determined by the eigenvalues of the closed-loop system matrix

$$\boldsymbol{F} = \boldsymbol{F} + \boldsymbol{g}\boldsymbol{c}^T \tag{D.35}$$

which are identical to the poles of the closed-loop transfer function of Equation (D.34).

While the system of Equation (D.27) has been discussed under the assumption that state variable feedback was in fact possible, in actual practice only the output function $y(t)$ is available for feedback purposes. The system design problem is to determine the transfer function $W(s)$, as indicated in Figure D-1, which will effect the same closed-loop system characteristics as the state variable feedback arrangement implied by Equation (D.34). The general concept is illustrated in Figure D-3, where the transfer function $W(s)$ is considered an operation on the output function $y(s)$ to produce the feedback function $-\boldsymbol{c}^T\boldsymbol{x}(s)$, as implied by Equation (D.28). The forward transfer function $V(s)$ is obtained from Equation (D.15).

From Figure D-3, the expression for $W(s)$ is

$$W(s) = -\frac{\boldsymbol{c}^T\boldsymbol{x}(s)}{y(s)} \tag{D.36}$$

or, upon substitution for $y(s)$,

$$W(s) = -\frac{\boldsymbol{c}^T\boldsymbol{x}(s)}{\boldsymbol{h}^T\boldsymbol{x}(s)}. \tag{D.37}$$

Further substitution for $\boldsymbol{x}(s)$ from Equation (D.15) yields

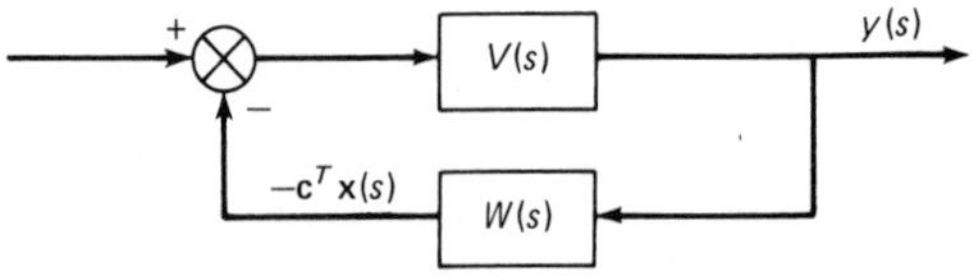

Figure D-3. Single input–single output equivalent of state variable feedback.

$$W(s) = -\frac{\boldsymbol{c}^T \boldsymbol{\Phi}(s) \boldsymbol{g} u(s)}{\boldsymbol{h}^T \boldsymbol{\Phi}(s) \boldsymbol{g} u(s)} \tag{D.38}$$

or, since $u(s)$ is a scalar function,

$$W(s) = -\frac{\boldsymbol{c}^T \boldsymbol{\Phi}(s) \boldsymbol{g}}{\boldsymbol{h}^T \boldsymbol{\Phi}(s) \boldsymbol{g}}. \tag{D.39}$$

This expression gives the transfer function $W(s)$ as a function of the open-loop resolvent matrix $\boldsymbol{\Phi}(s)$.

An important entity in classical control theory is the open-loop transfer function, defined as

$$V(s)\, W(s). \tag{D.40}$$

Using $V(s)$ given by Equation (D.15), and recognizing that all terms are scalar functions, one can obtain the open-loop transfer function immediately as simply

$$V(s)\, W(s) = -\boldsymbol{c}^T \boldsymbol{\Phi}(s) \boldsymbol{g}. \tag{D.41}$$

The open-loop transfer function is used extensively in the design techniques of classical control theory.

The equivalent of Equation (D.1), written in terms of the resolvent matrix $\boldsymbol{\Phi}(s)$, can now be obtained by substituting from Equations (D.15) and (D.41),

$$\frac{y(s)}{r(s)} = \frac{\boldsymbol{h}^T \boldsymbol{\Phi}(s) \boldsymbol{g}}{1 - \boldsymbol{c}^T \boldsymbol{\Phi}(s) \boldsymbol{g}} \tag{D.42}$$

This equation can be used to determine the closed-loop transfer function after the feedback gain vector $\boldsymbol{c}$ has been chosen, but it is not useful in the design process. The situation normally arising in system design is that the gain vector $\boldsymbol{c}$ has been selected to produce the desired closed-loop characteristics, and then the transfer function $W(s)$ is determined to yield the equivalent of implementing state variable feedback with the chosen gain vector $\boldsymbol{c}$. This approach should be compared with the use of observers as discussed in Chapter 5.

Alternate expressions in terms of the closed-loop resolvent matrix can be easily developed. They are

$$V(s) = \frac{\boldsymbol{h}^T \tilde{\boldsymbol{\Phi}}(s) \boldsymbol{g}}{1 - \boldsymbol{c}^T \tilde{\boldsymbol{\Phi}}(s) \boldsymbol{g}} \tag{D.43}$$

$$W(s) = -\frac{\boldsymbol{c}^T \tilde{\boldsymbol{\Phi}}(s) \boldsymbol{g}}{\boldsymbol{h}^T \tilde{\boldsymbol{\Phi}}(s) \boldsymbol{g}}$$

and

$$V(s)\,W(s) = -\frac{c^T\tilde{\Phi}(s)g}{1 - c^T\tilde{\Phi}(s)g}. \tag{D.44}$$

The expression for $W(s)$ in terms of $\tilde{\Phi}(s)$ above should be compared to the expression of Equation (D.39).

Using the results developed above, several conclusions can be reached relating to frequency domain characteristics of closed-loop systems designed by using the concept of state variable feedback. These results are presented here without further discussion, but their implications are obvious to anyone familiar with the classical feedback control concepts.

1. The poles of $W(s)$ are the zeros of $V(s)$.
2. $W(s)$ has $N - 1$ arbitrary zeros determined by the selection of gain vector $\mathbf{c}$.
3. The poles of the open-loop transfer function $V(s)\,W(s)$ are the poles of $V(s)$.
4. The zeros of the open-loop transfer function $V(s)\,W(s)$ are the zeros of $W(s)$.
5. The poles of the closed-loop transfer function $y(s)/r(s)$ are arbitrarily determined by selection of the control vector $\mathbf{c}$.
6. The zeros of the closed-loop transfer function $y(s)/r(s)$ are the zeros of $V(s)$.

Characteristic 6 is important to system realization and will be discussed further in Section D.2.3.4.

D.2.3.1. Realization Considerations. While $W(s)$ as determined from Equation (D.39) or (D.43) will indeed produce a closed-loop transfer function identical to that obtained with state variable feedback implemented using control vector $\mathbf{c}$, from the practical point of view problems may arise because $W(s)$ will have $N - 1$ zeros determined by selection of $\mathbf{c}$. These zeros represent differentiation of the output $y(t)$ and, while it is theoretically possible, differentiation should normally be avoided because it greatly enhances any noise which might be present in the system. If these zeros are a problem, one of three alternatives may be considered:

I. Reconstruction of $W(s)$ using minor loop equalization.
II. Use of series compensation.
III. Use of an observer.

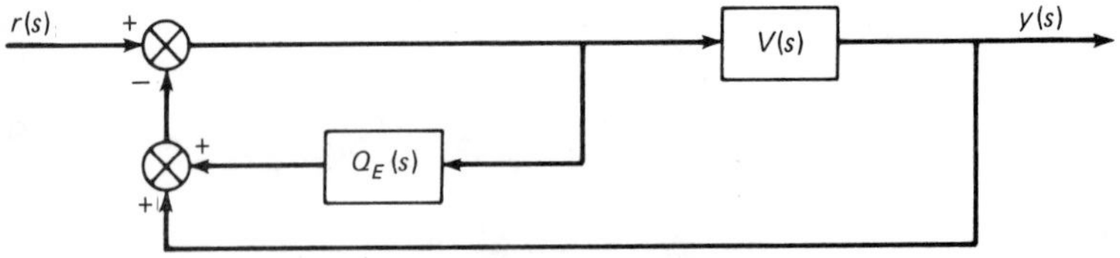

Figure D-4. Realization by minor loop equalization.

The first two alternatives are traditional classical control theory techniques. The third is a modern control theory result.

D.2.3.2. Minor Loop Equalization. To reduce the number of zeros (and hence differentiations) necessary to realize the closed-loop transfer function designed by state variable feedback, the alternative block diagram of Figure D-4 is considered. Here, the transfer function $Q_E(s)$ is to be selected so that the closed-loop transfer function is given by

$$\frac{y(s)}{r(s)} = \frac{V(s)}{1 + V(s)\,W(s)} \tag{D.45}$$

where $W(s)$ is the feedback transfer function determined from Equation (D.39). The closed-loop transfer function of this system is determined by standard techniques to be

$$\frac{y(s)}{r(s)} = \frac{V(s)}{1 + V(s) + Q_E(s)} \tag{D.46}$$

which, when compared to Equation (D.45), yields

$$Q_E(s) = V(s)[W(s) - 1]. \tag{D.47}$$

Substitution from Equations (D.15) and (D.39) yields an expression for $Q_E(s)$ in terms of the resolvent matrix $\mathbf{\Phi}(s)$:

$$Q_E(s) = -(\mathbf{c}^T + \mathbf{h}^T)\mathbf{\Phi}(s)\,\mathbf{g}. \tag{D.48}$$

Since the $\mathbf{c}$ and $\mathbf{h}$ vectors determine the poles and zeros of $W(s)$, either $Q_E(s)$ or $W(s)$ will normally be suitable for practical realization.

To illustrate these concepts, consider the system

$$\dot{\mathbf{x}}(t) = \mathbf{F}\mathbf{x}(t) + \mathbf{g}u(t)$$
$$y(t) = \mathbf{h}^T\mathbf{x}(t)$$

where

$$\mathbf{F} = \begin{bmatrix} 0 & 1 & 0 \\ 0 & -1 & 2 \\ 0 & 0 & -4 \end{bmatrix}; \quad \mathbf{g} = \begin{bmatrix} 0 \\ 1 \\ 1 \end{bmatrix}; \quad \mathbf{h} = \begin{bmatrix} 1 \\ 1 \\ 1 \end{bmatrix}.$$

The resolvent matrix is determined to be

$$\boldsymbol{\Phi}(s) = \begin{bmatrix} \frac{1}{s} & \frac{1}{s(s+1)} & \frac{2}{s(s+1)(s+4)} \\ 0 & \frac{1}{s+1} & \frac{2}{(s+1)(s+4)} \\ 0 & 0 & \frac{1}{s+4} \end{bmatrix}.$$

Suppose that state variable feedback analysis has determined that the gain vector $\boldsymbol{c}$ should be selected as

$$\boldsymbol{c} = \begin{bmatrix} -1 \\ -1 \\ 1 \end{bmatrix}$$

in order to produce the desired closed-loop characteristics. Then, $W(s)$ can be determined from Equation (D.39) to be

$$W(s) = -\frac{\boldsymbol{c}^T\boldsymbol{\Phi}(s)\boldsymbol{g}}{\boldsymbol{h}^T\boldsymbol{\Phi}(s)\boldsymbol{g}} = \frac{3}{s+3}.$$

This transfer function is suitable for realization, and no further development is necessary.

If the system is now altered so that the $\boldsymbol{c}$, $\boldsymbol{g}$, and $\boldsymbol{h}$ vectors are

$$\boldsymbol{c} = \begin{bmatrix} -1 \\ -1 \\ -1 \end{bmatrix}; \quad \boldsymbol{g} = \begin{bmatrix} 0 \\ 0 \\ 1 \end{bmatrix}; \quad \boldsymbol{h} = \begin{bmatrix} 1 \\ 0 \\ 0 \end{bmatrix}$$

then the feedback transfer function $W(s)$ is determined to be

$$W(s) = -\frac{\boldsymbol{c}^T\boldsymbol{\Phi}(s)\boldsymbol{g}}{\boldsymbol{h}^T\boldsymbol{\Phi}(s)\boldsymbol{g}} = \frac{1}{2}(s+2)(s+1).$$

This transfer function is obviously unsuitable for realization because it requires that the output function $y(t)$ be doubly differentiated to provide the feedback signal. To explore the possibility of minor loop equalization, the transfer function $Q_E(s)$ is determined from Equation (D.48) to be

$$Q(s) = -(\boldsymbol{c}^T + \boldsymbol{h}^T)\boldsymbol{\Phi}(s)\boldsymbol{g}$$

$$= \frac{s+3}{(s+1)(s+4)}$$

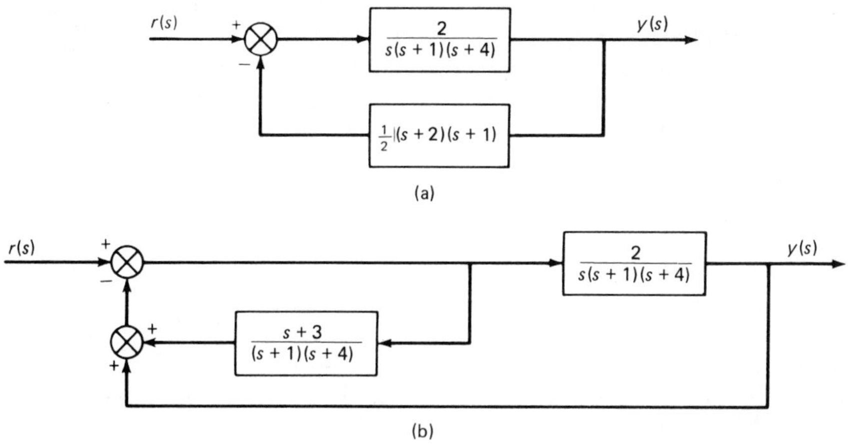

Figure D-5. State variable feedback realization: (a) with feedback transfer function; (b) with minor loop transfer function.

which is suitable for realization. The two realizations are shown in block diagram form in Figure D-5, where standard block diagram reduction techniques can verify that they both produce the closed-loop transfer function

$$\frac{y(s)}{r(s)} = \frac{2}{s^3 + 6s^2 + 7s + 2}.$$

In determining this transfer function, the expression for the system transfer function

$$V(s) = \boldsymbol{h}^T \boldsymbol{\Phi}(s) \boldsymbol{g}$$

from Equation (D.15) is used.

D.2.3.3. Series Compensation. Series compensation utilizes a form of feedback control in which the closed-loop system has unity feedback, as illustrated in Figure D-6. Straightforward application of block diagram manipulation produces the relationship between the minor loop equalization transfer function of the previous section and the series compensation transfer function $Q_C(s)$,

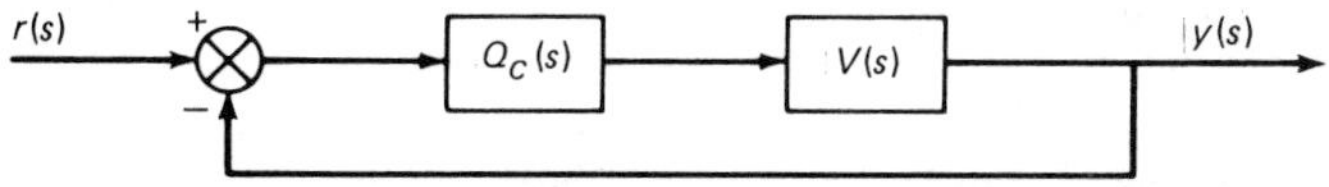

Figure D-6. Realization by series compensation.

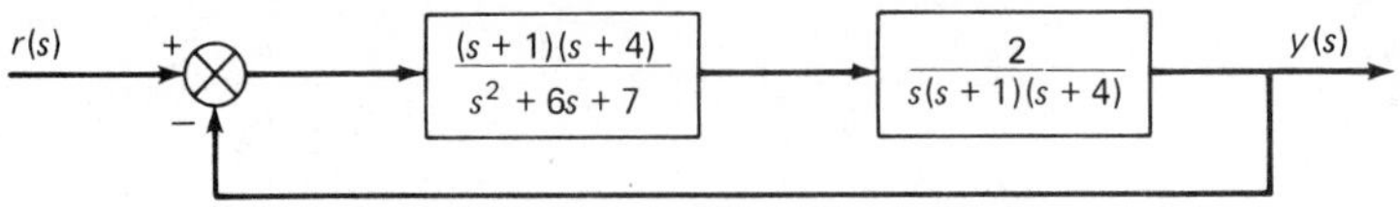

Figure D-7. State variable feedback realization by series compensation.

$$Q_C(s) = \frac{1}{1 + Q_E(s)}$$

$$= \frac{1}{1 - (\boldsymbol{c}^T + \boldsymbol{h}^T)\boldsymbol{\Phi}(s)\boldsymbol{g}} \tag{D.49}$$

where substitution from Equation (D.48) has been used. This equation permits the series compensation transfer function to be expressed directly in terms of the feedback gain vector $\boldsymbol{c}$, which presumably has been selected to produce the desired closed-loop characteristics.

This concept can be applied to the problem of the previous section to obtain the series compensation transfer function

$$Q_C(s) = \frac{(s + 1)(s + 4)}{s^2 + 6s + 7}$$

so that the block diagram of Figure D-7 is obtained. The closed-loop transfer function of this system can be determined to be

$$\frac{y(s)}{r(s)} = \frac{2}{s^3 + 6s^2 + 7s + 2}$$

which agrees with the result obtained from the two block diagrams of Figure D-5.

D.2.3.4. Use of Observers. The topic of observers is covered in Chapter 5, but their relationship to the synthesis techniques considered here must be noted. The use of an observer permits state variable design techniques to be used without regard to the actual availability of the state variables themselves for feedback purposes. Once the design is formulated, in the sense that the feedback vector $\boldsymbol{c}$ has been determined, an observer can be used to estimate the values of the state from the available measurement. The realization concepts considered here do the same thing, and the actual purpose of the transfer functions $W(s)$, $Q_E(s)$, and $Q_C(s)$ is merely to provide information on the state vector as gleaned from the output function $y(t)$. Both methods increase the order of the closed-loop system, usually by the same amount if a reduced-order observer is used.

For single input–single output systems, design by use of an observer has little advantage over the techniques considered here, because both provide a systematic approach once the feedback gain vector $\boldsymbol{c}$ has been determined. For systems with more than one output and/or more than one input, the approach presented here becomes more difficult and loses its systematic character. The use of observers, however, retains its general applicability in the design of multivariable systems.

An additional problem is encountered in single input–single output system synthesis by the frequency domain techniques considered here, resulting from characteristic 6 listed in Section D.2.3. It was stated there that the zeros of the closed-loop transfer function are determined by the zeros of the system transfer function $V(s)$, and therefore are not under the control of the designer. If the desired closed-loop transfer function has zeros different in location and/or number than those of $V(s)$, then additional compensation is required beyond that considered here. The determination of the required compensation is not systematic in nature, and one must resort to zero–pole cancellation techniques. On the other hand, the use of an observer is insensitive to the particular form of the feedback vector $\boldsymbol{c}$, and retains its systematic nature under this condition.

D.3. DISCRETE TIME SYSTEMS

Traditionally, difference equations arising from sampled data control systems have been treated by the z-transform method. This approach is in direct analogy to the use of Laplace transform techniques in the analysis of continuous time systems. Sampled data control systems are a more recent development in classical theory, but the true discrete time systems evolved with the rapid evolution of digital computer technology. Rather than considering sampled data systems in which analog signals are sampled, and continuous time systems subjected to pulselike input signals, one now speaks of discrete time systems in which elements of the system are truly digital in nature. Although the distinction is a fine point, it can be exemplified by comparing Figures D-8 and D-9 in the discussion to follow.

D.3.1. Single Input–Single Output Systems

When a discrete time single input–single output system is considered, its closed-loop response is given by an expression analogous to Equation (D.1), using transfer functions $V(z)$ and $W(z)$ based upon the z-transform. Then, if $r(z)$ and $y(z)$ are the transforms of the input and output time functions, the closed-loop transfer function $y(z)/r(z)$ is

$$\frac{y(z)}{r(z)} = \frac{V(z)}{1 + V(z)\,W(z)} \tag{D.50}$$

where the sampling is as indicated in Figure D-8. It must be noted that other sampling configurations may produce different forms of the closed-loop transfer function. Stability requires that the roots of the equation (D.50) lie within the unit circle.

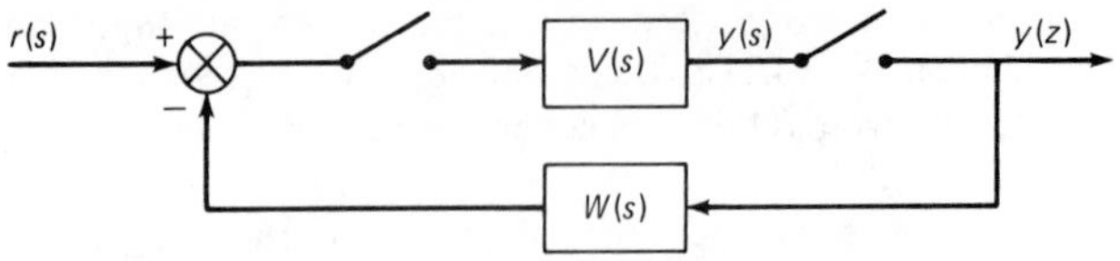

Figure D-8. Sampled data feedback system.

When viewed from the state space approach of modern control theory, the Nth-order difference equation represented by the $V(z)$ transfer function is expressed as N first-order difference equations in a manner completely analogous to the treatment of continuous time systems. Then the equations can be expressed as

$$\begin{aligned} \boldsymbol{x}(l+1) &= \boldsymbol{\Phi}\boldsymbol{x}(l) + \boldsymbol{\gamma} u(l) \\ y(l) &= \boldsymbol{h}^T \boldsymbol{x}(l). \end{aligned} \tag{D.51}$$

These equations may be transformed to produce, for $\boldsymbol{x}(0) = \boldsymbol{0}$,

$$\begin{aligned} z\boldsymbol{x}(z) &= \boldsymbol{\Phi}\boldsymbol{x}(z) + \boldsymbol{\gamma} u(z) \\ y(z) &= \boldsymbol{h}^T \boldsymbol{x}(z). \end{aligned} \tag{D.52}$$

From the first equation,

$$\boldsymbol{x}(z) = (\boldsymbol{I}z - \boldsymbol{\Phi})^{-1} \boldsymbol{\gamma} u(z) \tag{D.53}$$

and, by multiplying on the left by $\boldsymbol{h}^T$, one obtains

$$\boldsymbol{h}^T \boldsymbol{x}(z) = y(z) = \boldsymbol{h}^T [\boldsymbol{I}z - \boldsymbol{\Phi}]^{-1} \boldsymbol{\gamma} u(z). \tag{D.54}$$

From this equation, the transfer function $V(z)$ is obtained:

$$V(z) = \frac{y(z)}{r(z)} = \boldsymbol{h}^T (\boldsymbol{I}z - \boldsymbol{\Phi})^{-1} \boldsymbol{\gamma}. \tag{D.55}$$

For the system described by $V(z)$ to be stable, the roots of $V(z)$ must lie within the unit circle. Since the vectors $\boldsymbol{h}$ and $\boldsymbol{\gamma}$ are constants, and since the matrix inversion can be represented by the adjoint matrix divided by the determinant, the roots of $V(z)$ are the roots of the equation

$$\det (\boldsymbol{I}z - \boldsymbol{\Phi}) = 0. \tag{D.56}$$

Since this expression is just the eigenvalue equation, the requirement for stability is that the matrix have no eigenvalues which do not lie within the unit circle.

D.3.2. Closed-Loop Response

When a discrete time system uses linear state variable feedback of the form

$$\boldsymbol{c}^T\boldsymbol{x}(l) \tag{D.57}$$

and the input function $r(l)$ is considered, then the closed-loop system equations are seen to be

$$\begin{aligned} \boldsymbol{x}(l+1) &= [\boldsymbol{\Phi} + \boldsymbol{\gamma}\boldsymbol{c}^T]\boldsymbol{x}(l) + \boldsymbol{\gamma}r(l) \\ y(l) &= \boldsymbol{h}^T\boldsymbol{x}(l) \end{aligned} \tag{D.58}$$

Transforming these equations and solving for the closed-loop transfer function $y(z)/r(z)$ yield

$$\frac{y(z)}{r(z)} = \boldsymbol{h}^T(\boldsymbol{I}z - \tilde{\boldsymbol{\Phi}})^{-1}\boldsymbol{\gamma} \tag{D.59}$$

where the closed-loop transition matrix $\tilde{\boldsymbol{\Phi}}$ is defined as

$$\tilde{\boldsymbol{\Phi}} = \boldsymbol{\Phi} + \boldsymbol{\gamma}\boldsymbol{c}^T. \tag{D.60}$$

Relating this result to Equation (D.50) and previous discussions of stability implies that the closed-loop system response is determined by the eigenvalues of the closed-loop transition matrix.

The single input–single output system cannot provide state variable feedback, but the feedback transfer function $W(z)$ can be designed to effectively reconstruct the states, assuming that the system is observable. This concept is depicted in Figure D-9, where complete transition to a z-domain system description has been made for clarity. This transition can be obtained by adding redundant samplers to the system of Figure D-8. The problem, then, is to determine the transfer function $W(z)$ to provide an effective feedback control signal $\boldsymbol{c}^T\boldsymbol{x}(z)$.

From Figure D-9, an expression for $W(z)$ can be obtained:

$$W(z) = -\frac{\boldsymbol{c}^T\boldsymbol{x}(z)}{y(z)} = -\frac{\boldsymbol{c}^T\boldsymbol{x}(z)}{\boldsymbol{h}^T\boldsymbol{x}(z)}. \tag{D.61}$$

Substitution from Equation (D.53) then yields

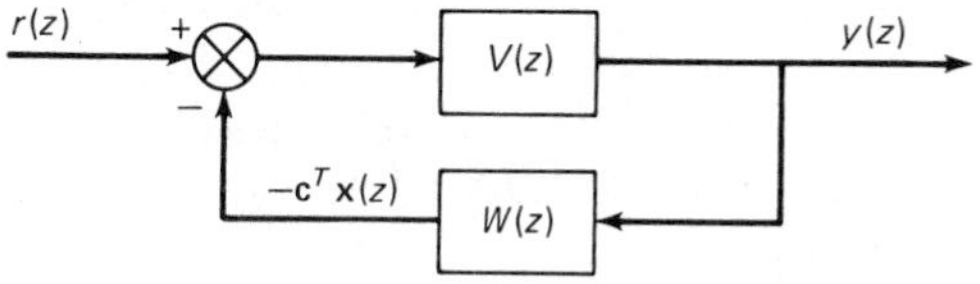

Figure D-9. Discrete time state variable feedback control.

$$W(z) = -\frac{\mathbf{c}^T(Iz - \mathbf{\Phi})^{-1}\gamma}{\mathbf{h}^T(Iz - \mathbf{\Phi})^{-1}\gamma} \tag{D.62}$$

in direct analogy to Equation (D.39). This expression gives the feedback transfer function $W(z)$ which will produce the same closed-loop transfer function as state variable feedback with feedback gain vector $\mathbf{c}$.

D.3.3. Multiple Input-Multiple Output Systems

Discrete time systems with multiple inputs and/or multiple outputs can be analyzed using a z-transformation matrix in a manner analogous to the treatment of continuous time systems discussed in Section D.2.2. The results of Equations (D.53) and (D.54) are easily generalized to yield

$$\begin{aligned} \mathbf{x}(z) &= (\mathbf{I}z - \mathbf{\Phi})^{-1}\mathbf{\Gamma}\mathbf{u}(z) \\ \mathbf{y}(z) &= \mathbf{H}(\mathbf{I}\mathbf{z} - \mathbf{\Phi})^{-1}\mathbf{\Gamma}\mathbf{u}(z). \end{aligned} \tag{D.63}$$

The second expression defines the transfer matrix $\mathbf{V}(z)$,

$$\mathbf{V}(z) = \mathbf{H}(\mathbf{I}\mathbf{z} - \mathbf{\Phi})^{-1}\mathbf{\Gamma} \tag{D.64}$$

where $\mathbf{H}$ and $\mathbf{\Gamma}$ are the observation and input matrices defined for the general discrete time system in Chapter 4.

D.4. SUMMARY

From the discussion presented in this appendix, it can be seen that the state space approach of modern control theory can be integrated with classical control concepts to present a systematic view of the basic fundamentals governing system behavior. While the classical theory is very effective in the analysis of single input–single output systems, it tends to become cumbersome when multivariable systems are considered. On the other hand, the modern techniques consider the multivariable system in general and treat the single input–single output system as a special case.

The primary distinction, of course, is that the classical methods deal with the system as defined in the frequency domain, while modern control theory utilizes the time domain. Before the advent of fast, inexpensive computation, time domain techniques were not a feasible alternative, and time response characteristics were inferred from frequency domain pole and zero location. Although this technique is still useful in the modern approach—as indicated by the use of the eigenvalues of the system matrix in estimating response characteristics—transition matrices and their direct indication of response characteristics can now be formulated.

The frequency domain approach of classical control theory deals specifically with only steady state solutions—although transient response is estimated as indicated above. The time domain approach of the modern theory, on the other hand, permits a unified approach which includes analysis of those systems in which the transient response is of paramount importance, as occurs with certain optimal strategies. Also, since the time domain approach of the modern theory permits use of very powerful techniques of system optimization, description of control systems in the time domain provides the necessary transition from design to realization.

In the area of time varying, linear systems, the modern theory is directly applicable in principle. The system matrix—and possibly the other matrices involved—become functions of time, and the state transition matrix is then a function of two arguments as discussed in Chapter 4. As a matter of fact, this situation is again the general case, with time invariant systems being a special consideration. Nonlinear systems, for which little of a general nature can be done by either approach, are still described in terms of the modern theory as the most general system, and certain results—such as Lyapunov stability criteria—have been developed for this most general case.

In summary, the time domain approach of modern control theory considers the general nonlinear, multivariable system. Progressively more severe restrictions reduce consideration to linear systems; to linear, time invariant systems; and finally to linear, time invariant, single input–single output systems. This last category is the domain of classical control theory, and even here transient response is not treated directly. This category of systems is a very important one however, and from the material in this appendix it can be seen that the two approaches can complement one another very effectively.

REFERENCES

1. D'Azzo, John J., and Houpis, Constantine H. *Linear Control System Analysis and Design.* New York: McGraw-Hill, 1981.
2. Katz, Paul. *Digital Control Using Microprocessors.* Englewood Cliffs, New Jersey: Prentice Hall International, 1981.
3. Kimura, Hidenori. A Further Result on the Problem of Pole Assignment by Output Feedback. *IEEE Transactions on Automatic Control,* **AC-22**: 458–463 (June 1977).
4. Kuo, Benjiman C. *Automatic Control Systems.* Englewood Cliffs, New Jersey: Prentice Hall, 1975.
5. Truxal, John G. *Control Systems Synthesis.* New York: McGraw-Hill, 1955.

DEVELOPMENTAL EXERCISES

D.1. Show that the expression for the closed-loop transfer function given by Equation (D.42) can also be expressed as

$$y(s)/r(s) = \boldsymbol{h}^T[s\boldsymbol{I} - (\boldsymbol{F} + \boldsymbol{g}\boldsymbol{c}^T)]^{-1}\boldsymbol{g}.$$

D.2. Prove the assertion of Equation (D.49).

D.3. Show that any single input–single output system can be represented by a feedback system with unity feedback.

PROBLEMS

D.1. The time domain representation of a second-order system is

$$\frac{y(s)}{u(s)} = \frac{10}{s(s+1)(s+2)}.$$

Find a state space representation of this equation.

D.2. Consider the third-order system described by the transfer function

$$\frac{y(s)}{u(s)} = \frac{2(s+4)}{s(s+1)(s+2)}.$$

(a) Find a resolvent matrix for this system.
(b) Find a state space representation of this system.

D.3. Consider a system with the transfer function

$$\frac{y(s)}{u(s)} = \frac{s+3}{s(s+1)(s+2)}.$$

Determine the state space representation of this system and find the state transition matrix.

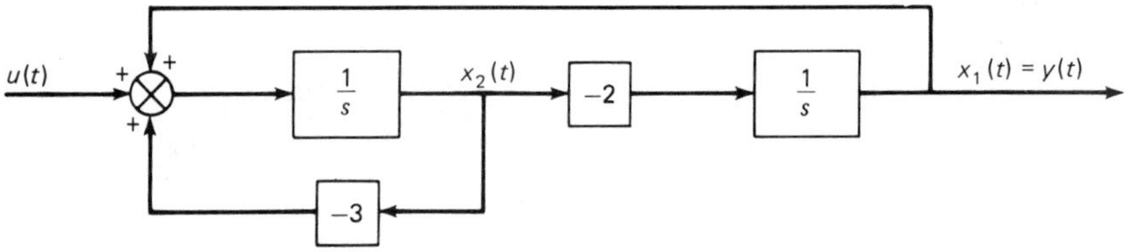

Figure PD-1.

D.4. Consider the system represented by the block diagram of Figure PD-1. Determine the state space representation, the resolvent matrix and the state transition matrix.

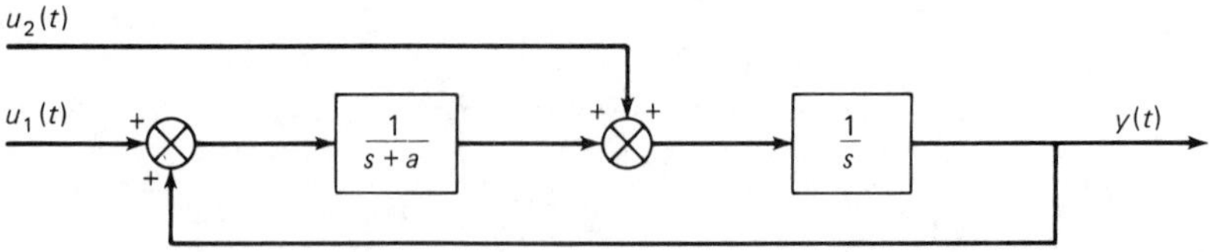

Figure PD-2.

D.5. A feedback control system is represented by the block diagram shown in Figure PD-2. Derive the state space representation for this system.

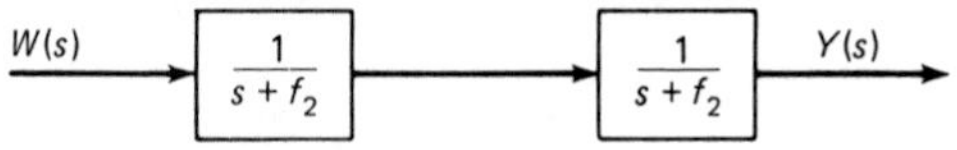

Figure PD-3.

D.6. Consider the second order system depicted in Figure PD-3. If the input function $W(t)$ is a white noise process with covariance kernel $q\delta(t - \sigma)$, find the covariance function of the output process $y(s)$. (*Hint:* Convert to state space representation and use concepts developed in Chapter 5.)

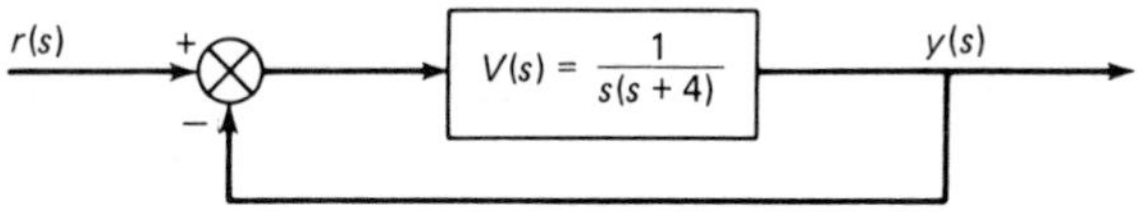

Figure PD-4.

D.7. Determine the state space representation of the feedback system shown in Figure PD-4.

D.8. Repeat Problem D.7 if

$$V(s) = \frac{1}{s^2(s + 10)}.$$

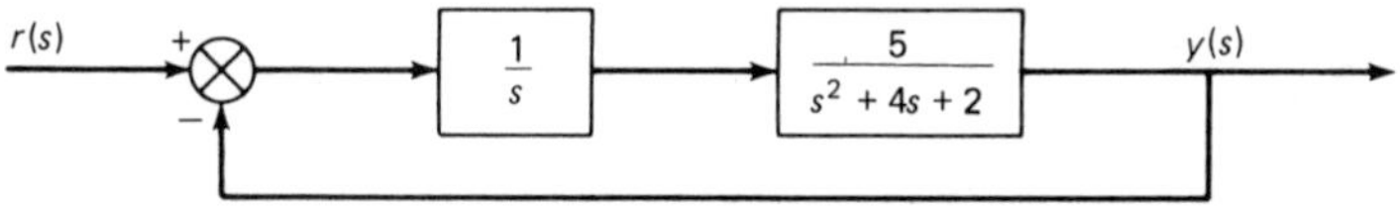

Figure PD-5.

D.9. Determine the state space representation of the feedback system shown in Figure PD-5.

D.10. A linear system has transfer function

$$V(s) = \frac{s^2 + 4s + 1}{s^3 + 9s^2 + 8s + 1} = \frac{y(s)}{u(s)}.$$

Find the state space representation of this system.

D.11. A linear system has transfer function

$$V(s) = \frac{2}{s^2(s^2 + s + 1)}$$

Find the state space representation of this system.

D.12. Consider the system described by the space equation

$$\dot{\boldsymbol{x}}(t) = \boldsymbol{F}(t)\boldsymbol{x}(t) + \boldsymbol{G}(t)u(t)$$

where

$$\boldsymbol{F} = \begin{bmatrix} 0 & 1 \\ 0 & -1 \end{bmatrix}; \qquad \boldsymbol{G} = \begin{bmatrix} 0 \\ 1 \end{bmatrix}.$$

(a) Find the resolvent matrix.
(b) Obtain the transfer function from the resolvent matrix.
(c) Find the transition matrix from the matrix expansion $e^{\boldsymbol{F}t}$.

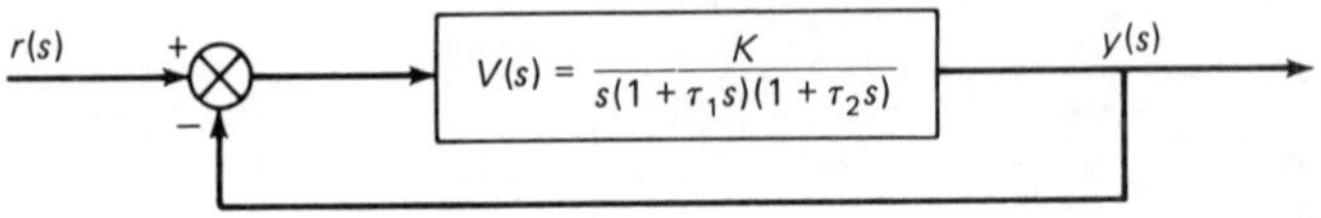

Figure PD-6.

D.13. A feedback system is shown in its frequency domain representation in Figure PD-6.
(a) Determine the closed-loop transfer function and the characteristic equation.
(b) Develop a state space representation of the system.
(c) Find the equation for the eigenvalues of the state transition matrix and compare with the result of part (a).

D.14. A system has transfer function

$$V(s) = \frac{5}{s^3 + 8s^2 + 9s + 1} = \frac{y(s)}{u(s)}.$$

Find the state space representation of this system.

D.15. A system has transfer function

$$V(s) = \frac{2}{s^2 + s + 2} = \frac{y(s)}{u(s)}.$$

Find the state space representation of this system.

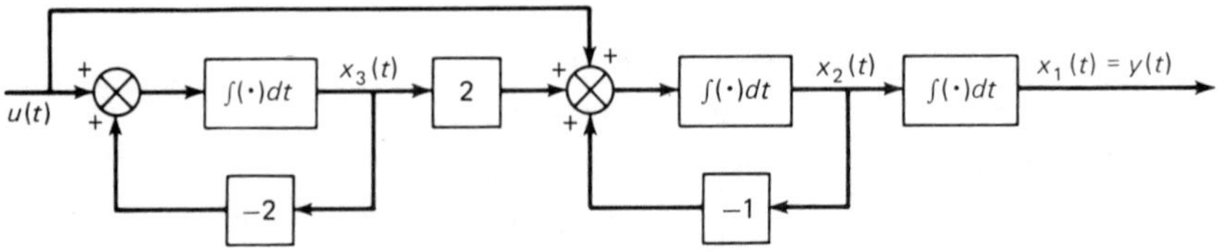

Figure PD-7.

D.16. A second-order, single input–single output system is described by the diagram of Figure PD-7. Use Equations (D.14) and (D.15) to find the transfer function of this system.

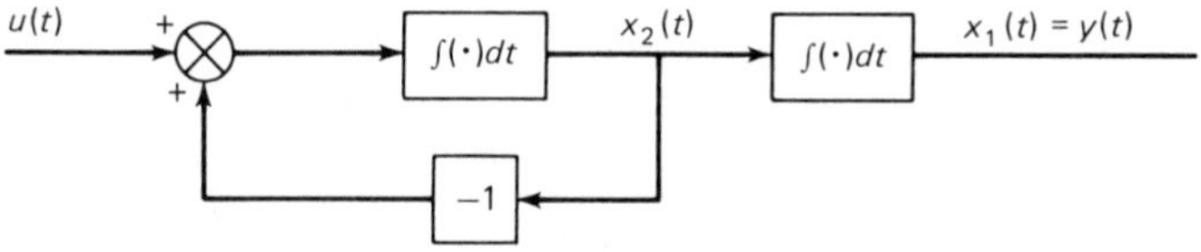

Figure PD-8.

D.17. Consider the second-order system described by the diagram of Figure PD-8.

(a) Use Equation (D.14) and (D.15) to find the transfer function.

(b) Design the optimal regulator using the performance index

$$J = \frac{1}{2} \int_0^{\infty} [x_1^2(t) + 0.1x_2^2(t) + u^2(t)]\, dt.$$

(c) Use Equation (D.39) to find the feedback transfer function $W(s)$ necessary to implement the optimal regulator.

(d) Use Equation (D.42) to find the closed-loop transfer function.

(e) If the result of (c) is not suitable, use some other method.

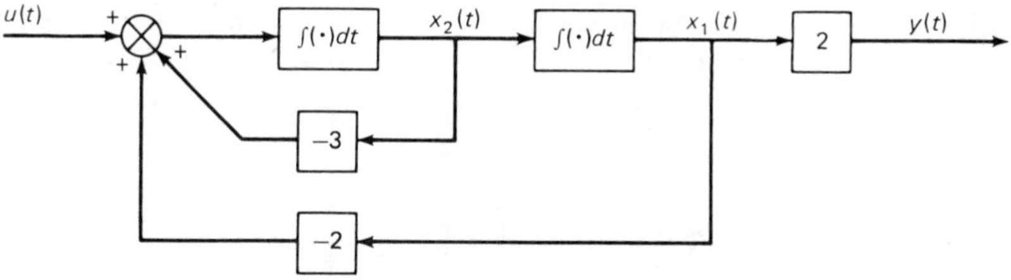

Figure PD-9.

D.18. Consider the second-order system described by the diagram of Figure PD-9.

(a) Use Equations (D.14) and (D.15) to find the transfer function.

(b) Determine the state variable feedback gains to produce closed-loop response characterized by critical damping with time constant 0.2.

(c) Use Equation (D.39) to determine the feedback transfer function necessary to accomplish the result of (b).

(d) Use Equation (D.42) to find the closed-loop transfer function.

(e) Use Equation (D.41) to find the open-loop transfer function.

(f) Repeat (d) and (e) using Equation (D.44).

(g) If the result of (d) is not suitable, use some other method.

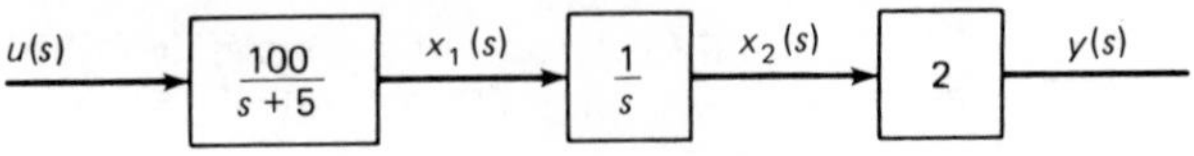

Figure PD-10.

D.19. Consider the system described by the block diagram shown in Figure PD-10.

(a) Determine the state space representation of this system.

(b) Find the state variable feedback gains necessary to produce closed-loop eigenvalues of $(10/\sqrt{2}) \pm [i(10/\sqrt{2})]$.

(c) Determine the closed-loop transfer function using Equation (D.42).

(d) Determine an alternative to the result of (c) using minor loop equalization.

(e) Determine an alternative to the result of (c) using series compensation.

D.20. Consider the discrete-time system described by the equations

$$x_1(l) = x_1(l-1) + 0.1x_2(l-1) + 0.005u(l-1)$$

$$x_2(l) = x_2(l) + 0.1u(l-1).$$

The measurement is $y(l) = x_1(l)$.

(a) Use Equation (D.55) to determine the transfer function $V(z)$.

(b) Find the state variable feedback gains to produce closed-loop response characterized by critical damping with eigenvalue 0.8.

(c) Use Equation (D.59) to determine the closed-loop transfer function.

(d) Find the feedback transfer function $W(z)$ from Equation (D.62). Comment on the suitability of this function.

D.21. Use the procedures of Section 5.2.7 to develop a reduced-order observer for the system of Figure D-5. The system is described by the matrices

$$\boldsymbol{F} = \begin{bmatrix} 0 & 1 & 0 \\ 0 & -1 & 2 \\ 0 & 0 & -4 \end{bmatrix}; \quad \boldsymbol{g} = \begin{bmatrix} 0 \\ 0 \\ 1 \end{bmatrix}; \quad \boldsymbol{h} = \begin{bmatrix} 1 \\ 0 \\ 0 \end{bmatrix}; \quad \boldsymbol{c} = \begin{bmatrix} -1 \\ -1 \\ -1 \end{bmatrix}.$$

Index